Lecture Notes in Computer Science 2045

Edited by G. Goos, J. Hartmanis and J. van Leeuwen

W0055465

Springer-Verlag Berlin Heidelberg GmbH

Birgit Pfitzmann (Ed.)

Advances in Cryptology – EUROCRYPT 2001

International Conference on the Theory
and Application of Cryptographic Techniques
Innsbruck, Austria, May 6-10, 2001
Proceedings

 Springer

Series Editors

Gerhard Goos, Karlsruhe University, Germany
Juris Hartmanis, Cornell University, NY, USA
Jan van Leeuwen, Utrecht University, The Netherlands

Volume Editors

Birgit Pfitzmann
Universität des Saarlandes, Fachrichtung Informatik
Postfach 15 11 50, 66041 Saarbrücken, Germany
E-mail: pfitzmann@cs.uni-sb.de

Cataloging-in-Publication Data applied for

Die Deutsche Bibliothek - CIP-Einheitsaufnahme

Advances in cryptology : proceedings / EUROCRYPT 2001, International
Conference on the Theory and Application of Cryptographic Techniques,
Innsbruck, Austria, May 6 - 10, 2001. Birgit Pfitzmann (ed.). - Berlin ;
Heidelberg ; New York ; Barcelona ; Hong Kong ; London ; Milan ;
Paris ; Singapore ; Tokyo : Springer, 2001
   (Lecture notes in computer science ; Vol. 2045)

CR Subject Classification (1998): E.3, G.2.1, D.4.6, K.6.5, F.2.1-2, C.2, J.1

ISSN 0302-9743
ISBN 978-3-540-42070-5   ISBN 978-3-540-44987-4 (eBook)
DOI 10.1007/978-3-540-44987-4

http://www.springer.de

© Springer-Verlag Berlin Heidelberg 2001

Originally published by Springer-Verlag Berlin Heidelberg New York 2001.

Typesetting: Camera-ready by author, data conversion by Steingräber Satztechnik GmbH, Heidelberg
Printed on acid-free paper   SPIN: 10781446   06/3142   5 4 3 2 1 0

EUROCRYPT 2001

May 6 – 10, 2001, Innsbruck (Tyrol), Austria

Sponsored by the
International Association for Cryptologic Research (IACR)
in cooperation with the
Austrian Computer Society (OCG)

General Chair

Reinhard Posch, Institute for Applied Information Processing and
Communications (IAIK), Austria

Program Chair

Birgit Pfitzmann, Saarland University, Saarbrücken, Germany

Program Committee

Josh Benaloh .... Microsoft Research, USA
Carlo Blundo .... Università di Salerno, Italy
Jan Camenisch ....IBM Zürich Research Laboratory, Switzerland
Matt Franklin ....UC Davis, USA
Shai Halevi ....IBM T. J. Watson Research Center, USA
Martin Hirt ....ETH Zürich, Switzerland
Thomas Johansson ....Lund University, Sweden
Neal Koblitz .... Univ. of Washington, USA
Hugo Krawcyk .... Technion, Israel
Kaoru Kurosawa ....Tokyo Institute of Technology, Japan
Arjen Lenstra ....Citicorp, USA
Willi Meier ....Fachhochschule Aargau, Switzerland
David Naccache ....Gemplus, France
Kaisa Nyberg ....Nokia, Finland
Torben Pryds Pedersen ....Cryptomathic, Denmark
Guillaume Poupard ....DCSSI Crypto Lab, France
Tal Rabin ....IBM T. J. Watson Research Center, USA
Vincent Rijmen ....K. U. Leuven, Belgium
Amit Sahai ....Princeton University, USA
Kazue Sako ....NEC, Japan
Louis Salvail ....BRICS, University of Århus, Denmark
Claus-Peter Schnorr ....University of Frankfurt, Germany
David Wagner ....UC Berkeley, USA
Michael Waidner ....IBM Zürich Research Laboratory, Switzerland

Preface

EUROCRYPT 2001, the 20th annual Eurocrypt conference, was sponsored by the IACR, the International Association for Cryptologic Research, see http://www.iacr.org/, this year in cooperation with the Austrian Computer Society (OCG). The General Chair, Reinhard Posch, was responsible for local organization, and registration was handled by the IACR Secretariat at the University of California, Santa Barbara.

In addition to the papers contained in these proceedings, we were pleased that the conference program also included a presentation by the 2001 IACR distinguished lecturer, Andrew Odlyzko, on "Economics and Cryptography" and an invited talk by Silvio Micali, "Zero Knowledge Has Come of Age." Furthermore, there was the rump session for presentations of recent results and other (possibly satirical) topics of interest to the crypto community, which Jean-Jacques Quisquater kindly agreed to run.

The Program Committee received 155 submissions and selected 33 papers for presentation; one of them was withdrawn by the authors. The review process was therefore a delicate and challenging task for the committee members, and I wish to thank them for all the effort they spent on it. Each committee member was responsible for the review of at least 20 submissions, so each paper was carefully evaluated by at least three reviewers, and submissions with a program committee member as a (co-)author by at least six. Final decisions, after intensive web discussions, were taken at a one-day face-to-face program committee meeting. The selection was based on originality, quality, and relevance to cryptology. In most cases, the reviewers provided extensive comments to the authors. Subsequently, the authors made a substantial effort to take these comments into account. I was pleased to see that the field is continuing to flourish and believe that we were able to select a varied and high-quality program. I wish to thank all the authors who submitted papers, thus making such a choice possible, and those of accepted papers for their cooperation in the timely production of revised versions.

Many thanks also go to the additional colleagues who reviewed submissions in their area of expertise: Joy Algesheimer, Seigo Arita, Giuseppe Ateniese, Olivier Baudron, Charles Bennett, Dan Boneh, Annalisa De Bonis, Wieb Bosma, Marco Bucci, Ran Canetti, Anne Canteaut, Suresh Chari, Philippe Chose, Christophe Clavier, Scott Contini, Don Coppersmith, Jean-Sébastien Coron, Ronald Cramer, Nora Dabbous, Ivan Damgård, Giovanni Di Crescenzo, Markus Dichtl, Yevgeniy Dodis, Paul Dumais, Serge Fehr, Marc Fischlin, Roger Fischlin, Matthias Fitzi, Pierre-Alain Fouque, Jun Furukawa, Pierre Girard, Clemente Gladi, Daniel Gottesman, Clemens Holenstein, Rosario Gennaro, Nick Howgrave-Graham, James Hughes, Yuval Ishai, Markus Jakobsson, Eliane Jaulmes, Antoine Joux, Olaf Keller, Ki Hyoung Ko, Reto Kohlas, Takeshi Koshiba, Eyal Kushilevitz, Yehuda Lindell, Helger Lipmaa, Anna Lysyanskaya, Subhamoy

Maitra, Tal Malkin, Daniel Mall, Barbara Masucci, Dominic Mayers, Alfred Menezes, Renato Menicocci, Daniele Micciancio, Markus Michels, Miodrag Mihaljevic, Phong Nguyen, Svetla Nikova, Satoshi Obana, Kazuo Ohta, Pino Persiano, David Pointcheval, Bartosz Przydatek, Michael Quisquater, Omer Reingold, Leonid Reyzin, Jean-Marc Robert, Pankaj Rohatgi, Alon Rosen, Ludovic Rousseau, Daniel Simon, Nigel Smart, Adam Smith, Othmar Staffelbach, Martijn Stam, Michael Steiner, Katsuyuki Takashima, Alain Tapp, Christophe Tymen, Shigenori Uchiyama, Frédéric Valette, Ramarathnam Venkatesan, Eric Verheul, Stefan Wolf, Akihiro Yamamura, Yuliang Zheng. I apologize for any inadvertent omissions.

The review process was greatly simplified by submission software written by Mihir Bellare and Chanathip Namprempre for Crypto 2000, and review software developed for EUROCRYPT 2000 by Bart Preneel, Wim Moreau, and Joris Claessens.

I am very grateful to André Adelsbach. Skillfully and patiently, he carried the main load of background work of the Program Chair, in particular in setting up the submission and review servers, providing technical help to the authors and committee members, and in the preparation of these proceedings. I would also like to thank Michael Steiner and Martin Wanke for technical support, Matthias Schunter for organizing the program committee meeting, and Mihir Bellare and Michael Waidner for advice.

March 2001   Birgit Pfitzmann

Table of Contents

Elliptic Curves

Commitments

Anonymity

Signatures and Hash Functions

XTR and NTRU

Assumptions

Multiparty Protocols

Block Ciphers

Primitives

Symmetric Ciphers

Author Index

A Memory Efficient Version of Satoh's Algorithm

Frederik Vercauteren*, Bart Preneel, and Joos Vandewalle

K.U. Leuven, Dept. Elektrotechniek-ESAT/COSIC,
Kasteelpark Arenberg 10, B-3001 Leuven-Heverlee, Belgium.
{Frederik.Vercauteren, Bart.Preneel, Joos.Vandewalle}@esat.kuleuven.ac.be

Abstract. In this paper we present an algorithm for counting points on elliptic curves over a finite field $\mathbb{F}_{p^n}$ of small characteristic, based on Satoh's algorithm. The memory requirement of our algorithm is $O(n^2)$, where Satoh's original algorithm needs $O(n^3)$ memory. Furthermore, our version has the same run time complexity of $O(n^{3+\varepsilon})$ bit operations, but is faster by a constant factor. We give a detailed description of the algorithm in characteristic 2 and show that the amount of memory needed for the generation of a secure 200-bit elliptic curve is within the range of current smart card technology.

Keywords: elliptic curve, finite field, order counting, Satoh's algorithm

1 Introduction

In 1985 Schoof [12] described a polynomial time algorithm for counting the number of points on an elliptic curve E defined over a finite field $\mathbb{F}_q$, with $q = p^n$. The run time of the algorithm is $O(\log^{5+\epsilon} q)$ bit operations using fast arithmetic and the memory requirements are $O(\log^2 q)$. Improvements by Elkies [6] and Atkin [1] led to the so called Schoof-Elkies-Atkin algorithm with a run time of $O(\log^{4+\epsilon} q)$ bit operations and further work by Couveignes [2,3] and Lercier [9] extended this SEA-algorithm to work in small characteristic. Csirik [4] implemented a reduced memory version of the algorithm. Recently Satoh [11] described a new algorithm for small characteristic $p \geq 5$ with run time $O(n^{3+\varepsilon})$ and memory complexity $O(n^3)$. Skjernaa [14] and Fouquet, Gaudry and Harley [7] independently extended Satoh's algorithm to characteristic 2.

In this paper we present a new version of Satoh's algorithm which still runs in $O(n^{3+\varepsilon})$ bit operations, but only needs $O(n^2)$ memory. The algorithm works for all small characteristics and is even faster than the original algorithm by a constant factor of about 1.5. Furthermore, the algorithm can be easily parallelized. We give a detailed description in the characteristic 2 case and present run times and memory usages of our implementation for elliptic curves in the range of interest to cryptography. The given data show that it now becomes feasible to compute the group order of a 200-bit elliptic curve on a smart card.

---

* F.W.O. research assistant, sponsored by the Fund for Scientific Research - Flanders (Belgium).

B. Pfitzmann (Ed.): EUROCRYPT 2001, LNCS 2045, pp. 1–13, 2001.

The remainder of the paper is organized as follows: after a brief review of Satoh's original algorithm in section 2, we outline our $O(n^2)$ memory version in its most general form in section 3. In section 4 we specialize this algorithm to the characteristic 2 case and give ready to implement pseudo-code. Section 5 discusses details of our implementation and contains run times and memory usages for field sizes relevant to cryptographical applications.

2   Satoh's Algorithm

Let E be an elliptic curve over $\mathbb{F}_q$, with $q = p^n$. The number of points $\#E(\mathbb{F}_q)$ satisfies the well known relation $\#E(\mathbb{F}_q) = q + 1 - t$, where t is the trace of the Frobenius endomorphism $F : E \longrightarrow E : (x, y) \mapsto (x^q, y^q)$. By Hasse's theorem [8] we have $|t| \leq 2\sqrt{q}$.

The basic idea of Satoh's algorithm is to lift both the curve E and the Frobenius endomorphism F to the valuation ring $\mathcal{R}$ of a degree n unramified extension $\mathcal{K}$ of the p-adic field $\mathbb{Q}_p$. Since this lifting is done in a canonical way, the trace of the lifted Frobenius $\mathcal{F}$ equals the trace of Frobenius t. However, the Frobenius endomorphism F itself is difficult to lift because it is inseparable. Therefore one actually works with the dual of the Frobenius endomorphism F, called the Verschiebung $\widehat{F}$. This Verschiebung is separable if and only if E is non-supersingular and can be lifted explicitly by lifting its kernel. Analyzing the action of the lift $\widehat{\mathcal{F}}$ of $\widehat{F}$ on the formal group of the canonical lift $\mathcal{E}$, we obtain an expression for the trace of $\widehat{\mathcal{F}}$ which equals the trace of Frobenius t.

2.1   The Canonical Lift of an Elliptic Curve

The main step in Satoh's algorithm is lifting the curve E and the Verschiebung $\widehat{F}$ to the valuation ring $\mathcal{R}$ of a degree n unramified extension $\mathcal{K}$ of $\mathbb{Q}_p$. Among the many possible lifts of E from $\mathbb{F}_q$ to $\mathcal{R}$ there is one which has particularly nice properties, called the canonical lift. The canonical lift $\mathcal{E}$ of a non-supersingular elliptic curve E over $\mathbb{F}_q$ is an elliptic curve over $\mathcal{K}$ which satisfies the following two properties: the reduction modulo p of $\mathcal{E}$ equals E and $\text{End}(E) \cong \text{End}(\mathcal{E})$ as a ring. Deuring [5] has shown that the canonical lift $\mathcal{E}$ always exists and is unique up to isomorphism. Furthermore, a theorem by Lubin, Serre and Tate [10] provides an effective, but slow algorithm to compute the j-invariant of $\mathcal{E}$ given the j-invariant of E.

Theorem 1 (Lubin-Serre-Tate) *Let E be a non-supersingular elliptic curve over $\mathbb{F}_q$ with j-invariant $j(E) \in \mathbb{F}_q \setminus \mathbb{F}_{p^2}$. Denote with Σ the Frobenius substitution on $\mathcal{R}$ and with $\Phi_p(X, Y)$ the p-th modular polynomial. Then the system of equations*

$$\Phi_p(X, \Sigma(X)) = 0 \quad \text{and} \quad X \equiv j(E) \bmod p, \tag{1}$$

has a unique solution $J \in \mathcal{R}$, which is the j-invariant of the canonical lift $\mathcal{E}$ of E.

Note that it is possible to solve the system of equations (1) directly, but this would lead to a slow algorithm because of the explicit computation of Σ. A detailed description of the Frobenius substitution Σ and its computation can be found in [13].

The hypothesis $j(E) \notin \mathbb{F}_{p^2}$ in Theorem 1 is necessary to ensure that a certain partial derivative of Φ_p does not vanish modulo p. This condition is necessary to guarantee the uniqueness of the solution of equation (1). The case $j(E) \in \mathbb{F}_{p^2}$ can be handled very easily using Weil's theorem: since $j(E) \in \mathbb{F}_{p^2}$ there exists an elliptic curve E' defined over $\mathbb{F}_{p^m}$ with $m = 1$ or $m = 2$, which is isomorphic to E over $\mathbb{F}_q$. Let $t_k = p^{mk} + 1 - \#E'(\mathbb{F}_{p^{mk}})$ then $t_{k+1} = t_1 t_k - p^m t_{k-1}$ with $t_0 = 2$ and therefore $\#E(\mathbb{F}_q) = p^n + 1 - t_{n/m}$. So in the remainder of the paper we can assume $j(E) \notin \mathbb{F}_{p^2}$ and in particular that E is non-supersingular.

Let $\sigma : E \longrightarrow E^\sigma : (x, y) \mapsto (x^p, y^p)$ be the p-th power Frobenius morphism, where E^σ is the curve obtained by raising each coefficient of E to the p-th power and let $\hat{\sigma}$ be the dual of σ. Repeatedly applying $\hat{\sigma}$ gives rise to the following cycle

$$E_0 \xrightarrow{\hat{\sigma}_0} E_1 \xrightarrow{\hat{\sigma}_1} \cdots \xrightarrow{\hat{\sigma}_{n-2}} E_{n-1} \xrightarrow{\hat{\sigma}_{n-1}} E_0 \ ,$$

with $E_{(n-i)} = E^{\sigma^i}$ and $\hat{\sigma}_i$ the dual of $\sigma_i : E_{i+1} \longrightarrow E_i : (x, y) \mapsto (x^p, y^p)$. Composing these, we see that $\widehat{F} = \hat{\sigma}_{n-1} \circ \hat{\sigma}_{n-2} \circ \ldots \circ \hat{\sigma}_0$. Instead of lifting E and $\widehat{F}$ directly, the crucial insight of Satoh was to lift the whole cycle $(E_0, E_1, \ldots, E_{n-1})$ simultaneously leading to the diagram

$$
\begin{array}{ccccccccc}
\mathcal{E}_0 & \xrightarrow{\widehat{\Sigma}_0} & \mathcal{E}_1 & \xrightarrow{\widehat{\Sigma}_1} & \cdots & \xrightarrow{\widehat{\Sigma}_{n-2}} & \mathcal{E}_{n-1} & \xrightarrow{\widehat{\Sigma}_{n-1}} & \mathcal{E}_0 \\
\uparrow & & \uparrow & & & & \uparrow & & \uparrow \\
E_0 & \xrightarrow{\hat{\sigma}_0} & E_1 & \xrightarrow{\hat{\sigma}_1} & \cdots & \xrightarrow{\hat{\sigma}_{n-2}} & E_{n-1} & \xrightarrow{\hat{\sigma}_{n-1}} & E_0,
\end{array}
\qquad (2)
$$

with $\mathcal{E}_i$ the canonical lift of E_i and $\widehat{\Sigma}_i$ the corresponding lift of $\hat{\sigma}_i$. The theorem of Lubin, Serre and Tate implies that the j-invariants of $\mathcal{E}_i$ satisfy

$$\Phi_p(j(\mathcal{E}_i), j(\mathcal{E}_{i+1})) = 0 \text{ and } j(\mathcal{E}_i) \equiv j(E_i) \bmod p, \qquad (3)$$

for $i = 0, \ldots, n - 1$. Define $\Theta : \mathcal{R}^n \longrightarrow \mathcal{R}^n$ by

$$\Theta(x_0, x_1, \ldots, x_{n-1}) = (\Phi_p(x_0, x_1), \Phi_p(x_1, x_2), \ldots, \Phi_p(x_{n-1}, x_0)), \qquad (4)$$

then clearly we have $\Theta(j(\mathcal{E}_0), j(\mathcal{E}_1), \ldots, j(\mathcal{E}_{n-1})) = (0, 0, \ldots, 0)$. Using a multivariate Newton iteration on Θ, we can lift the cycle $(j(E_0), j(E_1), \ldots, j(E_{n-1}))$ to $\mathcal{R}^n$ with arbitrary precision. The iteration step is given by

$$(J_0, J_1, \ldots, J_{n-1}) \leftarrow (J_0, J_1, \ldots, J_{n-1}) - ((D\Theta)^{-1}\Theta)(J_0, J_1, \ldots, J_{n-1}), \qquad (5)$$

with $D\Theta$ the Jacobian matrix

$$(D\Theta)(J_0, J_1, \ldots, J_{n-1}) = \begin{pmatrix} \frac{\partial \Phi_p}{\partial X}(J_0, J_1) & \frac{\partial \Phi_p}{\partial Y}(J_0, J_1) \cdots & & 0 \\ 0 & \frac{\partial \Phi_p}{\partial X}(J_1, J_2) \cdots & & 0 \\ \vdots & \vdots & & \vdots \\ 0 & 0 & \cdots & \frac{\partial \Phi_p}{\partial Y}(J_{n-2}, J_{n-1}) \\ \frac{\partial \Phi_p}{\partial Y}(J_{n-1}, J_0) & 0 & \cdots & \frac{\partial \Phi_p}{\partial X}(J_{n-1}, J_0) \end{pmatrix}.$$

(6)

The p-th modular equation satisfies the Kronecker relation

$$\Phi_p(X, Y) \equiv (X^p - Y)(X - Y^p) \bmod p \tag{7}$$

and since $j(E_i) \notin \mathbb{F}_{p^2}$ and $j(E_i) \equiv j(E_{i+1})^p \bmod p$, this leads to the following equations

$$\begin{cases} \frac{\partial \Phi_p}{\partial X}(j(E_i), j(E_{i+1})) \equiv j(E_{i+1})^{p^2} - j(E_{i+1}) \not\equiv 0 \bmod p, \\ \frac{\partial \Phi_p}{\partial Y}(j(E_i), j(E_{i+1})) \equiv j(E_{i+1})^p - j(E_{i+1})^p \equiv 0 \bmod p. \end{cases} \tag{8}$$

The above equations imply that the Jacobian matrix $(D\Theta)(J_0, J_1, \ldots, J_{n-1})$ is invertible over $\mathcal{R}$ and therefore we see $((D\Theta)^{-1}\Theta)(J_0, J_1, \ldots, J_{n-1}) \in \mathcal{R}^n$. Since Newton iteration has quadratic convergence, we can compute $J_i \equiv j(\mathcal{E}_i) \bmod p^N$ with $\log N$ iterations.

2.2   The Trace of Frobenius

The canonical lift $\mathcal{E}$ of a non-supersingular elliptic curve E over $\mathbb{F}_q$ has the property that $\mathrm{End}(E) \cong \mathrm{End}(\mathcal{E})$. Therefore we have $\mathrm{Tr}(F) = \mathrm{Tr}(\mathcal{F})$, where F is the Frobenius endomorphism on E and $\mathcal{F}$ the image of F under the ring isomorphism $\mathrm{End}(E) \cong \mathrm{End}(\mathcal{E})$. Furthermore, the trace of an endomorphism equals the trace of its dual, so $\mathrm{Tr}(F) = \mathrm{Tr}(\widehat{F}) = \mathrm{Tr}(\mathcal{F}) = \mathrm{Tr}(\widehat{\mathcal{F}})$. The following proposition by Satoh [11] gives a very simple relation between the trace of $\widehat{\mathcal{F}}$ and the leading coefficient of the endomorphism induced by $\widehat{\mathcal{F}}$ on the formal group of $\mathcal{E}$.

Proposition 1 (Satoh) *Let $\mathcal{E}$ be an elliptic curve over $\mathcal{K}$ and let $f \in \mathrm{End}_{\mathcal{K}}(\mathcal{E})$ be of degree d. Denote with τ the local parameter of $\mathcal{E}$ at $\mathcal{O}$ and assume that the reduction $\pi(f)$ of f modulo p is separable and that $f(\mathrm{Ker}(\pi)) \subset \mathrm{Ker}(\pi)$. Let $\tilde{f}(\tau) = c\tau + O(\tau^2)$ be the homomorphism induced by f on the formal group of $\mathcal{E}$, then $\mathrm{Tr}(f) = c + \frac{d}{c}$.*

Since the Frobenius endomorphism F is inseparable, we cannot apply the above proposition to $\mathcal{F}$. However, for a non-supersingular curve the Verschiebung $\widehat{F}$ is separable and we have $\mathrm{Tr}(F) = \mathrm{Tr}(\widehat{F}) = c + \frac{q}{c}$ with $\widetilde{\widehat{F}}(\tau) = c\tau + O(\tau^2)$. Diagram (2) shows that $\widehat{\mathcal{F}}$ can be written as $\widehat{\mathcal{F}} = \widehat{\Sigma}_{n-1} \circ \widehat{\Sigma}_{n-2} \circ \cdots \circ \widehat{\Sigma}_0$ and therefore we can compute c as the product of the leading coefficients of the morphisms induced by $\widehat{\Sigma}_i$. More precisely, let c_i be defined by $\tau_{i+1} \circ \widehat{\Sigma}_i = c_i \tau_i + O(\tau_i^2)$, with τ_i the local parameter of $\mathcal{E}_i$ at $\mathcal{O}$, then $c = \prod_{0 \leq i < n} c_i$. Since $\widehat{\mathcal{F}}$ is separable, c will be non-zero modulo p and we conclude

$$\mathrm{Tr}(F) \equiv \prod_{0 \leq i < n} c_i \bmod q. \tag{9}$$

The final step in Satoh's algorithm is to compute the coefficients c_i, based on the equations for $\mathcal{E}_i$ and $\mathcal{E}_{i+1}$ and the kernel of $\widehat{\Sigma}_i$, using Vélu's formulae [15]. The equations for $\mathcal{E}_i$ and $\mathcal{E}_{i+1}$ can be easily computed via a univariate Newton iteration, since we already know their j-invariants. The isogenies $\hat{\sigma}_i$ and $\widehat{\Sigma}_i$ are separable and of degree p, so $\hat{\sigma}_i$ can be explicitly lifted to $\widehat{\Sigma}_i$ by lifting its kernel. This kernel is a subgroup of the p-torsion group of E. The case $p \geq 5$ is discussed in [11] and proceeds by lifting a factor of the p-th division polynomial using a Hensel lift. The cases $p = 2, 3$ can be found in [7,14] and are handled by lifting a single non-trivial torsion point using a Newton iteration.

2.3   Complexity

According to Hasse's theorem we have $|t| \leq 2\sqrt{q}$. Therefore it suffices to lift all the data with precision $N \simeq n/2$. Since elements of $\mathcal{R}$ mod p^N can be represented as degree n polynomials with coefficients in $\mathbb{Z}/p^N\mathbb{Z}$ and since $N = O(n)$, every element will take $O(n^2)$ memory for fixed p. For each curve E_i with $0 \leq i < n$ we need $O(1)$ such elements, so the total memory required is $O(n^3)$. To lift the cycle of j-invariants with precision N, we need $\log N$ iterations. Working with the lowest possible precision in every iteration, the lifting of the cycle of j-invariants amounts to $O(nM(n^2))$ bit operations, where $M(m)$ is the time to multiply two m-bit objects. The computation of one coefficient c_i needs $O(1)$ multiplications, so to compute all c_i we also need $O(nM(n^2))$ bit operations. Therefore the total run time of Satoh's algorithm is $O(nM(n^2))$ bit operations or $O(n^{3+\varepsilon})$ using fast multiplication techniques.

3   An $O(n^2)$ Memory Algorithm

In this section we present a new version of Satoh's algorithm, which requires only $O(n^2)$ memory and still runs in $O(n^{3+\varepsilon})$ bit operations. The basic idea is very simple: the trace of Frobenius t can be computed as $t \equiv \prod_{0 \leq i < n} c_i \bmod q$ and the c_i only depend on $\mathcal{E}_i$ and $\mathcal{E}_{i+1}$. So the main problem of Satoh's original algorithm is that it lifts all j-invariants simultaneously, instead of lifting one j-invariant at a time. Note however that lifting all j-invariants simultaneously is exactly what makes Satoh's algorithm efficient, because this avoids slow Frobenius computations in $\mathcal{R}$. Thus if we would like our algorithm to run in $O(n^{3+\epsilon})$ bit operations and only use $O(n^2)$ memory, we have to find a method to lift one j-invariant without using Frobenius computations.

Our strategy is as follows: the j-invariants $j(\mathcal{E}_i)$ and $j(\mathcal{E}_{i+1})$ satisfy the following relations

$$\Phi_p(j(\mathcal{E}_i), j(\mathcal{E}_{i+1})) = 0, \quad j(\mathcal{E}_i) \equiv j(E_i) \bmod p \quad \text{and} \quad j(\mathcal{E}_{i+1}) \equiv j(E_{i+1}) \bmod p. \tag{10}$$

Suppose we have $J_{i+1} \equiv j(\mathcal{E}_{i+1}) \bmod p^N$ to our disposal, then we can compute $J_i \equiv j(\mathcal{E}_i) \bmod p^N$ using a univariate Newton iteration on $\Phi_p(X, J_{i+1})$. This iteration is given by

$$J_i \leftarrow J_i - \frac{\Phi_p(J_i, J_{i+1})}{\frac{\partial \Phi_p}{\partial X}(J_i, J_{i+1})}, \tag{11}$$

and we can use $j(E_i) \equiv j(\mathcal{E}_i) \bmod p$ as an initial approximation. Since $\Phi_p(X, Y)$ satisfies the Kronecker relation, $\frac{\partial \Phi_p}{\partial X}(J_i, J_{i+1})$ will be invertible in $\mathcal{R}$. Note that we are forced to walk backwards in the cycle, since $\frac{\partial \Phi_p}{\partial Y}(J_i, J_{i+1}) \equiv 0 \bmod p$. Applying this method repeatedly, one easily sees that it suffices to compute one j-invariant with precision N, e.g. $J_0 \equiv j(\mathcal{E}_0) \bmod p^N$. To solve this last problem, we analyze in detail the properties of a bivariate polynomial, which satisfies the same relations as $\Phi_p(X, Y)$.

Proposition 2 *Let $\mathcal{K}$ be an unramified extension of $\mathbb{Q}_p$ and denote with $\mathcal{R}$ its valuation ring. Let $g \in \mathcal{R}[X, Y]$ and assume $x_0, y_0 \in \mathcal{R}$ such that*

$$g(x_0, y_0) \equiv 0 \bmod p, \quad \frac{\partial g}{\partial X}(x_0, y_0) \not\equiv 0 \bmod p \quad \text{and} \quad \frac{\partial g}{\partial Y}(x_0, y_0) \equiv 0 \bmod p. \tag{12}$$

Then the following properties hold:

1. *For every $y \in \mathcal{R}$ with $y \equiv y_0 \bmod p$ there exists a unique $x \in \mathcal{R}$ such that $x \equiv x_0 \bmod p$ and $g(x, y) = 0$.*
2. *Let $y' \in \mathcal{R}$ with $y \equiv y' \bmod p^M$, $M \geq 1$ and let $x' \in \mathcal{R}$ be the unique element with $x' \equiv x_0 \bmod p$ and $g(x', y') = 0$. Then $x' \equiv x \bmod p^{M+1}$.*

Proof:

1. Define $h \in \mathcal{R}[X]$ by $h(X) = g(X, y)$. Then $h(x_0) \equiv 0 \bmod p$ and $h'(x_0) \equiv \frac{\partial g}{\partial X}(x_0, y_0) \bmod p$. Therefore, $h'(x_0) \not\equiv 0 \bmod p$ and Hensel's lemma guarantees the existence of a unique $x \in \mathcal{R}$ such that $h(x) = g(x, y) = 0$ and $x \equiv x_0 \bmod p$. Furthermore, given y, one can compute x with arbitrary precision using a univariate Newton iteration on $g(X, y)$ with $x_0 \bmod p$ as an initial approximation.

2. Define $\delta_x = x' - x$ and $\delta_y = y' - y$. Clearly $\delta_x \equiv \delta_y \equiv 0 \bmod p^M$. Writing out the Taylor series of $g(X, Y) = \sum_{i,j} g_{i,j} X^i Y^i$ leads to

$$
\begin{aligned}
0 = g(x', y') &= g(x + \delta_x, y + \delta_y) \\
&= \sum_{i,j} g_{i,j}(x + \delta_x)^i (y + \delta_y)^j \\
&= \sum_{i,j} g_{i,j}(x^i + ix^{i-1}\delta_x + \delta_x^2 R_x(x))(y^j + jy^{j-1}\delta_y + \delta_y^2 R_y(y)),
\end{aligned}
\tag{13}
$$

with R_x, R_y polynomials with coefficients in $\mathcal{R}$. Since $\delta_x^2 \equiv \delta_y^2 \equiv 0 \bmod p^{2M}$ and $M \geq 1$ we get

$$
0 \equiv \frac{\partial g}{\partial X}(x, y)(x - x') + \frac{\partial g}{\partial Y}(x, y)(y - y') \bmod p^{M+1}.
\tag{14}
$$

The above equation implies $x \equiv x' \bmod p^{M+1}$, since $\delta_y \equiv 0 \bmod p^M$, $\frac{\partial g}{\partial Y}(x, y) \equiv 0 \bmod p$ and $\frac{\partial g}{\partial X}(x, y) \not\equiv 0 \bmod p$.   □

Repeatedly applying Proposition 2 leads to a very simple iterative algorithm to compute $J_0 \equiv j(\mathcal{E}_0) \bmod p^N$. Starting with $J_{N-1} \equiv j(\mathcal{E}_{N-1}) \bmod p$, we compute $J_{N-2} \equiv j(\mathcal{E}_{N-2}) \bmod p^2$ using a Newton iteration on $\Phi_p(X, J_{N-1})$, similar to equation 11. More generally, given $J_{N-i+1} \equiv j(\mathcal{E}_{N-i+1}) \bmod p^{i-1}$, we determine $J_{N-i} \equiv j(\mathcal{E}_{N-i}) \bmod p^i$. After $N - 1$ steps we reach $J_0 \equiv j(\mathcal{E}_0) \bmod p^N$. Combining these ideas finally leads to algorithm Satoh_Low_Memory.

Algorithm 1 (Satoh_Low_Memory)

   IN: *A j-invariant $j \in \mathbb{F}_{p^n} \setminus \mathbb{F}_{p^2}$ of an elliptic curve E.*
OUT: *The trace of Frobenius $t = q + 1 - \#E(\mathbb{F}_q)$ of E.*

---

1. Compute $J \equiv j(\mathcal{E}) \bmod p^N$ with $N > n/2 + 1$ from $J_{N-1} = j(\mathcal{E}_{N-1}) \bmod p$ with $N - 1$ Newton iterations 11;

2. Set $c^2 = 1$;

3. For $i = 1$ To n Do

   3.1. Compute $J' \equiv j(\mathcal{E}_{n-i}) \bmod p^N$ using a Newton iteration 11 on $\Phi_p(X, J)$;

   3.2. Compute the square $c_{n-i}^2 \bmod p^N$ of coefficient $c_{n-i} \bmod p^N$;

   3.3. Set $c^2 = c^2 \times c_{n-i}^2$ and $J = J'$;

4. Compute $c \equiv \sqrt{c^2} \bmod p^N$ with the correct sign;

5. Return $t \equiv c \bmod p^N$.

The memory requirement of algorithm `Satoh_Low_Memory` is $O(n^2)$ for p fixed: every element in $\mathcal{R}$ mod p^N takes $O(n^2)$ memory, and the algorithm needs $O(1)$ such elements. Therefore, the total memory required is $O(n^2)$.

Lifting one j-invariant to precision N and computing one coefficient c_i can be done with $O(M(n^2))$ bit operations, so the loop in step 3 takes $O(nM(n^2))$ bit operations. Since the j-invariant in step 1 is computed using N Newton iterations with varying precision $i = 2, \ldots, N$, the total cost of step 1 is trivially bounded by $O(nM(n^2))$ bit operations. We therefore conclude that our version still runs in $O(nM(n^2))$ bit operations or $O(n^{3+\varepsilon})$ using fast arithmetic.

4   Algorithms in Characteristic 2

In this section we specialize the $O(n^2)$ memory algorithm of the previous section to the characteristic 2 case, which from a practical point of view is most important.

Let E be an elliptic curve over a finite field $\mathbb{F}_q$, with $q = 2^n$ and $j(E) \notin \mathbb{F}_4$. It is well known that either E or its quadratic twist is isomorphic over $\mathbb{F}_q$ with an elliptic curve given by an equation of the form $y^2 + xy = x^3 + a$, with $a \in \mathbb{F}_q^*$. Therefore, we can restrict ourselves to this case.

Let $\mathcal{K}$ be a degree n unramified extension of $\mathbb{Q}_2$ and $\mathcal{R}$ its valuation ring. Then $\mathcal{R}$ is isomorphic to $\mathbb{Z}_2[T]/(f(T))$, with $f \in \mathbb{Z}_2[T]$ a monic polynomial of degree n such that its reduction modulo 2 is irreducible in $\mathbb{F}_2[T]$. In practice all computations are carried out in the ring $\mathcal{R}$ mod 2^N, which can be represented as $(\mathbb{Z}/2^N\mathbb{Z})[T]/(f(T))$.

4.1   Lifting the j-Invariants

For $1 \le i < n$ define the elliptic curve E_i by the equation $y^2 + xy = x^3 + a^{2^{n-i}}$ and let $\mathcal{E}_i$ be the canonical lift of E_i. Using Proposition 2 we can compute $J_i \equiv j(E_i) \bmod 2^N$, starting from $J_{i+1} \equiv j(E_{i+1}) \bmod 2^{N-1}$, using a univariate Newton iteration on the polynomial $\Phi_2(X, J_{i+1})$, with

$$
\begin{aligned}
\Phi_2(X, Y) = &X^3 + Y^3 - X^2Y^2 + 1488(XY^2 + X^2Y) - 162000(X^2 + Y^2) \\
&+ 40773375XY + 8748000000(X + Y) - 157464000000000.
\end{aligned}
\tag{15}
$$

Algorithm `Lift_Previous_J_Invariant` computes coefficients $A, B, C \in \mathcal{R}$ mod 2^N, such that

$$
\Phi_2(X, J_{i+1}) \equiv X^3 + AX^2 + BX + C \bmod 2^N,
\tag{16}
$$

and then calls the recursive algorithm `Lift_Previous_J_Invariant_Rec` which performs the Newton iteration on the cubic polynomial $X^3 + AX^2 + BX + C$.

With every call of algorithm `Lift_Previous_J_Invariant` we gain 1 bit of precision, so if we would like to compute $J_0 \equiv j(\mathcal{E}_0) \bmod 2^N$ then it suffices to start with $j(E_{N-1}) \equiv j(\mathcal{E}_{N-1}) \bmod 2$ and iterate this algorithm $N - 1$ times, which immediately leads to algorithm `Lift_First_J_Invariant`.

Algorithm 2 (Lift_Previous_J_Invariant)

   IN: $J_{i+1} \in \mathcal{R} \mod 2^N$ *with* $J_{i+1} \equiv j(\mathcal{E}_{i+1}) \mod 2^{N-1}$ *and a precision* N.
OUT: $J_i \in \mathcal{R} \mod 2^N$ *with* $J_i \equiv j(\mathcal{E}_i) \mod 2^N$.

---

1. $A \equiv -J_{i+1}^2 + 1488 J_{i+1} - 162000 \mod 2^N$;

2. $B \equiv 1488 J_{i+1}^2 + 40773375 J_{i+1} + 8748000000 \mod 2^N$;

3. $C \equiv J_{i+1}^3 - 162000 J_{i+1}^2 + 8748000000 J_{i+1} - 157464000000000 \mod 2^N$;

4. $J_i = \text{Lift\_Previous\_J\_Invariant\_Rec}(J_{i+1}, A, B, C, N)$;

5. Return J_i.

Algorithm 3 (Lift_Previous_J_Invariant_Rec)

   IN: *Elements* $J_{i+1}, A, B, C \in \mathcal{R} \mod 2^N$ *with* $J_{i+1} \equiv j(\mathcal{E}_{i+1}) \mod 2^{N-1}$,
      $\Phi_2(X, J_{i+1}) \equiv X^3 + AX^2 + BX + C \mod 2^N$ *and a precision* N.
OUT: *An element* $J_i \in \mathcal{R} \mod 2^N$ *with* $J_i \equiv j(\mathcal{E}_i) \mod 2^N$.

---

1. If $N = 1$ Then

    1.1. $J_i = J_{i+1}^2 \mod 2$;

2. Else

    2.1. $N' = \left\lceil \frac{N}{2} \right\rceil$;

    2.2. $J_i = \text{Lift\_Previous\_J\_Inv\_Rec}(J_{i+1}, A, B, C, N')$;

    2.3. $J_i \equiv J_i - \dfrac{J_i^3 + AJ_i^2 + BJ_i + C}{3J_i^2 + 2AJ_i + B} \mod 2^N$;

3. Return J_i.

Algorithm 4 (Lift_First_J_Invariant)

   IN: *A j-invariant* $j_0 \in \mathbb{F}_{2^n} \setminus \mathbb{F}_4$ *and a precision* N.
OUT: $J_0 \in \mathcal{R} \mod 2^N$ *with* $J_0 \equiv j_0 \mod 2$ *and* $\Phi_2(J_0, \Sigma(J_0)) \equiv 0 \mod 2^N$.

---

1. $J_0 \equiv j_0^{2^{(n-N+1)}} \mod 2$;

2. For $i = 2$ To N Do

    2.1. $J_0 = \text{Lift\_Previous\_J\_Invariant}(J_0, i)$;

3. Return J_0.

4.2   Computing the Trace

In this section we give an explicit formula for the first coefficient c_i of the formal group expression of $\widehat{\Sigma}_i$. This suffices to compute the trace of Frobenius t, since $t \equiv \prod_{i=0}^{n-1} c_i \bmod q$.

The following proposition gives an expression for c_i^2 in terms of the j-invariant of $\mathcal{E}_i$ and the x-coordinate of the non-trivial point in $\mathrm{Ker}(\widehat{\Sigma}_i)$. Since $\widehat{\Sigma}_i$ is separable and of degree 2, its kernel is a subgroup of order 2 of the 2-torsion points and therefore contains exactly one non-trivial point. The proposition is adapted from [14]: the proof is exactly the same, but the given formulae have been simplified as much as possible.

Proposition 3 *Let* $\tau_i = -X/Y$ *be the local parameter of* $\mathcal{E}_i$ *at* $\mathcal{O}$ *and let* c_i *be defined as* $\tau_{i+1} \circ \widehat{\Sigma}_i = c_i \tau_i + O(\tau_i^2)$. *Denote the non-trivial point in* $\mathrm{Ker}(\widehat{\Sigma}_i)$ *by* $Q_i = (x_i, y_i)$ *and let* $z_i = x_i/2$ *and* $t_i = (12z_i^2 + z_i)(j(\mathcal{E}_i) - 1728) - 36$, *then*

$$c_i^2 = \frac{j(\mathcal{E}_i) - (504 + 12096 z_i)t_i}{j(\mathcal{E}_i) + 240 t_i}. \tag{17}$$

Algorithm 5 (Compute_Trace)

   IN: *A j-invariant $j \in \mathbb{F}_{2^n} \setminus \mathbb{F}_4$ of an elliptic curve E.*
   OUT: *The trace of Frobenius $t = q + 1 - \#E(\mathbb{F}_q)$ of E.*

---

1. $N = \lceil \frac{n}{2} \rceil + 13;\ M = N - 10;$

2. $J = \mathtt{Lift\_First\_J\_Invariant}(j,\ N);$

3. $CN = 1;\ CD = 1;$

4. For $i = 0$ To $n - 1$ Do

    4.1. $J' = \mathtt{Lift\_Previous\_J\_Invariant}(J,\ N);$

    4.2. $Z = -\dfrac{(J^2 + 195120J + 4095J' + 660960000)/2^{12}}{(J^2 + J(563760 - 512J') + 372735J' + 8981280000)/2^9};$

    4.3. $T = (12Z^2 + Z)(J' - 1728) - 36;$

    4.4. $CN = CN \times (J' - (504 + 12096Z)T);$

    4.5. $CD = CD \times (240T + J');$

    4.6. $J = J';$

5. $t = \mathtt{Sqrt}(CN/CD,\ 1,\ M) \bmod 2^{M-1};$

6. If $t > 2\sqrt{q}$ Then $t = t - 2^{M-1};$

7. Return t.

Thus to compute c_i^2, we need an expression for half the x-coordinate of the non-trivial point $Q_i \in \mathrm{Ker}(\widehat{\Sigma}_i)$. Again we follow [14], but considerably simplify the formula for z_i.

Proposition 4 *Let $Q_i = (x_i, y_i)$ be the non-trivial point in $\mathrm{Ker}(\widehat{\Sigma}_i)$ and let $z_i = x_i/2$, then*

$$z_i = -\frac{(j(\mathcal{E}_{i+1})^2 + 195120 j(\mathcal{E}_{i+1}) + 4095 j(\mathcal{E}_i) + 660960000)/2^{12}}{(j(\mathcal{E}_{i+1})^2 + j(\mathcal{E}_{i+1})(563760 - 512 j(\mathcal{E}_i)) + 372735 j(\mathcal{E}_i) + 8981280000)/2^9}. \tag{18}$$

Combining the above propositions we can compute $c^2 = \prod_{i=0}^{n-1} c_i^2$. Since the trace of Frobenius t satisfies $t \equiv c \bmod q$ and $|t| \le 2\sqrt{q}$, we have $t \equiv c \bmod 2^{\lceil \frac{n+4}{2} \rceil}$. The 2-adic square root can be found via a Newton iteration for the inverse square root, i.e. via a Newton iteration on $s(X) = c^2 X^2 - 1$. Clearly, we have $s(1/c) = 0$ and $s'(1/c) \equiv 0 \bmod 2$. Furthermore, $c \equiv 1 \bmod 4$, since E has a point of order 4 and thus $s'(1/c) \not\equiv 0 \bmod 4$. The vanishing of $s'(1/c)$ modulo 2 means that we lose exactly one bit of precision in the computation of the square root and therefore we need to compute c^2 modulo $2^{\lceil \frac{n+6}{2} \rceil}$. Substituting the expressions for z_i and t_i in c_i^2, we see that we have to determine the j-invariants $j(\mathcal{E}_i)$ with precision $2^{\lceil \frac{n}{2} \rceil + 13}$. This finally leads to the main algorithm `Compute_Trace`. In step 5 we use the function `Sqrt`, which computes the 2-adic square root of c^2 with precision M, such that $c \equiv 1 \bmod 4$.

5 Implementation

In this section we give practical run times and memory usages of both the original Satoh lifting-algorithm combined with the simplified formulae taken from [14] and our $O(n^2)$ memory version for elliptic curves in the range of interest to cryptography. Both algorithms have been implemented in the C programming language on a AMD Thunderbird 1 GHz PC with 384 MB of main memory, running Linux Redhat 6.2. All programs were compiled using the gcc compiler, version 2.7.2.3. Before giving the actual results we make some comments on our implementation.

Since efficiency was our main goal, we have written the basic operations on multiple precision integers in assembly. These include: addition and subtraction, shift left/right and the multiplication of a multi-precision integer by a word. To minimize the loop overhead for small multiple precision integers, i.e. integers which fit in four words or less, we implemented unrolled versions of the above-mentioned operations.

Elements of $\mathbb{F}_{2^n}$ are represented with respect to a standard polynomial basis, i.e. as polynomials over $\mathbb{F}_2$ modulo a degree n irreducible polynomial f. By choosing f as a trinomial or a pentanomial, reduction modulo f becomes very efficient. The same polynomial f is used to construct $\mathcal{R} \bmod 2^N$ as $(\mathbb{Z}/2^N\mathbb{Z})[T]/(f(T))$. Multiplication of two elements in $\mathcal{R} \bmod 2^N$ is implemented using Karatsuba's trick in the polynomial dimension and classical multiplication for the coefficients.

In Table 1 we compare the characteristic 2 version of Satoh's original algorithm with our $O(n^2)$ memory version for finite fields $\mathbb{F}_{2^n}$ relevant to cryptographical applications. The data in this table show that our algorithm is faster by a constant factor of about 1.5 and that the memory requirements are considerably lower than for Satoh's original algorithm. We note that our current implementation is more optimized towards speed than it is towards minimizing memory usage. Therefore it would be possible to lower the memory requirements by another 30%. Since a smart card typically has 32 KB of memory (in the near future this will be 64 KB), it becomes feasible to generate secure elliptic curves on a smart card.

Table 1. Run times and memory usage of Satoh's algorithm versus the $O(n^2)$ memory version on an AMD 1 GHz

Field size n	Original Satoh		$O(n^2)$ memory version	
	Time (s)	Memory (KB)	Time (s)	Memory (KB)
160	5.43	315	3.17	30
180	9.11	534	5.64	44
200	11.8	650	7.41	48
220	15.4	790	9.83	54
240	28.1	1162	15.8	73
260	36.0	1371	20.3	80
280	44.1	1574	25.1	86
300	64.3	2180	39.2	109
340	88.7	2790	55.3	125
380	133	4052	82.7	162
420	195	5643	123	197
460	244	6756	154	224
500	400	8964	225	275

6   Conclusion

In this paper we have presented a new version of Satoh's algorithm which only needs $O(n^2)$ memory, where the original algorithm needs $O(n^3)$ memory. Furthermore, we showed that our algorithm still runs in $O(n^{3+\varepsilon})$ bit operations, which equals the run time complexity of Satoh's original algorithm. Our version relies on univariate Newton iterations where Satoh also uses multivariate Newton iterations. In our implementation, this resulted in a speed-up of a factor of about 1.5. As a result of the $O(n^2)$ memory complexity, it now becomes feasible to generate secure elliptic curves on a smart card.

Acknowledgements

The authors are very grateful to Takakazu Satoh and Jan Denef for many interesting discussions and helpful remarks on this work.

References

1. A.O.L. Atkin. The number of points on an elliptic curve modulo a prime. *Series of e-mails to the* NMBRTHRY *mailing list*, 1992.
2. J.M. Couveignes. *Quelques calculs en théorie des nombres*. PhD thesis, Université de Bordeaux, 1994.
3. J.M. Couveignes. Computing *l*-isogenies with the *p*-torsion. *ANTS-II, Lecture Notes in Comp. Sci.*, 1122:59–65, 1996.
4. J.A. Csirik. Counting the number of points on an elliptic curve on a low-memory device. 1998. Preprint.
5. M. Deuring. Die Typen der Multiplikatorenringe elliptischer Funktionenkörper. *Abh. Math. Sem. Univ. Hamburg*, 14:197–272, 1941.
6. N. Elkies. Elliptic and modular curves over finite fields and related computational issues. *Computational Perspectives on Number Theory*, pages 21–76, 1998.
7. M. Fouquet, P. Gaudry, and R. Harley. On Satoh's algorithm and its implementation. *J. Ramanujan Math. Soc.*, 15:281–318, 2000.
8. H. Hasse. Beweis des Analogons der Riemannschen Vermutung für die Artinschen und F. K. Smidtschen Kongruenzzetafunctionen in gewissen elliptischen Fällen. *Ges. d. Wiss. Nachrichten. Math.-Phys. Klasse*, pages 253–262, 1933.
9. R. Lercier. *Algorithmique des courbes elliptiques dans les corps finis*. PhD thesis, L'École Polytechnique, Laboratoire D'Informatique, CNRS, Paris, June 1997.
10. J. Lubin, J.P. Serre, and J. Tate. Elliptic curves and formal groups. *Lecture notes prepared in connection with the seminars held at the Summer Institute on Algebraic Geometry, Whitney Estate, Woods Hole, Massachusetts*, 1964.
11. T. Satoh. The canonical lift of an ordinary elliptic curve over a finite field and its point counting. *J. Ramanujan Math. Soc.*, 15:247–270, 2000.
12. R. Schoof. Elliptic curves over finite fields and the computation of square roots mod *p*. *Math. Comput.*, 44:483–494, 1985.
13. J.P. Serre. *Local Fields*, volume 67 of *Graduate Texts in Mathematics*. Springer-Verlag, 1979.
14. B. Skjernaa. Satoh's algorithm in characteristic 2. *Preprint*, 2000.
15. J. Vélu. Isogénies entre courbes elliptiques. *C.R. Acad. Sc. Paris*, 273:238–241, 1971.

Finding Secure Curves
with the Satoh-FGH Algorithm
and an Early-Abort Strategy

Mireille Fouquet[1], Pierrick Gaudry[1], and Robert Harley[2]

[1] LIX, École polytechnique, 91128 Palaiseau Cedex, France
[2] ArgoTech, 26 ter rue Nicolaï, 75012 Paris, France

Abstract. The use of elliptic curves in cryptography relies on the ability to count the number of points on a given curve. Before 1999, the SEA algorithm was the only efficient method known for random curves. Then Satoh proposed a new algorithm based on the canonical p-adic lift of the curve for $p \geq 5$. In an earlier paper, the authors extended Satoh's method to the case of characteristics two and three. This paper presents an implementation of the Satoh-FGH algorithm and its application to the problem of finding curves suitable for cryptography. By combining Satoh-FGH and an early-abort strategy based on SEA, we are able to find secure random curves in characteristic two in much less time than previously reported. In particular we can generate curves widely considered to be as secure as RSA-1024 in less than one minute each on a fast workstation.

1 Introduction

Since elliptic curve cryptosystems were first proposed in the mid-eighties by Koblitz [Kob87] and Miller [Mil87], their efficiency and security have been the focus of intense study. In recent years, they have become widely accepted as an alternative to cryptosystems based on factorisation or discrete logarithms in finite fields, especially for constrained environments.

One of the initial steps in protocols based on elliptic curve cryptography is to generate a suitable curve defined over a finite field. To ensure that the system is secure, the curve must be chosen to have a number of points which is divisible by a large prime so that computing discrete logarithms on the curve is intractable using known attacks. Hence it is necessary to know the cardinality of the curve.

Among the elliptic curves defined over a given finite field, there are some classes of curves with particular properties that are useful for counting points or for accelerating arithmetic operations occurring in the protocols. However choosing such curves can be dangerous.

Perhaps the most striking example is trace 1 curves. The number of points over $\mathbb{F}_q$ is simply q. However Smart [Sma99], Satoh-Araki [SA98] and Semaev [Sem98] independently discovered a polynomial-time attack.

Another attack due to Menezes-Okamoto-Vanstone [MOV91], and generalised by Frey-Rück [FR94], reduces discrete logs on supersingular and trace

B. Pfitzmann (Ed.): EUROCRYPT 2001, LNCS 2045, pp. 14–29, 2001.

2 curves to discrete logs in a small-degree extension of $\mathbb{F}_q$. This yields an algorithm that runs in sub-exponential time.

A minor weakness is known for curves with many automorphisms [vOW99], [GLV], [DGM99] including curves defined over a small subfield, proposed by Koblitz, and some complex-multiplication curves. Attacks on these curves take less time than for generic curves, but remain in exponential time.

It has recently been shown by Gaudry-Hess-Smart [GHS00] that curves defined over composite extension fields are also weak in certain cases, using a reduction via hyperelliptic curves.

These results suggest that for maximum security one should avoid curves with special properties and instead choose a random curve whose number of points is divisible by a large prime, over a prime field or an extension of prime degree. This ideal procedure was made possible in practice by the SEA algorithm due to Schoof [Sch85], [Sch95], Elkies [Elk98], Atkin [Atk92] and others [Cou94] [Cou96], [Mor95], [Ler97a], [Mül95], [Dew98], etc. With this method, counting points on one given curve is reasonably fast.

However finding a cryptographically suitable curve requires testing many curves and this takes much more time. For instance, Johnson and Menezes [JM99] recently described this process as a "complicated and cumbersome task" requiring "a few hours on a workstation" for 200 bits.

Recently, a new algorithm for counting points on curves in small characteristic $p \geq 5$ was designed by Satoh [Sat00] and we extended it to characteristics two and three in [FGH00]. An independent extension to characteristic two is described by Skjernaa [Skj].

Satoh's algorithm is asymptotically superior to SEA for fixed p, requiring $O(\log^{3+\varepsilon} q)$ deterministic time, instead of $O(\log^{4+\varepsilon} q)$ under reasonable hypotheses. As demonstrated in [FGH00], the Satoh-FGH algorithm is much faster in practice in characteristic two. Indeed we were able to count points over much larger fields (up to 8009 bits) than had previously been possible, and could match the largest size reached with SEA (i.e. 1999 bits) in just three hours.

In the following we will describe a method for generating cryptographically suitable curves, over fields of 113 to 571 bits, using an implementation of the Satoh-FGH algorithm combined with an efficient early-abort strategy based on ideas from SEA. In this manner we reduce substantially the time required for curve-generation, finding suitable 200-bit curves in minutes rather than hours on a workstation, for instance.

In section 2, we recall some basic facts about elliptic curves defined over finite fields of characteristic two. Next we review some algorithms that can be used to compute the cardinality of a curve, and in particular we give a description of the Satoh-FGH algorithm. Section 4 gives the conditions that a curve must satisfy in order to be suitable for cryptographic applications. It also describes the early-abort strategy first used by Lercier in [Ler97a] for selecting good curves. Last but not least we describe our implementation and the results we obtained by combining a more aggressive early-abort strategy and the Satoh-FGH algorithm.

2   Elliptic Curves over Finite Fields of Characteristic Two

In this section, we recall some basic facts about elliptic curves defined over $\mathbb{F}_q$ where $q = 2^d$. We will only be concerned with characteristic two. For more informations on elliptic curves, the reader can refer to [Men93], [Sil86], [BSS99].

For our purposes, we can choose the equation of an elliptic curve E (with non-zero j-invariant) to be:

$$E : \ y^2 + xy = x^3 + a_6 \qquad \text{where } a_6 \in \mathbb{F}_q^*.$$

Its *twist* curve is:

$$E^* : \ y^2 + xy = x^3 + a_2 x^2 + a_6$$

where a_2 is some fixed element of trace 1.

An important invariant of the curve is its j-invariant $j(E) = 1/a_6$. In the following we assume $j(E) \notin \mathbb{F}_4$ and in particular that curves are ordinary i.e., not supersingular.

The set of points $E(\mathbb{F}_q)$ of the curve is:

$$E(\mathbb{F}_q) = \{(x, y) \in \mathbb{F}_q^2 | \ (x, y) \text{ satisfies the equation of } E\} \cup \{\mathcal{O}_E\},$$

where $\mathcal{O}_E$ is the *point at infinity*.

The Frobenius automorphism F is the map $x \mapsto x^q$ on $\mathbb{F}_q$. It can be extended to an endomorphism of E:

$$F : \quad E \to E$$
$$(x, y) \mapsto (x^q, y^q)$$

Its characteristic equation is of the form:

$$F^2 - cF + q = 0.$$

One can show that the number of points on E is

$$N = q + 1 - c, \quad \text{with} \quad |c| \leq 2\sqrt{q}$$

where c is the trace of Frobenius on E. The bound on c is due to Hasse [Has33]. Note that $4 \mid N$ since the point $(\sqrt[4]{a_6}, \sqrt{a_6})$ on E has order four. The number of points on E^* is $N^* = q + 1 + c$ and one has $2 \parallel N^*$.

The *little* Frobenius automorphism σ is the map $x \mapsto x^2$. It can be extended to an isogeny from E to the conjugate curve $E^\sigma : y^2 + xy = x^3 + a_6^2$ as follows:

$$\sigma : \quad E \to E^\sigma$$
$$(x, y) \mapsto (x^2, y^2).$$

3   Counting the Number of Points

3.1   The Schoof-Elkies-Atkin Algorithm

The first polynomial-time algorithm for counting points on elliptic curves over finite fields was described by Schoof in [Sch85]. The basic idea is to find the trace of the curve modulo small primes ℓ by studying the action of F on the ℓ–torsion part of E. Restricting the characteristic equation of F to the ℓ–torsion results in

$$(X^{q^2}, Y^{q^2}) - [q](X, Y) = [c_\ell](X^q, Y^q)$$

for each point (X, Y), where $c_\ell \equiv c \mod \ell$. This equality can be tested, for each candidate $c_\ell \in [0 \ldots \ell - 1]$, by doing polynomial arithmetic modulo the ℓ–division polynomial. Now, it suffices to compute c_ℓ for many small primes ℓ and then to recover the exact result using the Chinese Remainder Theorem. The time required for point-counting over $\mathbb{F}_q$ with this algorithm is $O(\log^{5+\varepsilon} q)$ using asymptotically fast methods for arithmetic (or $O(\log^8 q)$ using naïve arithmetic). The degree of the ℓ–division polynomial is $O(\ell^2)$, which grows quickly and causes this algorithm to be slow in practice.

In large characteristic, Elkies [Elk98] and Atkin [Atk92] improved Schoof's method yielding the so-called SEA algorithm (see [Sch95]) with run-time reduced to $O(\log^{4+\varepsilon} q)$ (or $O(\log^6 q)$) under reasonable hypotheses. Their idea is to construct a factor of degree $O(\ell)$ of the division polynomial and work with it instead. Such a factor can be found by factoring the modular polynomial to find eigenspaces of the Frobenius endomorphism F restricted to $E[\ell]$.

Further work by Morain [Mor95] and others led to practical implementations of SEA for prime fields. Couveignes extended SEA to work in small characteristic using the formal group [Cou94] or the p-torsion [Cou96] and Lercier found an efficient method for characteristic two [Ler97a].

3.2   The Satoh-FGH Algorithm

Here we present our adaptation of Satoh's algorithm to the case of characteristic two. The reader can find more details, including for odd characteristic, in [Sat00] and [FGH00].

The principal idea of this new algorithm is to lift E to a curve $\mathcal{E}$ over a 2–adic ring $\mathbb{Z}_q$ and to compute the trace of the Frobenius on $\mathcal{E}$.

Canonical Lift of the Curve. Just as $\mathbb{F}_q$ is obtained from $\mathbb{F}_2$ by taking an algebraic extension modulo an irreducible polynomial $f(x)$, one can obtain $\mathbb{Z}_q$ from the 2–adic integers $\mathbb{Z}_2$ by taking an extension modulo a polynomial $g(x)$ which reduces modulo 2 to $f(x)$. Thus we have $\mathbb{Z}_q = \mathbb{Z}_2[x]/(g(x))$. We represent this situation with the following figure.

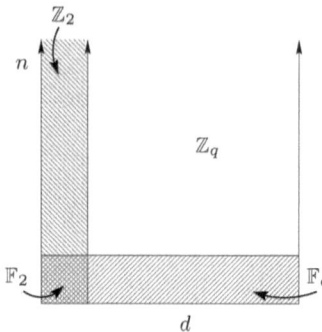

A Frobenius morphism $\mathcal{F}$ can also be defined on $\mathbb{Z}_q$. In this case it is not a simple q-th powering operation but something much more complicated. We do not define it explicitly since we will never have to compute it. Similarly, there exists a little Frobenius morphism Σ. For further details on $\mathbb{Z}_q$ and its Frobenius maps, see [Ser68].

A theorem of Lubin, Serre and Tate [LST64] guarantees the existence and uniqueness of a canonical lifted curve $\mathcal{E}$ over $\mathbb{Z}_q$ such that $\mathrm{End}(\mathcal{E}) = \mathrm{End}(E)$, via a canonical lift of the j-invariant. Indeed $J = j(\mathcal{E})$ is characterised by $J \equiv j(E)$ modulo 2 and $\Phi_2(J, \Sigma(J)) = 0$, where Φ_2 is the 2–modular polynomial.

A crucial part of Satoh's contribution is an efficient algorithm for lifting j-invariants. Instead of lifting $j(E)$ in isolation, he suggests lifting the whole cycle of conjugate j's simultaneously. He also proposes considering the duals $\hat{\Sigma}_i$ of the little Frobenius isogenies instead of Σ_i themselves. Indeed the duals are separable and hence are determined by their kernel. After having lifted the j-invariants using Satoh's method, we lift the coefficients of the curves and then compute the kernels by lifting a 2–torsion point on each conjugate curve, using the methods from [FGH00]. As a result, we compute the following diagram:

$$\mathcal{E}_0 \xrightarrow{\hat{\Sigma}_0} \mathcal{E}_1 \xrightarrow{\hat{\Sigma}_1} \cdots \xrightarrow{\hat{\Sigma}_{d-2}} \mathcal{E}_{d-1} \xrightarrow{\hat{\Sigma}_{d-1}} \mathcal{E}_0$$
$$\left\downarrow{\pi}\right. \qquad \left\downarrow{\pi}\right. \qquad\qquad\qquad \left\downarrow{\pi}\right.$$
$$E_0 \xrightarrow{\hat{\sigma}_0} E_1 \xrightarrow{\hat{\sigma}_1} \cdots \xrightarrow{\hat{\sigma}_{d-2}} E_{d-1} \xrightarrow{\hat{\sigma}_{d-1}} E_0$$

Here the top row is over $\mathbb{Z}_q$ to precision $O(2^{d/2+o(d)})$ and π is reduction modulo 2 down to $\mathbb{F}_q$.

Computing the Trace in $\mathbb{Z}_q$. Since traces are preserved by taking the dual and by canonical lifting, we have the equation:

$$\mathrm{Tr}(F) = \mathrm{Tr}(\hat{F}) = \mathrm{Tr}(\hat{\mathcal{F}}).$$

Moreover $\hat{\mathcal{F}}$ can be written as the composition

$$\hat{\mathcal{F}} = \hat{\Sigma}_{d-1} \circ \ldots \circ \hat{\Sigma}_1 \circ \hat{\Sigma}_0.$$

To find its trace we go to the formal groups of the curves. In formal groups, isogenies are represented by power series and composing isogenies is done by composing the power series. The first coefficient c_1 of the power series of $\hat{\mathcal{F}}$ is related to its trace as follows:

$$\operatorname{Tr} \hat{\mathcal{F}} = c_1 + \frac{q}{c_1}.$$

Therefore, computing the trace can be done by computing c_1, and the latter can be computed by composing all the power series of the $\hat{\Sigma}_i$. Only the first coefficients g_i of the $\hat{\Sigma}_i$ have to be determined, and this can be done with Vélu's formulae [Vél71]. More precisely, g_i^2 is given by an explicit formula involving the lifted curves and 2–torsion. Taking one of the square roots of $\prod g_i^2$ produces the trace to sufficient precision for it to be recovered exactly using Hasse's bound.

3.3   Description of the Algorithm

In this section, we give a synthetic description of the algorithm. For a more detailed one, we refer the reader to [FGH00]. The general procedure is:

Procedure MAINALGORITHM
Input: An elliptic curve E defined over $\mathbb{F}_q$, with $j(E) \notin \mathbb{F}_4$.
Output: The trace of the curve.

1. Compute the cycle of d curves E_i and their j-invariants j_i.
2. Lift all the j_i's simultaneously, yielding J_i.
3. Lift each curve by lifting its a_6 coefficient.
4. Lift the kernel of each $\hat{\Sigma}_i$.
5. Compute the trace from the lifted data.

In this procedure, points 2, 3 and 4 concern the lifting of the cycle of curves and of the kernels. We will detail these first. An essential ingredient is *Newton's iteration* for improving the (2–adic) precision of a root of a function.

Procedure LIFTCURVESAND2TORSION
Input: A cycle of d conjugate curves, and their j-invariants.
Output: The canonical lift of this cycle over $\mathbb{Z}_q$.

1. Lift the j-invariants simultaneously using an adaptation of the Newton iteration to the multivariate case. The function to be considered acts on a $1 \times d$ vector: $\Theta(x_0, \ldots, x_{d-1}) = (\Phi_2(x_0, x_1), \Phi_2(x_1, x_2), \cdots, \Phi_2(x_{d-1}, x_0))$ and the initial approximation of the root is the vector $(j_0, j_1, \ldots, j_{d-1})$ modulo 2.
2. Lift each curve E_i by lifting its a_6 coefficient, yielding A_i, using a Newton iteration with the function $f(x) = 1 + J(x + 432x^2)$ and the initial approximation $-1/J_i$ modulo 16.
3. Lift the 2–torsion point in the kernel of each $\hat{\Sigma}_i$ yielding (X_i, Y_i) on $\mathcal{E}_i$, using a Newton iteration based on the function $f(x) = 8x^3 + x^2 + A_i$ with initial approximation $1/J_{i+1}$ modulo 4.

With these algorithms, one can perform the lifting efficiently. Once this is done, it remains to compute the trace of $\hat{\mathcal{F}}$. The equations in the following algorithm are derived from Vélu's formulae.

Procedure COMPUTETRACE
Input: A cycle of d curves, given by A_i, and 2–torsion abscissae X_i.
Output: The trace of $\hat{\mathcal{F}}$.

1. Compute the square of the first coefficient of the expansion of each $\hat{\Sigma}_i$ in the formal group of $\mathcal{E}_i$ using Vélu's formulae. The result is:

$$g_i^2 = \frac{1 - 252X_i + 19008A_i}{(1 + 120(X_i + 6X_i^2))\,(1 + 864A_{i+1})}.$$

2. Compute $c^2 = \prod g_i^2$.
3. Compute c by computing a square root of c^2 and by determining the sign using $c \equiv 1 \mod 4$.

4 Good Elliptic Curves in Cryptography

The security of elliptic curve cryptosystems depends on the difficulty of solving the elliptic curve discrete logarithm (ECDL) problem. As mentioned in the introduction, there are several attacks against curves with special properties such as the one against trace 1 curves, or the MOV reduction for supersingular curves, etc.

For random curves, the chance that one of these methods can apply is vanishingly small. However there are other attacks that work for generic abelian finite groups.

The first is Pohlig-Hellman reduction [PH78]. When the group order N has all its prime factors small, discrete logs can be computed quickly by working in small subgroups. Thus for good security it is essential to pick a group whose order is divisible by a large prime.

The other attacks are algorithms that run in time $O(\sqrt{N})$. They include Shanks' baby-step giant-step algorithm (see [Coh96]) and Pollard's ρ method [Pol78]. In practice, the most difficult ECDL that has been computed is on a Koblitz curve over $\mathbb{F}_{2^{109}}$ using a distributed version of Pollard–ρ [Har00].

By extrapolating the work required to larger sizes and allowing safety margins for future increases in computing power, it is generally believed (see [FIPS186], [LV00], [P1363], [Sil00]) that a random curve whose order is divisible by a prime of at least 160 bits will offer reasonable security, comparable to 80-bit symmetric systems or 1024-bit RSA. For applications with the highest security requirements, one may take larger safety margins.

To find a secure curve, Lercier [Ler97a] proposed an early-abort strategy to use when computing the cardinality of the curve using SEA. The idea is to test on the fly if $q + 1 - c \equiv 0 \mod \ell$. If the test is true, then we throw away the curve and try again with another one. Since SEA computes $c \mod \ell$, this test is easy to implement and costs no extra run-time. In large characteristic where

Satoh-FGH does not apply this is still the best known method and we refer to the existing literature on the subject [LM95], [IKNY98], [MP98].

A difficulty that arises when designing an early-abort strategy to use with the Satoh-FGH algorithm is that $c \bmod \ell$ is not available (except for ℓ a power of p). Our solution is to implement a simplified version of SEA to determine whether the curve has a rational point of ℓ-torsion or not for the first few primes ℓ, as a preliminary step before launching Satoh-FGH. There is a trade-off to be made between the extra cost of these calculations and the benefit to be gained by avoiding an entire cardinality computation. In practice we found this strategy to be very worthwhile and obtained run-times lower than those previously reported in the literature.

5   Implementation and Results

5.1   Implementation Details

We wrote optimised implementations of the early-abort strategy and the Satoh-FGH algorithm for characteristic two, in the C programming language. This implementation of the early-abort strategy is independant of Lercier's one. For multiplication in $\mathbb{F}_q$ we used Karatsuba's algorithm; in $\mathbb{Z}_q$ we used Toom's algorithm. To ensure that modular reduction took very little time, we chose the irreducible polynomial to be a trinomial or pentanomial. For division we used the binary Euclidean algorithm in $\mathbb{F}_q$, and inversion by Newton iterations in $\mathbb{Z}_q$.

Most of our timing tests were run on a 750 MHz EV6 Alpha. In order to compare results with [Ler97a], we also ran some tests on a 266 MHz EV4 Alpha identical to the one Lercier used. Note that the difference between these processors is more than what we could think by just comparing the clock speeds: for usual applications, the gain is by a factor of about 15. Finally we timed curve generation for one small field on a 275 MHz StrongARM chip.

In the early-abort part, as explained below, the most time consuming parts are lazy factorizations of small-degree polynomials over $\mathbb{F}_q$. The most frequent operation is multiplication in $\mathbb{F}_q$. We give relevant timings obtained on the 750 MHz Alpha in Table 1.

Table 1. Cost of a multiplication in $\mathbb{F}_q$ on a 750 MHz EV6 Alpha.

Field size	163 bits	193 bits	239 bits	409 bits	571 bits
Cost of a multiplication in $\mathbb{F}_q$	0.488 μs	0.639 μs	0.917 μs	2.632 μs	4.685 μs

The most frequent operation in the point-counting part is multiplication in $\mathbb{Z}_q$. In Table 2, we give the time for one such operation at the highest 2–adic precision required i.e., $\lceil d/2 \rceil + 3$ bits, for various field sizes d. These measurements were also done on the 750 MHz Alpha.

Table 2. Cost of a multiplication in $\mathbb{Z}_q$ on a 750 MHz EV6 Alpha.

Base field size	163 bits	193 bits	239 bits	409 bits	571 bits
Maximal precision	85	100	123	208	289
Cost of a multiplication in $\mathbb{Z}_q$	0.19 ms	0.24 ms	0.36 ms	4.6 ms	8.0 ms

Table 3. Times for point-counting on a 266 MHz EV4 Alpha

Field size	SEA (timings from [Ler97b])			Satoh-FGH	Ratio
	Min	Max	Avg		
155 bits	58.8 s	132 s	86.5 s	36.3 s	2.4
196 bits	212 s	1029 s	308 s	68.8 s	4.5
300 bits	1519 s	3686 s	2434 s	408.4 s	6

5.2   Counting the Number of Points on One Curve

When computing the cardinality of a curve, one has to decide whether to use
SEA or Satoh. Two cases have to be dealt with differently: the case of large
characteristic and the case of small characteristic.

The complexity of Satoh's algorithm has a bad dependency in the characteristic p of the base field and when p is large, it is not efficient at all. This is due
to the use of the modular equation Φ_p for the lifting of the curves. This equation
has $O(p^2)$ coefficients that have to be known at least modulo $p^{(d/2)+O(1)}$. Hence
a complexity which is exponential in p appears to be unavoidable. On the other
hand, the SEA algorithm is polynomial-time independently of p. For instance,
Morain succeeded in counting the number of points of a curve over a field of size
$10^{499} + 153$ [Mor95].

However in small characteristic Satoh's algorithm is efficient. In particular in
characteristic two, Satoh-FGH is clearly faster than SEA in practice. To illustrate
the difference in speed between the two algorithms, we compare Lercier's results
[Ler97b] with the timings we get over the same fields, using an identical 266 MHz
Alpha. The results are given in Table 3. We do not give minimal or maximal
times for Satoh-FGH since the runtime of this algorithm is essentially constant
when treating different curves over the same field. These results show that the
bigger the field the greater the advantage for Satoh-FGH, as expected from the
asymptotics.

We give timings for point-counting on the 750 MHz Alpha in Table 4. Most
of the field sizes that we chose are recommended in cryptographic standards
(ANSI X9.63, IEEE P1363, IPSec, NIST, WAP).

Remark: In some cases, the SEA and Satoh-FGH algorithms can be combined
to speed-up point-counting. This works particularly well when the field size is
such that the maximum precision required in Satoh-FGH is a little more than
a multiple of the machine word-size. A good example is $q = 2^{251}$: the maximum
precision in the lifting calculations is $\lceil \frac{251}{2} \rceil + 3 = 129$ bits. In this case, computing

Table 4. Times for point-counting on a 750 MHz EV6 Alpha

Field size	Satoh-FGH	Field size	Satoh-FGH	Field size	Satoh-FGH
157 bits	2.39 s	197 bits	4.45 s	283 bits	26.5 s
163 bits	2.76 s	233 bits	6.57 s	409 bits	76.3 s
193 bits	4.10 s	239 bits	6.94 s	571 bits	257 s

the trace modulo 3 with the SEA algorithm allows the precision to be reduced to 128 bits which fits perfectly in a whole number of words. This approach could certainly be pushed further, although implementation complexity would appear to outweigh the moderate gain in speed.

5.3   Finding a Good Curve

The naïve strategy to find a curve suitable for cryptographic use is to count the number of points for many curves, until one with almost prime order is found. As mentioned before, if the SEA algorithm is used then many bad curves can be detected early; this nice property does not hold for the Satoh-FGH algorithm.

Hence, for small to medium sizes, the naïve strategy using Satoh-FGH is not better than the early-abort strategy with SEA. For instance over $\mathbb{F}_{2^{155}}$, Lercier [Ler97b] was able to select the good curves among a set of 1000 random ones in 14112 seconds. On the same computer, the Satoh-FGH method takes 36.5 seconds per curve, so that selecting the good ones would take 36500 seconds with the naive strategy, and would be worse by a factor 2.5. (For larger sizes, this phenomenon vanishes and Satoh-FGH is always better.)

To counter this, we take advantage of both methods: we first eliminate many candidate curves by an early-abort strategy based on SEA's techniques, and then run Satoh-FGH on the remaining ones.

Let E be a curve over $\mathbb{F}_q$. For a small prime ℓ, E is called ℓ-good if its order is coprime to ℓ, and ℓ-bad otherwise. Early-abort works as follows for each ℓ:

1. Compute the number of roots of $\Phi_\ell(X, j(E))$. It can be 0, 1, 2 or $\ell + 1$. (The cases 1 or $\ell + 1$ cannot occur unless q is a square modulo ℓ.)
2. If there are no roots, E is ℓ-good.
3. Otherwise, for each root of Φ_ℓ, build the corresponding factor of the ℓ-division polynomial and search for a root x of the factor. If there is such an x in $\mathbb{F}_q$ and a corresponding y too, then (x, y) is an ℓ-torsion point over $\mathbb{F}_q$ and E is ℓ-bad.
4. Otherwise E is ℓ-good.

The major cost in step 1 is that of computing X^q modulo $\Phi_\ell(X, j(E))$, which has degree $\ell + 1$. To accelerate the calculation, we replace Φ_ℓ by the *canonical* modular polynomial Φ_ℓ^c, which has the same degree but is sparser and involves lower powers of j. We refer to [Mor95] for the construction and the properties of these equations.

Table 5. Average runtime for checking if E is ℓ-good (EV6 – 750 MHz)

ℓ	$q = 2^{163}$		$q = 2^{239}$	
	Root finding of $\Phi_\ell^c(X, j)$	Average total time	Root finding of $\Phi_\ell^c(X, j)$	Average total time
3	0.17 ms	0.17 ms	0.28 ms	0.28 ms
5	0.34 ms	0.38 ms	0.61 ms	0.68 ms
7	0.34 ms	1.18 ms	0.56 ms	2.18 ms
11	4.42 ms	6.93 ms	9.14 ms	14.1 ms
13	1.07 ms	4.19 ms	1.94 ms	8.36 ms
17	3.71 ms	8.63 ms	7.34 ms	17.9 ms
19	4.97 ms	11.6 ms	10.1 ms	23.7 ms

Heuristically, in half of the cases there will be no root (in such a case ℓ is called an Atkin prime) and we are done. Otherwise, we have to continue to step 3. The factor of the division polynomial corresponding to a root of the modular polynomial is calculated using a system of formulae due to Lercier [Ler97a]. For small ℓ the solution to this system can be written explicitly, and the factor is obtained at almost no cost. (For larger ℓ the system could be solved efficiently by an algorithm also due to Lercier.) The cost of searching for a root is dominated by the computation of X^q modulo the factor, which has degree $(\ell - 1)/2$.

In Table 5 we give the run-time for this procedure, measured on the 750 MHz Alpha.

It is necessary to bound the maximum size of ℓ in order to balance the cost of early-abort against the gain obtained by avoiding point-counting. In theory, it would be beneficial to increase ℓ until the above early-abort procedure took approximately one ℓ-th of the time required for point-counting. Hence the maximum size of ℓ would grow with the field size.

However almost all of the advantage to be gained comes from using the first few primes and in practice we found $\ell \leq 19$ to be a good trade-off. For these primes it is not difficult to determine if curves are ℓ-good: Lercier's construction of isogenies is relatively easy, as in the search for ℓ-torsion points. Thus we were able to keep our code simple and reliable.

For comparison with Lercier's results reported in [Ler97b], we ran some further tests on the 266 MHz Alpha. We chose a similar early-abort strategy, searching for good curves with order $4p$ without considering the twist curves at all (but see below). The results can be found in Table 6. As a first step in the early-abort, we determine whether the order is divisible by 8. This can be decided very quickly by computing Tr a_6. Note that we measured our timings for 157 and 197 bits instead of 155 and 196 because composite extension fields may be weak in certain cases, as mentioned in the introduction.

Next, in order to maximise the performance of curve generation we decided to search simultaneously for twist curves with order $2p$ and this allowed us roughly to double the speed. As is clear from section 2, the cardinality of the twist can be found immediately from that of the curve itself. Furthermore, the early-abort

Table 6. Time to select good curves among 1000 (EV4 – 266 MHz)

Field size	SEA (from [Ler97b])	Satoh-FGH + early-abort
155 bits	14112 s	4490 s
196 bits	30254 s	7850 s

Table 7. Average time to find a good curve (EV6 – 750 MHz)

Field size (in bits)	Time for e.-a. on 10000 curves	Remaining curves	Time to count remaining curves	Good curves	Average time to find a good curve
157	21.1 s	435	17.3 min	45	23.6 s
163	23.1 s	473	21.7 min	55	24.1 s
193	25.1 s	402	27.5 min	33	50.7 s
197	30.8 s	415	30.8 min	43	43.6 s
233	40.3 s	402	44 min	29	92,4 s
239	43.5 s	435	50.3 min	29	105.6 s
283	122 s	418	3h 4 min	20	9.2 min
409	245 s	467	9h 54 min	22	27 min
571	524 s	375	26h 40 min	11	146 min

strategy can easily be adapted to take the twist into account since it has the same j-invariant and the same division polynomials. (This is because the curve and its twist are isomorphic over an algebraic closure and the isomorphism preserves the abscissae.)

One possibility would be to reject a pair consisting of a curve and its twist only when the early-abort strategy determines that both curves are cryptographically unsuitable. Alternatively one may pursue a more aggressive strategy by rejecting them both as soon as either one is found to be unsuitable, and immediately moving on to a new pair. Using the latter method for 10000 random curve pairs on the 750 MHz Alpha, we measured the timing results shown in Table 7.

Although the $O(d^3)$ space complexity of Satoh's algorithm grows quickly, the tricks described in [FGH00] keep the constant factor small. With these tricks, the largest key size we dealt with (571 bits) requires under 10 megabytes and for moderate key sizes the memory usage was only a few hundred kilobytes. We chose a different trade-off, using more memory in exchange for slightly higher speed.

To investigate the possibility of generating curves in constrained environments, we ran some tests at 113 bits on an ARM chip. This small key size is recommended for key-exchange in the Wireless Application Forum's WTLS standard (WAP) and can be used for short-term security at a level comparable to DES. The results can be seen in Table 8.

Table 8. Time to find a good WAP curve on an ARM chip

Field size	Frequency	Time to count one curve	Average time to find a good curve	RAM + ROM used
113 bits	275 MHz	5.9 s	38 s	240 KB + 136 KB

Table 9. New times for point-counting on a 750 MHz EV6 Alpha

Field size	Time	Field size	Time	Field size	Time
157 bits	0.50 s	197 bits	0.91 s	283 bits	6.32 s
163 bits	0.56 s	233 bits	1.39 s	409 bits	19.4 s
193 bits	0.84 s	239 bits	1.47 s	571 bits	58.2 s

6   Conclusion

The Satoh-FGH algorithm has proven to be the method of choice whenever one wants to compute the cardinality of a random elliptic curve defined over a finite field of characteristic two. But in spite of Satoh-FGH's excellent performance (see Table 4), the SEA algorithm should not be abandoned too quickly. In the case of large characteristic it is the only practical method available. Moreover the early-abort strategy, which is closely related to it, is valuable when looking for a curve for cryptographic use, even in small characteristic. By combining this technique and the Satoh-FGH algorithm, we obtain an efficient way of computing secure curves (see Table 7). We conclude that it is no longer necessary to use precomputed curves in cryptography since one can easily compute new curves as desired. Finding a curve with a security level comparable with RSA-1024 takes minutes or less. Curve generation for short-term security, with a level equivalent to DES, is feasible on a low-power chip. Finally, very high security levels similar to the highest AES level are now possible albeit in several hours.

Remark

We have recently implemented a new and quite different point-counting algorithm with lower memory requirements and a gain in speed by a factor ranging from 4 to 5 depending on key-size. For instance a secure 113-bit curve can be found in 8 seconds using 36 KB of RAM on the 275 MHz StrongARM. Repeating the calculations from Tables 4 and 7 gave the times in Table 9 and Table 10.

Acknowledgements

We would like to thank François Morain for his continuous support and many invaluable suggestions during this work.

We are also grateful to Rajit Manohar from Cornell Computer Systems Laboratory. He provided the computer resources needed for many of our calculations.

Table 10. New times to find good curves (EV6 – 750 MHz)

Field size (in bits)	Average time to find a good curve
157	5 s
163	5 s
193	10 s
197	10 s
233	21 s
239	22 s
283	138 s
409	7 min
571	34 min

References

FIPS186.   FIPS 186-2. Digital Signature Standard. Federal Information Processing Standards publication, january 2000. U.S. Departement of Commerce/National Institute of Standards and Technology. Available at `http://csrc.nist.gov/cryptval/dss.htm`.

P1363.   IEEE P1363. Standard specifications for public key cryptography. Available at `http://www.manta.ieee.org/groups/1363/`.

Atk92.   A. O. L. Atkin. The number of points on an elliptic curve modulo a prime. Series of e-mails to the NMBRTHRY mailing list, 1992.

BSS99.   I. Blake, G. Seroussi, and N. Smart. *Elliptic curves in cryptography*, volume 265 of *London Math. Soc. Lecture Note Ser.* Cambridge University Press, 1999.

Coh96.   H. Cohen. *A course in algorithmic algebraic number theory*, volume 138 of *Graduate Texts in Mathematics*. Springer–Verlag, 1996. Third printing.

Cou94.   J.-M. Couveignes. *Quelques calculs en théorie des nombres*. Thèse, Université de Bordeaux I, July 1994.

Cou96.   J.-M. Couveignes. Computing ℓ-isogenies using the p-torsion. In H. Cohen, editor, *Algorithmic Number Theory*, volume 1122 of *Lecture Notes in Comput. Sci.*, pages 59–65. Springer Verlag, 1996. Second International Symposium, ANTS II, Talence, France, May 1996, Proceedings.

Dew98.   L. Dewaghe. Remarks on the Schoof-Elkies-Atkin algorithm. *Math. Comp.*, 67(223):1247–1252, July 1998.

DGM99.   I. Duursma, P. Gaudry, and F. Morain. Speeding up the discrete log computation on curves with automorphisms. In Kwok Yan Lam, Eiji Okamoto, and Chaoping Xing, editors, *Advances in Cryptology – ASIACRYPT '99*, volume 1716 of *Lecture Notes in Comput. Sci.*, pages 103–121. Springer-Verlag, 1999. International Conference on the Theory and Applications of Cryptology and Information Security, Singapore, November 1999, Proceedings.

Elk98.   N. Elkies. Elliptic and modular curves over finite fields and related computational issues. In D.A. Buell and eds. J.T. Teitelbaum, editors, *Computational Perspectives on Number Theory*, pages 21–76. AMS/International Press, 1998. Proceedings of a Conference in Honor of A.O.L. Atkin.

FGH00.   M. Fouquet, P. Gaudry, and R. Harley. An extension of Satoh's algorithm and its implementation. *J. Ramanujan Math. Soc.*, 15:281–318, 2000.

FR94.   G. Frey and H.-G. Rück. A remark concerning m-divisibility and the discrete logarithm in the divisor class group of curves. *Math. Comp.*, 62(206):865–874, April 1994.

GHS00.   P. Gaudry, F. Hess, and N. Smart. Constructive and destructive facets of Weil descent on elliptic curves. Submitted to J. Crypt. and available at `http://www.cs.bris.ac.uk/~nigel/weil_descent.html`, 2000.

GLV.   R. Gallant, R. Lambert, and S. Vanstone. Improving the parallelized Pollard lambda search on binary anomalous curves. To appear in *Math. Comp.*

Har00.   R. Harley. `http://cristal.inria.fr/~harley/ecdl7/q`, 2000.

Has33.   H. Hasse. Beweis des Analogons der Riemannschen Vermutung für die Artinschen und F. K. Smidtschen Kongruenzzetafunktionen in gewissen elliptischen Fällen. *Ges. d. Wiss. Narichten. Math.-Phys. Klasse*, pages 253–262, 1933.

IKNY98.   T. Izu, J. Kogure, M. Noro, and K. Yokoyama. Efficient implementation of Schoof's algorithm. In K. Ohta and D. Pei, editors, *Advances in Cryptology – ASIACRYPT '98*, volume 1514 of *Lecture Notes in Comput. Sci.*, pages 66–79. Springer-Verlag, 1998. International Conference on the theory and application of cryptology and information security, Beijing, China, October 1998.

JM99.   D. Johnson and A. Menezes. The elliptic curve digital signature algorithm (ECDSA). Technical Report CORR 99-34, U. Waterloo, 1999. Available at `http://www.cacr.math.uwaterloo.ca/`.

Kob87.   N. Koblitz. Elliptic curve cryptosystems. *Math. Comp.*, 48(177):203–209, January 1987.

Ler97a.   R. Lercier. *Algorithmique des courbes elliptiques dans les corps finis*. Thèse, École polytechnique, June 1997.

Ler97b.   R. Lercier. Finding good random elliptic curves for cryptosystems defined over $\mathbb{F}_{2^n}$. In W. Fumy, editor, *Advances in Cryptology – EUROCRYPT '97*, volume 1233 of *Lecture Notes in Comput. Sci.*, pages 379–392. Springer-Verlag, 1997. International Conference on the Theory and Application of Cryptographic Techniques, Konstanz, Germany, May 1997, Proceedings.

LM95.   R. Lercier and F. Morain. Counting the number of points on elliptic curves over finite fields: strategies and performances. In L. C. Guillou and J.-J. Quisquater, editors, *Advances in Cryptology – EUROCRYPT '95*, volume 921 of *Lecture Notes in Comput. Sci.*, pages 79–94, 1995. International Conference on the Theory and Application of Cryptographic Techniques, Saint-Malo, France, May 1995, Proceedings.

LST64.   J. Lubin, J. P. Serre, and J. Tate. Elliptic curves and formal groups. In *Lecture notes prepared in connection with the seminars held at the Summer Institute on Algebraic Geometry, Whitney Estate, Woods Hole, Massachusetts, July 6-July 31, 1964*, 1964. Scanned copies available at `http://www.ma.utexas.edu/users/voloch/lst.html`.

LV00.   A. Lenstra and E. Verheul. Selecting cryptographic key sizes, January 2000. Presented at PKC2000.

Men93.   A. J. Menezes. *Elliptic curve public key cryptosystems*. Kluwer Academic Publishers, 1993.

Mil87.   V. Miller. Use of elliptic curves in cryptography. In A. M. Odlyzko, editor, *Advances in Cryptology – CRYPTO '86*, volume 263 of *Lecture Notes in*

	Comput. Sci., pages 417–426. Springer-Verlag, 1987. Proceedings, Santa Barbara (USA), August 11–15, 1986.
Mor95.	F. Morain. Calcul du nombre de points sur une courbe elliptique dans un corps fini : aspects algorithmiques. *J. Théor. Nombres Bordeaux*, 7:255–282, 1995.
MOV91.	A. Menezes, T. Okamoto, and S. A. Vanstone. Reducing elliptic curves logarithms to logarithms in a finite field. In *Proceedings 23rd Annual ACM Symposium on Theory of Computing (STOC)*, pages 80–89. ACM Press, 1991. May 6–8, New Orleans, Louisiana.
MP98.	V. Müller and S. Paulus. On the generation of cryptographically strong elliptic curves. Preprint, 1998.
Mül95.	V. Müller. *Ein Algorithmus zur Bestimmung der Punktanzahl elliptischer Kurven über endlichen Körpern der Charakteristik größer drei*. PhD thesis, University of Saarland, 1995.
PH78.	S. Pohlig and M. Hellman. An improved algorithm for computing logarithms over $GF(p)$ and its cryptographic significance. *IEEE Trans. Inform. Theory*, IT–24:106–110, 1978.
Pol78.	J. M. Pollard. Monte Carlo methods for index computation mod p. *Math. Comp.*, 32(143):918–924, July 1978.
SA98.	T. Satoh and K. Araki. Fermat quotients and the polynomial time discrete log algorithm for anomalous elliptic curves. *Comment. Math. Univ. St. Paul.*, 47:81–92, 1998.
Sat00.	T. Satoh. The canonical lift of an ordinary elliptic curve over a finite field and its point counting. *J. Ramanujan Math. Soc.*, 15:247–270, 2000.
Sch85.	R. Schoof. Elliptic curves over finite fields and the computation of square roots mod p. *Math. Comp.*, 44:483–494, 1985.
Sch95.	R. Schoof. Counting points on elliptic curves over finite fields. *J. Théor. Nombres Bordeaux*, 7:219–254, 1995.
Sem98.	I. A. Semaev. Evaluation of discrete logarithms in a group of p-torsion points of an elliptic curves in characteristic p. *Math. Comp.*, 67(221):353–356, January 1998.
Ser68.	J. P. Serre. *Corps locaux*. Hermann, 1968.
Sil86.	J. H. Silverman. *The arithmetic of elliptic curves*, volume 106 of *Graduate Texts in Mathematics*. Springer–Verlag, 1986.
Sil00.	R. Silverman. A cost-based security analysis of symmetric and assymetric key lengths. Bulletin Number 13 of RSA Security, April 2000.
Skj.	B. Skjernaa. Satoh's algorithm in characteristic 2. Copies available at `http://www.imf.au.dk/~skjernaa/`.
Sma99.	N. Smart. The discrete logarithm problem on elliptic curves of trace one. *J. Cryptology*, 12:193–196, 1999.
Vél71.	J. Vélu. Isogénies entre courbes elliptiques. *C. R. Acad. Sci. Paris Sér. I Math.*, 273:238–241, July 1971. Série A.
vOW99.	P. C. van Oorschot and M. J. Wiener. Parallel collision search with cryptanalytic applications. *J. of Cryptology*, 12:1–28, 1999.

How Secure Are Elliptic Curves over Composite Extension Fields?

Nigel P. Smart

Dept. Computer Science,
University of Bristol,
Merchant Venturers Building,
Woodland Road,
Bristol, BS8 1UB
nigel@cs.bris.ac.uk

Abstract. We compare the method of Weil descent for solving the ECDLP, over extensions fields of composite degree in characteristic two, against the standard method of parallelised Pollard rho. We give details of a theoretical and practical comparison and then use this to analyse the difficulty of actually solving the ECDLP for curves of the size needed in practical cryptographic systems. We show that composite degree extensions of degree divisible by four should be avoided. We also examine the elliptic curves proposed in the Oakley key determination protocol and show that with current technology they remain secure.

1 Introduction

Ever since its invention, in 1986 by Koblitz [9] and Miller [12], elliptic curve cryptography (ECC) has attracted considerable interest since it enables improved security, in the sense of greater perceived strength per bit of key, compared to conventional systems such as RSA, with the added benefit of smaller key sizes, less bandwidth and less computing power, see [4] for a complete treatment of ECC. Various standards bodies, both government sponsored and industry led (for example NIST [2] and SECG [3]), have standardised on elliptic curves defined over fields of the form $\mathbb{F}_{2^p}$ and $\mathbb{F}_p$, where p denotes a prime.

Despite this standardisation effort various people still propose using curves defined over so called composite extension fields, i.e. fields of the form $\mathbb{F}_{q^n}$ where q is some non-trivial power of the characteristic and $n > 1$. Composite extension fields are chosen because they provide greater computational efficiency for what at first glance appears to be the same security. The improved efficiency is particularly pronounced in characteristic two, where one chooses $q = 2^l$ and $n = 4$ or 5, in these later cases the use of look up tables to represent the subfield of degree 4 or 5 over $\mathbb{F}_2$ can significantly improve the efficiency of the resulting cryptographic scheme.

However, recent work of Frey, Galbraith, Gaudry, Hess and Smart, see [5], [6] and [8], has cast doubt on the claim that composite extension fields offer about the same security as those fields defined in the standards. This recent work is

B. Pfitzmann (Ed.): EUROCRYPT 2001, LNCS 2045, pp. 30–39, 2001.

based on the technique of Weil descent. Even though the work on Weil descent is now well known in the community there still appears to be a reluctance to drop composite extension fields in certain quarters.

In this paper we investigate in detail the security of such systems and try to quantify by how much the techniques based on Weil restriction weaken the cryptographic system. We shall concentrate solely on the case of characteristic two, which is important in applications. In section 2 we shall review the method of Weil restriction from [8]. In section 3 we examine in more detail Gaudry's method and explain a very efficient implementation of it. In section 4 we compare Gaudry's method for the hyperelliptic curves arising from Weil restriction to the method of Pollard rho on the original elliptic curve. In section 5 we discuss the curve over $\mathbb{F}_{2^{155}}$ proposed in the Oakley key determination protocol. Finally we give some conclusions.

2   The Method of Weil Restriction

Let $k = \mathbb{F}_q$ denote a finite field of characteristic two and let $K = \mathbb{F}_{q^n}$ denote an extension of degree $n \geq 4$. Suppose we are given an elliptic curve, defined over K,

$$E : Y^2 + XY = X^3 + \alpha X^2 + \beta$$

which is suitable for use in cryptography, i.e. $E(K)$ contains a large cyclic subgroup of prime order $s \approx q^n/2$. In particular this means that E must be defined over K and not over some proper subfield, since otherwise the order of $E(K)$ would not be almost prime, unless n were prime and $q = 2$. The elliptic curve discrete logarithm problem (ECDLP) for such curves is the following: Given $P, Q \in E(K)$ such that

$$[s]P = [s]Q = \mathcal{O}$$

find $\lambda \in (\mathbb{Z}/s\mathbb{Z})^*$ such that

$$Q = [\lambda]P.$$

Now let H denote a (imaginary quadratic) hyperelliptic curve of genus g, defined over k,

$$II : Y^2 + h(X)Y = f(X)$$

where $\deg h(X) \leq g$ and $\deg f(X) - 2g + 1$. The Jacobian of H has about q^g elements and one can also consider a hyperelliptic curve discrete logarithm problem (HCDLP) for such curves. We let the degree-zero divisor D_1 generate some large cyclic subgroup of $\mathrm{Jac}_k(H)$ and let $D_2 \in \langle D_1 \rangle$. The HCDLP is to find the integer λ such that

$$D_2 = [\lambda]D_1.$$

Further details on the hyperelliptic group law and the HCDLP can be found in [10] and [4].

The main result from [8] is the following: From an ECDLP in $E(K)$, i.e. $P_2 = [\lambda]P_1$ with $\lambda \in (\mathbb{Z}/s\mathbb{Z})^*$, one can construct a hyperelliptic curve H of genus g over k and two divisors D_1 and D_2 of order s in $\mathrm{Jac}_k(H)$ such that

- $g = 2^{m-1}$ or $2^{m-1} - 1$ where $1 \leq m \leq n$.
- $D_2 = [\lambda]D_1$.

We note that the construction of [8] is very fast and that the genus g is almost always equal to 2^{n-1} for curves of cryptographic interest. There is a small probability that the construction does not actually work in practice, but for real life examples this can usually be ignored.

Why this result is interesting is that it maps the discrete logarithm problem from a group $E(K)$ where the only known solution has exponential complexity in the size of q^n, to a group $\mathrm{Jac}_k(H)$ where the best known solution has subexponential complexity, albeit in the size of

$$q^g = q^{2^{n-1}}.$$

However, for fixed genus there is an algorithm due to Gaudry which solves the HCDLP in time $O(q^{2+\epsilon})$, which is much better than the algorithm for the equivalent ECDLP which takes time $O(q^{n/2}(\log q)^2)$. In [8] it is argued that for small fixed n, and hence essentially fixed g, this provides evidence for the weakness of the ECDLP on curves defined over composite extension fields, at least asymptotically. However, the asymptotic complexity hides a very bad dependence on g, and hence such a conclusion may not be able to be substantiated on curves over field sizes of cryptographic interest. In [8] a single experiment was reported on, involving an elliptic curve over a field of the form $\mathbb{F}_{q^4}$ which gave rise to a hyperelliptic curve of genus four. This experiment was conducted for an elliptic curve which is not typical of elliptic curves over fields of the form $\mathbb{F}_{q^4}$. Curves defined over $\mathbb{F}_{q^4}$ would usually give rise to a hyperelliptic curve of genus eight. It is this latter problem that we aim to address here.

3   Analysing and Implementing Gaudry's Method

We refer to [7] for a detailed explanation of Gaudry's method for the HCDLP. Essentially one takes a factor base of all the degree one prime divisors on H up to the equivalence

$$D_1 \equiv D_2 \text{ if } D_1 = -D_2.$$

This gives approximately $q/2$ such divisors, but one selects by some appropriate means (see [8]) a proportion, say $1/l$, of them. Hence, the total factorbase size is roughly

$$F = q/(2l).$$

Then one collects relations amongst the factor base elements by performing a random walk. Once $F + 1$ relations have been found one can solve the HCDLP by using a linear algebra technique for finding elements of the kernel of a large sparse matrix over $\mathbb{F}_s$, such as Lanczos [17].

We define the following estimates of the bit-complexity of certain algorithms:

- c_q = Maximum cost of an arithmetic operation in $\mathbb{F}_q$. For fields of crypto-graphic interest this is given by

$$c_q = (\log q)^2$$

- $c_{q,g}$ = Maximum cost of an arithmetic operation on a polynomial of degree g over $\mathbb{F}_q$. For fields of cryptographic interest and polynomials of degree $g \leq 32$ the actual methods used have cost, using a Karatsuba style multiplication,

$$c_{q,g} = g^{1.59} c_q.$$

- c_J = Cost of a doubling/addition in the Jacobian of H. By work of [15] this is given by

$$c_J = 22 c_{q,g}.$$

- c_s = Maximum cost of operation in $\mathbb{Z}/s\mathbb{Z}$, for values of s of cryptographic interest namely $s \approx q^n$ we have

$$c_s = (n \log q)^2.$$

Arguing as in [7] one can see that Gaudry's algorithm then takes around

$$F l^g g! c_J$$

bit operations to compute the matrix and then

$$F^2 c_s g$$

bit operations to actually compute an element in the kernel. Here we have assumed, as is born out by experiment, that the operations in the Jacobian dominate the time needed to compute the matrix.

The idea of the parameter l is to balance the time for finding the matrix with the time for solving the matrix. Assuming we have X times more computing power available to perform the relation finding, this gives the equation

$$2 l^{g+1} g! c_J = c_s g q / X.$$

In theory one should choose $X = 1$ but in practice a given organisation probably has more spare idle time available on desk top computers than on a single big server like that needed to run the matrix step. When $X = 1$ this means we should choose our proportion of good divisors as

$$l \approx \left(\frac{n^2 q}{44 g! g^{0.59}} \right)^{1/(g+1)} = \ell.$$

But since we must have $l \geq 1$, we shall choose $l \approx \min(1, \ell)$. In particular this means that the overall complexity of the attack on the ECDLP based on Weil descent, is given by

$$C = \frac{(q n \ln(q))^2 g}{4} \left(\frac{n^2 q g^{-0.59}}{44 g!} \right)^{(-2/(g+1))}$$

$$= (q n \ln(q))^2 2^{n-3} \left(\frac{n^2 q \left(2^{n-1}\right)^{-0.59}}{(44 \cdot 2^{n-1})!} \right)^{\left(-2/(2^{n-1}+1)\right)},$$

since $g \leq 2^{n-1}$. Therefore, for fixed n we obtain a complexity of

$$O(q^{2-\epsilon}(\log q)^2),$$

where $\epsilon = 2/(2^{n-1} + 1)$. For the purposes of extrapolating run times later we shall take this as $O(q^2(\log q)^2)$. We implemented Gaudry's algorithm with the following optimisations

- The field arithmetic in $\mathbb{F}_q$ was implemented using very fast hand coded loops for the particular finite fields we where interested in, namely $q = 2^i$ with $i \leq 31$. This on its own provided nearly a 200% improvement in performance.
- The polynomial code was also optimised heavily for the case where the polynomials have degree less than twenty, using Karatsuba type techniques.
- The linear algebra step was run using the code used in the McCurley challenge [18]. We thank T. Denny and D. Weber for allowing us to use this code. This was run on a machine with 6 processors and 8GB of RAM running HP-UX.

4   Comparison with Pollard Rho

To have something concrete to compare the method of Weil descent to we implemented the parallel version of Pollard's rho method [16] for the ECDLP. We used the method of distinguished points due to Wiener and van Oorschot [13] which has been used in recent years to solve various challenge ECDLP examples set by Certicom.

Since we are using elliptic curves defined over fields of the form $\mathbb{F}_{q^n}$ where $n = 4$ or 5 we implemented very efficient techniques for these fields, using lookup tables for the subfields of degree 4 or 5 over $\mathbb{F}_2$. In table 1 we give the time needed to solve an elliptic curve discrete logarithm problem on various elliptic curves over $\mathbb{F}_{q^4}$. This was for an implementation on a network of 80 Sun Sparc-5 and Sparc-10s, for comparison we also give the time to run the program on a single Sparc-10.

Table 1. Pollard rho for $E(\mathbb{F}_{q^4})$

q	2^7	2^{11}	2^{13}	2^{17}	2^{19}	2^{21}
80 Sparcs	00:00	00:00	00:06	38:32	$\approx$11d	$\approx$621d
Single Sparc	00:00	00:11	04:50	$\approx$38d	$\approx$3y	$\approx$71y

Times are given either in the format *hrs:mins* rounded to the nearest minute, or in the format xd or xy to denote a certain number of days or years. A $\approx$ in the table denotes an approximate run time deduced from running the program for a reasonable length of time and then calculating the expected run time from this empirical data. One should note that since the rho method is heuristic in nature the running times represent an average for the small values of q.

In tables 2 and 3 we give the run times for Gaudry's algorithm using the same set of 80 Sun Sparc-5 and Sparc-10s to compute the matrix, we also give the estimate of the time needed for a single Sparc-10 to compute the matrix. We also give the time needed for the matrix step using a HP-UX machine which had 8 GBytes of RAM. These times should be compared to the time needed to solve the equivalent problem on the elliptic curve using Pollard rho.

Table 2. Numerical data for $n = 4$ and $g = 4$

q	2^7	2^{11}	2^{13}	2^{17}	2^{19}	2^{21}
$\min(1, \ell)$	1	1.6	2.2	3.8	5.1	6.78
l used	1	2	2	4	4	8
$F = \#FB$	65	513	2049	16428	65537	131283
Time for relation step						
80 Sparcs	00:00	00:00	00:01	00:55	05:15	68:00
Single Sparc	00:00	00:02	00:10	16:50	70:00	$\approx$115d
Time for matrix step	00:00	00:00	00:01	00:06	02:10	13:00

Table 3. Numerical data for $n = 4$ and $g = 8$

q	2^7	2^{11}	2^{13}	2^{17}	2^{19}
$\min(1, \ell)$	1	1	1	1	1.03
l used	1	1	1	1	1
$\#F$	64	1024	4096	65536	262144
Time for relation step					
80 Sparcs	00:05	01:20	05:45	43:45	$\approx$8d
Single Sparc	01:30	19:20	95:10	$\approx$62d	$\approx$250d
Time for matrix step	00:00	00:00	00:02	31:00	$\approx$20d

We first examine the case of $n = 4$ and $g = 4$, this case occurs for around $1/q$ of all elliptic curves defined over the field $\mathbb{F}_{q^4}$. As can be seen from the table the method of Weil descent provides a far more efficient way of attacking such elliptic curves than the standard method of Pollard rho for all values of q.

For the case $n = 4$ and $g = 8$, which is the most common case for elliptic curve systems over fields of the form $\mathbb{F}_{q^4}$, we see that the cross over point between Pollard rho and the method of Weil descent occurs at a value of q just over 2^{17}. This, therefore, provides the missing evidence from [8] that all curves over fields of composite extension degree divisible by four should be avoided in cryptographic applications.

Hence, we now have a complete experimental treatment of the case $n = 4$ in the method of Weil descent. The next case to consider is $n = 5$, which in fact turns out to be the most interesting in practical applications. In the next section we turn to this case.

5   The Oakley 'Well-Known Groups' 3 and 4

In [1] two elliptic curve groups are proposed for use in a key agreement protocol used as part of the IPSEC set of protocols. These groups, denoted 'Well-Known Group' 3 and 'Well-Known Group' 4, are defined as elliptic curves over fields of composite degree over $\mathbb{F}_2$. The first group is defined over the field $\mathbb{F}_{2^{155}}$, whilst the second is defined over the field $\mathbb{F}_{2^{185}}$. Since the extension degree of these fields over $\mathbb{F}_2$ are composite it is an open question as to whether these curves should still be used within the IPSEC family of protocols. In this section we shall concentrate solely on group 3.

Group 3 is defined by the equation

$$Y^2 + XY = X^3 + \beta$$

where

$$\beta = \omega^{18} + \omega^{17} + \omega^{16} + \omega^{13} + \omega^{12} + \omega^9 + \omega^8 + \omega^7 + \omega^3 + \omega^2 + \omega + 1,$$

where $\omega^{155} + \omega^{62} + 1 = 0$. This has group order

$$E(\mathbb{F}_{2^{155}}) = 12 \cdot 3805993847215893016155463826195386266397436443.$$

We carried out a number of experiments on elliptic curves over fields of the form $\mathbb{F}_{q^5}$. For the Pollard rho method, using the various optimisations available in such fields, we obtained the times in Table 4. Extrapolating our experimental

Table 4. Pollard rho for $E(\mathbb{F}_{q^5})$

q	2^7	2^{11}	2^{13}	2^{17}	2^{19}
80 Sparcs	00:00	00:06	06:30	$\approx$376d	$\approx$41y
Single Sparc	00:00	02:05	$\approx$20d	$\approx$58y	$\approx$4000y

results on the Pollard rho algorithm to 'Well Known Group' 3 it would appear that we would require

$$10^{11} \text{ years}$$

to solve the discrete logarithm problem using our network of 80 Sparc 5 and Sparc 10 computers, or

$$10^{13} \text{ years}$$

using a single Sparc-10. Hence, it is clearly currently infeasible to attack this curve using the Pollard rho algorithm.

We now turn out attention to whether it is feasible to attack 'Well Known Group' 3 using techniques based on Weil descent. Applying the method of [8] to this curve we obtain the hyperelliptic curve

$$H : y^2 + y \left(\begin{array}{c} 1258097243x^{16} + 1177011841x^8 + 540379308x^4 \\ +1555798523x^2 + 613019365x \end{array} \right)$$
$$+558654746x^{33} + 1390366357x^{32} + 577010024x^{28}$$
$$+1211700991x^{26} + 2017104043x^{25} + 1674361774x^{24}$$
$$+993950732x^{22} + 1777282797x^{21} + 1982857394x^{20}$$
$$+144558341x^{19} + 693983331x^{18} + 1937134056x^{16}$$
$$+1947274294x^8 + 31687647x^4 + 1217310851x^2 + 493932675x$$

defined over the field $\mathbb{F}_{2^{31}}$, where $w^{31} + w^3 + 1 = 0$ and the curve H has genus 16. In the above equation to convert the decimal coefficients to field elements one should first convert the decimal to binary and then use the binary representation to define the polynomial in w which gives the corresponding field element. For example

$$1258097243 \equiv w^{30} + w^{27} + w^{25} + w^{23} + w^{22} + w^{21} + w^{20} + w^{19}$$
$$+w^{18} + w^{16} + w^{11} + w^9 + w^6 + w^4 + w^3 + w + 1$$

In our experiments using curves of genus 16 we found that it would take over three years for the network of 80 workstations to compute a single relation for a curve over a field of size 2^7. Hence, it makes very little sense to extrapolate from actual run times for Gaudry's algorithm. However, we can give a rough estimate as to how long it would take to perform the two steps for the curve over $\mathbb{F}_{2^{155}}$ considered above.

Firstly we note that for such a curve we would take $l = 1$ and hence the factor base would have size,

$$F \approx 2^{30}.$$

This on its own would imply that the matrix step would require around

$$10^7 \text{ years}$$

to process using the code used to produce the examples in the last section. To produce the matrix we estimate would take the network of 80 Sparcs over

$$10^{10} \text{ years.}$$

Hence, although the method of Weil descent would appear to produce a more efficient way to attack systems based on 'Well Known Group 3', it would appear that such curves are secure. However, this assumes there is no further algorithmic improvements in either the method of Weil descent or the method of Gaudry for solving HCDLP.

6 Conclusion

The 'Well Known Groups' 3 and 4 in IPSEC may still be considered secure, however, they are made less secure by the method of Weil descent. This does not pose an immediate threat, but future algorithmic improvements could render them insecure. It should be noted that since both Weil descent and Gaudry's algorithm are comparatively recent advances one cannot rule out further algorithmic improvements in the coming years.

For large genus the method of Gaudry will only be asymptotically better than Pollard rho, as $q \to \infty$, this is due to the bad dependence of the complexity estimate on g. For values of g where g is significantly larger than n the current techniques of Weil descent produce a major problem, namely the ECDLP is in a group of order q^n, whilst using Weil descent we have mapped it into a subgroup (of order q^n) of a group of order

$$q^g = q^{2^{n-1}}.$$

Hence, we seem to have made our problem more difficult. It may be that the best algorithm for the HCDLP in this setting may be the ones which have asymptotic complexity

$$O\left(L_{q^g}(1/2, c)\right) = O\left(\exp((c + o(1))\sqrt{(\log q^g)(\log \log q^g)})\right)$$

as q is fixed and $g \to \infty$. However, there has been little work on practical implementations of these methods, the only one in the literature being described in [14]. The algorithm in [14] does not appear practical for the curve which arose above when we considered the Oakley group.

We end by stating that for curves over characteristic two fields of size 2^p, where p is prime, the method of Weil descent does not apply. In [8] it was proved that for over fifty percent of all cryptographically interesting curves over $\mathbb{F}_{2^p}$ the method of Weil descent would not apply. Recently, Menezes and Qu [11] showed that the method did not apply to any cryptographically interesting curves over $\mathbb{F}_{2^p}$.

References

1. IETF. The Oakley Key Determination Protocol. *IETF RFC 2412*, Nov 1998.
2. NIST. FIPS PUB 186-2 : DIGITAL SIGNATURE STANDARD (DSS). *National Institute for Standards and Technology*, 2000.
3. SECG. SEC 2: Recommended Elliptic Curve Domain Parameters. *Standards for Efficient Cryptography Group*, 1999.
4. I.F. Blake, G. Seroussi and N.P. Smart. *Elliptic Curves in Cryptography*. Cambridge University Press, 1999.
5. G. Frey. How to disguise an elliptic curve. Talk at Waterloo workshop on the ECDLP, 1998. http://cacr.math.uwaterloo.ca/conferences/1998/ecc98/slides.html
6. S.D. Galbraith and N.P. Smart. A cryptographic application of Weil descent. *Cryptography and Coding, 7th IMA Conference*, Springer-Verlag, LNCS 1746, 191–200, 1999. The full version of the paper is *HP Labs Technical Report, HPL-1999-70*.
7. P. Gaudry. An algorithm for solving the discrete logarithm problem on hyperelliptic curves. In *Advances in Cryptology - EUROCRYPT 2000*, Springer-Verlag LNCS 1807, 19–34, 2000.

8. P. Gaudry, F. Hess and N.P. Smart. Constructive and destructive facets of Weil descent on elliptic curves. To appear *Journal Cryptology*.
9. N. Koblitz. Elliptic curve cryptosystems. *Math. Comp.*, **48**, 203–209, 1987.
10. N. Koblitz. Hyperelliptic cryptosystems. *J. Crypto.*, **1**, 139–150, 1989.
11. A. Menezes and M. Qu. Analysis of the Weil Descent Attack of Gaudry, Hess and Smart. To appear *Proceedings RSA 2001*, 2001.
12. V. Miller. Use of elliptic curves in cryptography. In *Advances in Cryptology, CRYPTO - '85*, Springer LNCS 218, 47–426, 1986.
13. P.C. van Oorschot and M.J. Wiener. Parallel collision search with cryptanalytic applications. *J. Crypto.*, **12**, 1–28, 1999.
14. S. Paulus. An algorithm of sub-exponential type computing the class group of quadratic orders over principal ideal domains. In *ANTS-2: Algorithmic Number Theory*, Springer-Verlag, LNCS 1122, 243–257, 1996.
15. S. Paulus and A. Stein. Comparing real and imaginary arithmetics for divisor class groups of hyperelliptic curves. In *ANTS-3: Algorithmic Number Theory*, Springer-Verlag, LNCS 1423, 576–591, 1998.
16. J.M. Pollard. Monte Carlo methods for index computation (mod *p*). *Math. Comp.*, **32**, 918–924, 1978.
17. J. Teitelbaum. Euclid's algorithm and the Lanczos method over finite fields. *Math. Comp.*, **67**, 1665-1678, 1998.
18. D. Weber and T. Denny. The solution of McCurley's discrete log challenge. In *Advances in Cryptology - CRYPTO '98*, Springer-Verlag LNCS 1462, 458–471, 1998.

Efficient and Non-interactive Non-malleable Commitment

Giovanni Di Crescenzo[1], Jonathan Katz[2], Rafail Ostrovsky[1], and Adam Smith[3]

[1] Telcordia Technologies, Inc.
{giovanni,rafail}@research.telcordia.com
[2] Telcordia Technologies and
Department of Computer Science, Columbia University.
jkatz@cs.columbia.edu
[3] Laboratory for Computer Science, MIT.
Work done while the author was at Telcordia Technologies.
asmith@theory.lcs.mit.edu

Abstract. We present new constructions of non-malleable commitment schemes, in the public parameter model (where a trusted party makes parameters available to all parties), based on the discrete logarithm or RSA assumptions. The main features of our schemes are: they achieve *near-optimal communication* for arbitrarily-large messages and are *non-interactive*. Previous schemes either required (several rounds of) interaction or focused on achieving non-malleable commitment based on general assumptions and were thus efficient only when committing to a single bit. Although our main constructions are for the case of perfectly-hiding commitment, we also present a communication-efficient, non-interactive commitment scheme (based on general assumptions) that is perfectly binding.

1 Introduction

Commitment protocols are one of the most fundamental cryptographic primitives, used as sub-protocols in such applications as zero-knowledge proofs (see Goldreich, Micali, and Wigderson [17] and Goldreich [15]), secure multi-party computation (see Goldreich, Micali, and Wigderson [16]), contract signing (see Even, Goldreich, and Lempel [13]), and many others. Commitment protocols can also be used directly; for example, in remote (electronic) bidding. In this setting, parties bid by committing to a value; once bidding is complete, parties reveal their bids by de-committing. In many of these settings, it is required that participants, upon viewing the commitment of one party, be unable to generate a commitment to a related value. For example, in the bidding scenario it is unacceptable if one party can generate a valid commitment to $x + 1$ upon viewing a commitment to x. Note that the value of the original commitment may remain unknown (and thus secrecy need not be violated); in fact, the second party may only be able to decommit his bid after viewing a decommitment of the first. Unfortunately, most known commitment protocols are easily susceptible to these types of attacks.

B. Pfitzmann (Ed.): EUROCRYPT 2001, LNCS 2045, pp. 40–59, 2001.

Two types of commitment schemes have been considered in the literature: perfectly-binding [19] and perfectly-hiding [21] (following [15] we refer to the former as *standard* and the latter as *perfect*). In a standard commitment scheme, each commitment is information-theoretically bound to only one possible (legal) decommitment value; on the other hand, the secrecy of the commitment is guaranteed only with respect to a computationally-bounded receiver. In a perfect commitment scheme, the secrecy of the commitment is information-theoretic, while the binding property guarantees only that a computationally-bounded sender cannot find a commitment which can be opened in two possible ways. The type of commitment scheme to be used depends on the application [15]; it may also depend on assumptions regarding the computational power of the participants. For example, in many protocols certain commitments are never opened; information-theoretic privacy ensures that the committed data will remain hidden indefinitely (for further discussion, see [23,21]).

Commitment size is an important parameter, particularly when committing to a very large message such as the contents of a database. Unfortunately, *standard* commitment schemes (even malleable ones) require commitment size at least $M + \omega(\log k)$, where M is the message size and k is the security parameter. *Perfect* commitment schemes, on the other hand, offer the opportunity to achieve much shorter commitment lengths. Indeed, the non-malleable, perfect commitment schemes presented here achieve commitment size only $3k$ for arbitrarily-large messages.

Previous Work. Non-malleability was first explicitly considered by Dolev, Dwork, and Naor [11], who define the notion in a number of different settings. They also provide the first construction of a standard commitment scheme which is provably non-malleable. Although their protocol is constructed from the minimal assumption of a one-way function (in particular, without assuming a public random string), it requires a non-constant number of rounds of interaction[1]. Assuming a public random string available to all participants, Di Crescenzo, Ishai, and Ostrovsky [9] construct a *non-interactive*, non-malleable standard commitment scheme. Interestingly, their construction can be modified to give a non-interactive, non-malleable *perfect* commitment scheme. Unfortunately, the resulting commitments are large (i.e., $\mathcal{O}(Mk)$), thus motivating the search for more efficient protocols.

Constructions of non-malleable public-key encryption schemes have also been proposed [11,6,25]. In some cases, these constructions give non-malleable standard commitment schemes, in the model where public parameters are published by a trusted party. We discuss this connection in more detail in Section 3.

Two efficient non-malleable commitment schemes, based on stronger (but standard) assumptions, have also been proposed. Like the construction of [9], these protocols both require publicly-available parameters generated by a trusted party (in some cases this can be reduced to the assumption of a public random

---

[1] Furthermore, their protocol allows an adversary to generate a different commitment to an identical value (unless user identities are assumed). Other protocols discussed in this paper (including our own) do not suffer from this drawback.

string). The first can be obtained from an adaptive chosen-ciphertext secure public-key encryption scheme proposed by Cramer and Shoup [6], whose security is based on the decisional Diffie-Hellman problem. More recently, non-malleable perfect commitment schemes based on the discrete logarithm and RSA assumptions were introduced by Fischlin and Fischlin [14]. Though efficient, these protocols require interaction between the sender and receiver.

Our Contribution. We present the first efficient constructions of non-interactive, non-malleable perfect commitment schemes. We work in the same setting as other efficient non-malleable commitment schemes, where public parameters are available to all participants [6,14] (our discrete logarithm construction can be implemented in the public random string model using standard techniques). Our constructions are based on the discrete logarithm or the RSA assumptions. Previous constructions are either for the case of standard commitment [11,9,6] or require interaction [11,14]. Our constructions allow efficient, perfectly-hiding commitment to arbitrarily-large messages. The schemes described in [14], while able to handle large messages, require modifications which render them less efficient and also result in statistical secrecy only.

Additionally, we discuss the case of non-interactive, non-malleable, standard commitment schemes and prove secure a folklore construction based on trapdoor permutations which is near-optimal in terms of commitment size. The large commitment size of this construction (though near-optimal) serves as motivation for our consideration of perfect commitment schemes. Indeed, for arbitrarily-large messages, our perfect commitment schemes require commitments of size $3k$, where k is the size of RSA or discrete log problems believed to be hard to solve (see Section 5 for improvements which reduce the commitment size even further). Our schemes require only $\mathcal{O}(k)$ bits of public information.

2   Definitions

We discuss the communication models in which we present our constructions, and recall the notions of commitment schemes, equivocable commitment schemes, and finally non-malleable commitment schemes.

Communication models. We will consider two models: the *public-random-string* model of [4,3], and a slight generalization of it, considered for instance by [14] in the context of commitment schemes, which we call the *public-parameter* model.

The former model was introduced in order to construct non-interactive zero-knowledge proofs (i.e., zero-knowledge proofs which consist of a single message sent from a prover to a verifier). In this model, all parties share a public *reference string* which is assumed to be uniformly distributed. The latter model generalizes the public random string model in the following sense: all parties still share a public reference string which is now defined as the output of an efficient algorithm (and may therefore have arbitrary distribution).

For a unified treatment, we present our definitions for the public-parameter model, keeping in mind that analogous definitions may be obtained for the public random string model if the algorithm generating the public reference string is replaced by an algorithm which chooses a uniformly distributed string.

Commitment schemes. A *commitment scheme* $(\mathcal{TTP}, \mathcal{S}, \mathcal{R})$ in the public-parameter model is a two-phase protocol between two probabilistic polynomial time parties $\mathcal{S}$ and $\mathcal{R}$, called the sender and the receiver, respectively, such that the following is true. In the first phase (the commitment phase), given the public reference string σ returned by the probabilistic polynomial time algorithm $\mathcal{TTP}$, $\mathcal{S}$ commits to bit b by computing a pair of keys (com, dec) and sending com (the commitment key) to $\mathcal{R}$. Given just σ and the commitment key, the polynomial-time receiver $\mathcal{R}$ cannot guess the bit with probability significantly better than $1/2$ (this is the *hiding property*). In the second phase (the decommitment phase) $\mathcal{S}$ reveals the bit b and the key dec (the decommitment key) to $\mathcal{R}$. Now $\mathcal{R}$ checks whether the decommitment key is valid; if not, $\mathcal{R}$ outputs a special string $\perp$, meaning that he rejects the decommitment from $\mathcal{S}$; otherwise, $\mathcal{R}$ can efficiently compute the bit b revealed by $\mathcal{S}$ and is convinced that b was indeed chosen by $\mathcal{S}$ in the first phase (this is the *binding property*).

We remark that the commitment schemes considered in the literature can be divided in two types, according to whether the hiding property holds with respect to computationally bounded adversaries or to unbounded adversaries. Commitment schemes of the first (resp., second) type have been shown to have applications to zero-knowledge proofs (resp., arguments) [17,21]. A computationally-hiding bit-commitment scheme has been constructed under the minimal assumption of the existence of pseudo-random generators [19]. A perfectly-hiding bit-commitment scheme has been constructed under the assumption of the existence of one-way permutations [21]. Both schemes have been designed in the interactive model (where no public reference string is available to parties); the former, however, can be adapted to run in the public parameter model.

Equivocable commitment schemes. Informally, an equivocable commitment scheme in the public parameter model is one for which there exists an efficient algorithm, substituting for the trusted third party $(\mathcal{TTP})$, which outputs a set of public parameters and a commitment such that: (a) the distribution of the generated public parameters, the commitment, and any decommitment is exactly equivalent to their distribution in a real execution of the protocol; and (b) the commitment can be opened in more than one possible way.

Definition 1. *Let* $(\mathcal{TTP}, \mathcal{S}, \mathcal{R})$ *be a perfectly-hiding commitment scheme in the public parameter model over message space* $\mathcal{M}$. *We say that* $(\mathcal{TTP}, \mathcal{S}, \mathcal{R})$ *is perfectly equivocable if there exists a probabilistic, polynomial time equivocable commitment generator* Equiv *such that:*

1. $\mathsf{Equiv}_1(1^k)$ *outputs* (σ, com, s) *(where* s *represents state information).*
2. *For all* $m \in \mathcal{M}$, $\mathsf{Equiv}_2(s, m)$ *outputs* dec *such that:*
   (a) $\mathcal{R}(\sigma, com, dec) = m$.
   (b) *The following two random variables are identically distributed:*

$$\{\sigma \leftarrow \mathcal{TTP}(1^k); (com, dec) \leftarrow \mathcal{S}(\sigma, m) : (\sigma, com, dec)\}$$
$$\{(\sigma, com, s) \leftarrow \mathsf{Equiv}_1(1^k); dec \leftarrow \mathsf{Equiv}_2(s, m) : (\sigma, com, dec)\}. \qquad \square$$

The notion of equivocable commitment was first discussed by Beaver [1]. In [9] it was shown that an adaptation of the commitment scheme in [19] is equivocable in the public random string model (this fact was used in the construction of the non-malleable commitment scheme of [9]). Other applications of such schemes include zero-knowledge protocols [10].

Non-malleable commitment schemes. Two definitions of non-malleable commitment have appeared in the literature, both seeking to capture the following intuition of security: if an adversary, after viewing a commitment to x, can produce a commitment to a related value y, then a simulator can perform at least as well without viewing a commitment to x. The difference is in the definition of "producing a commitment". In the original definition [11] (*non-malleability with respect to commitment*), generating a valid commitment of y is sufficient. Note that this definition does not apply to perfectly-hiding commitment schemes since for such schemes the value committed to by a commitment is not well-defined. In the definition of [9] (*non-malleability with respect to opening*), the adversary must also be able to give a (valid) decommitment to y after viewing the decommitment to x. Since our primary constructions are of perfectly-hiding commitment schemes (for which non-malleability with respect to opening is the appropriate notion), we present a formal definition of this variant, and refer the reader elsewhere [11,14] for definitions of non-malleability with respect to commitment.

Definition 2. *Let* $(\mathcal{TTP}, \mathcal{S}, \mathcal{R})$ *be a perfectly-hiding commitment scheme, and let k be a security parameter. We say that* $(\mathcal{TTP}, \mathcal{S}, \mathcal{R})$ *is ϵ-non-*malleable *(following [11]) with respect to opening if, for all $\epsilon > 0$ and every probabilistic, polynomial time algorithm $\mathcal{A}$, there exists a simulator $\mathcal{A}'$ running in* $\mathrm{poly}(k, 1/\epsilon)$ *time, such that for all poly-time computable, valid relations R (see note below), for all efficiently sampleable distributions $\mathcal{D}$, we have:*

$$\mathrm{Succ}^{\mathrm{NM}}_{\mathcal{A},\mathcal{D},R}(k) - \widetilde{\mathrm{Succ}}_{\mathcal{A}',\mathcal{D},R}(k) \leq \epsilon + \mathrm{negl}(k)$$

(for some negligible function negl*); where:*

$$\mathrm{Succ}^{\mathrm{NM}}_{\mathcal{A},\mathcal{D},R}(k) \stackrel{\mathrm{def}}{=}$$
$$\Pr\left[\sigma \leftarrow \mathcal{TTP}(1^k); m_1 \leftarrow \mathcal{D}; (\mathrm{com}_1, \mathrm{dec}_1) \leftarrow \mathcal{S}(\sigma, m_1); \mathrm{com}_2 \leftarrow \mathcal{A}(\sigma, \mathrm{com}_1); \right.$$
$$\mathrm{dec}_2 \leftarrow \mathcal{A}(\sigma, \mathrm{com}_1, \mathrm{dec}_1); m_2 \leftarrow \mathcal{R}(\sigma, \mathrm{com}_2, \mathrm{dec}_2) :$$
$$\left. \mathrm{com}_1 \neq \mathrm{com}_2 \wedge R(m_1, m_2) = 1\right]$$

$$\widetilde{\mathrm{Succ}}_{\mathcal{A}',\mathcal{D},R}(k) \stackrel{\mathrm{def}}{=}$$
$$\Pr\left[m_1 \leftarrow \mathcal{D}; m_2 \leftarrow \mathcal{A}'(1^k, \mathcal{D}) : R(m_1, m_2) = 1\right]. \qquad \square$$

DEFINITION OF NON-MALLEABILITY: The definition of security above allows for the possibility that the simulator may do *arbitrarily better than* the adversary. The reason for this is that the adversary may simply refuse to decommit, even when it would have otherwise succeeded[2]. In any case, if a simulator can do better than an adversary who gets to see a commitment to m_1, the scheme still satisfies our intuition of non-malleability.

VALID RELATIONS. In order for relation R to be valid, we impose the following restriction: for all $m \in \mathcal{M}$, we have $R(m, \perp) = 0$. This could also be taken into account by checking that $m_2 \neq \perp$ in the definitions of success, above; however, we find it easier to simply work with valid relations only.

MULTIPLE MESSAGES. The authors of [11] point out that a strictly stronger definition allows the adversary to produce *several* commitments $com_2^{(1)}, com_2^{(2)}, \ldots,$ and later several decommitments $dec_2^{(1)}, dec_2^{(1)}, \ldots$ to messages $m_2^{(1)}, m_2^{(2)}, \ldots.$ The simulator simply outputs messages $m_2^{(1)}, m_2^{(2)}, \ldots.$ The adversary (or simulator) succeeds when a relation $\mathcal{R}(m_1, m_2^{(1)}, m_2^{(2)}, \ldots)$ holds. For simplicity, we use the weaker definition in this paper. However, we stress that all the schemes in this paper are non-malleable with respect to this stronger definition.

HISTORY. The definition of [11] includes the possibility of giving the adversary $hist(m_1)$ (for any computable function $hist$) before he is required to generate his commitment. We note that the current proof of our perfect commitment schemes does not consider this property.

3   Computationally-Hiding Commitment Schemes

We first (briefly) examine the case of standard commitment schemes. Note that the size of a standard, non-interactive commitment (even for malleable schemes) must be at least $M + \omega(\log k)$, where M is the message length and k is the security parameter. Perfect binding implies that the size must be at least M, and semantic security requires, in particular, that each message have $\omega(\mathrm{poly}(k))$ possible commitments associated with it.

The lemma below indicates that we can achieve roughly this bound for standard non-malleable commitment, assuming the existence of trapdoor permutations[3] (in the model with public parameters). The commitment scheme is built from the following components: first, we use a cryptosystem that is secure indistinguishable under an adaptive-chosen-ciphertext attack. Such a scheme can be obtained using a construction in [11], and we denote this scheme by $\mathcal{E}_{\mathrm{pk}}(\cdot)$. Next, we use a symmetric-key cryptosystem (with secret key of length k) which is indistinguishable under adaptive chosen-ciphertext attack (which can be obtained using, e.g., the construction of [11]), and we denote this scheme by $\mathcal{E}_K^*(\cdot)$. The

---

[2] For any relation R, a simulator exists for R as well as for its complement $\bar{R}$, so one might think that this "problem" can be avoided. The difficulty is that there is an asymmetry here, in that both R and $\bar{R}$ must satisfy $R(*, \perp) = \bar{R}(*, \perp) = 0$ (see the note on valid relations).

[3] Recall that [9] achieves a non-interactive, non-malleable computationally-hiding commitment using only one-way functions. However, their scheme requires commitment size $\mathcal{O}(kM)$.

commitment scheme works as follows: public parameters consist of a public key pk for the public-key cryptosystem. Commitment is done by choosing a random secret key for the symmetric-key system, encrypting this secret key using the public key, and then encrypting the committed message using the secret key. A commitment to message m is then computed as:

$$\mathcal{E}_{\text{pk}}(K) \circ \mathcal{E}_K^*(m). \tag{1}$$

Decommitment consists of revealing m and the random bits used to form the commitment. Commitment verification is done in the obvious way.

Although the proof of the lemma is relatively straightforward (and is a "folk lemma" for the case of encryption), the result below was not widely known for the case of commitment. Indeed, there are some complications which require care to get right. A sketch of the proof can be found in Appendix A.

Lemma 1. *Assuming the existence of trapdoor permutations, there exists a computationally-hiding commitment scheme in the public parameter model that is non-malleable with respect to commitment and has commitment size $M + \text{poly}(k)$, where M is the size of the committed message and k is a security parameter.*

Note that this lemma immediately implies the security (under the decisional Diffie-Hellman assumption) of the above construction when using the efficient public-key cryptosystem of [6] for $\mathcal{E}$ and any adaptive chosen-ciphertext-secure private-key cryptosystem $\mathcal{E}^*$. Finally, we note that the security requirements for $\mathcal{E}$ and $\mathcal{E}^*$ can be relaxed. One can show that $\mathcal{E}$ is only required to be non-malleable under a chosen-plaintext attack (NM-CPA) and $\mathcal{E}^*$ need only be indistinguishable under a P0 plaintext attack and an adaptive chosen-ciphertext attack (IND-PO-C2); see [2,18] for formal definitions). This allows for much greater efficiency since NM-CPA-secure public-key cryptosystems can be constructed more efficiently than IND-CCA2 schemes [12] and IND-P0-C2-secure private-key schemes may be deterministic. We remark that the result in the lemma applies to the public random string model when so-called dense public-key encryption schemes [8,7] are used.

4   Perfectly-Hiding Commitment Schemes

The computationally-hiding commitment scheme presented in Section 3 achieves near-optimal commitment size $M + \text{poly}(k)$. We cannot hope to improve this by much (since computationally-hiding commitments have size at least M). In this section we present perfectly-hiding commitment schemes that improve significantly on the commitment length, achieving commitment size $3k$ for arbitrarily-large messages (see Section 5 for modifications allowing further reductions in the commitment size).

Both of our perfectly-hiding commitment schemes build on the paradigm established in [9], with changes which substantially improve the efficiency. A commitment consists of three components $\langle A, B, Tag \rangle$. The first component A is a commitment to parameters r_1 and r_2 for a one-time "message authentication code" (MAC) for B. The second component B contains the actual commitment to the message m, using public parameters which depend upon the first component A. Finally, $Tag = \text{MAC}_{r_1,r_2}(B)$. An adversary who wishes to generate a commitment to a related value has two choices: he can either re-use A or use a different A'. If he re-uses A, with high probability he will be unable to generate a correct Tag for a different B', since he does not know the values r_1, r_2. On the other hand, if he uses a different A', the public parameters he is forced to use for his commitment B' will be different from those used for the original commitment; thus, the adversary will be able to decommit in only one way, regardless of how the original B is decommitted. In particular, if it is possible to equivocate B for a particular choice of A, an adversary who uses a different A' will be unable to equivocate B' (without breaking some computational assumption). We refer the reader to [9] for further discussion.

In [9], the dependence (upon A) of the public parameters used for commitment B was achieved via a "selector function"[4], which results in public parameters of size dependent on the length of the committed message (as a consequence, the scheme can be efficient only in the case of commitment to a single bit). Here, we exploit algebraic properties to drastically reduce the size of the public parameters and obtain a more efficient scheme, even in the case of large messages.

4.1   Construction Based on the Discrete Logarithm Problem

The schemes discussed in this paper work over *any* group G of prime order for which extracting discrete logarithms is hard but multiplication is easy. However, for concreteness we will always assume that p, q are prime with $q|p - 1$ and the group $G \subseteq \mathbb{Z}_p^*$ is the set of elements of order q.

Our starting point is the perfect commitment scheme of Pedersen [24]. Let g, h be generators of G. To commit to a message $m \in \mathbb{Z}_q$, choose random $r \in \mathbb{Z}_q$ and output $com = g^m h^r$. This scheme achieves information-theoretic secrecy, since com is uniformly distributed in G; furthermore, it is computationally binding as long as the discrete logarithm problem is hard. Note that a simple extension of the scheme (which we refer to as *extended-Pedersen*) allows commitment to two messages: simply let g_1, g_2, g_3 be generators of G, and to commit to messages $m_1, m_2 \in \mathbb{Z}_q$, choose random r and output $com = g_1^{m_1} g_2^{m_2} g_3^r$. This scheme retains perfect secrecy; furthermore, computational binding of the extended-Pedersen scheme can be proved via a reduction to the standard Pedersen scheme (see [5]). Note further that the Pedersen and extended-Pedersen schemes are perfectly equivocable (one simply chooses public parameters with known discrete logarithms).

The public parameters, output by $\mathcal{TTP}(1^k)$, are primes p, q with $q|(p - 1)$ and $|p| = k$, along with random generators g_1, g_2, g_3 of G. Additionally, a random function H is chosen from a family of universal one-way hash (UOWH) functions [20]. Commitment is as shown in Figure 1.

---

[4] A different implementation of this technique first appeared in [11].

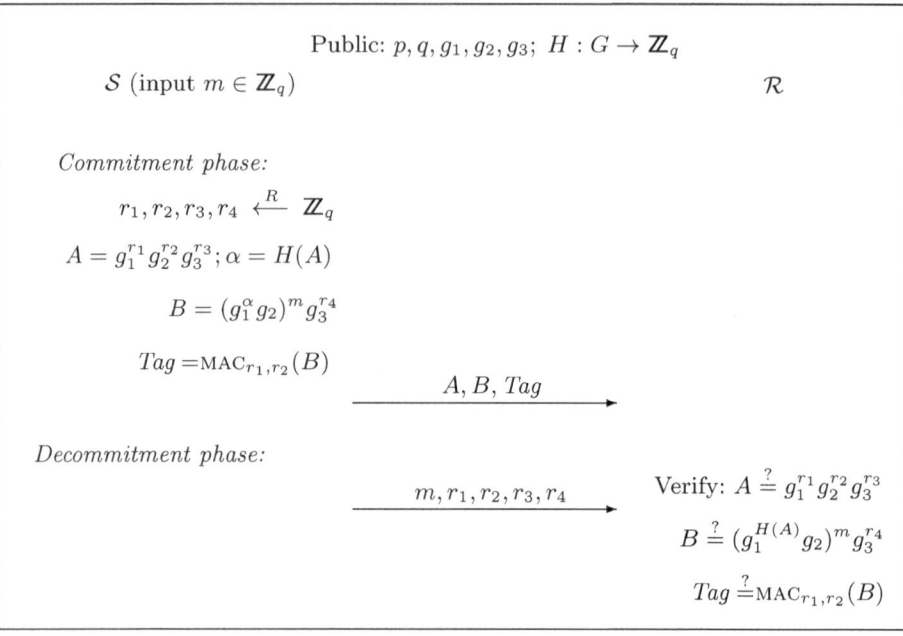

Fig. 1. DLog-based, NM perfect commitment scheme.

Theorem 1. *Assuming the hardness of the discrete logarithm problem in the underlying group, the protocol of Figure 1 is an ϵ-non-malleable perfectly-hiding commitment scheme in the public-parameter model.*

Proof   It is clear that the protocol is perfectly-hiding since B is uniformly distributed in group G independently from the distribution of every other component of the commitment. Computational binding of the protocol is also easy to show (proof omitted).

The proof of non-malleability is more involved; however, we provide some intuition here. As mentioned in Sec. 2, we prove non-malleability with respect to a single commitment output by the adversary; however, the same proof technique suffices to prove non-malleability with respect to multiple commitments. The simulator (which will do as well as the adversary without seeing the commitment) works as follows. First, it generates public parameters which are distributed identically to the real experiment, but for which the simulator knows some trapdoor information which allows it to perfectly equivocate its commitment (cf. Definition 1). The simulator generates a commitment com to a random message, gives this commitment to the adversary, and the adversary produces its commitment com_2. The simulator now tries to get the adversary to open com_2 (this will be the message output by the simulator). To do this, the simulator decommits com to a random message and gives the decommitment to the adversary, and repeats this step (rewinding the adversary each time) sufficiently many times until the adversary opens[5] com_2. Since the simulator can perfectly equivocate its commitment, the adversary's view is equivalent to its view in the original experiment. Furthermore, we show that the adversary itself is unable to equivocate its commitment com_2 (under the discrete logarithm assumption). A complete proof follows.

Assume an adversary $\mathcal{A}$ which, given commitment $\langle A, B, Tag \rangle$, generates commitment $\langle A', B', Tag' \rangle$. Given decommitment $\langle m, r_1, r_2, r_3, r_4 \rangle$, the adversary gives decommitment $\langle m', r_1', r_2', r_3', r_4' \rangle$. Following the proof structure of [9], we distinguish the following sub-cases:

CASE 1. $A' = A$. If this occurs, there are two possibilities: either $\langle r_1, r_2, r_3 \rangle = \langle r_1', r_2', r_3' \rangle$, or not. If they are equal, since r_1 and r_2 are information-theoretically hidden from the adversary when giving his commitment (and assuming the security of the MAC), the adversary will have been unable (except with negligible probability) to generate $B' \neq B$ and Tag' such that $Tag' =\mathrm{MAC}_{r_1, r_2}(B')$. If $\langle r_1, r_2, r_3 \rangle \neq \langle r_1', r_2', r_3' \rangle$, we can construct an adversary $\mathcal{C}$ which, given oracle access to $\mathcal{A}$, can violate the computational binding property of the extended-Pedersen scheme (via a standard reduction). Thus, the success probability of $\mathcal{A}$ in this case must be negligible.

CASE 2. $A' \neq A$ but $H(A') = H(A)$. If this happens, the security of the family of universal one-way hash functions is violated. Simply choose p, q along with random generators g_1, g_2, g_3. Then, select random m, r_1, r_2, r_3, r_4, generate the commitment $\langle A, B, Tag \rangle$, and output A. Upon being given a random member H from the UOWH family, run $\mathcal{A}$ on input the public parameters and the generated commitment. The first component of the commitment generated by $\mathcal{A}$ will then give the desired collision.

CASE 3. $A' \neq A$ and $H(A') \neq H(A)$. This is the most interesting case to consider. Fix ϵ, $\mathcal{D}$, and R, and assume adversary $\mathcal{A}$. Denote the process of selecting group parameters, as run by $\mathcal{TTP}$, by $p, q, G \leftarrow \mathcal{G}(1^k)$ (i.e., this selects primes p, q with $q|p-1$ and $|p| = k$). We describe an equivocable commitment generator Equiv which will be used as a subroutine of simulator $\mathcal{A}'$:

Equiv$_1(1^k)$
$\quad p, q, G \leftarrow \mathcal{G}(1^k)$
$\quad g_1, g_3 \leftarrow G; H \leftarrow \text{UOWH}$
$\quad r, s, t \leftarrow \mathbb{Z}_q$
$\quad A = g_1^r g_3^s; \alpha = H(A)$
$\quad g_2 = g_1^{-\alpha} g_3^t$
$\quad \sigma = \langle p, q, g_1, g_2, g_3, H \rangle$
$\quad r_2, u \leftarrow \mathbb{Z}_q$
$\quad r_1 = r + \alpha r_2; r_3 = s - t r_2$
$\quad B = g_3^u; Tag = \mathrm{MAC}_{r_1, r_2}(B)$
$\quad com = \langle A, B, Tag \rangle$
$\quad s = \langle r_1, r_2, r_3, t, u \rangle$
$\quad$ Output (σ, com, s)

Equiv$_2(\langle r_1, r_2, r_3, t, u \rangle, m)$
$\quad r_4 = u - tm$
$\quad dec = \langle m, r_1, r_2, r_3, r_4 \rangle$
$\quad$ Output dec

---
[5] If the adversary never opens its commitment, the simulator outputs $\perp$.

Note that Equiv satisfies Definition 1. In particular, the distributions of the public parameters output by Equiv and those of the real protocol are the same; they differ only in the "trapdoor" information stored by Equiv. Furthermore, note that p, q, g_1, g_3 can be chosen at random and given to Equiv; knowledge of $\log_{g_3} g_1$ is not necessary. This will be crucial for the proof of security. We now describe the simulator $\mathcal{A}'$:

$$\mathcal{A}'(1^k, \mathcal{D})$$
$$(\sigma, com, s) \leftarrow \mathsf{Equiv}(1^k)$$
Fix random coins ω
$$com_2 = \mathcal{A}(\sigma, com; \omega)$$
Repeat at most $2\epsilon^{-1} \ln 2\epsilon^{-1}$ times:
$$m_1 \leftarrow \mathcal{D}$$
$$dec = \mathsf{Equiv}(s, m_1)$$
$$dec_2 = \mathcal{A}(\sigma, com, dec)$$
$$m_2 = \mathcal{R}(\sigma, com_2, dec_2)$$
if $m_2 \neq \perp$ break
Output m_2

We show that the difference $\mathsf{Succ}^{\mathrm{NM}}_{\mathcal{A}, \mathcal{D}, R}(k) - \widetilde{\mathsf{Succ}}_{\mathcal{A}', \mathcal{D}, R}(k)$ (with terms as defined in Definition 2) is negligible. Straightforward manipulation, using the fact that Equiv is a perfectly equivocable commitment generator and $(\mathcal{TTP}, \mathcal{S}, \mathcal{R})$ is a perfect commitment scheme, gives:

$$\mathsf{Succ}^{\mathrm{NM}}_{\mathcal{A}, \mathcal{D}, R}(k) =$$
$$\Pr\left[\sigma \leftarrow \mathcal{TTP}(1^k); m_1 \leftarrow \mathcal{D}; \omega \leftarrow \Omega; r_1, r_2, r_3 \leftarrow \mathbb{Z}_q; \right.$$
$$(com_1, dec_1) \leftarrow \mathcal{S}(\sigma, m_1; r_1, r_2, r_3);$$
$$\left. m_2 = \mathcal{R}(\sigma, \mathcal{A}(\sigma, com_1; \omega), \mathcal{A}(\sigma, com_1, dec_1; \omega)) : R(m_1, m_2) = 1\right]$$

and

$$\widetilde{\mathsf{Succ}}_{\mathcal{A}', \mathcal{D}, R}(k) =$$
$$\Pr\left[\sigma \leftarrow \mathcal{TTP}(1^k); m_1 \leftarrow \mathcal{D}; \omega \leftarrow \Omega; r_1, r_2, r_3 \leftarrow \mathbb{Z}_q; \right.$$
$$(com_1, dec_1) \leftarrow \mathcal{S}(\sigma, m_1; r_1, r_2, r_3);$$
$$\left. m_2^* = \mathcal{R}(\sigma, \mathcal{A}(\sigma, com_1; \omega), \mathcal{A}(\sigma, com_1, dec^*; \omega)) : R(m_1, m_2^*) = 1\right].$$

The notation dec^* represents the fact that the decommitment given to $\mathcal{A}$ was produced according to algorithm $\mathcal{A}'$. In particular, dec^* represents either the first decommitment given to $\mathcal{A}$ which resulted in $m_2 \neq \perp$, or the $(2\epsilon^{-1} \ln 2\epsilon^{-1})^{\mathrm{th}}$ decommitment given to $\mathcal{A}$ (if all decommitments up to then had $m_2 = \perp$).

Define the tuple $(\sigma; \omega; r_1, r_2, r_3; com_1)$ as *good* if the following holds:

$$\Pr\left[m_1 \leftarrow \mathcal{D} : \mathcal{R}(\sigma, \mathcal{A}(\sigma, com_1; \omega), \mathcal{A}(\sigma, com_1, dec_1; \omega)) \neq \perp\right] \geq \epsilon/2,$$

(the above probability is over choice of m_1 only; note that once the tuple is fixed, choice of m_1 determines r_4, and hence dec_1). Furthermore, define event Good as occurring when the tuple generated by the experiment is *good*. We now have (for brevity, we denote generation of a random tuple by $\gamma \leftarrow \Gamma(1^k)$; also, we denote $m_2 = \mathcal{R}(\sigma, \mathcal{A}(\sigma, com; \omega), \mathcal{A}(\sigma, com, dec; \omega))$ by $m_2 = \mathcal{A}(\sigma, com, dec)$):

$$\mathsf{Succ}^{\mathrm{NM}}_{\mathcal{A},\mathcal{D},R}(k) - \widetilde{\mathsf{Succ}}_{\mathcal{A}',\mathcal{D},R}(k) =$$
$$\Pr\left[\gamma \leftarrow \Gamma(1^k); m_1 \leftarrow \mathcal{D}; m_2 = \mathcal{A}(\sigma, com_1, dec_1) : R(m_1, m_2) \wedge \mathsf{Good}\right]$$
$$+ \Pr\left[\gamma \leftarrow \Gamma(1^k); m_1 \leftarrow \mathcal{D}; m_2 = \mathcal{A}(\sigma, com_1, dec_1) : R(m_1, m_2) \wedge \overline{\mathsf{Good}}\right]$$
$$- \Pr\left[\gamma \leftarrow \Gamma(1^k); m_1 \leftarrow \mathcal{D}; m_2^* = \mathcal{A}(\sigma, com_1, dec^*) : R(m_1, m_2^*) \wedge \mathsf{Good}\right]$$
$$- \Pr\left[\gamma \leftarrow \Gamma(1^k); m_1 \leftarrow \mathcal{D}; m_2^* = \mathcal{A}(\sigma, com_1, dec^*) : R(m_1, m_2^*) \wedge \overline{\mathsf{Good}}\right],$$

from which we derive (by definition of event $\overline{\mathsf{Good}}$):

$$\mathsf{Succ}^{\mathrm{NM}}_{\mathcal{A},\mathcal{D},R}(k) - \widetilde{\mathsf{Succ}}_{\mathcal{A}',\mathcal{D},R}(k) \leq$$
$$\Pr\left[\gamma \leftarrow \Gamma(1^k); m_1 \leftarrow \mathcal{D}; m_2 = \mathcal{A}(\sigma, com_1, dec_1) : R(m_1, m_2) \wedge \mathsf{Good}\right]$$
$$+ \epsilon/2$$
$$- \Pr\left[\gamma \leftarrow \Gamma(1^k); m_1 \leftarrow \mathcal{D}; m_2^* = \mathcal{A}(\sigma, com_1, dec^*) : R(m_1, m_2^*) \wedge \mathsf{Good}\right].$$

But this, in turn, implies:

$$\mathsf{Succ}^{\mathrm{NM}}_{\mathcal{A},\mathcal{D},R}(k) - \widetilde{\mathsf{Succ}}_{\mathcal{A}',\mathcal{D},R}(k) \leq$$
$$\Pr\left[\gamma \leftarrow \Gamma(1^k); m_1 \leftarrow \mathcal{D}; m_2 = \mathcal{A}(\sigma, com_1, dec_1); m_2^* = \mathcal{A}(\sigma, com_1, dec^*) :\right.$$
$$\left. R(m_1, m_2) \wedge \overline{R(m_1, m_2^*)} \wedge \mathsf{Good}\right] + \epsilon/2,$$

which can be re-written as:

$$\mathsf{Succ}^{\mathrm{NM}}_{\mathcal{A},\mathcal{D},R}(k) - \widetilde{\mathsf{Succ}}_{\mathcal{A}',\mathcal{D},R}(k) \leq$$
$$\Pr\left[\gamma \leftarrow \Gamma(1^k); m_1 \leftarrow \mathcal{D}; m_2 = \mathcal{A}(\sigma, com_1, dec_1); m_2^* = \mathcal{A}(\sigma, com_1, dec^*) :\right.$$
$$\left. R(m_1, m_2) \wedge \overline{R(m_1, m_2^*)} \wedge m_2^* = \perp \wedge \mathsf{Good}\right] \qquad (2)$$
$$+ \Pr\left[\gamma \leftarrow \Gamma(1^k); m_1 \leftarrow \mathcal{D}; m_2 = \mathcal{A}(\sigma, com_1, dec_1); m_2^* = \mathcal{A}(\sigma, com_1, dec^*) :\right.$$
$$\left. R(m_1, m_2) \wedge \overline{R(m_1, m_2^*)} \wedge m_2^* \neq \perp \wedge \mathsf{Good}\right] \qquad (3)$$
$$+ \epsilon/2.$$

We now bound probabilities (2) and (3). First, notice that expression (2) is bounded from above by the probability that $m_2^* = \perp$. However, definition of event Good and a straightforward probability calculation show that:

$$\Pr\left[\gamma \leftarrow \Gamma(1^k); m_2^* \leftarrow \mathcal{A}(\sigma, com_1, dec^*) : m_2^* = \perp \wedge \mathsf{Good}\right] \leq$$
$$\Pr\left[\gamma \leftarrow \Gamma(1^k); m_2^* \leftarrow \mathcal{A}(\sigma, com_1, dec^*) : m_2^* = \perp \mid \mathsf{Good}\right] \leq \epsilon/2.$$

Finally, notice that for the event in expression (3) to occur, we must have $m_2 \neq \perp$ and $m_2 \neq m_2^*$. But this then gives a Pedersen commitment com_2 (using generators g_3 and $g_1^{\alpha'} g_2 = g_1^{(\alpha' - \alpha)} g_3^t$) which is decommited in two different ways. This would allow determination of $\log_{g_3} g_1$ (recall that $\alpha' \neq \alpha$ since we are dealing with Case 3). The experiment is as follows: choose random ω and r_1, r_2, r_3 and run Equiv using the given values g_1, g_3 to generate σ' and com_1 (recall that knowledge of $\log_{g_3} g_1$ is not necessary to run Equiv). The adversary $\mathcal{A}$ then produces a commitment com_2. Following the description of $\mathcal{A}'$, run $\mathcal{A}$ to obtain a decommitment to message m_2^*. Then, decommit once more to a randomly selected m_1 and give this as input to $\mathcal{A}$ to obtain a decommitment to m_2. If $m_2 \neq \perp$ and $m_2^* \neq \perp$ and $m_2 \neq m_2^*$ (which we call event Success), then $\log_{g_3} g_1$ can be calculated, as discussed above. But the probability of Success is bounded from below by expression (3); by assumption, however, the discrete logarithm problem is intractable and thus:

$$(3) \leq \Pr[\mathsf{Success}] \leq negl(k).$$

Putting everything together gives the desired result.   □

Note that the proof of non-malleability is exactly the same even if the message is hashed before commitment. Equiv can still perfectly equivocate to any (random) message M by first computing $m = \mathcal{H}(M)$ and then running the identical Equiv_2 algorithm. The simulator $\mathcal{A}'$ is also identical (messages will be longer, but this does not affect the analysis). The hash function must be collision resistant for the binding property to hold, but no other assumptions about the hash function are necessary, and the scheme is still perfectly secret[6]. The present scheme therefore gives a practical method for committing to arbitrarily long messages.

We remark that by making minor modifications to the above protocol, it can be proven secure in the public random string model as well.

We give an alternate proof of Theorem 1 in App. B. This proof, while more complicated than the proof given above, achieves a slightly stronger security guarantee by using a simulator which runs in expected polynomial time.

4.2   Construction Based on RSA

We have also developed an efficient non-interactive, non-malleable perfect commitment scheme based on the RSA assumption. Since the ideas underlying this construction, as well as the proof of security, are substantially similar to the scheme presented above, we defer details to the full version of this paper.

5   Extensions

There are extensions of our scheme which may be of practical value:

REDUCING THE COMMITMENT SIZE. Our schemes produce commitments $com = (A, B, Tag)$ of size $3k$, where k is the length of the string representing a group

---

[6] This can be compared to [14] which requires added complications when using an arbitrary hash function and achieves only statistical secrecy.

element. However, inspection of the proof of Thm. 1 reveals that one can replace this with any string that uniquely *binds* the sender to *com*. At least two modifications in this vein seem useful:

- Using a *collision-resistant* hash-function h, we can replace the commitment *com* with $h(com)$. The decommitment phase is the same as before. This does not increase the computational cost of the protocol by very much. The resulting commitment size is the output length of a hash function believed to be collision-resistant, e.g. SHA or MD5. In particular, this allows us to achieve optimal commitment size $\mathcal{O}(\omega(\log k))$, assuming an appropriate hash function. Note that this approach (hashing the commitment) does not seem to give provable security for general non-malleable commitment schemes, yet it *does* work (as can be seen by careful examination of the proof) for the particular construction given here.
- By adding one more public parameter and making appropriate (small) modifications to the scheme, we can set the commitment to the *product* of A, B and Tag (assuming Tag is computed as $B^{r_1}g_3^{r_2}$, which serves as an information-theoretically secure MAC). This reduces the commitment length to k. We defer a proof of security to the full version of the paper.

UNIQUE IDENTIFIERS. As mentioned in [11], in many situations there is a unique identifier (ID) associated to each user and using them can improve the efficiency of non-malleable primitives. This is also true of our scheme. If each user in the system has ID $id \in \mathbb{Z}_q$, we can simplify the scheme by replacing α with id. An adversary who attempts to generate related commitments must do so with respect to *his* identifier $id' \neq id$. The public parameters are p, q and three generators g_1, g_2, g_3. The commitment is $B = (g_1^{id}g_2)^m g_3^{r_3}$ (the components A and Tag are no longer needed, since their only role in the original protocol was to force an adversary to change α). The proof of non-malleability is the same as for the original scheme except there is no need to handle cases 1 and 2.

Acknowledgments

Thanks to Yuval Ishai for helpful discussions and to Marc Fischlin and Roger Fischlin for a pre-print of the full version of [14] and for close readings of preliminary drafts of this work. Thanks to Yehuda Lindell for pointing out that an NM-CPA scheme is sufficient in the construction of Section 3. We also appreciate the referees' comments, especially the suggestion (mentioned in Section 5) to extend our constructions to the setting in which users have unique IDs.

References

1. D. Beaver. Adaptive Zero-Knowledge and Computational Equivocation. FOCS '96.
2. M. Bellare, A. Desai, D. Pointcheval, and P. Rogaway. Relations Among Notions of Security for Public-Key Encryption Schemes. CRYPTO '98.
3. M. Blum, A. De Santis, S. Micali, and G. Persiano. Non-Interactive Zero-Knowledge. SIAM Journal of Computing, vol. 20, no. 6, Dec 1991, pp. 1084–1118.

4. M. Blum, P. Feldman, and S. Micali. Non-Interactive Zero-Knowledge and Applications. STOC '88.
5. S. Brands. Rapid Demonstration of Linear Relations Connected by Boolean Operators. Eurocrypt '97.
6. R. Cramer and V. Shoup. A Practical Public Key Cryptosystem Provably Secure Against Chosen Ciphertext Attack. CRYPTO '98.
7. A. De Santis, G. Di Crescenzo, and G. Persiano. Necessary and Sufficient Assumptions for Non-Interactive Zero-Knowledge Proofs of Knowledge for All NP Relations. ICALP '00.
8. A. De Santis and G. Persiano. Zero-Knowledge Proofs of Knowledge Without Interaction. FOCS '92.
9. G. Di Crescenzo, Y. Ishai, and R. Ostrovsky. Non-Interactive and Non-Malleable Commitment. STOC '98.
10. G. Di Crescenzo and R. Ostrovsky. On Concurrent Zero-Knowledge with Preprocessing. CRYPTO '99.
11. D. Dolev, C. Dwork, and M. Naor. Nonmalleable Cryptography. SIAM J. Comp. 30 (2) 391–437, 2000. A preliminary version appears in STOC '91.
12. C. Dwork. The Non-Malleability Lectures. Available from the author.
13. S. Even, O. Goldreich, A. Lempel. A Randomized Protocol for Signing Contracts. *Communications of the ACM* 28(6), 637–647, 1985.
14. M. Fischlin and R. Fischlin. Efficient Non-Malleable Commitment Schemes. CRYPTO 2000.
15. O. Goldreich. *Foundations of Cryptography, Fragments of a Book*, 1998.
16. O. Goldreich, S. Micali, and A. Wigderson. How to Play Any Mental Game or a Completeness Theorem for Protocols with Honest Majority. STOC '87.
17. O. Goldreich, S. Micali, and A. Wigderson. Proofs that Yield Nothing but their Validity or All Languages in NP have Zero-Knowledge Proof Systems. *J. ACM* 38(3): 691–729 (1991).
18. J. Katz and M. Yung. Complete Characterization of Security Notions for Probabilistic Private-Key Encryption. STOC '00.
19. M. Naor. Bit Commitment Using Pseudorandomness. J. Crypto. 4(2): 151–158 (1991).
20. M. Naor and M. Yung. Universal One-Way Hash Functions and Their Cryptographic Applications. STOC '89.
21. M. Naor, R. Ostrovsky, R. Venkatesan, and M. Yung. Perfect zero-knowledge arguments for NP can be based on general complexity assumptions. J. Cryptology, 11(2):87–108, 1998 (also CRYPTO '92).
22. T. Okamoto. Provable Secure and Practical Identification Schemes and Corresponding Signature Schemes. CRYPTO '92.
23. R. Ostrovsky, R. Venkatesan, and M. Yung. Fair games against an all-powerful adversary. AMS DIMACS Series in Discrete Mathematics and Theoretical Computer Science, Vol. 13 pp. 155-169, 1993.
24. T.P. Pedersen. Non-Interactive and Information-Theoretic Secure Verifiable Secret Sharing. CRYPTO '91.
25. A. Sahai. Non-Malleable Non-Interactive Zero-Knowledge and Adaptive Chosen-Ciphertext Security. FOCS '99.

A   Proof of Lemma 1

Sketch of Proof   First note that for (1) to be binding, we require that the decryption algorithms for both the public-key and symmetric-key systems have zero probability of decryption error[7]. Thus, revealing the randomness used to generate the commitment perfectly binds the sender to the message.

The proof of non-malleability with respect to commitment will imply that the scheme is semantically secure (this has been noted previously for the case of encryption [2,18], but a similar result holds for the case of commitment). Note that if we can prove that (1) constitutes a non-malleable (public-key) *encryption* scheme, we are done. Using the results of [2], it suffices to prove that (1) is secure under adaptive chosen-ciphertext attack.

Consider an adversary $\mathcal{A}$ who has non-negligible advantage in attacking (1) under an adaptive chosen-ciphertext attack. Define adversary $\mathcal{B}$ which uses $\mathcal{A}$ as a black box to break $\mathcal{E}_{\mathrm{pk}}$ under an adaptive chosen-ciphertext attack (the notation $\mathcal{D}(\cdot)$ means that $\mathcal{B}$ is given access to a decryption oracle for $\mathcal{E}_{\mathrm{pk}}$):

Algorithm $\mathcal{B}_1^{\mathcal{D}(\cdot)}(1^k, \mathrm{pk})$	Algorithm $\mathcal{B}_2^{\mathcal{D}(\cdot)}(y, (C, s))$
$(M_0, M_1, s) \leftarrow \mathcal{A}_1^{\widetilde{\mathcal{D}}(\cdot)}(1^k, \mathrm{pk})$	$b' \leftarrow \mathcal{A}_2^{\widetilde{\mathcal{D}}(\cdot)}(y \circ C, s)$
$K \leftarrow \{0,1\}^k$	if $b' = b$ return 1
$b \leftarrow \{0,1\}$	else return 0
$C \leftarrow \mathcal{E}_K^*(M_b)$	
return $(K, 0^k, (C, s))$	

The notation $\widetilde{\mathcal{D}}(\cdot)$ means that decryption oracle queries of $\mathcal{A}$ are handled by $\mathcal{B}$ in the following way: in the first stage, when $\mathcal{A}$ submits ciphertext $y' \circ C'$ to its decryption oracle, $\mathcal{B}$ submits y' to its decryption oracle for $\mathcal{E}_{\mathrm{pk}}$, receives key K', and then computes $M' := \mathcal{D}_{K'}(C')$. In the second stage, $\mathcal{B}$ answers as before except that $\mathcal{A}$ might submit a ciphertext $y \circ C'$. Note that $\mathcal{B}$ would *not* be allowed to submit y to its decryption oracle, since he cannot ask for decryption of the challenge ciphertext. Instead, $\mathcal{B}$ "assumes" that y is an encryption of K, and computes the response $M := \mathcal{D}_K(C)$. Adaptive chosen-ciphertext security of $\mathcal{E}_{\mathrm{pk}}$ implies that the advantage of $\mathcal{B}$ is negligible.

We now consider the following adversary which uses $\mathcal{A}$ as a black box to break $\mathcal{E}^*$ under an adaptive chosen-ciphertext attack. Here, the notation $\mathcal{D}(\cdot)$ means that $\mathcal{C}$ is given access to a decryption oracle for $\mathcal{E}_K^*$ (where K is some secret key unknown to $\mathcal{C}$). We let Gen denote the algorithm which selects public and private keys for $\mathcal{E}$.

Algorithm $\mathcal{C}_1^{\mathcal{D}(\cdot)}(1^k)$	Algorithm $\mathcal{C}_2^{\mathcal{D}(\cdot)}(C, (y, \mathrm{sk}, s))$
$(\mathrm{pk}, \mathrm{sk}) \leftarrow \mathrm{Gen}(1^k)$	$b' \leftarrow \mathcal{A}_2^{\widetilde{\mathcal{D}}(\cdot)}(y \circ C, s)$
$(M_0, M_1, s) \leftarrow \mathcal{A}_1^{\widetilde{\mathcal{D}}(\cdot)}(1^k, \mathrm{pk})$	return b'
$y \leftarrow \mathcal{E}_{\mathrm{pk}}(0^k)$	
return $(M_0, M_1, (y, \mathrm{sk}, s))$	

---

[7] This can be relaxed slightly, but since many commonly-used encryption schemes already have this property, we assume it here for simplicity of exposition.

Here, the notation $\widetilde{\mathcal{D}}(\cdot)$ means that decryption oracle queries of $\mathcal{A}$ are handled by $\mathcal{C}$ in the following way: in the first stage, when $\mathcal{A}$ submits ciphertext $y' \circ C'$ to its decryption oracle, $\mathcal{C}$ decrypts y' to get K' (it knows the secret key) and then computes $M' := \mathcal{D}_{K'}(C')$. In the second stage, however, $\mathcal{C}$ answers as before *unless* $\mathcal{A}$ submits a ciphertext $y \circ C'$. In this case, $\mathcal{C}$ submits C' to its decryption oracle for $\mathcal{E}_K^*$ and returns the result to $\mathcal{A}$. Adaptive chosen-ciphertext security of $\mathcal{E}^*$ implies that the advantage of $\mathcal{C}$ is negligible.

Informally, define the following probabilities of success:

$$p_{0,r} \stackrel{\text{def}}{=} \Pr[\mathcal{A}^{\mathcal{D}(\cdot)}(\mathcal{E}_{\text{pk}}(K) \circ \mathcal{E}_K^*(M_0)) = 0]$$
$$p_{0,f} \stackrel{\text{def}}{=} \Pr[\mathcal{A}^{\widetilde{\mathcal{D}}(\cdot)}(\mathcal{E}_{\text{pk}}(0^k) \circ \mathcal{E}_K^*(M_0)) = 0]$$
$$p_{1,r} \stackrel{\text{def}}{=} \Pr[\mathcal{A}^{\mathcal{D}(\cdot)}(\mathcal{E}_{\text{pk}}(K) \circ \mathcal{E}_K^*(M_1)) = 1]$$
$$p_{1,f} \stackrel{\text{def}}{=} \Pr[\mathcal{A}^{\widetilde{\mathcal{D}}(\cdot)}(\mathcal{E}_{\text{pk}}(0^k) \circ \mathcal{E}_K^*(M_1)) = 1].$$

$\mathcal{B}$'s advantage is given by $1/2 \{1/2(1 - p_{0,f}) + 1/2(1 - p_{1,f})\} + 1/4(p_{0,r} + p_{1,r})$. $\mathcal{C}$'s advantage is given by $1/2(p_{o,f} + p_{1,f})$. Note that these are both negligible, by the arguments advanced above. Finally, the advantage of $\mathcal{A}$ in the original experiment is given by $1/2(p_{0,r} + p_{1,r})$. Simple algebra implies that $\mathcal{A}$'s advantage must be negligible.

Note that adaptive-chosen-ciphertext-secure private-key encryption schemes can be constructed using a one-way function, while non-malleable public-key encryption schemes (with 0 probability of decryption error) are known to exist assuming trapdoor permutations [11,25]. This completes the proof.   □

B   Alternate Proof of Theorem 1

In this section, we present an alternate proof of Theorem 1, which in fact gives us a stronger security guarantee. First, notice that in the previous proof, the simulator had to cut off the simulation after $2\epsilon^{-1} \ln 2\epsilon^{-1}$ steps. This is because for some values of the initial setup γ, it is possible that the adversary would not decommit at all, and thus that the simulation would never terminate. This is an essential problem with the sort of simulation described above: even if the fraction of "bad setups γ" were barely noticeable, the expected running time of the simulation might be infinite!

Instead, we give a simulation which always runs in expected polynomial time, *provided that the adversary succeeds with noticeable probability*. To do so, we adapt the proof technique of DIO [9]. Unfortunately, one cannot apply their proof directly here since their proof relies on the fact that the DIO commitment scheme is statistically binding.

Let $p_{\mathcal{A}}$ be the success probability of the adversary in the original basic experiment for non-malleability with respect to opening, i.e. $p_{\mathcal{A}} = \text{Succ}_{\mathcal{A},\mathcal{D},R}^{\text{NM}}(k)$. For a given simulator A', let $\widetilde{p}_{\mathcal{A}'}$ denote the simulator's success probability, i.e. $\widetilde{p}_{\mathcal{A}'} = \widetilde{\text{Succ}}_{\mathcal{A}',\mathcal{D},R}(k)$. We will construct a simulator $\mathcal{A}'$ such that $p_{\mathcal{A}} - \widetilde{p}_{\mathcal{A}'} \leq$

$negl(k) \left(\frac{1}{p_A} \right)$ and the expected running time of $\mathcal{A}'$ is polynomial in $\frac{1}{p_A}$. Notice in particular that when the adversary's probability of success is noticeable, our simulation does (essentially) at least as well as the original adversary, and runs in expected polynomial time.

The simulator $\mathcal{A}'$ is simple: it runs the adversary in the basic non-malleability experiment until the adversary succeeds; it then outputs whatever m_2 the adversary succeeded with. We describe two equivalent formulations of this simulator below. The first simulation generates all its parameters honestly; the second simulation uses the equivocator of the previous section.

$\mathcal{A}'_1(1^k, \mathcal{D})$	$\mathcal{A}'_2(1^k, \mathcal{D})$
$\quad m_1, m_2 := \bot$	$\quad m_1, m_2 := \bot$
$\quad com_1, com_2 := 0$	$\quad com_1, com_2 := 0$
$\quad$ Repeat until $R(m_1, m_2) = 1$	$\quad$ Repeat until $R(m_1, m_2) = 1$
$\qquad$ and $com_1 \neq com_2$:	$\qquad$ and $com_1 \neq com_2$:
$\qquad \sigma \leftarrow \mathcal{TTP}(1^k)$	$\qquad (\sigma, com_1, s) \leftarrow \mathsf{Equiv}(1^k)$
$\qquad m_1 \leftarrow \mathcal{D}$	$\qquad$ Fix random coins ω
$\qquad (com_1, dec_1) \leftarrow \mathcal{S}(\sigma, m_1)$	$\qquad com_2 := \mathcal{A}(\sigma, com_1; \omega)$
$\qquad com_2 \leftarrow \mathcal{A}(\sigma, com_1)$	$\qquad m_1 \leftarrow \mathcal{D}$
$\qquad dec_2 \leftarrow \mathcal{A}(\sigma, com_1, dec_1)$	$\qquad dec_1 := \mathsf{Equiv}(s, m_1)$
$\qquad m_2 \leftarrow \mathcal{R}(\sigma, com_2, dec_2)$	$\qquad dec_2 := \mathcal{A}(\sigma, com_1, dec_1)$
	$\qquad m_2 := \mathcal{R}(\sigma, com_2, dec_2)$
$\quad$ Output m_2	$\quad$ Output m_2

From the point of view of the adversary, both of these simulations are equivalent, since the equivocator creates a public string σ and a commitment com which are from the same distribution as the "real" strings σ and com_1. Thus, the output distribution of the two simulations is the same, and hence so is their probability of success. Moreover, both simulations expect to make $\frac{1}{p_A}$ calls to $\mathcal{A}$, and thus their expected running times are essentially the same.

The only difference between the two simulations is that in the first simulation $\mathcal{A}'_1$, the simulator knows no more than the adversary about the relationship of the public parameters $g_1, g_2, g_3, \mathcal{H}$, and so all three of these values could come from an outside source. In the second simulation $\mathcal{A}'_2$, only g_1 and g_3 can come from an outside source; g_2 and the other parameters are carefully constructed.

As before, we now consider the three possible cases.

CASES 1 AND 2. Consider the first simulator $\mathcal{A}'_1$. As mentioned above, it does not choose the public parameters g_1, g_2, g_3, H, and so the analysis from the previous proof of cases 1 and 2 tells us that the probability of either of these cases is negligible. (Otherwise, the simulator would break either the computational binding of the Pedersen scheme or the intractability of finding collisions for the hash function).

Hence, we can assume *even in the second simulation* that whenever the adversary generates a new commitment to which he decommits, we have $H(A') \neq H(A)$.

CASE 3. As in the previous proof, we denote the generation of a tuple $\gamma = (\sigma, com, s, \omega)$ by the shorthand $\gamma \leftarrow \Gamma(1^k)$. Note that these variables uniquely determine both com_1 and com_2. Moreover, once the simulator chooses m_1, the decommitted message m_2 is completely determined by m_1 and γ. For conciseness we write simply $m_2 = \mathcal{A}(m_1, \gamma)$. By convention, we will take $\mathcal{A}(m_1, \gamma) = \perp$ whenever the adversary refuses to decommit or simply copies the commitment (i.e. $com_2 = com_1$).

On one hand, we can calculate the adversary's success probability, using the properties of the equivocator:

$$p_\mathcal{A} = \Pr\left[\gamma \leftarrow \Gamma(1^k); m_1 \leftarrow \mathcal{D} \ : \ m_2 = \mathcal{A}(m_1, \gamma) \text{ and } R(m_1, m_2)\right].$$

We can also calculate the success probability of the simulator (second formulation):

$$\widetilde{p}_{\mathcal{A}'} = \Pr\left[\gamma \leftarrow \Gamma(1^k); \ m_1, m_1' \leftarrow \mathcal{D} \ : \ R(m_1', \mathcal{A}(m_1, \gamma)) \ \middle| \ R(m_1, \mathcal{A}(m_1, \gamma))\right]$$

$$= \frac{\Pr\left[\gamma \leftarrow \Gamma(1^k); \ m_1, m_1' \leftarrow \mathcal{D} \ : \ R(m_1', \mathcal{A}(m_1, \gamma)) \text{ and } R(m_1, \mathcal{A}(m_1, \gamma))\right]}{p_\mathcal{A}}$$

The numerator in the last expression can be interpreted as the success probability of the following experiment:

Choose m_1 at random, and run the simulation to obtain a decommitment to a message m_2. Then pick a new message m_1' at random and see if both $R(m_1, m_2)$ and $R(m_1', m_2)$ hold.

Now intuitively, we expect that for any given γ the adversary can only decommit to one valid message. We want to use that intuition to show that the success probability of the experiment above is no worse than the following:

Choose m_1 and obtain m_2 as before. Now, for the same setup γ, pick a new message m_1' and run the simulation to get m_2'. Output a success if both $R(m_1, m_2)$ and $R(m_1', m_2')$ hold.

This intuition is captured in the following lemma:

Lemma 2. *Let $m_2 = \mathcal{A}(m_1, \gamma)$, and $m_2' = \mathcal{A}(m_1', \gamma)$, where γ, m_1, m_1' are chosen as in the previous discussion. Then we have:*

$$\Pr\left[R(m_1, m_2) \wedge R(m_1', m_2)\right] > \Pr\left[R(m_1, m_2) \wedge R(m_1', m_2')\right] - negl(k).$$

Proof For any two events A and B, we have $P(A) - P(B) \leq P(A \setminus B)$. Thus:

$$\Pr\left[R(m_1, m_2) \wedge R(m_1', m_2')\right] - \Pr\left[R(m_1, m_2) \wedge R(m_1', m_2)\right]$$

$$\leq \Pr\left[R(m_1, m_2) \wedge R(m_1', m_2') \wedge \overline{R(m_1', m_2)}\right].$$

Now this last event occurs only when m_2 and m_2' are different, yet both of them are valid messages. However, such an event allows extraction of the discrete log of g_3 with respect to g_1, even in the setting of the second simulation. Since $\mathcal{A}$ is polynomial time, this probability must be negligible in k. □

Using the lemma and the shorthand notation set up in the lemma, we get:

$$\widetilde{p}_{\mathcal{A}'} \geq \frac{\Pr\left[R(m_1, m_2) \wedge R(m'_1, m_2)\right]}{p_{\mathcal{A}}} \geq \frac{\Pr\left[R(m_1, m_2) \wedge R(m'_1, m'_2)\right] - negl(k)}{p_{\mathcal{A}}}.$$

Recall that once γ and m_1 are fixed, m_2 is also fixed. Similarly, m'_2 is fixed once γ and m'_1 are fixed. Thus, we can write:

$$\widetilde{p}_{\mathcal{A}'} \geq \frac{\left(\sum_\gamma \Pr\left[\gamma = \Gamma(1^k)\right] \cdot \Pr\left[R(m_1, m_2) \mid \gamma\right] \cdot \Pr\left[R(m'_1, m'_2) \mid \gamma\right]\right) - negl(k)}{p_{\mathcal{A}}}$$

$$= \frac{\left(\sum_\gamma \Pr\left[\gamma = \Gamma(1^k)\right] \cdot \Pr\left[R(m_1, m_2) \mid \gamma\right]^2\right) - negl(k)}{p_{\mathcal{A}}}.$$

For any random variable X we have $E(X^2) \geq (E(X))^2$. Applying this to the numerator we get:

$$\widetilde{p}_{\mathcal{A}'} \geq \frac{\left(\sum_\gamma \Pr\left[\gamma = \Gamma(1^k)\right] \cdot \Pr\left[R(m_1, m_2) \mid \gamma\right]\right)^2 - negl(k)}{p_{\mathcal{A}}}.$$

But the numerator is simply $(p_{\mathcal{A}})^2$. Thus $\widetilde{p}_{\mathcal{A}'} \geq p_{\mathcal{A}} - \frac{negl(k)}{p_{\mathcal{A}}}$. □

How to Convert the Flavor
of a Quantum Bit Commitment

Claude Crépeau[1], Frédéric Légaré[2], and Louis Salvail[3]

[1] School of Computer Science, McGill University[†], `crepeau@cs.mcgill.ca`
[2] ZKLabs[‡], Zero-Knowledge Systems Inc., `frederic@zeroknowledge.com`
[3] BRICS[§], Dept. of Computer Science, University of Århus, `salvail@brics.dk`

Abstract. In this paper we show how to convert a statistically binding but computationally concealing quantum bit commitment scheme into a computationally binding but statistically concealing QBC scheme. For a security parameter n, the construction of the statistically concealing scheme requires $O(n^2)$ executions of the statistically binding scheme. As a consequence, statistically concealing but computationally binding quantum bit commitments can be based upon any family of quantum one-way functions. Such a construction is not known to exist in the classical world.

1 Introduction

Finding the weakest computational assumptions from which the basic cryptographic primitives can be based upon is important for the theoretical foundations of cryptography. Protocols for secure 2-party computations are usually built from two basic and fundamental cryptographic primitives: Bit commitment and oblivious transfer. Classically, one-way functions are necessary and sufficient for secure bit commitment but not for oblivious transfer unless a major breakthrough is achieved in complexity theory [10,12]. This suggests that in classical cryptography, bit commitment is a weaker primitive than oblivious transfer. Bit commitments come in two main flavors: binding but computationally concealing and concealing but computationally binding. Informally, binding means that whatever the committer does, it is impossible to open both 0 and 1 with non-negligible probability of success (this is sometimes called statistically binding). Concealing means that the receiver cannot obtain more than a negligible amount of information about the committed bit (i.e. statistically concealing). The weakest known computational assumption from which bit commitment can be based upon depends on its flavor. Binding but computationally concealing bit commitments can be based upon any one-way function [17,11,7]. On the other hand, the weakest known assumption for concealing but computationally

---

[†] Part of this research was funded by Québec's Fonds FCAR and Canada's NSERC.
[‡] This research was done as part of the M.Sc. requirements at McGill University.
[§] Basic Research in Computer Science (www.brics.dk), funded by the Danish National Research Foundation.

B. Pfitzmann (Ed.): EUROCRYPT 2001, LNCS 2045, pp. 60–77, 2001.

binding commitments is the existence of one-way permutations [18]. It seems that in the classical world, concealing commitments are more difficult to achieve than binding ones. The two flavors allow for different cryptographic applications. For example, computational zero-knowledge proofs [8,9] can be constructed from binding commitments whereas perfect zero-knowledge arguments [4] use concealing commitments.

In quantum cryptography, computational assumptions are also required for bit commitment and oblivious transfer [15,16,14]. The standard computational assumptions for the quantum case are defined as in the classical case except that they must resist quantum inverters. A quantum one-way function is simply a classical function $f : \{0,1\}^n \rightarrow \{0,1\}^{l(n)}$ for which given any $x \in \{0,1\}^n$, $f(x)$ can be efficiently computed by a quantum computer but finding $x' \in f^{-1}(y)$ given $y := f(x)$, (when $x \in_R \{0,1\}^n$) is hard. In [6], a concealing quantum bit commitment scheme is built from any quantum one-way permutation. The resulting scheme, although improving the communication complexity of the known classical protocols, requires the same kind of assumption as in the classical case. In this paper, we show that the computational assumption for concealing quantum bit commitment schemes can be weakened compared to its classical counterpart. Our construction relies upon the QOT protocol for quantum 1-out-of-2 oblivious transfer of Crépeau [5]. The QOT protocol can be seen as a construction of quantum oblivious transfer from a black-box for bit commitment [5,19]. Therefore and unlike the classical case, there exists a black-box reduction of quantum oblivious transfer to bit commitment.

Our main contribution consists in showing how any statistically binding quantum bit commitment scheme can be transformed into a statistically concealing one. The construction is obtained by using the QOT protocol together with statistically binding but otherwise computationally concealing commitments (these commitments will be called *initial commitments* in the following). Using the QOT protocol that way, we construct a simple quantum commitment scheme that we show statistically concealing and computationally binding. The construction converts the flavor of the initial commitments after calling them $O(n^2)$ times for n a security parameter. As a byproduct, we show that the QOT protocol is an oblivious transfer that statistically hides one out of the two bits sent and computationally conceals the receiver's selection bit whenever it is used together with statistically binding but computationally concealing commitments instead of perfect commitments given as black-boxes. This extends the security result for the QOT protocol of [5,19] to the computational case. Our reduction of an adversary for the binding condition of the resulting commitment scheme to an adversary for the concealing condition of the initial commitment is expected polynomial-time black-box. Although quantum information has peculiar behaviors adding complexity to the security proofs of cryptographic protocols, we shall see that using quantum oblivious transfer as a primitive allows to return to an essentially classical situation. This might be of independent interest for the construction and analysis of complex quantum protocols.

One consequence of our result is that statistically concealing but computationally binding quantum commitment scheme can be based upon any quantum one-way function using Naor's construction [17] from pseudo-random bit generators. Only the ability to send and receive BB84[1] qubits is required in order to get the new flavor. The scheme can therefore be implemented using current technology. Our result gives more evidences that computational security in 2-party quantum cryptography enjoys different properties than its classical counterpart.

Paper's Organization. We introduce tools and definitions in Sect. 2. The protocol by which the flavor of an originally binding but computationally concealing commitment is transformed into a concealing but computationally binding commitment is described in Sect. 3. The security proof of our construction is given in Sect. 4 and Sect. 5. In Sect. 4, we show that the resulting commitment is computationally binding if the original one was computationally concealing. We then prove in Sect. 5 that if the initial commitment scheme is binding then the resulting one is concealing. We finally conclude in Sect. 6.

2 Preliminaries

2.1 Tools

Let $X \sim B(p)$ be a Bernoulli random variable with probability of success p (when $X = 1$). The following simple argument will be useful:

Hybrid Argument. Let $\mathcal{X} = \{X_1, X_2, \ldots, X_n\}$ be a set of independent random variables $X_i \sim B(p_i)$ for $1 \leq i \leq n$. Then, there exist $1 \leq k < n$ such that,

$$|p_{k+1} - p_k| \geq \frac{|p_n - p_1|}{n}. \tag{1}$$

The result also holds without the absolute values. Later, we shall be given $\mathcal{X}$ without the values of the p_i's but only circuits (quantum or classical) R_i for sampling in each $X_i \in \mathcal{X}$(i.e. $\mathrm{P}(\mathrm{R}_i = 0) = p_i$) and a guarantee that (1) holds for some k. In this scenario, we shall need an algorithm for estimating the p_i's and one for finding k' that satisfies a drop similar to (1).

Estimating the p_i's. Let R be a circuit for sampling in $B(p)$ where $p = q + \frac{1}{p(n)}$, $0 \leq q < 1$ is a known constant, and $p(n)$ is a positive polynomial. It is easy to devise an algorithm $\texttt{LowBound}(\mathrm{R}, q, n)$ that satisfies (see [13] for the proof and the algorithm):

Lemma 1. *For n sufficiently large, $\texttt{LowBound}(\mathrm{R}, q, n)$ returns $\frac{1}{g_n}$ such that $\frac{1}{n^2 p(n)} < \frac{1}{g_n} \leq \frac{1}{p(n)}$ except with probability $2^{-\alpha n}, \alpha > 0$ and after calling R an expected $O(n^5 p(n)^2)$ times.*

Finding a Drop. Let $\mathcal{D}_m(\frac{1}{p(n)}) = \{p_i\}_{i=0}^m$ be a family of Bernoulli distributions with unknown parameters $0 \leq p_i \leq 1$ for every $0 \leq i \leq m$ and such that $p_{k^*} - p_{k^*+1} \geq \frac{1}{p(n)}$ for some $0 \leq k^* < m$. Let S be a sampling circuit for $\mathcal{D}$ that given $0 \leq l \leq m$ runs R_m (i.e. $\mathrm{P}(\mathsf{S}(l) = 1) = 1 - \mathrm{P}(\mathsf{S}(l) = 0) = p_l$). We would like to find κ that exhibits a polynomial drop $p_\kappa - p_{\kappa+1}$ similar to $p_{k^*} - p_{k^*}$. It is not difficult to find an algorithm $\mathtt{FindDrop}$ that finds κ (using the sampling circuit S as a black-box) such that (see [13] for the proof and the algorithm):

Lemma 2. *Given a family of Bernoulli distributions* $\mathcal{D}_m(\frac{1}{p(n)}) = \{p_i\}_{i=1}^m$ *with sampling circuit* S *such that* $p_{k^*} - p_{k^*+1} \geq \frac{1}{p(n)}$ *for some* $0 \leq k^* \leq m - 1$, *algorithm* $\mathtt{FindDrop}(\mathsf{S}, \frac{1}{p(n)}, n)$ *returns* κ *such that* $p_\kappa - p_{\kappa+1} \geq \frac{1}{2p(n)}$ *except with negligible probability* $2^{-\alpha n}, \alpha > 0$ *and after calling* S *at most* $O(m^2 n p(n)^2)$ *times.*

2.2 Notations and Model of Computation

For simplicity, we shall often drop the security parameters associated with protocol executions. When protocols and adversaries are modeled as circuits they should be understood as infinite families of circuits, one circuit for each possible values of the security parameters. We write $poly(n)$ for the set of all positive polynomials.

Let $\mathcal{H}_n$ denote a n-dimensional Hilbert space, that is a complete inner product vector space over the complex numbers. The basis $\{|0\rangle, |1\rangle\}$ denotes the computational or rectilinear or "+" basis for $\mathcal{H}_2$. When the context requires, we write $|b\rangle_+$ to denote the bit b in the rectilinear basis. The diagonal basis, denoted "×", is defined as $\{|0\rangle_\times, |1\rangle_\times\}$ where $|0\rangle_\times = \frac{1}{\sqrt{2}}(|0\rangle + |1\rangle)$ and $|1\rangle_\times = \frac{1}{\sqrt{2}}(|0\rangle - |1\rangle)$. The states $|0\rangle, |1\rangle, |0\rangle_\times$ and $|1\rangle_\times$ are the four BB84 states. For any $x \in \{0,1\}^n$ and $\theta \in \{+, \times\}^n$, the state $|x\rangle_\theta$ is defined as $\otimes_{i=1}^n |x_i\rangle_{\theta_i}$ where $\otimes$ denotes the tensor product. An orthogonal (or von Neumann) measurement of a quantum state in $\mathcal{H}_m$ is described by a set of m orthogonal projections $\mathcal{M} = \{\mathbb{P}_i\}_{i=1}^m$ acting in $\mathcal{H}_m$ thus satisfying $\sum_i \mathbb{P}_i = \mathbb{1}_m$ where $\mathbb{1}_m$ denotes the identity operator in $\mathcal{H}_m$. Each projection or equivalently each index $i \in \{1, \ldots, m\}$ is a possible classical outcome for $\mathcal{M}$.

We model quantum algorithms by quantum circuits built out of a universal set of quantum gates $\mathcal{UG} = \{\mathtt{CNot}, \mathsf{H}, \mathsf{R}_\mathbb{Q}\}$, where $\mathtt{CNot}$ denotes the controlled-NOT, H the one qubit Hadamard gate, and $\mathsf{R}_\mathbb{Q}$ is an arbitrary one qubit non-trivial rotation specified by a matrix containing only rational numbers [2]. The time-complexity of a quantum circuit C is the number of elementary gates $\|C\|_{\mathcal{UG}}$ in C. In addition to the set of gates $\mathcal{UG}$, a quantum circuit is allowed to perform one kind of von Neumann measurement: $\mathcal{M}_+ = \{\mathbb{P}_0^+, \mathbb{P}_1^+\}$ where $\mathbb{P}_0^+ = |0\rangle\langle 0|$ and $\mathbb{P}_1^+ = |1\rangle\langle 1|$ are the two orthogonal projections of the computational basis. $\mathcal{M}_+$ is sometimes called the measurement in the *rectilinear* or *computational* basis. Another von Neumann measurement used by the receiver in the BB84 quantum coding scheme is the measurement in the *diagonal* basis $\mathcal{M}_\times = \{\mathbb{P}_0^\times, \mathbb{P}_1^\times\}$ for $\mathbb{P}_0^\times = \frac{1}{2}(|0\rangle + |1\rangle)(|0\rangle + |1\rangle)^\dagger$ and $\mathbb{P}_1^\times = \frac{1}{2}(|0\rangle - |1\rangle)(|0\rangle - |1\rangle)^\dagger$ where $\dagger$ denotes the transposed-complex conjugate operator. The Hadamard gate H is sufficient to build measurement $\mathcal{M}_\times \in \mathcal{UG}$ from $\mathcal{M}_+$ since $\mathcal{M}_\times = \{\text{H}\mathbb{P}_0^+\text{H}^\dagger, \text{H}\mathbb{P}_1^+\text{H}^\dagger\}$. For $x \in \{0,1\}^n$ and $\beta \in \{+, \times\}^n$ we write $\mathbb{P}_x^\beta \equiv \otimes_{i=1}^n \mathbb{P}_{x_i}^{\beta_i}$. If $|\Psi\rangle \in H_A \otimes H_B$ is a composite quantum state, we write $\mathbb{P}_x^A |\Psi\rangle$ (i.e. $\mathbb{P}_x^A \otimes \mathbb{1}^B |\Psi\rangle$) for the projector applied to the registers in H_A along the state $|x\rangle$ for $x \in \{0,1\}^{\text{Dim}(H_A)}$. The classical output $L(|\Psi\rangle)$ of circuit L is the classical outcomes of all von Neumann measurements $\mathcal{M}_+$ taking place during the computation $L|\Psi\rangle$. If the circuit L accepts two input states of the form $|\Psi_0\rangle \otimes |\Psi_1\rangle$ we may write similarly $L(|\Psi_0\rangle, |\Psi_1\rangle)$ for the classical output.

A 2-party quantum protocol is a pair of interactive quantum circuits (A, B) applied to some initial product state $|x_A\rangle^A \otimes |x_B\rangle^B$ representing A's and B's inputs to the protocol neglecting to write explicitly the states of A's and B's registers that do not encode their respective input to the protocol (thus all in initial states $|0\rangle$). Also, we shall often write $|x_A\rangle^A |x_B\rangle^B$ for the product state without explicitly writing the tensor product $\otimes$. Since communication takes place between A and B, the complete circuit representing one protocol execution may have quantum gates in A and B acting upon the same quantum registers. We write $A \odot B$ for the complete quantum circuit when A is interacting with B. The final composite state $|\Psi_{final}\rangle$ obtained after the execution is then written as $|\Psi_{final}\rangle = (A \odot B)|x_A\rangle^A |x_B\rangle^B$.

2.3   Cryptographic Primitives

The two relevant quantum primitives we shall use heavily in the following are quantum bit commitment and quantum oblivious transfer. They are defined as straightforward quantum generalizations of their classical counterparts.

Quantum Bit Commitment. A quantum bit commitment scheme is defined by two quantum protocols $((C^A, C^B), (O^A, O^B))$ where (C^A, C^B) is a pair of interactive quantum circuits for the committing stage and (O^A, O^B) is a pair of interactive quantum circuits for the opening stage (i.e. A being the committer and B the receiver). The committing stage generates the state $|\Psi_b\rangle = (C^A \odot C^B)|b\rangle^A |0\rangle^B$ upon which the opening stage is executed: $|\Psi_{final}\rangle = (O^A \odot O^B)|\Psi_b\rangle$. The binding condition of a quantum bit commitment is slightly more general than the usual classical definition. An adversary $\tilde{A} = (C^{\tilde{A}}, O^{\tilde{A}})$ is such that $|\tilde{\Psi}\rangle = (C^{\tilde{A}} \odot C^B)|0\rangle^{\tilde{A}}|0\rangle^B$ is generated during the committing stage. The dishonest opening circuit $O^{\tilde{A}}$ tries to open $b \in \{0,1\}$ given as an extra input bit $|b\rangle^{\tilde{A}}$. Given the final state $|\tilde{\Psi}_{final}\rangle = (O^{\tilde{A}} \odot O^B)|b\rangle^{\tilde{A}}|\tilde{\Psi}\rangle$ we define $s_b(n)$ as the probability to open b with success. More precisely, $s_b(n) = \|\mathbb{P}_{OK,b}^B |\tilde{\Psi}_{final}\rangle\|^2$ where $\mathbb{P}_{OK,b}^B$ is Bob's projection operator on the subspace leading to accept the opening of b. An adversary $\tilde{A}$ of the binding condition who can open $b = 0$ with probability at least $s_0(n)$ and open $b = 1$ with probability at least $s_1(n)$ will be called a $(s_0(n), s_1(n))$–*adversary against the binding condition*. We define the concealing and binding criteria similarly to [6]:

(computationally) binding: There exists no positive polynomial $p(n)$ and quantum $(s_0(n), s_1(n))$–adversary $\tilde{A}$ such that $s_0(n) + s_1(n) \geq 1 + \frac{1}{p(n)}$ for n sufficiently large. The scheme is *computationally binding* if we add the restriction that $\|\tilde{A}\|_{\mathcal{UG}} \in poly(n)$.

(computationally) concealing: For every interactive quantum circuit $\tilde{C}^B$ for the committing stage, all quantum circuits $L^{\tilde{B}}$ acting only upon $\tilde{B}$'s registers, all positive polynomials $p(n)$ and n sufficiently large, $\mathrm{P}\left(L^{\tilde{B}}((C^A \odot C^{\tilde{B}})|b\rangle^A|0\rangle^{\tilde{B}}) = b\right) < \frac{1}{2} + \frac{1}{p(n)}$ where the probabilities are taken over $b \in_R \{0,1\}$. The scheme is *computationally concealing* if we add the restriction $\|C^{\tilde{B}}\|_{\mathcal{UG}} + \|L^{\tilde{B}}\|_{\mathcal{UG}} \in poly(n)$.

Note that the concealing and binding conditions are statistical not perfect.

Quantum Oblivious Transfer. A 1–2 *quantum oblivious transfer protocol* [5] involves a sender Alice holding input bits (b_0, b_1) and a receiver Bob holding input $c \in \{0,1\}$. Alice sends (b_0, b_1) to Bob in such a way that Bob receives only b_c and Alice does not get to know c. The receiver must not be able to find $b_{\overline{c}}$ for at at least one $\overline{c} \in \{0,1\}$ and even given b_c. More precisely, a protocol (A, B) for 1–2 quantum oblivious is such that $|\Psi(b_0, b_1, c)\rangle = (A \odot B)|b_0 b_1\rangle^A|c\rangle^B$ allows Bob to recover b_c from applying $\mathcal{M}_+$ upon one of his registers. A protocol for 1–2 quantum oblivious transfer is *(computationally) secure* if it is both

(computationally) secure against the sender: For every quantum sender $\tilde{A}$, all quantum circuit $L^{\tilde{A}}$ acting only on $\tilde{A}$'s registers, all positive polynomials $p(n)$ and n sufficiently large, $\mathrm{P}\left(L^{\tilde{A}}((\tilde{A} \odot B)|00\rangle^{\tilde{A}}|c\rangle^B) = c\right) < \frac{1}{2} + \frac{1}{p(n)}$ where the probabilities are taken over $c \in_R \{0,1\}$. The security is *computational* if we add the restriction $\|L^{\tilde{A}}\|_{\mathcal{UG}} + \|\tilde{A}\|_{\mathcal{UG}} \in poly(n)$.

(computationally) secure against the receiver: For every quantum receiver $\tilde{B}$, all quantum circuits $L^{\tilde{B}}$ acting only on $\tilde{B}$'s registers, all positive polynomials $p(n)$ and n sufficiently large, there exists a random variable c with possible outcome 0 or 1 depending on $(A \odot \tilde{B})|b_0 b_1\rangle^A|0\rangle^{\tilde{B}}$ satisfying $\mathrm{P}\left(L^{\tilde{B}}((A \odot \tilde{B})|b_0 b_1\rangle^A|0\rangle^{\tilde{B}}, |b_{\overline{c}}\rangle^{\tilde{B}}) = b_{\overline{c}}\right) < \frac{1}{2} + \frac{1}{p(n)}$ where the probabilities are taken over $b_0, b_1 \in_R \{0,1\}$. The security is *computational* if we add the restriction $\|\tilde{B}\|_{\mathcal{UG}} + \|L^{\tilde{B}}\|_{\mathcal{UG}} \in poly(n)$.

As for bit commitment, the security is statistical not perfect.

3   The Protocols

In this section, we first describe the QOT protocol of [5] for 1-2 oblivious transfer. Then, we describe a simple quantum bit commitment scheme QBC, using QOT as a sub-protocol, that transforms any binding bit commitment scheme into a concealing one. Throughout this paper, we assume for simplicity that quantum transmission is error-free.

3.1   QOT Protocol

The QOT protocol [5] is based upon the BB84 quantum coding scheme [1]. If the receiver (Bob) of a random BB84 qubit $|s\rangle_\beta$, $s \in_R \{0,1\}$, $\beta \in_R \{+, \times\}$ measures it in basis $\hat{\beta} \in_R \{+, \times\}$ upon reception, then a noisy classical communication of bit s from Alice to Bob is implemented. Moreover, if later on Alice announces β, then Bob knows that he received s whenever $\beta = \hat{\beta}$ and an uncorrelated bit whenever $\beta \neq \hat{\beta}$. The QOT protocol amplifies this process in order to get a secure 1–2 oblivious transfer. In order to ensure that Bob measures the BB84 qubits upon reception, bit commitments are used. Bob commits upon each measurement basis[1] and measurement outcome right after the quantum transmission. Alice then verifies in random positions that Bob has really measured the transmitted qubits by testing that whenever $\beta = \hat{\beta}$ then Bob's classical outcome $r \in \{0,1\}$ is such that $r = s$.

In the following, we assume that Alice and Bob have access to some bit commitment scheme BBC in order for Bob to commit upon the measurement bases of the received qubits together with the outcomes. Since the two commitments are made together, we write $\mathrm{BBC}(x, y)$ where $x \in \{+, \times\}$ and $y \in \{0,1\}$ for the commitments of both the measurement basis and the measurement outcome. BBC may be given as a black-box for bit commitment or may be provided from some computational assumption. We denote by $\mathrm{Open\text{-}BBC}(x, y)$ the opening stage of $\mathrm{BBC}(x, y)$. Protocol $\mathrm{QOT}(b_0, b_1)(c)$ achieves the oblivious transfer of bit b_c.

Protocol 1 ($\mathrm{QOT}(b_0, b_1)(c)$)

  1: *For $1 \leq i \leq 2n$*
 - *Alice picks $s_i \in_R \{0,1\}$, $\beta_i \in_R \{+, \times\}$*
 - *Alice sends to Bob a qubit π_i in state $|s_i\rangle_{\beta_i}$*
 - *Bob picks a basis $\hat{\beta}_i \in_R \{+, \times\}$, measures π_i in basis $\hat{\beta}_i$, and obtains the outcome $r_i \in \{0,1\}$*

  2: *For $1 \leq i \leq n$*
 - *Bob runs $\mathrm{BBC}(\hat{\beta}_i, r_i)$ and $\mathrm{BBC}(\hat{\beta}_{n+i}, r_{n+i})$ with Alice*
 - *Alice picks $f_i \in_R \{0,1\}$ and announces it to Bob*
 - *Bob runs $\mathrm{Open\text{-}BBC}(\hat{\beta}_{nf_i+i}, r_{nf_i+i})$*
 - *Alice verifies that $\beta_{nf_i+i} = \hat{\beta}_{nf_i+i} \Rightarrow s_{nf_i+i} = r_{nf_i+i}$, otherwise she rejects the current execution.*
 - *if $f_i = 0$ then Alice sets $\beta_i \leftarrow \beta_{n+i}$ and $s_i \leftarrow s_{n+i}$ and Bob sets $\hat{\beta}_i \leftarrow \hat{\beta}_{n+i}$ and $r_i \leftarrow r_{n+i}$*

  3: *Alice announces her choices of bases $\beta_1, \beta_2, \ldots, \beta_n$ to Bob*

  4: *Bob chooses at random and announces two subsets of positions $J_0, J_1 \subset \{1, 2, \ldots, n\}$, $|J_0| = |J_1| = \frac{n}{3}$, $J_0 \cap J_1 = \emptyset$, and $\forall i \in J_c, \beta_i = \hat{\beta}_i$.*

  5: *Alice computes and announces $\hat{b}_0 = \bigoplus_{j \in J_0} s_j \oplus b_0$ and $\hat{b}_1 = \bigoplus_{j \in J_1} s_j \oplus b_1$*

  6: *Bob receives $\langle \hat{b}_0, \hat{b}_1 \rangle$ and computes $b_c = \bigoplus_{i \in J_c} r_i \oplus \hat{b}_c$*

---

[1] The bases $\{+, \times\}$ are encoded in $\{0,1\}$.

Known Security Results. The correctness and the security of the QOT protocol against the sender (Alice) has been reduced to the concealing property of BBC in [5]. The security against the receiver (Bob) has been provided by Yao in [19] given the commitment scheme BBC is perfectly binding. That is, given BBC is a perfect black-box for bit commitment then QOT is secure against any dishonest Bob irrespectively of his computing power.

3.2 QBC Protocol Using QOT

Given a binding but computationally concealing bit commitment scheme BBC in QOT the following simple commitment scheme will be shown concealing and computationally binding.

Protocol 2 (QBC(b))

1: QBC-COMMIT(b)
  – For $1 \leq j \leq n$
    • Alice prepares $a_{0j} \in_R \{0,1\}$ and $a_{1j} = a_{0j} \oplus b$
    • Bob prepares $c_j \in_R \{0,1\}$
    • Alice and Bob execute QOT(a_{0j}, a_{1j})(c_j) and Bob receives the result d_j

2: QBC-OPEN(b)
  • Alice announces b
  • For $1 \leq j \leq n$
    • Alice announces a_{0j} and a_{1j}
    • Bob verifies that $b = a_{0j} \oplus a_{1j}$ and $d_j = a_{c_j j}$

A commitment to bit b is done by sending through 1–2 oblivious transfers n pairs of bits $\{(a_{0j}, a_{1j})\}_{j=1}^n$ such that $a_{0j} \oplus a_{1j} = b$. The concealing condition relies on the security of QOT against a malicious receiver and the binding condition relies on the security against a malicious sender. Intuitively, QBC appears concealing since for $1 \leq j \leq n$ Bob cannot obtain information on more than one of the two bits (a_{0j}, a_{1j}) input in the j-th QOT and so, cannot determine $b = a_{0j} \oplus a_{1j}$. Similarly, QBC should be binding since for all $1 \leq j \leq n$ Alice needs to change the bit $a_{\bar{d}_j j}$ not selected by Bob in order to change her commitment.

More Notations. In the following we shall have to identify the variables generated during all calls to QOT in QBC. For that purpose, we use the following notation:

– π_i^j is the i-th qubit sent in the j-th call to QOT in QBC.
– $\beta_i^j \in \{+, \times\}$ is the basis β_i announced by Alice in the jth run of QOT in QBC. Note that a malicious Alice can send π_i^j other than $|0\rangle_{\beta_i^j}$ and $|1\rangle_{\beta_i^j}$.
– $\hat{\beta}_i^j \in \{+, \times\}$ is the basis used by Bob to measure π_i^j in the j-th call to QOT.
– $r_i^j \in \{0,1\}$ is the outcome of Bob's measurement of π_i^j in basis $\hat{\beta}_i^j$.
– $\hat{r}_i^j \in \{0,1\}$ is Carl's outcome for measurement of π_i^j in basis β_i^j.
– $J^j = (J_0^j, J_1^j)$ is the pair of sets announced by Bob in the jth run of QOT.

We denote by bold lowercases the values for all executions at one glance: $\boldsymbol{\beta} = \{\beta_i^j\}_{i,j}, \hat{\boldsymbol{\beta}} = \{\hat{\beta}_i^j\}_{i,j}, \boldsymbol{r} = \{r_i^j\}_{i,j}$, and $\hat{\boldsymbol{r}} = \{\hat{r}_i^j\}_{i,j}$. We denote by $\hat{\boldsymbol{b}}_0 = \hat{b}_0^1, \ldots, \hat{b}_0^n$ and $\hat{\boldsymbol{b}}_1 = \hat{b}_1^1, \ldots, \hat{b}_1^n$ the bits announced by Alice at step 5 of each call to QOT. Similarly, we denote by $\boldsymbol{a} = (\boldsymbol{a}_0, \boldsymbol{a}_1) = (a_{01}, a_{11}), (a_{02}, a_{12}), \ldots, (a_{0n}, a_{1n}) \in \{0,1\}^{2n}$ Alice's announcements during the opening stage. We also denote $\boldsymbol{J}_0 = J_0^1, \ldots, J_0^n$ and $\boldsymbol{J}_1 = J_1^1, \ldots, J_1^n$ all sets announced by Bob and we write $\boldsymbol{J} = (\boldsymbol{J}_0, \boldsymbol{J}_1)$. Let $\boldsymbol{c} = c_1, \ldots, c_n$ be all selection bits used by Bob and let $\boldsymbol{d} = d_1, \ldots, d_n$ be all bits received by QOT. We write $\boldsymbol{J}_{\boldsymbol{c}} = J_{c_1}^1, J_{c_2}^2, \ldots, J_{c_n}^n$ for all set of positions corresponding to qubits measured by Bob in bases announced by Alice.

4 The Binding Condition

In the following section, we show that QBC is secure against any Alice (the sender) who cannot break the concealing condition of the initial commitment scheme BBC. BBC is used in the calls to QOT in order for Bob to commit on his measurements and outcomes.

Simplified Version of QOT. In our analysis of the binding condition of QBC, we shall assume that the opening of half of the commitments in step 2 of QOT doesn't occur. The opening of the commitments allows Alice to make sure that Bob measured the qubits received in QOT upon reception. This test is not relevant to the binding condition of QBC.

Protocol 3 (QOT*$(b_0, b_1)(c)$)

 1: ...*step 1 of protocol 2*
 2: *For* $1 \le i \le n$
  – *Bob runs* BBC$(\hat{\beta}_i, r_i)$ *and* BBC$(\hat{\beta}_{n+i}, r_{n+i})$ *with Alice*
  – *Alice picks* $f_i \in_R \{0,1\}$ *and announces it to Bob*
  – *if* $f_i = 0$ *then Alice sets* $\beta_i \leftarrow \beta_{n+i}$ *and* $s_i \leftarrow s_{n+i}$ *and Bob sets* $\hat{\beta}_i \leftarrow \hat{\beta}_{n+i}$ *and* $r_i \leftarrow r_{n+i}$
 3–6: ...*as steps 3 to 6 in protocol 2.*

We omit the proof of the following simple lemma:

Lemma 3. *If* QOT* *is secure against the sender then* QOT *is secure against the sender.*

Throughout Sect. 4, we shall assume implicitly calls to QOT* in QBC instead of calls to QOT. This simplifies the analysis and according to Lemma 3, it can be done without loss of generality.

4.1 How to Prove the Binding Condition

In order to show that QBC is computationally binding, we introduce intermediary protocols that will allow us to bridge the security of QBC with the known security of QOT given black-boxes for bit commitments. Let's consider the following four modified protocols:

U-QOT: Protocol QOT except that in step 2, Bob commits to random values. In other words, for $1 \leq i \leq n$, Bob runs BBC(u_{0i}, u_{1i}) and BBC(u_{2i}, u_{3i}) with $u_{0i}, u_{2i} \in_R \{+, \times\}$ and $u_{1i}, u_{3i} \in_R \{0, 1\}$.

M-QOT: The same as U-QOT but a third party named Carl, for $1 \leq i \leq n$, intercepts the i-th qubit π_i sent by Alice in step 1, measures in basis β_i (announced by Alice in step 3) and sends the resulting state to Bob.

U-QBC: Protocol QBC using U-QOT.

M-QBC: Protocol QBC using M-QOT.

The security against any dishonest sender in U-QOT and M-QOT is a direct consequence of the analysis in [5]. Since the commitments upon measurements do not carry any information about Bob's measurement, Alice cannot obtain any information about his selection bit c. The security is information-theoretic, no complexity assumption on Alice's computing power is required.

We reduce the security of the binding condition of QBC to the security of the concealing condition of BBC in two steps:

1. Using Lemmas 4 and 5, we conclude in Lemma 6 that U-QBC is binding. The modified protocol M-QBC is used for reducing the security of U-QBC to the security of U-QOT. Carl's presence allows one to reduce the analysis to an essentially classical argument which becomes simpler than working from U-QBC directly.
2. Theorem 1 establishes the desired result using the fact that an adversary for the binding condition of QBC cannot be an adversary of U-QBC (Lemma 6). It is shown how to construct an adversary for the concealing condition of BBC given an adversary for the binding condition of QBC.

4.2   U-QBC Is Binding

In this section, we show that U-QBC is binding (Lemma 6) using Lemmas 4 and 5 as intermediary steps.

First, we show that an adversary against the binding condition of U-QBC can be transformed into an adversary against the binding condition of M-QBC.

Lemma 4. *If there exists a $(s_0(n), s_1(n))$-adversary $\tilde{A}$ against the binding condition of U-QBC there also exists a $(s_0(n), s_1(n))$-adversary A^* against the binding condition of M-QBC.*

Proof. We observe first that $\tilde{A}$'s announcement of β at step 3 of U-QOT commutes with step 2. That is, since only commitments to random values are received, $\tilde{A}$ can determine β without Bob's commitments. Moreover, $\tilde{A}$ could simulate the commitments on her own and then determine β before the qubits are sent to Bob at step 1. Let A^* be the quantum adversary that does that. If $\tilde{A}$ provides a $(s_0(n), s_1(n))$–advantage in U-QBC then so it is for A^*. We now show that A^* is also an adversary for the binding condition of M-QBC.

Now assume for simplicity and without loss of generality that, Bob in U-QBC or Bob and Carl in M-QBC wait until after Alice announces $\boldsymbol{a} = (\boldsymbol{a}_0, \boldsymbol{a}_1)$ before measuring all qubits received. It is easy to verify that this can always be done since nothing in the committing stage of U-QBC or M-QBC relies on those measurements' outcomes (i.e. since the commitments are made to random values). Clearly, postponing measurements do not influence Alice's probability of success at the opening stage.

Let $V = (\boldsymbol{\beta}, \boldsymbol{J}, \hat{\boldsymbol{b}}_0, \hat{\boldsymbol{b}}_1, \boldsymbol{c}, \boldsymbol{a})$ be the partial view in U-QBC or in M-QBC up to Alice's announcement of $\boldsymbol{a}$ (and b since for all $1 \leq j \leq n$, $a_{j0} \oplus a_{j1} = b$) in the opening stage. Let $\boldsymbol{V}_U$ and $\boldsymbol{V}_M$ be the random variable for the partial view in U-QBC and M-QBC respectively. By construction we have that for all $V = (\boldsymbol{\beta}, \boldsymbol{J}, \hat{\boldsymbol{b}}_0, \hat{\boldsymbol{b}}_1, \boldsymbol{c}, \boldsymbol{a})$, $p(V) = \mathrm{P}(\boldsymbol{V}_U = V) = \mathrm{P}(\boldsymbol{V}_M = V)$. Moreover, we have that for all partial views V, the joint states $|\Psi_U(V)\rangle$ for U-QBC and $|\Psi_M(V)\rangle$ for M-QBC satisfy $|\Psi_U(V)\rangle = |\Psi_M(V)\rangle$. Let $\mathcal{V}_b = \{(\boldsymbol{\beta}, \boldsymbol{J}, \hat{\boldsymbol{b}}_0, \hat{\boldsymbol{b}}_1, \boldsymbol{c}, \boldsymbol{a}) | (\forall 1 \leq j \leq n)[a_{j0} \oplus a_{j1} = b]\}$ be the set of partial views corresponding for Alice to open bit b. Given V, Bob's test will succeed if he gets $\boldsymbol{d} = \boldsymbol{a}_c = a_{1c_1}, a_{2c_2}, \ldots, a_{nc_n}$ after measuring the qubits in positions in $\boldsymbol{J}_c$ using Alice's bases β_i^j for all $i \in J_{c_j}^j$ and $j \in \{1, \ldots, n\}$. Let $\mathcal{M}_{test}(V) = \{\mathbb{Q}_{ok}^V, \mathbb{1} - \mathbb{Q}_{ok}^V\}$ be the measurement allowing Bob to test Alice's announcement when she unveils b given partial view $V \in \mathcal{V}_b$. $\mathbb{Q}_{ok}^V$ is the projection for the state of all qubits received in positions in $\boldsymbol{J}_c$ into the subspace corresponding to parity $d_j = a_{jc_j}$ for all $j \in \{1, \ldots, n\}$. More precisely, $\mathbb{Q}_{ok}^V = \bigotimes_{j=1}^n \sum_{x \in T(V,j)} \mathbb{P}_x^{\boldsymbol{\beta}(V,j)}$ where $T(V, j) = \{x \in \{0,1\}^{|J_{c_j}^j|} | \oplus_i x_i = a_{jc_j} \oplus \hat{b}_{c_j}^j\}$ and $\boldsymbol{\beta}(V, j) = \{\beta_i^j | i \in J_{c_j}^j\}$ for all $j \in \{1, \ldots, n\}$. Let $s_b'(n)$ be the probability of success when A^* opens b in M-QBC. We get that

$$s_b(n) = \sum_{V \in \mathcal{V}_b} p(V) \| \mathbb{Q}_{ok}^V | \Psi_U(V)\rangle \|^2 = \sum_{V \in \mathcal{V}_b} p(V) \| \mathbb{Q}_{ok}^V \mathbb{Q}_{ok}^V | \Psi_M(V)\rangle \|^2 = s_b'(n) \quad (2)$$

since the only difference between U-QBC and M-QBC is that in the former case both Carl and Bob measure the qubits in positions in $\boldsymbol{J}_c$ with the same measurement $\mathcal{M}_{test}$ (this is why we have $\mathbb{Q}_{ok}^V \mathbb{Q}_{ok}^V = \mathbb{Q}_{ok}^V$ in (2)). Carl's measurements for positions in $\boldsymbol{J}_{\bar{c}}$ are irrelevant to the success probability. The result follows. $\square$

Next, we reduce the binding condition of M-QBC to the security against the sender in M-QOT. We show that from any successful adversary against the binding condition of M-QBC one can construct an adversary able to extract non-negligible information about Bob's selection bit in M-QOT. Carl's measurements in M-QBC allows one to use a classical argument for most of the reduction thus simplifying the proof that U-QBC is binding.

Lemma 5. *If there exists a $(s_0(n), s_1(n))$-adversary $\tilde{A} = (C^{\tilde{A}}, O^{\tilde{A}})$ against the binding condition of M-QBC with $s_0(n) + s_1(n) \geq 1 + \frac{1}{p(n)}$ for some positive polynomial $p(n)$, then there also exists a cheating sender A^* for M-QOT.*

Proof. Let a_{j0}' and a_{j1}' be the two input bits for the j-th call to M-QOT computed according to Carl's outcomes $\hat{\boldsymbol{r}}$. Let $\boldsymbol{V}$ be the random variable for the joint view

$(\boldsymbol{a}, \boldsymbol{a}', \boldsymbol{d}, \boldsymbol{c})$ for an execution of the committing and the opening stages of M-QBC between $\tilde{A}$ and an honest receiver B and where $\tilde{A}$ is opening a random bit $b \in_R \{0, 1\}$. Without loss of generality, we assume the announcements made by $\tilde{A}$ to be consistent, that is $a_{0i} \oplus a_{1i} = b$ for $1 \leq i \leq n$ when she opens bit b. Given $V = (\boldsymbol{a}, \boldsymbol{a}', \boldsymbol{d}, \boldsymbol{c})$, we define the ordered set $S(V) = \{j | a'_{j0} \oplus a'_{j1} \neq a_{j0} \oplus a_{j1}\} \subseteq \{1, \ldots, n\}$ of calls to M-QOT for which given view V Alice's announcement of $\boldsymbol{a}$ disagree with Carl's outcomes $\boldsymbol{a}'$. Given the ordered set $S(V) = \{\sigma_1, \sigma_2, \ldots, \sigma_s\}$, let $X_j(V) \in \{0, 1\}$ for $1 \leq j \leq s$ be defined as

$$X_j(V) = \begin{cases} 0 \text{ if } d_{\sigma_j} \neq a_{\sigma_j c_{\sigma_j}} \\ 1 \text{ if } d_{\sigma_j} = a_{\sigma_j c_{\sigma_j}}. \end{cases}$$

We let $X(V) = X_1(V), \ldots, X_{l(V)}(V)$ for $l(V) = \min(|S(V)|, \lceil \frac{n}{2} \rceil)$. Clearly, for $\tilde{A}$ to open with success given V, we must have $X(V) = 1^{l(V)}$. Note that $P\left(|S(\boldsymbol{V})| \geq \frac{n}{2}\right) \geq \frac{1}{2}$ since for at least one choice of b, $|S(\boldsymbol{V})| \geq \frac{n}{2}$ given that $\boldsymbol{V}$ always describes a consistent opening. We easily get that

$$P\left(X(\boldsymbol{V}) = 1^{\lceil \frac{n}{2} \rceil}\right) = P\left(X(\boldsymbol{V}) = 1^{l(\boldsymbol{V})}\right) - P\left(X(\boldsymbol{V}) = 1^{l(\boldsymbol{V})} \wedge l(\boldsymbol{V}) < \frac{n}{2}\right)$$

$$\geq \frac{1}{2}(s_0(n) + s_1(n)) - \frac{1}{2}P\left(X(\boldsymbol{V}) = 1^{l(\boldsymbol{V})} \mid l(\boldsymbol{V}) < \frac{n}{2}\right) \geq \frac{1}{2p(n)}. \quad (3)$$

Since $\sum_{x \in \{0,1\}^{\lceil \frac{n}{2} \rceil}} P\left(X(\boldsymbol{V}) = x\right) \leq 1$, for n sufficiently large there exists a string $\hat{y}^0 \in \{0, 1\}^{\lceil \frac{n}{2} \rceil}$ such that $P\left(X(\boldsymbol{V}) = \hat{y}^0\right) \leq \frac{1}{4p(n)}$. Let ρ be the number of zeros in $\hat{y}^0$ and $R(\hat{y}^0) = \{r_1, r_2, \ldots, r_\rho\} \subseteq \{1, \ldots, \lceil \frac{n}{2} \rceil\}$ be the ordered set of positions $1 \leq r \leq \lceil \frac{n}{2} \rceil$ where $\hat{y}_r^0 = 0$. We now define for $1 \leq j \leq \rho$ the hybrid strings $\hat{y}^j = \hat{y}_1^j \hat{y}_2^j \ldots \hat{y}_{\lceil \frac{n}{2} \rceil}^j$ between $\hat{y}^0$ and $1^{\lceil \frac{n}{2} \rceil}$:

$$\hat{y}_i^j = \begin{cases} 1 \text{ if } i = r_k \text{ for } k \leq j \\ \hat{y}_i^0 \text{ Otherwise.} \end{cases}$$

Hence, $P\left(X(\boldsymbol{V}) = \hat{y}^\rho = 1^n\right) - P\left(X(\boldsymbol{V}) = \hat{y}^0\right) \geq \frac{1}{4p(n)}$ and we conclude by an hybrid argument that there exist $1 \leq k^* \leq \rho$ such that

$$P\left(X(\boldsymbol{V}) = \hat{y}^{k^*}\right) - P\left(X(\boldsymbol{V}) = \hat{y}^{k^*-1}\right) \geq \frac{1}{\rho 4p(n)} \geq \frac{1}{2(n+1)p(n)}. \quad (4)$$

Note that $\hat{y}^{k^*}$ and $\hat{y}^{k^*-1}$ differs only by the bit in position r_{k^*} where they respectively have a 1 and a 0.

A^* uses $\tilde{A}$ and $B = (C^B, O^B)$ in the following way: after choosing $h \in_R \{1, \ldots, n\}$, it makes $\tilde{A}$ interact with a simulated honest receiver B for M-QBC except for the h-th execution of M-QOT for which $\tilde{A}$ interacts with the targeted receiver for M-QOT. Let $V = (\boldsymbol{a}, \boldsymbol{a}', \boldsymbol{d}, \boldsymbol{c})$ be the view generated during the execution. Given A^*'s view, algorithm L^{A^*} produces a guess $\tilde{c}$ for Bob's selection bit $c = c_h$ in M-QOT as follows:

- If $|S(V)| \geq \lceil \frac{n}{2} \rceil$, $h = \sigma_{r_{k*}}$ and $\forall i \in \{1, \ldots, \lceil \frac{n}{2} \rceil\} \setminus \{r_{k*}\}$, $X_i(V) = \hat{y}_i^{k^*}$, then $\tilde{c} \in \{0, 1\}$ is defined such that $a_{h\tilde{c}} = a'_{h\tilde{c}}$ (which necessarily exists since $h \in S(V)$),
- Otherwise, $\tilde{c} \in_R \{0, 1\}$.

Let $\mathcal{T}(V)$ be the event of a successful test in the previous computation. Since independently $|S(V)| \geq \frac{n}{2}$ with probability at least $\frac{1}{2}$, $h = \sigma_{r_{k*}}$ with probability $\frac{1}{n}$, and $\forall i \in \{1, \ldots, \lceil \frac{n}{2} \rceil\} \setminus \{r_{k*}\}$, $X_i(V) = \hat{y}_i^{k^*}$ with probability $P\left(X(V) = \hat{y}^{k^*}\right) + P\left(X(V) = \hat{y}^{k^*-1}\right)$, we have that

$$P\left(\mathcal{T}(V)\right) \geq \frac{P\left(X(V) = \hat{y}^{k^*}\right) + P\left(X(V) = \hat{y}^{k^*-1}\right)}{2n}. \tag{5}$$

Given $\mathcal{T}(V)$, the guess $\tilde{c}$ is the only value for Bob's selection bit c that would lead to $X(V) = \hat{y}^{k^*}$ instead of $X(V) = \hat{y}^{k^*-1}$ (the two strings are the only possible given $\mathcal{T}(V)$). We get that

$$P\left(\tilde{c} = c | \mathcal{T}(V)\right) = \frac{P\left(X(V) = \hat{y}^{k^*}\right)}{P\left(X(V) = \hat{y}^{k^*}\right) + P\left(X(V) = \hat{y}^{k^*-1}\right)}. \tag{6}$$

It follows that (A^*, L^{A^*}) is a cheating sender for M-QOT since

$$P\left(\tilde{c} = c\right) = \frac{1}{2}(1 - P\left(\mathcal{T}(V)\right)) + P\left(\mathcal{T}(V)\right) P\left(\tilde{c} = c | \mathcal{T}(V)\right)$$

$$\geq \frac{1}{2} + \frac{1}{8n(n+1)p(n)}. \tag{7}$$

$\square$

Using Lemmas 3, 4 and 5 together with the fact that M-QOT is unconditionally secure against the sender [5], we get the desired result:

Lemma 6. *Protocol* U-QBC *is binding.*

As we shall see next, Lemma 6 helps a great deal in proving that QBC is computationally binding.

4.3   QBC Is Binding when BBC Is Concealing

In the following, we conclude that QBC is computationally binding whenever BBC is computationally concealing. We use the fact that U-QBC is binding (Lemma 6) in order to use any adversary against the binding condition of QBC as a distinguisher between random (U-QBC) and real (QBC) commitments for some hybrids between U-QBC and QBC.

Theorem 1. *If there exists a $(s_0(n), s_1(n))$-adversary $\tilde{A} = (C^{\tilde{A}}, O^{\tilde{A}})$ against the binding condition of QBC with $s_0(n) + s_1(n) \geq 1 + \frac{1}{p(n)}$ for a positive polynomial $p(n)$, then there exists a quantum receiver $C^{\tilde{B}}$ in BBC and a quantum algorithm $L^{\tilde{B}}$ such that $P\left(L^{\tilde{B}}((C^A \odot C^{\tilde{B}})|b\rangle^A|0\rangle^{\tilde{B}}) = b\right) \geq \frac{1}{2} + \Omega(\frac{1}{n^4 p(n)})$ whenever $b \in_R \{0, 1\}$ and where $C^{\tilde{B}}$ calls $\tilde{A}$ an expected $O(n^5 p(n)^2)$ times.*

Proof. Let $B = (C^B, O^B)$ be the circuits for the honest receiver in QBC and let $\mathcal{A}$ be an honest committer in BBC. Given $\tilde{A}$, we construct a receiver $C^{\tilde{B}}$ in BBC from which a bias for $\mathcal{A}$'s committed bit can be extracted. Remember that the only difference between U-QBC and QBC is that a honest receiver commits to random bits instead of his measurements and outcomes. There are $4n$ calls to Commit-BBC per QOT (U-QOT) for a total of $4n^2$ during the committing stage of QBC (U-QBC). Let's note as *significant* the committed bits specified by the protocol QOT (to measurements and outcomes) and as *random* the ones specified by the protocol U-QOT (to random bits). We describe hybrids in between QBC and U-QBC by letting the number of significant and random commitments vary. Let QBC^k be protocol QBC but where the first k commitments out of $4n^2$ are made to random values. We have that U-QBC $\equiv$ QBC$^{4n^2}$ is binding whereas $\tilde{A}$ is a $(s_0(n), s_1(n))$–adversary for the binding condition of QBC$^0 \equiv$ QBC. Let $s_b^k(n)$ be the probability that $\tilde{A}$ succeeds when opening $b \in \{0, 1\}$ in QBC^k for $0 \leq k \leq 4n^2$. Defining $\hat{s}^k(n) = (s_0^k(n) + s_1^k(n))/2$, we get that $\hat{s}^0(n) \geq \frac{1}{2} + \frac{1}{2p(n)}$ and from Lemma 6, $\hat{s}^{4n^2}(n) < \frac{1}{2} + \frac{1}{e(n)}$ where $e(n) > p(n)$ for all $p(n) \in poly(n)$ and n sufficiently large. By the hybrid argument, there exists $0 \leq k^* \leq 4n^2 - 1$ such that for n sufficiently large,

$$\hat{s}^{k^*}(n) - \hat{s}^{k^*+1}(n) \geq \frac{1}{9n^2 p(n)}. \tag{8}$$

Hence, $\mathcal{D}_{4n^2}(\frac{1}{9n^2 p(n)}) = \{\hat{s}^i(n)\}_{i=0}^{4n^2}$ is a family of Bernoulli distributions that satisfies the condition of Lemma 2. The sampling circuit S is easy to construct given $\tilde{A}$ and B. Upon classical input $|l\rangle$ for $0 \leq l \leq 4n^2$, S runs $\tilde{A}$ and B except that the first l commitments sent from B to $\tilde{A}$ (using BBC) are made to random values instead of the measurements $\hat{\beta}$ and the outcomes r. $\tilde{A}$ then opens a random bit $b \in_R \{0, 1\}$. If B accepts the opening of b then $S(|l\rangle) = 1$ otherwise it returns $S(|l\rangle) = 0$. Circuit S is therefore a sampling circuit for $\mathcal{D}_{4n^2}(\frac{1}{9n^2 p(n)})$ such that $\|S\|_{\mathcal{U}\mathcal{G}} \in O(\|\tilde{A}\|_{\mathcal{U}\mathcal{G}})$ assuming without loss of generality that $\|B\|_{\mathcal{U}\mathcal{G}} \in O(\|\tilde{A}\|_{\mathcal{U}\mathcal{G}})$.

We now construct the adversary $C^{\tilde{B}}$ for the concealing condition of BBC given $\tilde{A}$. In order to use algorithm FindDrop (defined in Sect. 2.1), $C^{\tilde{B}}$ must first determine a lower bound $\frac{1}{p'(n)}$ for the drop $\frac{1}{9n^2 p(n)}$. This is done by finding a lower bound $\tilde{p}(n)$ for $\frac{1}{2p(n)}$ and then setting $p'(n) = \frac{5n^2}{\tilde{p}(n)}$. $C^{\tilde{B}}$ computes $\tilde{p}(n) =$ LowBound(S$_0$, $\frac{1}{2}$, n) (defined in Sect. 2.1) where S$_0$ is the circuit S with the input bits fixed to $|0\rangle$. From Lemma 1, LowBound returns $\tilde{p}(n)$ such that $\frac{1}{2n^2 p(n)} \leq \tilde{p}(n) \leq \frac{1}{2p(n)}$ except with negligible probability and after an expected $O(n^5 p(n)^2)$ calls to S$_0$.

Now $C^{\tilde{B}}$ can use FindDrop(S, $\frac{1}{p'(n)}$, n) with the family of distributions $\mathcal{D}_{4n^2}(\frac{1}{p'(n)}) = \{\hat{s}^i(n)\}_{i=0}^{4n^2}$ which exhibits a drop $\frac{1}{p'(n)}$ except with negligible probability. From Lemma 2, $C^{\tilde{B}}$ gets $0 \leq \kappa \leq 4n^2 - 1$ such that

$$\hat{s}^{\kappa}(n) - \hat{s}^{\kappa+1}(n) \geq \frac{1}{2p'(n)} \tag{9}$$

except with negligible probability. The value of κ is obtained after calling S (including the calls to S_0 in LowBound) an expected $O(n^5 p(n)^2)$ times.

$C^{\tilde{B}}$ then uses κ for attacking the concealing condition of BBC in the following way: It makes $\tilde{A}$ and B interact (where $\tilde{A}$ opens $b \in_R \{0,1\}$) as in $\text{QBC}^{\kappa+1}$ except that the $(\kappa+1)$-th random commitment is provided by the committer $\mathcal{A}$ in BBC. Let $b \in \{0,1\}$ be the bit committed by $\mathcal{A}$. Let V be the random variable for the view generated during the interaction between $\tilde{A}$ and B when $\tilde{A}$ opens the random bit. Let $c_{\kappa+1}(V) \in \{0,1\}$ be the bit that B would have committed if the $(\kappa+1)$-th commitment was significant. The distinguisher $L^{\tilde{B}}$ (which is classical given the view V) returns the guess $\tilde{b}$ for b the following way:

- If V is a successful opening then $\tilde{b} = c_{\kappa+1}(V)$,
- Otherwise, $\tilde{b} \in_R \{0,1\}$.

Let $\mathcal{V}_{ok}^{\kappa+1}$ be the set of views for $\text{QBC}^{\kappa+1}$ resulting in a successful opening and let $\mathcal{G}$ be the set of values κ for which (9) holds. We have $\hat{s}^{\kappa}(n) = P\left(V \in \mathcal{V}_{ok}^{\kappa+1} | c_{\kappa+1}(V) = b\right)$ and $\hat{s}^{\kappa+1}(n) = \frac{1}{2}\left(P\left(V \in \mathcal{V}_{ok}^{\kappa+1} | c_{\kappa+1}(V) \neq b\right) + P\left(V \in \mathcal{V}_{ok}^{\kappa+1} | c_{\kappa+1}(V) = b\right)\right)$ which, using (9), leads to

$$P\left(V \in \mathcal{V}_{ok}^{\kappa+1} \wedge c_{\kappa+1}(V) \neq b\right) \leq P\left(V \in \mathcal{V}_{ok}^{\kappa+1} \wedge c_{\kappa+1}(V) = b\right) - \frac{1}{2p'(n)}.$$

Since we also have $P\left(V \in \mathcal{V}_{ok}^{\kappa+1}\right) = P\left(V \in \mathcal{V}_{ok}^{\kappa+1} \wedge c_{\kappa+1}(V) \neq b\right) + P\left(V \in \mathcal{V}_{ok}^{\kappa+1} \wedge c_{\kappa+1}(V) = b\right)$, we get

$$P\left(\tilde{b} = b | \kappa \in \mathcal{G}\right) = P\left(V \in \mathcal{V}_{ok}^{\kappa+1} \wedge c_{\kappa+1}(V) = b\right) + \frac{1}{2}\left(1 - P\left(V \in \mathcal{V}_{ok}^{\kappa+1}\right)\right)$$

$$\geq \frac{1}{2}\left(1 + \frac{1}{2p'(n)}\right).$$

Since $P\left(\tilde{b} = b\right) \geq P\left(\kappa \in \mathcal{G}\right) P\left(\tilde{b} = b | \kappa \in \mathcal{G}\right)$ and $P\left(\kappa \in \mathcal{G}\right) \geq 1 - 2^{-\alpha n}, \alpha > 0$ (Lemma 1) we finally get that $(C^{\tilde{B}}, L^{\tilde{B}})$ is an adversary for the concealing condition of BBC providing a bias in $\Omega(\frac{1}{p'(n)}) = \Omega(\frac{1}{n^4 p(n)})$ after calling $\tilde{A}$ an expected $O(n^5 p(n)^2)$ times. $\qquad\square$

5   The Concealing Condition

We now reduce the concealing condition of QBC to the security of QOT against a malicious receiver.

Lemma 7. *If there exists a quantum circuit $C^{\tilde{B}}$ for the receiver in* Commit-QBC *and a quantum algorithm $L^{\tilde{B}}$ acting only on $\tilde{B}$'s registers such that $P\left(L^{\tilde{B}}((C^A \odot C^{\tilde{B}})|b\rangle^A |0\rangle^{\tilde{B}}) = b\right) \geq \frac{1}{2} + \frac{1}{p(n)}$ for some positive polynomial $p(n)$ and an honest committing circuit C^A for $b \in_R \{0,1\}$, then there also exists a cheating receiver (B^*, L^{B^*}) for QOT.*

Proof. For the receiver $C^{\tilde{B}}$ and C^A described in the statement, we have

$$P\left(L^{\tilde{B}}((C^A \odot C^{\tilde{B}})|1\rangle^A|0\rangle^{\tilde{B}}) = 1\right) -$$

$$P\left(L^{\tilde{B}}((C^A \odot C^{\tilde{B}})|0\rangle^A|0\rangle^{\tilde{B}}) = 1\right) \geq \frac{2}{p(n)}.$$

Let's define a modification of an honest committing circuit for QBC, noted $C^{\tilde{A}}$, which is the same as C^A but takes a string $\hat{f} \in \{0,1\}^n$ instead of a bit b and sends in the i-th call to QOT the bits $a_{0i} \in_R \{0,1\}$ and $a_{1i} = a_{0i} \oplus \hat{f}_i$ for $1 \leq i \leq n$. The circuit C^A with input b is equivalent to $C^{\tilde{A}}$ with input b^n. Once again, by an hybrid argument, there exists $1 \leq k^* \leq n$ such that for

$$P\left(L^{\tilde{B}}((C^{\tilde{A}} \odot C^{\tilde{B}})|1^{k^*}0^{n-k^*}\rangle^{\tilde{A}}|0\rangle^{\tilde{B}}) = 1\right) -$$

$$P\left(L^{\tilde{B}}((C^{\tilde{A}} \odot C^{\tilde{B}})|1^{k^*-1}0^{n-k^*+1}\rangle^{\tilde{A}}|0\rangle^{\tilde{B}}) = 1\right)$$

$$\geq \frac{2}{np(n)}.$$

With such value k^*, B^* cheats an honest sender A' for $\text{QOT}(e_0, e_1)(0)$ in the following way: it makes $C^{\tilde{B}}$ interact with $C^{\tilde{A}}$ with input $(1^{k^*-1}?0^{n-k^*})$ for Commit-QBC except for the k^*-th call to QOT where it makes $C^{\tilde{B}}$ interact with the targeted sender A' with inputs $e_0, e_1 \in_R \{0,1\}$. Then, knowing e_c for $c \in \{0,1\}$, we take the output of $L^{\tilde{B}}$, b' say, and compute a guess $e_c \oplus b'$ for $e_{\bar{c}}$. For this algorithm L^{B^*} we have

$$P\left(L^{B^*}((A' \odot B^*)|e_0 e_1\rangle^A|0\rangle^{B^*}, |e_c\rangle^{B^*}) = e_{\bar{c}}\right) = P\left(b' = e_0 \oplus e_1\right)$$

$$\geq \frac{1}{2} + \frac{1}{np(n)}$$

where the probabilities are taken over $e_0, e_1 \in_R \{0,1\}$. $\square$

From Yao's result [19] and Lemma 7 it is straightforward to conclude that QBC is concealing.

6   Conclusion and Open Questions

Having shown in Theorem 1, that a computationally concealing BBC results in a computationally binding QBC and, from Lemma 7 together with Yao's result [19], that no adversary against the concealing condition of QBC exists, we conclude with our main result:

Theorem 2. *If* BBC *is binding and computationally concealing then* QBC *is concealing and computationally binding.*

For security parameter n, the reduction of an adversary $(C_n^{\tilde{B}}, L_n^{\tilde{B}})$ for the concealing condition of BBC to an adversary $\tilde{A}_n$ for the binding condition of QBC is expected polynomial-time black-box. The adversary $\{(C_n^{\tilde{B}}, L_n^{\tilde{B}})\}_{n>0}$ is a uniform family of quantum circuits whenever $\{\tilde{A}_n\}_{n>0}$ is uniform. It is an interesting open problem to find an exact polynomial-time black-box reduction.

One consequence of Theorem 2 is that concealing commitment schemes can be built from any quantum one-way function. We first observe that Naor's commitment scheme [17] is also secure against the quantum computer if the pseudo-random bit generator (PRBG) it is based upon is secure against the quantum computer. This follows from the fact that any quantum circuit able to distinguish between commitments to 0 and 1 is also able to distinguish a truly random sequence from a pseudo-random one. To complete the argument, we must make sure that given a quantum one-way function one can construct a PRBG resistant to quantum distinguishers. A tedious but not difficult exercise allows to verify that the classical construction of [11] results in a PRBG secure against quantum distinguishers given it is built from quantum one-way functions. We get the following corollary which is not known to hold in the classical case:

Corollary 1. *Both binding but computationally concealing and concealing but computationally binding quantum bit commitments can be constructed from quantum one-way functions.*

It would be interesting to find a concealing quantum bit commitment scheme directly constructed from one-way functions which improves the complexity of our construction. Is it possible to find a non-interactive concealing commitment scheme from the same complexity assumption or are such constructions inherently interactive? It is also unclear whether or not perfectly concealing schemes can be based upon any quantum one-way function.

Although we assumed in this paper a perfect quantum channel, our construction should also work with noisy quantum transmission [3]. It would be nice to provide the analysis for this general case.

References

1. C. H. Bennett and G. Brassard. Quantum cryptography: Public key distribution and coin tossing. In *Proceedings of IEEE International Conference on Computers, Systems, and Signal Processing*, pages 175–179. IEEE, 1984.
2. BARENCO, A., C. H. BENNETT, R. CLEVE, D. P. DIVINCENZO, N. MARGOLUS, P. SHOR, T. SLEATOR, J. SMOLIN and H. WEINFURTER, "Elementary Gates for Quantum Computation", *Physical Review A*, vol. 52, no 5, November 1995, pp. 3457–3467.
3. BENNETT, C. H., G. BRASSARD, C. CRÉPEAU and M.-H. SKUBISZEWSKA, "Practical Quantum Oblivious Transfer", *Advances in Cryptology : CRYPTO '91 : Proceedings*, Lecture Notes in Computer Science, vol. 576, Springer-Verlag, August 1992, pp. 362–371.
4. BRASSARD, G., D. CHAUM and C. CRÉPEAU, "Minimum Disclosure Proofs of Knowledge", *Journal of Computing and System Science*, vol. 37 , 1988, pp. 156–189.

5. CRÉPEAU, C., "Quantum Oblivious Transfer", *Journal of Modern Optics*, vol. 41, no 12, December 1994, pp. 2445–2454. A preliminary version of this work appeared in CRÉPEAU, C. and J. KILIAN, "Achieving oblivious transfer using weakened security assumptions", *Proceedings of 29th IEEE Symposium on the Foundations of Computer Science*, October 1988, pp. 42–52.

6. DUMAIS, P., D. MAYERS, and L. SALVAIL, "Perfectly Concealing Quantum Bit Commitment From Any Quantum One-Way Permutation", *Advances in Cryptology : EUROCRYPT '00 : Proceedings*, Lecture Notes in Computer Science, vol. 1807, Springer-Verlag, 2000, pp. 300–315.

7. GOLDREICH, O., and L. LEVIN, "A Hard-Core Predicate for Any One-Way Function", *Proceedings of the 21st ACM Symposium on Theory of Computing*, 1989, pp. 25–32.

8. GOLDWASSER, S., S. MICALI and C. RACKOFF, "The Knowledge Complexity of Interactive Proof Systems", *SIAM Journal on Computing*, vol. 18, 1989, pp. 186–208.

9. GOLDREICH, O., S. MICALI, and A. WIGDERSON, "Proofs that Yield Nothing but their Validity or All Language in NP Have Zero-Knowledge Proof Systems", *Journal of the ACM*, vol. 38, no 1, 1991, pp. 691–729.

10. GOLDREICH, O., S. MICALI, and A. WIGDERSON, "How to play any mental game or a completeness theorem for protocols with honest majority", *Proceedings of the 19th ACM Symposium on Theory of Computing*, 1987, pp. 218–229.

11. HÅSTAD, J., R. IMPAGLIAZZO, L. LEVIN and M. LUBY "A pseudo-random generator from any one-way function", *SIAM Journal on Computing*, vol. 28, no 4, 1999, pp. 1364–1396.

12. IMPAGLIAZZO, R. and S. RUDICH, "Limits on Provable Consequences of One-Way Permutations", *Advances in Cryptology : CRYPTO '88 : Proceedings*, Lecture Notes in Computer Science, vol. 403, Springer-Verlag, 1989, pp. 2–7.

13. LÉGARÉ, F., "Converting the flavor of a quantum bit commitment", M.Sc. thesis, School of Computer Science, McGill University, 2001. Supervised by C. Crépeau. Thesis available at `http://www.cs.McGill.ca/~crepeau/students.html`.

14. LO, H.–K. and H. F. CHAU, "Is quantum Bit Commitment Really Possible?", *Physical Review Letters*, vol. 78, no 17, April 1997, pp. 3410–3413.

15. MAYERS, D., "The Trouble With Quantum Bit Commitment", available at `http://xxx.lanl.gov/abs/quant-ph/9603015`.

16. MAYERS, D., "Unconditionally Secure Quantum Bit Commitment is Impossible", *Physical Review Letters*, vol. 78, no 17, April 1997, pp. 3414–3417.

17. NAOR, M., "Bit Commitment Using Pseudo-Randomness", *Journal of Cryptology*, vol. 4, 1991, pp. 151–158.

18. NAOR, M., R. OSTROVSKY, R. VENTKATESAN, and M. YOUNG, "Perfect Zero-Knowledge Arguments For NP Using Any One-Way Permutation", *Journal of Cryptology*, vol. 11, no 2, 1998, pp. 87–108.

19. YAO, A. C., "Security of Quantum Protocols Against Coherent Measurements", *Proceedings of the 27th ACM Symposium on Theory of Computing*, 1995, pp. 67–75.

Cryptographic Counters
and Applications to Electronic Voting

Jonathan Katz[1], Steven Myers[2], and Rafail Ostrovsky[3]

[1] Telcordia Technologies and
Department of Computer Science, Columbia University.
`jkatz@cs.columbia.edu`
[2] Department of Computer Science, University of Toronto[†]
`myers@cs.toronto.edu`
[3] Telcordia Technologies, Inc., 445 South Street, Morristown, NJ 07960.
`rafail@research.telcordia.com`

Abstract. We formalize the notion of a *cryptographic counter*, which allows a group of participants to increment and decrement a cryptographic representation of a (hidden) numerical value privately and robustly. The value of the counter can only be determined by a trusted authority (or group of authorities, which may include participants themselves), and participants cannot determine any information about the increment/decrement operations performed by other parties.

Previous *efficient* implementations of such counters have relied on fully-homomorphic encryption schemes; this is a relatively strong requirement which not all encryption schemes satisfy. We provide an alternate approach, starting with any encryption scheme homomorphic over the additive group $\mathbb{Z}_2$ (i.e., 1-bit XOR). As our main result, we show a general and efficient reduction from any such encryption scheme to a general cryptographic counter. Our main reduction does not use additional assumptions, is efficient, and gives a novel implementation of a general counter. The result can also be viewed as an efficient construction of a general n-bit cryptographic counter from any 1-bit counter which has the additional property that counters can be added securely.

As an example of the applicability of our construction, we present a cryptographic counter based on the quadratic residuosity assumption and use it to construct an efficient voting scheme which satisfies universal verifiability, privacy, and robustness.

1 Introduction

1.1 Cryptographic Counters

In this paper we present an efficient and secure protocol for calculating the sum of integers, where each integer is held privately by a single participant. Although it is clear that this can be achieved via the completeness results for multi-party computation (see [14] for a complete review of multi-party computation and

---

[†] Work done while the author was at Telcordia Technologies.

B. Pfitzmann (Ed.): EUROCRYPT 2001, LNCS 2045, pp. 78–92, 2001.
© Springer-Verlag Berlin Heidelberg 2001

related results), such constructions are only of theoretical interest as they are too inefficient to be of practical use. In order to construct our secure addition protocol, we introduce an abstraction we call a *cryptographic counter* that may be of independent interest. In particular, such counters may have a variety of applications, especially as subroutines in larger multi-party computations. We give a formal definition of cryptographic counters, and provide a construction based on any encryption scheme homomorphic over the additive group $\mathbb{Z}_2$.

Informally, a cryptographic counter is a public string which can be viewed as an encryption of a value such that the value is hidden from all participants except a trusted authority (who holds some secret key). Only the trusted authority can decrypt and thereby determine the value of the counter, whereas all participants have the ability to increment or decrement (*update*) the counter by an arbitrary amount. Information about updates (e.g., whether the counter was incremented or decremented) is kept hidden from all other participants. We also consider *restricted* cryptographic counters for which the set of legal update operations is constrained in some publicly-known way.

Previous constructions of cryptographic counters (in the context of voting schemes) have relied on what we call *fully-homomorphic* encryption. Informally, this is an encryption scheme for which, for any $n_0 > 0$, there is some choice of the security parameter such that the resulting encryption is homomorphic over (the additive group) $\mathbb{Z}_n$, where $n \geq n_0$. It is clear how a cryptographic counter can be constructed given this strong property (the difficult aspects of previous constructions were providing efficient proofs of validity and achieving threshold decryption). In this paper, we provide a construction of an n-bit cryptographic counter based on any 1-bit cryptographic counter that also allows secure addition (mod 2) of multiple counters. This immediately implies a construction from any encryption scheme homomorphic over $\mathbb{Z}_2$. As a concrete example, we present an *efficient* n-bit counter based only on the quadratic residuosity assumption.

Addition is a useful function to compute privately, as many of the currently-proposed applications of secure multi-party computation rely heavily on summing secret values held by different individuals. It has particular relevance to the problem of secure electronic voting, in which each participant holds a vote which is either 0 or 1, and the participants wish to determine the tally without revealing individual votes. As an example of the applicability of cryptographic counters, we use them to build a secure voting scheme and compare it to previously-proposed solutions. In particular, ours is the first efficient construction of a voting scheme which is not based on fully-homomorphic encryption.

1.2   Secure Electronic Voting

An electronic voting scheme is a protocol allowing voters to cast a vote by interacting with a set of authorities who collect the votes, tally them, and publish the final result. There are a variety of properties which may be desired of an electronic voting scheme; however, the cryptographic literature has traditionally focused on the following three requirements:

Privacy ensures that an individual's vote is kept hidden from (any reasonably-sized coalition of) other voters and even the authorities themselves.

Universal Verifiability means that any party, including a passive observer, can be convinced that all votes cast were valid and that the final tally was computed correctly.

Robustness guarantees that the final tally can be correctly computed even in the presence of faulty behavior of a number of parties.

It is furthermore desirable to minimize the interaction between parties. In particular, voters should not have to interact with each other to cast a vote or (ideally) to prove validity of votes, and the authorities should be able to remain off-line until the election is concluded. Other features are not considered in the present work. For example, information-theoretic privacy is sometimes required [8], while we only require computational privacy. Receipt-freeness [2] and preventing vote-duplication can be achieved by other means (see, for example, [17]) and are not considered here.

Many voting schemes meeting the above requirements have been proposed [6,3,4,8,9,23,10]. However, all previously-known schemes achieving universal verifiability rely on *fully-homomorphic* encryption schemes, where the homomorphism is over additive group $\mathbb{Z}_n$ and n is larger than the number of voters (our use of the term "fully-homomorphic" is explained above). One typical paradigm is as follows: say voter i wishes to cast vote v_i, where, for a valid vote, we have $v_i \in \{0, 1\}$. To vote, voter i publicly posts[1] $\mathcal{E}_{pk}(v_i)$, the encryption of v_i under some public key established by the set of authorities. When everyone has voted, the authorities compute the product of the encryptions (which can be publicly computed) and decrypt the result; this gives the correct final tally since:

$$\mathcal{D}_{sk}\left(\mathcal{E}_{pk}(v_1)\cdots\mathcal{E}_{pk}(v_N)\right) = v_1 + \cdots + v_N,$$

where equality holds by the homomorphic properties of the encryption scheme. Depending on the level of trust in the authorities, they may also provide a (publicly verifiable) proof that decryption was done correctly. In this way, everyone is assured that all votes were correctly counted.

Many examples of fully-homomorphic encryption schemes are known (for example: [12,6,21]). The voting schemes of [6,3,4] are based on the r-th residuosity assumption, those of [8,9,23] are based on the discrete logarithm assumption in prime groups, and the scheme of [10] is based on hardness of deciding residue classes in $\mathbb{Z}_{N^2}^*$. Even so, it is interesting to determine the minimal assumptions under which an efficient voting protocol can be constructed.

We show how privacy and universal verifiability can be achieved without fully-homomorphic encryption. Our construction uses an n-bit counter which, in turn, is constructed from any encryption scheme homomorphic over $\mathbb{Z}_2$ (i.e., the

---

[1] This might be accompanied by a proof of validity, but for simplicity we focus here on that portion of the protocol which relies on the homomorphic properties of the encryption.

Table 1. Efficiency of some voting schemes. L is the number of voters, M is the number of authorities, k_1 is a security parameter, and 2^{-k_2} is a bound on the probability of cheating (in [8,9], the probability of cheating is 2^{-k_1}). Computation is measured in bitwise operations, assuming multiplication of k-bit numbers requires $\mathcal{O}(k^2)$ operations.

	Size of Vote + Proof	Voter Computation	Authority Computation
[8]	$\mathcal{O}(k_1 M)$	$\mathcal{O}(k_1^3 M)$	$\mathcal{O}(k_1^3 L)$
[9]	$\mathcal{O}(k_1)$	$\mathcal{O}(k_1^3)$	$\mathcal{O}(k_1^3 L)$
Present work	$\mathcal{O}(k_1 k_2 \log L)$	$\mathcal{O}(k_1^2 k_2 \log L)$	$\mathcal{O}(k_1^2 \log L + L)$

1-bit XOR operation). Using as a specific example the well-studied encryption scheme based on the hardness of deciding quadratic residuosity [16], we show how to achieve robustness as well.

Often, basing a result on a weaker assumption results in an impractical scheme. However, our resulting voting scheme is efficient enough to be practical. A comparison of the efficiency of our construction with those of [8,9] appears in Table 1. Our simplest solution, while being both size- and computation-efficient, requires sequential execution and hence $\mathcal{O}(L)$ rounds (as compared with previous solutions which require $\mathcal{O}(1)$ rounds). We discuss ways of dealing with this issue in Section 5.

2 Definitions

In this section we formalize the notion of a cryptographic counter. Although related notions have been folklore in the cryptographic community (particularly in the context of electronic voting), a formal definition has, to the best of our knowledge, not previously appeared.

COUNTERS. In order to more easily define a *cryptographic* counter, we first need a formal definition of a counter.

Definition 1. *An n-counter consists of a set S along with a pair of algorithms (D, T) in which:*

- $S = \{s_1, \ldots\}$ *represents the set of states of the counter.*
- D, *the decoding algorithm, is a deterministic algorithm which takes as input a state $s \in S$ and returns a number $i \in \mathbb{Z}_n$. This defines a mapping from states in S to numbers in the range $[0, n - 1]$.*
- T, *the transition algorithm, is a probabilistic algorithm which takes as input a state $s \in S$ and an integer $i \in \mathbb{Z}_n$ and returns a state $s' \in S$. This function defines legal update operations on the counter.*

We require that for all $s \in S$ and $i \in \mathbb{Z}_n$, if $s' \leftarrow T(s, i)$, then $D(s') = D(s) + i \bmod n$.

Note that subtraction of integer i can be done by simply computing the inverse of i in $\mathbb{Z}_n$ and adding $-i$ using the transition algorithm.

CRYPTOGRAPHIC COUNTERS. We now turn to the definition of a cryptographic counter. We first define its components, and follow this with definitions of security against two types of adversaries: honest-but-curious and malicious. All algorithms are assumed to run in time polynomial in the security parameter k, and n is fixed independently of k.

Definition 2. *A cryptographic n-counter is a triple of algorithms $(\mathcal{G}, D, T)$ in which:*

- *$\mathcal{G}$, the key generation algorithm, is a probabilistic algorithm that on input 1^k outputs a public key/secret key pair (pk,sk) and a string s_0. The secret key, in turn, implicitly defines[2] an associated set of states S_{sk}. It is the case that $s_0 \in S_{sk}$.*
- *D, the decryption algorithm, is a deterministic algorithm that takes as input a secret key sk and a string s. If $s \in S_{sk}$, then D outputs an integer $i \in \mathbb{Z}_n$. Otherwise, D outputs $\perp$.*
- *T, the transition algorithm, is a probabilistic algorithm that takes as input the public key pk, a string s, and an integer $i \in \mathbb{Z}_n$ and outputs a string s'.*

For any (pk, sk) output by $\mathcal{G}(1^k)$, define $D' = D(sk, \cdot)$ and $T' = T(pk, \cdot, \cdot)$. Then we require that the set S_{sk} along with algorithms (D', T') define an n-counter. Furthermore, we require that $D'(s_0) = 0$ (this represents initialization of the counter to 0).

SECURITY (HONEST-BUT-CURIOUS). We briefly describe the attack scenario before giving the formal definition. Adversary A is given the public key and the initial state s_0. The adversary then outputs[3] a sequence of integers $i_1, \ldots, i_\ell \in \mathbb{Z}_n$. The state is updated accordingly; that is, the transition algorithm T is run ℓ times, generating $s_1, \ldots, s_\ell$. All intermediate states are given to the adversary, who then outputs $x_0, x_1 \in \mathbb{Z}_n$. A bit b is selected at random, and the counter is incremented by x_b to give state s^*. The adversary, given s^*, must then guess the value of b.

Definition 3. *We say that cryptographic n-counter $(\mathcal{G}, D, T)$ is secure against honest-but-curious adversaries if, for all poly-time adversaries A, the following is negligible (in k):*

$$\left| \Pr \left[\begin{array}{l} (pk, sk, s_0) \leftarrow \mathcal{G}(1^k) \\ (i_1, \ldots, i_\ell) \leftarrow A(1^k, pk, s_0) \\ s_1 \leftarrow T(pk, s_0, i_1); \ldots; s_\ell \leftarrow T(pk, s_{\ell-1}, i_\ell) \\ (x_0, x_1) \leftarrow A(s_1, \ldots, s_\ell) \\ b \leftarrow \{0, 1\} \\ s^* \leftarrow T(pk, s_\ell, x_b) \\ b' \leftarrow A(s^*) \end{array} : b' = b \right] - 1/2 \right|.$$

---

[2] Note that membership in S_{sk} may not be efficiently decidable when given only pk. We require, however, that membership *is* efficiently decidable, given sk.

[3] These integers may be chosen adaptively, but for simplicity we present the non-adaptive case here. Note that the construction of Section 3.2 achieves security against an adaptive adversary as well.

SECURITY (MALICIOUS). An honest-but-curious adversary is restricted to having the increment operations (which he must distinguish between) performed on a state distributed according to the output of the transition algorithm T. A malicious adversary, in contrast, is allowed to select the state to be incremented freely. In fact, we allow the adversary to select any *string* to be incremented by T; this allows us to deal with the case in which there is no efficient way to determine whether a string s is a valid state (i.e., whether $s \in S_{sk}$).

Definition 4. *We say that cryptographic n-counter* $(\mathcal{G}, D, T)$ *is secure against malicious adversaries if, for all poly-time adversaries A, the following is negligible (in k):*

$$\left| \Pr \left[\begin{array}{l} (pk, sk, s_0) \leftarrow \mathcal{G}(1^k) \\ (s, x_0, x_1) \leftarrow A(1^k, pk, s_0) \\ b \leftarrow \{0,1\} \\ s^* \leftarrow T(pk, s, x_b) \\ b' \leftarrow A(s^*) \end{array} : b' = b \right] - 1/2 \right|.$$

VERIFIABLE COUNTERS. It may sometimes be useful to verify whether transitions were indeed computed correctly. For example, when using a counter for voting, it should be publicly verifiable that each voter acted in a correct manner. We therefore define the notion of a *verifiable cryptographic counter* as follows:

Definition 5. *A verifiable cryptographic n-counter is a tuple* $(\mathcal{G}, D, T, V)$ *such that:*

- $(\mathcal{G}, D, T)$ *is a cryptographic n-counter.*
- V, the verification algorithm, *is a probabilistic algorithm satisfying completeness and soundness for all (pk, sk) output by $\mathcal{G}$, as follows:*
  1. *(Completeness) For all $s \in S_{sk}$, if $s' \leftarrow T(pk, s, i)$ for some $i \in \mathbb{Z}_n$, then:*

     $$V(pk, s, s') = 1.$$

     (Note that V does not require i as input.)
  2. *(Soundness) For all s and all strings s' such that for all i, s' is not in the range of $T(pk, s, i)$, the following probability is negligible (in k):*

     $$\Pr[V(pk, s, s') = 1].$$

RESTRICTED COUNTERS. Definitions 1, 2, and 5 may be modified to allow for the possibility that although the counter can store values in $\mathbb{Z}_n$, update operations are restricted to some subset of $\mathbb{Z}_n$. We call counters with this property *restricted*. An illustrative example is a counter used in a voting scheme. Although the counter needs to be able to store values up to L (the number of voters), it may be required to restrict update operations to the set $\{0, 1\}$ (representing a yes/no vote). Modifications to the definitions are straightforward.

ADDITIVE COUNTERS. The transition algorithms described above take an old state s and an integer i and output a new state s' which represents the old value incremented by i. However, definitions 1 and 2 may be modified such that the transition algorithm takes an old state s and a second state s' and then outputs a new state s'' which represents the old value incremented by the value stored in s'. Such counters are termed *additive*. Note that additive cryptographic n-counters include the case of homomorphic encryption over $\mathbb{Z}_n$; yet, the former are more general since the transition algorithm need not be multiplication. Definitions 3 and 4 can be modified for the case of additive counters in the natural way.

3   Constructing Cryptographic Counters

In Sections 3.1 and 3.2, we describe the construction of a cryptographic n-counter based on any 1-bit additive cryptographic counter. We also discuss the extension to the case of verifiable cryptographic counters. In Section 3.4, using as a particular example the encryption scheme based on quadratic residuosity [16] (see Appendix A), which is homomorphic over $\mathbb{Z}_2$, we give an efficient construction of a verifiable cryptographic n-counter where update operations are restricted to $\{0, 1\}$. This provides a natural foundation for a voting protocol; we discuss this connection further in Section 4.

3.1   Linear Feedback Shift Registers

Before presenting our main result, we provide an introduction to the theory of linear feedback shift registers; a more comprehensive treatment can be found in [20,19]. Let $r_1, r_2, \ldots \in \{0, 1\}$ be a sequence of elements (called *registers*) satisfying the k-th order linear recurrence relation:

$$r_{j+k} = b_k r_{j+k-1} + \cdots + b_1 r_j, \tag{1}$$

where $b_i \in \{0, 1\}$ (throughout this section, addition is over the field $\mathbb{Z}_2$). The sequence $r_1, r_2, \ldots$ is called a *linear recurring sequence*. Once the terms $r_1, \ldots, r_k$ have been fixed, the rest of the sequence is uniquely determined. Define the j-th state of this sequence to be the vector $(r_j, \ldots, r_{j+k-1})$. Equation (1) defines transitions between these states: given state $s = (r_1, \ldots, r_k)$, the next state $s' = (r'_1, \ldots, r'_k)$ can be computed as follows:

$$r'_i = \begin{cases} r_{i+1} & 1 \leq i < k \\ f(r_1, \ldots, r_k) & i = k \end{cases},$$

where the function f is given by (1) as:

$$f(r_1, \ldots, r_k) = b_k r_k + \cdots + b_1 r_1.$$

This sequence of states defines a linear feedback shift register (LFSR). For the present application, it is important to note that f can be computed using XOR operations only.

Since an LFSR has a finite set of states, the sequence of states eventually repeats. The number of states which appear before the first state repeats (and the sequence begins again) is called the *period*. Clearly, an LFSR with period n can be used to count from 0 to $n - 1$: choose an arbitrary initial state giving rise to a sequence of period n, label this initial state "0", and label every succeeding state by one more than the label of its predecessor.

It is possible to associate with every LFSR (whose underlying recurrence relation is given by Equation (1)) the characteristic polynomial $g(x) = x^k - b_k x^{k-1} - \cdots - b_1$. The period of an LFSR is related to the order of its characteristic polynomial. In particular, if the characteristic polynomial of an LFSR is *primitive*[4], then the LFSR has maximum possible period $2^k - 1$ (assuming the initial state of the LFSR is not the zero vector) [20,19]. Primitive polynomials can be generated efficiently using a probabilistic algorithm [22]. It is thus possible to efficiently construct an LFSR which counts from 0 to $n - 1$ using the minimum possible $\lceil \log_2 n \rceil$ registers (each representing a single bit).

Given a state s of an LFSR (and assuming knowledge of the initial state), it is easy to decode the state and determine the number it represents by either counting down from s to the initial state, or counting up from the initial state until state s is reached. This requires time $\mathcal{O}(n)$. This procedure is fast, however, even for large[5] n, since each state transition consists of only simple, *bitwise* manipulations (shifts and XORs). More efficient approaches are mentioned in Section 3.3.

3.2   General Construction of a Cryptographic Counter

Theorem 1. *An additive cryptographic 2-counter secure against honest-but-curious (resp. malicious) adversaries implies the existence of a cryptographic n-counter secure against honest-but-curious (resp. malicious) adversaries, for all n of the form $n = 2^x - 1$.*

Sketch of Proof   An encryption scheme homomorphic over (the additive group) $\mathbb{Z}_2$ is an example of an additive cryptographic 2-counter secure against honest-but-curious adversaries. For ease of exposition, we describe the construction of a cryptographic n-counter using an encryption scheme $(\mathcal{G}, \mathcal{E}, \mathcal{D})$ which is homomorphic over $\mathbb{Z}_2$; it should be clear, however, that a substantially-similar construction yields a cryptographic n-counter starting from *any* additive cryptographic 2-counter.

We show how to use the encryption scheme as a building block to construct a cryptographic n-counter. First, note that an LFSR (as described in Section 3.1) is an n-counter. The idea behind the construction is as follows: since only XOR operations are needed to effect transitions, the encryption scheme allows

---

[4] A polynomial $g \in \mathbb{Z}_2[x]$ of degree k is primitive if the smallest integer N for which $g|(x^N - 1)$ is $N = 2^k - 1$.

[5] For a typical voting scheme, n will be on the order of the number of voters. So, even for the U.S. election, we have n only (roughly) 10^8.

a participant to change the counter without leaking any information about the transition. Below is a complete description of the protocol (here, $\ell = \lceil \log_2 n \rceil$):

Key Generation Algorithm $\mathcal{G}'(1^k)$:

1. Run $\mathcal{G}(1^k)$ to generate public key pk_0 and secret key sk_0.
2. Generate a primitive polynomial $g \in \mathbb{Z}_2[x]$ of degree ℓ using [22].
3. Set $r_1 = \mathcal{E}_{pk_0}(1)$ and $r_2 = \mathcal{E}_{pk_0}(0), \ldots, r_\ell = \mathcal{E}_{pk_0}(0)$.
4. Set $s_0 = (r_1, \ldots, r_\ell)$, $sk = (sk_0, g)$, and $pk = (pk_0, g)$. Output pk, sk, and s_0.

Transition Algorithm [defined for $i \in \mathbb{Z}_n$] $T((pk_0, g), (r_1, \ldots, r_\ell), i)$:

1. Polynomial g defines (nonzero) $f(r_1, \ldots, r_\ell) = b_\ell r_\ell + \cdots + b_1 r_1$ (see Section 3.1).
2. Repeat the following procedure i times[6]:
   (a) Set $r_1' = r_2; \ldots; r_{\ell-1}' = r_\ell$.
   (b) Set $r_\ell' = \prod_{i=1}^{\ell} r_i^{b_i}$.
   (c) Set $r_1 = r_1'; \ldots; r_\ell = r_\ell'$.
3. Set $r_i' = r_i \cdot \mathcal{E}_{pk_0}(0)$, for $1 \leq i \leq \ell$. Output $s' = (r_1', \ldots, r_\ell')$.

Decryption Algorithm $D(sk = (sk_0, g), s = (r_1, \ldots, r_\ell))$:

1. Let $r_i^* = \mathcal{D}_{sk_0}(r_i)$, for $1 \leq i \leq \ell$.
2. Let $s^* = (r_1^*, \ldots, r_k^*)$
3. Increment the LFSR defined by polynomial g, beginning with initial state $(1, 0, \ldots, 0)$, until reaching state s^*. Let t be the number of transitions made. Output t.

The protocol described above is a cryptographic n-counter secure against an honest-but-curious adversary. To see this, fix n. The size of the LFSR, ℓ, is thus a constant (independent of the security parameter). A simple hybrid argument shows that an adversary cannot distinguish between random representations of any two states of the counter. Therefore, an adversary cannot gain any information about the current value of the counter, nor about transitions made. We leave a formal proof to the full version of the paper.

Note that if we start with a cryptographic 2-counter secure against malicious adversaries, the above construction is also secure against malicious adversaries. When using an arbitrary encryption scheme homomorphic over $\mathbb{Z}_2$, the above construction is secure against malicious adversaries if it can be efficiently determined (given pk) whether a string represents a valid ciphertext[7]; in this case, the transition algorithm must first check whether every register in s represents a valid ciphertext before computing s' (if this is not true, it aborts).   □

---

[6] This algorithm can be made significantly more efficient to run in time polynomial in $\log n$. This is discussed briefly in Section 3.3.

[7] For example, in the case of encryption using quadratic residuosity, it is possible to tell whether a string C is a valid ciphertext by checking that the Jacobi symbol of C is 1.

In order to make the above construction verifiable, only a few changes are needed. First, we include a random string τ in the public key. Additionally, we change the transition algorithm so that after s' has been output, we append a non-interactive zero-knowledge proof (NIZK) [5] using random string τ that the transition from s to s' was valid. The verification algorithm V runs the proof-verification algorithm for the NIZK proof. If the proof verification succeeds, the verification algorithm outputs 1; otherwise, it outputs 0. A verifiable, restricted n-counter can be constructed in a similar way.

3.3   Observations on the Cryptographic Counter Construction

Linear feedback shift registers have an algebraic interpretation: the state of an ℓ-bit LFSR represents an element of $GF^*(2^\ell)$. Incrementing the counter corresponds to multiplication of the state by a generator, g, of the multiplicative group in $GF^*(2^\ell)$. This allows for two important gains in efficiency, which are highlighted below.

First, the counter may be efficiently updated by values larger than 1. In particular, the counter may be incremented by value i in only $\mathcal{O}(\ell^2 \log i)$ steps, as opposed to the $\mathcal{O}(\ell \cdot i)$ steps used in the transition function of Section 3.2.

Next, note that the state of the LFSR can be viewed as an element of the form g^j in $GF^*(2^\ell)$. Therefore, one can use algorithms for solving the discrete logarithm problem to determine the value represented by the state of the LFSR,. In particular, it is relatively straightforward to determine the value of an ℓ-bit LFSR in time $\sqrt{2^\ell}$, and an algorithm due to Coppersmith [7] allows decoding in time $\mathcal{O}(2^{\ell^{1/3}(\log^{2/3} \ell)})$.

3.4   An Efficient Cryptographic Counter

The well-known encryption scheme based on quadratic residuosity [16] (see Appendix A) is homomorphic over $\mathbb{Z}_2$. Application of Theorem 1 (see also footnote 7) shows that the construction outlined there results in a cryptographic counter secure against malicious adversaries when instantiated with this encryption scheme. If we are interested in verifiability, however, the generic construction of Section 3.2 will be impractical unless there exists an efficient NIZK proof that the transition algorithm was executed correctly. In the case of quadratic residuosity, we show that efficient NIZK proofs are possible. Since we are interested in eventual applications to electronic voting, we focus on the case of a restricted counter where transitions are limited to either no change in the counter (a 0 vote) or incrementing the counter by 1 (a 1 vote).

Consider the cryptographic counter protocol of Section 3.2, instantiated with encryption based on quadratic residues. Let N be a Blum integer which is part of the associated public key. The string $s = (r_1, \ldots, r_\ell)$ (with $r_i \in \mathbb{Z}_N^{+1}$) is a cryptographic representation of some state of the LFSR, but this underlying state cannot be determined unless one knows the secret key. However, following a transition to $s' = (r'_1, \ldots, r'_\ell)$, there are two possibilities: either

$$\mathcal{QR}_N(r'_i) = \mathcal{QR}_N(r_i), \text{ for } 1 \leq i \leq \ell, \tag{2}$$

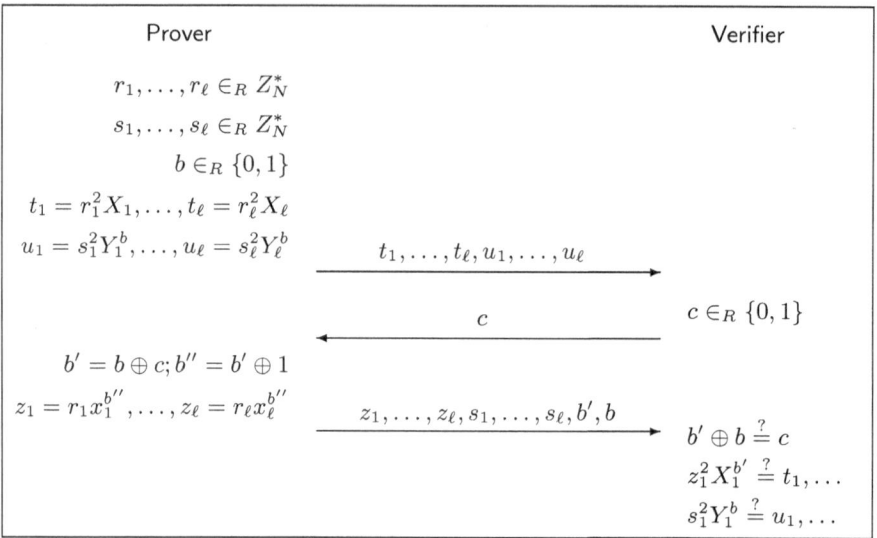

Fig. 1. Proof of validity for a counter transition.

which represents a 0 vote, or

$$QR_N(r_i') = QR_N(r_{i+1}), \text{ for } 1 \le i < \ell \text{ and } QR_N(r_\ell') = QR_N(\prod_{i=1}^{\ell} r_i^{b_i}), \qquad (3)$$

(with b_i as defined in Section 3.2), which represents a 1 vote. We seek an NIZK proof that either condition (2) or condition (3) holds. Note that these conditions are equivalent to the following: either

$$QR_N(r_i' \cdot r_i) = 0, \text{ for } 1 \le i \le \ell, \qquad (4)$$

or else

$$QR_N(r_i' \cdot r_{i+1}) = 0, \text{ for } 1 \le i < \ell \text{ and } QR_N(r_\ell' \cdot \prod_{i=1}^{\ell} r_i^{b_i}) = 0. \qquad (5)$$

Therefore, an NIZK proof that one of (4) or (5) holds is sufficient.

In Figure 1 we describe a protocol which takes as input two sequences $X_1, \ldots, X_\ell$ and $Y_1, \ldots, Y_\ell$, and proves the following statement:

$$((QR_N(X_1)=0) \wedge \cdots \wedge (QR_N(X_\ell)=0)) \vee ((QR_N(Y_1)=0) \wedge \cdots \wedge (QR_N(Y_\ell)=0)). \qquad (6)$$

By the arguments of the previous paragraph, this is sufficient for our application. The prover knows the square roots of every element of at least one of these sequences[8] (for someone who honestly increments the counter by either 0 or 1,

---

[8] Without loss of generality, we assume the prover knows the square roots for the first input sequence; thus, in Figure 1, we assume the prover knows $\{x_i\}$ such that $x_i^2 = X_i$, for $1 \le i \le \ell$.

this will be the case); these are the witnesses that these elements are quadratic residues.

By repeating this protocol k_2 times, the probability of cheating is reduced to 2^{-k_2}. This protocol can be made non-interactive using the Fiat-Shamir heuristic [13], by which the challenge of the verifier is replaced by applying a hash function (viewed as a random oracle [1]) to the statement to be proved and the first message of the prover. Let $\mathcal{H}$ be a suitable hash function. The prover need only send $z_1, s_1, \ldots, z_\ell, s_\ell, b', b$ as his proof. The verifier can compute $t_i = z_i^2 X_i^{b'}$ and $u_i = s_i^2 Y_i^b$ and then verify whether $b' \oplus b = \mathcal{H}(X_1, Y_1, t_1, u_1, \ldots, X_\ell, Y_\ell, t_\ell, u_\ell)$.

Theorem 2. *Take the cryptographic counter as described in Theorem 1, instantiated with encryption based on quadratic residuosity. An update of the counter now includes a non-interactive proof (as outlined in Figure 1 and using the Fiat-Shamir heuristic) for statement (6). This then constitutes a verifiable, restricted cryptographic n-counter (for all n of the form $n = 2^x - 1$) which is secure against malicious adversaries.*

Sketch of Proof   The protocol given in Figure 1 constitutes an honest-verifier perfect zero knowledge proof with soundness probability $1/2$. The proof of this fact follows from techniques outlined in [11]; we refer the reader there for discussion and a complete proof. Repeating the proof k_2 times (non-interactively, using the Fiat-Shamir heuristic) reduces the probability of cheating to 2^{-k_2}, and is a non-interactive zero-knowledge proof (in the random oracle model). The counter is thus *restricted* in that updates are limited to adding an integer from $\{0, 1\}$, and *verifiable* in that updates can be publicly verified as being in this range.

The security of the construction against a malicious adversary follows from Theorem 1 and the zero-knowledge properties of the above protocol.   □

3.5   Distributed Decryption of the Counter

We mention that robustness with respect to the trusted authorities can be achieved via distributed generation of the secret key along with threshold decryption of the final counter (which can always be achieved via general multi-party techniques [15]). For the particular case when encryption is done using quadratic residuosity, we are able to achieve efficient distributed key generation and threshold decryption [18]. As this is not the focus of this work, we defer a complete discussion until the full version of the paper.

4   Voting with Cryptographic Counters

We briefly discuss the application of cryptographic counters to the problem of electronic voting. The discussion will be kept as general as possible. For efficient implementation, we have outlined above how it is possible to build an efficient scheme using the encryption scheme based on quadratic residuosity.

We follow the model introduced by Benaloh, et al. [6,3,4]. The parties participating in the election consist of a set of voters $V_1, \ldots, V_L$ and a set of authorities

$A_1, \ldots, A_M$, which need not be disjoint. We assume that everyone has access to a *bulletin board* to which all voters will post their messages. Messages are authenticated, and the identity of a sender cannot be forged, nor can messages to the bulletin board be tampered with. Messages are listed in order of arrival (or, equivalently, every message includes the time it was sent), and no one can erase anything from the bulletin board once posted. Note that we do not assume any private channels between voters and the authorities. We now give a high-level description of a voting protocol based on a restricted cryptographic counter; this proves the following theorem:

Theorem 3. *A voting scheme satisfying universal verifiability, privacy, and robustness can be efficiently constructed from any (robust) verifiable, restricted cryptographic counter secure against malicious adversaries (where votes are restricted to the set $\{0, 1\}$).*

Sketch of Proof   We describe the voting protocol assuming the existence of a verifiable, restricted cryptographic n-counter (where votes are restricted to the set $\{0, 1\}$) secure against malicious adversaries. Robustness (with respect to the authorities) follows if the counter itself is robust (as described in Section 3.5).

SYSTEM SETUP. The authorities run the key generation algorithm for the cryptographic n-counter. Here, n is chosen to be equal to the total number of voters (or an upper bound on the number of voters if the exact number is unknown). If robustness is desired, and/or if some voters are also authorities, the key generation may be done in a robust manner as outlined in Section 3.5. The public key pk and the initial state s_0 are announced to all voters. The key generation step may be the most expensive part of the entire protocol, but it is only a one-time operation which can be done months before the election takes place.

VOTING. The counter always holds the current vote total. The current counter value is always defined as the most recently posted (valid) counter value. Denote the counter after the ith vote by s_i. The $(i + 1)^{\text{th}}$ vote is cast as follows: a voter looks at the current counter and computes new state s_{i+1} using the transition function, the previous state s_i, the desired vote $v \in \{0, 1\}$, and the public key pk. The voter publishes this updated state s_{i+1} which then becomes the current state (since it is the most recently posted counter). This proceeds for L rounds until every voter has voted once (see Section 5 for ways to reduce the number of rounds).

Universal verifiability (and hence vote correctness) follows from verifiability of the counter, and voter privacy follows from the definition of security against a malicious adversary. Robustness with respect to the authorities follows from the (robust) distributed key generation and decryption.

TALLYING. When the election is complete, the authorities determine the final tally by decrypting the last (valid) counter. If there is more than one trusted authority, threshold decryption (see Section 3.5) will be necessary. It may also be desirable to have the authorities prove correctness of the decryption; note that it is not acceptable to just publish the secret key, since this would allow determination of every voter's vote retroactively. In the particular case where encryption is done via quadratic residues, the authorities can easily prove that decryption was done correctly by publishing an x for each encrypted value y such that $y = \pm x^2$. $\qquad\square$

5 Conclusion

For small-scale elections, the voting scheme outlined here (when based on the encryption scheme using quadratic residuosity) is efficient enough to be practical (cf. Table 1). The required computation and vote size are quite reasonable. One drawback to this scheme is the number of rounds required for voting to take place. When a single cryptographic counter is used, the number of rounds is equal to the number of voters, L. However, by using k cryptographic counters, assigning each voter to one of k groups, and allowing voting to take place in parallel, the number of rounds can be reduced to L/k. Even in a national election, such an approach may be acceptable; for example, by assigning a set of counters to each voting district.

From a theoretical point of view, the approach outlined in this paper is especially interesting since it was previously unclear whether voting could be done efficiently without using fully-homomorphic encryption.

References

1. M. Bellare and P. Rogaway. Random Oracles are Practical: A Paradigm for Designing Efficient Protocols. ACM CCCS 1993.
2. J. Benaloh and D. Tuinstra. Receipt-Free Secret-Ballot Elections. STOC 1994.
3. J. Benaloh and M. Yung. Distributing the Power of a Government to Enhance the Privacy of Voters. PODC 1986.
4. J. Benaloh. Verifiable Secret-Ballot Elections. PhD thesis, Yale University, Department of Computer Science, New Haven, CT, 1987.
5. M. Blum, P. Feldman, and S. Micali. Non-Interactive Zero-Knowledge and its Applications. STOC 1988.
6. J. Cohen and M. Fischer. A Robust and Verifiable Cryptographically Secure Election Scheme. FOCS 1985.
7. D. Coppersmith. Fast Evaluation of Logarithms in Fields of Characteristic Two. IEEE Transactions on Information Theory, Vol. 30, pp. 587-594, 1984
8. R. Cramer, M. Franklin, B. Schoenmakers, and M. Yung. Multi-Authority Secret-Ballot Elections with Linear Work. Eurocrypt 1996.
9. R. Cramer, R. Gennaro, and B. Schoenmakers. A Secure and Optimally Efficient Multi-Authority Election Scheme. Eurocrypt 1997.
10. I. Damgård and M. Jurik. Efficient Protocols Based on Probabilistic Encryption Using Composite Degree Residue Classes. Manuscript, May 2000.
11. A. De Santis, G. Di Crescenzo, G. Persiano, and M. Yung. On Monotone Formula Closure of SZK. FOCS 1994.
12. T. ElGamal. A Public-Key Cryptosystem and a Signature Scheme Based on Discrete Logarithms. IEEE Trans. Info. Theory, 31(4): 469–472, 1985.

13. A. Fiat and A. Shamir. How to Prove Yourself: Practical Solution to Identification and Signature Problems. CRYPTO 1986.
14. O. Goldreich. Secure Multi-Party Computation (Working Draft, Version 1.1). Manuscript, 1998.
15. O. Goldreich, S. Micali, and A. Wigderson. How to Play Any Mental Game, or a Completeness Theorem for Protocols with an Honest Majority. STOC '87.
16. S. Goldwasser and S. Micali. Probabilistic Encryption. JCSS 28(2): 270–299, 1984.
17. M. Hirt and K. Sako. Efficient Receipt-Free Voting Based on Homomorphic Encryption. Eurocrypt 2000.
18. J. Katz and M. Yung. Threshold Cryptosystems with Distributed Prime Factors. Manuscript.
19. R. Lidl and H. Niederreiter. *Introduction to Finite Fields and their Applications (Revised Edition)*, Cambridge University Press, 1994.
20. R. Lidl and G. Pilz. *Applied Abstract Algebra (Second Edition)*, Springer, 1997.
21. P. Pallier. Public-Key Cryptosystems Based on Composite Degree Residue Classes. Eurocrypt 1999.
22. J. Rifa and J. Borrell. Improving the Time Complexity of the Computation of Irreducible and Primitive Polynomials in Finite Fields. Applied Algebra, Algebraic Algorithms, and Error-Correcting Codes, 1991.
23. B. Schoenmakers. A Simple Publicly Verifiable Secret Sharing Scheme and its Application to Electronic Voting. CRYPTO 1999.

A   The Quadratic Residuosity Assumption

These definitions are standard [16,11]. We say $y \in \mathbb{Z}_N^*$ is a *quadratic residue* modulo N iff there exists an $x \in \mathbb{Z}_N^*$ such that $y = x^2 \bmod N$; otherwise, y is a *quadratic non-residue* modulo N. Define the predicate $\mathcal{QR}_N(y)$ to be 0 iff y is a quadratic residue modulo N, and 1 otherwise. For p prime, the problem of deciding quadratic residuosity is equivalent to computing the Legendre symbol. In fact, the Legendre symbol of y modulo p is defined by $\mathcal{L}_p(y) = +1$ iff y is a quadratic residue, and -1 otherwise.

Now, let $p, q \equiv 3 \bmod 4$ be primes and let $N = pq$ (such N are known as *Blum integers*). No efficient algorithm is known for deciding quadratic residuosity modulo a Blum integer whose factorization is not known. Some information is given by the *Jacobi symbol*, which extends the Legendre symbol as $\mathcal{J}_N(y) = \mathcal{L}_p(y)\mathcal{L}_q(y)$. Despite the way the Jacobi symbol is defined, it is well-known that it can be computed in polynomial time without knowledge of the factors of N. Application of the Chinese Remainder Theorem shows that if $\mathcal{J}_N(y) = -1$, then y cannot be a quadratic residue modulo N. On the other hand, if $\mathcal{J}_N(y) = +1$, no polynomial-time algorithm is known for computing $\mathcal{QR}_N(y)$ if the factorization of N is unknown.

Define $\mathbb{Z}_N^{+1}$ as the set of elements of $\mathbb{Z}_N^*$ with Jacobi symbol 1. It is easy to generate a random $y \in \mathbb{Z}_N^{+1}$ which is a quadratic residue: choose random $r \in \mathbb{Z}_N^*$ and set $y = r^2 \bmod N$. It is equally easy to generate a random quadratic non-residue: choose random $r \in \mathbb{Z}_N^*$ and set $y = -r^2 \bmod N$. This suggests the following semantically secure encryption scheme [16]: the public key is a Blum integer N, and the secret key is the prime factors of N. To encrypt a 0, send a random quadratic residue; to encrypt a 1, send a random quadratic non-residue. This can be extended to n-bit messages in the obvious way, by concatenating n single-bit encryptions.

When $y_1, y_2 \in \mathbb{Z}_N^{+1}$, it is easily verified that $\mathcal{QR}_N(y_1 y_2) = \mathcal{QR}_N(y_1) \oplus \mathcal{QR}_N(y_2)$. This shows that the above encryption scheme is homomorphic over addition in its message space $\mathbb{Z}_2$.

An Efficient System
for Non-transferable Anonymous Credentials
with Optional Anonymity Revocation

Jan Camenisch[1] and Anna Lysyanskaya[2,*]

[1] IBM Research
Zurich Research Laboratory
CH–8803 Rüschlikon
jca@zurich.ibm.com
[2] MIT LCS
545 Technology Square
Cambridge, MA 02139 USA
anna@theory.lcs.mit.edu

Abstract. A credential system is a system in which users can obtain credentials from organizations and demonstrate possession of these credentials. Such a system is anonymous when transactions carried out by the same user cannot be linked. An anonymous credential system is of significant practical relevance because it is the best means of providing privacy for users. In this paper we propose a practical anonymous credential system that is based on the strong RSA assumption and the decisional Diffie-Hellman assumption modulo a safe prime product and is considerably superior to existing ones: (1) We give the first practical solution that allows a user to unlinkably demonstrate possession of a credential as many times as necessary without involving the issuing organization. (2) To prevent misuse of anonymity, our scheme is the first to offer optional anonymity revocation for particular transactions. (3) Our scheme offers separability: all organizations can choose their cryptographic keys independently of each other. Moreover, we suggest more effective means of preventing users from sharing their credentials, by introducing *all-or-nothing* sharing: a user who allows a friend to use one of her credentials once, gives him the ability to use all of her credentials, i.e., taking over her identity. This is implemented by a new primitive, called *circular encryption*, which is of independent interest, and can be realized from any semantically secure cryptosystem in the random oracle model.

Keywords. Privacy protection, credential system, pseudonym system, e-cash, blind signatures, circular encryption, key-oblivious encryption.

* This research was carried out while the author was visiting IBM Zürich Research Laboratory.

B. Pfitzmann (Ed.): EUROCRYPT 2001, LNCS 2045, pp. 93–118, 2001.

1   Introduction

As information becomes increasingly accessible, protecting the privacy of individuals becomes a more challenging task. To solve this problem, an application that allows the individual to control the dissemination of personal information is needed. An anonymous credential system (also called pseudonym system), introduced by Chaum [18], is the best known idea for such a system. In this paper, we propose a new efficient anonymous credential system, considerably superior to previously proposed ones. The communication and computation costs of our solution are small, thus introducing almost no overhead to realizing privacy in a credential system.

An anonymous credential system [18,19,21,25,34] consists of users and organizations. Organizations know the users only by pseudonyms. Different pseudonyms of the same user cannot be linked. Yet, an organization can issue a credential to a pseudonym, and the corresponding user can prove possession of this credential to another organization (who knows her by a different pseudonym), without revealing anything more than the fact that she owns such a credential. Credentials can be for unlimited use (these are called *multiple-show* credentials) and for one-time use (these are called *one-show* credentials). Possession of a multi-show credential can be demonstrated an arbitrary number of times; these demonstrations cannot be linked to each other.

Basic desirable properties. It should be impossible to forge a credential for a user, even if users and other organizations team up and launch an adaptive attack on the organization. Each pseudonym and credential must belong to some well-defined user [34]. In particular, it should not be possible for different users to team up and show some of their credentials to an organization and obtain a credential for one of them that that user alone would not have gotten. Systems where this is not possible are said to have *consistency of credentials*. As organizations are autonomous entities, it is desirable that they be separable, i.e., be able to choose their keys themselves and independently of other entities, so as to ensure security of these keys and facilitate the system's key management.

The scheme should also provide user privacy. An organization cannot find out anything about a user, apart from the fact of the user's ownership of some set of credentials, even if it cooperates with other organizations. In particular, two pseudonyms belonging to the same user cannot be linked [8,18,19,21,25,34].

Finally, it is desirable that the system be efficient. Besides requiring that it be based on efficient protocols, we also require that each interaction involve as few entities as possible, and the rounds and amount of communication be minimal. In particular, if a user has a multiple-show credential from some organization, she ought to be able to demonstrate it without getting the organization to reissue credentials each time.

Additional desirable properties. It is an important additional requirement that the users should be discouraged from sharing their pseudonyms and credentials with other users. The previously known way of discouraging the user from doing this was by *PKI-assured non-transferability*. That is, sharing a credential implies also sharing a particular, valuable secret key from *outside* the system (e.g., the secret key that gives access to the user's bank account) [26,32,34]. However, such a valuable key does not always exist. Thus we introduce an alternative, novel way of achieving this: *all-or-nothing* non-transferability. Here, sharing just one pseudonym or credential implies sharing all of the user's other credentials and pseudonyms in the system, i.e., sharing *all* of the user's secret keys *inside* the system. These two methods of guaranteeing non-transferability are different: neither implies the other, and both are desirable and can in fact be combined.

In addition, it may be desirable to have a mechanism for discovering the identity of a user whose transactions are illegal (this feature, called *global anonymity revocation*, is optional); or reveal a user's pseudonym with an issuing organization in case the user misuses her credential (this feature, called *local anonymity revocation*, is also optional). It can also be beneficial to allow *one-show* credentials, i.e., credentials that should only be usable once and should incorporate an *off-line* double-spending test. It should be possible to encode attributes, such as expiration dates, into a credential.

Related work. The scenario with multiple users who, while remaining anonymous to the organizations, manage to transfer credentials from one organization to another, was first introduced by Chaum [18]. Subsequently, Chaum and Evertse [19] proposed a solution that is based on the existence of a semi-trusted third party who is involved in all transactions. However, the involvement of a semi-trusted third party is undesirable.

The scheme later proposed by Damgård [25] employs general complexity-theoretic primitives (one-way functions and zero-knowledge proofs) and is therefore not applicable for practical use. Moreover, it does not protect organizations against colluding users. The scheme proposed by Chen [21] is based on discrete-logarithm-based blind signatures. It is efficient but does not address the problem of colluding users. Another drawback of her scheme and the other practical schemes previously proposed is that to use a credential several times, a user needs to obtain several signatures from the issuing organization.

Lysyanskaya, Rivest, Sahai, and Wolf [34] propose a general credential system. While their general solution captures many of the desirable properties, it is not usable in practice because their constructions are based on one-way functions and general zero-knowledge proofs. Their practical construction, based on a non-standard discrete-logarithm-based assumption, has the same problem as the one due to Chen [21]: a user needs to obtain several signatures from the issuing organization in order to use unlinkably a credential several times.

Other related work is that of Brands [8] who provides a certificate system in which a user has control over what is known about the attributes of a pseudonym. Although a credential system with one-show credentials can be inferred from his framework, obtaining a credential system with multi-show credentials is not immediate and may in fact be impossible in practice. Another inconvenience of these and the other discrete-logarithm-based schemes mentioned above is that all the users and the certification authorities in these schemes need to share the same discrete logarithm group.

The concept of revocable anonymity is found in electronic payment systems (e.g., [9,37]) and group signature and identity escrow (e.g., [2,14,20,33] schemes.

Prior to our work, the problem of constructing a practical system with multiple-use credentials eluded researchers for some time [8,21,25,34]. We solve it by extending ideas found in the constructions of strong-RSA-based signature schemes [23,30] and group signature schemes [2].

Our contribution. In Section 2 we present our definitions for a credential system with the basic properties. Although not conceptually new and inspired by the literature on multi-party computation [15,16] and reactive systems [36], these definitions are of interest, as our treatment is more formal than the one usually encountered in the literature on credential and electronic cash systems. We omit formal definitions for a credential system satisfying the additional desirable properties and instead refer the reader to the full version of this paper [12].

Our basic credential system, presented in Section 4, provably satisfies the basic properties listed above under the strong RSA assumption and the decisional Diffie-Hellman assumption modulo a strong prime product. Our basic solution is practical. When using an RSA modulus n of 1024 bits, a credential-pseudonym pair is about 4K bits, and the most expensive operation of proving possession of a credential requires about 22 exponentiations in $\mathbb{Z}_n^*$ for both parties and can be done in three rounds.

Our extended credential system, presented in Sections 6 and 7, describes how to incorporate additional desirable properties into the basic credential system. These are also efficient, except the one with all-or-nothing non-transferability: when using RSA moduli of length 1024 bits, establishing a pseudonym is somewhat less efficient: it takes about 200 exponentiations in $\mathbb{Z}_n^*$ for both parties, but batch-verification techniques [4] could be applied to reduce this, and organizations have to store about 25K bits per user (here computation complexity could be traded against storage).

All-or-nothing non-transferability is based on a new primitive we call *circular encryption*, discussed in Section 6.1; we implement this primitive in the random oracle model; it is a challenging open problem whether this primitive can be realized outside the random oracle model.

This work is the first to introduce one-show credentials with an off-line double-spending test, similar to known e-cash schemes (in fact, our one-show credentials can be used as anonymous coins). These one-show credentials are described in Section 7.1. More precisely, our double-spending test mechanism together with the all-or-nothing property ensures that, if a user presents such a credential more than once, then the verifying entity gets the ability to demonstrate possession of all the pseudonyms and credentials of that user — a strong incentive to users not to double-spend. From a technical point of view, it might be interesting that the anonymity of our one-show credentials is not obtained by blind signatures but by an alternative mechanism.

Another innovation of this work is the possibility of anonymity revocation, described in Section 7.2. We stress that this feature is entirely optional. Moreover, for each transaction, the user has the freedom of specifying under which conditions the anonymity can be revoked (maybe subject to conditions of the other parties involved in the transaction). The user may also choose unconditional anonymity, and then his identity will not be retrievable under any circumstances. Yet another innovation is separability for organizations.

2   Formal Definitions and Requirements

A basic credential system has *users*, *organizations*, and *verifiers* as types of players. Users are entities that receive credentials. The set of users in the system may grow over time. Organizations are entities that grant and verify the credentials of the users. Each organization grants a unique (for simplicity of exposition) type of credential. Finally, verifiers are entities that verify credentials of the users.

Variations of such a system allow a single organization to issue different types of credentials. For the purposes of non-transferability, we can add a *CA* to the model who verifies that the users entering the system possess an external public and secret key. This *CA* will be trusted to do his job properly. For PKI-assured non-transferability, this will make sure that access to a user's pseudonym or credential is sufficient to obtain this user's secret key from an external PKI. For all-or-nothing non-transferability, this will make sure that access to one pseudonym or credential of a user is sufficient to obtain access to all of them. To allow revocable anonymity, an anonymity revocation manager can be added. This entity will be trusted not to use his ability to find out a user's identity or pseudonym unless dictated to do so. As the trusted parties perform tasks that are not required frequently, these parties can be implemented in a distributed fashion to weaken the trust assumptions. Finally, a credential may include an attribute, such as an expiration date. These variations are simple to handle in the model; for simplicity of exposition of the model, however, we do not discuss them here. The extended solution we propose already incorporates some of them, and can be easily adapted to incorporate others.

We first give a specification for an ideal credential system that relies on a trusted party T as an intermediator; we then explain what it means for a cryptographic system to conform to this specification.

Initialization: To initialize the system, the organizations create a public file which, for each organization O, describes the type of credential that the organization grants.

Ideal communication: All communication is routed through T. If the sender of a message wishes to be anonymous, he requests T not to reveal his identity to the recipient. Also, a sender of a message may request that a session be established between him and the recipient. This session then gets a session id *sid*.

Events in the system: Each transaction between players is an event in the system. Events in the system can be triggered through external processes, some of which may be controlled by an adversary. An external process can trigger some particular event between a particular user and organization; or may trigger a set of events; or may cause some probability distribution on the events.

Input of the players: The players are interactive Turing machines. Initially, they have no input; then as transactions are triggered, they obtain inputs and act accordingly.

Output of the players: In the end of the system's lifetime, each user outputs a list of the transactions she participated in, complete with the pseudonym used in each transaction, session ids of the transactions, and transaction outcomes. Organizations and verifiers output a list of transaction identifiers for transactions in which they participated, the pseudonym involved (in case of organizations), and the outcome of the transaction.

The system supports the following transactions:

FormNym(U, O): This protocol is a session between a user U and an organization O. The user U contacts T with a request to establish a pseudonym between herself and organization O. She further specifies the login name L_U by which T knows her and the corresponding authenticating key K_U. If she does not have an account with T yet, she first establishes it by providing to T a login name L_U and obtaining K_U in return. She further specifies N_1. Then T verifies the validity of (L_U, K_U) and, if K_U is the authenticating key corresponding to login name L_U, contacts O and tells it that some user wants to establish a pseudonym with it with prefix N_1. The organization either accepts or rejects. If it accepts, it sends a pseudonym suffix N_2, so the pseudonym becomes $N_{(U,O)} := N_1 \| N_2$ ($\|$ denotes concatenation). T forwards the resulting $N_{(U,O)}$ to U in case of acceptance, or notifies U of rejection.

GrantCred(N, O): This protocol is a session between a user U and an organization O. U approaches T, and submits her login name L_U, her authenticating key K_U, the pseudonym N, and the name of organization O. If K_U is not a valid authenticating key for L_U, or if N is not U's pseudonym with O, then T replies with a "Fail" message. Otherwise, T contacts O. If O accepts, then T notifies the user that a credential has been granted, otherwise it replies with "Reject."

VerifyCred(V, N, O): This protocol is a session between a user U and a verifier V. A user approaches T and gives it her login L_U, her authenticating key K_U, a name of the verifier V, a pseudonym N and a name of a credential-granting organization O. If K_U is a valid authenticating key for L_U, and N is U's pseudonym with organization O, and a credential has been granted by O to N, then T notifies V that the user talking to V in the current session *sid* has a credential from O. Otherwise, T replies with a "Fail" message.

VerifyCredOnNym(V, N_V, N_O, O): This protocol is a session between a user U and an verifier V. U approaches T and gives it her login name L_U, her authenticating key K_U, a name of the verifier V, pseudonyms N_V and N_O, and a name of a credential-granting organization O. If K_U is a valid authenticating key for L_U, and N_V is U's pseudonym with V while N_O is U's pseudonym with organization O, and a credential has been granted by O to N_O, then T notifies V that the user with pseudonym N_V has a credential from O.

This ideal system captures the intuitive requirements, such as unforgeability of credentials, anonymity of users, unlinkability of credential showings, and consistency of credentials. Ideal operations that allow additional desirable features can be implemented as well.

Let us briefly illustrate the use of the credential system by a typical example. Consider a user U who wants to get a credential from organization O. Organization O requires the possession of credentials from organizations O_1 and O_2 as a prerequisite to get a credential from O. Assume that U possesses such credentials. Then U can get a credential from O as follows: she first establishes a pseudonym with O by executing FormNym(U, O) and then shows O her credentials from O_1 and O_2 be executing VerifyCredOnNym(O, N_O, N_{O_1}, O_1) and VerifyCredOnNym(O, N_O, N_{O_2}, O_2). Now O knows that the user it knows under N_O possesses credentials from O_1 and O_2 and will grant U a credential, i.e., U can execute GrantCred(N_O, O). We remark that the operation VerifyCred(V, N, O) exists for efficiency reasons. This operation can be used by U if she wants to show a party only a single credential, e.g., to access a subscription-based service.

The ideal-world (resp., real-world) adversary. The ideal-world (resp., real-world) adversary is a probabilistic polynomial-time machine that gets control over the corrupted parties in the ideal world (resp., real world). He receives, as input, the number of honest users and organizations, as well as all the public information of the system. The adversary can trigger an event as described above.

Definition 1. *Let the ideal credential system described above be denoted ICS. Let a cryptographic credential system without T be denoted CCS. Let $V = \text{poly}(k)$ be the number of players in the system with security parameter k. By $ICS(1^k, E)$ (resp., $CCS(1^k, E)$) we denote a credential system with security parameter k and event scheduler E for the events that take place in this system. As events are scheduled adversarially, E schedules them according to the adversary's wishes, therefore we will write $E^{\mathcal{A}}$. By $Z_i(1^k)$, we denote the output of party i in the credential system. If $\{A_1(1^k), \ldots, A_V(1^k)\}$ is a list of the players' outputs, then we denote these players' outputs by $\{A_1(1^k), \ldots, A_l(1^k)\}^{CS(1^k, E)}$ when all of them, together, exist within a credential system CS. CCS is secure if there exists a simulator $\mathcal{S}$ (ideal-world adversary) such that the following holds, for all interactive probabilistic polynomial-time machines $\mathcal{A}$ (real-world adversary), for all sufficiently large k:*

1. In the ICS, $\mathcal{S}$ controls the ideal-world players corresponding to the real-world players controlled by $\mathcal{A}$.

2. For all event schedulers $E^{\mathcal{A}}$

$$\{\{Z_i(1^k)\}_{i=1}^V, \mathcal{A}(1^k)\}^{CCS(1^k, E)} \overset{c}{\approx} \{\{Z_i(1^k)\}_{i=1}^V, \mathcal{S}^{\mathcal{A}}(1^k)\}^{ICS(1^k, E)} \ ,$$

where $\mathcal{S}$ is given black-box access to $\mathcal{A}$. ("$D_1(1^k) \overset{c}{\approx} D_2(1^k)$" denotes computational indistinguishability of the distributions D_1 and D_2.)

3   Protocol Notation

By $\text{neg}(k)$ we denote any function that vanishes faster than any inverse polynomial in k. By $\text{poly}(k)$ we denote a function bounded by a polynomial in k.

In the description of our scheme, we use the notation introduced by Camenisch and Stadler[14] for various proofs of knowledge of discrete logarithms and proofs of the validity of statements about discrete logarithms. For instance,

$$PK\{(\alpha, \beta, \gamma) : y = g^\alpha h^\beta \ \wedge \ \tilde{y} = \tilde{g}^\alpha \tilde{h}^\gamma \ \wedge \ (u \le \alpha \le v)\}$$

denotes a *"zero-knowledge Proof of Knowledge of integers α, β, and γ such that $y = g^\alpha h^\beta$ and $\tilde{y} = \tilde{g}^\alpha \tilde{h}^\gamma$ holds, where $v < \alpha < u$,"* where $y, g, h, \tilde{y}, \tilde{g}$, and $\tilde{h}$ are elements of some groups $G = \langle g \rangle = \langle h \rangle$ and $\tilde{G} = \langle \tilde{g} \rangle = \langle \tilde{h} \rangle$. The convention is that Greek letters denote quantities the knowledge of which is being proved, while all other parameters are known to the verifier. Using this notation, a proof-protocol can be described by just pointing out its aim while hiding all details.

In the random oracle model, such protocols can be turned into signature schemes using the Fiat-Shamir heuristic [28]. We use the notation $SPK\{(\alpha) : y = g^\alpha\}(m)$ to denote a signature obtained in this way.

It is important that we use protocols that are *concurrent* zero-knowledge. They are characterized by remaining zero-knowledge even if several instances of the same protocol are run arbitrarily interleaved. In the public key model, Damgård [24] shows a general technique for making the so-called Σ-protocols (these include all the proofs of knowledge used here) composable under concurrent composition without incurring a penalty in communication or round complexity. All the proofs of knowledge we use in this paper incorporate this technique.

In this paper we apply such PK's and SPK's to the group of quadratic residues modulo a composite n, i.e., $G = QR_n$. This choice for the underlying group has some consequences. First, the protocols are proofs of knowledge under the strong RSA assumption [29]. Second, the largest possible value of the challenge c must be smaller that the smallest factor of G's order. Third, soundness needs special attention in the case that the verifier is not equipped with the factorization of n because then deciding membership in QR_n is believed to be hard. Thus the prover needs to convince the verifier that the elements he presents are indeed quadratic residues, i.e., that the square roots of the presented elements exist. This can in principle be done with a protocol by Fiat and Shamir [28]. However, often it is sufficient to simply execute $PK\{(\alpha) : y^2 = (g^2)^\alpha\}$ instead of $PK\{(\alpha) : y = g^\alpha\}$. The quantity α is defined as $\log_{g^2} y^2$, which is the same as $\log_g y$ in case y is in QR_n.

For the an explanation of how the PK's used in the paper can be realized efficiently, we refer to the full version of this paper [12].

4   The Basic Anonymous Credential System

The basic system comprises protocols for a user to join the system, register with an organization, obtain multi-show credentials, and show such credentials.

Throughout we assume that the users and organizations are connected by perfectly anonymous channels. Furthermore, we assume that for each protocol an organization authenticates itself to the user and that they establish a secure channel between them for each session. For any protocol we describe, we implicitly assume that if some check or sub-protocol (e.g., some proof of knowledge PK) fails for some party, it informs the other participants of this and stops.

4.1 High-Level Description

In our system, each organization O will have, in its public key PK_O, an RSA modulus n_O, and five elements of QR_{n_O}: $(a_O, b_O, d_O, g_O, h_O)$. Each user U will have her own master secret key x_U. A pseudonym of user U with organization O, denoted $N_{(U,O)}$, is just a name by which the user is known to the organization, and consists of a user-generated part N_1 and an organization-generated part N_2. The pseudonym $N_{(U,O)} = N_1 \| N_2$ will be tagged with a value $P_{(U,O)}$. This *validating tag* is of the form $P_{(U,O)} = a_O^{x_U} b_O^{s_{(U,O)}}$, where $s_{(U,O)}$ is a short random string to which the user and organization contribute randomness, but of which only the user knows its value. An appropriate choice of parameters for the length of x_U and $s_{(U,O)}$ ensures that the resulting $P_{(U,O)}$ is statistically independent of the user's key x_U and of any other validating tags formed by the same user with other organizations.

A credential issued by O to a pseudonym $N_{(U,O)}$ is a tuple $(e_{(U,O)}, c_{(U,O)})$ where $e_{(U,O)}$ is a sufficiently long prime and $c_{(U,O)}{}^{e_{(U,O)}} \equiv P_{(U,O)} d_O$. Under the strong RSA assumption, such tuples cannot be existentially forged for correctly formed tags even by an adaptive attack (Theorem 2).

To protect the user's privacy in our system, proof of possession of a credential is realized by a proof of knowledge of a correctly formed tag $P_{(U,O)}$ and a credential on it. This is done by publishing statistically secure commitments to both the validating tag and the credential, and proving relationships between these commitments. It can also include a proof that the underlying secret key is the same in both the committed validating tag (corresponding to the pseudonym formed with the issuing organization) and the validating tag with the verifying organization. This ensures consistency of credentials, e.g., guarantees that even users that fully trust each other cannot pool their credentials.

4.2 System Parameter and Key Generation

We name some common system parameters: the length of all the RSA moduli ℓ_n, the integer intervals $\Gamma = \ ] - 2^{\ell_\Gamma}, 2^{\ell_\Gamma} [$, $\Delta = \ ] - 2^{\ell_\Delta}, 2^{\ell_\Delta} [$, $\Lambda = \ ] 2^{\ell_\Lambda}, 2^{\ell_\Lambda + \ell_\Sigma} [$ such that $\ell_\Delta = \epsilon(\ell_\Lambda + \ell_n) + 1$, where $\epsilon > 1$ is a security parameter, and $\ell_\Lambda > \ell_\Sigma + \ell_\Delta + 4$.

Each organization O_i chooses random $\ell_n/2$-bit primes p'_{O_i}, q'_{O_i} such that $p_{O_i} = 2p'_{O_i} + 1$ and $q_{O_i} = 2q'_{O_i} + 1$ are prime and sets modulus $n_{O_i} = p_{O_i} q_{O_i}$. It also chooses random elements $a_{O_i}, b_{O_i}, d_{O_i}, g_{O_i}, h_{O_i} \in QR_{n_{O_i}}$. It stores $SK_{O_i} := (p_{O_i}, q_{O_i})$ as its secret key and publishes $PK_{O_i} := (n_{O_i}, a_{O_i}, b_{O_i}, d_{O_i}, g_{O_i}, h_{O_i})$ as its public key. In the public-key model, we assume that there is a special entity that verifies, through a zero-knowledge protocol with O_i, that n_{O_i} is the product of two safe primes (see [13] for how this can be done efficiently) and that the elements $a_{O_i}, b_{O_i}, d_{O_i}, g_{O_i}, h_{O_i}$ are indeed in $QR_{n_{O_i}}$ (see, for example, Goldwasser et al. [31]). Alternatively, this can be carried out in the random oracle model using the Fiat-Shamir heuristic [28]. The parameter ℓ_Λ should be chosen such that computing discrete logarithms in $QR_{n_{O_i}}$ with ℓ_Λ-bits exponents is hard.

4.3   Generation of a Pseudonym

We now describe how a user U establishes a pseudonym $N_{(U,O)}$ and its validating tag $P_{(U,O)}$ with organization O. Let $x_U \in \Gamma$ be U's master secret. The protocol below assures that the pseudonym's validating tag is of the right form, i.e., $P_{(U,O)} = a_O^{x_U} b_O^{s_{(U,O)}}$, with $x_U \in \Gamma$ and $s_{(U,O)} \in \Delta$. The value $s_{(U,O)}$ is chosen jointly by O and U without O learning anything about either x_U or $s_{U,O}$. Note that this protocol does not force U to use the same x_U as with other organizations; this is taken care of later in Protocol 4.

Protocol 1

1. *U chooses a value $N_1 \in \{0,1\}^k$, and values $r_1 \in_R \Delta$ and $r_2, r_3 \in_R \{0,1\}^{2\ell_n}$. U sets $C_1 := g_O^{r_1} h_O^{r_2}$, $C_2 := g_O^{x_U} h_O^{r_3}$. U sends N_1, C_1, and C_2 to O.*
2. *To prove that C_1 and C_2 are formed correctly, U serves as the prover to verifier O in*

$$PK\{(\alpha, \beta, \gamma, \delta) : C_1^2 = (g_O^2)^\alpha (h_O^2)^\beta \wedge C_2^2 = (g_O^2)^\gamma (h_O^2)^\delta\} .$$

3. *O chooses a random $r \in_R \Delta$ and a value N_2 and sends r, N_2 to U.*
4. *U sets her pseudonym $N_{(U,O)} := N_1 \| N_2$. U computes $s_{(U,O)} = (r_1 + r \bmod (2^{\ell_\Delta+1} - 1)) - 2^{\ell_\Delta} + 1$, ($s_{(U,O)}$ is the sum of r_1 and r, adjusted appropriately so as to fall in the interval Δ). U then sets her validating tag $P_{(U,O)} := a_O^{x_U} b_O^{s_{(U,O)}}$ and sends $P_{(U,O)}$ to O.*
5. *Now, U must show that $P_{(U,O)}$ was formed correctly. To that end, she computes $\tilde{s} = \lfloor \frac{r_1 + r}{2^{\ell_\Delta+1} - 1} \rfloor$ ($\tilde{s}$ is the value of the carry resulting from the computation of $s_{(U,O)}$ above) and chooses $r_4 \in_R \{0,1\}^{\ell_n}$, sets $C_3 := g_O^{\tilde{s}} h_O^{r_4}$, and sends C_3 to O. Furthermore, U proves to O that the values in step 4 were chosen correctly by executing*

$$PK\{(\alpha, \beta, \gamma, \delta, \varepsilon, \zeta, \vartheta, \xi) : C_1^2 = (g_O^2)^\alpha (h_O^2)^\beta \wedge C_2^2 = (g_O^2)^\gamma (h_O^2)^\delta \wedge$$
$$C_3^2 = (g_O^2)^\varepsilon (h_O^2)^\zeta \wedge \frac{C_1^2 (g_O^2)^{(r - 2^{\ell_\Delta} + 1)}}{(C_3^2)^{(2^{\ell_\Delta+1} - 1)}} = (g_O^2)^\vartheta (h_O^2)^\xi \wedge$$
$$P_{(U,O)}^2 = (a_O^2)^\gamma (b_O^2)^\vartheta \wedge \gamma \in \Gamma \wedge \vartheta \in \Delta\} .$$

6. *O stores $N_{(U,O)}$, $P_{(U,O)}^2$ and $P_{(U,O)}$.*
7. *U stores $N_{(U,O)}$, $P_{(U,O)}^2$, $P_{(U,O)}$, and $s_{(U,O)}$.*

4.4   Generation of a Credential

A credential on (N, P) issued by O is a pair $(c, e) \in \mathbb{Z}_{n_O}^* \times \Lambda$ such that $P_{(U,O)} d_O = c^e$. To generate a credential on a previously established pseudonym $N_{(U,O)}$ with validity tag $P_{(U,O)}$, organization O and user U carry out the following protocol:

Protocol 2

1. U sends $(N_{(U,O)}, P_{(U,O)})$ to O and authenticates herself as its owner by executing
$$PK\{(\alpha, \beta) : P_{(U,O)}^2 = (a_O^2)^\alpha (b_O^2)^\beta\} \ .$$

2. O makes sure $(N_{(U,O)}, P_{(U,O)})$ is in its database, chooses a random prime $e_{(U,O)} \in_R \Lambda$, computes $c_{(U,O)} = (P_{(U,O)} d_O)^{1/e_{(U,O)}} \bmod n_O$, sends $c_{(U,O)}$ and $e_{(U,O)}$ to U and stores $(c_{(U,O)}, e_{(U,O)})$ in its record for $N_{(U,O)}$.

3. U checks if $c_{(U,O)}{}^{e_{(U,O)}} \equiv P_{(U,O)} d_O \pmod{n_O}$ and stores $(c_{(U,O)}, e_{(U,O)})$ in its record with organization O. The tuple $(P_{(U,O)}, c_{(U,O)}, e_{(U,O)})$ is called a credential record.

Step 1 can be omitted if Protocol 2 takes place in the same session as some other protocol where U already proved ownership of $N_{(U,O)}$.

4.5   Showing a Single Credential

Assume a user U wants to prove to a verifier V the possession of a credential issued by O, i.e., possession of values $(P_{(U,O)} = a_O^{x_U} b_O^{s_{(U,O)}}, c_{(U,O)}, e_{(U,O)})$, where $c_{(U,O)}^{e_{(U,O)}} = d_O P_{(U,O)}$. U and verifier V engage in the following protocol:

Protocol 3

1. U chooses $r_1, r_2 \in_R \{0, 1\}^{2\ell_n}$, computes $A = c_{(U,O)} h_O^{r_1}$ and $B = h_O^{r_1} g_O^{r_2}$, and sends A, B to V.

2. U engages with V in

$$PK\{(\alpha, \beta, \gamma, \delta, \varepsilon, \zeta, \xi) : \ d_O^2 = (A^2)^\alpha (\frac{1}{a_O^2})^\beta (\frac{1}{b_O^2})^\gamma (\frac{1}{h_O^2})^\delta \ \wedge$$

$$B^2 = (h_O^2)^\varepsilon (g_O^2)^\zeta \ \wedge \ 1 = (B^2)^\alpha (\frac{1}{h_O^2})^\delta (\frac{1}{g_O^2})^\xi \ \wedge$$

$$\beta \in \Gamma \ \wedge \ \gamma \in \Delta \ \wedge \ \alpha \in \Lambda\} \ .$$

The PK in step 2 proves that U possesses a credential issued by O on some pseudonym registered with O. We refer to the proof of Lemma 2 in the full version of this paper [12] for more details about this PK.

4.6   Showing a Credential with Respect to a Pseudonym

Assume a user U wants to prove possession of a credential record $(P_{(U,O_j)} = a^{x_U} b^{s_{(U,O_j)}}, c_{(U,O_j)}, e_{(U,O_j)})$ to organization O_i with whom U has established a pseudonym $(N_{(U,O_i)}, P_{(U,O_i)})$. That means O_i not only wants to be assured that U owns a credential by O_j but also that the pseudonym connected with this credential is based on the same master secret key as $P_{(U,O_i)}$.

Protocol 4

1. U chooses random $r_1, r_2, r_3 \in_R \{0,1\}^{2\ell_n}$, computes $A = c_{(U,O_j)} h_{O_j}^{r_1}$ and $B = h_{O_j}^{r_1} g_{O_j}^{r_2}$, and sends $N_{(U,O_i)}, A, B$ to O_i.
2. U engages with O_i in

$$PK\{(\alpha, \beta, \gamma, \delta, \varepsilon, \zeta, \xi, \eta) : \; d_{O_j}^2 = (A^2)^\alpha (\frac{1}{a_{O_j}^2})^\beta (\frac{1}{b_{O_j}^2})^\gamma (\frac{1}{h_{O_j}^2})^\delta \; \wedge$$

$$B^2 = (h_{O_j}^2)^\varepsilon (g_{O_j}^2)^\zeta \; \wedge \; 1 = (B^2)^\alpha (\frac{1}{h_{O_j}^2})^\delta (\frac{1}{g_{O_j}^2})^\xi \; \wedge$$

$$P_{(U,O_i)}^2 = (a_{O_i}^2)^\beta (b_{O_i}^2)^\eta \; \wedge \; \beta \in \Gamma \; \wedge \; \gamma \in \Delta \; \wedge \; \alpha \in \Lambda\} \; .$$

The first three equations of this proof of knowledge are the same as Protocol 3. The fourth equation proves that the same master secret key is used in $P_{(U,O_i)}$ and in the validating tag to the pseudonym established with O_j.

In the random oracle model, the verifier (or verifying organization) can obtain the receipt from a showing transaction by turning step 2 of Protocol 4 (or Protocol 3, respectively) into the corresponding *SPK* on the description of the transaction. This step will add efficiency and also will enable a user to sign an agreement with a verifier using her credential as a signature public key. This could, for instance, be useful if possessing a credential means being allowed to sign on behalf of the issuing organization (cf. group signatures).

5   Proof of Security for the Basic Credential System

The following technical lemmas about the protocols described above are stated here without proof; their proofs can be found in the full version of this paper [12].

Lemma 1. *Under the strong RSA assumption and the decisional Diffie-Hellman assumption modulo a safe prime product, step 5 of Protocol 1 (the protocol for establishing a pseudonym) is a statistical zero-knowledge proof of knowledge of the correctly formed values x_U, $s_{(U,O)}$ that correspond to a pseudonym validating tag $P_{(U,O)}$.*

Lemma 2. *Under the strong RSA assumption and the decisional Diffie-Hellman assumption modulo a safe prime product, step 2 of Protocol 3 (the protocol for showing a single credential) is a statistical zero-knowledge proof of knowledge of the values $x \in \Gamma$, $s \in \Delta$, $e \in \Lambda$, and c such that x, s correspond to a pseudonym validating tag $P = a_O^x b_O^s$, and $c^e = P d_O \bmod n_O$.*

Lemma 3. *Under the strong RSA assumption and the decisional Diffie-Hellman assumption modulo a safe prime product, step 2 of Protocol 4 (the protocol for showing a credential corresponding to a given validating tag $P_{(U,O_i)}$) is a statistical zero-knowledge proof of knowledge of the values $x \in \Gamma$, $s_1, s_2 \in \Delta$, $e \in \Lambda$, and c such that $P_{(U,O_i)} = a_{O_i}^x b_{O_i}^{s_1} \bmod n_{O_i}$, x, s_2 correspond to a validating tag $P = a_{O_j}^x b_{O_j}^{s_2}$ and $c^e = Pd_{O_j} \bmod n_{O_j}$ holds.*

5.1 Description of the Simulator

We now describe the simulator $\mathcal{S}$ for our scheme and then in Section 5.2 show that it satisfies Definition 1.

Setup. For the organizations not controlled by the adversary, the simulator sets up their secret and public keys as dictated by the protocol. For each organization, the simulator creates an archive where it will record the credentials issued by this organization to the users controlled by the adversary. It also initializes a list of the users controlled by the adversary.

Generation of a pseudonym. If a user controlled by the adversary establishes a pseudonym from an honest organization, the simulator uses the knowledge extractor of Lemma 1 to discover the user's underlying key x and the value s. If no user with key x is present in the list of dishonest users, $\mathcal{S}$ creates a new user U with login name L_U, and runs $\mathsf{FormNym}(U, O)$ to create a pseudonym $N_{(U,O)}$ for this user, and to obtain a key K_U for further interactions of this user with T. The simulator stores the record $(U, L_U, x, K_U, N_{(U,O)}, s)$ in its list of users controlled by the adversary. If some user U with key x is already present, the simulator runs $\mathsf{FormNym}(U, O)$ to create a pseudonym $N_{(U,O)}$ for this user, and adds $(N_{(U,O)}, s)$ to U's record.

If an honest user, through T, establishes a pseudonym with an organization controlled by the adversary, our simulator will use the zero-knowledge simulator from Lemma 1 to furnish the adversary's view of the protocol.

Generate a credential. If a user controlled by the adversary requests a credential from an honest organization O, then, upon receiving a message from T to that effect, the simulator runs the knowledge extractor for the proof of knowledge of step 1 of Protocol 2. It determines the values x and s. The simulator looks at its list of the pseudonyms of users controlled by the adversary. If it does not find a record with x and s, then it refuses to grant a credential (as an organization would). If it finds that there is a record containing these x and s and pseudonym N, then the simulator runs $\mathsf{GrantCred}(N, O)$ with T. Upon hearing from T that the user may have a credential, the simulator runs the organization's side of the rest of the Protocol 2, and issues the correct e and c. It stores the values (x,s,e,c) in the archive for organization O.

If an honest user, through T, requests a credential from an organization controlled by the adversary, then the simulator will run the zero-knowledge simulator for step 1 of Protocol 2, and execute the rest of the user's side of it. If the user accepts, then the simulator informs T that the credential was granted.

Showing a single credential. This part of the simulator can easily be inferred from the part for *Showing a credential with respect to a pseudonym* that follows.

Showing a credential with respect to a pseudonym. If a user controlled by the adversary wants to show a credential from an honest organization O_j to an honest organization O_i with whom it has pseudonym $N_{(U,O_i)}$, then the simulator runs O_i's part of Protocol 4, and extracts the user's values $(x, s_{(U,O_i)}, s_{(U,O_j)}, e, c)$ with the knowledge extractor of Lemma 3. If O_i's side of Protocol 4 accepts, while $(x, s_{(U,O_j)}, e, c)$ is not in the archive of O_j, then $\mathcal{S}$ rejects. Otherwise, it finds the user U with key x, the user's corresponding key K and pseudonym $N_{(U,O_j)}$ and runs $\mathsf{VerifyCred}(O_i, N_{(U,O_i)}, N_{(U,O_j)}, O_j)$.

   If a dishonest user wants to prove to an honest organization O_i that he has a credential from a dishonest organization O_j, then the simulator runs O_i's side of Protocol 4, with the knowledge extractor of Lemma 3 to obtain the values $(x, s_{(U,O_i)}, s, e, c)$. If O_i's side of the protocol rejects, it does nothing. Otherwise: (1) It checks if there exists a user with key x in O_j's archive. If so, denote this user by U. If not, let U be the user with key x. Next it runs $\mathsf{FormNym}(U, O_j)$ to get $N_{(U,O)}$. (2) It checks if U has a credential record in O_j's archive. If not, it runs $\mathsf{GrantCred}(N_{(U,O_j)}, O_j)$. (3) It runs $\mathsf{VerifyCredOnNym}(O_i, N_{(U,O_i)}, N_{(U,O_j)}, O_j)$.

   If an honest user (through T) wants to prove to organization O_i controlled by the adversary, that he has a credential from an honest organization O_j, then the simulator runs the zero-knowledge simulator of Lemma 3 to do that.

5.2   Proof of Successful Simulation

We show that our simulator fails with negligible probability only. As a first step, we show in Theorem 2 that a tuple (x, s, e, c) the knowledge of which is essential for proving possession of a credential, is unforgeable even under an adaptive attack. For this we rely on the following theorem due to Ateniese et al. [2]:

Theorem 1. *Suppose an ℓ_n-bit RSA modulus $n = pq = (2p' + 1)(2q' + 1)$ is given, where p, q, p', and q' are primes. Let $\Delta' =] - 2^{\ell_{\Delta'}}, 2^{\ell_{\Delta'}}[$, $\Pi' =] - T, T[$ and $\Lambda' =]2^{\ell_{\Lambda'}}, 2^{\ell_{\Lambda'} + \ell_{\Sigma'}}[$ with $2^{\ell_{\Delta'}} \geq T > 2^{2\ell_n}$ and $\ell_{\Lambda'} > \ell_{\Sigma'} + \ell_{\Delta'} + 3$. Suppose random $b, d \in_R QR_n$ are given. Further, suppose we have access to an oracle which, on the i-th query outputs tuples (y_i, e_i, c_i) such that $y_i \in_R \Pi'$, $e_i \in_R \Lambda'$ is a prime, and $c_i^{e_i} = b^{y_i} d \bmod n$. Under the strong RSA assumption, it is hard, upon seeing the oracle output for $1 \leq i \leq K$, K polynomial in ℓ_n, to produce a tuple (y, e, c) such that for all $1 \leq i \leq K$, $(y, e) \neq (y_i, e_i)$, and $y \in \Delta'$, $e \in \Lambda'$, and $c^{2e} = (b^y d)^2$.*

Theorem 2. *Suppose an ℓ_n-bit RSA modulus $n = pq = (2p' + 1)(2q' + 1)$ is given, where p, q, p', and q' are primes. Suppose random $a, b, d \in_R QR_n$ are given. Further, suppose we have access to an oracle $\mathcal{O}$ which, on the i-th query with a value $x_i \in \Gamma$, outputs a tuple (s_i, e_i, c_i) such that that $s_i \in_R \Delta$, $e_i \in_R \Lambda$ is a prime, and $c_i^{e_i} = a^{x_i} b^{s_i} d$. Under the strong RSA assumption and the discrete logarithm assumption modulo a safe prime product, it is hard, upon seeing the* *oracle output for $1 \leq i \leq K$, K polynomial in ℓ_n, to produce a tuple (x, s, c, e) such that for all $1 \leq i \leq K$, $(x, s, e, c) \neq (x_i, s_i, c_i, e_i)$, and $x \in \Gamma$, $s \in \Delta$, $e \in \Lambda$, and $c^{2e} = (a^x b^s d)^2 \bmod n$.*

Proof. We will prove our theorem by exhibiting a reduction to Theorem 1. The reduction has access to a forger $\mathcal{A}$ that forges a tuple (x, s, e, c) under conditions stated in the theorem. Using $\mathcal{A}$, the reduction will forge a tuple (y, e, u) under conditions stated in Theorem 1. This, in turn, contradicts the strong RSA assumption.

The reduction will take, as input, the public parameters (n, b, d) as in Theorem 1. Then it will define the public parameters for the setting of the theorem: b and d are as given, and to form a, pick $\alpha \in_R [0, n/4]$ and set $a := b^\alpha$.

Then, the reduction makes K queries and obtains a set of tuples $\{(y_i, e_i, c_i)\}$. Now the reduction proceeds as follows: upon receiving a query $x_i \in \Gamma$, set $s_i := y_i - \alpha x_i$. Setting the parameter T of Theorem 1 to $2^{\ell_\Delta} - 2^{\ell_\Lambda + \ell_n}$ will assure that $s_i \in \Delta$. Setting $\ell_\Delta = \epsilon(\ell_\Lambda + \ell_n) + 1$ with $\epsilon > 1$ (cf. Section 4.2) assures that $T > 2^{\ell_n}$ and also that s_i will be distributed statistically close to uniformly from Δ. Further, note that $a^{x_i} b^{s_i} d = b^{\alpha x_i + s_i} d = a^{y_i} d = c_i^{e_i}$. Setting $\ell_{\Sigma'} = \ell_\Sigma$ and $\ell_{\Lambda'} = \ell_\Lambda$, the tuple (s_i, e_i, c_i) is distributed statistically close to the distribution induced by the actual oracle $\mathcal{O}$.

After answering the K queries, the reduction receives from the forger a tuple (x, s, e, c) such that for all $1 \leq i \leq K$, $(x, s, e, c) \neq (x_i, s_i, e_i, c_i)$, $x \in \Gamma$, $s \in \Delta$, $e \in \Lambda$, and $c^{2e} = (a^x b^s d)^2 \bmod n$. Compute $y = s + \alpha x$. Setting ℓ'_Λ of Theorem 1 to $\ell_\Lambda + 1$ gives us the condition $\ell_\Lambda = \ell_{\Lambda'} > \ell_{\Sigma'} + \ell_{\Delta'} + 3 = \ell_\Sigma + \ell_\Delta + 4$ (cf. Section 4.2). With these settings, the triple (y, e, c) constitutes a forgery for Theorem 1, provided that $(y, e) \neq (y_i, e_i)$ for all i.

Suppose the probability that $(y, e) = (y_i, e_i)$ for some i is non-negligible. Then we can use $\mathcal{A}$ to break discrete logarithm modulo n. Suppose $(g, h) \in QR_n$ are given. It is known that finding (α_1, β_1) and a distinct (α_2, β_2) such that $g^{\alpha_1} h^{\beta_1} = g^{\alpha_2} h^{\beta_2}$ is hard if factoring is hard and computing discrete logarithms modulo a safe prime product is hard.

The reduction takes, as input, the modulus n, and the values (g, h). Then it selects K random primes $\{e_i \in_R \Lambda\}_{i=1}^K$, chooses a random $v \in_R QR_n$, and sets $d = v^{\prod_{i=1}^k e_i}$, $a = g^{\prod_{i=1}^k e_i}$, $b = h^{\prod_{i=1}^k e_i}$.

On input (x_i, V_i), do the following: select an $s_i \in_R \Delta$. Compute the value $E_i - \prod_{j=1, j \neq i}^K e_j$. Set $u_i := (g^{x_i} h^{s_i} v)^{E_i}$. Note that by construction, $c_i^{e_i} = a^{x_i} b^{s_i} d$. Then output (s_i, e_i, c_i). With non-negligible probability, obtain a forgery (x, s, e, c) from the forger such that $(x, s, e, c) \neq (x_i, s_i, e_i, c_i)$ for all i, and yet for some i, $(a^{x_i} b^{s_i} d)^2 = (a^x b^s d)^2$. Because a, b, and d are quadratic residues, it follows that $a^{x_i} b^{s_i} d = a^x b^s d$. From here, we either break the discrete logarithm problem or factor n.

Lemma 4. *Under the strong RSA assumption and the decisional Diffie-Hellman assumption modulo a safe prime product, the simulator rejects with only negligible probability.*

Proof. (Sketch) Note that the only time when the simulator rejects is when a dishonest user makes the verifier accept in Protocol 3 or in Protocol 4, and yet the tuple (x, s, e, c) extracted by the simulator was not given to the adversary by the simulator itself. Under the appropriate assumptions, by Lemmas 2 and 3 knowledge extraction succeeds with probability $1 - \mathsf{neg}(k)$. Then if we are given an adversary that can make the simulator reject non-negligibly often, we can use this adversary to create a forgery to contradict Theorem 2.

The statistical zero-knowledge property of the underlying protocols gives us Lemma 5 which in turn implies Theorem 3.

Lemma 5. *The view of the adversary in the real protocol is statistically close to his view in the simulation.*

Theorem 3. *Under the strong RSA assumption, the decisional Diffie-Hellman assumption modulo a safe prime product, and the assumption that factoring is hard, the credential system described above is secure.*

6   All-or-Nothing and PKI-Based Non-transferability

The protocols described in Section 4 ensure consistency of credentials, i.e., credential pooling is not possible. However, credential (or pseudonym) lending is still possible. More precisely, revealing to a friend the secrets x_U and $s_{(U,O_i)}$ attached to some credential does not mean that the friend obtains some other valuable secret of the user or can use any of the user's other credentials. This section provides protocols to obtain PKI-based non-transferability and all-or-nothing non-transferability to discourage users from credential lending.

The idea of the former is that the user provides the *CA* with a (verifiable) encryption of some valuable external secret that can be decrypted with x_U.

The idea for achieving the latter is similar, i.e., the user provides each organization with a (verifiable) encryption of the secrets underlying her validating tag. This approach raises some technical problems:

First, the approach requires that each user encrypts each of her secret keys D_i under one of her public keys E_j, thereby creating "circular encryptions". However, the canonical definitions [35] of secure encryption do not provide security for such encryptions. Moreover, it is not known whether circular security is possible under general assumptions. Nevertheless, we introduce in this section a new cryptographic primitive called *circular encryption* which is an encryption scheme that provides security for circular encryptions. Given any semantically secure encryption scheme, we provide a generic construction of such a scheme and prove its security in the random oracle model

Second, the encryptions made by a user must not reveal the public key this encryption was made with, i.e., we require that the encryption scheme be *key-oblivious*. We provide a formal definition of this and show that our circular encryption scheme satisfies it.

Third, the encryption must be verifiable. To this end we review the verifiable encryption protocol due to Camenisch and Damgård [10] and adapt it to suit our needs. Specifically, we want to enable verification without revealing the public key. We provide a verification method involving a *committed* public key, so that by inspecting this verifiable encryption, an adversary would not be able to discover the underlying public key.

Independently of and concurrently with our work, Black et al. [5] proposed symmetric encryption schemes for key-dependent messages (which is what we call circular symmetric encryption) and Bellare et al. [3] studied key-private encryption (which is what we call key-oblivious encryption).

6.1   Circular Encryption

Definition 2. *Let* $n, m \in \text{poly}(k)$. *A semantically secure encryption scheme* $\mathcal{G} = (\mathcal{E}, \mathcal{D})$ *is circular-secure if*

1. *There exists a message, denoted by* 0, *such that for all* $E \in \mathcal{E}(1^k)$, 0 *is in the message space of* E.
2. *For all* $E_1 \in \mathcal{E}(1^k)$, $D_2 \in \mathcal{D}(1^k)$, *the message space of* E_1 *includes* D_2.
3. *For all* n-*node directed graphs* G *with* m *edges, given* n *randomly chosen public keys,* $\{E_i\}_{i=1}^n$, *we have:* $\{E_i(D_j)\}_{(i,j) \in E(G)} \overset{c}{\approx} \{E_i(0)\}_{(i,j) \in E(G)}$.

The idea here is that having access to encryptions of the secret keys does not help the adversary in breaking the security of the system. Note that if, in the definition above, we had limited our attention to acyclic graphs, then any semantically secure cryptosystem would be enough to satisfy such a definition. As the definition can only be more powerful if we include graphs that have cycles, we call this notion of security "circular security."

Let us present a cryptosystem that satisfies this definition in the random oracle model. Suppose the length of a secret key is $p(k)$. Let $\mathcal{H} : \{0,1\}^* \rightarrow \{0,1\}^{p(k)}$ be a random oracle, and let $\oplus$ denote the bitwise XOR operation. Let $\mathcal{G} = (\mathcal{E}, \mathcal{D})$ be a semantically secure cryptosystem with a sufficiently large message space. Construct $\mathcal{G}' = (\mathcal{E}', \mathcal{D}')$ as follows: generate (E, D) according to $\mathcal{G}$. To encrypt a message $m \in \{0,1\}^{p(k)}$, E' picks a random $r \in_R \{0,1\}^\ell$ and sets $E'(m) := (E(r), \mathcal{H}(r) \oplus m)$. To decrypt a tuple (a, b), D' computes $\tilde{m} := \mathcal{H}(D(a)) \oplus b$. For this construction, the following theorem holds (the proof can be found in the full version of this paper [12]).

Theorem 4. *If* $\mathcal{G}$ *is semantically secure,* $\mathcal{G}'$ *is circular-secure.*

As a basis for our circular encryption scheme, we use the ElGamal encryption [27] in some $G = \langle g \rangle$. It is easy to see that the ElGamal cryptosystem is semantically secure under the decisional Diffie-Hellman assumption. Let $P = g^x$ be a public key. The resulting circular encryption scheme is as follows. To encrypt a message $m \in \{0,1\}^k$, choose a random element $r_1 \in G$ and a random integer $r_2 \in \{0,1\}^{2\ell}$, and compute the encryption $(u, v, z) := (P^{r_2} r_1, g^{r_2}, \mathcal{H}(r_1) \oplus m)$. Decryption works by computing $\mathcal{H}(u/v^x) \oplus z$. We denote this encryption scheme by CEIG.

6.2   Verifiable Encryption with a Committed Public Key

Verifiable encryption [1,10], is a protocol between a prover and a verifier such that as a result of the protocol, on input public key E and value v, the verifier obtains an encryption e of some value s under E such that $(s,v) \in \mathcal{R}$. Here $\mathcal{R}$ is a relation such as, e.g., $\{(s,g^s)|s \in \mathbb{Z}_q\} \subset \mathbb{Z}_q \times G$. More formally,

Definition 3. *Let* $(\mathcal{E}, \mathcal{D})$ *be a semantically secure encryption scheme. A two-party protocol between a prover* $\mathcal{P}(\mathcal{R}, E, s, v)$ *and a verifier* $\mathcal{V}(\mathcal{R}, E, v)$ *is a verifiable encryption protocol with respect to public keys* $\mathcal{E}$ *for a polynomial-time verifiable relation* $\mathcal{R}$ *if*

- *For all* $(E, D) \in \mathcal{G}(1^k)$ *and for all* $(s, v) \in \mathcal{R}$, *if* $\mathcal{P}$ *and* $\mathcal{V}$ *are honest then* $\mathcal{V}_{\mathcal{P}(\mathcal{R}, E, s, v)}(\mathcal{R}, E, v) \neq \perp$.
- *There is an efficient extractor algorithm* C *such that for all sufficiently large* k, *and* $\forall (E, D) \in (\mathcal{E}, \mathcal{D})(1^k)$

$$\Pr[(C(D, e), v) \in \mathcal{R} \mid e = \mathcal{V}_{\widetilde{\mathcal{P}}(\mathcal{R}, E, s, v)}(\mathcal{R}, E, v) \wedge e \neq \perp] = 1 - \mathbf{neg}(k) \ .$$

- *There is a black-box simulator* $\mathcal{S}$ *such that* $\forall \ \widetilde{\mathcal{V}}, \ \forall (s, v) \in \mathcal{R}$ *we have* $\mathcal{S}^{\widetilde{\mathcal{V}}(\mathcal{R}, E, v)}(\mathcal{R}, E, v) \overset{c}{\approx} \widetilde{\mathcal{V}}_{\mathcal{P}(\mathcal{R}, E, s, v)}(\mathcal{R}, E, v)$, *where the probability "hidden" in the* $\overset{c}{\approx}$ *notation is over the choice of* E *and the random cointosses of* $\widetilde{\mathcal{V}}$.

Note that e is not a single message from the prover, but the verifier's entire transcript of the protocol. Furthermore, C does not necessarily extract the same s that was the additional input to the prover. It could extract some other value $s' \neq s$, but only if $(s', v) \in \mathcal{R}$.

It is clear that an (inefficient) way of implementing verifiable encryption would be for the prover to encrypt s under the public key E, and then carry out a zero-knowledge proof that the encrypted value satisfies relation $\mathcal{R}$ with respect to v. But this is not satisfactory, because it is important that verifiably encryption be executed efficiently enough to be useful in practice. Generalizing the protocol of Asokan et al. [1], Camenisch and Damgård [10] provide a practical verifiable encryption scheme for all relations $\mathcal{R}$ that have an honest-verifier zero-knowledge three-move proof of knowledge where the second message is a random challenge and the witness can be computed from two transcripts with the same first message but different challenges. This includes most known proofs of knowledge, and all proofs about discrete logarithms considered in this paper. Their construction is secure with respect to any public key for a semantically secure cryptosystem.

We use similar notation for verifiable encryption as for the PK's and denote by, e.g., $e := VE(\mathsf{ElGamal}, (g, y))\{(\xi) : a = b^\xi\}$ the verifiable encryption protocol for the ElGamal scheme, whereby $\log_b a$ is encrypted in e under public key (y, g).

For guaranteeing the all-or-nothing non-transferability, we need to have each user verifiable encrypt all of her secret information under a public key that corresponds to her secret key. However, revealing this public key will leak information about the user. Therefore, we need to realize verifiable encryption in such a manner that the public key corresponding to the resulting ciphertext cannot be linked to the verifier's view, i.e., a verifiable encryption scheme must be key-oblivious:

Definition 4. *Let $(\mathcal{P}, \mathcal{V})$ be a verifiable encryption scheme with respect to public keys $\mathcal{E}$, for a polynomial-time verifiable relation $\mathcal{R}$. We say that this scheme is key-oblivious if for all polynomially bounded $\tilde{\mathcal{V}}$, for all $E, E' \in \mathcal{E}(1^k)$ and $\forall (s, v) \in \mathcal{R}$ we have $\tilde{\mathcal{V}}_{\mathcal{P}(\mathcal{R}, E, s, v)}(\mathcal{R}, v, E, E') \overset{c}{\approx} \tilde{\mathcal{V}}_{\mathcal{P}(\mathcal{R}, E', s, v)}(\mathcal{R}, v, E, E')$, where the probability "hidden" in the $\overset{c}{\approx}$ notation is over the random cointosses of $\tilde{\mathcal{V}}$.*

In case the verifier does not know the public key under which the encryption is carried out, previously known constructions do not work, as they require that the verifier be able to check that a given ciphertext is an encryption of a given value. Thus we propose a new construction, based on the circularly secure variant of the ElGamal cryptosystem described above. Here we assume that the prover $\mathcal{P}$ knows the secret key of the encryption; this is not the general case, but it works for our construction. Let $P = g^x$ serve as a public key, and x as the corresponding secret key. Let $C = Ph^r$ be a commitment to P, where h is another generator of $G = \langle g \rangle$, and let $(u, v, z) = (P^{r_2} r_1, g^{r_2}, \mathcal{H}(r_1) \oplus m)$ be an encryption of m as above. To convince the verifier that (u, v, z) is an encryption of m under the public key committed to by C, the prover reveals r_1 and engages with the verifier in $PK\{(\alpha, \beta, \gamma) : C = g^\alpha h^\beta \ \wedge \ v = g^\gamma \ \wedge \ u/r_1 = v^\alpha\}$. The verifier further needs to check if $z = \mathcal{H}(r_1) \oplus m$. By using techniques developed by Camenisch and Damgård [10], a key-oblivious verifiable encryption scheme is obtained.

In the sequel, we write, e.g., $Com\text{-}VE(\mathsf{CEIG}, (\mathcal{H}, g, h, C))\{(\xi) : a = b^\xi\}$ for this key-oblivious verifiable encryption with respect to a committed public key. The proof of the following lemma uses standard techniques and is given in the full version of this paper [12]:

Lemma 6. *Under the decisional Diffie-Hellman assumption, the verifiable encryption scheme described above is key-oblivious.*

6.3   All-or-Nothing Non-transferability

As already mentioned, all-or-nothing non-transferability is achieved by ensuring that if a user U gives away her master secret x_U, then she will also reveal the secret keys underlying her validating tag with O. More precisely, U has to supply O a verifiable encryption of these secrets w.r.t. the secret key x_U. This is done in the following protocol, which U and O should carry out as part of Protocol 1. A prerequisite of the protocol is that during the setup of the system, a group $G = \langle g \rangle = \langle h \rangle$ of prime order $q > 2^{\ell_r}$ is chosen such that $\log_g h$ is unknown.

Protocol 5

1. *U chooses $r \in_R \mathbb{Z}_q$, sets $C := g^{x_U} h^r$, and sends C to O. U proves to O that C is a commitment to her public key by carrying out*

$$PK\{(\gamma, \vartheta, \varphi) : P^2_{(U,O)} = (a_O^2)^\gamma (b_O^2)^\vartheta \ \wedge \ C = g^\gamma h^\varphi\} \ .$$

2. U and O engage in the verifiable encryption protocol

$$w_{P_{(U,O)}} = \text{Com-VE}(\text{CEIG}, (\mathcal{H}, g, h, C))\{(\alpha, \beta) : P^2_{(U,O)} = (a_O^2)^\alpha (b_O^2)^\beta\} \ .$$

3. O publishes $N_{(U,O)}$ and $w_{P_{(U,O)}}$.

However, publishing $N_{(U,O)}$ and $w_{P_{(U,O)}}$ is not sufficient for using U's credential with O even when knowing x_U. Therefore, the organizations must publish all related information together with the verifiable encryption. Hence, at the end of Protocol 2, O must publish $(c_{(U,O)}, e_{(U,O)})$ together with $N_{(U,O)}$. Thus, we obtain all-or-nothing transferability: whenever a user's friend gets to know x_U, he can look at the organizations' public records to obtain all information needed to use all the user's credentials.

6.4   PKI-Assured Non-transferability

We assume that the user possesses some external valuable public key PK_U. Then PKI-assured non-transferability is achieved by having the CA ask for this public key, check whether it is indeed the user's public key (e.g., via some external certificate), and require the user to verifiably encrypt the corresponding secret key SK_U with respect to x_U. This verifiable encryption is then published by the CA. Now, if the user ever gives x_U away to her friend, then her friend, by reading the CA's public records, will recover the verifiable encryption of SK_U, and will succeed in decrypting it.

The technical realization is similar to the one for all-or-nothing non-transferability. The main difference is that we do not need circular encryption and thus can use regular ElGamal. We give an example for what this protocol looks like when U's external public key Y_U is discrete-logarithm based, i.e., $Y_U = g^x$ for some generator g in some group G. Other cases are similar. A prerequisite of the protocol is that during the setup of the system, a group $G = \langle g \rangle = \langle h \rangle$ of prime order $q > 2^{\ell_r}$ is chosen such that $\log_g h$ is unknown.

Protocol 6

1. U sends Y_U, g, and the certificate on Y_U of the external PKI to CA who checks their validity.
2. U chooses $r \in_R \mathbb{Z}_q$, sets $C := g^{x_U} h^r$, and sends C to CA. U proves to CA that C is a commitment to her public key by carrying out

$$PK\{(\gamma, \vartheta, \varphi) : P^2_{(U,O_0)} = (a_{O_0}^2)^\gamma (b_{O_0}^2)^\vartheta \ \wedge \ C = g^\gamma h^\varphi\} \ .$$

3. U and CA engage in

$$w_{PKI} = \text{Com-VE}(\text{ElGamal}, (\mathcal{H}, g, h, C))\{(\alpha) : Y_U = g^\alpha\} \ .$$

4. CA publishes (w_{PKI}, PKI).

7   One-Show Credentials and Revocation

This section describes how the basic credential scheme can be extended to allow for global and local revocation as well as to enable organizations to issue one-show credentials.

7.1   One-Show Credentials

The credentials we considered so far can be shown an unlimited number of times. However, for some services it might be required that a credential can only be used once (e.g., when it represents money). Of course, one possibility would be that a user just reveals the credential to the verifier. This, however, would mean that the user is not fully anonymous any more as the verifier and the organization then both know the credential and thus can link the transaction to the user's pseudonym. Traditionally, this problem has been solved using so-called blind signatures [17]. Here, we provide a novel and alternative way to approach this problem, i.e., instead of blinding the signer we blind the verifier. In the sequel we describe the general idea, the changes to the protocols that need to be made, and provide a protocol for showing one-show credentials.

Addition to key generation. Each organization O publishes an additional generator $z_O \in QR_{n_O}$.

Changes to Protocol   1. The validating tag $P_{(U,O)}$ on a user's pseudonym $N_{(U,O)}$ is formed slightly differently: $P_{(U,O)} = a_O^{x_U} b_O^{s_{(U,O)}} z_O^{r_{(U,O)}}$, where $r_{(U,O)}$ is chosen by O and U together in the same way as $s_{(U,O)}$ is. (Credentials, however, are issued in the same way as before, i.e., U obtains $c_{(U,O)}$ and $e_{(U,O)}$ such that $c_{(U,O)}^{e_{(U,O)}} \equiv P_{(U,O)} d_O \pmod{n_O}$ holds.)

Showing a one-show credential. When proving possession of a one-show credential issued by O (with respect to a pseudonym or not), the user provides to verifier V (which might be an organization) the value $H_{(U,O)} = h_O^{r_{(U,O)}}$ and proves that it is formed correctly w.r.t. to the pseudonym U established with O. Of course, the various proofs of knowledge in the respective protocols have to be adapted to reflect the different form of the pseudonym U holds with O. These adaptions, however, are immediate and we do not describe them here.

Now, different usages of the same credential can be linked to each other but not to the user's pseudonym with the issuing organization. This allows to prevent users from using the same credential several times, if the verifier checks with the issuing organization whether $H_{(U,O)}$ was already used or not, similar as it is done for anonymous on-line e-cash.

Off-line checking could be done as well. As here double usage can only be detected but not prevented, a mechanism for identifying double-users is required. This could for instance be achieved using revocation as described in the previous section, or using similar techniques that are used in for anonymous off-line e-cash (e.g., [7]).

We now describe how the latter can be done such that using a one-show credential twice would expose the user's secret keys connected with the corresponding pseudonym. Together with (any kind of) non-transferability this would be quite a strong incentive for the users not to use one-show credentials twice. The main idea is that the verifying entity chooses some random challenge c from a suitably large set, say $\{0,1\}^{\ell_c}$ with $\ell_c = 60$, and the user replies with $r = cx_U + s_{(U,O)}$ and proves correctness of this result. To assure that r hides x_U statistically, we must have that $\ell_\Delta > \epsilon(\ell_\Gamma + \ell_c)$ because $x_U \in \Gamma$ and $s_{(U,O)} \in \Delta$. However, when a user uses the same credential twice, one can compute x_U from the the different replies the user provides. We present the resulting protocol for showing a single credential (cf. Protocol 3).

Protocol 7

1. U chooses $r_1, r_2 \in_R \{0,1\}^{2\ell_n}$, computes $A = c_{(U,O_i)} h_O^{r_1}$ and $B = h_O^{r_1} g_O^{r_2}$, and sends $A, B, H_{(U,O)}$ to V.
2. V chooses $c \in_R \{0,1\}^{\ell_c}$ and sends c to U.
3. U replies with $r = cx_U + s_{(U,O)}$ (computed in $\mathbb{Z}$).
4. U engages with V in

$$PK\{(\alpha, \beta, \gamma, \varphi, \delta, \varepsilon, \zeta, \xi) : \ d_O^2 = (A^2)^\alpha (\frac{1}{a_O^2})^\beta (\frac{1}{b_O^2})^\gamma (\frac{1}{z_O^2})^\varphi (\frac{1}{h_O^2})^\delta \ \wedge$$

$$B^2 = (h_O^2)^\varepsilon (g_O^2)^\zeta \ \wedge \ 1 = (B^2)^\alpha (\frac{1}{h_O^2})^\delta (\frac{1}{g_O^2})^\xi \ \wedge$$

$$H_{(U,O)} = h_O^\varphi \ \wedge \ g_O^r = (g_O^c)^\beta g_O^\gamma \ \wedge \ \beta \in \Gamma \ \wedge \ \gamma \in \Delta \ \wedge \ \varphi \in \Delta \ \wedge \ \alpha \in \Lambda \} \ .$$

The adaption of Protocol 4 to implement one-show credentials with built-in anonymity revocation is similar.

7.2   Local and Global Revocation

For simplicity we assume a single revocation manager R who is responsible for local and global revocation (extending the scheme to one revocation manager per organization is easy). Given the transcript of a protocol where some user proved possession of a credential from organization O_i, R will have the task of providing information that allows the organization to identify the pseudonym of the user in case of local revocation, or allows the CA to retrieve the identity of the user.

In the sequel we describe how the protocols for proving possession of a credential must be adapted such that local revocation is possible using Cramer-Shoup encryption [22]. We then discuss global revocation. We remark that it can be decided at the time when the possession of a credential is proved whether local and/or global revocation shall be possible for the transaction at hand.

Additions to key generation. The revocation manager R chooses a group $G = \langle g \rangle = \langle h \rangle$ of prime order $q > 2^{\ell_\Gamma}$. The he chooses five secret keys $x_1, \ldots, x_5 \in_R$

$\mathbb{Z}_q$ and computes $(y_1, y_2, y_3) := (g^{x_1} h^{x_2}, g^{x_3} h^{x_4}, g^{x_5})$ as his public key. Each organization O publishes an additional generator $v_O \in QR_{n_O}$.

Changes to Protocol 1. A validating tag $P_{(U,O)}$ on a user's pseudonym $N_{(U,O)}$ is formed slightly differently: $P_{(U,O)} = a_O^{x_U} b_O^{s_{(U,O)}} v_O^{x_{(U,O)}}$, where $x_{(U,O)}$ is chosen from Γ by U. However, credentials are issued in the same way as before, i.e., U obtains $c_{(U,O)}$ and $e_{(U,O)}$ such that $c_{(U,O)}{}^{e_{(U,O)}} \equiv P_{(U,O)} d_O \pmod{n_O}$ holds.

If Protocol 1 is carried out with the CA, it is extended by the following steps.

8. U computes $Y_U = g^{x_U}$ and sends Y_U to CA.
9. U engages with CA in

$$PK\{(\alpha, \beta, \gamma) : P_{(U,CA)}^2 = (a_{CA}^2)^\alpha (b_{CA}^2)^\beta (v_O^2)^\gamma \wedge Y_U = g^\alpha \wedge \gamma \in \Gamma\} .$$

10. Both CA and U store Y_U with $P_{(U,CA)}$.

In case Protocol 1 is carried out with an organization O different from the CA, it is extended by the following steps.

8. U computes $Y_{(U,O)} = g^{x_{(U,O)}}$ and sends $Y_{(U,O)}$ to O.
9. U engages with O in

$$PK\{(\alpha, \beta, \gamma) : P_{(U,O)}^2 = (a_O^2)^\alpha (b_O^2)^\beta (v_O^2)^\gamma \wedge Y_{(U,O)} = g^\gamma \wedge \gamma \in \Gamma\} .$$

10. Both O and U store $Y_{(U,O)}$ with $P_{(U,CA)}$.

Changes to Protocols 3 and 4. Suppose Protocol 3 (resp., Protocol 4) is being executed. Suppose the user U and the verifying organization V agree upon text m that describes under what conditions V can find out U's identifying information. Specifically, m describes the conditions under which V may find out U's pseudonym with the issuing organization O, as well as the conditions under which V may find out U's identity. The text of m can also include part of the communication transcript of the current protocol. The former mode of anonymity revocation is called *local* revocation, while the latter is called *global* revocation. We provide the two protocols to be executed as sub-routines of Protocol 3 (resp., Protocol 4) in order to get local and/or global revocation, respectively, where the user proves possession of a credential issued by organization O.

Protocol 8 (Global Revocation)

1. U chooses $r_2 \in_R \mathbb{Z}_n$ and computes $w_1 := g^{r_2}$, $w_2 := h^{r_2}$, $w_3 := y_3^{r_2} Y_U$, and $w_4 := y_1^{r_2} y_2^{r_2 \mathcal{H}(w_1, w_2, w_3, m_0)}$ and sends $w_{(U,R)} = (w_1, w_2, w_3, w_4)$ to V.
2. U and V engage in

$$PK\{(\alpha, \beta, \gamma, \delta, \varepsilon, \xi) : d_O^2 = (A^2)^\alpha (\frac{1}{a_O^2})^\beta (\frac{1}{b_O^2})^\gamma (\frac{1}{v_O^2})^\xi \frac{1}{h_O^2})^\delta \wedge w_1 = g^\varepsilon \wedge$$
$$w_2 = h^\varepsilon \wedge w_3 = g^\beta y_3^\varepsilon \wedge w_4 = (y_1 y_2^{\mathcal{H}(w_1, w_2, w_3, m_0)})^\varepsilon\} .$$

Protocol 9 (Local Revocation)

1. U chooses $r_1 \in_R \mathbb{Z}_q$ and computes $w_1 := g^{r_1}$, $w_2 := h^{r_1}$, $w_3 := y_3^{r_1} Y_{(U,O)}$, and $w_4 = y_1^{r_1} y_2^{r_1 \mathcal{H}(w_1, w_2, w_3, m_j)}$ and sends $w_{(U,R_j^l)} = (w_1, w_2, w_3, w_4)$ to V.
2. U and V engage in

$$PK\{(\alpha, \beta, \gamma, \delta, \varepsilon, \xi) : \ d_O^2 = (A^2)^\alpha (\frac{1}{a_O^2})^\beta (\frac{1}{b_O^2})^\gamma (\frac{1}{v_O^2})^\xi (\frac{1}{h_O^2})^\delta \ \wedge \ w_1 = g^\varepsilon \ \wedge$$

$$w_2 = h^\varepsilon \ \wedge \ w_3 = g^\xi y_3^\varepsilon \ \wedge \ w_4 = \left(y_1 y_2^{\mathcal{H}(w_1, w_2, w_3, m_j)}\right)^\varepsilon \} \ .$$

Revocation. Upon presentation of an encryption $w = (w_1, w_2, w_3, w_4)$ and a revocation condition m, stemming from Protocol 8 or 9, the revocation manager checks whether $w_4 = w_1^{x_1 + x_3 \mathcal{H}(w_1 \| w_2 \| w_3 \| m)} w_2^{x_2 + x_4 \mathcal{H}(w_1 \| w_2 \| w_3 \| m)}$ and whether m is satisfied. If these checks succeed, he returns $\hat{Y} := w_3 / w_1^{x_5}$. In case of local revocation, $\hat{Y}$ will allow retrieval of the user's pseudonym with the organization that issued the credential of which that user proved possession. In case of global revocation, $\hat{Y}$ will allow the CA to retrieve the identity of the user.

7.3 Encoding Expiration Dates and Other Personal Attributes

Expiration dates and other attributes of credentials can be encoded in the exponent $e_{(U,O)}$ as this is the organization's choice. We need to divide the interval Λ into subintervals. Then, if a user is required to prove certain attributes of her credential, she proves that the exponent lies in the subinterval instead of proving that it lies in Λ.

Acknowledgements

The authors are grateful to Ron Rivest for fruitful discussions and comments. We thank the anonymous referees for their helpful and detailed remarks. The second author acknowledges the support of an NSF graduate fellowship and of the Lucent Technologies GRPW program.

References

1. N. Asokan, V. Shoup, and M. Waidner. Optimistic fair exchange of digital signatures. *IEEE Journal on Selected Areas in Communications*, 18(4):591–610, 2000.
2. G. Ateniese, J. Camenisch, M. Joye, and G. Tsudik. A practical and provably secure coalition-resistant group signature scheme. In *CRYPTO 2000*, vol. 1880 of *LNCS*, pp. 255–270. Springer Verlag, 2000.
3. M. Bellare, A. Boldyreva, A. Desai, and D. Pointcheval. Key-privacy in public-key encryption. Manuscript, 2001.
4. M. Bellare, J. A. Garay, and T. Rabin. Fast batch verification for modular exponentiation and digital signatures. In *EUROCRYPT '98*, vol. 1403 of *LNCS*, pp. 236–250. Springer Verlag, 1998.

5. J. Black, P. Rogaway, and T. Shrimpton. Encryption scheme security in the presence of key-dependent messages. Manuscript, 2001.
6. F. Boudot. Efficient proofs that a committed number lies in an interval. In *EUROCRYPT 2000*, vol. 1807 of *LNCS*, pp. 431–444. Springer Verlag, 2000.
7. S. Brands. Untraceable Off-line Cash in Wallets With Observers. In *CRYPTO '93*, vol. of *LNCS*. pp. 302–318. Springer Verlag, 1993.
8. S. Brands. *Rethinking Public Key Infrastructures and Digital Certificates; Building in Privacy*. PhD thesis, Eindhoven Institute of Technology, the Netherlands, 1999.
9. E. Brickell, P. Gemmel, and D. Kravitz. Trustee-based tracing extensions to anonymous cash and the making of anonymous change. In *Proc. ACM-SIAMs*, pp. 457–466. ACM press, 1995.
10. J. Camenisch and I. Damgård. Verifiable encryption and applications to group signatures and signature sharing. Technical Report RS-98-32, BRICS, Departement of Computer Science, University of Aarhus, December 1998.
11. J. Camenisch and A. Lysyanskaya. Efficient non-transferable anonymous multishow credential system with optional anonymity revocation. Technical Report Research Report RZ 3295, IBM Research Division, 2000.
12. J. Camenisch and A. Lysyanskaya. An Efficient Non-transferable Anonymous Credential System with Optional Anonymity Revocation. http://eprint.iacr.org/2001.
13. J. Camenisch and M. Michels. Proving in zero-knowledge that a number n is the product of two safe primes. In *EUROCRYPT '99*, vol. 1592 of *LNCS*, pp. 107–122.
14. J. Camenisch and M. Stadler. Efficient group signature schemes for large groups. In *CRYPTO '97*, vol. 1296 of *LNCS*, pp. 410–424. Springer Verlag, 1997.
15. R. Canetti. *Studies in Secure Multiparty Computation and Applications*. PhD thesis, Weizmann Institute of Science, Rehovot 76100, Israel, 1995.
16. R. Canetti. Security and composition of multi-party cryptographic protocols. *Journal of Cryptology*, 13(1):143–202, 2000.
17. D. Chaum. Blind signatures for untraceable payments. In *CRYPTO '82*, pp. 199–203. Plenum Press, 1983.
18. D. Chaum. Security without identification: Transaction systems to make big brother obsolete. *Communications of the ACM*, 28(10):1030–1044, 1985.
19. D. Chaum and J.-H. Evertse. A secure and privacy-protecting protocol for transmitting personal information between organizations. In *CRYPTO '86*, vol. 263 of *LNCS*, pp. 118–167. Springer-Verlag, 1987.
20. D. Chaum and E. van Heyst. Group signatures. In *EUROCRYPT '91*, vol. 547 of *LNCS*, pp. 257–265. Springer-Verlag, 1991.
21. L. Chen. Access with pseudonyms. In *Cryptography: Policy and Algorithms*, vol. 1029 of *LNCS*, pp. 232 243. Springer Verlag, 1995.
22. R. Cramer and V. Shoup. A practical public key cryptosystem provably secure against adaptive chosen ciphertext attack. In *CRYPTO '98*, vol. 1642 of *LNCS*, pp. 13–25, 1998, Springer Verlag.
23. R. Cramer and V. Shoup. Signature schemes based on the strong RSA assumption. In *Proc. 6th ACM CCS*, pp. 46–52. ACM press, 1999.
24. I. Damgård. Efficient concurrent zero-knowledge in the auxiliary string model. In *EUROCRYPT 2000*, vol. 1807 of *LNCS*, pp. 431–444. Springer Verlag, 2000.
25. I. Damgård. Payment systems and credential mechanism with provable security against abuse by individuals. In *CRYPTO '88*, vol. 403 of *LNCS*, pp. 328–335.
26. C. Dwork, J. Lotspiech, and M. Naor. Digital signets: Self-enforcing protection of digital information. In *Proc. 28th STOC*, 1996.

27. T. ElGamal. A public key cryptosystem and a signature scheme based on discrete logarithms. In *CRYPTO '84*, vol. 196 of *LNCS*, pp. 10–18. Springer Verlag, 1985.
28. A. Fiat and A. Shamir. How to prove yourself: Practical solution to identification and signature problems. In *CRYPTO '86*, vol. 263 of *LNCS*, pp. 186–194, 1987.
29. E. Fujisaki and T. Okamoto. Statistical zero knowledge protocols to prove modular polynomial relations. In *CRYPTO '97*, vol. 1294 of *LNCS*, pp. 16–30, 1997.
30. R. Gennaro, S. Halevi, and T. Rabin. Secure hash-and-sign signatures without the random oracle. In *EUROCRYPT '99*, vol. 1592 of *LNCS*, pp. 123–139, 1999.
31. S. Goldwasser, S. Micali, and C. Rackoff. The knowledge complexity of interactive proof systems. In *Proc. 27th FOCS*, pages 291–304, 1985.
32. O. Goldreich, B. Pfitzman, and R. Rivest. Self-delegation with controlled propagation—or—what if you lose your laptop. In *CRYPTO '98*, vol. 1642 of *LNCS*, pp. 153–168, 1998.
33. J. Kilian and E. Petrank. Identity escrow. In *CRYPTO '98*, vol. 1642 of *LNCS*, pp. 169–185, Springer Verlag, 1998.
34. A. Lysyanskaya, R. Rivest, A. Sahai, and S. Wolf. Pseudonym systems. In *Selected Areas in Cryptography*, vol. 1758 of *LNCS*. Springer Verlag, 1999.
35. S. Micali, C. Rackoff, and B. Sloan. The notion of security for probabilistic cryptosystems. *SIAM Journal on Computing*, 17(2):412–426, 1988.
36. B. Pfitzmann and M. Waidner. Composition and integrity preservation of secure reactive systems. In *Proc. 7th ACM CCS*, pp. 245–254. ACM press, 2000.
37. M. Stadler, J.-M. Piveteau, and J. Camenisch. Fair blind signatures. In *EUROCRYPT '95*, vol. 921 of *LNCS*, pp. 209–219. Springer Verlag, 1995.

Priced Oblivious Transfer:
How to Sell Digital Goods

Bill Aiello, Yuval Ishai, and Omer Reingold

[1] AT&T Labs – Research, 180 Park Ave., Florham Park, NJ 07932,
aiello@research.att.com
[2] DIMACS and AT&T Labs – Research,
yuval@dimacs.rutgers.edu
[3] AT&T Labs – Research, 180 Park Ave., Florham Park, NJ 07932,
omer@research.att.com

Abstract. We consider the question of protecting the privacy of customers buying digital goods. More specifically, our goal is to allow a buyer to purchase digital goods from a vendor without letting the vendor learn *what*, and to the extent possible also *when* and *how much*, it is buying. We propose solutions which allow the buyer, after making an initial deposit, to engage in an unlimited number of *priced oblivious-transfer* protocols, satisfying the following requirements: As long as the buyer's balance contains sufficient funds, it will successfully retrieve the selected item and its balance will be debited by the item's price. However, the buyer should be unable to retrieve an item whose cost exceeds its remaining balance. The vendor should learn nothing except what must inevitably be learned, namely, the amount of interaction and the initial deposit amount (which imply upper bounds on the quantity and total price of all information obtained by the buyer). In particular, the vendor should be unable to learn what the buyer's current balance is or when it actually runs out of its funds.

The technical tools we develop, in the process of solving this problem, seem to be of independent interest. In particular, we present the first one-round (two-pass) protocol for oblivious transfer that does not rely on the random oracle model (a very similar protocol was independently proposed by Naor and Pinkas [21]). This protocol is a special case of a more general "conditional disclosure" methodology, which extends a previous approach from [11] and adapts it to the 2-party setting.

1 Introduction

Consider a scenario where a buyer wishes to purchase digital goods from a vendor without disclosing *what* it is buying, or even *when* exactly it is buying. For instance, the buyer may wish to subscribe to a pay-per-view service, where different costs are associated with different channels, or get an up-to-date information on its stock portfolio. In both cases buyers may wish to hide from vendors what items they are buying, or even whether at a given moment they are buying anything at all.

B. Pfitzmann (Ed.): EUROCRYPT 2001, LNCS 2045, pp. 119–135, 2001.
© Springer-Verlag Berlin Heidelberg 2001

In the realm of physical goods, it is inherently impossible to hide from the vendor what, when, and how much it is selling. Being bounded to a limited inventory, the vendor must keep track of how many items of each kind it has in stock. However, unlike physical goods, digital goods are typically of unlimited supply. The purpose of this paper is to exploit the difference between the physical and the digital worlds in order to obtain privacy of buyers in the following electronic commerce scenario. Assume that a buyer first deposits a pre-payment at the hands of a vendor.[1] The buyer should then be able to engage in a virtually unlimited number of interactions with the vendor in order to obtain digital goods (also referred to as *items*) at a total cost which does not exceed its initial deposit amount. After spending all of its initial credit, the buyer should be unable to obtain any additional items before depositing an additional pre-payment. This paper provides efficient ways to implement this, rather standard, e-commerce task with the added requirement of *maintaining the buyer's privacy*. That is, the vendor should learn nothing except what must inevitably be learned: the amount of interaction and the initial deposit amount (which imply upper bounds on the quantity and total price of all information obtained by the buyer). In particular, the vendor should be unable to learn what the buyer's current balance is or when it actually runs out of its funds..

Traditional approaches for protecting the privacy of buyers, such as anonymous digital payments (e.g., [7,8]), do not address the problem of hiding which goods are being bought and when. This information, possibly combined with additional information from other sources (such as traffic analysis), may facilitate attacks on the privacy of individual buyers.[2] Moreover, strong anonymity is not only difficult to implement and prone to various types of attacks [2], but in some contexts it is also undesirable [26]. We stress that our solutions do not require anonymity of buyers and do not attempt to achieve this property. On the contrary, our work provides an *alternative* approach for protecting individual buyers engaging in e-commerce activities, which promises a different (and in a sense stronger) type of security. This approach is most beneficial when anonymity is insufficient, undesirable, or difficult to achieve.

Priced Oblivious Transfer. The well-known *oblivious transfer* primitive [25,10,4,15] provides a partial solution to our problem. If all items are identically priced, then the buyer's initial deposit determines the number of items it is entitled to obtain. In this case, the vendor may allow the buyer to retrieve just the right number of items using multiple invocations of oblivious transfer. However, this solution is not applicable in the realistic scenario where the items are not identically priced. Moreover, coping with differently priced items may be highly beneficial even in the case that all "real" items have the same price. By adding a single dummy item with price 0, the buyer has the option of "buying" this item an arbitrary number of times for the sole purpose of hiding when it is buying *real* items. This

---

[1] By having the buyer pay to a third party, the vendor may be initialized with an *encryption* of the buyer's deposit and therefore not even learn the deposit amount.

[2] One may argue that without any such information, the vendor can hardly optimize the offered goods. However, marketing-related information can still be *voluntarily* provided to the vendor by potential buyers.

added privacy feature is impossible to achieve with a standard use of oblivious transfer, unless the buyer is willing to pay for all the dummy items it retrieves.

Obtaining a complete solution to our problem requires a more general protocol that we call *priced oblivious transfer*. Assume that at the beginning of each phase of interaction the vendor holds an encryption of the buyer's current balance. A phase of interaction (also referred to as a *transaction*) should allow the buyer to *privately* retrieve a *single* item. This in itself is an oblivious transfer protocol. However, in this case we have the following additional requirements: (1) The buyer can only retrieve an item if its current balance is larger than the item's price; (2) The price of the item the buyer retrieves should be decreased from the buyer's (encrypted) balance.

Broadcast Encryption. A prime motivating example for priced oblivious transfer is as follows. A vendor is broadcasting n different data streams. The data streams may be video, audio, or text and the content may be news, entertainment, technical and professional information, etc. To accomplish private buying in this setting, the vendor encrypts each of the n streams with a different key. The buyer and vendor then engage in a priced oblivious transfer protocol where the keys are the items being transferred. The buyer is then able to decrypt the data stream that it paid for, but as it does not have knowledge of the other keys, it is unable to gain access to the content of the other data streams.

Subscriptions. An important extension to enabling the purchase of a single digital good per transaction is to allow subscriptions. In a subscription scenario, the vendor changes the database periodically. Denote the ith data item at time j as x_j^i. The sequence of the ith data items over time, x_0^i, $x_1^i, \ldots$, is called the ith *channel* or channel stream. For example, a channel may be a daily financial white paper or a daily decryption key for a broadcast stream as above. In this setting the buyer is allowed to *subscribe* to a channel. As with a single data item from a static database, the channel to which a buyer subscribes should remain private. While the buyer is subscribed to a channel, it receives the sequence of data items of the channel and its balance is deducted by the appropriate amount each time period of the channel. The buyer remains subscribed to the channel until it explicitly *unsubscribes* or until its balance becomes negative. It is clear that the operation of subscribing to a channel can be simulated by repeated operations of purchasing an item. The issue however is one of efficiency and in particular it is a question of the communication pattern: While buying inherently requires some non-trivial interaction, maintaining a subscription should ideally require only efficient one-way communication from the vendor to the buyer. Allowing an efficient subscription implementation (with one-way communication) seems to be vital in many of the applications we have in mind. We therefore extend our solutions to handle this additional requirement.

A NOTE CONCERNING EFFICIENCY. The main goals of this work are to put forward a new problem, establish a "practical feasibility" result for this problem, and in the process develop some useful general tools. We do not attempt at minor optimizations which would complicate the presentation. Our solution should be mainly viewed as a feasible framework which may be the basis for further optimizations.

Additional Contributions. Several ingredients of our construction seem to be of independent interest. In particular, we obtain the first implementation of a 1-round oblivious transfer protocol satisfying a "reasonable" security definition and provably secure under a "reasonable" security assumption. The security of our protocol can be based on the decisional Diffie-Hellman (DDH) assumption. A similar protocol has been independently obtained by Naor and Pinkas [21]. The oblivious transfer protocol follows from a more general *conditional disclosure* methodology, which can be used in some contexts as a light-weight alternative to zero-knowledge proofs. In this we extend an "information-theoretic" technique from [11] (see Section 2.3) and adapts it to the 2-party setting. In the course of addressing the case of subscriptions, we propose efficient solutions for the problem of privately retrieving a chosen *prefix* of a long stream of information.

Related Work. General techniques for secure 2-party computation [28,13] may be used to solve our problem. However, similarly to most other works in this area, our goal is to use the specific structure of the problem at hand for providing far more efficient solutions than those obtained via general techniques.

The current work has been greatly inspired by previous works on *specific* secure computation tasks such as private information retrieval (PIR) and oblivious transfer. In Section 2.3 we describe some relevant techniques from these works which we rely on or extend. A restricted "off-line" variant of our problem may be viewed as a special case of a *generalized oblivious transfer* primitive studied in [14]. In a distributed multi-vendor setting, an off-line variant of our problem has been considered in [11]. Adapting the solutions from [14,11] to our setting would result in very inefficient protocols. We stress that unlike the PIR-related context of [11], where the main concern is that of minimizing the asymptotic complexity as a function of the number of data items, most aspects of our problem are equally interesting even when the number of items is as small as 2.

Organization. The remainder of the paper is organized as follows. In Section 2 we specify the problem and its security requirements, and review the tools we will use. In Section 3 we describe our basic protocol and its properties. We also discuss some efficiency improvements. In Section 4 we discuss an extension to the subscription scenario. Finally, in Section 5 we present the one-round OT protocol which is a special case of our methodology.

2   Preliminaries

2.1   Problem Specification

As discussed in the introduction, our goal is to construct an "on-line" protocol between a buyer B and a vendor V which allows the buyer and the vendor to engage in multiple transactions. Both the buyer and vendor are allowed to store a (short) state information between transactions. Before specifying the security aspects of the protocol, we will first describe its desired functionality.

Initialization: At time 0, the buyer initializes its balance with a pre-payment to the vendor.

Main Protocol: At time t, $t = 1, 2, \ldots$

- The vendor may choose a database $\mathbf{x} = (x^0, x^1, \ldots, x^{n-1})$ of n items for sale and some public information $\mathbf{P}$ concerning the identity of these items. $\mathbf{P}$ contains a price list $\mathbf{p} = (p^0, p^1, \ldots, p^{n-1})$. By convention, x^0 is a dummy item with $p^0 = 0$.
- The buyer may then decide either to:
  - *Buy* the i-th item, where $0 \leq i < n$; if the buyer's remaining balance is sufficiently large (i.e., the combined price of all items previously received and the current price p_i does not exceed the initial deposit), the buyer receives x_i.
  - *Subscribe* to the i-th channel; by subscribing, the buyer indicates that it wishes to continue buying the i-th item until overriding the subscription with a new request. We assume that throughout the subscription, the buyer is charged the price p^i effective when initiating the subscription (even though $\mathbf{p}$ may change).
  - *Unsubscribe*, i.e., terminate a previous "subscribe" request.
  - *Do nothing*, i.e., maintain its default subscription if such exists, and otherwise keep idle.

2.2   Security Requirements

Efficiency considerations dictate some compromises we make in comparison to full-fledged simulation-based definitions for secure computation (e.g., those of [6,12]). Nonetheless, our solutions are provably secure under standard security assumptions. Our formal security requirements, which are only sketched below, can be found in the full version.

Both $\mathcal{B}$ and $\mathcal{V}$ are modeled by efficient randomized algorithms, and are initially given a security parameter 1^κ and a number of items 1^n as inputs. We assume that subsequent "inputs" are dynamically chosen by $\mathcal{B}, \mathcal{V}$ as the protocol proceeds. The protocol is assumed to terminate after a polynomial number of transactions. An *honest* buyer is restricted to choose items such that their total price does not exceed the initial deposit amount $b^{(0)}$. We first address a default scenario which only allows the buyer to issue "buy" requests. A protocol $(\mathcal{B}, \mathcal{V})$ as above is considered *secure* if it satisfies the following requirements:

CORRECTNESS. If both $\mathcal{B}$ and $\mathcal{V}$ are honest, then $\mathcal{B}$ outputs the correct item x^i at the end of each transaction.

BUYER'S SECURITY. A malicious vendor should not learn the choices made by an honest buyer. More formally, the view of any efficient (and possibly malicious) $\mathcal{V}^*$ in the interaction $(\mathcal{B}, \mathcal{V}^*)(1^\kappa)$ can be efficiently simulated. We note that this requirement is weaker than that of general security definitions in that it does not address the effect $\mathcal{V}^*$ may have on the output of $\mathcal{B}$. In particular, $\mathcal{V}^*$ does not need to "know" a database x which is effectively determined by its strategy in a given transaction. This is consistent with other definitions of related primitives (such as PIR, see Section 2.3, or even some definitions of oblivious transfer).

VENDOR'S SECURITY. A malicious buyer should not obtain more information than what its initial deposit allows. This is formalized by requiring that the interaction of $\mathcal{B}^*$ with an honest vendor $\mathcal{V}$ could be efficiently simulated in the natural idealized model.

Our security definitions for the general case, where the buyer may take any of the four actions, are more subtle. In a nutshell, the vendor's security requirement remains unchanged, and can be defined as above. The buyer's security in this setting, may also be defined similarly to the above. However, such a definition will only be satisfied when the buyer's action type is oblivious to the received items, i.e. depends only on public data (yet its specific selections i may also depends on received items). The reader is referred to the full version of the paper for a more detailed discussion.

Finally, while we do not explicitly address issues of robustness or recovery from faults, our protocols can be extended in a straightforward manner to deal with these issues.

2.3   Tools

Homomorphic Encryption. Our constructions rely on the widely used tool of *homomorphic encryption*. Loosely speaking, an encryption scheme is said to be homomorphic if: (1) The plaintexts are taken from a group $(H, +)$; (2) From encryptions of group elements h_1, h_2 it is possible to *efficiently* compute a *random* encryption of $h_1 + h_2$. A useful consequence is that given an encryption of a group element h and an integer c in binary representation, one can *efficiently* compute a random encryption of $c \cdot h$. This is done in a similar fashion to the repeated squaring procedure for modular exponentiation.

In what follows H will always be a group of a (large) prime order Q. It is important to note that by "+" we denote an abstract group operation. Hence, our notation applies both in a case where $H = Z_Q$ is an *additive* group, and where $H \subset Z_P^*$ is a *multiplicative* group. A useful example of a multiplicative homomorphic encryption is the *El-Gamal* scheme. (We refer the reader to, e.g., [22] for relevant definitions.) In this case, H is a subgroup of Z_P^*, where Q is a prime of length κ that divides $P - 1$.

We prefer an additive notation over a multiplicative one due to its more intuitive nature in our context. However, our protocols can be instantiated with both types of encryption. We note that *all* of our constructions can be based on the El-Gamal encryption (whose security is equivalent to the DDH assumption, cf. [22]) and most on any other homomorphic encryption scheme candidate, e.g. [18,23,24]. An additional property enjoyed by the El-Gamal encryption, which explains the above distinction, is discussed below.

Verifiability. It is sometimes required to verify the validity of a public key k and the validity of a ciphertext c relative to a valid k. Luckily, the latter verification task is typically easy, and we can therefore assume it as part of our default requirements. However, in most encryption schemes the validity of the public key itself is difficult to verify. To this end a special zero-knowledge proof procedure may be employed during the initialization stage of our protocols. This step, however, is not always needed. A useful added feature of the El-Gamal scheme is that its public keys are easily verifiable: to verify that (P, Q, g, h) constitutes a valid public key, it is enough to verify that P, Q are prime, Q divides $P - 1$, and $g^Q \equiv h^Q \equiv 1 \pmod{Q}$.

PIR. A Private Information Retrieval (PIR) protocol [9] allows a user to retrieve a selected item from a database while hiding the identity of this item from the server holding the database. PIR only requires the protection of the user, and makes no requirement on the privacy of the database. Thus, a naive solution to the PIR problem is to send the entire database to the user. When the database is large, this solution is very expensive in terms of communication. The main goal of PIR-related research has been to minimize the communication complexity of PIR, which is measured by default as the cost of retrieving one out of n bits. The current state of the art can be briefly summarized as follows. Assuming either a general homomorphic encryption [16,17,27] or a stronger number theoretic assumption [5], the asymptotic communication complexity of PIR can be made very small. In practice, however, the naive solution is still preferable when the database does not contain too many items. Thus, when we use PIR as a building block in our protocols, one should always keep in mind that the naive solution can be used in a case where the number of items is small.

Naor-Pinkas Pseudo-random Sequence. A variant of PIR where the user is restricted to learn no more than a single data item has been referred to in the literature as *symmetrically* private information retrieval (SPIR) [11].[3] In [19] (followed by [20]), Naor and Pinkas suggested the following reduction from SPIR to PIR. Suppose that there is an efficient method allowing the user to retrieve exactly one out of n *pseudo-random* items $(r^0, \ldots, r^{n-1})$ chosen by the server. Then, SPIR can be solved by applying such a procedure and concurrently applying PIR on $(x^0 \oplus r^0, \ldots, x^{n-1} \oplus r^{n-1})$. The pseudo-random sequence $(r^0, \ldots, r^{n-1})$ is created in the following way. Represent i as a length-ℓ binary string (in this case, $\ell = \log n$). Let $(s_1^0, s_1^1), (s_2^0, s_2^1), \ldots, (s_\ell^0, s_\ell^1)$ be ℓ pairs of independent keys to a pseudo-random function f, and define $r^i = \oplus_{j=1}^{\ell} f_{s_j}(i)$ where $s_j = s_j^{i_j}$. By letting the user choose *one* key from each pair (s_j^0, s_j^1), the user can learn any selected r^i, but no more than one r^i. A SPIR protocol constructed via the above method keeps all but a single data item x^i semantically secure from the user. More precisely, it is possible to simulate the view of a user, whose $\log n$ selections define an index i, based on x^i alone (up to *computational indistinguishability*).

Conditional Disclosure of Secrets. Motivated by the problem of constructing efficient SPIR protocols in the multi-server setting, Gertner et al. [11] suggested the following *conditional disclosure* primitive. An input string y to a public Boolean predicate C is *partitioned* among k servers, such that no server knows the entire string y. In addition, one of the servers holds a secret s. The goal of the servers is to each send a single message to a user, who knows y, such that the user will learn s if $C(y) = 1$ and otherwise will learn no information on s. To make this possible, the servers have a common random input r which is unknown to the user. In [11], the problem is reduced to *linear secret-sharing*. It is shown that the communication complexity of conditional disclosure as above is linear in the span program size of C (and in particular in the *formula size* of C). If the user is allowed to "help" the servers by secret-sharing a *witness* to the validity of $F(y)$

---

[3] This problem is very similar to $\binom{n}{1}$-OT, except for a different multi-server model and the focus on sublinear communication.

between them (without letting individual servers learn additional information on y), the communication can be made linear in the *circuit size* of C. Moreover, these solutions were efficiently extended to the non-Boolean case, where y is a string over a large field, and the condition C tests whether y satisfies some linear equation over F (or more complicated predicates over such atomic conditions).

A main ingredient of our protocol is an almost exact adaptation of the above conditional disclosure scenario to the single-server setting. In our setting, y will always be viewed as a vector over a large field $F = Z_Q$. Instead of partitioning $y = (y_1, \ldots, y_m)$ among several servers, a single server holds a public key k, the *encryptions* $E_k(y_1), \ldots, E_k(y_m)$, and a secret $s \in F$. The user holds both y and the secret key corresponding to k. An important observation regarding the solutions to the multi-server conditional disclosure problem mentioned above is that the joint messages sent by the servers may be expressed as a random *linear* function of (y, s), where the distribution of this linear function depends only on C. Therefore, if the encryption scheme E is homomorphic, the server may compute an encryption of these messages from $E_k(y)$. Instead of formulating our solutions in a general complexity-theoretic terminology, we will solve the required instances along the way in an intuitive way.

3   Solving the Problem

In this section we describe our solutions for the priced oblivious transfer problem. For the sake of presentation, we develop our solutions gradually and improve their efficiency along the way. In particular, the only operation we consider at first is 'buy'. We deal with subscription operations in Section 4.

Establishing a Public-Key Meta Structure. As described in introduction, during the entire run of our protocol the vendor will maintain an encryption of the buyer's current balance (using the public key of the buyer). Let E, D and G be the encryption, decryption and key-generation algorithms respectively. In the *initialization phase* of the protocol (time 0), the buyer applies G to sample a public-key, secret-key pair (k, sk) and sends the public-key k to the vendor. The vendor needs to verify that k is indeed a valid public-key and that the buyer knows a private-key sk that corresponds to k. Therefore, the buyer also proves in zero-knowledge that it knows an input of G that generates the public key k.[4] Finally, the vendor sets the current balance $b^{(0)}$ to the initial deposit of the buyer and creates an encryption $E_k(b^{(0)})$ of the balance.

The first challenge in designing our protocol is that, at each transaction, the vendor needs to update the encrypted balance $E_k(b)$ by some value p *without knowing either b or p*. It should not be surprising that in order to do so it is useful to let E be a *homomorphic encryption*. Recall that we assume that the plaintexts are taken from a group G_Q of order Q, where Q is a prime of length κ. Under our additive notation, it is convenient to view G_Q as the *field* $F = Z_Q$.

---

[4] Suppose that the encryption scheme enjoys the *verifiability* property discussed in Section 2.3. In this case, if the vendor is willing to settle on a somewhat weaker notion of security, it can verify the validity of k on its own instead of letting the buyer prove this validity as above. We use this fact in our one-round oblivious transfer (where we cannot afford the additional rounds required for the zero-knowledge proof).

Representations. We assume for simplicity that the length of each data item x^i is smaller than the security parameter κ. Even if this is not the case, our problem can be reduced to that of selling *keys* which encrypt the actual data. We take $B = 2^\ell$ to be an upper bound on the initial balance, where $B < Q$. This allows to represent prices and balances as elements of F by identifying (in the natural way) each integer i in the interval $[B - Q, B - 1]$ with the corresponding element of F. Thus, the elements $0, 1, \ldots, B - 1 \in F$ will be referred to as non-negative, and $B, \ldots, Q - 1$ as negative. In all of our protocols we will view a positive balance as being valid, and a negative balance as being invalid. If the buyer's balance is negative, it should not be allowed to learn any additional information.

3.1   Basic Solution

We present a solution where each transaction (here, a single 'buy' operation) requires two passes of communications: (1) A message from the buyer; (2) The vendor's reply. This is optimal since even without privacy the buyer still needs to specify the item it wants to retrieve and the vendor needs to send this item.

Assume without loss of generality that all item prices are distinct. (This assumption can be easily dispensed with at a moderate efficiency cost, e.g. by replacing each price p^i by $B'p^i - i$ for a sufficiently large B', and scaling the initial deposit by a factor of B'.) The most essential part of the *buyer's message* is an encryption $E_k(p)$ where p is the price of the item it wants to retrieve. The vendor needs to perform two operations: (1) Update the balance; (2) Send back (in some encrypted form) the item x^i such that $p = p^i$.

Updating the Balance. Since the vendor has an encryption of the current balance $E_k(b)$ and it received an encryption $E_k(p)$ of the retrieved item's price, it seems that updating the balance is not a problem. Simply create an encryption $E_k(b-p)$ of the new balance using the homomorphism of E. However, we should be careful: By setting p to be negative (e.g. $b - B + 1$), the buyer can arbitrarily increase its balance (this is of course undesirable, regardless of whether in this specific transaction the buyer gains any information).

One way to prevent the buyer from cheating in this manner is to require it to prove in a zero-knowledge fashion that $0 \le p \le b$. Such a solution requires more passes of interaction than desired. A better solution *in this respect* is for the buyer to use *non-interactive* zero-knowledge proofs of this claim (for that the buyer and vendor can agree upon a random string in the initialization phase of the protocol). However, non-interactive zero-knowledge proofs are usually very inefficient and we therefore give in Section 3.3 an alternative (more efficient) solution. Jumping ahead, the vendor in the revised protocol will not try to *verify* that p is in the right range but will rather make sure that any such violation on the part of the buyer will *cripple all future interactions*. We note that [3] gives an efficient zero-knowledge proof to a related problem, of proving that a *committed* number lies in a an interval. However, the problem we solve (and hence our machinery for solving it) is easier.

Sending an Item. We now assume that $0 \le p \le b$ and that the balance was updated by the vendor. All that is left is for the vendor to "send" an item x^i such that $p = p^i$ (if such an item exists).

The vendor's message is composed of n (parallel) messages $m^0, m^1, \ldots, m^{n-1}$. For every j, the message m^j allows the buyer to compute x^j in case $p = p^j$ and gives the buyer no information if $p \neq p^j$. Note that for a fixed j, what we have is in a sense an instance of *conditional disclosure in a computational setting*. The vendor wants to disclose the value x^j conditioned on $p = p^j$. For this simple condition (equality) the solution is very simple: For every j, the vendor uniformly samples $\alpha^j \in Z_Q$ and sets m^j to be a (random) encryption $E(\beta^j)$ of $\beta^j = \alpha^j(p - p^j) + x^j$. It is immediate that $\beta^j = x^j$ in case $p = p^j$ and is random in Z_Q if $p \neq p^j$ (therefore, in this case the buyer gets no information on x^j in an information theoretic sense).

Adapting the conditional disclosure methodology of [11] to the computational setting is one of the main tools of our solution. In addition to the example above, it is used extensively in Sections 3.3 and 4.

3.2   Reducing the Communication

The protocol of Section 3.1 has the disadvantage that the vendor's message is of linear length as a function of n (the number of items). This in itself is a non-trivial task and for some applications may be sufficient. We now give a simple method for reducing the communication. In Section 3.5 we provide a method for reducing the communication which is superior in most settings of the parameters (but is slightly more involved).

The main observation for reducing the communication is simple: If the buyer wants to retrieve item x^i then the only part of the vendor's message it needs is the value m^i (in fact, the rest of the message is useless). Therefore, instead of getting the entire sequence, $m^0, m^1, \ldots, m^{n-1}$, the buyer can just retrieve m^i using a PIR protocol (where we view the vendor's message as a database of n records). Note that in this case PIR is sufficient since security is preserved even if the buyer learns the entire sequence.

3.3   Avoiding Zero-Knowledge Proofs

In the protocol of Section 3.1, the buyer sends an encryption $E_k(p)$ and proves in zero-knowledge that $0 \leq p \leq b$. This was important for two reasons: (1) To prevent the buyer from learning x^i with $p^i > b$ *in the current transaction*; (2) To prevent the buyer from increasing its balance (in order to gain additional information *in future transactions*). However, as discussed above, both interactive and non-interactive zero-knowledge proofs are not efficient enough for our needs, and are in a sense an overkill. We now show how to replace zero-knowledge proofs with *conditional disclosures*. In these solutions, the vendor will not be able to *detect* a value p that is outside of the range $[0, b]$. Nevertheless, each such violation will prevent the buyer from learning any additional information.

The idea is simple. At the t-th transaction, the vendor will sample a random mask v^t and a random receipt u^t. The vendor will disclose v^t and u^t under the condition that $0 \leq p \leq b$. The value v^t will be used to mask the interaction in the current transaction (i.e. instead of retrieving x^i the buyer will retrieve $x^i + v^t$). The value u^t will be used as a receipt for future interaction – knowing u^t implies that the buyer behaved correctly until now.

A naive way to use the receipt u^t is to require the buyer to send it at the beginning of the next transaction. As it turns out, this solution may compromise the privacy of the buyer against a malicious vendor. We therefore use a chaining technique: at the t-th transaction the buyer will also send an encryption $E_k(u)$. The vendor will disclose v^t and u^t under the condition $(0 \le p \le b) \wedge (u = u^{t-1})$. We note that other methods of chaining are possible in this scenario. However, we find this particular solution appealing, both from a conceptual point of view and because it allows to maintain *statistical* vendor's security.

One may view this kind of chaining as an ongoing proof of the buyer that it behaves correctly, where the proof never gets to its conclusion (i.e. convincing the vendor). This kind of a technique may be useful in other scenarios.

It remains to show how to perform the more involved conditional disclosure needed here. We already saw how to perform conditional disclosure for equality. This also implies a recursive way to perform conditional disclosure under any condition that can be described as a monotone formula where the leaves are equalities: Assume we know how to perform conditional disclosure under the conditions A_1 and A_2. To perform conditional disclosure of x under $(A_1 \vee A_2)$, just perform two independent conditional disclosures of x — One under A_1 and the other under A_2. To perform conditional disclosure of x under $(A_1 \wedge A_2)$, sample a random mask r, disclose r under A_1 and $x + r$ under A_2.

To perform a conditional disclosure under the condition $(0 \le p \le b) \wedge (u = u^{t-1})$ it is enough to describe the condition $0 \le p \le b$ by a small monotone formula as above. For this purpose we will need some help from the buyer. Recall that $B = 2^\ell$ is an upper bound on a valid balance. In its message, the buyer will send separate encryptions of the bits $b_{\ell-1}, \ldots, b_0$ and $p_{\ell-1}, \ldots, p_0$ where $b_{\ell-1} \ldots b_0$ is supposed to be the binary representation of the current balance b and $p_{\ell-1} \ldots p_0$ defines the price p (i.e. $p = \sum_j p_j 2^j$). Note that the vendor can create an encryption of p from the encryptions of the bits p_i. The condition $0 \le p \le b$ is implied by the conjunction of the following conditions: (1) $b = \sum_j b_j 2^j$; (2) $b_{\ell-1}, \ldots, b_0$ and $p_{\ell-1}, \ldots, p_0$ are all bits; (3) $p \le b$ when p and b are viewed as integers. It is well known (and rather simple) that (3) can be represented as a monotone formula of size $O(\ell)$ with leaves that are equalities (in the bits $b_{\ell-1}, \ldots, b_0, p_{\ell-1}, \ldots, p_0$ and the constants 0 and 1). We may therefore conclude that $0 \le p \le b$ can also be represented as such a monotone formula of size $O(\ell)$.

3.4 Putting the Pieces Together

The ideas presented so far already combine into a protocol that satisfies the specification of Section 2.1, has the desired communication pattern, and is relatively efficient. This protocol is still not the most efficient we propose (significant improvements are described in Section 3.5) and it does not handle subscriptions (which are dealt with in Section 4). Nevertheless, since most of the ideas already appear in this solution, we now give a short summary of the protocol and informally discuss its properties.

The Protocol.

Initialization. The buyer applies the key generator G to sample a public-key, secret-key pair (k, sk) and sends the public-key k to the vendor. The buyer also proves in zero-knowledge that it knows an input of G that generates the public key k. The vendor creates an encryption $E_k(b^{(0)})$ of the initial balance $b^{(0)}$. Finally, both set u^0 to be some predefined string (e.g. the all zero string).

Buyer (Time $t > 0$). The buyer's message is composed of (1) $E_k(u)$ (u is supposed to be u^{t-1}); (2) $E_k(b_{\ell-1}), \ldots, E_k(b_0)$ and $E_k(p_{\ell-1}), \ldots, E_k(p_0)$, where $b_{\ell-1} \ldots b_0$ is supposed to be the binary representation of the current balance $b^{(t-1)}$ and $p_{\ell-1} \ldots p_0$ the binary representation of the price p^i; (3) A PIR query q for the index i.

Vendor. The vendor computes an encryption of $p = \sum_j p_j 2^j$ and creates an encryption of the new balance $b^{(t)} = b^{(t-1)} - p$. It samples two keys v^t and u^t uniformly at random in F and discloses both under the condition $(b = \sum_j b_j 2^j) \wedge (0 \le p \le b) \wedge (u = u^{t-1})$. For every j, the vendor computes m^j which is the conditional disclosure of $x^j + v^t$ under the condition $p^j = p$. Finally, the vendor answers with the PIR answer to the query q for the database $(m^0, \ldots, m^{n-1})$.

Buyer's Output. The buyer retrieves m^i and computes x^i (which is its output for this transaction). In addition, the buyer recovers and stores u^t for future interaction and also remembers the new balance.

Properties.

Correctness. For honest buyer and vendor is straightforward.

Buyer's Security. Follows from the semantic security of E since all the (even malicious) vendor sees at each transaction is a fixed number of encryptions. That is, a simulator for the view of $\mathcal{V}^*$ may first simulate the initialization stage, and then produce an appropriate number of encryptions for each transaction.

Vendor's Security. For any buyer $\mathcal{B}^*$, even malicious *and unbounded*, there exists an efficient simulator $\mathcal{S}$, with black-box access to $\mathcal{B}^*$, that produces an output which is *statistically close* to the view of $\mathcal{B}^*$. The simulator invokes $\mathcal{B}^*$ and simulates its conversation with $\mathcal{V}$. The first step is to extract (using the zero-knowledge extractor) the secret-key sk that corresponds to k. Given this information, the rest of the simulation is fairly trivial. The only point that needs arguing is that starting at the first time t for which the condition $(b = \sum_j b_j 2^j) \wedge (0 \le p \le b) \wedge (u = u^{t-1})$ is violated it will be violated at all subsequent transactions (with overwhelming probability).

Efficiency. Excluding the PIR protocol, the buyer performs $O(\ell)$ public-key operations and its message consists of $O(\ell)$ encryptions. The vendor however is much less efficient — it performs $O(n)$ public-key operations (to create the messages m^j). The vendor's message consists of a PIR reply for the database containing the strings m^j. This in itself already seems optimal: Any solution to our problem will in particular give a PIR protocol for the database $x^0, x^1, \ldots, x^{n-1}$. Therefore, we cannot expect to have communication which is smaller than that of a PIR protocol. However, here the strings m^j can be significantly longer than the strings x^j, which may result in a communication blowout. In Section 3.5 we show how to achieve savings in both the communication and work on the part of the vendor.

3.5   Additional Improvements

We now describe a modification of the protocol of Section 3.4 that typically improves its performance. The alternative approach is especially natural in the case where the vendor only sells *keys* encrypting the data, and the encrypted data is accessed by other means (e.g., via broadcast, or a PIR protocol). We assume that these keys are refreshed at each transaction (in particular, we would not like the buyer to get all values of x^i after buying it once) and describe the modification in this setting.

The keys that the vendor will sell are a carefully chosen subsequence of the Naor-Pinkas pseudo-random sequence (see Section 2.3). Let ℓ be as above (i.e. the length of the binary representation of prices). Let $(s_0^0, s_0^1), (s_1^0, s_1^1), \ldots, (s_{\ell-1}^0, s_{\ell-1}^1)$ be ℓ pairs of independent keys to a pseudo-random function f, and let $\{k^z\}_{z \in \{0,1\}^\ell}$ be the Naor-Pinkas sequence that is generated by these ℓ key pairs.

The idea is the following. Let the j-th *key* that the vendor sells be the element of the Naor-Pinkas sequence *indexed by the price p^j of this key* (i.e. the element k^{p^j}). This slightly unusual choice (the more natural choice seems to be taking the j-th key to simply be k^j) is the main observation of the revised protocol. To make it even more compatible with our solution, we let the j-th key at time t be $k^{p^j} + v^t$ (recall that the sequence $\{k^z\}_{z \in \{0,1\}^\ell}$ is refreshed at each transaction). We can now consider the following adjustment in the protocol.

The buyer sends almost the same message as before (there is no need to send the PIR query). Recall that as part of its message, the buyer sends encryptions of the bits of the price $E_k(p_{\ell-1}), \ldots, E_k(p_0)$. The vendor updates the balance and discloses v^t and u^t as before. In addition, for every $0 \le j < \ell$ and $\sigma \in \{0,1\}$, the vendor discloses s_j^σ conditioned on $p_j = \sigma$. Recall that given the ℓ keys $\langle s_{\ell-1}^{p_{\ell-1}} \ldots s_0^{p_0} \rangle$, the buyer can compute k^p whereas the rest of the sequence (i.e. k^z for $z \ne p$) remains pseudo-random. This implies the security of the protocol.

As for efficiency, we have that the $O(n + \ell)$ public-key operations of the previous protocol are reduced to $O(\ell)$ public-key operations, plus at most $n\ell$ private-key operations. The above excludes the computational cost of PIR, which depends on its specific implementation. In terms of communication, both the buyer and the vendor need only to send $O(\ell)$ encryptions, and in addition to invoke a PIR protocol on a database which is now of an optimal size (since here each item is masked with a pseudo-random string of the same size).

4   Subscription

Recall that our motivation for letting the buyer issue a "subscribe" request is to allow efficient one-way communication from the vendor to the buyer. In this abstract we will sketch a relatively simple solution to this problem. A more efficient solution, whose details can be found in the full version, will be briefly discussed at the end of this section.

Subscribing. As in the previous protocol, $\mathcal{B}$ sends to $\mathcal{V}$ encryptions of the bits of $p = p^i$ and b. In addition, $\mathcal{B}$ picks a value τ, $0 \le \tau < 2^\ell$, which is assumed to be a length of a prefix of the i-th channel it is *entitled* to buy, and sends encryptions of the ℓ bits of τ. An honest buyer can let $\tau = \lfloor b/p \rfloor$, regardless of the intended subscription length. $\mathcal{V}$ discloses a mask v and a receipt u subject to the condition $(0 \leq)\tau \cdot p <= b$, and a key k^p encrypting the future contents of the channel indexed by p. An efficient implementation of the former disclosure, which requires some additional help from $\mathcal{B}$, will be described later. As before, k^p and v will be used to encrypt the received data during the current subscription, and the receipt to cripple future transactions in a case of cheating.

Maintaining a Subscription. At the t-th transaction following a subscription, each channel will further be masked with a key v_t, which will be disclosed subject to the condition $t \leq \tau$. Note that this does not require the help of $\mathcal{B}$, since the encrypted bits of τ are given to the vendor during the initialization.

Unsubscribing. If $\mathcal{B}$ unsubscribes after T transactions, $\mathcal{V}$ deducts from its balance the amount $T \cdot p$ (note that this can be done efficiently from the public value T and the encrypted values of p, b). If the buyer's balance turns negative (by failing to unsubscribe before depleting its balance), all its future transactions will automatically be crippled.[5]

It remains to describe the implementation of the conditional disclosure in the subscription procedure described above, namely a disclosure subject to the condition $\tau \cdot p \leq b$. The fact that the underlying field F is large allows to obtain much greater efficiency than that obtained by emulating a Boolean multiplication circuit. The disclosure procedure proceeds as follows. $\mathcal{B}$ will provide, as additional help, encryptions of $a_{\ell-1} = (\tau_{\ell-1}2^{\ell-1}) \cdot p, \ldots, a_0 = (\tau_0 2^0) \cdot p$. If $\mathcal{B}$ acts honestly, these should sum up to the product $\tau \cdot p$. To guarantee that each a_j is valid, observe that $\mathcal{V}$ can compute the two possible valid values of a_j, and disclose to $\mathcal{B}$ a mask subject to the condition that a_j is indeed consistent with the value of τ_j. That is, the j-th conjunct in the condition is of the form: $(\tau_j = 0 \wedge a_j = 0) \vee (\tau_j = 1 \wedge a_j = 2^j p)$. Finally, using the methods of the previous section, an additional mask will be disclosed subject to the condition $\sum a_j \leq p$ (note that an encryption of $\sum a_j$ can be computed by the vendor alone). As before, the latter conditional disclosure will require $\mathcal{B}$ to send the encrypted bit representation of the sum. This concludes the description of the conditional disclosure procedure, and thus of the entire subscription protocol.

Efficiency. Both initializing a subscription and each subsequent transaction require $O(\ell)$ public-key operations, with communication consisting of $O(\ell)$ encryptions. In comparison to the implementation of a "buy" operation from the previous section, initializing a subscription is more expensive, but maintaining it is significantly cheaper.

A More Efficient Protocol. In a typical case where subscriptions are more frequently maintained than initialized, it is important to optimize the efficiency of the procedure for maintaining a subscription. In particular, it is desirable to avoid public-key operations altogether. In the full version of this paper we describe an implementation which achieves the above goal. In the core of this solution is an efficient subprotocol, performed during the subscription initialization stage, which allows $\mathcal{B}$ to effectively learn a prefix of length τ from a pseudo-random key sequence of length 2^ℓ. This subprotocol may be of independent interest.

---

[5] We assume here that $|F|/2^\ell$ is sufficiently large to make a balance wraparound infeasible. This assumption holds for any reasonable choice of parameters.

5   One-Round Oblivious Transfer

Oblivious Transfer (OT) [25,10,4] may be viewed as the simplest atomic building block for general secure computation [15]. OT is a 2-party protocol between Alice and Bob. In its most common variant, also known as $\binom{2}{1}$-OT, Alice holds a selection bit b and Bob holds a pair of secrets x^0, x^1. At the end of the protocol, Alice should output x^b and learn no information on x^{1-b}, and Bob should output and learn nothing.

As a special case of our general methodology, we obtain an efficient 1-round OT protocol which satisfies a *reasonable* security definition. Unlike a previous construction of [1] which is not known to be secure under a standard computational assumption (i.e. without using the random oracle methodology), our construction can be based on the standard DDH assumption. A similar construction (and definition) has been independently proposed by Naor and Pinkas [21]. For lack of space in this extended abstract, we only briefly describe the protocol and discuss its security features.

Our $\binom{2}{1}$-OT protocol naturally extends into a more general $\binom{n}{1}$-OT protocol (where Alice retrieves one of n secrets held by Bob). We therefore directly describe our solution in this setting.

5.1 $\binom{n}{1}$-OT Protocol

Each transaction of a priced oblivious transfer protocol trivially implies an OT protocol. However, in our one-round implementations of such a transaction we assumed an *initialization phase*, which is not part of the setting in a standalone OT protocol. In fact, one part of the initialization phase will also be part of our OT protocol: Alice still needs to sample a public-key, secret-key pair (k, sk) and send the public-key k to Bob. Moreover, Bob still needs to verify that k is valid. However, in this case Alice cannot prove that k is valid (there is just not enough interaction). We therefore assume that the underlying homomorphic encryption scheme enjoys the *verifiability* property discussed in Section 2.3, as is the case for the El-Gamal scheme. For such an encryption scheme, Bob can verify on its own that k has a corresponding secret key sk (although Alice may not know this key). We can now define our basic $\binom{n}{1}$-OT protocol:

Alice invokes G to sample a public-key, secret-key pair (k, sk). She then sends to Bob the public-key k and a random encryption $c = E_k(i)$ of i.

Bob verifies that k is a valid public key and c is a valid encryption. In such a case, for every $j \in [n]$, Bob computes m^j which is the conditional disclosure of x^j conditioned on $j = i$ (i.e. m^j is a random encryption of $\alpha^j(i - j) + x^j$ for a uniformly distributed element α^j of F). Bob sends $m^0, \ldots, m^{n-1}$.

Alice decrypts $m_i = E(x^i)$ and outputs x^i.

Security. Various definitions of security for OT have been proposed. The most widely accepted are those relying on a general framework for defining secure two-party computation (cf., [6,12]). We are unable to obtain this level of security while preserving the minimal number of rounds in our protocol. In a nutshell, the security definition satisfied by the above protocol relaxes the simulation-based definition of [6,12] in two ways. First, the simulator for Alice is allowed to be computationally unbounded (yet its simulation quality is perfect or statistical rather than computational). This may be interpreted as saying that Bob's security is *purely* information theoretic. Second, the simulator for Bob should simulate Bob's view alone, without considering its correlation with Alice's output. In particular, we do not require that a cheating Bob *knows* the input to which Alice's selection effectively applies. We feel however that the notion of security we achieve is perfectly suitable for OT, either as a standalone application, or in more general "information-retrieval" contexts such as the one studied in this work. Next we analyze the security of the above protocol.

The view of a possibly cheating Bob only contains a random public-key and a (random) encryption. Therefore, the semantic security of E implies that this view can be simulated. The view of a possibly cheating Alice (even an unbounded one) can be perfectly emulated by an *unbounded* simulator. The simulator first computes the private key sk that corresponds to k (if such a key does not exist, Bob would refuse to interact with Alice). Note that this requires the simulator to be unbounded. Now there exists at most a single i for which $c = E_k(i)$. If such an i exists the simulator queries for x^i and defines m^i to be a random encryption of x^i. For all other j, the simulator defines m^j to be a random encryption of a random element. It is easy to verify that this is a perfect simulation.[6]

Efficiency and Improvements. Alice's work consists of sampling a key and a constant number of public-key operations. Bob performs $O(n)$ public-key operations and its message contains n encryptions. However, the improvements in efficiency that are described in Sections 3.2 and 3.5 apply also in the contexts of the OT protocol. We omit the details in this preliminary version.

Acknowledgments

We wish to thank Avi Rubin for discussions which initiated this work, and Moni Naor and Benny Pinkas for helpful discussions and pointers. We also thank the anonymous referees for their comments.

References

1. M. Bellare and S. Micali. Non-interactive oblivious transfer and applications. CRYPTO '89, pp. 547-557, 1989.
2. O. Berthold, H. Federrath, and M. Kohntopp. Project "Anonymity and Unobservability in the Internet". Conference on Freedom and Privacy, pp. 57-65, 2000.
3. F. Boudot. Efficient proofs that a committed number lies in an interval. EUROCRYPT 2000, pp. 431-444. Springer-Verlag, 2000.
4. G. Brassard, C. Crépeau, and J.-M. Robert. All-or-nothing disclosure of secrets. CRYPTO '86, pp. 234-238, 1987.
5. C. Cachin, S. Micali, and M. Stadler. Computationally private information retrieval with polylogarithmic communication. EUROCRYPT '99, pp. 402-414, 1999.

---

[6] In this we assume that the verification process of E is never wrong (as is the case for the El-Gamal encryption). We also assume that the secrets are group elements (otherwise the simulated view can only be made statistically close to the actual view).

6. R. Canetti. Security and composition of multiparty cryptographic protocols. *J. of Cryptology*, 13(1), 2000.
7. D. Chaum. Security without identification: Transaction systems to make big brother obsolete. *Communications of the ACM*, 28(10):1030-1044, 1985.
8. D. Chaum, A. Fiat, and M. Naor. Untraceable electronic cash. CRYPTO '88, pp. 319-327
9. B. Chor, O. Goldreich, E. Kushilevitz, and M. Sudan. Private information retrieval. FOCS '95, pp. 41-51, 1995.
10. S. Even, O. Goldreich, and A. Lempel. A randomized protocol for signing contracts. *C. ACM*, 28:637-647, 1985.
11. Y. Gertner, Y. Ishai, E. Kushilevitz, and T. Malkin. Protecting data privacy in private information retrieval schemes. *JCSS*, 60(3):592-629, 2000. Preliminary version in STOC '98.
12. O. Goldreich. Secure multi-party computation. http://philby.ucsb.edu/cryptolib/ BOOKS, February 1999.
13. O. Goldreich, S. Micali, and A. Wigderson. How to play any mental game. STOC '87, pp. 218-229, 1987.
14. Y. Ishai and E. Kushilevitz. Private simultaneous messages protocols with applications. STOC '97, pp. 174-183, 1997.
15. J. Kilian. Basing cryptography on oblivious transfer. STOC '98, pp. 20-31, 1988.
16. E. Kushilevitz and R. Ostrovsky. Replication is not needed: Single database, computationally-private information retrieval. FOCS '97, pp. 364-373, 1997.
17. E. Mann. Private access to distributed information. Master's thesis, Technion - Israel Institute of Technology, Haifa, 1998.
18. D. Naccache and J. Stern. A new public key cryptosystem. EUROCRYPT '97, pp. 27-36, 1997.
19. M. Naor and B. Pinkas. Oblivious transfer and polynomial evaluation. STOC '99, pp. 245-254.
20. M. Naor and B. Pinkas. Oblivious transfer with adaptive queries. CRYPTO '99, pp. 573-590.
21. M. Naor and B. Pinkas. Efficient oblivious transfer protocols. SODA 2001.
22. M. Naor and O. Reingold. Number theoretic constructions of efficient pseudo-random functions. FOCS '97.
23. T. Okamoto and S. Uchiyama. A new public key cryptosystem as secure as factoring. EUROCRYPT '98, pp. 308-318, 1998.
24. P. Pallier. Public-key cryptosystems based on composite degree residuosity classes. EUROCRYPT '99, pp. 223-238, 1999.
25. M. O. Rabin. How to exchange secrets by oblivious transfer. Technical Report TR-81, Harvard Aiken Computation Laboratory, 1981.
26. S. von Solms and D. Naccache. On blind signatures and perfect crimes. *Computers and Security*, 11(6):581-583,1992.
27. J. P. Stern. A new and efficient all-or-nothing disclosure of secrets protocol. ASI-ACRYPT '98, pp. 357-371, 1998.
28. A. C. Yao. How to generate and exchange secrets. FOCS '86, pp. 162-167, 1986.

A Secure Three-Move Blind Signature Scheme for Polynomially Many Signatures

Masayuki Abe

NTT Laboratories
1-1 Hikari-no-oka, Yokosuka-shi, 239-0847 Japan
abe@isl.ntt.co.jp

Abstract. Known practical blind signature schemes whose security against adaptive and parallel attacks can be proven in the random oracle model either need five data exchanges between the signer and the user or are limited to issue only logarithmically many signatures in terms of a security parameter. This paper presents an efficient blind signature scheme that allows a polynomial number of signatures to be securely issued while only three data exchanges are needed. Its security is proven in the random oracle model. As an application, a provably secure solution for double-spender-traceable e-cash is presented.

1 Introduction

Blind signatures are a key part of some information systems that offer both user privacy and data authenticity. Such systems include anonymous electronic cash and electronic voting as typical examples. The notion of blind signatures was first introduced by Chaum in [12] with the first scheme based on RSA. Later, some discrete-log based signature schemes were turned into blind signatures [24,10,21]. For some applications, extra functionalities, such as partial blindness [2,1,3] and revocability [6,11,9], were added. A secure blind signature scheme should be one-more unforgeable against adaptive and parallel attacks. Namely, users should not be able to produce more signatures than legitimately issued.

There are some theoretical results on the security of blind signatures [14,25,22]. In [22], a formal security definition and a secure scheme were introduced, though the scheme was rather impractical compared to ordinary signature schemes in real use. In [27,29], Pointcheval and Stern proved that one type of efficient blind signature schemes, which includes Okamoto-Schnorr [23] and Okamoto-Guillou-Quisquater [20] signatures, to be secure in the random oracle model [4] as long as a logarithmic number of signatures were issued. Later, [26] introduced a generic adaptation that renders logarithmically secure blind signature schemes into secure ones with polynomially many signatures. Its cost is two additional data transfers. As the underlying schemes require three data transfers, the resulting schemes need five moves of data between the signer and a user. In [30], Schnorr and Jakobsson argued the security of the Schnorr blind signature in the random oracle model with a strong assumption; the attacker is generic, i.e., restricted to use the group operation only. In [17], Fischlin pointed out some pitfalls that

B. Pfitzmann (Ed.): EUROCRYPT 2001, LNCS 2045, pp. 136–151, 2001.

could be found between the generic adversary plus random oracle model and the reality.

This paper presents a blind signature scheme that needs only three data moves and provides polynomial security, i.e., one-more unforgeable even if polynomially many signatures are issued in an adaptive and concurrent manner. The security is proven in the random oracle model. The scheme remains practical as it requires only three to four times more computation than the original Schnorr signatures [31].

Another advantage of our scheme is its potential support of protocols that need additional functionality. By following the idea of [3], one can easily extend our scheme to be partially blind schemes. Furthermore, it is shown that a variant of our scheme gives a provably secure solution for double-spender-traceable electronic cash systems. Note that such e-cash schemes in the literature, e.g. [6,7,18], rely on a variant of blind signatures called restrictive blind signatures [7], whose security has been proved only under non-standard and strong assumptions and only against certain restricted attacks [8] while our solution withstands the most general attacks.

2   Security Definitions

Blind signature schemes have two aspects of security; blindness and one-more unforgeability. Let $(\mathcal{G}, \mathcal{S}, \mathcal{U}, \mathcal{V})$ be a blind signature scheme where $\mathcal{G}$ is the key generation algorithm, $\mathcal{S}$ and $\mathcal{U}$ are a signer and a user, respectively, and $\mathcal{V}$ is a verification algorithm (refer to [22] for a formal definition of blind signature schemes).

Definition 1. *(Blindness) Let $\mathcal{S}^*$ and $\mathcal{D}^*$ be a signer and a distinguisher. Let $view_0$ and $view_1$ be views of $\mathcal{S}^*$ during executions of the signature issuing protocol where honest user $\mathcal{U}$ obtains valid signature-message pairs (Σ_0, msg_0) and (Σ_1, msg_1), respectively. Given $(view_0, view_1, \Sigma_b, msg_b)$ for $b \in_U \{0, 1\}$, $\mathcal{D}^*$ outputs $b' \in \{0, 1\}$. A signature scheme is blind if, for all polynomial-time $\mathcal{S}^*$ and $\mathcal{D}^*$, $b' = b$ happens with probability at most $1/2 + 1/n^c$ for sufficiently large n and some constant c. The probability is taken over the coin flips of $\mathcal{G}$, $\mathcal{S}^*$, $\mathcal{D}^*$ and $\mathcal{U}$.*

Note that our scheme provides *computational* blindness defined as above while some of the previously known schemes achieve *perfect* blindness where the success probability of unbound $\mathcal{D}^*$ is exactly $1/2$.

Definition 2. *(One-more unforgeability) A blind signature scheme is $(\ell, \ell+1)$ unforgeable if, for any probabilistic polynomial-time algorithm $\mathcal{U}^*$, $\mathcal{U}^*$ outputs $\ell + 1$ valid signatures with probability at most $1/n^c$ for sufficiently large n and some constant c after interacting with legitimate signer $\mathcal{S}$ at most ℓ times in an adaptive and concurrent manner. The probability is taken over the coin flips of $\mathcal{G}$, $\mathcal{S}$, and $\mathcal{U}^*$.*

In the random oracle model, these success probabilities also depend on the choice of random oracles.

3 The Proposed Scheme

3.1 Underlying Idea

The proposed scheme is based on the partially blind signature scheme of [3]. Roughly, their scheme is a witness indistinguishable variant of the Schnorr signature scheme where the signer uses two public keys $y(= g^x)$ and $z(= g^w)$, which we call the *real public key* and the *tag public key*, respectively, in such a way that the signature can be issued only with real secret key x but no one can distinguish which secret key, i.e., x or w, was used. Their scheme then allows the signer to sign with several different tag public keys to achieve partial blindness. It was proven that the same tag key could be used only for logarithmically many signatures but the signer could use polynomially many tag keys. Accordingly, if the signer generates a one-time tag key each time he signs, it achieves polynomial security, though the blindness is lost.

Our scheme follows the above approach with additional ideas to retain blindness. It allows the user to blind the tag public key so that the resulting signature can be verified with the real public key provided by the signer and the blinded tag public key provided by the user. However, if the blinding is *perfectly* done and the resulting tag public key just looks like a random public key, the user could himself generate such a signature by arbitrarily creating the tag key and exploiting witness indistinguishability. Accordingly, we restrict the blinding so that the resulting blinded tag key maintains a link to the original one but the link is computationally hidden. Namely, our scheme provides *computational* blindness. The main idea to realize this property is to use a pair of tag public-keys, say (z, z_1), in such a way that z is fixed and z_1 is changed for every signature. The user blinds them into $(\zeta, \zeta_1) = (z^\gamma, z_1^\gamma)$ with random factor γ so that $\log_z z_1 = \log_\zeta \zeta_1$ holds. Accordingly, (ζ, ζ_1) preserves the relation that underlies (z, z_1). The blindness is now provided if the signer cannot decide whether (z, z_1, ζ, ζ_1) is in such relation or not. Some more tricks are added to force the user follow the blinding procedure to get valid signatures.

This restrictive blinding stealthily preserves the link between each valid signature to a particular execution of the issuing protocol. Thus, if $\ell + 1$ signatures are generated after ℓ executions of the signing protocol, there exists an execution that yields at least two signatures. Accordingly, we only need to consider the possibility of yielding two signatures from one issuing, which results in more efficient reduction than the previous results.

3.2 Construction

Let $\mathcal{G}$ be a probabilistic polynomial-time algorithm that takes security parameter n and outputs (p, q, g) where p, q are large primes that satisfy $q|p - 1$, and g is an element of $\mathbb{Z}_p^*$ whose order is q. By $\langle g \rangle$, we denote a prime subgroup in $\mathbb{Z}_p^*$ generated by g. Let $\mathcal{H}_1 : \{0, 1\}^* \to \langle g \rangle$, $\mathcal{H}_2 : \{0, 1\}^* \to \langle g \rangle$, and $\mathcal{H}_3 : \{0, 1\}^* \to \mathbb{Z}_q$ be hash functions. We assume that it is hard to compute the discrete log of the outputs of $\mathcal{H}_1$ and $\mathcal{H}_2$. Such hash functions may be constructed in practice as $\mathrm{SHA}(\mathrm{str})^{(p-1)/q} \bmod p$ allowing negligibly small error probability [3].

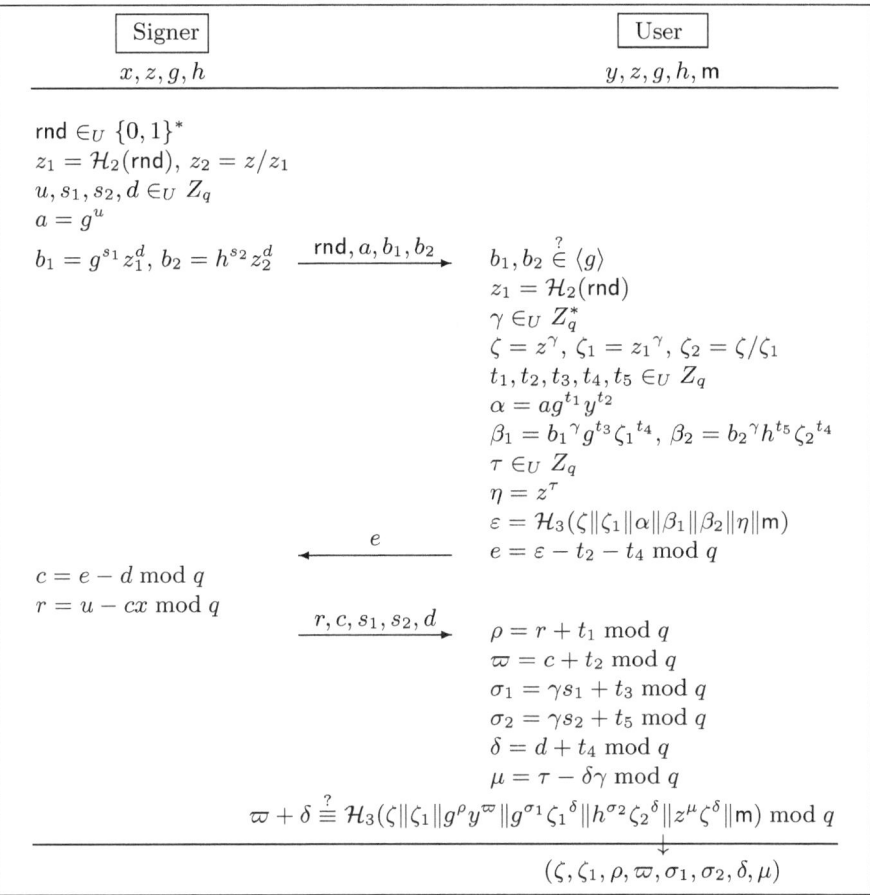

Fig. 1. The signature issuing protocol. The user aborts if any of the checks $(\overset{?}{\equiv}, \overset{?}{\in})$ fails.

[Key Generation]

The signer executes $(p, q, g) \leftarrow \mathcal{G}(1^n)$, and selects $h \in_U \langle g \rangle$, $x \in_U \mathbb{Z}_q$. It then computes real public-key y and fixed tag key z as $y = g^x \bmod p$ and $z = \mathcal{H}_1(p\|q\|g\|h\|y)$, respectively. If $z = 1$, abandon the key and retry. The public key is (p, q, g, h, y, z), and the private key is x.

[Signature Issuing]

Here we overview the signature issuing protocol at a higher level. The details are illustrated in Figure 1. Hereafter, all arithmetic operations are done in $\mathbb{Z}_p$ unless otherwise noted.

Signer $\mathcal{S}$: $\mathcal{S}$ generates a random string rnd and a one-time tag key $z_1 = \mathcal{H}_2(\text{rnd})$. Sending rnd convinces $\mathcal{U}$ that $\log_g z_1$ is not known to $\mathcal{S}$. Then z_2 is computed so that $z = z_1 \cdot z_2$ holds. The rest of the issuing protocol consists of two parts:

- **y-side:** Proof of knowledge x of $y = g^x$, and
- **z-side:** Proof of knowledge (w_1, w_2) of $z_1 = g^{w_1}$, $z_2 = h^{w_2}$.

Since z-side witness is not known to $\mathcal{S}$, the z-side proof is done by simulation as illustrated in Figure 1 by using the OR-proof technique of [13]. Accordingly, $\mathcal{S}$ can complete the protocol only with y-side witness x.

User $\mathcal{U}$: $\mathcal{U}$ blinds and converts the y-side proof into a signature in the same way as done in Schnorr blind signatures [24,10]. For z-side, $\mathcal{U}$ blinds z, z_1, z_2 into ζ, ζ_1, ζ_2 by raising them with random factor γ. The proofs for z_1, z_2 given from $\mathcal{S}$ are also blinded, and then converted into signatures in the standard way with adjustment for the effect of γ. $\mathcal{U}$ then creates an additional Schnorr signature that proves $\zeta = z^\gamma$.

The resulting signature Σ is 8-tuple $\Sigma = (\zeta, \zeta_1, \rho, \varpi, \sigma_1, \sigma_2, \delta, \mu)$ that proves the knowledge of $\log_g y \lor (\log_g \zeta_1 \land \log_h(\zeta/\zeta_1) \land \log_z \zeta)$.

[Signature Verification]
A signature message pair (Σ, m) is valid if it satisfies $\zeta \not\equiv 1$ and

$$\varpi + \delta \equiv \mathcal{H}_3(\zeta \| \zeta_1 \| g^\rho y^\varpi \| g^{\sigma_1} \zeta_1^{\delta} \| h^{\sigma_2}(\zeta/\zeta_1)^{\delta} \| z^\mu \zeta^{\delta} \| \mathsf{m}) \bmod q.$$

4   Security Proofs

4.1   Correctness

Theorem 1. *If the signer and the user follow the issuing protocol, the resulting signature satisfies the verification predicates with provability 1.*

Proof. Observe that the following holds.

$$\varpi + \delta = c + t_2 + d + t_4 = e + t_2 + t_4 = \varepsilon \pmod{q}$$
$$g^\rho y^\varpi = g^{r+t_1} y^{c+t_2} = g^{r+cx} g^{t_1} y^{t_2} = a g^{t_1} y^{t_2} = \alpha$$
$$g^{\sigma_1} \zeta_1^\delta = g^{\gamma s_1 + t_3} \zeta_1^{d+t_4} = (b_1 z_1^{-d})^\gamma g^{t_3} \zeta_1^{d+t_4} = b_1^\gamma g^{t_3} \zeta_1^{t_4} = \beta_1$$
$$h^{\sigma_2}(\zeta/\zeta_1)^\delta = h^{\gamma s_2 + t_5} \zeta_2^{d+t_4} = (b_2 z_2^{-d})^\gamma h^{t_5} \zeta_2^{d+t_4} = b_2^\gamma h^{t_5} \zeta_2^{t_4} = \beta_2$$
$$z^\mu \zeta^\delta = z^{\tau - \delta\gamma} \zeta^\delta = z^\tau = \eta$$

Furthermore, $\zeta \not\equiv 1$ holds as $\gamma \neq 0$ when the user is honest.   □

4.2   Blindness

Theorem 2. *The proposed scheme is blind if the decision Diffie-Hellman problem is intractable and H_1, H_2, H_3 are random oracles.*

Proof. (sketch) Suppose that $(\mathcal{S}^*, \mathcal{D}^*)$ is successful in breaking blindness with probability $1/2 + \epsilon$ where ϵ is not negligible. Let t_s be the maximum running time of $\mathcal{D}^*$, which is also polynomially bound. We show that $\mathcal{S}^*$ can be used to solve the DDH problem. Define $\mathcal{DH} = \{(X_1, X_2, X_3, X_4) \in \langle g \rangle^4 | \log_{X_1} X_2 = \log_{X_3} X_4\}$ and $\mathcal{R} = \{(X_1, X_2, X_3, X_4) \in \langle g \rangle^4\}$. Let $(A, B, C, D) \in \langle g \rangle^4$ be a DDH instance, which is taken from $\mathcal{DH}$ or $\mathcal{R}$ with equal probability. Given such an instance, first define H_1 so that $z = A$. Select $b \in_U \{0, 1\}$ and engage in the issuing protocol with $\mathcal{S}^*$ twice. Label the executions run_0 and run_1. Define H_2 so that $z_1 = B$ in run_b, and $z_1 \in_U \langle g \rangle$ in run_{1-b}. Follow the protocol in both run. Then, generate a signature-message pair (Σ, m) that includes $(\zeta, \zeta_1) = (C, D)$. Other variables in Σ are generated by using the standard zero knowledge simulation technique; randomly choose $\rho, \varpi, \sigma_1, \sigma_2, \delta, \mu$, and then define H_3 so that it looks consistent. Given (Σ, m) and views from $\mathcal{S}$, distinguisher $\mathcal{D}^*$ outputs b'. If $b' = b$, we conclude that the instance is in $\mathcal{DH}$. It is in $\mathcal{R}$, otherwise.

Observe that if $(A, B, C, D) \in \mathcal{DH}$, Σ is a valid signature that can be produced in run_b, since $\log_z z_1 = \log_A B = \log_C D = \log_\zeta \zeta_1$ and there exist blinding factors t_1, t_2, t_3, t_4, t_5 that convert the view of run_b into Σ [1]. On the other hand, Σ cannot be produced from run_{1-b} since $\log_z z_1 \neq \log_\zeta \zeta_1$ except for negligible probability. Therefore, given Σ, $\mathcal{D}^*$ outputs correct b with probability $1/2 + \epsilon$. Next, observe that if $(A, B, C, D) \in \mathcal{R}$, Σ cannot be produced in either run_0 and run_1 since $\log_z z_1 \neq \log_\zeta \zeta_1$ for both runs except for negligible probability. Hence, b is independent of Σ, and $b' = b$ happens with probability $1/2$. Thus, the success probability in DDH problem is $1/2(1/2 + \epsilon) + 1/2(1/2) = 1/2 + \epsilon/2$, which contradicts to the DDH assumption when ϵ is not negligible. Note that $\mathcal{D}^*$ may not terminate in time t_s if the instance is in $\mathcal{R}$. However, this is also to our advantage since we can see that Σ is not a proper input to $\mathcal{D}^*$ and the instance is in $\mathcal{R}$.

Finally, note that if $\mathcal{S}^*$ chooses the same rnd in both executions, the resulting signatures are perfectly indistinguishable as there exist consistent blinding factors for any combination of the views and signatures.   □

Note that the blindness relies on the decision Diffie-Hellman assumption over the public key of the signer. This suggests that an adversarial signer could choose p, q, g so that the DDH problem could be solved with those parameters. However, as we shall show in the next section, one-more unforgeability is based on the discrete logarithm assumption. Therefore, choosing weak parameters to violate blindness could result in the loss of one-more unforgeability unless DL is strictly harder than DDH. Nevertheless, it is beneficial for the users to verify that the public keys are generated and the hash functions are chosen so that those assumptions are likely to hold. There are several practical solutions for this matter. An inexpensive solution would be to use a widely believed secure hash function like SHA-1, and plug it into the source of randomness of $\mathcal{G}$ so that the users can believe that there is no room for the adversarial signer to control the resulting parameters. It is also needed to check if y is in $\langle g \rangle$ and z

---

[1] This is why $b_1, b_2 \overset{?}{\in} \langle g \rangle$ has to be checked. Without this check, wrong b_1, b_2 could produce a valid signature if γ is a lucky choice. This results in a nonuniform distribution of γ while the one that underlies the simulated signature follows the uniform distribution.

is correctly made. In practice these could be examined by a certificate authority at registration on behalf of the users.

4.3   One-More Unforgeability

Theorem 3. *The proposed scheme is $(\ell, \ell + 1)$-unforgeable for polynomially bound ℓ if the discrete logarithm problem is intractable and H_1, H_2, H_3 are random oracles.*

The proof is structured as follows. We first observe that the scheme is witness indistinguishable [15] (Lemma 1), which helps us to simulate the signer with either y-side or z-side witness(es) to extract the witness of the other side. It is then proven that the user can blind (z, z_1) into (ζ, ζ_1) only in such a way that $\log_z \zeta = \log_{z_1} \zeta_1$ to obtain a valid signature (Lemma 2). We then show that creating a valid signature without engaging in the issuing protocol with the legitimate signer is infeasible (Lemma 3). From Lemma 2 and 3, one can see that if the user engages in the signature issuing protocol ℓ times and outputs $\ell + 1$ signatures, there exist at least two valid signatures linked to a particular run of the issuing protocol. So the rest is to prove that such a forger who is successful in producing two signatures from a single protocol run can be used to solve the discrete logarithm problem.

Lemma 1. *The signature issuing protocol is witness indistinguishable.*

The above lemma holds immediately according to [13]. Indeed, it is not hard to see that the issuing protocol can be completed if the signer knows either y-side witness x, or z-side witness $(w_1, w_2) = (\log_g z_1, \log_h z_2)$.

Hereafter, let run_i denote the label of i-th execution of the issuing protocol. We define z-side witness in run_i as (w_{1i}, w_{2i}).

Lemma 2. *(Restrictive Blinding) Let $\mathcal{U}_0^*$ be a user that engages in the signature issuing protocol ℓ times, and outputs a valid message-signature pair, $(m, \zeta, \zeta_1, \rho, \varpi, \sigma_1, \sigma_2, \delta, \mu)$. Let z_{1i} denote z_1 used by $\mathcal{S}$ in run_i. For polynomially bound ℓ and for all polynomial-time $\mathcal{U}_0^*$, the probability that $\log_z \zeta \neq \log_{z_{1i}} \zeta_1$ holds for all i is negligible if the discrete logarithm problem is intractable and H_1, H_2, H_3 are random oracles.*

Proof idea: Suppose that $\log_g h$ is not known. We assign $z = g^{w_1} h^{w_2}$ and $(z_{1J}, z_{2J}) = (g^{w_1}, h^{w_2})$ for $J \in_U \{1, \ldots, \ell\}$ by defining $\mathcal{H}_1$ and $\mathcal{H}_2$ so. Since the signature contains proofs of $\zeta = z^\gamma$, $\zeta_1 = g^{w_1'}$, $\zeta_2 = h^{w_2'}$, we may be capable of extracting (γ, w_1', w_2') by rewinding the user in the random oracle model. Once it is done, the condition $\log_z \zeta \neq \log_{z_{1J}} \zeta_1$ guarantees that we obtain two different representations of z, i.e., $z = g^{w_1} h^{w_2} = g^{w_1'/\gamma} h^{w_2'/\gamma}$, which allows us to compute $\log_g h$. For this to be done, we need to simulate $\mathcal{S}$ that issues ℓ signatures without knowing $\log_g h$. We do this with y-side witness x by exploiting witness indistinguishability. The problem is that, due to witness indistinguishability, the rewinding may result in extracting y-side witness x, which is already known. So we first flip a coin to decide with which witness, y-side or z-side, the simulation is performed, and expect that one of the following happens.

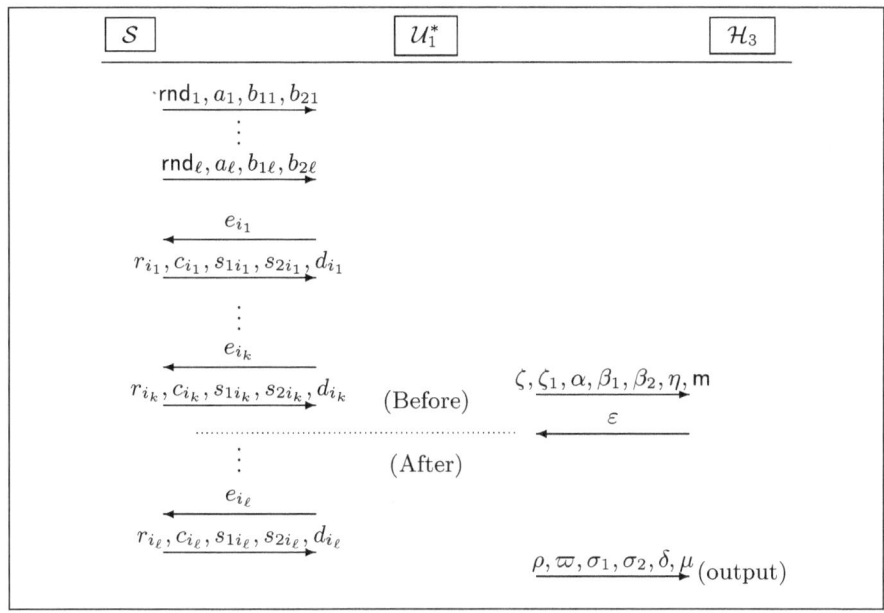

Fig. 2. The interaction among signer $\mathcal{S}$, adversary $\mathcal{U}_1^*$, and random oracle $\mathcal{H}_3$.

- Simulation is done with y-side witness (and z-side witness in run$_J$). Then another z-side witness is extracted by rewinding. This solves $\log_g h$.
- Simulation is done with z-side witnesses. Then y-side witness is extracted by rewinding. This solves $\log_g y$.

Proof. Assume that, having at most q_h accesses to $\mathcal{H}_3$ and asking at most ℓ signatures to $\mathcal{S}$, $\mathcal{U}_0^*$ outputs signature $(\zeta, \zeta_1, \rho, \varpi, \sigma_1, \sigma_2, \delta, \mu)$ that satisfies $\log_z \zeta_1 \neq \log_{z_{1i}} \zeta_1$ for all i with probability ϵ_0 which is not negligible in n. Here, q_h and ℓ are bound by a polynomial of security parameter n. We randomly fix an index $Q \in \{1, \ldots, q_h\}$ and regard $\mathcal{U}_0^*$ as successful only if the resulting signature corresponds to the Q-th query to $\mathcal{H}_3$. (If it does not correspond to any query, $\mathcal{U}_0^*$ is successful only with negligible probability due to the randomness of $\mathcal{H}_3$.) Accordingly, it is equivalent to assuming an adversary, say $\mathcal{U}_1^*$, that asks $\mathcal{H}_3$ only once and succeeds with probability $\epsilon_1 \geq \epsilon_0/q_h$. Figure 2 illustrates the interaction among the signer $\mathcal{S}$, adversarial user $\mathcal{U}_1^*$, and random oracle $\mathcal{H}_3$. Given $\mathcal{U}_1^*$, we construct machine $\mathcal{M}_1$ that solves the discrete-log problem by simulating the interaction. Let $(\mathbf{p}, \mathbf{q}, \mathbf{g}, \mathbf{Y})$ be an instance to solve $\log_{\mathbf{g}} \mathbf{Y}$ in $\mathbb{Z}_{\mathbf{q}}$.

Reduction Algorithm: $\mathcal{M}_1$ first sets $(p, q, g) := (\mathbf{p}, \mathbf{q}, \mathbf{g})$. It then flips a coin $\chi \in_U \{0, 1\}$ to select either $y := \mathbf{Y}$ (case $\chi = 0$) , or $h := \mathbf{Y}$ (case $\chi = 1$).

Case $y = \mathbf{Y}$: (Extracting y-side witness)

1. $\mathcal{M}_1$ selects $w, w_0 \in_U \mathbb{Z}_q$ and sets $h := g^w$ and $z := \mathcal{H}_1(p\|q\|g\|y) = g^{w_0}$.
2. $\mathcal{M}_1$ runs $\mathcal{U}_1^*$ simulating $\mathcal{S}$ with z-side witnesses as follows.

(a) Select $c_i, r_i \in_U \mathbb{Z}_q$ and compute $a_i := g^{r_i} y^{c_i}$.

(b) Select $\mathsf{rnd}_i \in_U \{0,1\}^*$ and $w_{1i} \in_U \mathbb{Z}_q$ and define $\mathcal{H}_2(\mathsf{rnd}_i)$ as $g^{w_{1i}}$. Then compute $w_{2i} := (w_0 - w_{1i})/w \bmod q$. (Accordingly, $z_{1i} = g^{w_{1i}}$ and $z_{2i} = h^{w_{2i}}$.)

(c) Compute $b_{1i} := g^{u_{1i}}$ and $b_{2i} := h^{u_{2i}}$ with $u_{1i}, u_{2i} \in_U \mathbb{Z}_q$.

(d) Send $\mathsf{rnd}_i, a_i, b_{1i}, b_{2i}$ to $\mathcal{U}_1^*$.

(e) Given e_i from $\mathcal{U}_1^*$, compute $d_i := e_i - c_i \bmod q$, $s_{1i} := u_{1i} - d_i w_{1i} \bmod q$, and $s_{2i} := u_{2i} - d_i w_{2i} \bmod q$.

(f) Send $r_i, c_i, s_{1i}, s_{2i}, d_i$ to $\mathcal{U}_1^*$.

$\mathcal{M}_1$ simulates $\mathcal{H}_3$ by returning $\varepsilon \in_U \mathbb{Z}_q$.

3. $\mathcal{U}_1^*$ outputs a signature, say $(\zeta, \zeta_1, \rho, \varpi, \sigma_1, \sigma_2, \delta, \mu)$, that corresponds to ε.

4. Reset and restart $\mathcal{U}_1^*$ with the same setting. $\mathcal{M}_1$ simulates $\mathcal{H}_3$ with $\varepsilon' \in_U \mathbb{Z}_q$.

5. $\mathcal{U}_1^*$ outputs a signature, say $(\zeta, \zeta_1, \rho', \varpi', \sigma_1', \sigma_2', \delta', \mu')$, that corresponds to ε'.

6. If $\varpi \neq \varpi'$, $\mathcal{M}_1$ outputs $x := (\rho - \rho')/(\varpi' - \varpi) \bmod q$. The simulation fails, otherwise.

Case $h = \mathbf{Y}$: (Extracting z-side witness)

1. $\mathcal{M}_1$ selects $x \in_U \mathbb{Z}_q$ and sets $y := g^x$. It also selects $w_1, w_2 \in_U \mathbb{Z}_q$ and sets $z := \mathcal{H}_1(p\|q\|g\|y) = g^{w_1} h^{w_2}$.

2. $\mathcal{M}_1$ selects $I \in_U \{0, \ldots, \ell\}$ and $J \in_U \{1, \ldots, \ell\}$.

3. $\mathcal{M}_1$ runs $\mathcal{U}_1^*$ simulating as follows.

(a) For $i \neq J$, $\mathcal{M}_1$ follows the protocol with y-side witness, x. $\mathcal{H}_2$ is simulated by returning random choices from $\langle g \rangle$.

(b) For $i = J$, $\mathcal{M}_1$ engages in the issuing protocol using both y-side witness x and z-side witness (w_1, w_2) as follows.

  i. Define $\mathcal{H}_2(\mathsf{rnd}_J)$ so that $z_{1J} = g^{w_1}$ and $z_{2J} = h^{w_2}$.

  ii. Compute $a_J = g^{u_J}, b_{1J} = g^{u_{1J}}, b_{2J} = h^{u_{2J}}$ with $u_J, u_{1J}, u_{2J} \in_U \mathbb{Z}_q$.

  iii. Send $(\mathsf{rnd}_J, a_J, b_{1J}, b_{2J})$ to $\mathcal{U}_1^*$.

  iv. Given e_J from $\mathcal{U}_1^*$, choose $d_J \in_U \mathbb{Z}_q$ and compute $c_J := e_J - d_J \bmod q$, $r_J := u_J - c_J x \bmod q$, $s_{1J} := u_{1J} - d_J w_1 \bmod q$, and $s_{2J} := u_{2J} - d_J w_2 \bmod q$.

  v. Send $(r_J, c_J, s_{1J}, s_{2J}, d_J)$ to $\mathcal{U}_1^*$.

$\mathcal{M}_1$ simulates $\mathcal{H}_3$ by returning $\varepsilon \in_U \mathbb{Z}_q$.

4. $\mathcal{U}_1^*$ outputs a signature, say $(\zeta, \zeta_1, \rho, \varpi, \sigma_1, \sigma_2, \delta, \mu)$, that corresponds to ε.

5. Rewind and restart $\mathcal{U}_1^*$ with the same setting.

  – If $I = 0$, $\mathcal{M}_1$ simulates $\mathcal{H}_3$ by returning $\varepsilon' \in_U \mathbb{Z}_q$. Otherwise, set $\varepsilon' = \varepsilon$.

  – If $I \neq 0$ and run_J has not yet been completed before the query to $\mathcal{H}_3$ is sent, $\mathcal{M}_1$ simulates the execution by using both y-side and z-side witnesses as above choosing $d_J' \in_U \mathbb{Z}_q$. Otherwise, $\mathcal{M}_1$ simulates only with y-side witness choosing $d_J' = d_J$.

6. $\mathcal{U}_1^*$ outputs a signature, say $(\zeta, \zeta_1, \rho', \varpi', \sigma_1', \sigma_2', \delta', \mu')$, that corresponds to ε'.

7. If $\delta \neq \delta'$, $\mathcal{M}_1$ computes $w_1' = (\sigma_1 - \sigma_1')/(\mu - \mu') \bmod q$, $w_2' = (\sigma_2 - \sigma_2')/(\mu - \mu') \bmod q$, and outputs $w = (w_1 - w_1')/(w_2' - w_2) \bmod q$. Simulation fails, otherwise.

Evaluation of success probability:
In Figure 2, observe that independent variables given to $\mathcal{U}_1^*$ are p, q, g, h, y, $\mathcal{H}_1$, $\mathcal{H}_2$, rnd_i, a_i, b_{1i}, b_{2i}, d_i for all i, and ε and the random tape of $\mathcal{U}_1^*$. All other variables are uniquely determined by these independent variables. Note that e_i's are also determined by the random tape of $\mathcal{U}_1^*$ and the variables that appeared so far. We wrap all these independent variables into Λ, except for $\{\varepsilon, d_{i_{k+1}}, \dots, d_{i_\ell}\}$, which is defined as D_ε. Let D denote $D_\varepsilon \setminus \{\varepsilon\}$.

Let S be the set of all (Λ, D_ε) that leads $\mathcal{U}_1^*$ to a success, i.e., $\Pr_{\Lambda, D_\varepsilon}[(\Lambda, D_\varepsilon) \in S] \geq \epsilon_1$. According to Lemma 4, with probability at least $\epsilon_1/2$, randomly selected Λ satisfies $\Pr_{D_\varepsilon}[(\Lambda, D_\varepsilon) \in S] \geq \epsilon_1/2$. Once Λ is fixed, δ is uniquely determined by D_ε. By $\delta \leftarrow D_\varepsilon$, we denote the map from (Λ, D_ε) in S to δ. If $(\Lambda, D_\varepsilon) \notin S$, we denote $\perp \leftarrow D_\varepsilon$.

We consider how sensitive δ is to D_ε. Define function ψ as

$$\psi(\delta) = \Pr_{D_\varepsilon}[\delta \leftarrow D_\varepsilon].$$

Let δ_{max} be the value of δ that maximizes $\psi(\delta)$. That is, δ_{max} is the value of δ that is most likely to appear in a successful output of $\mathcal{U}^*$. Let $\psi_{max} = \psi(\delta_{max})$. We consider two cases.

Case 1 (ψ_{max} is not negligible) :
In this case, δ is not likely to change even if D_ε changes, so we perform the rewinding simulation with z-side witnesses choosing D_ε and D'_ε uniformly. By the definition of ψ_{max}, uniformly chosen D_ε and D'_ε yield δ_{max} with probability greater than ψ_{max}^2, which is not negligible. Since ε differs in D_ε and D'_ε with overwhelming probability, we have $\varpi + \delta_{max} = \varepsilon \neq \varepsilon' = \varpi' + \delta_{max} \pmod{q}$. Thus, we obtain $\varpi \neq \varpi'$ with which y-side witness can be extracted as written in Step-6 of Case $y = \mathbf{Y}$.

Case 2 (ψ_{max} is negligible) :
In this case, δ tends to change if D_ε changes. We first observe that there exists at least one element in D_ε whose change impacts δ. Hereafter, we treat ε in D_ε as d_0, so the elements in D_ε are suffixed as $(0, i_{k+1}, \dots, i_\ell)$. Define $Id = (0, i_{k+1}, \dots, i_\ell)$. Let D_ε^{-i} for $i \in Id$ denote a sequence obtained by removing d_i from D_ε. Observe that $\Pr_{D_\varepsilon}[\delta \leftarrow D_\varepsilon] \leq \psi_{max}$ holds for any δ by the definition of ψ_{max}. Suppose that D_ε is uniformly chosen and δ is produced as $\delta \leftarrow D_\varepsilon$. Then, according to Corollary 1, there exists $J \in Id$ such that randomly chosen D_ε^{-J} satisfies

$$\Pr_{d_J}[\delta \leftarrow D_\varepsilon^{-J} \cup \{d_J\}] > 1 - \psi_{max}$$

with probability $< \psi_{max}$. We can correctly guess such index J with probability at least $1/(\ell+1)$ by randomly taking it from $\{0, \dots, \ell\}$. Taking the complement of the above, we see that randomly chosen D_ε^{-J} satisfies

$$\Pr_{d_J}[\delta \nleftarrow D_\varepsilon^{-J} \cup \{d_J\}] \geq \psi_{max}$$

with probability $\geq 1 - \psi_{max}$. Now suppose that D'_ε is made from D_ε by choosing $d_J \in_U \mathbb{Z}_q$, and δ' is produced as $\delta' \leftarrow D'_\varepsilon$. From the above observation, $\{\delta' \neq \delta\}$

$\vee\ \{(\Lambda, D'_\varepsilon) \notin S\}$ happens with probability not negligible in n. According to Lemma 4, with probability $\epsilon_1/4$, uniformly chosen D_ε^{-J} satisfies

$$\Pr_{d_J}[(\Lambda, D_\varepsilon^{-J} \cup \{d_J\}) \in S] \geq \epsilon_1/4.$$

Thus, with probability not negligible in n, such D_ε and D'_ε are in S and result in $\delta' \neq \delta$. From this collision, z-side witness $\log_g h$ can be extracted as shown in Step-7 of Case $h = \mathbf{Y}$. The simulation with such D_ε and D'_ε can be done if the simulator has y-side witness and z-side witness of run_J since they differ at only one index J.

The probability distribution over these cases depends on Λ and the strategy of $\mathcal{U}_1^*$. Note that the distribution of Λ does not depend on the choice of χ as the protocol is witness indistinguishable and the public key are generated so that it distributes uniformly. Accordingly, the coin flip of χ turns the simulation to the proper case with probability $1/2$.   □

Lemma 3. *Any poly-time adversary $\mathcal{U}_3^*$ outputs a valid signature without interacting with $\mathcal{S}$ only with negligible probability if the discrete logarithm problem is intractable and H_1, H_2, H_3 are random oracles.*

Proof. (sketch) This is equivalent to proving the security of the ordinary (i.e., non-blind) version of the signature scheme against key-only attack [19]. Thus it can be done by the rewinding simulation in the random oracle model in a similar way as done in [28]. Given $\mathbf{Y} \in_U \langle g \rangle$, we construct a machine, $\mathcal{M}_2$, that finds $\log_g \mathbf{Y}$ in $\mathbb{Z}_q$. $\mathcal{M}_2$ first selects w, ξ randomly and sets $y = \mathbf{Y}$, $h = g^w$, $z = \mathbf{Y}g^\xi$. (Since $\mathcal{M}_2$ does not need to simulate signer $\mathcal{S}$, it can put $\mathbf{Y}$ into both y and z.) $\mathcal{M}_2$ then invokes $\mathcal{U}_3^*$ twice with the same initial settings and different ε and ε' as answers of $\mathcal{H}_3$. Let the resulting signatures be $(\zeta, \zeta_1, \rho, \varpi, \sigma_1, \sigma_2, \delta, \mu)$ and $(\zeta, \zeta_1, \rho', \varpi', \sigma_1', \sigma_2', \delta', \mu')$. Since $\varpi + \delta = \varepsilon \neq \varepsilon' = \varpi' + \delta'$, at least either $\varpi \neq \varpi'$ or $\delta \neq \delta'$ happens. If $\varpi \neq \varpi'$, $\mathcal{M}_2$ computes $\log_g \mathbf{Y} = \log_g y = (\rho - \rho')/(\varpi' - \varpi) \bmod q$. For the case $\delta \neq \delta'$, $\mathcal{M}_2$ computes $\gamma = \log_z \zeta = (\mu - \mu')/(\delta' - \delta) \bmod q$, $w_1 = \log_g \zeta_1 = (\sigma_1 - \sigma_1')/(\delta' - \delta) \bmod q$, $w_2 = \log_g \zeta_2 = (\sigma_2 - \sigma_2')/(\delta' - \delta) \bmod q$, and $\log_g \mathbf{Y} = \log_g z - \xi = (w_1 + w_2/w)/\gamma - \xi \bmod q$.   □

Proof of Theorem 3. Suppose that there exists an adversary $\mathcal{U}_4^*$ that outputs $\ell + 1$ valid signatures with probability ϵ_4 not negligible in n after interacting with $\mathcal{S}$ at most ℓ times. The case of $\ell = 0$ has been proven by Lemma 3. We consider $\ell \geq 1$.

Due to Lemma 2 and 3, among the $\ell + 1$ signatures, there exist at least two signature-message pairs which contains (ζ, ζ_1) and $(\tilde{\zeta}, \tilde{\zeta}_1)$ such that $\log_\zeta \zeta_1 = \log_{\tilde{\zeta}} \tilde{\zeta}_1 = \log_z z_{1I}$ holds for z_{1I} used in run_I for some I in $\{1, \ldots, \ell\}$. Now, there exist two queries to $\mathcal{H}_3$ that correspond to those signatures. In a similar way as used in the proof of Lemma 2, we guess the indexes of these queries and regard $\mathcal{U}_4^*$ as being successful only if the guess is correct. Accordingly, this is equivalent to an adversary, say $\mathcal{U}_5^*$, that asks $\mathcal{H}_3$ only twice and succeeds with probability $\epsilon_5 = \epsilon_4/\binom{q_h}{2}$ in producing two signatures in the expected relation.

We construct a machine $\mathcal{M}_3$ that, given $(\mathbf{p}, \mathbf{q}, \mathbf{g}, \mathbf{Y})$, solves $\log_g \mathbf{Y}$ in $\mathbb{Z}_q$ by using $\mathcal{U}_5^*$.

Reduction algorithm: $\mathcal{M}_3$ sets $(p, q, g) := (\mathbf{p}, \mathbf{q}, \mathbf{g})$. It then flips a coin, $\chi \in_U \{0, 1\}$, to select either $y := \mathbf{Y}$ (case $\chi = 0$) , or $y := g^x$ with randomly chosen x (case $\chi = 1$) .

1. $\mathcal{M}_3$ selects $w, w_0 \in_U \mathbb{Z}_q$ and sets $h := g^w$ and $z := g^{w_0}$ by defining $\mathcal{H}_1$ so.
2. $\mathcal{M}_3$ selects $I \in_U \{1, \ldots, \ell\}$ and $J \in_U \{1, 2\}$.
3. $\mathcal{M}_3$ runs $\mathcal{U}_5^*$ simulating $\mathcal{S}$ as follows.
   - For run_i ($i \neq I$), $\mathcal{M}_3$ simulates with z-side witness in the same way as shown in Step-2 of Case $y = \mathbf{Y}$ in the proof of Lemma 2.
   - For run_I,
     - if $\chi = 0$, $\mathcal{M}_3$ simulates with z-side witness as above, or
     - if $\chi = 1$, it defines $z_{1I} := \mathcal{H}_2(\mathsf{rnd}_I) = \mathbf{Y}$ and follows the issuing protocol by using y-side witness.

   $\mathcal{M}_3$ simulates $\mathcal{H}_3$ by returning random values, say ε_1 and ε_2.
4. $\mathcal{U}_5^*$ outputs two signatures.
5. $\mathcal{M}_3$ rewinds and restarts $\mathcal{U}_5^*$ with the same setting. $\mathcal{M}_3$ answers J-th query to $\mathcal{H}_3$ with $\varepsilon'_J \in_U \mathbb{Z}_q$.
6. $\mathcal{U}_5^*$ outputs two signatures.
7. Let $(\zeta, \zeta_1, \rho, \varpi, \sigma_1, \sigma_2, \delta, \mu)$ and $(\zeta, \zeta_1, \rho', \varpi', \sigma'_1, \sigma'_2, \delta', \mu')$ be the resulting signatures that correspond to ε_J and ε'_J respectively. (If any of the resulting signatures does not correspond to the hash value, $\mathcal{M}_3$ fails.) If $\chi = 0$ and $\varpi \neq \varpi'$, $\mathcal{M}_3$ outputs $\log_g y = \log_g \mathbf{Y} = (\rho - \rho')/(\varpi' - \varpi) \bmod q$. If $\chi = 1$ and $\delta \neq \delta'$, it outputs $\log_g z_{1I} = \log_g \mathbf{Y} = (\sigma_1 - \sigma'_1)/(\mu - \mu') \bmod q$. $\mathcal{M}_3$ fails, otherwise.

Evaluation of success probability: (sketch)

The probability that $\mathcal{U}_5^*$ is successful and the obtained twin signatures are correlated to run_I is at least ϵ_5/ℓ. The probability is taken over the coin flips of $\mathcal{G}$, $\mathcal{S}$, $\mathcal{U}_5^*$ and the choices of $\mathcal{H}_1, \mathcal{H}_2, \mathcal{H}_3$.

According to Lemma 4, we can find, with probability at least $\epsilon_5/2\ell$, a convenient random tapes of $\mathcal{G}, \mathcal{S}, \mathcal{U}_5^*$ and $\mathcal{H}_1, \mathcal{H}_2$ that lead $\mathcal{U}_5^*$ to output twin signatures that corresponds to run_I with probability $\geq \epsilon_5/2\ell$. The success probability of $\mathcal{U}_5^*$ is now taken over the choice of $\mathcal{H}_3$, i.e., ε_1 and ε_2. We show that the standard rewinding simulation works to extract the witness of the desired side with probability not negligible in the security parameter. (The rest of the proof is actually the same as that in [3], so we give only a brief sketch below.) By $\boldsymbol{\varepsilon}$, we denote $(\varepsilon_1, \varepsilon_2)$ hereafter. Note that the number of all possible $\boldsymbol{\varepsilon}$ is q^2. Define $Succ$ as a set of $\boldsymbol{\varepsilon}$ with which $\mathcal{U}_5^*$ succeeds. Then, there exists a many-to-one mapping from $\boldsymbol{\varepsilon} \in Succ$ to e_I, which is the challenge from $\mathcal{U}_5^*$ used in run_I. Since $\epsilon_5/2\ell$ is not negligible in n, $\#Succ > q$ holds for infinitely many values of n. Thus, there exist $\boldsymbol{\varepsilon}$ and $\boldsymbol{\varepsilon}'$ in $Succ$ that result in the same e_I. Let tr_i denote a transcript obtained in run_i. That is, $tr_i = \{(\mathsf{rnd}_i, a_i, b_{1i}, b_{2i}), e_i, d_i\}$ (excluding dependent variables, r_i, w_i, s_{1i}, s_{2i}). For such $\boldsymbol{\varepsilon}$ and $\boldsymbol{\varepsilon}'$, the sequences of the transcriptions are identical with regard to run_I, that is, $(tr_1, \cdots, tr_I, \cdots, tr_\ell)$ and $(tr'_1, \cdots, tr_I, \cdots, tr'_\ell)$.

Since the issuing protocol is witness indistinguishable, the distribution of tr_I does not depend on the choice of χ. The same is true for other tr_i and tr_i' as they are produced by z-side witnesses selected independently from χ. Thus, if $\mathcal{U}_5^*$ is run twice with such ε and ε', $\mathcal{U}_5^*$ produces a collision that results in exposing either z-side witness or y-side witness independently from χ. It is successful if y-side witness is extracted when $\chi = 0$, or z-side witness, which contains $w_1 = \log_g z_1 = \log_g \mathbf{Y}$, is extracted when $\chi = 1$. These successful cases happen with probability $1/2$ due to the random choice of χ. The difficulty is that we rarely find such ε and ε'. So we consider what happens if ε and ε'' that result in different e_I and e_I' are chosen in the simulation. In this case, tr_I and tr_I' differ and may reflect the choice of χ so that they only yield a useless witness that we already have. We can, however, prove that such useless result cannot occur all the time. Suppose that $\chi = 0$ and ε and ε' yield y-side witness as desired, but ε and ε'' only yield useless z-side witness. This means that $\varpi \neq \varpi'$ and $\varpi = \varpi''$. Thus, $\varpi' \neq \varpi''$ and desired y-side witness can be extracted if ε' and ε'' are chosen. Following this observation, [3] estimated the probability of finding such a convenient pair of ε and concluded that it was not negligible in the security parameter n.   □

5   Application to Double-Spender-Traceable E-cash

Here we apply the proposed blind signature scheme to create a secure anonymous e-cash scheme that provides double-spender traceability.

The withdrawal protocol is exactly the same as the signature issuing protocol. A coin is 7-tuple $\mathsf{coin} = (\zeta, \zeta_1, \rho, \varpi, \sigma_1, \sigma_2, \delta)$, which omits μ from the signature described in the previous section. The user stores the coin together with τ and γ. To pay, the user releases the coin and (ε_p, μ_p) where $\varepsilon_p = \mathcal{H}_4(z^\tau \| \mathsf{coin} \| \mathsf{desc})$ and $\mu_p = \tau - \varepsilon_p \gamma \bmod q$. Here $\mathcal{H}_4$ is a hash function $\mathcal{H}_4 : \{0,1\}^* \to \mathbb{Z}_q$ and desc is the unique description of the transaction. The shop accepts if

$$\zeta \not\equiv 1,$$
$$\varpi + \delta \equiv \mathcal{H}_3(\zeta \| \zeta_1 \| g^\rho y^\varpi \| g^{\sigma_1} \zeta_1{}^\delta \| h^{\sigma_2}(\zeta/\zeta_1)^\delta \| z^{\mu_p} \zeta^{\varepsilon_p}) \bmod q, \text{ and}$$
$$\varepsilon_p \equiv \mathcal{H}_4(z^{\mu_p} \zeta^{\varepsilon_p} \| \mathsf{coin} \| \mathsf{desc}) \bmod q.$$

It is not hard to see that a double payment using different desc and desc' with the same coin yields (ε_p, μ_p) and (ε_p', μ_p') which allows the bank to extract blinding factor γ as $\gamma = (\mu_p' - \mu_p)/(\varepsilon_p - \varepsilon_p') \bmod p$. Since we can prove that Lemma 2 also applies to this variant, $\zeta^{1/\gamma}$ should expose z_1 used in a particular withdrawal session invoked by an authenticated user.

6   Conclusion

We presented an efficient three-move blind signature scheme. It provides one-more unforgeability with polynomially many signatures. From a practical point of view, the scheme is less efficient than known logarithmically-secure schemes but remains practical as it costs only a few times more than the Schnorr blind signature scheme.

The unforgeability was proven under the discrete-log assumption in the random oracle model. Computing the exact reduction cost in the style of [5] seems hard due to the intricate reduction algorithm. Accordingly, the success probability was argued in a classical style, i.e., it was shown that the success probability of the reduction is not negligible with regard to the security parameter.

We also have presented a secure double-spender-traceable e-cash scheme to demonstrate the suitability of our scheme. The scheme is the first single-term scheme whose security against parallel withdrawals can be proven only under the discrete-log and the random oracle assumption.

Acknowledgments

The author wishes to thank Jan Camenisch and Eiichiro Fujisaki for their helpful comments. Early discussions with Miyako Ohkubo helped simplify the scheme.

References

1. M. Abe and J. Camenisch. Partially blind signatures. In the 1997 Symposium on Cryptography and Information Security, 1997.
2. M. Abe and E. Fujisaki. How to date blind signatures. In *Asiacrypt '96*, LNCS 1163, pp. 244–251. Springer-Verlag, 1996.
3. M. Abe and T. Okamoto. Provably secure partially blind signatures. In *Crypto 2000*, LNCS 1880, pp. 271–286. Springer-Verlag, 2000.
4. M. Bellare and P. Rogaway. Optimal asymmetric encryption. In *Eurocrypt '94*, LNCS 950, pp. 92–111. Springer-Verlag, 1995.
5. M. Bellare and P. Rogaway. The exact security of digital signatures – how to sign with RSA and Rabin. In *Eurocrypt '96*, LNCS 1070, pp. 399–416. Springer-Verlag, 1996.
6. S. Brands. Untraceable off-line cash in wallet with observers. In *Crypto '93*, LNCS 773, pp. 302–318. Springer-Verlag, 1993.
7. S. Brands. Restrictive binding of secret-key certificates. In *Eurocrypt '95*, LNCS 921, pp. 231–247. Springer-Verlag, 1995.
8. S. Brands. Restrictive binding of secret-key certificates. Tech. report, CWI, 1995.
9. J. Camenisch. *Group Signature Schemes and Payment Systems Based on the Discrete Logarithm Problem*. PhD thesis, ETH Zürich, 1998.
10. J. Camenisch, J.-M. Piveteau, and M. Stadler. Blind signatures based on the discrete logarithm problem. In *Eurocrypt '94*, LNCS 950, pp. 428–432. Springer-Verlag, 1995.
11. J. Camenisch, J.-M. Piveteau, and M. Stadler. Fair blind signatures. In *Eurocrypt '95*, LNCS 921, pp. 209–219. Springer-Verlag, 1995.
12. D. Chaum. Blind signatures for untraceable payments. In *Crypto '82*, pp. 199–204. Prenum Publishing Corporation, 1982.
13. R. Cramer, I. Damgård, and B. Schoenmakers. Proofs of partial knowledge and simplified design of witness hiding protocols. In *Crypto '94*, LNCS 839, pp. 174–187. Springer-Verlag, 1994.

14. I. Damgård. A design principle for hash functions. In *Crypto '89*, LNCS 435, pp. 416–427. Springer-Verlag, 1990.
15. U. Feige and A. Shamir. Witness indistinguishable and witness hiding protocols. In 21st *STOC*, pp. 416–426, 1990.
16. A. Fiat and A. Shamir. How to prove yourself: Practical solutions to identification and signature problems. In *Crypto '86*, LNCS 263, pp. 186–199. Springer-Verlag, 1986.
17. M. Fischlin. A note on security proofs in the generic model. In *Asiacrypt 2000*, LNCS 1976, pp. 458–469. Springer-Verlag, 2000.
18. Y. Frankel, Y. Tsiounis, and M. Yung. "Indirect discourse proofs": Achieving efficient fair off-line e-cash. In *Asiacrypt '96*, LNCS 1163, pp. 286–300. Springer-Verlag, 1996.
19. S. Goldwasser, S. Micali, and R. Rivest. A digital signature scheme secure against adaptive chosen-message attacks. *SIAM Journal of Computing*, 17(2):281–308, April 1988.
20. L. C. Guillou and J.-J. Quisquater. A practical zero-knowledge protocol fitted to security microprocessor minimizing both transmission and memory. In *Eurocrypt '88*, LNCS 330, pp. 123–128. Springer-Verlag, 1988.
21. H. Horster, M. Michels, and H. Petersen. Meta-message recovery and meta-blind signature schemes based on the discrete logarithm problem and their applications. In *Asiacrypt '92*, LNCS 917, pp. 224–237. Springer-Verlag, 1992.
22. A. Juels, M. Luby, and R. Ostrovsky. Security of blind digital signatures. In *Crypto '97*, LNCS 1294, pp. 150–164. Springer-Verlag, 1997.
23. T. Okamoto. Provably secure and practical identification schemes and corresponding signature schemes. In *Crypto '92*, LNCS 740, pp. 31–53. Springer-Verlag, 1993.
24. T. Okamoto and K. Ohta. Divertible zero knowledge interactive proofs and commutative random self-reducibility. In *Eurocrypt '89*, LNCS 434, pp. 134–149. Springer-Verlag, 1990.
25. B. Pfitzmann and M. Waidner. How to break and repair a "probably secure" untraceable payment system. In *Crypto '91*, LNCS 576, pp. 338–350. Springer-Verlag, 1992.
26. D. Pointcheval. Strengthened security for blind signatures. In *Eurocrypt '98*, LNCS, pp. 391–405. Springer-Verlag, 1998.
27. D. Pointcheval and J. Stern. Provably secure blind signature schemes. In *Asiacrypt '96*, LNCS 1163, pp. 252–265. Springer-Verlag, 1996.
28. D. Pointcheval and J. Stern. Security proofs for signature schemes. In *Eurocrypt '96*, LNCS 1070, pp. 387–398. Springer-Verlag, 1996.
29. D. Pointcheval and J. Stern. Security arguments for digital signatures and blind signatures. *Journal of Cryptology*, 2000.
30. C. Schnorr and M. Jakobsson. Security of discrete log cryptosystems in the random oracle and generic model. Tech. report, University Frankfurt and Bell Labs., 1999.
31. C. P. Schnorr. Efficient signature generation for smart cards. *Journal of Cryptology*, 4(3):239–252, 1991.

Appendix

The following Lemma is known as the Heavy-row Lemma [16] or the Splitting Lemma [28,29]. Let $X \times Y$ be a product space and A its subset. Let (x, y) denote an element in $X \times Y$.

Lemma 4. *Let A be $\Pr[(x,y) \in A] \geq \epsilon$ for some ϵ, and B be $B = \{x \in X \mid \Pr_{y \in Y}[(x,y) \in A] \geq \epsilon/2\}$. Then, $\Pr_{x \in X}[x \in B] \geq \epsilon/2$.*

The following lemma is the reverse of the above in some sense.

Lemma 5. *Let A be $\Pr[(x,y) \in A] < \epsilon$ for $\epsilon \leq 1/3$. Define*

$$B = \{x \in X \mid \Pr_{y \in Y}[(x,y) \in A] > 1 - \epsilon\}, \text{ and}$$
$$C = \{y \in Y \mid \Pr_{x \in X}[(x,y) \in A] > 1 - \epsilon\}.$$

Then, either $\Pr[x \in B] < \epsilon$ or $\Pr[y \in C] < \epsilon$ holds.

Proof. By contradiction. Assume that $\Pr[x \in B] \geq \epsilon$ and $\Pr[y \in C] \geq \epsilon$. Let

$$BY = \{(x,y) \in A \mid x \in B\}, \text{ and}$$
$$CX = \{(x,y) \in A \mid y \in C\}.$$

Observe that $|CX| > (1-\epsilon)|X| \cdot \epsilon|Y|$ and $|BY| > \epsilon|X| \cdot (1-\epsilon)|Y|$. Let CX' and BY' denote minimal subsets of CX and BY, which, respectively, can be considered as $(1-\epsilon)|X| \times \epsilon|Y|$ and $\epsilon|X| \times (1-\epsilon)|Y|$ squares over plain $X \times Y$. Since $1 - \epsilon > \epsilon$, the maximum overlap of those squares is $\epsilon|X| \times \epsilon|Y|$. So, $|CX' \cap BY'| \leq \epsilon^2|X||Y|$. Since $|A| > |CX'| + |BY'| - |CX' \cap BY'|$, we have

$$\epsilon|X||Y| > (1-\epsilon)|X| \cdot \epsilon|Y| + \epsilon|X| \cdot (1-\epsilon)|Y| - \epsilon^2|X||Y|,$$
$$\epsilon > 1/3,$$

which is a contradiction. $\square$

   Lemma 5 can be generalized in the following way by repeatedly applying itself. Let $(x_1, \ldots, x_k)$ denote an element of product space X^k. Let $(x_1, \ldots, x_k)^j$ denote removal of the j-th element, i.e., $(x_1, \ldots, x_{j-1}, x_{j+1}, \ldots, x_k)^j$.

Corollary 1. *Let A be $\Pr[(x_1, \ldots, x_k) \in A] < \epsilon$ for $\epsilon \leq 1/3$. Then, there exists j such that $\Pr[(x_1, \ldots, x_k)^j \in B_j] < \epsilon$ where*

$$B_j = \{(x_1, \ldots, x_k)^j \mid \Pr_{x_j}[(x_1, \ldots, x_k) \in A] > 1 - \epsilon\}.$$

Practical Threshold RSA Signatures
without a Trusted Dealer

Ivan Damgård and Maciej Koprowski

BRICS*, Aarhus University

Abstract. We propose a threshold RSA scheme which is as efficient as
the fastest previous threshold RSA scheme (by Shoup), but where two
assumptions needed in Shoup's and in previous schemes can be dropped,
namely that the modulus must be a product of safe primes and that a
trusted dealer generates the keys. The robustness (but not the unforgeability) of our scheme depends on a new intractability assumption, in
addition to security of the underlying standard RSA scheme.

1 Introduction

In a threshold public-key system we have a standard public key (for the RSA
system, for instance), while the private key is shared among a set of servers,
in such a way that by collaborating, these servers can apply the private key
operation to a given input, to decrypt it or sign it, as the case may be. If there
are l servers, such schemes typically ensure that even if an active adversary
corrupts less than $l/2$ servers, he will not learn additional information about
the private key, and will be unable to force the network to compute incorrect
results. Thus threshold cryptography is an important concept because it can
improve substantially the reliability and security of applications in practice of
public-key systems.

Threshold schemes based on the discrete log problem are relatively straightforward to build, and have been known for a long time. It is even possible to
make efficient schemes where also the key generation phase is done by the servers
in a distributed way [13,9]. This way we can completely avoid assuming any fixed
trusted parties in the system.

Basing threshold schemes on RSA is technically more difficult because we
have to work in a group of non-prime and unknown order (Z_n^* rather than a
prime order subgroup of Z_p^* for a prime p). Nevertheless RSA-based schemes
have been known for some time, see [10,14] for the first reasonably efficient and
robust solutions. However, due to the technical difficulties mentioned, they tend
to be more complex and less efficient in comparison to the discrete log schemes.
One concrete reason is that they use secret sharing "in two levels", i.e. server i
knows a number d_i, such that $\sum_i d_i = d$, the secret RSA exponent. In addition,
each d_i is a verifiable secret shared among the servers. In such a scenario, testing

---

* Basic Research in Computer Science,
  Centre of the Danish National Research Foundation.

B. Pfitzmann (Ed.): EUROCRYPT 2001, LNCS 2045, pp. 152–165, 2001.

if servers have behaved correctly is more complex than in the discrete log case, and if faults do occur, interaction between the servers is necessary to recover.

Recently, however, Shoup[16] proposed a threshold RSA signature scheme which is essentially as efficient as possible: the scheme uses only one level of secret sharing, to sign a message, each server simply sends a single response to a signature request, and must do work that is equivalent up to a constant factor to computing a single RSA signature. No further interaction is needed to recover from faults. Unfortunately, that scheme - like any previous efficient RSA-based scheme - needs to assume a trusted dealer to generate keys. This is caused by the fact that it relies on a special property for the RSA modulus, namely it must be the product of two so called *safe primes* (i.e. the modulus n is the product of primes p, q, where $p' = (p - 1)/2, q' = (q - 1)/2$ are also prime). The problem now is that although reasonably efficient distributed RSA key generation protocols are known[1,6], none of these protocols can ensure that the modulus is a product of safe primes[1]. One attempt to overcome this was made by Miyazaki et al. [11], who build a threshold RSA scheme that can use the key generation protocol from [6]. Unfortunately that scheme is significantly less efficient than Shoup's. It uses two-level secret sharing and needs interaction between servers for each message signed, even if no faults occur. Fouque and Stern [8] present independently of our work a distributed threshold RSA scheme in which they modify the distributed generation of RSA keys from [1],[6] and combine this with Shoup's scheme [16]. The security of their protocol is based only on the underlying standard RSA scheme, but is less efficient than ours by a factor of $\Omega(k^2)$, where k is the security parameter (Fouque and Stern estimate a factor of 30 for a realistic setting of the parameters).

In this paper, we overcome the problem in a more efficient way by constructing a new threshold RSA scheme which may be seen as a generalization of Shoup's, is essentially as efficient as that scheme, follows the same communication pattern, but does not need the assumption about safe primes. As we shall see, this implies that the distributed RSA key generation protocol from [6] can be used to generate keys for our scheme. Note that there may be good reasons to avoid safe primes, other than the distributed key generation issue: first, we do not even know if there are infinitely many safe primes, and second it may turn out to be the case that safe primes are not "safe" at all: although it currently looks as if safe prime products are as hard to factor as RSA moduli in general, this may eventually turn out to be false, indeed most experts agree that choosing the primes as randomly as possibly gives the best security.

On the technical level, one difficulty that arises when safe primes are not assumed, relates to the efficient zero-knowledge protocols used in [16] to verify the behavior of servers. These protocols seem to fail if safe primes are not used, primarily because the group we are working in is no longer cyclic, and may have small prime factors in its order. We get around this by showing that with

---

[1] Of course, generic multiparty computation methods could be used to generate and share such keys in a distributed fashion, but this would be extremely inefficient and completely unsatisfactory in practice

small modifications to the protocols and under an appropriate intractability assumption, the adversary will not be able to exploit the "deficiencies" of the group. Concretely, we show that zero-knowledge proofs of equality of discrete logarithms over a general RSA modulus can be done very efficiently (i.e. without resorting to binary challenge proofs) as long as the prover does not know the factorization. This may be of independent interest, and was previously only known if the modulus was a safe prime product.

Following Shoup, we describe and prove our scheme assuming the random oracle model, however, we rely on it only for robustness of the scheme (and not for unforgeability). At the expense of an additional round of interaction when signing a message, we can avoid using random oracles. The details of this are are omitted since they are standard and straightforward.

To prove the security of our scheme, we need an intractability assumption in addition to the standard RSA assumption. Informally speaking, we assume that given the public key n, e:

- The adversary cannot compute an element $a \neq 1, -1 \bmod n$ such that a has "extremely small order". More precisely the adversary cannot compute an $a \neq 1, -1 \bmod n$ whose order is not divisible by q, where q is the largest prime factor in $\phi(n)$.
- The adversary cannot distinguish a random square modulo n from a random square of maximal order.

As evidence in favor of this assumption, we first note that it is well known that computing the order of a random element is equivalent to factoring. Specifically w.r.t. the first item, for a random RSA modulus n, there is overwhelming heuristic evidence that the prime q will be large (superpolynomial) with overwhelming probability. And so a suitable a cannot be found by choosing randomly. Indeed, it seems that one would need to raise a randomly chosen element to the q'th power to find such an a, however, guessing q is very unlikely to be feasible if factoring is difficult at all. Also, we note that if $n = pq$ is chosen such that $(p-1)/2$ and $(q-1)/2$ have no prime factors less than some number B, then finding $a \neq 1, -1$ of order less than B is as hard as factoring n, since the only possibilities for a are the two non-trivial square roots of 1. In [8], Fouque and Stern show a distributed protocol for generating such RSA moduli efficiently when B is (essentially) a constant (so this does not quite suffice to show that our assumption is equivalent to factoring for such n). The second item can be seen as a generalization of the Quadratic Residuosity Assumption, which can be interpreted as stating that it is difficult to decide if a given element has a maximal power of 2 dividing its order. Our conjecture makes a similar statement for other prime factors.

For the version of our scheme we describe here, we actually need that this assumption holds, even if the adversary is given an oracle for RSA signatures, i.e. , the adversary can specify an M and will be given the e'th root modulo n of $\tilde{H}(M)$, where $\tilde{H}$ is a secure hash function. While this extra condition does not seem to help the adversary in computing orders of elements, it can be removed completely if we are willing to assume that $\tilde{H}$ can be modelled as a random oracle.

In [4] and [7], Damgård/Jurik and Fouque et al. construct threshold versions for (generalizations of) Paillier's probabilistic public key system [12] using the basic techniques from Shoups scheme. These protocols all assume a trusted dealer. Using similar constructions, but starting from our scheme instead of Shoup's, threshold versions of Paillier's scheme without a trusted dealer are easily obtained.

2   Model

Here we describe the model for threshold signature schemes we use, rather informally, due to space limitations. In the type of schemes we consider there are l *servers*. In the *generation phase* on input a security parameter k the public key pk and *secret key shares* $s_1, ..., s_l$ are created, where s_i belongs to server number i. There is a *signing protocol* defined for the servers which takes a message M as input and outputs (publically) a signature σ. Finally, there is a *verification predicate* V, which is efficiently computable, takes pk, message M and signature σ as inputs, and returns *accept* or *reject*. Both the signing protocol and the verification predicate may make use of a random oracle.

To define security, we assume a polynomially bounded static and active adversary $\mathcal{A}$, who corrupts initially $t < l/2$ of the l servers. Thus, the adversary always learns pk and the s_i's of corrupted servers. As the adversary's algorithm is executed, he may issue two types of requests:

- An *oracle request*, where he queries the random oracle used, he is then given the oracle's answer to the query he specified.
- A signature request, where the adversary specifies a message M. This causes the signing protocol to be executed on input M, where the adversary controls the behaviour of corrupted servers (and will of course see whatever information is made public by honest servers).

At the end, $\mathcal{A}$ outputs a message M_0 and a signature σ_0.

We say that $\mathcal{A}$ *wins*, if any of the signing requests resulted in an invalid signature being output, or if he produced a forged signature on a new message, i.e. M_0 was not used in a previous signature request, and $V(pk, M_0, \sigma_0) = accept$.

We say that the scheme is *secure*, if every adversary wins with probability negligible in k^2.

3   The Honest Dealer Scheme

In this section we first describe our scheme assuming an honest dealer that will generate and distribute the keys. The algorithm we specify for the dealer looks

---

[2] unlike the definition in [10], we treat robustness and unforgeability together - this does not make any essential difference.

rather strange, taken by itself. However, the dealer is designed in such a way that the information he distributes matches the output that can be generated by the distributed RSA key generation protocol of Frankel et al. [6]. Therefore, once we prove the security of the honest dealer scheme, a secure (and efficient) scheme without an honest dealer follows easily. We return to this issue in Section 5 [3].

In the threshold scheme to be described, an RSA public key n, e will be selected. We will assume that there exists some method, represented by a function $\tilde{H}$ for mapping an input message M to an element $\tilde{H}(M) \in Z_n^*$. Then the signature we will compute is just the standard RSA signature $\tilde{H}(M)^d \bmod n$, where d is the private exponent corresponding to e. We will refer to this as *the underlying RSA scheme*. The function $\tilde{H}$ can be a hash function, a redundancy scheme, or a combination of both, our construction will work fine in any case. But we do assume throughout *that the underlying RSA signature scheme is secure* - more precisely that it is not existentially forgeable under a chosen message attack. This is clearly a necessary assumption, since without it, no threshold scheme we build from the underlying scheme can be secure. Note that, assuming $\tilde{H}$ can be modelled as a random oracle, security of RSA signatures using $\tilde{H}$ follow from only the standard RSA assumption.

The dealer. Let $\Delta = l!$. The dealer chooses at random $p_1, \ldots, p_l, q_1, \ldots, q_l \in_R [2^{k-1}, 2^k]$ until $p = (p_1 + \cdots + p_l)$ and $q = (q_1 + \cdots + q_l)$ are prime numbers and $gcd((p-1)/2, \Delta) = gcd((q-1)/2, \Delta) = 1$. The RSA modulus is $n = pq$. The dealer also chooses a public exponent e as a prime $e > l$. The public key is $pk = (n, e)$.

Next the dealer executes generation of private keys from [6] to compute $d\Delta^2 = d_1 + \cdots + d_{t+1} \in \mathbb{Z}$ such that $de \equiv 1 \bmod \Phi(n)$, $\Delta | d_1, \ldots, \Delta | d_{t+1}$ and $|d_1| < Cl^{l+1}\Delta^{11}n^2, \ldots, |d_{t+1}| < Cl^{l+1}\Delta^{11}n^2$ for some constant $C > 1$. [4]

The dealer performs secret sharing over the integers, which was introduced in [5] and presented in a modified version in [6]. For $1 \le i \le t+1$ a random polynomial $f_i(x) = \sum_{j=0}^t f_{i,j}x^j$ is chosen such that $f_{i,0} = d_i$ and for $1 \le j \le t$ we have $f_{i,j} \in_R \{0, \Delta, 2\Delta, \ldots, \Delta^{10}n^2 \cdot \Delta\}$. We define a polynomial $f(x) = f_1(x) + \cdots + f_{t+1}(x)$. We can observe that $f(0) = d\Delta^2$ and $f(x)$ is a multiple of Δ for all integers x.

For $1 \le i \le l$, the dealer computes $s_i = f(i) = f_1(i) + \cdots + f_{t+1}(i)$, which is a secret key of server i. If we define $\alpha(k, l) = 4k + (12l + 4)\log l$, it is easy to verify that $0 \le s_i < 2^{\alpha(k,l)}$.

The dealer chooses $v \in Z_n^*$ as a random square. For $1 \le i \le l$, the dealer computes verification key $v_i = v^{s_i \Delta^2}$ of server i.

---

[3] We note that the description in [6] in some places leaves open alternatives for how details in their key generation protocol are executed. Choosing different options lead to minor differences in the output distribution. We stick to one option here for simplicity. Any of the other options could easily be accommodated here by adjusting the description of the honest dealer.

[4] In case of a small public exponent, the protocol from [6] instead generates the private exponent $d\Delta^2$ as a sum of l shares. Our construction could also be based on this method. The protocol and the proofs would be analogous.

Signing protocol.

1. When a message M is requested to be signed, we set $x = \tilde{H}(M)$, where $\tilde{H}$ is a hash function, a redundancy scheme, or a combination of both, and use the scheme to compute $x^d \bmod n$.
2. We define the signature share x_i of server i by

$$x_i = x^{2\Delta^2 s_i}.$$

3. The server i can prove that the discrete logarithm of x_i^2 to base $\tilde{x} = x^{4\Delta^2}$ is the same as the discrete logarithm of v_i to the base v^{Δ^2}.
   We construct the proof of correctness. Let H be a hash function modelled as a random oracle, whose output is an L_1-bit integer, where L_1 is a secondary security parameter (e.g. $L_1 = 128$).
   Each server i chooses at random a number $r \in \{0, \ldots, 2^{\alpha(k,l)+2L_1} - 1\}$. Let

$$c = H(v, \tilde{x}, v_i, x_i^2, v^{r\Delta^2}, x^{4r\Delta^2}), z = s_i c + r.$$

   The proof of correctness produced by server i is (z, c).
4. To verify this proof of correctness, one should check that

$$c = H(v, \tilde{x}, v_i, x_i^2, v^{z\Delta^2} v_i^{-c}, \tilde{x}^z x_i^{-2c}).$$

5. Suppose that valid shares were generated by honest servers from a set $S = \{i_1, \ldots, i_{t+1}\} \subset \{1, \ldots, l\}$.
   For all $j \in S$ we define the Lagrange coefficients multiplied by Δ:

$$\lambda_{0,j}^S = \Delta \cdot \prod_{i \in S \setminus \{j\}} \frac{i}{i - j}.$$

   Clearly the coefficients $\lambda_{0,j}^S$ are integers and

$$d\Delta^3 = f(0)\Delta = \sum_{j \in S} \lambda_{0,j}^S f(j) = \sum_{j \in S} \lambda_{0,j}^S s_j.$$

   Therefore to combine shares, we can compute

$$\omega = \prod_{j \in S} x_j^{2\lambda_{0,j}^S} = x^{4\Delta^2 \sum_{j \in S}\left(s_j \lambda_{0,j}^S\right)} = x^{4\Delta^5 d}.$$

6. We can observe that

$$\omega^e = x^{4\Delta^5}.$$

   Since e is prime to $4\Delta^5$, we can obtain such integers a and b from the extended Euclidian algorithm that $a4\Delta^5 + be = 1$. Finally we have a signature $y = \omega^a x^b$, because

$$y^e = \left(\omega^a x^b\right)^e = x.$$

This shows that y is obtained if the signature shares x_i are computed by honest servers only. In real life, we will only know that the x_i's are values that allow the servers to produce acceptable proofs of correctness. We will later show that this (with overwhelming probability) is sufficient.

4   Proof of Security for the Honest Dealer Scheme

We start by stating our intractability assumption more formally. To this end, we define a *signing oracle* $O(n, e, \tilde{H})$ to be an oracle that on input a message M returns the signature $\tilde{H}(M)^d \bmod n$.

Conjecture 1.   – Consider any probabilistic polynomial time algorithm who gets as input n, e (as chosen by the honest dealer on input k), gets access to a signing oracle $O(n, e, \tilde{H})$, and outputs a number a. For any such algorithm, the probability that $a \neq 1, -1 \bmod n$ and q does not divide the order of a, where q is the largest prime factor in $\phi(n)$, is negligible in k.
  – Let $D = \{D(k) | \ k = 1, 2..\}$ be the family of distributions where $D(k)$ is the distribution of n, e, v generated by our honest dealer on input k. Define D' to be the same, except that v is chosen as a random square of maximal order. Then D and D' are polynomial time indistinguishable, where distinguishers are given access to a signing oracle $O(n, e, \tilde{H})$.

This assumption was already discussed in the introduction. Note that if we are willing to assume that $\tilde{H}$ can be modelled as a random oracle, then the signing oracles can be removed from the conjecture by a standard argument[5].

A number of preliminary observations:

Lemma 1. *The proofs of correctness for signature shares produced by honest servers can be simulated with a statistically close distribution, given the public key and the message to be signed.*

Proof. We construct a simulator which can simulate the proof of correctness generated by server i without knowing the value of secret s_i. Recall that we invoke the random oracle for the hash function H. The simulator controls the random oracle. Whenever the adversary queries the random oracle, if it has not been defined yet at the given point, the simulator picks a random value and sends it to the adversary. When a honest server is expected to produce a proof of correctness for given x, x_i, the simulator picks random $c' \in \{0, \ldots, 2^{L_1} - 1\}$ and $z' \in \{0, \ldots, 2^{\alpha(k,l) + 2L_1} - 1\}$. We declare the value of the random oracle at the point $(v, \tilde{x}, v_i, x_i^2, v^{z'}{}^{\Delta^2} v_i^{-c'}, \tilde{x}^z x_i^{-2c'})$ to be c'. With overwhelming probability, the random oracle has not been defined at this point before. The simulated proof is (z', c'). The only difference to a real proof (z, c) is that in a real execution, we have $z = r + cs_i$, where r is a random $\alpha(k, l) + 2L_1$-bit number. But since r and z' are L_1 bits longer than cs_i, the distance between the distributions of z and z' is exponentially small in L_1. $\square$

Lemma 2. *Let q be the largest prime factor in $\phi(n)$, and consider a signature share x_i (for an input x) produced by a corrupt server. Assume that the element v produced by the honest dealer has maximal order (among all squares modulo n),*

---

[5] Since under this assumption, a signing oracle is easy to implement: if the adversary wants to see a signature on message M, choose a random $\sigma \in Z_n^*$, define the output of $\tilde{H}$ on input M to be $\sigma^e \bmod n$, so that σ now is the signature on M

and that x_i is incorrect, i.e., $x_i^2 \neq (x^{4\Delta^2})^{s_i} \bmod n$. Then, either q does not divide the order of $x_i^2 \cdot (x^{4\Delta^2})^{-s_i} \bmod n$, or the probability that the adversary can construct an acceptable proof of correctness for x_i is negligible. Furthermore, a correct signature can be computed from $t+1$ correct signature shares.

Proof. Let (z, c) be an acceptable proof produced by a corrupt server i. Therefore

$$c = H(v, \tilde{x}, v_i, x_i^2, v^{z\Delta^2} v_i^{-c}, \tilde{x}^z x_i^{-2c}).$$

We can reinterpret this proof as an application of the following interactive protocol, where the verifier is replaced by a call to the random oracle:

Let G be a group of squares in $\mathbb{Z}_n^*$. We have elements $\tilde{v}, w = v^s \in G$, where $\tilde{v}$ has maximal order in G and the prover knows s. The prover P makes elements α, β, guaranteed to be in G as well, and wants to convince us that $\alpha^s = \beta$.

So $\alpha, \beta, \tilde{v}$ correspond to $\tilde{x}, x_i^2, v^{\Delta^2}$ above. Note that if v has maximal order, so does v^{Δ^2}, since n was chosen such that G has order prime to Δ.

The prover performs the following steps:

1. P chooses r in some large enough interval and sends $a = \tilde{v}^r, b = \alpha^r$.
2. P gets a random challenge c from the verifier.
3. P replies by sending $z = r + cs$
4. To check the proof, one verifies that $\tilde{v}^z = aw^c$ and $\alpha^z = b\beta^c$.

We can always write $G = G_1 \times .. \times G_u$, where the order of G_j is a power of q_j and $q_1, \ldots, q_u$ are the distinct prime factors in the order of G. So then we can think of α as a u-tuple, $\alpha = (\alpha_1, ..., \alpha_u)$, $\alpha_j \in G_j$, and similarly for the other group elements. Now, of course, $\alpha^s = \beta$ iff $\alpha_j^s = \beta_j$ for all j.

Claim. If for some j, $\alpha_j^s \neq \beta_j$, then for any initial message (a, b) in the protocol, there is at most one value of $c \bmod q_j$, for which a satisfactory reply z to c exists.

To prove this, there are two cases we must look at, depending on whether $\beta_j \in < \alpha_j >$ or $\beta_j \notin < \alpha_j >$.

Assume first that $\beta_j \notin < \alpha_j >$. Suppose that for some initial message a, the prover can answer both c and c', where $c \neq c' \bmod q_j$. This means that the prover can send z and z' such that $\alpha_j^z = b_j\beta_j^c$ and $\alpha_j^{z'} = b_j\beta_j^{c'}$. Dividing one equation by the other we get $\alpha_j^{z'-z} = \beta_j^{c'-c}$. Since we assumed that $\beta_j \notin < \alpha_j >$, it must be the case that $< \beta_j > \cap < \alpha_j >$ is a proper subgroup of $< \beta_j >$. Hence the order of $\beta_j^{c'-c}$ must be strictly smaller than the order of β_j, but this is a contradiction since $c - c'$ is relatively prime to q_j by assumption.

Next, assume that $\beta_j = \alpha_j^{\tilde{s}}$ for some $\tilde{s}$, but nevertheless $\beta_j \neq \alpha_j^s$. So $\tilde{s} \neq s \bmod ord(\alpha_j)$, where $ord(\alpha_j)$ is some power of q_j. If we let $q_j^{v_j}$ be the order of $\tilde{v}_j$, we have $ord(\alpha_j) \leq q_j^{v_j}$ because $\tilde{v}$ has maximal order in G. Assume again that given some initial message a, the prover can answer both c and c', where $c \neq c' \bmod q_j$, by sending responses z, z'. From the equations the verifier checks, we get

$$\tilde{v}_j^{z'-z} = w_j^{c'-c}, \quad \alpha_j^{z'-z} = \beta_j^{c'-c}$$

Now, $c' - c$ is relatively prime to q_j, we can set $d = (c' - c)^{-1} \bmod q_j^{\nu_j}$ and raise both equations to the d'th power. Since the order of α_j - and hence of β_j - is at most $q_j^{\nu_j}$, this gives us

$$\tilde{v}_j^{d(z'-z)} = w_j, \quad \alpha_j^{d(z'-z)} = \beta_j$$

Hence $d(z - z') = s \bmod ord(\tilde{v}_j)$ and also $d(z - z') = \tilde{s} \bmod ord(\alpha_j)$, which implies $s = \tilde{s} \bmod ord(\alpha_j)$, a contradiction.

This finishes the proof of the claim.

We now return to the situation where we have given an incorrect signature share x_i. Recall that we defined q to be the largest prime factor in the order of Z_n^*, and hence in the order of G, so q is one of the q_j's, say $q = q_1$. Let ϕ be the natural homomorphism from G to G_1. We may assume that $\phi(x_i^2) \neq \phi((x^{4\Delta^2})^{s_i})$, i.e., x_i is "incorrect in G_1", since otherwise q does not divide the order of $x_i^2(x^{4\Delta^2})^{-s_i}$. It then follows from the claim we just proved that for each oracle call the adversary makes where x_i occurs as signature share, the probability that this results in an acceptable proof is at most $1/q$. (note that if the adversary attempts to make a proof without calling the oracle it is clear that it will be accepted with probability at most 2^{-L_1}). It follows from the first part of conjecture 1 that $1/q$ must be negligible, since otherwise a small order element could be found by guessing at random. Since the adversary can only make a polynomial number of oracle calls, it follows that the probability that he can make an acceptable proof for such an x_i is negligible.

Combining shares. Assume that we have $t + 1$ correct signature shares $x'_{i_1}, \ldots,$ $x'_{i_{t+1}}$. For $1 \leq j \leq t + 1$ the signature shares satisfy a property

$$x'_{i_j} = \tilde{x}^{s'_{i_j}}, \quad \text{where } s'_{i_j} \equiv s_{i_j} \bmod ord(v^{\Delta^2}).$$

Since v is an element of maximal order in the group of squares in Z_n^* and $\tilde{x} = x^{4\Delta^2}$, we have

$$x'_{i_j} = \tilde{x}^{s'_{i_j}} = \tilde{x}^{s_{i_j}} \bmod n.$$

Therefore $t + 1$ correct signature shares allow us to compute a correct signature. $\square$

Lemma 3. *Let n, e, distributed as the honest dealer chooses them, be given. Based on this, the information the adversary learns from the honest dealer initially can be simulated with a statistically close distribution.*

Proof. Suppose that the adversary corrupted t servers $i_1, \ldots, i_t$.

We choose at random $r \in Z_{ln}$ and distribute $r\Delta^2$ randomly as a sum $r\Delta^2 = r_1 + \cdots + r_{t+1}$, where $\Delta | r_1, \ldots, \Delta | r_{t+1}$ and $|r_1| < Cl^{l+1}\Delta^{11}n^2, \ldots, |r_{t+1}| < Cl^{l+1}\Delta^{11}n^2$. We perform secret sharing over the integers to share r. For $1 \leq i \leq l$ a random polynomial $g_i(x) = \sum_{j=0}^{t} g_{i,j}x^j$ is chosen such that $g_{i,0} = r_i$ and for $1 \leq j \leq t$ we have $g_{i,j} \in_R \{0, \Delta, \ldots, \Delta^{10}n^2 \cdot \Delta\}$. We define a polynomial $g(x) = g_1(x) + \cdots + g_{t+1}(x)$. We can observe that $g(0) = r\Delta^2$.

The function g gives us a polynomial sharing over the integers of r, which was generated like in the sum-to-poly protocol [5] and by Lemma 3 from [5] it is almost t-wise independent. Since the adversary learns t shares $s_{i_1} = (f_1 + \cdots + f_{t+1})(i_1), \ldots, s_{i_t} = (f_1 + \cdots + f_{t+1})(i_t)$, he can not distinguish these shares from random values and from the shares generated for him by the honest dealer.

Let w be a random square in $\mathbb{Z}_n^*$. We define the verification key $v = w^e \bmod n$.

The verification key of a corrupted server i is $v_i = v^{s_i \Delta^2}$. For an uncorrupted server i, we define set $S = \{0, i_1, \ldots, i_t\}$. We can take the normal Lagrange coefficients and multiply them by Δ so they become integers. The results are called $\lambda_{i,j}^S$ and we have

$$\Delta f(i) = \sum_{j \in S} \lambda_{i,j}^S f(j).$$

Since the adversary can not distinguish our secret d from r, we can compute

$$v_i = v^{s_i \Delta^2} = v^{\Delta(d\Delta^2 \lambda_{i,0}^S + \lambda_{i,i_1}^S s_{i_1} + \cdots + \lambda_{i,i_t}^S s_{i_t})}$$
$$= w^{\Delta(\Delta^2 \lambda_{i,0}^S + e(\lambda_{i,i_1}^S s_{i_1} + \cdots + \lambda_{i,i_t}^S s_{i_t}))} \bmod n.$$

The adversary's view consists from $n, e, s_{i_1}, \ldots, s_{i_t}, v, v_1, \ldots, v_l$. Since it was generated on the basis of the adversary's shares $s_{i_1}, \ldots, s_{i_t}$, which were statistically indistinguishable from the adversary's shares produced by the honest dealer, the adversary can not distinguish this view from the one given by the honest dealer. $\square$

Lemma 4. *Assume we are given a set of values distributed by the honest dealer to the adversary, i.e., $n, e, v, v_1, v_2, \ldots, v_l$ and the s_i's sent to the corrupt servers. Let also a message M, and the signature $\tilde{H}(M)^d \bmod n$ be given. Based on this, the contributions from honest servers in the protocol where M is signed can be simulated with the correct distribution.*

Proof. Let $\{i_1, \ldots, i_t\}$ be the set of corrupted servers. Let $y \equiv \tilde{H}(M)^d \bmod n$ and $x \equiv y^e \equiv \tilde{H}(M) \bmod n$.

We define set $S = \{0, i_1, \ldots, i_t\}$. We can take the normal Lagrange coefficients and multiply them by Δ so they become integers. We can easily compute $x_i = x^{2\Delta^2 s_i}$ for an uncorrupted player i as

$$x_i = y^{2\Delta(\Delta^2 \lambda_{i,0}^S + e(\lambda_{i,i_1}^S s_{i_1} + \cdots + \lambda_{i,i_t}^S s_{i_t}))} \bmod n.$$

$\square$

4.1   Proof Assuming v Has Maximal Order

As a first step, we prove:

Lemma 5. *Modify the honest dealer scheme described above such that the honest dealer chooses v to be a random element of maximal order (among the squares modulo n). Then the resulting scheme is secure under Conjecture 1 and assuming the underlying standard RSA signature scheme is secure.*

Proof. Assume we are given an adversary $\mathcal{A}$ that breaks the scheme, with probability at least $1/p(k)$, for some polynomial $p()$. We will then build an algorithm that with approximately the same probability either breaks the first part of Conjecture 1 or the underlying RSA scheme. So our algorithm gets n, e as input, and also gets a chosen message attack on the underlying RSA scheme, i.e., access to an oracle which on input M returns $\tilde{H}(M)^d \bmod n$. The algorithm now behaves as follows:

1. Invoke Lemma 3 to generate from n, e a simulation of the honest dealer (note that this produces a random square v which does not necessarily have maximal order - we deal with this problem below). Send the data produced to $\mathcal{A}$.
2. For every oracle request $\mathcal{A}$ issues, check if the input value to the oracle that $\mathcal{A}$ specified has been asked for before. If so, return the same answer that was returned earlier. Otherwise, return a fresh random value as an answer and record this value.
3. For every signature request (say, on message M) $\mathcal{A}$ issues, call the oracle to obtain the signature $\tilde{H}(M)^d \bmod n$. Use this and the data generated in step 1 to invoke Lemma 4 and compute the contributions from honest servers in the signing protocol where M is the input. Invoke Lemma 1 to simulate the proofs of correctness from honest players. Send all data produced in this step to $\mathcal{A}$, and receive signature shares x_i and proofs for the corrupt servers from $\mathcal{A}$.
4. If $\mathcal{A}$ produces an incorrect signature share x_i and an acceptable proof for this share, stop and output $x_i^2 \cdot (x^{4\Delta^2})^{-s_i} \bmod n$ (where $x = \tilde{H}(M)$ and M is the message that was signed).
5. If $\mathcal{A}$ stops and outputs M_0, σ_0, output this pair and stop.

To analyze this algorithm, note first that the simulation of the honest dealer in step 1 produces v as a random square, where the honest dealer we have assumed in this subsection chooses v as a random square of maximal order. However, for any prime p, there is a non-negligible probability that a randomly chosen number modulo p has maximal order, namely $p - 1$ (see [15]). This (and the Chinese Remainder Theorem) implies that if we let GOOD be the event that v is a square of maximal order, there is a non-negligible probability that GOOD occurs. It will therefore be sufficient to show that the probability that our algorithm breaks one of the two assumptions, given that GOOD occurs, is non-negligible.

Under this assumption, step 3 simulates our honest dealer with a statistically close distribution. Therefore, the simulations of the signing protocols are also statistically close to the real life distributions (by Lemma 4 and 1). The simulation of the random oracle is trivially perfectly indistinguishable from the real thing. It follows that the probability that $\mathcal{A}$ breaks the threshold signature scheme during our simulation is equal to the probability with which this happens in real life except for a negligible amount, and certainly is at least $1/p'(k)$ for some polynomial.

However, assume first that $\mathcal{A}$ does this by producing an incorrect signature share x_i and a valid proof for it (by Lemma 2 this is necessary to make the signing protocol output a bad signature). By Lemma 2, this means that $x_i^2 \cdot (x^{4\Delta^2})^{-s_i} \bmod n$ has order not divisible by q, except with negligible probability, and so we have broken the first part of Conjecture 1. On the other hand, if M_0 did not occur in any of $\mathcal{A}$'s signature requests, it did not occur in any of ours either, so if also $\sigma_0 = \tilde{H}(M_0)^d \bmod n$, i.e., is a valid signature, we have broken the underlying RSA signature scheme.

4.2   Proof in General

We are now ready for the main result:

Theorem 1. *Consider the original honest dealer scheme described above where the honest dealer chooses v to be a random square modulo n. This scheme is secure under Conjecture 1 and assuming the underlying standard RSA signature scheme is secure.*

Proof. Assume the result is false, i.e. there exists an adversary $\mathcal{A}$ that breaks the scheme with significant probability. We will then argue that this leads to a contradiction with the second part of Conjecture 1. So let us assume that we are given values n, e, v. We know that n, e are chosen as the honest dealer would choose them, and we will show how to use the assumed adversary $\mathcal{A}$ to decide if v is a random square or a square of maximal order.

Note first that we may as well try to decide if $v^e \bmod n$ is random or of maximal order, since raising to the e'th power preserves order and is a 1-1 mapping. So by replacing v by $v^e \bmod n$, we see that we may assume without loss of generality that we know the e'th root of v. With this in mind, a trivial modification of Lemma 3 shows how the honest dealer can be simulated given n, e and v (and the e'th root of v).

We now run the simulation algorithm that appears in the proof of Lemma 5, with two changes:

- In step 1, we run the modified simulation of the honest dealer we just described.
- Having finished, we output v *is random* if $\mathcal{A}$ broke the threshold signature scheme, and v *has maximal order* otherwise.

It is evident from this description that if v has maximal order, we will be producing a simulation that is statistically close to the view of A attacking the scheme with maximal order v, and similarly for random v. It now follows that if v is in fact random, we will output v *is random* with probability at least $1/p(k)$ for some polynomial $p()$, by assumption on $\mathcal{A}$, while this happens with negligible probability if v has maximal order, by Lemma 5.

5   Removing the Honest Dealer

By inspection, it is straightforward to check that the output data from the distributed key generation protocol of [6] matches the data we have assumed that the trusted dealer generates, with one exception : we have required that n is such that $\phi(n)/4$ is not divisible by any prime less than l, and this condition is not automatically satisfied using [6].

This is easily handled, however: the protocol from [6] contains a test division step where each candidate p for a prime factor in n is testdivided by small prime factors. At this point, p is shared additively among the players, so it is trivial to obtain an additive sharing of $p - 1$, and testdivide $p - 1$ by all primes less than l. This will of course slow down the protocol because more candidates will be rejected, however, by Mertens' theorem the cost will only be a factor proportional to $\log l$.

To show security of the combined scheme, we assume (for concreteness) that the protocol from [6] according to the definition of Canetti [2] is a secure protocol for computing the function F, which on input the security parameter k outputs to all players the values $n, e, v, \{v^{s_i} \bmod n\}$, and s_i as private output to server i[6]. Security of the entire combined scheme now follows from Canetti's composition theorem, provided we show that our protocol is secure given an "ideal implementation" of F, i.e., an oracle that on input k outputs to all players a set of output values for F chosen according to the correct distribution. But since such an oracle is equivalent to an honest dealer, the required proof is precisely what we have given in the previous sections [7].

6   Efficiency Analysis

It is straightforward to check that the number of bits sent by each server in order to sign a message, as well as the number of modular multiplications the server needs to perform, is proportional to the bit length of its share s_i of the secret key. From the estimates on s_i in Section 3 it therefore follows that the communication complexity per server is $O(k + l \log l)$ bits and the computation is $O(k + l \log l)$ modular multiplications, where l is the number of servers and k is the length of the modulus.

This is more than in Shoups[16] scheme which has complexity $O(k)$, however, in practice k must be 1000 or more for security reasons, while l is going to be much smaller, so this difference is hardly significant in practice. In the hidden constants, the main difference is a factor of 2 in Shoup's favor. As a concrete example, for a 1000 bit modulus and 32 servers, Shoups scheme will have shares of size 1 Kbit while our shares will be about 4 Kbits.

---

[6] [6] does not directly reference the definition of [2]. Nevertheless, the simulation based security proof they give fits with Canetti's definition

[7] Note that we allow the adversary to do a chosen message attack after seeing the public key. Strictly speaking, the model from [2] does not permit this because contributions from corrupted players must be chosen initially when the adversary is static. However, recent work by Canetti [3] does allow taking this into account

References

1. D. Boneh and M. Franklin *Efficient generation of shared RSA keys*, Proc. of Crypto' 97, Springer-Verlag LNCS series, nr. 1233.
2. R. Canetti, *Security and Composition of Multiparty Cryptographic Protocols*, Journal of Cryptology, vol.13, 2000. On-line version at http://philby.ucsd.edu/cryptolib/1998/98-18.html.
3. R. Canetti, *A unified framework for analyzing security of protocols* , Cryptology Eprint archive 2000/67, http://eprint.iacr.org/2000/067.ps
4. Damgård and Jurik: *A Generalization and some Applications of Paillier's Probabilistic Public-key System*, to appear in Public Key Cryptography 2001.
5. Yair Frankel, Peter Gemmell, Philip D. MacKenzie and Moti Yung *Optimal-Resilience Proactive Public-Key Cryptosystems* Proc. of FOCS 97.
6. Yair Frankel, Philip D. MacKenzie and Moti Yung *Robust Efficient Distributed RSA-Key Generation*, Proc. of STOC 98.
7. P. Fouque, G. Poupard, J. Stern: *Sharing Decryption in the Context of Voting or Lotteries*, Proceedings of Financial Crypto 2000.
8. Pierre-Alain Fouque and Jacques Stern: *Fully Distributed Threshold RSA under Standard Assumptions*, IACR Cryptology ePrint Archive: Report 2001/008, February 2001
9. Gennaro, Jarecki, Krawczyk and Rabin: *Secure Distributed Key Generation for Discrete-Log Based Cryptosystems*, Proc. of EuroCrypt 99, Springer Verlag LNCS series, nr. 1592.
10. Gennaro, Rabin, Jarecki and Krawczyk: *Robust and Efficient Sharing of RSA Functions*, J.Crypt. vol.13, no.2.
11. Shingo Miyazaki, Kouichi Sakurai and Moti Yung *On Threshold RSA-Signing with no Dealer*, Proc. of ICISC 1999, Springer Verlag LNCS series, nr.1787.
12. P.Pallier: *Public-Key Cryptosystems based on Composite Degree Residue Classes*, Proceedings of EuroCrypt 99, Springer Verlag LNCS series, pp. 223-238.
13. Pedersen: *A Threshold cryptosystem without a trusted third party*, proc. of EuroCrypt 91, Springer Verlag LNCS nr. 547.
14. T.Rabin: *A Simplified Approach to Threshold and Proactive RSA*, proc. of Crypto 98, Springer Verlag LNCS 1462.
15. J. B. Rosser and L. Schoenfeld: *Approximate formulas for some functions of prime numbers*, Ill. J. Math. 6 (1962), 64–94.
16. Victor Shoup *Practical Threshold Signatures*, Proceedings of EuroCrypt 2000, Springer Verlag LNCS series nr. 1807.

Hash Functions:
From Merkle-Damgård to Shoup

Ilya Mironov

Computer Science Department, Stanford University, Stanford, CA 94305[*]
mironov@cs.stanford.edu

Abstract. In this paper we study two possible approaches to improving existing schemes for constructing hash functions that hash arbitrary long messages. First, we introduce a continuum of function classes that lie between universal one-way hash functions and collision-resistant functions. For some of these classes efficient (yielding short keys) composite schemes exist. Second, we prove that the schedule of the Shoup construction, which is the most efficient composition scheme for universal one-way hash functions known so far, is optimal.

1   Introduction

In the pursuit of efficient and provably secure constructions of practical cryptosystems several basic primitives have emerged as useful building blocks. Two of them are collision-resistant hash functions (CRHFs) and universal one-way hash functions (UOWHFs). In the complexity-theoretic sense UOWHF is a strictly weaker primitive than CRHF, because the latter is also the former but there is an oracle relative to which UOWHFs exist but not CRHFs [Si98]. Therefore it might be reasonable to base practical cryptosystems on a weaker primitive, which can be easier to construct. Also, since no unconditionally secure UOWHFs are known, the assumption that a particular family of functions is a UOWHF can be more plausible than the assumption of its collision-resistance.

A UOWHF is a collection of keyed compressing functions $\{h_k\}_{k \in K}$ such that winning the following game is infeasible: The adversary chooses x, then receives a key $k \in K$ picked at random and wins if he can find y such that $h_k(x) = h_k(y)$.

A CRHF is a set of keyed compressing functions $\{f_k\}_{k \in K}$ such that for a random $k \in K$ it is infeasible to find x and y that satisfy $f_k(x) = f_k(y)$.

In many applications it is convenient to have a family of UOWHFs or CRHFs, i.e., a collection of functions that map bit strings of different lengths into fixed length strings. The problem is to construct such a family given a single UOWHF or CRHF, which is typically the case when one begins with an off-the-shelf function, for instance, MD5 or SHA-1. For CRHFs a widely used, provably secure and efficient method is the Merkle-Damgård construction [D89, M89]. Surprisingly, this construction does not apply to UOWHFs ([BR97] gave a concrete example of a UOWHF on which the Merkle-Damgård construction fails). For building UOWHF families the best method known so far is due to Shoup [Sh00].

---

[*] Supported by NSF contract #CCR-9984259

B. Pfitzmann (Ed.): EUROCRYPT 2001, LNCS 2045, pp. 166–181, 2001.
© Springer-Verlag Berlin Heidelberg 2001

In this paper we study applicability of the Merkle-Damgård construction, introducing a continuum of primitives that lie between CRHF and UOWHF. Then we give an alternative proof of the Shoup construction and prove that this construction is optimal in some restricted model of computations. The optimality result is the major contribution of the paper.

2 Motivation: Key Length of Different Composition Schemes

The first application of UOWHFs in [NY89] was to use them as a tool for constructing a signature scheme secure under the most general attack. However, most practical signature schemes that follow "hash-and-sign" paradigm use UOWHFs or CRHFs in a different way. They take a message M of an arbitrary length and hash it to obtain a constant length string, which is then fed into a signing algorithm. Many schemes use CRHF families to hash M, but as it was first noted in [BR97] a UOWHF suffices for that purpose. Indeed, if $\{h_k\}_{k \in K}$ is a UOWHF, then $(k, h_k(M))$, where k chosen at random, can be signed and still be as secure as the underlying signature algorithm. If the key length varies with the length of a message, the signing algorithm is applied to $(h_{K'}(k), h_k(M))$, where K' is part of the signer's public key. Here function $h_{K'}$ can be replaced by any second-preimage resistant function, because its input is random and chosen by the signer. Since messages can be very long, hashing speed is a crucial factor. Again, because a UOWHF is a weaker primitive than a CRHF, we may hope to find a more efficient algorithm that implements a UOWHF, thus speeding up the signature scheme.

A closer look at this approach reveals that the key k must be part of the signature so the receiver can recompute the hash. Therefore the shorter the key the better. This is our motivation for studying different composition schemes that yield hash functions with a short key.

The problem of composing a family of UOWHFs does not exist in case of CRHFs, since the Merkle-Damgård construction does not increase the key size. Ironically, if we consider two competing algorithms one implementing a CRHF and a more efficient one, which is supposedly a UOWHF, a signature scheme based on the CRHF can outperform a scheme that uses a family of UOWHFs.

Among several composition schemes for UOWHFs [BR97, Sh00] the one with the smallest key expansion is due to Shoup [Sh00]. Characteristics of the Shoup construction are the following. Suppose that the starting point is a UOWHF that has key length l and compresses n bits to m bits. The composition scheme yields a family of UOWHFs such that a function that compresses N bits to m bits is keyed by $m \cdot \log_2\lceil N/(n-m)\rceil + l$ bits. The key length grows logarithmically with the length of a message. Schemes in [BR97, NY89] have the same asymptotics but a bigger constant factor.

In Table 1 we give a concrete example of the signature length on messages of various sizes if we couple 1024-bit modulus RSA with either a CRHF or a

UOWHF. The UOWHF as in [Sh00] results from the Shoup construction applied to a keyed SHA-1 compression function, which hashes 672 bits to 160 bits.

Table 1. Length of RSA signatures with 1024-bit modulus.

Message length	CRHF	UOWHF
$|M| = $ 1Kb	$|S| = $ 1Kb	$|S| = $ 1.81Kb
1Mb	1Kb	3.22Kb
1Gb	1Kb	4.87Kb

3   Between CRHF and UOWHF

The condition imposed on the round function by the Merkle-Damgård composition theorem can be relaxed. We consider the Merkle-Damgård construction as a useful test that can be applied to function classes filling the gap between CRHFs and UOWHFs. We define these classes in the next section.

3.1   Definitions

CRHFs and UOWHFs enjoy different types of collision-resistance and their constructions base on different assumptions. This adds to the impression that these two primitives have nothing in common. In fact, the only difference between them is in the degree of freedom that the adversary has in choosing one of the colliding elements. In case of a UOWHF, the adversary commits to x before he knows the key, while to defeat a CRHF the adversary is free to choose x afterwards. This difference can be easily quantified by specifying how many bits of x the adversary commits to before he knows the key. Qualitative differences between several variations of hash functions were demonstrated in [ZMI90]. We shall see that the Merkle-Damgård construction may be extended to a class of functions that lie between CRHF and UOWHF.

Definition 1 (class $CR_{\ell_i}(n_i \to m_i)$). Let $\{(n_i, m_i, \ell_i)\}_{i \in \mathbb{N}}$ be a sequence of non-repeating triplets of integer numbers such that $0 < m_i < n_i$ and $0 \leq \ell_i \leq n_i$ for any i. We say that a collection of keyed functions $h_k^i \colon \{0,1\}^{n_i} \to \{0,1\}^{m_i}$, where $k \in K_i$, belongs to **class $CR_{\ell_i}(n_i \to m_i)$** if no adversary can win the following game for infinitely many i in time $\text{poly}(n_i)$ with probability at least $1/\text{poly}(n_i)$:
1. The adversary selects some $x_0 \in \{0,1\}^{n_i - \ell_i}$.
2. Key k is chosen at random from K_i.
3. The adversary selects $x_1 \in \{0,1\}^{\ell_i}$ and $y \in \{0,1\}^{n_i}$ such that $h_k^i(x_1 \| x_0) = h_k^i(y)$.

We call the ℓ_i bits that the adversary is free to choose the *flexibility* of a class.

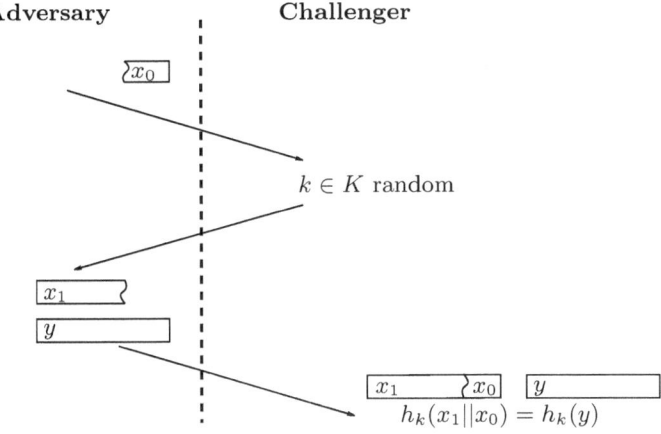

Fig. 1. Function h_k from $\mathrm{CR}_\ell(n \to m)$.

This definition subsumes the definitions of UOWHFs and CRHFs. The class of functions with zero flexibility, $\mathrm{CR}_0(n_i \to m_i)$, is the class of UOWHFs, where the adversary must choose in its entirety one of the colliding elements before he knows the key. On the other hand, functions with full flexibility, $\mathrm{CR}_{n_i}(n_i \to m_i)$, constitute the class of CRHFs, since the adversary commits to nothing ahead of time.

We may omit the index parameter i for the sake of notation brevity. It does not imply that we consider a single triple (n, m, ℓ) (our asymptotic definition is inept in this setting), but that the subsequent arguments can be uniformly applied to the whole family of $\{(n_i, m_i, \ell_i)\}_{i \in \mathbb{N}}$. For example, we can formulate and prove the following propositions without utilizing the index variable.

Proposition 1. $\mathrm{CR}_{\ell_1}(n \to m) \subseteq \mathrm{CR}_{\ell_2}(n \to m)$ if $\ell_1 \geq \ell_2$.

Proof. Because higher flexibility gives more power to the adversary, any set of functions that qualifies as $\mathrm{CR}_{\ell_1}(n \to m)$ also belongs to $\mathrm{CR}_{\ell_2}(n \to m)$.   □

Proposition 2. *A collision for a function from* $\mathrm{CR}_\ell(n \to m)$ *can be found in* $O(2^{\max(m-\ell, m/2)})$ *evaluations of this function.*

Proof. Consider the birthday attack that applies to the flexible part of the input.   □

3.2 Merkle-Damgård Construction Applies to $\mathbf{CR}_\ell(n \to m)$, where $\ell \geq m$

Suppose we have a family of functions $\{h_k\}_{k \in K} \in \mathrm{CR}_\ell(n \to m)$, where $\ell \geq m$. *Merkle-Damgård construction with variable IV and r rounds* (Merkle-Damgård construction for short) is an operator that takes a function h_k and transforms it into a function $\mathbf{MD}^r h_k \colon \{0,1\}^{r \cdot (n-m)+m} \mapsto \{0,1\}^m$. This function is built according to this rule:

Merkle-Damgård construction with variable IV and r rounds
1. Input x formatted as $(x_0, x_1, \ldots, x_r)$ such that $|x_0| = m$, $|x_1| = \cdots = |x_r| = n - m$.
2. Chaining variable C_0 is initialized as x_0.
3. For $i = 1$ to r let $C_i = h_k(C_{i-1}, x_i)$.
4. Output of the function $\mathbf{MD}^r h_k(x)$ is C_r.

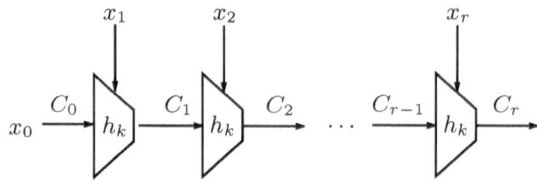

Fig. 2. r-round Merkle-Damgård construction.

This is the usual Merkle-Damgård construction except that the initializing value (IV) is also part of the input. The following theorem proves that the construction works correctly on functions from the class $\mathrm{CR}_\ell(n \to m)$, where $\ell \geq m$.

Theorem 1. *Suppose we have a family of functions $\{h_k^i\}_{i \in \mathbb{N}, k \in K} \in \mathrm{CR}_{\ell_i}(n_i \to m_i)$, where $\ell_i \geq m_i$, and a sequence $\{r_i\}_{i \in \mathbb{N}}$ such that $r_i < \mathrm{poly}(n_i)$. Then $\{\mathbf{MD}^{r_i} h_k^i\}_{i \in \mathbb{N}, k \in K}$ is in the class $\mathrm{CR}_{\ell_i}(r_i \cdot (n_i - m_i) + m_i \to m_i)$.*

Proof. Suppose that $\ell_i = m_i$. The case of $\ell_i > m_i$ is treated analogously. Assume for contradiction that there is an algorithm $\mathcal{A}$ that wins the game described in Definition 1 for infinitely many i. We build an algorithm $\mathcal{B}$ that contradicts the fact that $\{h_k^i\}_{i \in \mathbb{N}, k \in K} \in \mathrm{CR}_{\ell_i}(n_i \to m_i)$.

Fix some i. Denote the value $\mathcal{A}$ commits to on the first step of this game by $x = (x_1, \ldots, x_r)$. Choose $0 < j \leq r_i$ at random. Commit to x_j in the game played by $\mathcal{B}$.

Key k is chosen at random. Let $\mathcal{A}$ find x_0 and $y = (y_0, y_1, \ldots, y_r)$ that make a collision. Now $\mathbf{MD}^{r_i} h_k^i(x_0 \| x) = \mathbf{MD}^{r_i} h_k^i(y)$. With probability at least $1/r$

we have a collision on the jth application of the function h_k^i. It means that $h_k^i(C_{j-1}||x_j) = h_k^i(C'_{j-1}||y_j)$, but $C_{j-1}||x_j \neq C'_{j-1}||y_j$, where C_j and C'_j are the chaining variables. If it is the case, $\mathcal{B}$ outputs a colliding pair $C_{j-1}||x_j$ and $C'_{j-1}||y_j$. $\qquad\square$

The fact that $\ell_i \geq m_i$ is crucial for the proof. Because the flexible part of the hashes' input is longer than their output, the adversary does not need to commit to the chaining variable if he wants to use $\mathcal{A}$ to find a collision. The value of the chaining variable depends on the key and cannot be predicted ahead of time. It is where the proof breaks if we want to apply it to the case when $\ell_i < m_i$.

Note 1. In practice we want to have a CRHF or a UOWHF that takes as input any string of some bounded length and maps it to a fixed-length string. This is stronger primitive than a collection of functions each taking a fixed-length input, because it must be collision resistant (resp. be a UOWHF) across inputs of different length. It turns out that a collection of fixed-input length functions can be strengthened to allow a variable input length. We assume that the message is padded to a length divisible by the block size ($n - m$ in case of the Merkle-Damgård construction) and the last block of the message uniquely encodes the message length before padding. This preprocessing stage was proposed in [M89] and discussed in [LM92] along with a definition of a free-start attack. Theorem 1 can be generalized to hold for this strengthened construction ([LM92] proved the theorem for families of pure CRHFs and UOWHFs).

3.3   Boundary

Because there is a complexity-theoretic jump between CRHFs and UOWHFs, we may expect to observe at least one such a jump in the sequence of classes $CR_0(n \to m) \supseteq \cdots \supseteq CR_n(n \to m)$. The following theorems show that this is indeed the case and there are two classes of complexity-theoretic equivalence. The boundary between them coincides with the limit of validity of the Merkle-Damgård construction (Section 3.2).

We recall that UOWHFs are one-way functions and can be built from a family of one-way functions [Ro90]. This also implies existence (via black box construc tions) of other cryptographic primitives such as secure signature schemes [NY89], pseudo-random generators [HILL99], telephone coin flipping [B82] and bit com mitment protocols [N91]. [Si98] proved that there is no black-box (relativizing) construction of a CRHF based on a UOWHF. In our terminology it means that there is no unconditional construction of $CR_n(n \to m)$ given access to $CR_0(n \to m)$ as a black-box.

Theorem 2. $CR_{m-O(\log n)}(n \to m)$ *is non-empty if and only if CRHFs exist.*

Proof. The adversary playing the game from Definition 1 may choose values for $O(\log n)$ bits randomly. His probability of success drops in this case by a factor of $2^{O(\log n)} = \text{poly}(n)$. Therefore $CR_{m-O(\log n)}(n \to m) = CR_m(n \to m)$.

Since $\mathrm{CR}_m(n \to m) \supseteq \mathrm{CR}_n(n \to m)$ by Proposition 1, the "if" part is trivial.

Suppose we have a $\{h_k\}_{k \in K} \in \mathrm{CR}_m(n \to m)$. Define $g_k\colon \{0,1\}^{m+1} \mapsto \{0,1\}^m$ for any $k \in K$ as follows. Suppose that g_k takes two arguments—a single bit b and x, which is m-bit long. Let

$$g_k(b, x) = h_k(x|| \underbrace{b \ldots b}_{n-m \text{ times}}).$$

We claim that $\{g_k\}_{k \in K} \in \mathrm{CR}_{m+1}(m+1 \to m)$, i.e., it is a CRHF family.

Assume the opposite. There is an efficient algorithm $\mathcal{A}$ that for a random k finds a collision $g_k(b_0, x) = g_k(b_1, y)$. Wlog we may assume that $b_0 = 0$ with probability at least $1/2$. We want to show that $\{h_k\}_{k \in K} \notin \mathrm{CR}_m(n \to m)$. In order to prove it we build algorithm $\mathcal{B}$ that wins the game from Definition 1 as follows:

step 1. Commit to 0^{n-m}.

step 2. Get $k \in K$.

step 3. Run $\mathcal{A}$ to find a collision $g_k(b_0, x) = g_k(b_1, y)$. If $b_0 = 0$ proceed to the next step, otherwise the algorithm fails.

step 4. Output $(x||0^{n-m}, y||b_1^{n-m})$ as a collision for h_k.

The output of $\mathcal{B}$ is indeed a collision, since

$$h_k(x||0^{n-m}) = g_k(x, 0) = g_k(x, b_0) = g_k(y, b_1) = h_k(y||b_1^{n-m})$$

and, because $b_0||x \neq b_1||y$, these two elements of the domain of h_k are different. The success probability of $\mathcal{B}$ is at least one half of the success probability of $\mathcal{A}$.

Notice that $g_k(0, x)$ and $g_k(1, x)$ is a pair of claw-free pseudo-injections (see [Ru95]). $\qquad\square$

Theorem 3. $\mathrm{CR}_{m-m^{\Omega(1)}}(n \to m)$ *is not empty if and only if UOWHFs exist.*

Formally, if some $\mathrm{CR}_{\ell_i}(n_i \to m_i)$ is not empty, then UOWHFs exist. If UOWHFs exist, then for any $\ell(m)\colon \mathbb{N} \mapsto \mathbb{N}$, such that $m - \ell(m) > m^c$ for some $0 < c < 1$, a non-empty class $\mathrm{CR}_{\ell(m_i)}(n_i \to m_i)$ exists.

Proof. Since $\mathrm{CR}_0(n \to m) \supseteq \mathrm{CR}_\ell(n \to m)$, the "only if" part is trivial.

Suppose UOWHFs exist. Take $\{h_k\}_{k \in K} \in \mathrm{CR}_0(n \to m)$. Then for any $\ell < \mathrm{poly}(n)$ we may define $g_k\colon \{0,1\}^{n+\ell} \mapsto \{0,1\}^{m+\ell}$ as

$$g_k(x, y) = y || h_k(x),$$

where $|y| = \ell$ and $|x| = n$. We claim that $\{g_k\}_{k \in K} \in \mathrm{CR}_\ell(n + \ell \to m + \ell)$.

Indeed, if there is a collision $g_k(x_0, y_0) = g_k(x_1, y_1)$, then $y_0 = y_1$ and $h_k(x_0) = h_k(x_1)$. Note that y_0 is the flexible part of the input and x_0 is the part that the adversary commits to before he knows the key. The adversary works poly-time in $n + \ell$. Since $\ell < \mathrm{poly}(n)$, the adversary's running time is also polynomial in n. Therefore the same adversary can be used to break UOWHF-ness property of $\{h_k\}_{k \in K}$.

Suppose we are given a family of UOWHFs $\{h_k^i\}_{i\in\mathbb{N}, k\in K} \in \mathrm{CR}_0(n_i' \to m_i')$. For every m_i' there is some m, such that $m^c/2 < m_i' < m^c \le m - \ell(m)$. The construction above with $\ell = m - m_i' < (2m_i')^{1/c} < (2n_i')^{1/c} = \mathrm{poly}(n_i')$ yields a collection of functions from $\mathrm{CR}_\ell(n_i' + \ell \to m_i' + \ell) = \mathrm{CR}_\ell(n_i' + \ell \to m_i) \subseteq \mathrm{CR}_{\ell(m_i)}(n_i' + \ell \to m_i)$. The last inclusion is because $\ell = m - m_i' > \ell(m)$ and by Proposition 1. $\qquad\square$

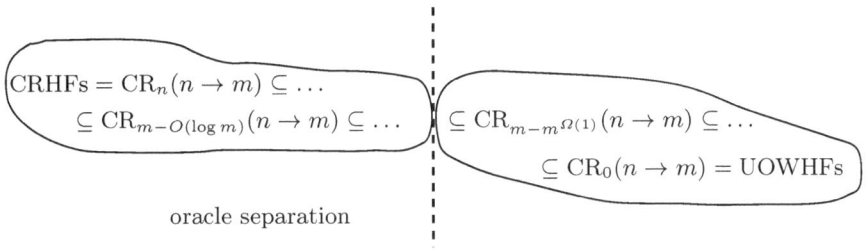

oracle separation

Fig. 3. Hierarchy of classes.

Theorems 2 and 3 show that there are two classes of complexity-theoretic equivalence of classes $\mathrm{CR}_\ell(n \to m)$ (Figure 3). One contains CRHFs and all $\mathrm{CR}_\ell(n \to m)$ for $\ell \ge m - O(\log n)$, the other one spans classes between UOWHFs and $\mathrm{CR}_\ell(n \to m)$ for $\ell < m - m^{\Omega(1)}$. The following note eliminates the gap between them.

Note 2. The claim of Theorem 3 can be improved if we assume that $n < \mathrm{poly}(m)$ and $\mathrm{CR}_0(n \to m)$ has "ideal" security $\Omega(2^m)$ as in Proposition 2. With these assumptions there is no gap between the two theorems and there are only two classes of equivalence.

4   Optimality of the Shoup Construction

[BR97] gave an example of a UOWHF on which the two-round Merkle-Damgård construction fails. Since we want to build UOWHFs the same way we build families of CRHFs, i.e., starting with a keyed function that has fixed-length input, other constructions have to be studied. The most efficient among different composition schemes that have appeared in the literature is the Shoup construction. We give an alternative proof of its correctness, which is technically simpler than in [Sh00] and conceptually better matches our main result, the proof of its optimality.

4.1 Shoup Construction

The Shoup construction (see Figure 4) can be viewed as an extended Merkle-Damgård construction, where the chaining variable is XORed with some mask on each iteration. Because these masks are reused, the key length grows logarithmically with the size of the message.

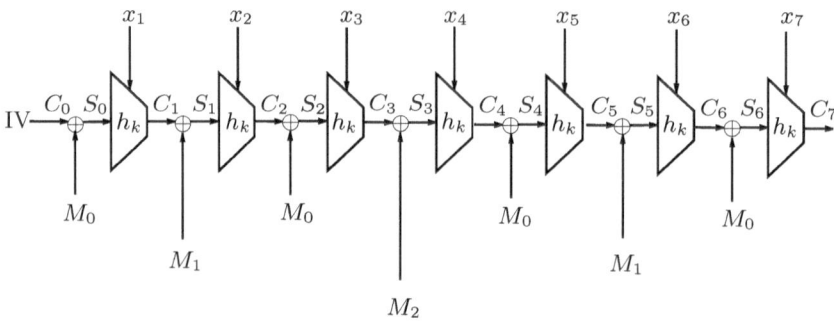

Fig. 4. 7-round Shoup construction.

Formally, *the r-round Shoup construction* is an operator that takes function $h_k : \{0,1\}^n \mapsto \{0,1\}^m$, bit-vector M with length $Lm > m\lfloor \log r \rfloor$, which is formatted as L masks $M = (M_0, \ldots, M_{L-1})$, and transforms it into a function $\mathbf{S}^{r,M} h_k : \{0,1\}^{r \cdot (n-m)} \mapsto \{0,1\}^m$. This function is built according to this rule:

r-round Shoup construction

1. Input x formatted as $(x_1, \ldots, x_r)$ such that $|x_1| = \cdots = |x_r| = n - m$.
2. Chaining variable C_0 is initialized as IV.
3. For $i = 1$ to r let $C_i = h_k(C_{i-1} \oplus M_{\nu(i)}, x_i)$, where the auxiliary function $\nu(i)$ is the highest power of 2 that divides i.
4. Output of the function $\mathbf{S}^{r,M} h_k(x)$ is C_r.

We defer the proof of correctness of this construction to Section 4.3.

4.2 Optimality of the Shoup Construction

The Shoup construction achieves its short, compared to other constructions, key length of the composite scheme by reusing the bit-masks. A legitimate question is whether the masks can be reused even more. In this section we give a negative answer to this question. We prove that the Shoup construction really reuses masks as much as possible in the strongest sense.

Definition 2. A **generalized r-round Shoup construction** is the Shoup construction as described in Section 4.1 but with function ν, which selects a mask to use on every iteration of the construction, being any function that maps $[1, \ldots, r]$ to $[0, \ldots, L-1]$. Function ν is the **schedule** of the construction. The construction is **valid** if it transforms any UOWHF family into a UOWHF family.

The Shoup construction instantiates the function $\nu(i) = \max\{j: 2^j | i\}$, so $L = \lfloor \log r \rfloor + 1$. A schedule is optimal if L is minimal for a fixed r. The following theorem says the Shoup scheduling is optimal for all r.

Theorem 4. *For any valid generalized r-round Shoup construction $r < 2^L$.*

Proof. The proof consists of two steps. First, we show that any schedule of a valid construction must be even-free (defined below). Second, we prove that any schedule with $r \geq 2^L$ is not even-free.

Definition 3 (even-freeness property). We say that a schedule ν is **even-free** if for any a and b, such that $1 \leq a \leq b \leq r$, there is some $0 \leq \eta < L$, such that the number of times $\nu(c)$ takes value η for $a \leq c \leq b$ is odd. In other words, there is no sub-interval that contains every mask an even number of times.

Lemma 1. *Any valid schedule ν is even-free.*

Proof. Suppose there is a non even-free schedule ν of some valid r-round generalized Shoup construction. We build a UOWHF family on which this construction fails, thus contradicting validity of the construction.

Assume that $g_k: \{0,1\}^n \mapsto \{0,1\}^m$ is a UOWHF, $2m + 2 < n$ and $k \in K = \{0,1\}^m$. Of course, if UOWHFs do not exist, then every construction is valid but the problem itself is moot. If we have *some* UOWHF family, by adding an additional argument to its input that gets replicated to the output we can ensure that $2m + 2 < n$. The size of the key space can be adjusted similarly. We define function $h_k: \{0,1\}^n \mapsto \{0,1\}^{2m+1}$ as follows:

$$h_k(y, z, b, x) = \begin{cases} g_k(y, z, b, x) || z || 1 & \text{if } x \neq 0^l \text{ and } z \neq k \\ g_k(y, z, b, x) || k || 1 & \text{if } x = 0^l \text{ and } z \neq k \\ 0^{2m+1} & \text{if } z = k, \end{cases}$$

where $|y| = |z| = |k| = m$, b is a bit and $|x| = l = n - 2m - 1 > 1$. As usual, we omit the index i of the family of UOWHFs, assuming that the construction of h_k and the proof below apply uniformly to all functions of the family.

We claim that h_k is a UOWHF. Indeed, a collision $h_k(y, z, x) = h_k(y', z', x')$ also yields a collision $g_k(y, z, x) = g_k(y', z', x')$ unless $z = k$ and $z' = k$. Probability that the adversary hits $z = k$ *before* he knows k is negligible.

If ν is not even-free, there is a sub-interval $[a, b]$ that contains each mask an even number of times. We exploit this property to find a collision with a previously committed value.

The composite scheme takes as its input $x = (x_1, \ldots, x_r)$, where $|x_i| = l = n - 2m - 1$ for all i. We claim that the following x and x' collide:

$$x = \underbrace{0^l || \ldots || 0^l}_{a \text{ times}} \, || \, \underbrace{1^l || \ldots || 1^l}_{r - a \text{ times}},$$

$$x' = \underbrace{0^l || \ldots || 0^l}_{a - 1 \text{ times}} \, || 1^{l-1} 0 || \, \underbrace{1^l || \ldots || 1^l}_{r - a \text{ times}}.$$

Denote the input of h_k on the ith iteration of the composite scheme by (y_i, z_i, b_i, x_i) and (y'_i, z'_i, b'_i, x'_i) for the inputs x and x' respectively. Format masks $M_i = (M_i^{(1)}, M_i^{(2)}, M_i^{(3)})$, where $|M_i^{(1)}| = |M_i^{(2)}| = m$, $|M_i^{(3)}| = 1$. By definition of h_k we may compute $z_a, \ldots, z_b$ and $z'_a, \ldots, z'_b$ as follows:

$$z_a = k \quad \oplus M_{\nu(a)}^{(2)} \qquad\qquad z'_a = k \quad \oplus M_{\nu(a)}^{(2)}$$

$$z_{a+1} = z_a \quad \oplus M_{\nu(a+1)}^{(2)} \qquad\qquad z'_{a+1} = z'_a \quad \oplus M_{\nu(a+1)}^{(2)}$$

$$\cdots \qquad\qquad\qquad \cdots$$

$$z_b = z_{b-1} \oplus M_{\nu(b)}^{(2)} \qquad\qquad z'_b = z'_{b-1} \oplus M_{\nu(b)}^{(2)}.$$

Therefore,

$$z_b = z'_b = k \oplus M_{\nu(a)}^{(2)} \oplus M_{\nu(a+1)}^{(2)} \cdots \oplus M_{\nu(b)}^{(2)}.$$

Since every mask appears between a and b an even number of times, all masks XOR themselves out and

$$z_b = z'_b = k.$$

If $C_i(x)$ and $C_i(x')$ are the ith chaining variable of the composite scheme evaluated on x and x',

$$C_b(x) = h_k(y_b, k, b_b, x_b) = 0^{2m+1},$$
$$C_b(x') = h_k(y'_b, k, b'_b, x'_b) = 0^{2m+1}$$

by the third case of the definition of h_k.

Since x and x' agree after their a^{th} component, the output of the composite scheme will be the same on both inputs. $\square$(Lemma 1)

Lemma 2. *If schedule is even-free, then $r < 2^L$.*

Proof. Assume the opposite. There is an even-free schedule ν with $r \geq 2^L$.

Let $\#_{a,b}(i)$ be a function that counts the number of appearances of the ith mask between $\nu(a)$ to $\nu(b)$ inclusive.

Define a sequence of bit-vectors $d^i = (\#_{1,i}(0) \bmod 2, \ldots, \#_{1,i}(L-1) \bmod 2)$. Each vector has length L. Because the schedule is even-free, none of these vectors is $(0, \ldots, 0)$. Therefore, there are $r \geq 2^L$ L-bit vectors and one of 2^L possible values is not available. By the pigeonhole principle there are two equal vectors $d^a = d^b$ among them. Consider their difference.

$$
\begin{aligned}
d^b - d^a &= (0, \ldots, 0) \\
&= ((\#_{1,b}(0) - \#_{1,a}(0)) \bmod 2, \ldots, (\#_{1,b}(L-1) - \#_{1,a}(L-1)) \bmod 2) \\
&= (\#_{a+1,b}(0) \bmod 2, \ldots, \#_{a+1,b}(L-1) \bmod 2).
\end{aligned}
$$

Because of the interval $[a+1, b]$ the schedule function is not even-free. $\square$(Lemma 2)

From lemmas 1 and 2 the theorem follows. $\square$

Note 3. It is instructive to see why the Shoup scheduling $\nu(i) = \max\{j : 2^j | i\}$ is even-free. In any interval $[a, b]$ there is a *unique* element c that maximizes ν. Indeed, if there were two such elements c_1 and c_2, necessarily $\nu(c_1) = \nu(c_2)$. But then the element $c = (c_1 + c_2)/2$ would be divisible by a higher power of 2. Existence of an element that appears only once, i.e., an odd number of times, in every interval is enough for even-freeness.

Note 4. What if one uses addition modulo 2^m instead of XOR to mingle a mask and a chaining variable? If this operation is commutative and has an efficiently computable inverse, then the proof goes through with minor modifications. Having an inverse is required for our proof of the Shoup construction (below, Theorem 5), but being commutative is not necessary.

4.3   Correctness of the Shoup Construction

In this section we give an alternative proof of the Shoup construction. It is different from [Sh00] in presentation of the key reconstruction algorithm.

Theorem 5. *If $\{h_k\}_{k \in K}$ is a UOWHF, so is $\{\mathbf{S}^{r,M} h_k\}_{k \in K, |M| = m(\lfloor \log r \rfloor + 1)}$ for $r < \mathrm{poly}(n)$.*

Proof. Suppose that there is an adversary $\mathcal{A}$ that finds a collision of $\mathbf{S}^{r,M} h_k$ with a non-negligible probability over the key of the composite scheme. We build an algorithm $\mathcal{B}$ that finds a collision in $\{h_k\}_{k \in K}$. Let $S_j(x) = C_{j-1} \oplus M_{\nu(j)}$—first m bits of the input of h_k on the jth iteration of the scheme on input x.

Algorithm $\mathcal{B}$
1. Run $\mathcal{A}$. $\mathcal{A}$ commits to some $x = (x_1, \ldots, x_r)$.
2. Choose randomly j from $\{1, \ldots, r\}$ and an m-bit string $C \in \{0, 1\}^m$. Commit to $C \| x_j$.
3. Receive a key $k \in K$.
4. Run the key reconstruction algorithm (described below) that will output M such that $S_j(x) = C$.
5. Feed k and M to $\mathcal{A}$.

6. If $\mathcal{A}$ finds a collision $x' = (x'_1, \ldots, x'_r)$, check if $S_j(x)||x_j \neq S_j(x')||x'_j$ and $h_k(S_j(x)||x_j) = h_k(S_j(x')||x'_j)$. If so, output $S_j(x')||x'_j$ that collides with $C||x_j$.

Suppose that the output of the key reconstruction algorithm on uniformly distributed C with fixed x, j and k also has the uniform distribution. Then the probability that $\mathcal{B}$ finds a collision is $1/r$ of the success probability of $\mathcal{A}$. Indeed, if $\mathcal{A}$ finds a collision, there is at least one $i \in \{1, \ldots, r\}$ such that $S_i(x)||x_i \neq S_i(x')||x'_i$ but $h_k(S_i(x)||x_i) = h_k(S_i(x')||x'_i)$ (consider the output of each iteration of the scheme going backward). This contradicts the assumption that $\{h_k\}_{k \in K}$ is a UOWHF. From this the claim of the theorem follows.

Now all we need is to show and prove the key reconstruction algorithm.

Key reconstruction algorithm

Input: $x = (x_1, \ldots, x_r)$, $k \in K$, $C \in \{0, 1\}^m$.
Output: $M = (M_0, \ldots, M_{L-1})$, such that $S_j(x) = C$.
1. Label all masks $M_0, \ldots, M_{L-1}$ as "undefined."
2. Repeat the following steps while $j > 0$. If $j = 0$, randomly define all undefined masks and quit.
3. Let $i = j - 2^{\nu(j)}$.
4. Pick D at random from $\{0, 1\}^m$.
5. Randomly define all yet undefined masks from the list $M_{\nu(i+1)}, \ldots, M_{\nu(j-1)}$.
6. If $i = 0$, let $C_0 = IV$, otherwise let $C_i = h_k(D, x_i)$. Compute $C_{i+1} = h_k(M_{\nu(i+1)} \oplus C_i, x_{i+1}), \ldots, C_{j-1} = h_k(M_{\nu(j-1)} \oplus C_{j-2}, x_{j-1})$.
7. Let $M_{\nu(j)} = C_{j-1} \oplus C$.
8. Assign $C \leftarrow D$, $j \leftarrow i$ and go to step 2.

First, note two invariants of the algorithm.
Invariant 1. $\nu(i) > \nu(j)$.
Invariant 2. $\nu(j) > \nu(l)$ for any $i < l < j$.

Both invariants follow from the fact that $j \equiv 2^{\nu(j)} \pmod{2^{\nu(j)+1}}$.

To prove the correctness of the algorithm we need to show that a mask is never *redefined*. Masks are defined in three steps of the algorithm. In steps 2 and 5 only undefined masks are assigned random values. By Invariant 2 their numbers are less than $\nu(j)$. In step 7 mask $M_{\nu(j)}$ is defined. Because $\nu(j)$ always increases (by Invariant 1) and masks that have been defined have numbers less than $\nu(j)$, before execution of this step $M_{\nu(j)}$ was not defined.

Since $C_i, \ldots, C_{j-1}$ computed in step 6 of the algorithm are indeed the values of the corresponding chaining variables, $S_j(x) = C_{j-1} \oplus M_{\nu(j)} = C_{j-1} \oplus C_{j-1} \oplus C = C$ as required. It completes our proof of correctness of the key reconstruction algorithm.

As the last step toward the proof of the theorem we have to show that the bit-vectors M output by the algorithm have the uniform distribution. Write down all the "decisions" (strings chosen at random) done during the execution of the algorithm (including its input C). It is a list of type C, $S_{j_1}(x)$, $M_{\nu(j_2)}$, $\ldots$, $M_{\nu(j_3)}$, $S_{j_4}(x)$, $M_{\nu(j_5)}$, $\ldots$, $M_{\nu(j_6)}, \ldots$ that contains exactly $L + 1$ m-bit string (because all strings are equally long, and every mask must be defined in steps 2, 5, or 7). Since the algorithm never stops without yielding a result, there is a mapping from the set of these strings into the set of possible outputs, which has the same cardinality. This mapping is an injection, because a preimage of an output value can be uniquely determined (it suffices to compute $S_j(x)$ for all corresponding j given M, k and x, which is trivial). Therefore, if the "decisions" are uniformly distributed, so are the outputs of the algorithm.   □

Shoup proved the security of the construction above. Our proof makes the key reconstruction algorithm more explicit and one-pass, thus giving a more efficient reduction, and requires less from the function ν (we do not need Fact 2 in [Sh00]). The operation performed by the key reconstruction algorithm in step 2 brings in Theorem 5. The mask $M_{\nu(i)}$ for $i = j - 2^{\nu(j)}$ is the only mask that appears an odd number of times between $M_{\nu(j)}$ and its previous appearance (if there exists one) at $M_{\nu(j-2\cdot 2^{\nu(j)})} = M_{\nu(j)}$.

We stress that our result of optimality of the Shoup scheduling does not rule out existence of a composite scheme with a shorter key. Even more important, our result implies that there must be at least $1 + \lfloor \log r \rfloor$ *different* masks, but says nothing about their *independence*. However, the proof of validity of the Shoup construction does need full independence of masks. There is an apparent gap between these two proofs.

We may try to reduce the key length by letting the masks be the output of a pseudo-random generator initialized with a short seed. Unfortunately, once the seed is exposed we cannot suppose anything about the output of the generator unless we resort to the random-oracle model. But in this model one could assume existence of CRHFs in the first place and our construction would be of no use in this world.

5   Conclusion and Open Problem

Recent attacks on MD4, MD5 and a flaw in the first version of SHA demonstrate that practical CRHFs are hard to construct. The oracle separation result due to [Si98] backed up this empirical fact by proving that CRHFs cannot be constructed from one-way permutations. Though UOWHFs, an alternative to CRHFs, have been known for years, their deployment in practical cryptosystems was hindered by lack of efficient composite schemes. While a family of CRHFs can be based on a single compression function, similar constructions for UOWHFs can only yield families of functions with variable key length. A variable key-length hash function stands out from all cryptographic primitives we use in practice and this annoying property can propagate to higher levels of construction (see [SS00] for an example).

We may approach this problem from two directions. First, it is possible that there exists a class of functions that are weaker than CRHFs, at least as strong as UOWHFs and for which an efficient composite scheme exists. We introduce a continuum of function classes that lie between CRHFs and UOWHFs and characterized by the degree of freedom the adversary has in choosing one of the colliding elements. From the complexity-theoretic point of view the hierarchy almost collapses to two large classes. The Merkle-Damgård construction, which yields fixed length-key families of functions, applies to one class of functions and not to the other.

Another approach is to improve existing composite schemes for UOWHFs. We take the Shoup construction, which is the most efficient (key length-wise), and prove that the scheme is optimal in respect to its mask scheduling. We also give a simplified proof of the Shoup construction.

An open problem is whether there exists a lower bound on the key length of a family of UOWHFs built via a black-box construction out of one-way functions. The upper bound given by numerous schemes from [BR97, Sh00] is $O(\log n)$, where n is the length of the input to a particular function. Such a lower bound would complement the line of research of [KST99, GT00] on *efficiency* of black-box constructions for UOWHFs.

Acknowledgement

The author is grateful to Victor Shoup for motivational discussions on this problem and Michael Waidner for his hospitality rendered through IBM Zurich. Additional thanks go to Dan Boneh and anonymous referees for their valuable comments.

References

[B82]   M. Blum, "Coin flipping by telephone," CRYPTO 81, pp. 11–15, 1981.

[BR97]   M. Bellare, P. Rogaway, "Collision-resistant hashing: towards making UOWHFs practical," Proc. of CRYPTO 97, pp. 470–484, Full version of this paper is available from http://www-cse.ucsd.edu/users/mihir/, 1997.

[D89]   I. Damgård, "A design principle for hash functions," Proc. of CRYPTO 89, pp. 416–427, 1989.

[GT00]   R. Gennaro, L. Trevisan, "Lower bounds on the efficiency of generic cryptographic constructions," Proc. of FOCS'00, pp. 305–313, 2000.

[HILL99] J. Hastad, R. Impagliazzo, L. Levin, M. Luby, "A pseudo-random generator from any one-way function," SIAM J. Computing, 28(4):1364–1396, 1999.

[KST99]   J.H. Kim, D. Simon, P. Tetali, "Limits on the efficiency of one-way permutation-based hash functions," Proc. of FOCS'99, pp. 535–542, 1999.

[LM92]   X. Lai, J. Massey, "Hash function based on block ciphers," Proc. of EUROCRYPT 92, pp. 55–70, 1992.

[M89]   R. Merkle, "One way hash functions and DES," Proc. of CRYPTO 89, pp. 428–446, 1989.

[N91]   M. Naor, "Bit commitment using pseudorandomness," J. Cryptology, 4(2): 151–158, 1991.

[NY89]   M. Naor, M. Yung, "Universal one-way hash functions and their cryptographic applications," Proc. of STOC'89, pp. 33–43, 1989.

[Ro90]   J. Rompel, "One-way functions are necessary and sufficient for secure signatures," Proc. of STOC'90, pp. 387–394, 1990.

[Ru95]   A. Russell, "Necessary and sufficient condtions for collision-free hashing," J. of Cryptology 8(2), pp. 87–100, 1995.

[SS00]   T. Schweinberger, V. Shoup, "ACE: The Advanced Cryptographic Engine," Manuscript. Available from `http://www.shoup.net`, 2000.

[Sh00]   V. Shoup, "A composite theorem for universal one-way hash functions," Proc. of EUROCRYPT 2000, pp. 445–452, 2000.

[Si98]   D. Simon, "Finding collisions on a one-way street: Can secure hash functions be based on general assumptions?" Proc. of EUROCRYPT 98, pp. 334-345, 1998.

[ZMI90]   Y. Zheng, T. Matsumoto, H. Imai, "Structural properties of one-way hash functions," Proc. of Crypto 90, pp. 285–302, 1990.

Key Recovery and Message Attacks on NTRU-Composite

Craig Gentry

DoCoMo Communications Laboratories USA, Inc.
181 Metro Dr., San Jose, CA 95110, USA
cgentry@dcl.docomo-usa.com

Abstract. NTRU is a fast public key cryptosystem presented in 1996 by Hoffstein, Pipher and Silverman of Brown University. It operates in the ring of polynomials $\mathbb{Z}[X]/(X^N - 1)$, where the domain parameter N largely determines the security of the system. Although N is typically chosen to be prime, Silverman proposes taking N to be a power of two to enable the use of Fast Fourier Transforms. We break this scheme for the specified parameters by reducing lattices of manageably small dimension to recover partial information about the private key. We then use this partial information to recover partial information about the message or to recover the private key in its entirety.

1 Introduction

NTRU is a fast public key cryptosystem that operates in the ring of truncated polynomials given by $\mathbb{Z}[X]/(X^N - 1)$, where the domain parameter N largely determines the security of the system. Typically N is chosen to be a prime number (not for security reasons, but because having N prime maximizes the probability that the private key has an inverse with respect to a specified modulus [11]). Recently, however, Silverman has proposed taking N to be a power of two to allow the use of Fast Fourier Transforms when computing the convolution product of elements in the ring [13].

In this paper, we present lattice-based attacks that are especially effective when N is composite. We show how to use low-dimensional lattices to find a folded version of the private key, where the folded private key has d coefficients with d dividing N. This folded private key can be used either to obtain a folding of the plaintext message, or as partial information to help us recover the entire private key. Using this attack, we were able to recover entire private keys for the NTRU-256 scheme proposed by Silverman in an average of about 3 minutes.

2 Notation

We denote the ring of integers by $\mathbb{Z}$, and the ring of integers modulo q by $\mathbb{Z}_q$, which are taken in the interval $(-\frac{q}{2}, \frac{q}{2}]$. The polynomial ring $\mathbb{Z}_q[X]/(X^n - 1)$ contains all polynomials with degree less than n and coefficients in $\mathbb{Z}_q$. The

B. Pfitzmann (Ed.): EUROCRYPT 2001, LNCS 2045, pp. 182–194, 2001.

inverse of a polynomial $f \in \mathbb{Z}_q[X]/(X^n - 1)$ is denoted by f_q^{-1}. A polynomial may be described as a row vector:

$$f = (f_0, f_1, \ldots, f_{n-1}) = \sum_{i=0}^{n-1} f_i X^i .$$

Concatenation of f and g is denoted by (f, g). The convolution $f * g$ of two vectors is analogous to ordinary polynomial multiplication over $\mathbb{Z}[X]/(X^n - 1)$:

$$(f * g)_k = \sum_{i+j=k \bmod n} f_i g_j .$$

When d divides n, the d-dimensional folded version of f is defined by:

$$f_{(d)} = (\sum_{\substack{0 \leq i < n \\ i=0 \bmod d}} f_i, \sum_{\substack{0 \leq i < n \\ i=1 \bmod d}} f_i, \ldots, \sum_{\substack{0 \leq i < n \\ i=d-1 \bmod d}} f_i) .$$

In algebraic terms, $f_{(d)}$ may be described as the image of f under the canonical mapping from $\mathbb{Z}[X]/(X^n - 1)$ to $\mathbb{Z}[X]/(X^d - 1)$. The ith term of $f_{(d)}$ will be denoted $f_{(d),i}$. The circulant matrix associated with f is given by F, where:

$$F_{ij} = f_{j-i \bmod n} .$$

F_i will denote the ith row vector of F. $F_{(d)}$ will denote the circulant associated with $f_{(d)}$, and $F_{(d),i}$ will denote its ith row vector. $I_{(d)}$ will refer to the d-dimensional identity matrix.

3   The NTRU Cryptosystem

Public Parameters. The basic objects of the NTRU Cryptosystem are polynomials from the ring $\mathbb{Z}[X]/(X^N - 1)$, where N is a public parameter. Also public are two moduli, p and q, with g.c.d.$(p, q) = 1$ and $p \ll q$. For example, $(N, p, q) = (167, 3, 128)$ has been proposed as a high security parameter set [7]. Additional public parameters include S_f, S_g, S_m, and S_ϕ, which describe the space of allowable polynomials for private keys f and g, the plaintext message m, and a random polynomial ϕ that the sender uses in encrypting the message. These spaces are designed to limit f, g, m, and ϕ to vectors that have short Euclidean length (in practice, less than $\sqrt{N}$) and that typically are also very short in the l_∞-norm — i.e., the magnitudes of the individual coefficients are typically very small in relation to q. For example, in NTRU-167, S_f might limit f to those polynomials having exactly 61 coefficients equal to 1, 60 coefficients equal to -1, and 46 coefficients equal to 0 [7]. S_m always restricts the coefficients of m to $\mathbb{Z}_p$.

Key Creation. Choose random $f \in S_f$ and $g \in S_g$. Compute f_q^{-1} and publish the polynomial

$$h = f_q^{-1} * g \pmod{q}$$

as the public key. Both f and g are private, with f serving as the private key.

Encryption. Choose random ϕ in S_ϕ and compute the ciphertext:

$$e = m + p\phi * h \pmod{q} .$$

Decryption. Compute

$$f * e = f * m + p\phi * f * h \pmod{q}$$
$$= f * m + p\phi * g \pmod{q} ,$$

where the second equality follows from the definition of h. Assuming S_f, S_m, S_ϕ, and S_g are chosen wisely, such that the coefficients of f, m, ϕ, and g are very small in relation to q, then, with high probability, we get

$$f * m + p\phi * g \pmod{q} = f * m + p\phi * g ,$$

which is to say that reduction modulo q has no effect. This is because the co-efficients of $f * m + p\phi * g$, with high probability, already lie in $(-\frac{q}{2}, \frac{q}{2}]$ before reduction modulo q. Possessing the unreduced value of $f * m + p\phi * g$, we can compute

$$f * m + p\phi * g \pmod{p} = f * m \pmod{p} ,$$

and then

$$f_p^{-1} * f * m \pmod{p} = m \pmod{p} .$$

4   Previous Lattice Attacks on NTRU

Lattice attacks on NTRU (including our attack) have focused primarily on the following "key recovery" problem: find the private key f using only the public key h and public information about how f and g are chosen (S_f and S_g).[1] By the definition of h, we know that $f * h = g \pmod{q}$, but this information alone is clearly insufficient to recover f. Indeed, the set of pairs $u, v \in \mathbb{Z}^N$ that satisfy $u * h = v \pmod{q}$ is an additive abelian group of infinite cardinality. Even if we limit ourselves to pairs $u, v \in \mathbb{Z}_q^N$, we are still left with q^N distinct (u, v) pairs corresponding to the q^N distinct values that u can assume. How do we find the pair (f, g) from among these q^N possibilities?

We know that, to enable error-free decoding, the coefficient vectors of f and g each have short Euclidean length (less than $\sqrt{N}$ in current NTRU implementations). They are considerably shorter than the typical "random" N-dimensional vector with coefficients in $\mathbb{Z}_q$, which has an expected length of more than $\frac{q}{4}\sqrt{N}$.

---

[1] Non-lattice-based cryptanalysis of NTRU includes a meet-in-the-middle attack found by Odlyzko [12] and a chosen-ciphertext attack presented by Jaulmes and Joux [5], which exploited NTRU's inappropriate use of OAEP-like padding. We understand that NTRU now uses the hybridization method presented by Fujisaki and Okamoto [2] to obviate chosen-ciphertext attacks.

Also, one can show that it is extremely unlikely that the abelian group generated by a "randomly" chosen h' has a (u, v) pair as short as (f, g).[2] These facts may lead us to hypothesize that (f, g) is, in fact, the shortest nonzero vector $(u, v) \in \mathbb{Z}^{2N}$ such that $u * h = v \pmod q$. If this hypothesis is true, and if we can find an efficient way to find the shortest vector belonging to the group of (u, v) pairs, we can recover the private key. This provides the motivation for representing the group of (u, v) pairs as a "lattice," and then using "lattice basis reduction."

A "lattice" is a discrete additive subgroup of $\mathbb{R}^n$. For example, $\mathbb{Z}^n$ is a lattice. Also, the set of (u, v) pairs is a lattice, being an additive subgroup of $\mathbb{Z}^{2N}$. An equivalent, but more concrete, definition is that a lattice L consists of all integer linear combinations of some set of m linearly independent vectors $B = \{b_0, b_1, \ldots, b_{m-1}\}$, $b_i \in \mathbb{R}^n$. Here, m is the "dimension" of L, and B is called a "basis" of L. The basis B can be compactly represented by an $m \times n$ matrix where the ith row is the "basis vector" b_i, in which case L consists of the vectors that can be expressed as integer linear combinations of the rows of B. Bases for a lattice are not unique, but are related by unimodular transformations — i.e., if U is an integral $m \times m$ matrix with determinant ± 1, then UB is an equally valid basis for L. Typically, the goal of "lattice basis reduction" is to find a basis B' for L in which the basis vectors are as short as possible (usually in the Euclidean sense), with the basis vector b_0' being the shortest nonzero vector in the entire lattice (allowing for possible ties).

Coppersmith and Shamir [1] give us the following explicit basis for the lattice of (u, v) pairs (recall that H denotes the circulant matrix corresponding to the public key h):

$$L_{CS} = \begin{bmatrix} I_{(N)} & H \\ 0 & qI_{(N)} \end{bmatrix}.$$

To see that any pair $(u, v) \in \mathbb{Z}^{2N}$ for which $u * h = v \pmod q$ is contained in the lattice generated by L_{CS}, let $a \in \mathbb{Z}^N$ be such that $u * h = v + qa$. Then, if we left-multiply L_{CS} by $(u, -a)$, we obtain (u, v). As a consequence, the private key pair (f, g) is an integer linear combination of the rows of L_{CS}. If (f, g) is actually the shortest vector in the generated lattice, as we have reason to believe,[3] then an "SVP-oracle" — a magical device which gives us the answer to the "shortest vector problem" *in a reasonable (polynomial) amount of time* — would give us the private key when given L_{CS} as input.

For the attacker, the problem is that actual lattice basis reduction algorithms, such as LLL and its variants, do not behave like SVP-oracles. The original LLL algorithm terminates in time polynomial in the dimension n of the lattice, but it is only guaranteed to find a vector that is no more than $2^{(n-1)/2}$ times — $2^{(2N-1)/2}$ times, in the case of L_{CS} — as long as the shortest vector. Obviously, such an algorithm is useless to us, considering that it is trivial to find vectors only about $\frac{q}{4}$ times as long as (f, g), as suggested above, and even these are far too long to be useful for decryption. Variants of LLL exist that find shorter

---

[2] See Appendix A.1.
[3] See Appendix A.1.

vectors, but they naturally have greater time-complexity. In particular, Schnorr defines a family of LLL-variants whose performances depend on a parameter called the "blocksize." Little is known about the average-case complexity of these variants, but it appears, based on numerous experiments by the authors of NTRU using Shoup's NTL library [8], that the time necessary to find (f, g) in the lattice grows at least exponentially in N (because the block size required for LLL to find (f, g) grows roughly linearly in N [3], and the running time of LLL is exponential in the block size [10]). The authors of NTRU estimate that, for $N > 90$, it takes current lattice reduction algorithms $e^{.2002N-7.608}$ seconds to find (f, g) on a 400 MHz machine, which translates into 4.607×10^{14} MIPS-years to break NTRU-263 [10].[4]

5   Cryptanalysis of NTRU-Composite

The problem with previous lattice-based attacks is that the dimension of the lattices involved is too high, given that the running time of LLL to return the target vector of these lattices is empirically exponential in the lattice dimension. Ideally, we would like to construct much smaller (and more easily reduceable) lattices whose shortest vectors contain at least some useful cryptanalytic information. We can do this if N is composite.

Theorem 1. *Let N be composite, and d be a nontrivial divisor. The mapping $\theta : \mathbb{Z}[X]/(X^N - 1) \to \mathbb{Z}[X]/(X^d - 1)$ given by*

$$\theta(f) = f_{(d)}$$

is a ring homomorphism.

Although this is a basic algebraic result, arising from the fact that $(X^d - 1)$ divides $(X^N - 1)$ when d divides N, we prove multiplication in a concrete fashion.

Proof.

$$
\begin{aligned}
g_{(d),k} &= \sum_{\substack{i=k \bmod d}}^{0 \leq i < N} g_i \\
&= \sum_{\substack{i=k \bmod d}}^{0 \leq i < N} \left(\sum_{\substack{x+y=i \bmod N}}^{0 \leq x,y < N} f_x h_y \right) \\
&= \sum_{\substack{x+y=k \bmod d}}^{0 \leq x,y < N} f_x h_y \\
&= \sum_{\substack{v+w=k \bmod d}}^{0 \leq v,w < d} \left(\left(\sum_{\substack{x=v \bmod d}}^{0 \leq x < N} f_x \right) \left(\sum_{\substack{y=w \bmod d}}^{0 \leq y < N} h_y \right) \right)
\end{aligned}
$$

[4] Refinements to L_{CS} by May [6] have made it possible to recover an NTRU-107 private key in 12 to 24 hours on a single 400 MHz machine [10], but do not seriously affect the security estimates for higher security levels, such as NTRU-167 [9].

$$= \sum_{\substack{v+w=k \bmod d}}^{0\le v,w<d} f_{(d),v}h_{(d),w} \ .$$

This gives us $f_{(d)} * h_{(d)} = g_{(d)}$, which is what we wanted. □

5.1   A Smaller Version of L_{CS}

With the equation $f_{(d)} * h_{(d)} = g_{(d)}$ in mind, we construct the following $2d$-dimensional analog of L_{CS} (recall that $H_{(d)}$ is the circulant corresponding to $h_{(d)}$):

$$L_{(d)} = \begin{bmatrix} I_{(d)} & H_{(d)} \\ 0 & qI_{(d)} \end{bmatrix} \ .$$

This lattice contains the vector $(f_{(d)}, g_{(d)})$. Notice that if N/d is not too large, then the smallness of the coefficients of f and g ensures that the coefficients of $f_{(d)}$ and $g_{(d)}$, each of which is a summation of N/d coefficients of f and g respectively, are also small. Assuming $(f_{(d)}, g_{(d)})$ is the shortest vector in $L_{(d)}$, we can find it using lattice reduction. We can then recover significant partial information about the private key by reducing a lattice whose dimension is only a fraction of the dimension of the lattice generated by L_{CS}.

In Appendix A.2, we give a tight upper bound on the length of $(f_{(d)}, g_{(d)})$ and show that, assuming f and g are "random" in a specified way, the expected length of $(f_{(d)}, g_{(d)})$ is equal to the length of (f, g) (once certain modifications are made to these vectors). This leads us to conclude that, at least when $d > \sqrt{N}$, $(f_{(d)}, g_{(d)})$ is almost certainly the shortest vector in $L_{(d)}$ for the same reasons that (f, g) is almost certainly the shortest vector in L_{CS}.

Remark: In the discussion above, we have limited our focus to homomorphisms of the form $\theta : \mathbb{Z}[X]/(X^N - 1) \to \mathbb{Z}[X]/(X^d - 1)$ and the folded lattices derived therefrom, but this need not be the case. More generally, we could consider homomorphisms of the form $\alpha : \mathbb{Z}_q[X]/(X^N - 1) \to \mathbb{Z}_q[X]/s(X)$ given by $\alpha(f) = f + < s(X), q >$, where $s(X)t(X) = (X^N - 1) \pmod{q}$ for some $t(X)$. However, such homomorphisms appear to be useful only when $(\alpha(f), \alpha(g))$ is a short vector that can be found using lattice basis reduction, and $(\alpha(f), \alpha(g))$ is always short only if $s(X)$ is an *extremely* short vector, preferably with a minimum of high degree coefficients (e.g., $(X^d - 1)$). Useful alternative homomorphisms therefore appear to be rare.

5.2   Message Attacks

Once we find $f_{(d)}$, we can make immediate use of it to recover the folded plaintext. Since folding is a ring homomorphism, we get:

$$f_{(d)} * e_{(d)} = f_{(d)} * m_{(d)} + p\phi_{(d)} * f_{(d)} * h_{(d)} = f_{(d)} * m_{(d)} + p\phi_{(d)} * g_{(d)} \pmod{q}.$$

We then proceed through the steps of decryption in the usual way until we obtain $m_{(d)}$. If $N/d = 2$, for example, knowing $m_{(d)}$ is tantamount to knowing

$m_i + m_{i+d}$ for $0 \leq i < d$, where the m_i are coefficients from the original plaintext. This could be useful information.

However, folding entails an increased likelihood of decryption errors, since the expected magnitudes of the coefficients of $f_{(d)} * m_{(d)} + p\phi_{(d)} * g_{(d)}$ are larger than those of $f * m + p\phi * g$ by a factor of $\sqrt{N/d}$. So, this message attack appears to be practical only for very small values of N/d.

5.3   Key Recovery Attacks

Alternatively, we can use $f_{(d)}$ to help us recover f. The basic concept behind this attack is the Chinese Remainder Theorem, which tells us, for example, that f is completely determined by the values of $f \pmod{X^d - 1}$ and $f \pmod{(X^N - 1)/(X^d - 1)}$.[5] Instead of using the lattice corresponding to $f \pmod{(X^N - 1)/(X^d - 1)}$, however, we use a different lattice with a shorter target vector.

Supposing, for example, that $N/d = 2$, we obtain linear equations of the form $f_{i+d} = f_{(d),i} - f_i$, so that we have

$$f = (f_0, f_1, \ldots, f_{d-1}, f_{(d),0} - f_0, f_{(d),1} - f_1, \ldots, f_{(d),d-1} - f_{d-1}).$$

Recall that in the lattice generated by L_{CS}, the target vector

$$(f, g) = \sum_{i=0}^{N-1} f_i(I_i, H_i) \pmod{q} \, ,$$

where (I_i, H_i) denotes the concatenation of the ith rows of the identity matrix and the circulant H. Using the dependencies in f, we obtain

$$(f, g) = \sum_{i=0}^{d-1} f_i(I_i, H_i) - \sum_{i=0}^{d-1} f_i(I_{i+d}, H_{i+d}) + \sum_{i=0}^{d-1} f_{(d),i}(I_{i+d}, H_{i+d}) \pmod{q}$$

$$= \sum_{i=0}^{d-1} f_i(I_i - I_{i+d}, H_i - H_{i+d}) + \sum_{i=0}^{d-1} f_{(d),i}(I_{i+d}, H_{i+d}) \pmod{q} \, .$$

Notice that we already know all of the terms in the second summation; let (s, t) be this known vector. If we denote by u the d-dimensional vector with coefficients equal to the first d coefficients of f, then (u, g) is in the following $(N + d + 1) \times (N + d)$ lattice:

$$L_{ug} = \begin{bmatrix} 0 & t \\ I_{(d)} & H_{(N),i} - H_{(N),i+d} \\ 0 & qI_{(N)} \end{bmatrix} ,$$

---

[5] More generally, f is determined by $\{f \pmod{s_1}, f \pmod{s_2}, \ldots, f \pmod{s_z}\}$, $0 \neq s_i \in \mathbb{Z}[X]$, when $(k(X))(X^N - 1) = L.C.M.(s_1, s_2, \ldots, s_z) \pmod{q}$ for some $k(X) \in \mathbb{Z}[X]$.

where $H_{(N),i} - H_{(N),i+d}$ is a $d \times N$ matrix formed by pairing rows, and the top third of the lattice consists only of a single row. Now, if we wish, we may discard the last d columns, obtaining a $(2d + 1) \times (2d)$ lattice with target vector (u, v), where v consists of the first d coefficients of g. Clearly, (u, v) is a short vector, most likely the shortest vector in this lattice. Once we obtain (u, v), this information can be combined with $(f_{(d)}, g_{(d)})$ to completely recover (f, g). We thereby obtain the private key without ever having to reduce a $2N$-dimensional lattice. When $N/d > 2$, we can decrease the dimension of L_{CS} by about $2d$ in a similar fashion. This $2d$ reduction in lattice dimension should reduce LLL's running time by a factor exponential in $2d$.

6   NTRU-256

Silverman [13] proposes choosing N to be a power of 2, because then convolution products can be computed rapidly using Fast Fourier Transforms. In particular, he suggests $(N, p, q) = (256, 2, 127)$ as an advantageous choice of parameters. We found that an NTRU-256 private key can be recovered in about 3 minutes using the folding technique described above.

In our experiments, we used a three-staged approach in recovering the private key. First, we recovered $(f_{(64)}, g_{(64)})$ by reducing the lattice generated by $L_{(64)}$ - a 128-dimensional matrix with $H_{(64)}$ in upper right quadrant. Second, we recovered $(f_{(128)}, g_{(128)})$ by reducing the 129×128 lattice constructed as described above. Finally, we took advantage of the modulo p structure of the private keys to create an over-defined system of linear equations. For example, upon computing that $f_{(128),i} = 0$, we know that $f_i = f_{i+128} = 0$, because $f_i \in [0, 1]$ for all i. Similarly, $f_{(128),i} = 2$ implies $f_i = f_{i+128} = 1$. When $p = 2$, this trick most likely results in more than half of g's coefficients being known, and less than half of f's coefficients being unknown, so that we may solve for the unknown coefficients in f. We thereby recover the entire private key (f, g) using lattices one-fourth the size of L_{CS}.

Since the S_f and S_g parameters were not specified for NTRU-(256,2,127), we used those for NTRU-(263,2,127) - specifically, f and g both have 35 1's, the rest 0's [13]. Using the NTL's implementation of LLL with a block size of 10 for both reductions, the 3 stages took an average of 40, 43 and 3 seconds, respectively.[6] Out of 20 trials, the correct key was recovered every time. We also tested the case when f has 75 1's and g has 65 1's, which is more challenging cryptanalytically, both in terms of the lattice reduction (since the target vector is longer) and the linear system (since there are more equations to solve). To avoid errors, the block size for the second reduction was increased to 12. The

---

[6] For these particular values of S_f and S_g, we could even have begun by recovering $(f_{(32)}, g_{(32)})$, which, heuristically, is recoverable even though N/d is somewhat large. After finding $(f_{(64)}, g_{(64)})$, it is probable that at least half of the coefficients of each of $f_{(64)}$ and $g_{(64)}$ are zero, and that, consequently, over half of the coefficients of each of f and g are known to be zero. If such is the case, we can proceed directly to the third stage without having to reduce a lattice larger than 65×64.

three stages took an average of 84, 94 and 12 seconds, respectively. Out of 10 trials, the correct key was recovered 9 times.

Some gimmicks were required to pick the target vector from among other short, but useless, vectors. For example, in the first reduction, we searched for rather than $(f_{(64)}^{\perp}, g_{(64)}^{\perp})$ rather than $(f_{(64)}, g_{(64)})$, where $f_{(64)}^{\perp}$ is the projection of f orthogonal to 1^N, the vector having every coefficient equal to 1. This technique, originated by Coppersmith and Shamir [1], prevents LLL from returning the short, but cryptanalytically useless vector $(1^N, 0^N)$. Also in the first reduction, LLL would often return other trivial vectors. For example, $(1^{\pm}, 1^{\pm})$ is a trivial vector that can arise in $L_{(64)}$, where $1^{\pm}$ is the vector consisting of alternating 1's and -1's. We simply skipped over these trivial vectors manually, continuing on to the next vector in the lattice until the desired nontrivial one was obtained. This process certainly could have been automated.

7   Remarks on NTRU-Prime

As we noted in the introduction, the domain parameter N is typically chosen to be prime not, apparently, for security reasons, but because it maximizes the probability that a randomly chosen private key f has inverses modulo q and p, these inverses being necessary for public key generation and decryption, respectively [11]. In terms of security, NTRU's typical use of a prime domain parameter appears to be merely fortuitous.

Folding does not work when N is prime — e.g., $f_{(d)} * h_{(d)}$ does not equal $g_{(d)}$ when d does not divide N. However, we can say that if $f * h = g$, and if h has a period of c in the sense that $h_i = h_{i+c}$ for $0 = i < N - c$, then the first N coefficients of $g' = (f, 0) * h'$ (where $(f, 0)$ is f followed by c zeros and h' is h followed by the c coefficients of h's period) are precisely the N coefficients of g. (Proof omitted.) For example, suppose $N = 263$ and h has a period of 37 — i.e., $h_0 = h_{37}, \ldots, h_{225} = h_{262}$. Then, we obtain $(f, 0)$ by appending 37 zeros to f and obtain the last c coefficients of h' by the relations $h_{263} = h_{226}, \ldots, h_{289} = h_{262}$. This will give us $(f, 0) * h' = (g, w)$, where the coefficients of w are not necessarily small. Since $(f, 0)$, h' and (g, w) have dimension $263 + 37 = 300$, a composite number, we can fold them to dimension, say, 100, obtaining the relation $(f, 0)_{(100)} * h'_{(100)} = (g, w)_{(100)}$. All of the coefficients of $(f, 0)_{(100)}$ will be small, and 100 - 37 = 63 of the coefficients of $(g, w)_{(100)}$ will be small, so we can construct a lattice of dimension 163 that may give us partial information about f and g. Although this approach works well in the rare case that h has a small period, it does not appear to lead to an attack against NTRU-prime that works in general.

Circulant lattices have a rather interesting property - namely, given an n-dimensional lattice generated by circulant matrix C, one can construct an $\lfloor \frac{n+1}{2} \rfloor$-dimensional lattice C_{half} that contains a C-vector no more than twice as long as the shortest vector in C. This is a consequence of the fact that for any vectors b and c, $\|b * c\| = \|b_{rev} * c\|$, where b_{rev} is the reverse of b given by $b_{rev,i} = b_{n-i}$. Now, if we suppose that the matrix C consists of cyclic rotations of the vector

c, and that the shortest vector in the lattice generated by C is obtained through left-multiplication by b — i.e., the shortest vector is $b * c$ — we get

$$\|(b + b_{rev}) * c\| \leq \|b * c\| + \|b_{rev} * c\| = 2\|b * c\|.$$

Since $(b + b_{rev})$ is a palindrome — i.e., $(b + b_{rev})_i = (b + b_{rev})_{n-i}$ for all i — $(b + b_{rev})$ has at most $\lfloor \frac{n+1}{2} \rfloor$ distinct coefficients, so that the rows of C can be paired together. This property of circulant lattices does not appear to allow one to cut the dimension of L_{CS} (a *block*-circulant lattice) in half, unless h is a palindrome.

8   Summary and Conclusion

We have shown that choosing N to be a composite number, especially one with a small factor, significantly reduces the security of the NTRU cryptosystem. Also, we have shown that it is possible to recover entire NTRU private keys using lattices of much smaller dimension than was previously thought. To avoid the presented attacks, N should be chosen to be prime, or to have only large nontrivial factors.

Acknowledgements

The author would like to thank Yiqun Lisa Yin, Satomi Okazaki and Atsushi Takeshita for their numerous helpful comments and suggestions.

References

1. D. Coppersmith and A. Shamir. Lattice Attacks on NTRU. In Proc. of Eurocrypt '97, volume 1233 of LNCS, pages 52-61. Springer-Verlag, 1997.
2. E. Fujisaki and T. Okamoto. Secure Integration of Asymmetric and Symmetric Encryption Schemes. In Proc. of Crypto '99, volume 1666 of LNCS, pages 537-554. Springer-Verlag, 1999.
3. J. Hoffstein, D. Lieman, J. Pipher, and J.H. Silverman. NTRU: A Public Key Cryptosystem. Submission to IEEE P1363 (1999). Available at http://www.manta.iccc.org/groups/1363/StudyGroup/NewFam.html.
4. J. Hoffstein, J. Pipher, and J.H. Silverman. NTRU: A Ring Based Public Key Cryptosystem. In Proc. of ANTS III, volume 1423 of LNCS, pages 267-288. Springer-Verlag, 1998. Available at http://www.ntru.com.
5. E. Jaulmes and A. Joux. A Chosen-Ciphertext Attack against NTRU. In Proc. of Crypto '00, volume 1880 of LNCS, pages 20-35. Springer-Verlag, 2000.
6. A. May. Cryptanalysis of NTRU. Preprint, February 1999. Available at http://www.informatik.uni-frankfurt.de/~alex/crypto.html.
7. NTRU Cryptosystems. The NTRU Public Key Cryptosystem. Available at http://www.ntru.com/technology/tutorials/pkcstutorial.htm.
8. V. Shoup. Number Theory C++ Library (NTL) version 3.9. Available at http://www.shoup.net/ntl.

9. J.H. Silverman. Dimension-Reduced Lattices, Zero-Forced Lattices, and the NTRU Public Key Cryptosystem. NTRU Cryptosystems Technical Report No.13 (1999). Available at http://www.ntru.com.
10. J.H. Silverman. Estimated Breaking Times for NTRU Lattices. NTRU Cryptosystems Technical Report No.12 (1999). Available at http://www.ntru.com.
11. J.H. Silverman. Invertibility in Truncated Polynomial Rings. NTRU Cryptosystems Technical Report No.9 (1999). Available at http://www.ntru.com.
12. See J.H. Silverman. A Meet-in-the Middle Attack on an NTRU Private Key. NTRU Cryptosystems Technical Report No.4 (1997). Available at http://www.ntru.com.
13. J.H. Silverman. Wraps, Gaps, and Lattice Constants. NTRU Cryptosystems Technical Report No.11 (1999). Available at http://www.ntru.com.

A   Appendix

A.1   Probability of Very Short (u, v) Pair

Assume we choose h' from $\mathbb{Z}_q^N$ with uniform distribution; what is the expected length of the shortest nonzero pair (f', g'), $f', g' \in \mathbb{Z}_q^N$, that satisfies $f' * h' = g'$ (mod q)? We begin with the (admittedly heuristic) observation that the choices for h' essentially partition the set of (u, v) pairs according to whether $u * h' = v$ (mod q). In other words, it is rare that the set of (u, v) pairs for h'_1 and those for h'_2 overlap such that the equalities $u' * h'_1 = v'$ (mod q) and $u' * h'_2 = v'$ (mod q) are simultaneously satisfied for some (u', v'). This notion could be made more precise, but we will make do with the heuristic observation.

Assuming that the (u, v) pairs are, in fact, partitioned among the choices for h', the probability that a randomly chosen h' has a vector of length less than R is less than or equal to $V(R)/q^N$, where $V(R)$ is the volume of a $2N$-dimensional ball of radius R, and q^N is the number of (u, v) pairs that belong to h'. Using Stirling's Formula for the volume of an n-dimensional ball, we find that the probability that h' has a (u, v) pair shorter than $\sqrt{Nq/2\pi e}$ is negligibly small.

Since the probability of a random h' having a (u, v) pair as short as (f, g) is extremely small, we have some basis for concluding that it is very unlikely that h has a (u, v) pair unrelated to (f, g) that is as small as (f, g). This conclusion comes with caveats, the most important being that the trivial vector $(1^N, 0^N)$, where 1^N is the vector having all N elements equal to 1, is typically a (u, v) pair for h, and may very well be shorter than (f, g). This is not a serious problem, because, as shown in Appendix A.2, the group of (u, v) pairs can be slightly modified to exclude this trivial vector.

A.2   Length of $f_{(d)}$

We can establish an upper bound on the Euclidean norm of $(f_{(d)}, g_{(d)})$ as follows:

Theorem 2. $\|(f_{(d)}, g_{(d)})\| \le \sqrt{N/d}\|(f, g)\|$.

Proof. Let $b = (1 + X^d + \cdots + X^{(N/d-1)d})$, $v_f = f * b$ and $v_g = g * b$. Since $\|(f * X^i, g * X^i)\| = \|(f,g)\|$ for all i, we have

$$\|(v_f, v_g)\| \leq N/d\|(f,g)\|$$

by the triangle equality, with equality holding when $f = f * X^d = \cdots = f * X^{(N/d-1)d}$ and $g = g * X^d = \cdots = g * X^{(N/d-1)d}$. Notice that the ith coefficient of v_f is equal to $f_{(d),i \bmod d}$. In other words, the coefficients of v_f are precisely the coefficients of $f_{(d)}$, repeated N/d times. The same goes for v_g. Thus,

$$\|(v_f, v_g)\| = \sqrt{N/d}\|(f_{(d)}, g_{(d)})\| \ ,$$

from which the desired inequality follows. $\square$

For the expected length of $(f_{(d)}, g_{(d)})$, recall that we have $\|(f_{(d)}, g_{(d)})\| = \sqrt{N/d}\|(f,g)\|$ only when

$$f = f * X^d = \cdots = f * X^{(N/d-1)d}, g = g * X^d = \cdots = g * X^{(N/d-1)d}$$

— i.e., when the coefficients of f and g have a period of d. Of course, parameters S_f and S_g can be chosen to require f and g to be periodic, or nearly periodic, but this would reduce the keyspace and invite other attacks.

We can use ideas of Coppersmith and Shamir to obtain a better approximation of the length of the target vector of $L_{(d)}$ when f and g behave like random vectors. Let $f^\perp$ denote the projection of f orthogonal to 1^N, the vector in which all N elements are equal to 1. We find that $v_f^\perp$ — i.e., the projection of v_f orthogonal to 1^N — is equal to $f^\perp * b$. Then, following Coppersmith and Shamir, we get:

$$\|v_f^\perp\|^2 = \sum_k (f^\perp * b)_k^2$$

$$= \left(\sum_i (f_i^\perp)^2\right)\left(\sum_l b_l^2\right) + \sum_{j \neq 0}\left(\sum_i f_i^\perp f_{i+j}^\perp\right)\left(\sum_l b_l b_{l+j}\right)$$

$$= (N/d)\|f^\perp\|^2 + \sum_{j \neq 0}\left(\sum_i f_i^\perp f_{i+j}^\perp\right)\left(\sum_l b_l b_{l+j}\right) \ .$$

Since b_l is nonzero only when $l = 0 \pmod d$, each term $b_l b_{l+j}$ must be zero, and thus the entire rightmost summation must be zero, unless $j = 0 \pmod d$. When $j = 0 \pmod d$, the rightmost summation is equal to N/d. Thus, we obtain:

$$\|v_f^\perp\|^2 = (N/d)\|f^\perp\|^2 + \sum_{j \neq 0, j = 0 \bmod d} (N/d)\left(\sum_i f_i^\perp f_{i+j}^\perp\right) \ .$$

If f behaves like a random vector, then, for each j, we would expect $\sum_i f_i^\perp f_{i+j}^\perp$ to be less than $\sum_i f_i^\perp f_i^\perp = \|f^\perp\|^2$ by a factor of about $1/\sqrt{N}$. Since the terms have random sign, we would also expect some cancellation to occur. Thus, if N/d is not too large (there are $N/d - 1$ terms in the first summation), such as when $d > \sqrt{N}$, then we can expect that

$$\|(v_f^\perp, v_g^\perp)\| \approx \sqrt{N/d}\|(f^\perp, g^\perp)\| .$$

Since, as before, the coefficients of $v_f^\perp$ are precisely the coefficients of $f_{(d)}^\perp$ repeated N/d times, where $f_{(d)}^\perp$ denotes the projection of $f_{(d)}$ orthogonal to 1^d, we get:

$$\|(f_{(d)}^\perp, g_{(d)}^\perp)\| \approx \|(f^\perp, g^\perp)\| .$$

Coppersmith and Shamir have shown that $(f^\perp, g^\perp)$ is the optimal target vector for L_{CS} (ignoring their additional "balancing constant" refinement"). Similarly, $(f_{(d)}^\perp, g_{(d)}^\perp)$ is the optimal target vector for $L_{(d)}$. Thus, when N/d is not too large, we can expect the target vector of $L_{(d)}$ to be about the same length as the target vector of L_{CS}.

Applying the techniques used in Appendix A.1, we find that that while $\|(f_{(d)}^\perp, g_{(d)}^\perp)\| \approx \|(f^\perp, g^\perp)\|$, the expected length of the shortest vector in $L_{(d)}$ is less than that of L_{CS} by a factor of $\sqrt{N/d}$. One might think that this tightening of the ratio between the expected length of the shortest vector to the length of the target vector would make the target vector more difficult for LLL to find, but, empirically, the small reduction in this ratio does not even come close to offsetting the exponential reduction in running times obtained by decreasing the lattice dimension.

Evidence that XTR Is More Secure than Supersingular Elliptic Curve Cryptosystems

Eric R. Verheul

PricewaterhouseCoopers, GRMS Crypto group,
P.O. Box 85096, 3508 AB Utrecht, The Netherlands
eric.verheul@[nl.pwcglobal.com, pobox.com]

Abstract. We show that finding an efficiently computable injective homomorphism from the XTR subgroup into the group of points over $GF(p^2)$ of a particular type of supersingular elliptic curve is at least as hard as solving the Diffie-Hellman problem in the XTR subgroup. This provides strong evidence for a negative answer to the question posed by S. Vanstone and A. Menezes at the Crypto 2000 Rump Session on the possibility of efficiently inverting the MOV embedding into the XTR subgroup. As a side result we show that the Decision Diffie-Hellman problem in the group of points on this type of supersingular elliptic curves is efficiently computable, which provides an example of a group where the Decision Diffie-Hellman problem is simple, while the Diffie-Hellman and discrete logarithm problem are presumably not. The cryptanalytical tools we use also lead to cryptographic applications of independent interest. These applications are an improvement of Joux's one round protocol for tripartite Diffie-Hellman key exchange and a non refutable digital signature scheme that supports escrowable encryption. We also discuss the applicability of our methods to general elliptic curves defined over finite fields.

1   Introduction

XTR is an efficient and compact method to work with order $p^2 - p + 1$ subgroups of the multiplicative group $GF(p^6)^*$ of the finite field $GF(p^6)$. It was introduced in [10], followed by several practical improvements in [11] and [12].

Throughout this paper we let $p, q > 3$ denote prime numbers. In the context of XTR we further demand that $p \equiv 2 \bmod 3$ and that q divides $p^2 - p + 1$. Let g be a generator of the order q subgroup μ_q of $GF(p^6)^*$. In [10] it is shown that elements of μ_q, the *XTR subgroup*, can conveniently be represented by their so-called trace over $GF(p^2)$, and it is shown in [10] how this representation can efficiently be computed. Any familiar cryptosystem based on the XTR subgroup (like Diffie-Hellman, ElGamal, DSA) can be easily transformed using this representation, yielding both efficient and compact cryptosystems. Moreover, it is shown in [10] that the security of these transformed systems is equivalent to the ones started with, that is, the security of the discrete logarithm problem in the multiplicative group of the finite field $GF(p^6)^*$. We refer to the group of order $p^2 - p + 1$ of

B. Pfitzmann (Ed.): EUROCRYPT 2001, LNCS 2045, pp. 195–210, 2001.

$\mathrm{GF}(p^6)^*$ as the *XTR supergroup*. It is widely believed that the Diffie-Hellman and discrete logarithm problem in these XTR groups is hard.

At the Crypto 2000 Rump Session, [16], the following comparison was presented, suggesting that XTR is nothing else than an elliptic curve cryptosystem in disguise. As is well known, the number of points over $\mathrm{GF}(p^2)$ (including the point at infinity) on an elliptic curve defined over $\mathrm{GF}(p^2)$ takes the form $p^2 - t + 1$ for some integer called the Frobenius trace number $t \in [-2p, 2p]$. There exist elliptic curves over $\mathrm{GF}(p^2)$ of such order equal to $p^2 - p + 1$. These curves are actually characterized in [14] as **C**lass **T**hree supersingular elliptic curves over $\mathrm{GF}(p^2)$ with **P**ositive parameter t, namely $t = p$ (as opposed to $t = -p$). This is why we call these curves simply the *CTP curves* for short. Moreover, there exist efficiently computable (i.e., in polynomial time and space in length of input), injective homomorphisms of such curves onto the XTR supergroup. The Menezes-Okamoto-Vanstone (MOV) imbedding [15], provides an example of such a homomorphism.

It seems like a plausible hypothesis (cf. [16]) that the inverses of such homomorphisms might be efficiently computable too. Under this hypothesis the XTR (sub)group is just an instance of an elliptic curve (sub)group and so an attack affecting the security of elliptic curve cryptosystems would affect the security of the XTR cryptosystem. Or in other words, under this hypothesis the security of XTR cryptosystems is not better than that of elliptic curve cryptosystems.

In this paper we show that the hypothesis mentioned above is unlikely to be correct, as we show that under this hypothesis, we can solve several problems that are widely believed to be hard. The Diffie-Hellman problem in the XTR subgroup is an example of such a problem. As a side result we show that the Decision Diffie-Hellman problem in many supersingular elliptic curves is efficiently computable. The results presented in this paper are specifically geared towards XTR, to counter the suggestion that XTR is nothing else than an elliptic curve cryptosystem in disguise. We did not attempt to fully generalize them to other classes of (supersingular) elliptic curves, although we expect they can be (cf. Section 4). The results in this paper should therefore be interpreted in a broader context. Namely, they provide evidence that the multiplicative group of a finite field provides essentially more, and in any case not less, security than the group of points of a supersingular elliptic curve of comparable size.

The CTP curves take the form $y^2 = x^3 + a$ where $a \in \mathrm{GF}(p^2)$ is a square but not a cube in $\mathrm{GF}(p^2)$, cf. [8]. We denote the CTP curves by C_a. Actually, in the category of elliptic curves over $\mathrm{GF}(p^2)$ only two such curves exist; all others are isomorphic under an efficiently computable isomorphism. Compare Lemma 1. The set of points over $\mathrm{GF}(p^2)$ (including the point at infinity) on C_a is denoted by C_{a,p^2} and the subgroup thereof of order l is denoted by $C_{a,p^2}[l]$. It is important to consider the elliptic curve $y^2 = x^3 + a$ over the extension field $\mathrm{GF}(p^6)$ as well, respectively subgroups of order l therein. These are denoted by respectively C_{a,p^6} and $C_{a,p^6}[l]$. For further reference, we formulate the hypothesis mentioned above as follows:

X2C There exists an efficiently computable element $s \in \mathrm{GF}(p^2)$ and an efficiently computable, injective group homomorphism from the XTR subgroup into $C_{s,p^2}[q]$.

A similar problem is posed by N. Koblitz in [9, p.328]. Note that **X2C** is more general than only assuming that (a restriction of) an MOV embedding is efficiently invertible. It actually follows from our results (Theorem 10) that under the **X2C** hypothesis, (restrictions of) MOV embeddings are efficiently invertible.

Outline of the paper
In Section 2 we explore the structure of CTP curves. We introduce a so-called distortion map on CTP curves which is of crucial importance for our results, and we prove a more convenient formulation of the **X2C** hypothesis. In Section 3 we present and prove our main results and in Section 4 we briefly discuss some possible extensions of our results. In Section 5 we discuss some practical applications of distortion maps, including a more computational and communicational efficient variant of the one round protocol for tripartite Diffie-Hellman key exchange described in [5] and a non refutable digital signature scheme that supports escrowable encryption. Finally, we summarize our results in Section 6.

2   Group Isomorphisms between CTP Curves

We recall that any isomorphism between two elliptic curves defined over a field K induces a group isomorphism between the points on the elliptic curves over K, but not vice versa. See [14], [18]. This distinction is important in the following lemma.

Lemma 1 *Let C_a and C_b be CTP curves (in particular, a, b are squares in $\mathrm{GF}(p^2)$ but not cubes), then the following hold:*

1. *The map $S : C_{a,p^2} \to C_{a^p,p^2} : (x,y) \to (x^p, y^p)$ is an efficiently computable group isomorphism.*
2. *The equation $u^6 = b/a$ has its solutions in $\mathrm{GF}(p^6)$ and for any such solution u, the map $R_u : C_a \to C_b : (x,y) \to (u^2x, u^3y)$ is an isomorphism in the category of elliptic curves over $\mathrm{GF}(p^6)$ and induces in particular an efficiently computable group isomorphism $C_{a,p^6} \to C_{b,p^6}$.*
3. *The map R_u is an isomorphism in the category of elliptic curves over $\mathrm{GF}(p^2)$ iff b/a is a cube in $\mathrm{GF}(p^2)$.*
4. *If b/a is not a cube in $\mathrm{GF}(p^2)$, then b/a^p is a cube in $\mathrm{GF}(p^2)$. Also the equation $w^6 = b/a^p$ has its solutions w in $\mathrm{GF}(p^2)$ and the composite map $R_w \circ S$ is an efficiently computable group isomorphism from C_{a,p^2} to C_{b,p^2}.*

Proof: The first part of the lemma is well known and easily verified. That the equation mentioned in the second part of the lemma has a solution in $\mathrm{GF}(p^6)$ follows as b/a is a square in $\mathrm{GF}(p^2)$. The remainder of the second part of the lemma follows for instance from [14, Theorem 2.2]. The third part also follows from this result combined with the observation that $u^6 = b/a$ has all its solutions u in $GF(p^2)$ iff b/a is a cube in $GF(p^2)$. For a proof of the fourth part, let α be a generator of the multiplicative group of $GF(p^2)$. As $p > 3$ it follows that $p^2 - 1 \equiv 0 \bmod 3$, so the element $x = \alpha^j$ is a cube in $GF(p^2)^*$ iff j is divisible by three. Now write $a = \alpha^k$ and $b = \alpha^l$. If b/a is not a cube in $GF(p^2)$, then $k \bmod 3$ and $l \bmod 3$ are different. As $k, l \bmod 3$ are non-zero, it follows from $p \equiv 2 \bmod 3$ that $k \cdot p \bmod 3$ and $l \bmod 3$ are equal. That is, b/a^p is a cube in $GF(p^2)$. The remainder of the proof of the fourth part of the lemma now follows from the first and third part.   □

From Lemma 1 it follows that the CTP curves split into two equivalence classes under the equivalence relation $C_a \simeq C_b$ iff b/a is a third power in $GF(p^2)$. From [14, Theorem 3.2] it follows that there are exactly two isomorphism classes of supersingular elliptic curves over $GF(p^2)$ of order $p^2 - p + 1$. We conclude that the CTP curves provide a complete representation of such curves.

From the previous result we immediately deduce the following.

Theorem 2 *All CTP groups C_{a,p^2} are efficiently computable group isomorphic. Moreover, we can reformulate* **X2C** *as:*

X2C *For each CTP subgroup $C_{a,p^2}[q]$ there exists an efficiently computable, injective homomorphism from the XTR subgroup into $C_{a,p^2}[q]$.*

Let C_a be a CTP curve. We recall some facts on elliptic curves which can all be found in [14]. For a divisor l of $p^2 - p + 1$, the l-th *torsion group* of C_a is the collection of all points of order dividing l on the curve $y^2 = x^3 + a$ over the algebraic closure of the field $GF(p^2)$. The torsion group is isomorphic to $\mathbf{Z}_l \oplus \mathbf{Z}_l$, which is a non-cyclic, abelian group. In addition, as C_a is a so-called Class III supersingular curve, the l-th torsion group of C_a is just the collection of all points of order dividing l over $GF(p^6)$ (including the point at infinity) on the curve $y^2 = x^3 + a$. That is, the l-th torsion group of C_a is equal to $C_{a,p^6}[l]$ and is hence a subset of $GF(p^6) \times GF(p^6)$.

Before formulating the theorem that is crucial to our results, we need a definition.

Definition 3 *Let H be an abelian group, then two elements g_1, g_2 are called independent, provided that $g_1 \notin \langle g_2 \rangle$ and $g_2 \notin \langle g_1 \rangle$.*

This definition becomes relevant when the group H is not cyclic itself, which is typically the situation in torsion groups. Before coming to our next result we remark that it is easily verified that the two points in C_{a,p^2} that have a zero first coordinate, augmented with the point at infinity, that is $\{(0, w), (0, -w), \mathcal{O}\}$ with $w^2 = a$, constitutes a subgroup of order three. We denote this group by G_3.

Theorem 4 *Let C_a be a CTP curve and let $P \neq \mathcal{O}$ be a point on C_{a,p^2}. Then, using the notation from Lemma 1, the following hold:*

1. *The equation $u^6 = a/a^p$ has its solutions u in $GF(p^6) \setminus GF(p^2)$ and for any such solution u, the map $D : C_{a,p^6} \overset{S}{\to} C_{a^p,p^6} \overset{R_u}{\to} C_{a,p^6}$ is a group automorphism which takes the form $(x, y) \to (u^2 x^p, u^3 y^p)$.*

2. $\langle P \rangle \cap \langle D(P) \rangle = \mathcal{O}$ if the order of P is not divisible by 3 and $\langle P \rangle \cap \langle D(P) \rangle = G_3$ otherwise.
3. The point P is independent from its image under $D(.)$ iff P has an order different from 1 or 3.

Proof: For a proof of the first part of the theorem, it easily follows (cf. the proof of Lemma 1) that a/a^p is not a cube in $\mathrm{GF}(p^2)$. Now the proof follows from the last part of Lemma 1. For a proof of the second part of the theorem: the first coordinate of the value $(u^2 x^p, u^3 y^p)$ under $D(.)$ of a point $Q = (x, y)$ is clearly not an element of $\mathrm{GF}(p^2)$ when x is non-zero. That is, apart from the point at infinity, the only points that can belong to $\langle P \rangle \cap \langle D(P) \rangle$ have a zero first coordinate. As $\langle P \rangle \cap \langle D(P) \rangle$ is a group it is either equal to $\{\mathcal{O}\}$ or G_3. In the latter case it follows that the order of P must be divisible by 3. For a proof of the last part, as $D(.)$ is a group automorphism, the orders of P and $D(P)$ coincide. So if these points are dependent it follows from the second part that either P or $D(P)$ is an element of G_3, i.e., of order 1 or 3.   □

For convenience we refer to the map $D(.)$ introduced in Theorem 4 as the *distortion* map. In Figure 1 a few pages below we have depicted the property of $D(.)$ with $K = \mathrm{GF}(p^2)$ and $K' = \mathrm{GF}(p^6)$. Related to the l-th torsion group of C_a, i.e., $C_{a,p^6}[l]$, is the Weil pairing, a function

$$e_l : C_{a,p^6}[l] \times C_{a,p^6}[l] \to \mu_l,$$

where μ_l is the subgroup of $\mathrm{GF}(p^6)^*$ of order l. Hence, μ_q is equal to the XTR subgroup. In the setting of supersingular curves, the Weil pairing can be computed efficiently. The Weil pairing satisfies the Identity rule, i.e., $e_l(P, P) = 1$, and is bilinear. From the latter property it follows that $e_l(a*P, b*Q) = e_l(P, Q)^{ab}$. This formula is particularly useful when $e_l(P, Q)$ is a generator of μ_l, as the map $< P > \mapsto \mu_l : x \to e_l(x, Q)$ is then a group isomorphism. Actually, this is the MOV embedding mentioned in the introduction. We finally mention that two points P, Q in the torsion group $C_{a,p^6}[l]$ are dependent, iff $e_l(P, Q) = 1$, see [14, p.70].

The following corollary describes the order of a value of the Weil pairing.

Corollary 5 *Let l dividing $p^2 - p + 1$ be a power of a prime number r and let P be a point on C_{a,p^2} of order l. Then, letting $D(.)$ denote the distortion map from Theorem 4, the following hold:*

1. *If $r \neq 3$, then the element $e_l(P, D(P))$ is of order l in $\mathrm{GF}(p^6)^*$.*
2. *If $r = 3$, then the element $e_l(P, D(P))$ is of order at least $l/3$ in $\mathrm{GF}(p^6)^*$.*

Proof: First note that the point $D(P)$ is of order l as $D(.)$ is a group automorphism. For a proof of the first statement, suppose to the contrary that we have $e_l(P, D(P))^{l/r} = 1$. Then it follows that $e_l(P, l/r \cdot D(P)) = 1$, that is, P and $l/r \cdot D(P)$ are dependent. Hence either, $P \in \langle l/r \cdot D(P) \rangle$ or $l/r \cdot D(P) \in \langle P \rangle$. The first option is ruled out as it implies that the order of P is divisible by l/r. So,

$$l/r \cdot D(P) \in \langle P \rangle \cap \langle D(P) \rangle = \{\mathcal{O}\},$$

where the last equality follows from Theorem 4. That is, $l/r \cdot D(P) = \mathcal{O}$ contradicting that the order of $D(P)$ is equal to l. For a proof of the second statement, we may assume without loss of generality that $l \geq 3^2$. If we assume to the contrary that $e_l(P, D(P))^{l/9} = 1$ and reasoning in a similar way as in the proof of the first part, we conclude that

$$l/9 \cdot D(P) \in \langle P \rangle \cap \langle D(P) \rangle = G_3,$$

where the last equality follows from Theorem 4. This contradicts that the order of $l/9 \cdot D(P)$ is nine. $\square$

3   Hardness of the X2C Hypothesis

Before coming to our main results, we recall some general notions. Let $G = \langle \gamma \rangle$ be any cyclic, multiplicative group of order l, generated by an element γ. The security of the Diffie-Hellman key agreement protocol with respect to γ lies in the *Diffie-Hellman problem* of computing the values of the function $DH(\gamma^x, \gamma^y) = \gamma^{xy}$. Two other problems are related to the DH problem. The first one is the *Decision Diffie-Hellman* (DDH) problem with respect to γ: given $\alpha, \beta, \delta \in G$ decide whether $\delta = DH(\alpha, \beta)$ or not. The DH problem is at least as difficult as the DDH problem. The second related problem is the *discrete logarithm* (DL) problem in G with respect to γ: given $\alpha = \gamma^x \in G$, with $0 \leq x < l$ then find $x = DL(\alpha)$. The DL problem is at least as difficult as the DH problem. It is widely assumed that if the DL problem G is hard, then so are the other two. In [5], Joux notes that Decision Diffie-Hellman type of problems in extensions of supersingular elliptic curves are often efficiently computable. We use Joux's reasoning in the proof of the next result, which in particular provides an example of a supersingular elliptic curve where the Decision Diffie-Hellman problem is efficiently computable, while the discrete logarithm problem is presumably hard.

Theorem 6 *The Decision Diffie-Hellman problem in any supersingular elliptic curve over* $\mathrm{GF}(p^2)$ *of order* $p^2 - p + 1$ *is efficiently computable.*

   Proof: We can restrict ourselves to curves of type C_a. Write $p^2 - p + 1 = t \cdot v$ where t is a power of three and v is relatively prime with three. By virtue of the Pohlig-Hellman algorithm [17], the DDH problem in C_{a,p^2} can be reduced to the DDH problem in the subgroups of order t and v. As one can easily solve the discrete logarithm related to the first subgroup, one can efficiently the Decision Diffie-Hellman problem for this subgroup too.
   Now, let P be a generator of the subgroup $C_{a,p^2}[v]$ and suppose that points $X = x * P, Y = y * P, Z = z * P$ in $C_{a,p^2}[v]$ are given. To solve the Decision Diffie-Hellman problem in $C_{a,p^2}[v]$, we need to determine whether $z = x * y \bmod v$. By the Identity property of the Weil pairing, its bilinearity and Corollary 5, the Weil pairing $e_v(P, D(P))$ is a v-th root of unity of $\mathrm{GF}(p^6)$. So on the one hand, $e_v(X, D(Y)) = e_v(P, D(P))^{xy}$ and on the other hand $e_v(P, D(Z)) = e_v(P, D(P))^z$. That is $z = x * y \bmod v$ iff $e_v(X, D(Y))$ is equal to $e_v(P, D(Z))$, which is an efficiently computable condition. $\square$

There are several cryptographic protocols whose security depends on the difficulty of the Decision Diffie-Hellman problem, like the publicly verifiable voting system in [2] and the Cramer-Shoup [3] public key cryptosystem that is provable secure against adaptive chosen ciphertext attacks. Theorem 6 shows that these protocols should not be based on (CTP) supersingular elliptic curves, even with the "appropriate" key sizes. We now obtain our first evidence that the **X2C** hypothesis is not valid.

Corollary 7 *Under the* **X2C** *hypothesis, the Decision Diffie-Hellman problem in the XTR subgroup is efficiently computable.*

Proof: This follows immediately from Theorem 6.   □

Next we show an even stronger consequence of the **X2C** hypothesis, namely that the Diffie-Hellman problem in the XTR subgroup is efficiently computable. It is convenient to first introduce three variants of the Diffie-Hellman problem. To this end, again let $G = \langle \gamma \rangle$ be any cyclic, multiplicative group of (known) order l, generated by the (known) element γ. Then the *weak DH problem* with respect to γ is the problem of finding any generator κ, such that for all $0 \leq x, y < l$ determining κ^{xy} can be efficiently done on basis of γ^x and γ^y. That is, κ is only dependent of γ and not of x, y. The *strong DH problem* with respect to γ is the problem of efficiently determining ξ^{xy} on basis of γ^x and γ^y, for all $0 \leq x, y < l$ and *any* generator ξ of G. Finally, the *DH problem with respect to the group* G is the problem of efficiently determining ξ^{xy} on basis of α^x and α^y for all $0 \leq x, y < l$ and *any* generators ξ, α of G. Note that this notion is independent of the choice of a particular generator γ of G.

Lemma 8 *In the setting above, the weak, conventional and strong Diffie-Hellman problem w.r.t. γ and the Diffie-Hellman problem w.r.t. G are equivalent.*

Proof: We first show equivalence of the first three problems. Clearly, if one can solve the strong Diffie-Hellman problem, one can solve the conventional Diffie-Hellman problem. Moreover, if one can solve the conventional Diffie-Hellman problem then by taking $\kappa = \gamma$ one can solve the weak Diffie-Hellman problem. To show that these three problems are equivalent, it suffices to show that if one can solve the weak Diffie-Hellman problem, one can solve the strong Diffie-Hellman problem. To this end, let γ, κ be as described in the definition of weak Diffie-Hellman problem and let ξ be any generator of G. Also, let the function $WDH(.,.)$ be defined by $\kappa^{xy} = WDH(\gamma^x, \gamma^y)$. Then by hypothesis $WDH(.,.)$ is efficiently computable. We only prove the lemma in the case that l is a prime number which is important to us and leave the general case to the reader.

We can write $\kappa = \gamma^s$ and $\xi = \gamma^t$ for some $0 \leq s, t < l$, which are unknown. We first claim that we can efficiently compute $\gamma^{(s^n)}$ for any $n \geq 1$. To this end, for any $i \geq 1$ define

$$T(i) = (\gamma^{(s^{i-1})}, \gamma^{(s^i)}).$$

Note that $T(1) = (\gamma, \kappa)$ is efficiently computable. Also note that if $T(i) = (A, B)$ is given, then $T(2i)$ is equal to $(WDH(A, A), WDH(A, B))$ and $T(2i + 1)$ is equal to $(WDH(A, B), WDH(B, B))$. This means that we can compute $T(n)$ in $2 \cdot \log_2(n)$ calls to the function $WDH(., .)$ using repeated squaring and multiplication (cf. [10, Algorithm 2.3.7]). That is, we can efficiently compute $\gamma^{(s^n)}$ for any $n \geq 1$. In particular, we can efficiently compute the element $D = \gamma^{(s^{l-4})}$.

We now are ready to prove that we can solve the strong Diffie-Hellman problem with respect to γ. To this end, let $A = \gamma^x$ and $B = \gamma^y$ be given. Then, first of all,

$$
\begin{aligned}
E = WDH(D, WDH(A, B)) &= WDH(\gamma^{(s^{l-4})}, WDH(\gamma^x, \gamma^y)) \\
&= WDH(\gamma^{(s^{l-4})}, \kappa^{xy}) \\
&= WDH(\gamma^{(s^{l-4})}, \gamma^{xys}) \\
&= \kappa^{(s^{l-4}xys)} = \kappa^{(xys^{l-3})} \\
&= \gamma^{s(xys^{l-3})} = \gamma^{(xys^{l-2})} \\
&= \gamma^{(xys^{-1})}
\end{aligned}
$$

Here we have used that $s^{l-1} \equiv 1 \bmod l$ for any prime number l (i.e., Fermat's little theorem). Now,

$$
WDH(E, \xi) = WDH(\gamma^{(xys^{-1})}, \gamma^t) = \kappa^{xys^{-1}t} = \gamma^{s(xyts^{-1})} = \gamma^{xyt} = \xi^{xy}.
$$

As we can efficiently compute $E = WDH(D, WDH(A, B))$ and $WDH(E, \xi)$ we can efficiently compute ξ^{xy} on basis on γ^x and γ^y. That is, we have solved the strong Diffie-Hellman problem with respect to γ.

We are left with showing the equivalence between the first three properties mentioned in the lemma and the last one. To this end, let ξ, α be generators of G and suppose that α^x, α^y are given for some $0 \leq x, y < l$. Write $\alpha = \gamma^a$ and $\xi = \gamma^t$ for some $0 \leq a, t < l$. First of all, we can efficiently determine $\gamma^{(a^2)}$ from α, which is a conventional Diffie-Hellman problem w.r.t. γ. Secondly, from the latter result one can efficiently determine $\gamma^{(a^{-2})}$ by using the techniques described above. Finally, from the latter result and ξ, we can efficiently determine $\delta = \gamma^{(a^{-2}t)}$ which is again a conventional Diffie-Hellman problem w.r.t. γ. Now, if we present α^x, α^y to the efficient algorithm solving the strong Diffie-Hellman problem with respect to γ and δ it returns $\delta^{(a^2xy)}$ which is equal to $\gamma^{(a^{-2}ta^2xy)} = \gamma^{txy} = \xi^{xy}$. We conclude that we have solved the Diffie-Hellman problem with respect to α and ξ.   □

Lemma 9 *Let G, Γ be two isomorphic, cyclic groups and let $i : G \to \Gamma$ and $j : \Gamma \to G$ be two efficiently computable, injective homomorphisms. We assume that the order l of G and Γ and some generators are known. Then, the Diffie-Hellman problem with respect to G is efficiently computable iff it is with respect to Γ. Moreover, under this condition, the inverses of $i(.)$ and $j(.)$ are efficiently computable too.*

Proof: It easily follows that if one can solve the Diffie-Hellman problem in one of G or Γ, then one can solve the weak Diffie-Hellman problem in the other one. So the first part of the lemma follows from Lemma 8. For a proof of the second part of the lemma, we show that $i^{-1}(.)$ is efficiently computable by efficiently computing $i^{-1}(\omega)$ for any element ω of Γ. To this end, let g be a generator of G and let $\gamma = i(g)$ and $g_2 = j(\gamma)$. One can easily verify that the algorithm solving the Diffie-Hellman problem with respect to g_2 and g yields $i^{-1}(\omega)$ when presented g_2 and $j(\omega)$. $\qquad\square$

Theorem 10 *Under the* **X2C** *hypothesis, the following problems are efficiently computable:*

1. *The Diffie-Hellman problem in the XTR subgroup.*
2. *The Diffie-Hellman problem in the group of points of order q on a supersingular elliptic curve over $\mathrm{GF}(p^2)$ of order $p^2 - p + 1$.*
3. *Inverting any efficiently computable embedding (e.g., based on the MOV embedding) from the group of points of order q on a supersingular elliptic curves over $\mathrm{GF}(p^2)$ of order $p^2 - p + 1$ into the XTR subgroup.*

Proof: Suppose that $H(.)$ is an efficiently computable injective homomorphism from the XTR subgroup into some $C_{a,p^2}[q]$. We first prove the first part of the theorem. Consider any generator g of the XTR subgroup. We construct another generator h in the XTR subgroup satisfying the definition of the weak DH problem. To this end, let $h = e_q(H(g), D(H(g)))$ where $e_q(.,.)$ denotes the Weil pairing on the q-th torsion group of C_{a,p^2} and $D(.)$ denotes the distortion map from Theorem 4. It also follows from this theorem that the order of h is equal to q.

To break the weak Decision Diffie-Hellman problem, with respect to g, h, suppose that $X = g^x, Y = g^y$ are given. Then:

$$e_q(H(X), D(H(Y))) = e_q(x * H(g), y * D(H(g))) = e_q(H(g), D(H(g)))^{xy} = h^{xy}.$$

That is, by computing $e_q(H(X), D(H(Y)))$, which can be done efficiently, we have solved the weak DH problem with respect to g, h. The result now follows from Lemma 8. The second and third part of the theorem follow from the first part and Lemma 9. $\qquad\square$

The last part of Theorem 10 states that to prove the validity of the **X2C** hypothesis, one can concentrate on efficiently inverting any MOV embedding into the XTR subgroup.

4   Extensions

4.1   Other Extension Field Based Public Key Systems

Two other public key cryptosystems exist that are based on the discrete logarithm problem in the extension field $\mathrm{GF}(p^6)^*$, or actually subfields thereof. The

LUC cryptosystem, [19] and [13], is based on the order $p+1$ subgroup of $GF(p^2)^*$. The variant by Gong & Harn of LUC is based on the $p^2 + p + 1$ subgroup of $GF(p^3)^*$, where as in the XTR setting $p = 2 \bmod 3$. For both subgroups one can find supersingular elliptic curves (cf. [14]) and efficiently computable, isomorphisms from these curves onto these subgroups, based on the Weil pairing. That is, for each of the two cryptosystems one can formulate an hypothesis similar to **X2C**. We remark that there do not exist elliptic curves defined over $GF(p^2)$ with $p^2 + p + 1$ or $p^2 - p + 1$ points over $GF(p^2)$ if $p = 1 \bmod 3$, as the number of isomorphism classes is equal to $1 - \left(\frac{-3}{p}\right)$ (cf. [14, Theorem 3.2]), which is equal to zero if $p = 1 \bmod 3$ and equal to two if $p = 2 \bmod 3$.

With respect to the Gong and Harn variant of LUC, one could call the related curves *CTN curves*: **C**lass **T**hree supersingular elliptic curves defined over $GF(p^2)$ with **N**egative parameter t, namely $t = -p$ (as opposed to $t = p$). Provided $p \equiv 2 \bmod 3$, it follows that these elliptic curves take the form $y^2 = x^3 + a$ where $a \in GF(p^2)$ is neither a square nor a cube in $GF(p^2)$. This means that the difference with CTP curves lies in the fact that a is a non-quadratic residue. However, it easily follows that this property is not of significance in the proofs in this paper and all results for CTP curves generalize to CTN elliptic curves. More in particular, the map $(x, y) \to (u^2 x^p, u^3 y^p)$ where u is a solution of $u^6 = a/a^p$ is an appropriate distortion map on these types of curves. As there exists no point on such curves with first coordinates equal to zero, all points different from the point at infinity on the curve over $GF(p^2)$ are mapped to points outside the curve over $GF(p^2)$. It follows that the existence of any efficiently computable, injective homomorphism from the Gong & Harn group in any supersingular elliptic curve over $GF(p^2)$ of order $p^2 + p + 1$ implies that we can solve the Diffie-Hellman problem in the Gong & Harn subgroup of $GF(p^3)^*$ as well as in the related elliptic curve group of points. Moreover, it follows that the Decision Diffie-Hellman problem in these elliptic curve groups is always efficiently computable, irrespective of additional hypotheses.

Our techniques do not completely generalize, at least not in a straightforward fashion, to disprove this hypothesis for the LUC cryptosystem. This is partly due to the fact that we are not aware of a full representation of all isomorphism classes of the corresponding supersingular elliptic curves, i.e., curves over $GF(p)$ of trace zero. However, our techniques do generalize to two particular subclasses of such elliptic curves over $GF(p)$, as one can easily find the appropriate distortion maps. These classes of curves and distortion maps are:

1. $y^2 = x^3 - bx$ with $p = 3 \bmod 4$ and a any non-zero element in $GF(p)$. Here an appropriate distortion map is given by $(x, y) \to (-x, i \cdot y)$ where $i \in GF(p^2) \setminus GF(p)$ satisfies $i^2 = -1$.
2. $y^2 = x^3 + a$ with $p = 2 \bmod 3$ and a any non-zero element in $GF(p)$. Here an appropriate distortion map is given by $(x, y) \to (x, w \cdot y)$ where $w \in GF(p^2) \setminus GF(p)$ satisfies $w^3 = 1$.

It follows in particular that the Decision Diffie-Hellman problem in the group of points over $GF(p^2)$ on these curves is efficiently computable. Recently, A. Joux and K. Nguyen, [6], have constructed examples of supersingular elliptic curves, of the type described above that have the additional property that the Diffie-Hellman problem and the discrete logarithm problem are equivalently difficult.

4.2   Possible Generalizations

In this section we discuss the applicability our techniques to general elliptic curves, e.g., non-supersingular ones. To this end, let $E : y^2 + a_1 xy + a_3 y = x^3 + a_2 x^2 + a_4 x + a_6$ be an elliptic curve defined over a finite field $K = \mathrm{GF}(p^n)$ and let P be a point on E over K of prime order q. As usual, we refer to the points on the curve E over a field L (including the point at infinity) as $E(L)$. Now, a *distortion* map with respect to P is an endomorphism defined over the completion $\overline{K}$ of K that maps P to a point $D(P)$ independent from P (cf. Figure 1). As $D(.)$ is a group homomorphism, it follows that $D(P)$ is an element

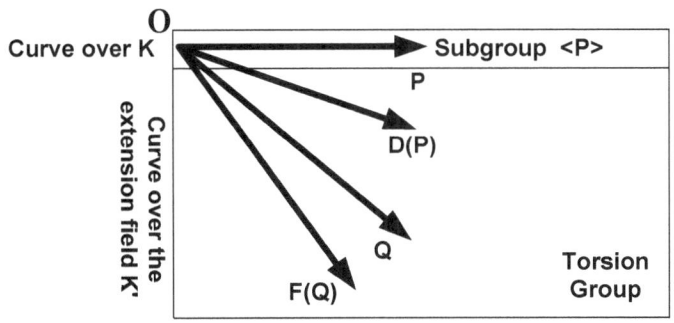

Fig. 1. Distortion maps

of the q-th torsion points $E[q]$ of E. Suppose that the set of q-th torsion points $E[q]$ of E is contained in $E(K')$ for some extension field $K' = \mathrm{GF}(p^{nk})$ of K of degree k, the so-called *MOV degree*. It it known (cf., [14]) that if the degree k is of polynomial size in $\log_2(\#(K))$ then computing the Weil pairing $E_q(.,.)$ can be done in probabilistic polynomial time in $\log_2(\#(K))$ too.

Under this condition it directly follows from the techniques employed in Section 3 that the existence of a distortion map implies that the Decision Diffie-Hellman problem in the group $\langle P \rangle$ is efficiently computable then. Now the following question arises: under what conditions can we expect that distortion maps exist? As pointed out to us by A. Joux, it is a consequence of [18, Ch. III, Th.9.5] that the endomorphism group of a supersingular elliptic curve is so large, that distortion maps always exist in these circumstances, with only a finite number of exceptions. As in this situation the degree k is either 1, 2, 3, 4 or 6, it also follows that the Decision Diffie-Hellman problem is efficiently computable in subgroups on such curves.

With respect to the case of "ordinary", i.e., non supersingular elliptic curves there is one prominent example of a distortion map, namely the Frobenius map with respect to K $F : (x, y) \rightarrow (x^{(p^n)}, y^{(p^n)})$. The Frobenius map acts as a $GF(q)$-linear mapping on $E[q]$ (considered as a two dimensional linear space over $GF(q)$) and its characteristic equation is $\lambda^2 - t\lambda + p^n$ (cf. [14]). The eigenvalues of F with respect to $E[q]$ are one (with corresponding eigenspace $\langle P \rangle$) and $t - 1 \bmod q$. If $t \neq 2 \bmod q$, then the eigenvectors corresponding with the eigenvalue $t - 1 \bmod q$ are not elements of the curve over K. That is, they are outside the original curve and really lie on the extension of the curve over K'. Now if we consider any subgroup $\langle Q \rangle$ of the q-th torsion group different from these eigenspaces, we see that Q and $F(Q)$ are independent. Compare Figure 1 above. So if the MOV degree k is polynomial in $\log_2(\#(K))$ then the Decision Diffie-Hellman problem is efficiently computable in such subgroups. In [1] it is shown that for general elliptic curves over basic fields K it is unlikely that $k < \log_2(\#(K))^2$. We are not aware of general results concerning the case that k is not polynomial in $\log_2(\#(K))$. Moreover, we observe that our techniques do not really require to actually compute values of Weil pairings: only the ability to compare them suffices. For this the efficient calculation of a one bit predicate of a Weil pairing is probably sufficient. It is not a priori clear that it can be excluded that this is possible in polynomial time even if k is not polynomially bounded.

Of course, this does not settle the existence of distortion maps in the original group $\langle P \rangle$. This is very relevant from a practical, cryptographic point of view, as such existence would make the Decision Diffie-Hellman problem in practically used elliptic curve subgroups (possibly) efficiently computable. In discussions with numerous knowledgeable colleagues, it emerged that distortion maps in such elliptic curve subgroups do not exist. The following elegant proof of this was presented to us by Ruud Pellikaan.

Theorem 11 *Let E be a non-supersingular curve and let $P \in E(K)$ be of order q. If q is relatively prime to p and the q-th torsion group is not contained in $E(K)$ then there can not exist a distortion map $D(.)$ w.r.t. P. Moreover, the second condition is implied by the condition that q^2 does not divide $\#(E(K))$.*

Proof: Suppose, at the contrary, that such a distortion map D exists. Notice that $Q = D(P)$ is not a point on $E(K)$ as this implies that the q-th torsion group is contained in $E(K)$. The crux of the proof is that the endomorphism ring of a non-supersingular elliptic curve is abelian. This follows for instance from the fact that this ring is an order in a quadratic imaginary field (cf. [18, Ch. V, Theorem 3.1]). As before, let F be the K Frobenius map. Now,

$$Q = D(P) = D(F(P)) = F(D(P)) = F(Q),$$

where the second equality follows as $P \in E(K)$. But this means that Q is an element of $E(K)$ and we arrive at a contradiction. The last part of the result easily follows.   □

As elliptic curve subgroups used in practical cryptosystems, satisfy the conditions of Theorem 11, we conclude that in such circumstances distortion maps do not exist. It seems like an interesting problem to find out if distortion maps can exist in the situation that the q-torsion group is contained in $E(K)$, but that no point of order q is contained in $E(K_0)$ for any genuine subfield K_0 of K.

5   Applications

Distortion maps on (supersingular) elliptic curves can not only be used as cryptanalytical tools, but also as building blocks in actual applications.

5.1   A One Round Protocol
for Tripartite Diffie-Hellman Key Exchange

In [5] A. Joux proposes schemes for a three participants variation of the Diffie-Hellman protocol. One of his schemes is based on a subgroup of prime order q of a supersingular elliptic curve over a field $GF(p^n)$. Two points P, Q of order q are chosen, such that P is an element of the elliptic curve over $GF(p^n)$ and Q is an element of the q-th torsion group that is independent from P. A simple way to establish this, is to choose the element Q of order q so that it is not on the curve itself, but it is is on the curve over the extension field $GF(p^{nk})$ of $GF(p^n)$. Here k is called the MOV degree, which is either $1, 2, 3, 4$ or 6. It follows in particular that the Weil pairing $e_q(P, Q)$ is a q-th root of unity in $GF(p^{nk})$. It is assumed that taking discrete logarithms in the groups $\langle P \rangle$ and $\langle Q \rangle$ is not practically possible.

Now in the tripartite Diffie-Hellman protocol, three parties A, B, C want to establish a shared key, whereby each party only exchanges one message with another party. That is, at most 6 messages are exchanged. Joux proposes the following protocol. Each i-th participant $(i = 1, 2, 3)$ generates a random $0 \leq x_i < q$, forms $(A_i, B_i) = (x_i \cdot P, x_i \cdot Q)$, and sends this to the other participants. Now the shared key is the element $e_q(P, Q)^{x_1 \cdot x_2 \cdot x_3}$. To illustrate that each participant can compute the shared key, the first participant can do so by determining:

$$e_q(A_2, B_3)^{x_1} = e_q(x_2 \cdot P, x_3 \cdot Q)^{x_1} = e_q(P, Q)^{x_1 \cdot x_2 \cdot x_3}.$$

We now describe the possible application of distortion maps. To this end, let P be a point on an elliptic curve E of order q such that taking discrete logarithms in $\langle P \rangle$ is not practically possible and assume there exists a distortion map $D(.)$ on the curve that maps P to a point $D(P)$ independent from P.

Now if, in our variant of the tripartite Diffie-Hellman protocol, three parties A, B, C want to establish a shared key then, each i-th participant $(i = 1, 2, 3)$ generates a random $0 \leq x_i < q$, forms the point $x_i \cdot P$, and sends this to the other participants. The shared key is the element $e_q(P, D(P))^{x_1 \cdot x_2 \cdot x_3}$. It is a simple verification to see that each participant can compute this key. Compared with the original tripartite Diffie-Hellman protocol in the curve E, this variant only requires two thirds of the number of exponentiations and half the number of bits exchanged.

If one can solve the Diffie-Hellman problem with respect to P or $e_q(P, Q)$ then one can break this protocol. We are not aware of reverse results.

5.2 Supporting Non-repudiation and Escrowable Encryption with Only Public Key

To fully support non-repudiation of digital signatures it is common practice not to escrow the related private keys. To prevent loss of information resulting from loss of private key material, or to comply with legal requirements end-users will typically be issued two (or even three) certificates: one for non-repudiation services and others for different services.

The use of distortion mappings make it possible to employ one public key (and hence certificate) for a non-repudiation service as well as for an encryption service, in such a way that the private signing key is not escrowed, while the encryption service is recoverable. To describe this scheme, once again let P be a point on an elliptic curve E over a finite field $GF(p^n)$ such that taking discrete logarithms in $\langle P \rangle$ is not practically possible. Assume there exists a distortion map $D(.)$ on the curve that maps P to a point $D(P)$ independent from P in the q-th torsion group contained in the elliptic curve over the extension field $GF(p^{nk})$. We assume that the Weil pairing is efficiently computable on $\langle P \rangle \times \langle D(P) \rangle$. Denote the q-th root of unity $e_q(P, D(P))$ in $GF(p^{nk})$ by g.

In our scheme an end-user A chooses its private signing key $0 \le x < q$ randomly. Its public key (for both the non-repudiation and the encryption service) is the element $y = g^x$ in $GF(p^{nk})^*$. The user's certificate is based on this public key and also references to (or contains) the system parameters, e.g., the elliptic curve E, the group order q, the point P on it and the element g. To make the encryption service recoverable, the user also forms the point $Y = x \cdot P$ and escrows this at a trusted third party. Now, the end-user could employ any discrete logarithm based digital signature scheme, like Schnorr, ElGamal or DSA thereby using the g, y and the private key x. The encryption service supported, is the following variant of the ElGamal [4] encryption scheme:

1. The sender generates a random $0 \le k < q$ and symmetrically encrypts the information for end-user A using y^k as a session key.
2. The sender forms the point $K = k \cdot P$ on the curve E and sends both the encrypted information and the point K to end-user A.

Now, there are essentially two ways for the end-user A to decrypt information encrypted this way. The first way is to first calculate $e_q(K, D(P)) = e_q(k \cdot P, D(P)) = e_q(P, D(P))^k = g^k$ and then secondly calculate $(g^k)^x = y^k$ which enables the end-user to decrypt the symmetrically encrypted information. Note that no secret information is required to determine g^k, so this information could in fact be sent along by the sender, avoiding that the end-user needs to calculate a Weil pairing. The second way to decrypt this information is to directly calculate $e_q(K, D(Y)) = e_q(k \cdot P, D(x \cdot P)) = e_q(k \cdot P, x \cdot D(P)) = g^{kx} = y^k$ on basis of Y. Note that this operation does not require the private key x but that the escrowed value Y suffices. Hence, if the end-user retrieves a copy of Y from his escrow agent then he is able to decrypt his messages when he loses his private x. However, the end-user is not able to make new digital signatures as determining the private key x from $Y = x \cdot P$ requires one to solve a discrete logarithm problem in the elliptic curve, which we assumed is not practically possible.

For an indication of security, suppose that an attacker can compute Y on basis of y, then as y is chosen randomly by the end-user, the attacker has found an computable injective homomorphism from $\langle g \rangle$ to $\langle P \rangle$. It follows from the arguments in Section 3 that the attacker is then also able to solve the Diffie-Hellman problem in both these groups. We are not aware of more rigorous security proofs. We finally remark that there exists a more general but less efficient variant of this scheme that does not require a distortion map and whereby one uses two independent points P, Q. We leave the details, which are straightforward, to the reader.

6   Conclusion

We have shown that the existence of any efficiently computable, injective homomorphism from the XTR subgroup in the group of points over $\mathrm{GF}(p^2)$ on a supersingular elliptic curve over $\mathrm{GF}(p^2)$ of order $p^2 - p + 1$ implies that we can solve several problems that are widely believed to be hard. The Diffie-Hellman problem in the XTR subgroup is an example of such a problem. We have also shown that the Decision Diffie-Hellman problem in such elliptic curve groups is efficiently computable and that our results can be extended to other supersingular elliptic curve groups. The results in this paper therefore provide evidence that the multiplicative group of a finite field provides essentially more, and in any case not less, security than the group of points of a supersingular elliptic curve of comparable size. In addition to this, we have discussed generalizations to tackle the Decision Diffie-Hellman problem in certain groups of points on non-supersingular elliptic curves over finite fields. Finally, we have shown that the tools we used in our cryptanalysis (distortion maps) can also be used as building blocks in new cryptographic applications. We have illustrated that with two examples: an improvement of Joux's one round protocol for tripartite Diffie-Hellman key exchange and a non refutable digital signature scheme that supports escrowable encryption.

Acknowledgments

I want to thank Scott Vanstone for posing the **X2C** hypothesis on the Crypto 2000 Rump Session. I am grateful to Antoine Joux, David Kohel, Arjen Lenstra, Ruud Pellikaan and Joe Silverman for the stimulating discussions I had with them. Antoine is specifically thanked for his observation that my initial proof techniques could be elegantly formulated with distortion maps. Arjen is specifically thanked for challenging and helping me to show that the weak Diffie-Hellman problem is equivalent to the conventional one. Ruud is specifically thanked for his permission to include Theorem 11.

References

1. R. Barasubramanian, N. Koblitz, *The improbability that an elliptic curve has subexponential discrete log problem under the MOV algorithm*, J. of Cryptology, vol 11, 141-145, 1999.
2. R. Cramer, R. Gennaro, B. Schoenmakers, *A Secure and Optimally Efficient Multi-Authority Election Scheme* Advances in Cryptology - EUROCRYPT '97 Proceedings, Springer-Verlag, 1997, 103-118.
3. R. Cramer, V. Shoup, *A practical public key cryptosystem provably secure against adaptive chosen ciphertext attack*, Proceedings of Crypto 1998, LNCS 1462, Springer-Verlag, 1998, 13-25.
4. T. ElGamal, *A Public Key Cryptosystem and a Signature scheme Based on Discrete Logarithms*, IEEE Transactions on Information Theory 31(4), 1985, 469-472.
5. A. Joux, *A one round protocol for tripartite Diffie-Hellman*, 4th International Symposium, Proceedings of ANTS, LNCS 1838, Springer-Verlag, 2000, 385-394.
6. A. Joux, K. Nguyen, *Seperating Decision Diffie-Hellman from Diffie-Hellman in cryptographic groups*, in preparation. Available from eprint.iacr.org.
7. G. Gong, L. Harn, *Public key cryptosystems based on cubic finite field extensions*, IEEE Trans. on I.T., November 1999.
8. N. Koblitz, The 4th workshop on Elliptic Curve Cryptography (ECC 2000), Essen, October 4-6 2000.
9. N. Koblitz, *An Elliptic Curve Implementation of the Finite Field Digital Signature Algorithm*, Proceedings of Crypto '98, LNCS 1462, Springer-Verlag, 1998, 327-337.
10. A.K. Lenstra, E.R. Verheul, *The XTR public key system*, Proceedings of Crypto 2000, LNCS 1880, Springer-Verlag, 2000, 1-19; available from www.ecstr.com.
11. A.K. Lenstra, E.R. Verheul, *Key improvements to XTR*, Proceedings of Asiacrypt 2000, LNCS 1976, Springer-Verlag, 2000, 220-223; available from www.ecstr.com.
12. A.K. Lenstra, E.R. Verheul, *Fast irreducibility and subgroup membership testing in XTR*, Proceedings of the 2001 Public Key Cryptography conference, LNCS 1992, Springer-Verlag, 2001, 73-86; available from www.ecstr.com.
13. R. Lidl, W.B. Müller, *Permutation Polynomials in RSA-cryptosystems*, Crypto '83 Proceedings, Plemium Press, 1984, 293-301.
14. A. Menezes, *Elliptic Curve Public Key Cryptosystems*, Kluwer Academic Publishers, Boston 1993.
15. A. Menezes, T. Okamoto, S.A. Vanstone *Reducing elliptic curve logarithms to a finite field*, IEEE Trans. Info. Theory, 39, 1639-1646, 1993.
16. A. Menezes, S.A. Vanstone, *ECSTR (XTR): Elliptic Curve Singular Trace Representation*, Rump Session of Crypto 2000.
17. S.C. Pohlig, M.E. Hellman, *An improved algorithm for computing logarithms over GF(p) and its cryptographic significance*, IEEE Trans. on IT, 24 (1978), 106-110.
18. J. Silverman, *The Arithmetic on Elliptic Curves*, Springer-Verlag, New York, 1986.
19. P. Smith, C. Skinner, *A public-key cryptosystem and a digital signature system based on the Lucas function analogue to discrete logarithms*, Asiacrypt '94 proceedings, Springer-Verlag, 1995, 357-364.

NSS: An NTRU Lattice-Based Signature Scheme

Jeffrey Hoffstein, Jill Pipher, and Joseph H. Silverman

NTRU Cryptosystems, Inc., 5 Burlington Woods,
Burlington, MA 01803 USA,
jhoff@ntru.com, jpipher@ntru.com, jhs@ntru.com

Abstract. A new authentication and digital signature scheme called the NTRU Signature Scheme (NSS) is introduced. NSS provides an authentication/signature method complementary to the NTRU public key cryptosystem. The hard lattice problem underlying NSS is similar to the hard problem underlying NTRU, and NSS similarly features high speed, low footprint, and easy key creation.

Keywords: digital signature, public key authentication, lattice-based cryptography, NTRU, NSS

Introduction

Secure public key authentication and digital signatures are increasingly important for electronic communications and commerce, and they are required not only on high powered desktop computers, but also on SmartCards and wireless devices with severely constrained memory and processing capabilities. The importance of public key authentication and digital signatures is amply demonstrated by the large literature devoted to both theoretical and practical aspects of the problem, see for example [1,2,6,7,9,11,12,15,16,17].

At CRYPTO '96 the authors introduced a highly efficient new public key cryptosystem called NTRU. (See [4] for details.) Underlying NTRU is a hard mathematical problem of finding short vectors in certain lattices. In this note we introduce a complementary fast authentication and digital signature scheme that uses public and private keys of the same form as those used by the NTRU public key cryptosystem. We call this new algorithm NSS for NTRU Signature Scheme.

In the original version of this paper for Eurocrypt 2001, we both introduced NSS and optimized it for maximum efficiency and minimum signature length. As a result the underlying ideas and security analysis were less transparent than they might have been. To alleviate this problem and attempt to address some of the concerns of the referees, the present paper takes the following form. We first present a complete version of NSS and a set of parameters optimized to provide security comparable to RSA 1024 along with high efficiency. We then describe the properties of an implementation of this system at these parameters. The version of this paper originally submitted to Eurocrypt then provided a security analysis tailored specifically to these parameters. In the current version

B. Pfitzmann (Ed.): EUROCRYPT 2001, LNCS 2045, pp. 211–228, 2001.
© Springer-Verlag Berlin Heidelberg 2001

we eliminate some details of the security analysis of the optimized version in order to include a discussion of the less efficient version. In this way we hope to elucidate the main ideas underlying NSS and thereby make this paper easier to read. Complete details of the analysis of the optimized version are available on our website at <www.ntru.com/technology/tech.technical.htm>.

We also note that the signature scheme described in this paper differs in some respects from the scheme described by Jeff Hoffstein at the CRYPTO 2000 rump session. In order to optimize NSS, the rump session version used disparate sized coefficients whose existence was concealed by allowing p to divide q, which led to a statistical weakness. (This weakness was independently noted by Mironov [10].) The use of uniform coefficients and relatively prime values for p and q makes NSS more closely resemble the original NTRU public key cryptosystem, a system that has withstood intense scrunity since its introduction at CRYPTO '96.

The authors would like to thank Phil Hirschhorn for much computational assistance and Don Coppersmith for substantial help in analyzing the security of NSS. Any remaining weaknesses or errors in the signature scheme described below are, of course, entirely the responsibility of the authors.

1   A Brief Description of NSS

In this section we briefly describe NSS, the NTRU Signature Scheme. In order to avoid excessive duplication of exposition, we assume some familiarity with [4], but we repeat definitions and concepts when it appears useful. Thus this paper should be readable without reference to [4].

The basic operations occur in the ring of polynomials

$$R = \mathbb{Z}[X]/(X^N - 1)$$

of degree $N - 1$, where multiplication is performed using the rule $X^N = 1$. The coefficients of these polynomials are then reduced modulo p or modulo q, where p and q are fixed integers.

There are five integer parameters associated to NSS,

$$(N, p, q, D_{\min}, D_{\max}).$$

There are also several sets of polynomials $\mathcal{F}_f, \mathcal{F}_g, \mathcal{F}_w, \mathcal{F}_m$ having small coefficients that serve as sample spaces. For concreteness, we mention the choice of integer parameters

$$(N, p, q, D_{\min}, D_{\max}) = (251, 3, 128, 55, 87), \tag{1}$$

which appears to yield a secure and practical signature scheme. See Section 2 for futher details.

Remark 1. For ease of exposition we often assume that $p = 3$. We further assume that polynomials with mod q coefficients are chosen with coefficients in the range $-q/2$ to $q/2$.

The public and private keys for NSS are formed as follows. Bob begins by choosing two polynomials f and g having the form

$$f = f_0 + pf_1 \quad \text{and} \quad g = g_0 + pg_1. \tag{2}$$

Here f_0 and g_0 are fixed universal polynomials (e.g., $f_0 = 1$ and $g_0 = 1 - 2X$) and f_1 and g_1 are polynomials with small coefficients chosen from the sets $\mathcal{F}_f$ and $\mathcal{F}_g$, respectively. Bob next computes the inverse f^{-1} of f modulo q, that is, f^{-1} satisfies

$$f^{-1} * f \equiv 1 \pmod{q}.$$

Bob's public verification key is the polynomial

$$h \equiv f^{-1} * g \pmod{q}.$$

Bob's private signing key is the polynomial f.

Before describing exactly how NSS works, we would like to explain the underlying idea. The coefficients of the polynomial h have the appearance of being random numbers modulo q, but Bob knows a small polynomial f (i.e., f has coefficients that have small absolute value compared to q) with the property that the product $g \equiv f * h \pmod{q}$ also has small coefficients. Equivalently (see Section 4.2), Bob knows a short vector in the NTRU lattice generated by h. It is a difficult mathematical problem, starting from h, to find f or to find some other small polynomial F with the property that $G \equiv F * h \pmod{q}$ is small. Bob's signature s on a digital document D will be linked to D and will demonstrate to Alice that he knows a decomposition $h \equiv f^{-1} * g \pmod{q}$ without giving Alice information that helps her to find f. The mechanism by which Bob shows that he knows f without actually revealing its value lies at the heart of NSS and is described in the next section.

1.1   NSS Key Generation, Signing, and Verifying

We now describe in more detail the steps used by Bob to sign a document and by Alice to verify Bob's signature. The key computation involves the following quantity.

Definition 1. *Let $a(X)$ and $b(X)$ be two polynomials in R. First reduce their coefficients modulo q to lie between $-q/2$ to $q/2$, then reduce their coefficients modulo p to lie in the range between $-p/2$ and $p/2$. If*

$$\bar{a}(X) = \bar{a}_0 + \cdots + \bar{a}_{N-1}X^{N-1} \quad \text{and} \quad \bar{b}(X) = \bar{b}_0 + \cdots + \bar{b}_{N-1}X^{N-1}$$

are the reductions of a and b, respectively, then the deviation *of a and b is*

$$\text{Dev}(a, b) = \#\{i : \bar{a}_i \neq \bar{b}_i\}.$$

Intuitively, $\text{Dev}(a, b)$ is the number of coefficients of $a \bmod q$ and $b \bmod q$ that differ modulo p.

Key Generation: This was described above, but we briefly repeat it for convenience. Bob chooses two polynomials f and g having the appropriate form (2). He computes the inverse f^{-1} of f modulo q. Bob's public verification key is the polynomial $h \equiv f^{-1} * g \bmod q$ and his private signing key is the pair (f, g).

Signing: Bob's document is a polynomial m modulo p. (In practice, m must be the hash of a document, see Section 4.9.) Bob chooses a polynomial $w \in \mathcal{F}_w$ of the form

$$w = m + w_1 + pw_2,$$

where w_1 and w_2 are small polynomials whose precise form we describe later, see Section 2.1. He then computes

$$s \equiv f * w \pmod{q}.$$

Bob's signed message is the pair (m, s).

Verification: In order to verify Bob's signature s on the message m, Alice checks that $s \neq 0$ and then verifies the following two conditions:

(A) Alice compares s to $f_0 * m$ by checking if their deviation satisfies

$$D_{\min} \leq \mathrm{Dev}(s, f_0 * m) \leq D_{\max}.$$

(B) Alice uses Bob's public verification key h to compute the polynomial $t \equiv h * s \pmod{q}$. She then checks if the deviation of t from $g_0 * m$ satisfies

$$D_{\min} \leq \mathrm{Dev}(t, g_0 * m) \leq D_{\max}.$$

If Bob's signature passes tests (A) and (B), then Alice accepts it as valid.

The check by Alice that $s \neq 0$ is done to eliminate the small possibility of a forgery via the trivial signature. This is described in more detail in [5] We defer until Section 3 below a detailed explanation of why NSS works. However, we want to mention here the reason for allowing s and t to deviate from $f_0 * m$ and $g_0 * m$, respectively. This permits us to take w_1 to be nonzero and to allow a significant amount of reduction modulo q to occur in the products $f * w$ and $g * w$. This makes it difficult for an attacker to find the exact values of $f * w$ or $g * w$ over $\mathbb{Z}$, which in turn means that potential attacks via lattice reduction require lattices of dimension $2N$ rather than N.

This is the key difference between the optimized version of NSS presented in the next section and a somewhat less efficient version. If we take $D_{\min} = D_{\max} = 0$, i.e., if we allow no deviations, then a transcript will reveal $f * w$ and $g * w$ exactly. Lattices of dimension N can be reduced faster than lattices of dimension $2N$. Consequently, for a secure version of NSS assuming no deviations we require a larger value of N. We will show that if N is chosen greater than about 700 this still gives a fast and equally secure signature scheme, albeit with somewhat larger key and signature sizes than the optimized version of NSS described in this note.

This concludes our overview of how NSS works. In the next section we suggest a parameter set and explain why we believe that it provides a level of security comparable to RSA 1024. Table 1 compares the efficiency of NSS to other systems. In the following sections we provide a security analysis, although due to space constraints, we refer the reader to [5] for some details, especially for the optimized version with $D_{min}, D_{max} > 0$.

2   A Practical Implementation of NSS

The following parameter selection for NSS appears to create a scheme with a breaking time of at least 10^{12} MIPS years:

$$(N, p, q, D_{min}, D_{max}) = (251, 3, 128, 55, 87). \tag{3}$$

This leads to the following key and signature sizes for NSS:

Public Key: 1757 bits   Private Key: 502 bits   Signature: 1757 bits

We take $f_0 = 1$ and $g_0 = 1 - 2X$, where recall that $f = f_0 + pf_1$ and $g = g_0 + pg_1$. In order to describe the sample spaces, we let

$$\mathcal{T}(d) = \{F(X) \in R : F \text{ has } d \text{ coefs} = 1 \text{ and } = -1, \text{ with the rest } 0\}.$$

Then the sample spaces corresponding to the parameter set (3) are

$$\mathcal{F}_f = \mathcal{T}(70), \quad \mathcal{F}_g = \mathcal{T}(40), \quad \mathcal{F}_m = \mathcal{T}(32).$$

Note that m is a hash of the digital document D being signed. Thus the users must agree on a method (e.g., using SHA1) to transform D into a list of 64 distinct integers $0 \le e_i < 251$, and then $m = \sum_{i=1}^{32} X^{e_i} - \sum_{i=33}^{64} X^{e_i}$.

The polynomial w has the form $w = m + w_1 + pw_2$, so we also must explain how to choose the polynomials w_1 and w_2. This must be done carefully so as to prevent an attacker from either lifting to a lattice over $\mathbb{Z}$ (see Section 4.4) or gaining information via a reversal averaging attack (see Section 4.6). Roughly, the idea is to choose random w_2, compute $s' \equiv f * (m + pw_2) \pmod{q}$ and $t' \equiv g * (m + pw_2) \pmod{q}$, choose w_1 to cancel all of the common deviations of $(s', f_0 * m)$ and $(t', g_0 * m)$ and to exchange some of the noncommon deviations, and finally to alter w_2 to move approximately $1/p$ of the nonzero coefficients of $m + w_1$. For the parameter set (3) given above, the polynomial w_1 has up to 25 nonzero coefficients and w_2 is initially chosen at random from the set $\mathcal{T}(32)$. The precise prescription for creating w is described in Section 2.1.

We have implemented NSS in C and run it on various platforms. Table 1 describes the performance of NSS on a desktop machine and on a constrained device and gives comparable figures for RSA and ECDSA signatures.

Table 1. Speed Comparison of NSS, RSA, and ECDSA

	Pentium	Palm
NSS Sign	0.35 ms	0.33 sec
RSA Sign	66.56 ms	36.13 sec
ECDSA Sign	1.18 ms	1.79 sec
NSS Verify	0.29 ms	0.25 sec
RSA Verify	1.23 ms	0.73 sec
ECDSA Verify	1.70 ms	3.26 sec

Notes for Table 1.

1. NSS speeds from the NERI implementation of NSS by NTRU Cryptosystems.
2. RSA and ECDSA speeds presented by Alfred Menezes [8] at CHES 2000.
3. RSA 1024 bit verify uses a small verification exponent for increased speed.
4. ECDSA 163 bit uses a Koblitz curve for increased speed. Time is approximately doubled if a random curve over $\mathbb{F}_{2^{163}}$ is used.

2.1   Selection of the Masking Polynomial w

The polynomial $w = m + w_1 + pw_2$ has two purposes. First, it includes the message digest m and is thus the means by which m is attached to the signature s. Second, it contains polynomials w_1 and w_2 that introduce variability into the signature and prevent an attacker from gaining useful information that might be used to find the private key f or to directly forge a signature.

There are two principle areas that must be addressed when selecting w. First, in the optimized version we must ensure that an attacker cannot lift the values of $s \equiv f * w \pmod{q}$ and $t \equiv g * w \pmod{q}$ to the exact values of $f * w$ or $g * w$ in $\mathbb{Z}[X]$. Second, we must ensure that the attacker cannot use averages formed from long transcripts of signatures to deduce information about f or g.

The first item is addressed by selecting w_1 so as to alter many of the coefficients of $f * (m + pw_2)$ and $g * (m + pw_2)$ that lie outside the range from $-q/2$ to $q/2$. This has the effect of masking the coefficients that have suffered nontrivial reduction modulo q and prevents the attacker from undoing the reduction. The second item is handled by changing $1/p$ of the coefficients of w_2; this has the effect of forcing all second moment transcript averages to converge to 0. We now describe exactly how w_1 and w_2 are created. For ease of exposition, we assume that $p = 3$. For further details of why this procedure protects against lifting and averaging attacks, see [5].

The first step is to choose a random polynomial $w_2 \in \mathcal{T}(d_{w_2})$. That is, w_2 has a specified number of 1's and -1's. For example, the parameter set (3) takes $w_2 \in \mathcal{T}(32)$. The next step is to compute preliminary signature polynomials

$$s' \equiv f * (m + pw_2) \pmod{q} \qquad \text{and} \qquad t' \equiv g * (m + pw_2) \pmod{q}. \quad (4)$$

Next we choose w_1. We start with $w_1 = 0$. We let $i = 0, 1, 2, \ldots, N - 1$ and run through the coefficients s_i' and t_i' of s' and t', performing the following steps. [The quantity $\mathtt{w_1\text{-}Limit}$ used below is a prespecified parameter. For the parameter set (3), its value is 25.]

- If $s'_i \not\equiv m_i \pmod{p}$ and $t'_i \not\equiv m_i \pmod{p}$ and $s'_i \equiv t'_i \pmod{p}$,
  then set $w_{1,i} \equiv m_i - s'_i \pmod{p}$.
- If $s'_i \not\equiv m_i \pmod{p}$ and $t'_i \not\equiv m_i \pmod{p}$ and $s'_i \not\equiv t'_i \pmod{p}$,
  then set $w_{1,i} = 1$ or -1 at random.
- If $s'_i \not\equiv m_i \pmod{p}$ and $t'_i \equiv m_i \pmod{p}$,
  then with probability 25%, set $w_{1,i} \equiv m_i - s'_i \pmod{p}$.
- If $s'_i \equiv m_i \pmod{p}$ and $t'_i \not\equiv m_i \pmod{p}$,
  then with probability 25%, set $w_{1,i} \equiv m_i - t'_i \pmod{p}$.
- If $i = N - 1$ or if $w_1(X)$ has more than w₁-Limit nonzero coordinates, the
  construction of w_1 is complete.

Finally, we need to make some alterations to w_2 to prevent the averaging of long transcripts of signatures. This is done by taking each coefficient $w_{2,i}$, $0 \le i < N$, and with probability $1/3$, replacing it with with $w_{2,i} - m_i - w_{1,i}$.

This completes the description of how w_1 and w_2 are chosen.

3   Completeness of NSS

A signature scheme is deemed to be complete if Bob's signature, created with the private signing key f, will be accepted as valid. Thus we need to check that Bob's signed message (m, s) passes the two tests (A) and (B).

3.1   The Norm of a Polynomial

In order to analyze the two verification conditions we briefly digress to discuss norms of polynomials.

Let
$$a(X) = a_0 + a_1 X + a_2 X^2 + \cdots + a_{N-1} X^{N-1}$$

be a polynomial with integer coefficients and let μ be the average of the coefficients. We define the *centered Euclidean Norm* and the *Sup Norm* of a, denoted respectively $\|a\|$ and $\|a\|_\infty$, by the formulas

$$\|a\| = \sqrt{(u_0 - \mu)^2 + \cdots + (u_{N-1} - \mu)^2} \quad \text{and} \quad \|a\|_\infty - \max\{|a_0|, \ldots, |a_{N-1}|\}.$$

In our examples, μ will be close to or equal to zero.

We require certain facts about polynomials with small coefficients. For random polynomials with small coefficients such as f and w, it is generally true that

$$\|f * w\| \approx \|f\| \cdot \|w\| \quad \text{and} \quad \|f * w\|_\infty \approx \gamma \|f\| \cdot \|w\|, \tag{5}$$

where $\gamma < 0.15$ for $N < 1000$. The NTRU cryptosystem relies on these properties of small polynomials, which are discussed in [4]. (Note that the infinity norm defined in [4] is actually twice the infinity norm defined here.)

With this background we now easily check the completeness of NSS.

Test (A): The polynomial s that Alice tests is congruent to the product

$$s \equiv f * w \pmod{q}$$
$$\equiv (f_0 + pf_1)(m + w_1 + pw_2) \pmod{q}$$
$$\equiv f_0 * m + f_0 * w_1 + pf_0 * w_2 + pf_1 * w \pmod{q}.$$

We see that the ith coefficients of s and $f_0 * m$ will agree modulo p unless one of the following situations occurs:

- The ith coefficient of $f_0 * w_1$ is nonzero.
- The ith coefficient of $f * w$ is outside the range $(-q/2, q/2]$, so differs from the ith coefficient of s by some multiple of q.

The estimates in (5) tell us that before reduction modulo q, the absolute value of the coefficents of $f * w$ is bounded above by $\gamma \|f\| \cdot \|w\|$. As long as this quantity does not greatly exceed $q/2$, little reduction modulo q will take place. If the parameters and sample spaces are chosen properly (e.g., as in Section 2) then there will be at least $D_{\min}$ and at most $D_{\max}$ deviations between $s \bmod p$ and $m \bmod p$. Alternatively, if $\|f\|$ and $\|w\|$ are sufficiently small, then no reduction modulo q will take place and one can set $D_{\min} = D_{\max} = 0$. Thus Bob's signature will pass test (A).

Test (B): The polynomial t is given by

$$t \equiv h * s \equiv (f^{-1} * g) * (f * w) \equiv g * w \pmod{q}.$$

Since g has the same form as f, the same reasoning as for test (A) shows that t will pass test (B).

Remark 2. We have indicated why, for appropriate choices of parameters, Bob's signature will probably be accepted by Alice. Note that when Bob creates his signature, he should check to make sure that it is a valid signature. For the parameters $(N, p, q, D_{\min}, D_{\max}) = (251, 3, 128, 55, 87)$ from Section 2, we see from Table 2 that the probability that $\text{Dev}(s, f_0 * m)$ is valid is approximately 87.33% and the probability that $\text{Dev}(t, g_0 * m)$ is valid is approximately 90.92%. Thus Bob's signature will be valid about 79.40% of the time. Of course, if it is not valid, he simply chooses a new random polynomial w_2 and tries again. In practice it will not take very many tries to find a valid signature. The timings given in Table 1 take this factor into account.

4   Security Analysis of NSS

It was shown in Section 3 that given a message m, Bob can produce a signature s satisfying the necessary requirements. In this section we discuss various ways in which an observer Oscar might try to break the system. There are many attacks that he might try. For example, he might attempt to discover the private key f or a useful imitation, either directly from the public key h or from a long transcript

Table 2. Deviations Between $f_0 * m$ and s and Between $g_0 * m$ and t

Range	$\mathrm{Dev}(s, f_0 * m)$	$\mathrm{Dev}(t, g_0 * m)$
32 to 39	0.02%	0.08%
40 to 47	0.38%	0.99%
48 to 55	3.53%	6.98%
56 to 63	14.21%	26.32%
64 to 71	27.58%	37.79%
72 to 79	28.51%	21.22%
80 to 87	17.03%	5.58%
88 to 95	6.54%	0.90%
96 to 103	1.74%	0.11%
104 to 158	0.46	0.02%

$$(N, p, q) = (251, 3, 128)\text{---}10^6 \text{ Trials}$$

of valid signatures. He might also try to forge a signature on a message without first finding the private key. We describe the hard lattice problems that underlie some of these attacks and examine the success probabilities of other attacks that rely on random searches. In all cases we explain why the indicated attacks are infeasible for an appropriate choice of parameters such as those given in Section 2. Due to space constraints, we must refer the reader to [5] for many of the technical details related to the analysis of the optimized parameter set.

4.1 Random Search for a Valid Signature on a Given Message

Given a message m, Oscar must produce a signature s satisfying:

(A) $D_{\min} \leq \mathrm{Dev}(s, f_0 * m) \leq D_{\max}$.
(B) $D_{\min} \leq \mathrm{Dev}(t, g_0 * m) \leq D_{\max}$, where $t \equiv s * h \pmod{q}$.

If $D_{\min} = D_{\max} = 0$ these conditions become:

(A') $s \equiv f_0 * m \pmod{p}$.
(B') $t \equiv h * s \pmod{q}$ satisfies $t \equiv g_0 * m \pmod{p}$.

The most straightforward approach for Oscar is to choose s at random satisfying condition (A), which is obviously easy to do, and then to hope that t satisfies condition (B). If it does, then Oscar has successfully forged Bob's signature, and if not, then Oscar can try again with a different s. Thus we must examine the probability that a randomly chosen s satisfying (A) will yield a t that satisfies (B).

The condition (A) on s has no real effect on the end result t, since t is formed by multiplying $s * h$ and reducing the coefficients modulo q, and the coefficients of h are essentially uniformly distributed modulo q. Thus we are really asking for the probability that a randomly chosen polynomial t with coefficients between $-q/2$ and $q/2$ will satisfy condition (B). This is easily computed using elementary probability theory.

The coefficients of a randomly chosen t can be viewed as N independent random variables taking values uniformly modulo q. The coefficients of m are fixed target values modulo p. We need to compute the probability that a randomly chosen N-tuple of integers modulo q has at least $D_{\min}$ and no more than $D_{\max}$ of its coordinates equal modulo p to fixed target values. Assuming that q is significantly larger than p, this probability is approximately

$$\mathrm{Prob}(D_{\min} \le \mathrm{Dev}(t, g_0 * m) \le D_{\max}) \approx \frac{1}{p^N} \sum_{d=D_{\min}}^{D_{\max}} \binom{N}{d} (p-1)^d.$$

(Notice that for condition (B'), the probability is p^{-N}, since all N "random" coefficients of t (mod p) must match $g_0 * m$.) Table 3 gives this probability for $(N, p) = (251, 3)$ and several values of $D_{\min}$ and $D_{\max}$. For example, the table shows that for $D = 87$, the probability of a successful forgery using a randomly selected s is approximately $2^{-80.95}$.

Table 3. Probability Random t Satisfies $D_{\min} \le \mathrm{Dev}(t, g_0 * m) \le D_{\max}$

$D_{\min}$	$D_{\max}$	Probability
55	82	$2^{-90.86}$
55	87	$2^{-80.95}$
55	92	$2^{-71.66}$
55	98	$2^{-61.32}$

4.2 NTRU Lattices and Lattice Attacks on the Public Key

Oscar can try to extract the private key f from the public key h with or without a long transcript of genuine signatures. Alternatively, he can try to forge a signature without knowledge of f, using only h and a transcript. In this section we discuss attempts by Oscar to obtain the private key from the public key by lattice reduction methods. As is the case with the NTRU cryptosystem, recovery of the private key by this means is equivalent to solving a certain class of shortest or closest vector problems.

We begin with a brief exposition of our approach to the analysis of lattice reduction problems. We have perfomed a large number of computer experiments to quantify the effectiveness of current lattice reduction techniques. This has given us a strong empirical foundation for analyzing and quantifying the vulnerability of several general classes of lattices to lattice reduction attacks. The following analysis and heuristics applies to the lattices discussed in this paper. (See also the lattice material in the papers [3,4,6,7].)

Let L be a lattice of determinant d and dimension n. Let v_0 denote a given fixed vector, possibly the origin. Let r denote a given radius and consider the problem of locating a vector $v \in L$ such that $\|v - v_0\| < r$. The difficulty of solving this problem for large n is related to the quantity

$$\kappa = \kappa(L, r) = \frac{r}{d^{1/n}\sqrt{n/(2\pi e)}}. \tag{6}$$

Here the denominator is the length that the gaussian heuristic predicts for the shortest expected vector in L. See [4] for a similar analysis.

If $\kappa < 1$, then the gaussian heuristic says that a solution, if one exists at all, will probably be unique (or unique up to obvious symmetries of the lattice). The closer that κ is to 0, the easier it will be to find the unique solution using lattice reduction methods. As κ gets close to 1, lattice reduction methods become less effective.

For example, let $(L_n, r_n, v_{0,n})$ be a sequence of lattices, radii, and target vectors of increasing dimension n that contain a target vector $v_n \in L_n$ (i.e., satisfying $|v_n - v_{0,n}| < r_n$) and whose κ values satisfy

$$\kappa_n = \kappa(L_n, r_n) = c/\sqrt{n} \tag{7}$$

for a constant c. Then our experiments suggest that the time necessary for lattice reduction methods to find the target vector v_n grows like $e^{\alpha n}$ for a value of α that is roughly proportional to c. Similarly, if $\kappa \geq 1$, then a solution will probably not be unique, but it becomes progressively harder to find a solution as κ approaches 1.

We must stress here that the above statements are not intended to be a proof of security or to convey any assurance of security. They merely supply a conceptual framework that we have found useful for formulating working parameter sets. The lattices associated to these parameter sets are then subjected to extensive experimental testing.

Recall from (2) that the public key has the form $h \equiv f^{-1} * g \pmod{q}$, where $f = f_0 + pf_1$ and $g = g_0 + pg_1$. As this is very similar to the form of an NTRU public key, a $2N$-dimensional lattice attack based on the shortest vector can be used to try to derive f and g from h. See [4,13] for details on the NTRU lattice and the use of lattice reduction methods to compute the shortest expected vector.

If we identify polynomials with their vector of coefficients, then the $2N$-dimensional NTRU lattice L^{NT} consists of the linear combinations of the $2N$ vectors in the set

$$\left\{(X^i, X^i * h) : 0 \leq i < N\right\} \cup \left\{(0, qX^i) : 0 \leq i < N\right\}.$$

Equivalently, L^{NT} is the set of all vectors $(F(X), F(X) * h(X))$, where $F(X)$ varies over all N-dimensional vectors and the last N coordinates are allowed to be changed by arbitrary multiples of q. It is not hard to see that the vector (f, g) is contained in L^{NT} and will be shorter than the expected shortest vector of L^{NT} (i.e., $\kappa < 1$). Thus in principle, (f, g) should be essentially unique and findable by lattice reduction methods.

A more effective attack is to use the knowledge of f_0, g_0 to set up a closest vector attack on f_1, g_1 in the same $2N$-dimensional lattice The object is to search for the vector in L^{NT} that is closest to the vector $(0, (g_0 - f_0 * h)p')$, where $pp' \equiv 1 \pmod{q}$. If successful, this attack produces a small F such that

$G \equiv F * h - (g_0 - f_0 * h)p'$ (mod q) is also small. Then $(f_0 + pF, g_0 + pG)$ is either the original key or a useful substitute. With this approach, after balancing the lattice as in [4], we obtain the following estimate for the constant c in equation (7):

$$c > 2\sqrt{\pi e} \|f_1\| \|g_1\| / q. \tag{8}$$

Experimental evidence shows that if L runs through a sequence of NTRU type lattices of dimension $2N$ with $N > 80$ and $q \approx N/2$ and if the constant c of (7) satisfies $c > 3.7$, then the time T (in MIPS-years) necessary for the LLL reduction algorithm to find a useful solution to the closest vector problem satisfies

$$\log T \geq 0.1707N - 15.82. \tag{9}$$

Thus if $N = 251$ and $c = 3.7$, one has $T > 5 \cdot 10^{11}$ MIPS-years.

For the optimized version of NSS presented in Section 2, we have $N = 251$ and $c > 5.3$. Since larger c values in (7) yield longer LLL running times, we see that the time to find the target vector should be at least 10^{12} MIPS-years, and is probably considerably higher. In general, we obtain this lower bound provided that $N, \mathcal{F}_f, \mathcal{F}_g$ are chosen so that $\|f_1\|, \|g_1\|$ give a large enough value for c in (8).

4.3   Lattice Attacks on Transcripts

Another potential area of vulnerability is a transcript of signed messages. Oscar can examine a list of signatures $s, s', s'' \ldots$, which means that he has at his disposal the lists

$$fw, fw', fw'', \ldots \bmod q \qquad \text{and} \qquad gw, gw', gw'', \ldots \bmod q. \tag{10}$$

If Oscar can determine any of the w values, then he can easily recover f and g. Using division, Oscar can obtain $w^{-1}w' \bmod q$ and other similar ratios, so he can launch an attack on the pair (w, w') identical to that described in the preceding section. As long as $\|w\|, \|f\|$, and $\|g\|$ are about the same size, the value of κ will remain the same or increase, leading to no improvement in the breaking time.

Oscar can also set up a kN-dimensional NTRU type lattice using the ratios of signatures $w^{(1)}/w^{(1)}, w^{(2)}/w^{(1)}, \ldots, w^{(k)}/w^{(1)}$. The target is $(w^{(1)}, \ldots, w^{(k)})$. With this approach the value of κ decreases as k increases, giving the attacker a potential advantage, but the increasing dimension more than offsets any advantage gained. With the parameters given in Section 2, the optimal value of k for the attacker is $k = 10$, giving $\kappa = 4.87/\sqrt{10N}$. This is a bit better than the $c > 5.3$ coming from the original $2N$ dimensional lattice, but still considerably worse than the $c = 3.7$ that gave us the original lower bound of 10^{12} MIPS-years.

There are several other variations on the lattice attacks described in this and the previous section, but none appears to be stronger then the closest vector attack on the public key given in Section 4.2.

4.4   Lifting a NSS Signature Lattice to $\mathbb{Z}$

Recall that an attacker Oscar is presumed to have access to a transcript of signed messages such as given in (10). Various ways in which he might try to exploit this mod q information are described in Sections 4.3. In this section we are concerned with the possibility that Oscar might lift the transcript information (10) and recover the values of $f * w, f * w', \ldots$ exactly over $\mathbb{Z}$.

This is the primary area where the signature scheme with zero deviations differs from the optimized scheme. If the signatures can be recovered over $\mathbb{Z}$, as they can be if $D_{\min} = D_{\max} = 0$, then two additional lattice attacks are made possible. In the optimized scheme of Section 2, we ensure that a lift back to $\mathbb{Z}$ is impractical by making the number of possible liftings greater than 2^{80}. This leaves Oscar with only the lattice attacks described in Sections 4.2 and 4.3 and allows us to take $N = 251$ while maintaing a breaking time in excess of 10^{12} MIPS years.

We now investigate the attacks that are possible if such a lifting can be accomplished. This analysis, irrelevant for the optimized parameters, allows us to set parameters for a simpler variant of NSS with $D_{\min} = D_{\max} = 0$.

Suppose that Oscar forms the lattice L' generated by $X^i * f * w$ with $0 \leq i < N$ and a few different values of w (or similarly for $X^i * g * w$). It is highly likely that the shortest vectors in L' are the rotations of f. Essentially, Oscar is searching for a greatest common divisor of the products $f * w$, though the exponentially large class number of the underlying cyclotomic field greatly obstructs the search. Although it is still not easy to find very short vectors in the lattice L' using lattice reduction, the fact that $\dim(L') = N$, as compared to the NTRU lattice L^{NT} of dimension $2N$, means that L' is easier to reduce than L^{NT}.

The difficulty of finding a solution to the shortest vector problem for the lattice L' appears to be related, as one might expect, to the magnitude of the norm of f. For example, if one considers a sequence of lattices L' of dimension N formed with f satisfying $\|f\| \approx \sqrt{2N/3}$, then our experiments have shown that the extrapolated time necessary for the LLL reduction algorithm to locate f is at least T MIPS years, where T is given by the formula

$$\log T = 0.1151N - 7.9530. \tag{11}$$

As the norm of f is reduced, the time goes down. For example, if we take $\|f\| \approx \sqrt{0.068N}$, then our experiments show that the breaking time is greater than the T given by the formula

$$\log T = 0.0785N - 6.2305. \tag{12}$$

One further lattice attack of dimension $2N$ is enabled if a lifting to $\mathbb{Z}$ is possible. One can view it as an alternative attack on the gcd problem. Given two products $f * w$ and $g * w$, one can reduce these modulo any integer Q and then take the ratio, obtaining $f^{-1} * g$ modulo Q. This is very similar to the original problem of finding the private key from the public key, but there is an important difference. The integer Q can be chosen as large as desired, which has the effect of decreasing the value of κ. As a result, it becomes easier to reduce the lattice. The advantage of making Q larger does not continue indefinitely, and the ultimate result is to reduce the effective dimension of the lattice from $2N$ to N. Experiments have shown that when f and g satisfying $\|f\| = \|g\| = \sqrt{2N/3}$ are used to generate these lattices and an optimal value of Q is chosen for each N, the extrapolated time necessary for the LLL reduction algorithm to locate f is at least T MIPS years, where T is given by the formula

$$\log T = 0.0549N + 1.7693. \tag{13}$$

This third approach seems to be the strongest attack, yielding a lower bound of 10^{12} MIPS years when $N > 680$. As with the N-dimensional lattice, decreasing the norms of f and g does not seem to lower the slope of the line very much, while increasing the norms increases the slope somewhat. A closest vector attack on (f_1, g_1) might decrease this lower bound a bit, but should not alter it substantially.

4.5   Forgery via Lattice Reduction

The opponent, Oscar, can try to forge a signature s on a given message m by means of lattice reduction. We show in this section that an ability to accomplish this implies an ability to consistently locate a very short vector in a large class of $(2N + 1)$-dimensional lattices.

First consider the case that $D_{\min} = D_{\max} = 0$, so Oscar must find a polynomial s satisfying $s \equiv f_0 * m \pmod{p}$ and such that $t \equiv h * s \pmod{q}$ satisfies $t \equiv g_0 * m \pmod{p}$. Let m_s and m_t be the polynomials with coefficients between $-p/2$ and $p/2$ satisfying $m_s \equiv f_0 * m \bmod p$ and $m_t \equiv g_0 * m \bmod p$, respectively. Consider the $(2N + 1)$-dimensional lattice L_m generated by

$$\left\{ (X^i, X^i * h, 0) : 0 \le i < N \right\} \cup \left\{ (0, qX^i, 0) : 0 \le i < N \right\} \cup \left\{ (m_s, m_t, 1) \right\}.$$

Then L_m contains the vector $\tau = (s - m_s, t - m_t, -1)$. The norm of τ can be estimated by assuming that its coordinates are more-or-less randomly distributed in the interval $[-q/2, q/2]$. This yields $\|\tau\| \approx q\sqrt{N/6}$.

The vector τ is also contained in the lattice $L_p = (p\mathbb{Z})^{2N} \oplus \mathbb{Z}$. Let $L_{m,p} = L_m \cap L_p$ be the intersection. In other words, letting I_N denote the N-by-N identity matrix and H the N-by-N circulant matrix formed from the coefficients of the public key h, the lattice $L_{m,p}$ is the intersection of the lattices generated by the rows of the following matrices:

$$L_{m,p} = \begin{bmatrix} I_N & H & 0 \\ 0 & qI_N & 0 \\ m_s & m_t & 1 \end{bmatrix} \cap \begin{bmatrix} pI_N & 0 & 0 \\ 0 & pI_N & 0 \\ 0 & 0 & 1 \end{bmatrix}.$$

Then $L_{m,p}$ has determinant equal to $(\det L)p^{2N}$. Referring to (6) we see that

$$\kappa \approx \sqrt{\pi eq/6p^2}.$$

For example, $(N, p, q) = (719, 3, 359)$ gives $\kappa \approx 7.5$. This means that the construction of a signed message is equivalent to finding a vector in $L_{m,p}$ that is about 7.5 times longer than the expected shortest vector. It follows that if Oscar is able to forge messages with a reasonable probability, then with reasonable probability he can also find vectors within a factor of 7.5 of the shortest vector. Experiments have indicated that for $N \approx 700$, it requires far in excess of 10^{12} MIPS-years to find such a vector in the $(2N + 1)$-dimensional lattice $L_{m,p}$. We note also that the probability that such a vector would have all of its coefficients bounded in absolute value by $q/2$ is extremely low.

The case of the optimized parameters of Section 2 is similar. Oscar's best strategy is probably to simply choose m_s at random having the correct properties (i.e., with $\text{Dev}(m_s, f_0 * m)$ in the allowable range) and to choose

$$m_t \equiv g_0 * m \bmod p$$

exactly. The optimized parameters $(N, p, q) = (251, 3, 128)$ lead to a 503-dimensional lattice with $\kappa = 4.5$. Oscar must first try to find a vector no more than 4.5 times longer than the shortest vector. He must then refine his search so that the first N coordinates of his vector have absolute value less than $q/2$ and so that the second N coordinates have at least 55 and no more than 87 coordinates with absolute value greater than $q/2$. The norm condition alone requires about 10^5 MIPS years for LLL to produce a candidate. Experiments indicate that if the necessary additional constraints are placed on the sup norms of the vectors, then the required time will significantly exceed 10^{12} MIPS years.

Another, less efficient, forgery attack requiring a $3N$-dimensional lattice is described in detail in [5].

In conclusion, forgery solutions probably exist in both the general and the optimized versions of NSS, but the time required to find a forgery is sufficiently large so as to preclude a successful attack based on this approach.

4.6   Transcript Averaging Attacks

As mentioned previously, examination of a transcript (10) of genuine signatures gives the attacker a sequence of polynomials of the form

$$s = f * w \equiv (f_0 + pf_1)(m + w_1 + pw_2) \pmod{q}$$

with varying w_1 and w_2. A similar sequence is known for g. Because of the inherent linearity of these expressions, we must prevent Oscar from obtaining useful information via a clever averaging of long transcripts.

The primary tool for exploiting such averages is the *reversal* of a polynomial $a(X) \in R$ defined by $\rho(a) = a(X^{-1})$. Then the average of $a * \rho(a)$ over a sequence of polynomials with uncorrelated coefficients will approach the constant $\|a\|^2$, while the average of $a' * \rho(a)$ over uncorrelated polynomials will converge to 0. If m, w_1, and w_2 were essentially uncorrelated, then Oscar could obtain useful information by averaging expressions like $s * \rho(m)$ over many signatures. Indeed, this particular expression would converge to $f\|m\|^2$, and thus would reveal the private key f.

There is an easy way to prevent all second moment attacks of this sort. Briefly, after m, w_1, and a preliminary w_2 are chosen, Bob goes through the coefficients of $m + w_1$ and, with probability $1/p$, subtracts that value from the corresponding coefficient of w_2. This causes averages of the form $a * \rho(b)$ created from signatures to equal 0. For further details on this attack and the defense that we have described, see [5]. We also mention that it might be possible to compute averages that yield the value of $f * \rho(f)$ and averages that use fourth power moments, but the former does not appear to be useful for breaking the scheme and the latter, experimentally, appears to converge much too slowly to be useful. Again we refer to [5] for details.

4.7   Forging Messages to Known Signatures

Another possible attack is to take a list of one or more valid signatures (s, t, m), generate a large number of messages m', and try to find a signature in the list that validly signs one of the messages. It is important to rule out attacks of this sort, since for example, one might take a signature in which m says "IOU \$10" and try to find an m' that says "IOU \$1000". Note that this attack is different from the attack in Section 4.1 in which one chooses an m and an s with valid $\text{Dev}(s, m)$ and hopes that $t \equiv h*s \pmod{q}$ has a valid $\text{Dev}(t, g_0*m)$. The fact that (s, t, m) is already a valid signature implies some correlation between s and t, which may make it more likely that (s, t) also signs some other m'.

In the case of zero deviations, if signature encoding is used as suggested in Section 4.9 then it is quite clear that the probability of a successful attack by this method is negligible.

In the case of the optimized parameters the situation is somewhat harder to analyze, but a conservative probabilistic estimate shows that the possibility of a successful forgery is less than 2^{-67}. For added security, one can reduce the value of $D_{\max}$ to 81. This makes it only a little harder to produce a valid signature while reducing the above probability to less than 2^{-82}. See [5] for details.

4.8   Soundness of NSS

A signature scheme is considered sound if it can be proved that the ability to produce several valid signatures on random messages implies an ability to recreate the secret key. We can not prove this for the parameters given in Section 2, which have been chosen to maximize efficiency. Instead, the preceding sections on security analysis make a strong argument that forgery is not feasible without the private key, and that it is not feasible to recover the private key from either a transcript of valid signatures or the public key.

We can, however, make a probabilistic argument for soundness under certain assumptions. For example, recall from Section 4.5 that the existence of a signed message (m, s) implies the existence of a vector in a lattice which is a factor of $\kappa = \sqrt{\pi eq/(6p^2)}$ times larger than the expected smallest vector. We have chosen $p = 3$ for efficiency, but if p is somewhat larger, for fixed N, then κ will be less than 1. This implies that the existence of such a vector by random chance is extremely unlikely, and that such a vector is probably related to a genuine product $f * w$. If we assume the ability of Oscar to produce such products on demand, given an input m, with a somewhat larger p it is not too hard to see that Oscar can probably recover f_1.

4.9 Signature Encoding

In practice, it is important that the signature be encoded (i.e., padded and transformed) so as to prevent a forger from combining valid signatures to produce new valid signatures. For example, let s_1 and s_2 be valid signatures on messages m_1 and m_2, respectively. Then there is a nontrivial possibility that the sum $s_1 + s_2$ will serve as a valid signature for the message $m_1 + m_2$. This and other similar sorts of attacks are easily thwarted by encoding the signature. For example, one might start with the message M (which is itself probably the hash of a digital document) and concatenate it with a time/date stamp D and a random string R. Then apply an all-or-nothing transformation to $M\|D\|R$ to produce the message m to be signed using NSS. This allows the verifier to check that m has the correct form and prevents a forger from combining or altering valid signatures to produce a new valid signature.

This is related to the more general question of whether or not Oscar can create any valid signature pairs (m, s), even if he does not care what the value of m is. When encoding is used, the probability that a random m will have a valid form can easily be made smaller than 2^{-80}.

References

1. E.F. Brickell and K.S. McCurley. *Interactive Identification and Digital Signatures*, AT&T Technical Journal, November/December, 1991, 73–86.

2. L.C. Guillou and J.-J. Quisquater. *A practical zero-knowledge protocol fitted to security microprocessor minimizing both transmission and memory*, Advances in Cryptology—Eurocrypt '88, Lecture Notes in Computer Science 330 (C.G. Günther, ed.), Springer-Verlag, 1988, 123–128.

3. J. Hoffstein, B.S. Kaliski, D. Lieman, M.J.B. Robshaw, Y.L. Yin, *Secure user identification based on constrained polynomials*, US Patent 6,076,163, June 13, 2000.

4. J. Hoffstein, J. Pipher, J.H. Silverman, *NTRU: A new high speed public key cryptosystem*, in Algorithmic Number Theory (ANTS III), Portland, OR, June 1998, Lecture Notes in Computer Science 1423 (J.P. Buhler, ed.), Springer-Verlag, Berlin, 1998, 267–288.

5. J. Hoffstein, J. Pipher, J.H. Silverman, *NSS: A Detailed Analysis of the NTRU Lattice-Based Signature Scheme*, <www.ntru.com>.

6. J. Hoffstein, D. Lieman, J.H. Silverman, *Polynomial Rings and Efficient Public Key Authentication*, in Proceeding of the International Workshop on Cryptographic Techniques and E-Commerce (CrypTEC '99), Hong Kong, (M. Blum and C.H. Lee, eds.), City University of Hong Kong Press.

7. J. Hoffstein, J.H. Silverman, *Polynomial Rings and Efficient Public Key Authentication II*, in Proceedings of a Conference on Cryptography and Number Theory (CCNT '99), (I. Shparlinski, ed.), Birkhauser.

8. A.J. Menezes, *Software Implementation of Elliptic Curve Cryptosystems Over Binary Fields*, presentation at CHES 2000, August 17, 2000.

9. A.J. Menezes and P.C. van Oorschot and S.A. Vanstone. *Handbook of Applied Cryptography*, CRC Press, 1996.

10. I. Mironov, *A note on cryptanalysis of the preliminary version of the NTRU signature scheme*, IACR preprint server, <http://eprint.iacr.org/2001/005/>

11. T. Okamoto. *Provably secure and practical identification schemes and corresponding signature schemes*, Advances in Cryptology—Crypto '92, Lecture Notes in Computer Science 740 (E.F. Brickell, ed.) Springer-Verlag, 1993, 31–53.

12. C.-P. Schnorr. *Efficient identification and signatures for smart cards*, Advances in Cryptology—Crypto '89, Lecture Notes in Computer Science 435 (G. Brassard, ed), Springer-Verlag, 1990, 239–251.

13. J.H. Silverman. *Estimated Breaking Times for NTRU Lattices*, NTRU Technical Note #012, March 1999, <www.ntru.com>.

14. J.H. Silverman. *Almost Inverses and Fast NTRU Key Creation*, NTRU Technical Note #014, March 1999, <www.ntru.com>.

15. J. Stern. *A new identification scheme based on syndrome decoding*, Advances in Cryptology—Crypto '93, Lecture Notes in Computer Science 773 (D. Stinson, ed.), Springer-Verlag, 1994, 13–21.

16. J. Stern. *Designing identification schemes with keys of short size*, Advances in Cryptology—Crypto '94, Lecture Notes in Computer Science 839 (Y.G. Desmedt, ed), Springer-Verlag,1994, 164–173.

17. D. Stinson, *Cryptography: Theory and Practice*. CRC Press, 1997.

The Bit Security of Paillier's Encryption Scheme and Its Applications*

Dario Catalano[1], Rosario Gennaro[2], and Nick Howgrave-Graham[2]

[1] Dipartimento di Matematica e Informatica
Università di Catania. Viale A. Doria 6, 95125 Catania.
catalano@dmi.unict.it.
[2] IBM T.J.Watson Research Center
PO Box 704, Yorktown Heights, New York 10598, USA.
{rosario,nahg}@watson.ibm.com

Abstract. At EuroCrypt'99, Paillier proposed a new encryption scheme based on higher residuosity classes. The new scheme was proven to be one-way under the assumption that *computing N-residuosity classes in $Z^*_{N^2}$ is hard*. Similarly the scheme can be proven to be semantically secure under a much stronger *decisional* assumption: given $w \in Z^*_{N^2}$ it is hard to decide if w is an N-residue or not.

In this paper we examine the bit security of Paillier's scheme. We prove that, if computing residuosity classes is hard, then given a random w it is impossible to predict the least significant bit of its class significantly better than at random. This immediately yields a way to obtain semantic security without relying on the decisional assumption (at the cost of several invocations of Paillier's original function).

In order to improve efficiency we then turn to the problem of simultaneous security of many bits. We prove that Paillier's scheme hides $n - b$ (up to $O(n)$) bits if one assumes that computing the class c of a random w remains hard even when we are told that $c < 2^b$. We thoroughly examine the security of this stronger version of the intractability of the class problem.

An important theoretical implication of our result is the construction of the first trapdoor function that hides super-logarithmically (up to $O(n)$) many bits. We generalize our techniques to provide sufficient conditions for a trapdoor function to have this property.

1 Introduction

At EuroCrypt'99 Paillier [10] proposed a new encryption scheme based on higher residuosity classes. It generalized previous work by Okamoto and Uchiyama [9]. Both works are based on the problem of computing high-degree residuosity classes modulo a composite of a special form (in Paillier's the modulus is N^2 where N is a typical RSA modulus, while in [9] the modulus is $N = p^2q$ where p, q are large primes.)

---

* The first author's research was carried out while visiting the Computer Science Department of Columbia University.

B. Pfitzmann (Ed.): EUROCRYPT 2001, LNCS 2045, pp. 229–243, 2001.

The mathematical details are described below, but for now let us sketch the basics of Paillier's scheme. It can be shown that $Z_{N^2}^*$ can be partitioned into N equivalence classes generated by the following equivalence relationship: $a, b \in Z_{N^2}^*$ are equivalent iff ab^{-1} is an N-residue in $Z_{N^2}^*$. The N-residuosity class of $w \in Z_{N^2}^*$ is the integer $c = Class(w)$ such that w belongs to the cth residuosity class (in a well specified ordering of them). The conjectured hard problem is: given a random w, compute c. It can be shown that computing $c = Class(w)$ is possible if the factorization of N is known.

Thus Paillier suggests the following encryption scheme: To encrypt a message $m \in Z_N$, the sender sends a random element $w \in Z_{N^2}^*$ such that $Class(w) = m$ (this can be done efficiently as it is shown later). The receiver who knows the factorization of N, given w can compute m.

If we assume that computing residuosity classes is hard, then this scheme is simply one-way. Indeed even if computing the whole of m is hard, it is possible that partial information about m can be leaked.

What we would like to have is instead a *semantically secure* scheme. Semantic security (introduced by Goldwasser and Micali in [7]) basically says that to a polynomial time observer the encryption of a message m should look indistinguishable from the encryption of a different message m'. Paillier's scheme is semantically secure if we assume a stronger *decisional* assumption: given a random element $w \in Z_{N^2}^*$ it is impossible to decide efficiently if w is an N-residue or not.

HARD-CORE BITS. The concept of hard-core bits for one-way functions was introduced by Blum and Micali in [4].

Given a one-way function $f : \{0,1\}^n \to \{0,1\}^n$ we say that $\pi : \{0,1\}^n \to \{0,1\}$ is a hard-core predicate for f if given $y = f(x)$ it is hard to guess $\pi(x)$ with probability significantly higher than $1/2$. Another way of saying this is that if x is chosen at random then $\pi(x)$ looks random (to a polynomial time observer) even when given $y = f(x)$.

Blum and Micali showed the existence of a hard-core predicate for the discrete logarithm function. Later a hard-core bit for the RSA/Rabin functions was presented in [1]. Goldreich and Levin in [6] show that any one-way function has a hard-core predicate.

The concept can be generalized to many hard bits. We say that k predicates $\pi_1, \ldots, \pi_k$ are *simultaneously* hard-core for f if given $f(x)$ the collection of bits $\pi_1(x), \ldots, \pi_k(x)$ looks random to a polynomial time observer.

OUR RESULT: In this paper we investigate the hard core bits of Paillier's new trapdoor scheme. We first prove that the least significant bit of the $c = Class(w)$ is a hard-core bit if we assume computing residuosity classes is hard. In other words we show that given a random $w \in Z_{N^2}^*$, if one can guess $lsb(Class(w))$ better than at random, then one can compute the whole $Class(w)$ efficiently.

Let $n = |N|$. The result above can be generalized to the simultaneous hardness of the least $O(\log n)$ bits using standard techniques. We then show that by slightly strengthening the assumption on computing residuosity classes we are able to extract many more simultaneously hard-core bits. More precisely, for any

$\omega(\log n) \leq b < n$ we show that Paillier's scheme hides the $n - b$ least significant bits, if we assume that computing residuosity classes remain hard even if we are told that the class is smaller than 2^b.

The residuosity class problem seems to remain hard even in this case. Actually we see *no* way to exploit knowledge of the bound (i.e. the fastest known algorithm to compute c even in this case is to factor N). We discuss this further in section 3.4.

An interesting feature of our construction is that the number of bits hidden by the function is related to the underlying complexity assumption that one is willing to make. The smaller the bound is (i.e. the stronger the assumption), the more bits one can hide.

A Theoretical Implication. If f is a trapdoor permutation that simultaneously hides k bits, then we can securely encrypt k bits with a single invocation of f (as originally described in [7]).

However, for all previously known trapdoor functions (like RSA) $f:\{0,1\}^n \to \{0,1\}^n$ we only know how to prove that $k = O(\log n)$ bits are simultaneously hard-core. Thus to securely encrypt m bits one needs to invoke the function $\Omega(m/\log n)$ times.

Another way to look at our result is that we show a candidate trapdoor function that hide up to $O(n)$ bits. To our knowledge this is the first example of trapdoor problems with a super-logarithmic number of hard-core predicates.

We also generalize our construction to a large class of trapdoor functions by giving sufficient conditions for a trapdoor function to hide super-logarithmically many bits[1].

Decisional Assumptions. As we mentioned earlier, the scheme of Paillier [10] can also be proven to be semantically secure under a decisional problem involving residuosity classes. In other words if assuming that deciding N-residuosity is hard, then his scheme hide *all* n input bits.

Notice however the difference with our result. We prove that these two schemes hide many bits, under a *computational* assumption, about computing residuosity class.

Decisional assumptions are very strong. Basically a decisional problem is a true/ false question which we assume the adversary is not able to solve. Conversely computational assumptions (only) require that the adversary cannot compute the *full* solution of a computational problem. Thus, whenever possible, computational assumptions should be preferred to decisional ones.

The goal of this paper is to show example of trapdoor functions that hides several bits without resorting to true/false questions.

---

[1] The above discussion implicitly rules out *iterated* functions. Indeed [4] shows that if $f(x)$ is a one-way function and $\pi(x)$ is a hard-core predicate for it, then the iterated function $f^k(x)$ is clearly also one-way and it simultaneously hide the following k bits: $\pi(x), \pi(f(x)), ..., \pi(f^{k-1}(x))$. We are interested in functions that hide several bits in a *single* iteration.

APPLICATIONS. The main application of our result is the construction of a new semantically secure encryption scheme based on Paillier's scheme. Assuming that Paillier's function securely hides k bits, we can then securely encrypt an m-bit message using only $O(m/k)$ invocations; k is of course a function of n, the security parameter of the trapdoor function. We can do this without resorting to the decisional assumption about N-residuosity, but simply basing our security on the hardness of computing residuosity classes.

Today we can assume that $n = 1024$. Also in practice public-key cryptography is used to exchange keys for symmetric encryption. Thus we can then assume that $m = 128$. With a reasonable computational assumption we can encrypt the whole 128-bit key with a *single* invocation of Paillier's scheme. The assumption is that computing the class is hard even when we are promised that $c < N^{.875}$.

We discuss this new scheme and make comparisons with existing ones in Section 5.

1.1   Related Work

Computing high-degree residuosity classes is related to the original work of Goldwasser and Micali [7] who suggested quadratic residuosity in Z_N^* as a hard trapdoor problem (where N is an RSA modulus). Later Benaloh [2] generalized this to deciding s-residuosity where s is a small prime dividing $\phi(N)$. In Benaloh's scheme, s is required to be small (i.e. $|s| = O(\log n)$) since the decryption procedure is exponential in s. By changing the structure of the underlying field, Okamoto-Uchiyama in [9] and Paillier in [10] were able to lift this restriction and consider higher degree residuosity classes.

The idea of restricting the size of the input space of a one-way function in order to extract more hard bits goes back to Hastad *et al.* [8]. They basically show that the ability to invert $f(x) = g^x \bmod N$ when x is a random integer $x < O(\sqrt{N})$ is sufficient to factor N. Then they show that discrete log modulo a composite must have $n/2$ simultaneously hard bits, otherwise the above restricted-input function can be inverted (i.e. we could factor N). [8] shows the first example of one-way function with a superlogarithmic number of hard-core bits. No such examples was known for *trapdoor* function.

Building on ideas from [8], Patel and Sundaram in [11] show that if one assumes that $f(x) = g^x \bmod p$ (with p prime) remains hard to invert even when $x < B$, then discrete logarithm simultaneously hide $k - b$ bits ($k = |p|, b = |B|$). In their case, as in ours, one must make an explicit computational assumption about the hardness of inverting the function with small inputs. There is an important difference between [11] and our computational assumption though. In [11] we know that there exist algorithms to find $x < B$ given $y = g^x$, which run in $O(\sqrt{B})$ steps. In our case , as discussed in section 3.4, an attack with a similar complexity is not known.

1.2   Paper Organization

In Section 3 we describe in detail the scheme based on Paillier's function. In Section 4 we generalize our result to a larger class of trapdoor functions, giving sufficient conditions for a trapdoor function to hide super-logarithmically many bits. We then discuss applications to public-key encryption and comparisons to other schemes in Section 5. Our work raises some interesting open problems which we list at the end in Section 6.

2   Definitions

In the following we denote with $\mathbf{N}$ the set of natural numbers and with $\mathbf{R}^+$ the set of positive real numbers. We say that a function $\mathsf{negl} : \mathbf{N} \to \mathbf{R}^+$ is *negligible* iff for every polynomial $P(n)$ there exists a $n_0 \in \mathbf{N}$ s.t. for all $n > n_0$, $\mathsf{negl}(n) \leq 1/P(n)$. We denote with $\mathcal{PRIMES}(k)$ the set of primes of length k.

If A is a set, then $a \leftarrow A$ indicates the process of selecting a at random and uniformly over A (which in particular assumes that A can be sampled efficiently).

TRAPDOOR PERMUTATIONS. Let $f_n : \{0,1\}^n \to \{0,1\}^n$ be a family of permutations. We say that f_n is a *trapdoor family* if the following conditions hold:

- f_n can be computed in polynomial time (in n)
- f_n can be inverted in polynomial time only if given a description of f_n^{-1}. I.e. for any probabilistic polynomial time Turing Machine $\mathcal{A}$ we have that

$$\Pr[x \leftarrow \{0,1\}^n; \mathcal{A}(f_n, f_n(x)) = x] = \mathsf{negl}(n)$$

The above notion can be generalized to *probabilistic* functions where each $f_n : \{0,1\}^n \times \{0,1\}^r \to \{0,1\}^{n+r}$ is a permutation, but we look at the second argument as a random string and we assume that given $y \in \{0,1\}^{n+r}$ we cannot compute the first argument, i.e. for any probabilistic polynomial time Turing Machine $\mathcal{A}$ we have that

$$\Pr[x \leftarrow \{0,1\}^n; s \leftarrow \{0,1\}^r; \mathcal{A}(f_n, f_n(x,s)) = x] = \mathsf{negl}(n)$$

HARD-CORE BITS. A Boolean predicate π is said to be *hard* for a function f_n if no efficient algorithm $\mathcal{A}$, given $y - f(x)$ guesses $\pi(x)$ with probability substantially better than $1/2$. More formally for any probabilistic polynomial time Turing Machine $\mathcal{A}$ we have that

$$\left| \Pr[x \leftarrow \{0,1\}^n; \mathcal{A}(f_n, f_n(x)) = \pi(x)] - \frac{1}{2} \right| = \mathsf{negl}(n)$$

For one-way functions f_n, a possible way to prove that a predicate π is hard is to show that any efficient algorithm $\mathcal{A}$ that on input $y = f_n(x)$ guesses $\pi(x)$ with probability bounded away from $1/2$ can be used to build another algorithm $\mathcal{A}'$ that on input y computes x with non-negligible probability.

SIMULTANEOUSLY HARD BITS. A collection of k predicates $\pi_1, \ldots, \pi_k$ is called simultaneously hard-core for f_n if, given $y = f_n(x)$, the whole collection of bits $\pi_1(x), \ldots, \pi_k(x)$ looks "random". A way to formalize this (following [14]) is to say that it is not possible to guess the value of the jth predicate even after seeing $f_n(x)$ and the value of the previous $j - 1$ predicates over x. Formally, for every $j = 1, \ldots, k$, for every probabilistic polynomial time Turing Machine $\mathcal{A}$ we have that:

$$\left| \Pr[x \leftarrow \{0,1\}^n; \mathcal{A}(f_n, f_n(x), \pi_1(x), \ldots, \pi_{j-1}(x)) = \pi_j(x)] - \frac{1}{2} \right| = \mathsf{negl}(n)$$

Here too, a proof method for simultaneously hard-core bits is to show that an efficient algorithm $\mathcal{A}$ contradicting the above equation can be used to build another efficient algorithm $\mathcal{A}$ which inverts f_n with non-negligible probability.

3   Bit Security of Paillier's Scheme

In this section we present our candidate trapdoor function which is based on work by Paillier [10]. Readers are referred to [10] for details and proofs which are not given here.

PRELIMINARIES. Let $N = pq$ be an RSA modulus, i.e. product of two large primes of roughly the same size. Consider the multiplicative group $Z_{N^2}^*$.

Let $g \in Z_{N^2}^*$ be an element whose order is a non zero multiple of N. Let us denote with $\mathcal{B}$ the set of such elements. It can be shown that g induces a bijection

$$\mathcal{E}_g : Z_N \times Z_N^* \rightarrow Z_{N^2}^*$$
$$\mathcal{E}_g(x, y) = g^x y^N \bmod N^2$$

Thus, given g, for an element $w \in Z_{N^2}^*$ there exists an unique pair $(c, z) \in Z_N \times Z_N^*$ such that $w = g^c z^N \bmod N^2$. We say that c is the *class* of w relative to g. We may also denote this with $Class_g(w)$.

We define the *Computational Composite Residuosity Class Problem* as the problem of computing c given w and assume that it is hard to solve.

Definition 1. We say that computing the function $Class_g(\cdot)$ is hard if, for every probabilistic polynomial time algorithm $\mathcal{A}$, there exists a negligible function $\mathsf{negl}()$ such that

$$\Pr \left[\begin{array}{l} p, q \leftarrow \mathcal{PRIMES}(n/2); \ N = pq; \\ g \leftarrow Z_{N^2}^* \text{ s.t. } ord(g) > N; \\ c \leftarrow Z_N; \ z \leftarrow Z_N^*; \ w = g^c z^N \bmod N^2; \\ \mathcal{A}(N, g, w) = c \end{array} \right] = \mathsf{negl}(n)$$

It can be shown that if the factorization of N is known then one could solve this problem: indeed let $\lambda = \mathrm{lcm}(p - 1, q - 1)$ then

$$Class_g(w) = \frac{L(w^\lambda \bmod N^2)}{L(g^\lambda \bmod N^2)} \bmod N \tag{1}$$

where L is defined as the integer[2] $L(u) = (u-1)/N$.

An interesting property of the class function is that it is homomorphic: for $x, y \in Z_{N^2}^*$

$$Class_g(xy \bmod N^2) = Class_g(x) + Class_g(y) \bmod N$$

It is also easy to see that $Class_g(\cdot)$ induces an equivalence relationship (where elements are equivalent if they have the same class) and thus for each c we have N elements in $Z_{N^2}^*$ with class equal to c.

3.1   The Least Significant Bit of *Class* Is Hard

As we said in the introduction, Goldreich and Levin [6] proved that any one-way function has a hard-core bit. Clearly their result applies to Paillier's scheme as well. Here, however, we present a direct and more efficient construction of a hard-core bit.

Consider the function $Class_g(\cdot)$ defined as in the previous section. We show that, given $w = g^c y^N \bmod N^2$, for some $c \in Z_N$ and $y \in Z_N^*$, computing the predicate $lsb(c)$ is equivalent to computing $Class_g(w)$, i.e. $lsb(c)$ is hard for $Class_g$. We start with the following Lemma.

Lemma 1. *Let N be a random n-bit RSA modulus, $y \in Z_N^*$, c an even element of Z_N and g an element in $\mathcal{B}$. Then, denoting $z = 2^{-1} \bmod N$,*

$$(g^c y^N)^z = g^{\frac{c}{2}} y'^N \bmod N^2$$

for some $y' \in Z_N^*$

Proof. Since $z = 2^{-1} \bmod N$, there exist an integer k such that $2z = 1 + kN$. Now

$$(g^c y^N)^z = g^{2z \frac{c}{2}} y^{zN} \bmod N^2 = g^{\frac{c}{2}} (g^{\frac{ck}{2}} y^z)^N \bmod N^2$$

Observe that, being the group Z_{N^2} isomorphic to $Z_N^* \times Z_N$ (for $g \in \mathcal{B}$) [10], this is enough to conclude the proof.   □

Theorem 1. *Let N be a random n-bit RSA modulus, and let the functions $\mathcal{E}_g(\cdot, \cdot)$ and $Class_g(\cdot)$ be defined as above. If the function $Class_g(\cdot)$ is hard (see Definition 1), then the predicate $lsb(\cdot)$ is hard for it.*

Proof. The proof goes by *reductio ad absurdum*: we suppose the given predicate not to be hard, and then we prove that if some oracle $\mathcal{O}$ for $lsb(\cdot)$ exists, then this oracle can be used to construct an algorithm that computes the assumed intractable function, in probabilistic polynomial time. In other words, given $w \in Z_{N^2}^*$ such that $w = \mathcal{E}_g(c, y)$, and an oracle $\mathcal{O}(g, w) = lsb(c)$, we show how to compute, in probabilistic polynomial time, the whole value $c = Class_g(w)$.

For the sake of clarity we divide the proof in two cases, depending on what kind of oracle is given to us. In the first case we suppose to have access to a

---

[2] It is easy to see that both w^λ and g^λ are $\equiv 1 \bmod N$.

perfect oracle, that is an oracle for which $Pr_w[\mathcal{O}(g, w) = lsb(c)] = 1$. Then we will show how to generalize the proof for the more general case in which the oracle is not perfect, but has some non negligible advantage in predicting the required bit. In this last case we will suppose $Pr_w[\mathcal{O}(g, w) = lsb(c)] \geq \frac{1}{2} + \epsilon(n)$ where $\epsilon(n) > \frac{1}{p(n)}$, for some polynomial $p(\cdot)$. For convenience we will denote $\epsilon(n)$ by simply ϵ in the following analysis.

THE PERFECT CASE. The algorithm computes c, bit by bit starting from $lsb(c)$. Denote $c = c_n \ldots c_2 c_1$ the bit expansion of c. It starts by querying $\mathcal{O}(g, w)$ which by assumption will return $c_1 = lsb(c)$. Once we know c_1 we can "zero it out" by using the homorphic properties of the function $Class$. This is done by computing $w' = w \cdot g^{-c_1}$. Finally we use Lemma 1 to perform a "bit shift" and position c_2 in the lsb position. We then iterate the above procedure to compute all of c. A detailed description of the algorithm follows(where () denotes the empty string and $\alpha|\beta$ is the concatenation of the bit strings α and β):

$ComputeClass(\mathcal{O}, w, g, N)$
1. $z = 2^{-1} \bmod N$
2. $c = ()$
3. **for** $i = 0$ **to** $n = |N|$
4. $x = \mathcal{O}(g, w)$
5. $c = c|x$
6.   **if** (x==1) **then**
7. $w = w \cdot g^{-1} \bmod N^2$ (bit zeroing)
8. $w = w^z \bmod N^2$ (bit shifting)
9. **return** c

THE IMPERFECT ORACLE. In this case the above algorithm does not work, because we are not guaranteed that x is the correct bit during any of the iterations. We need to use randomization to make use of the statistical advantage of the oracle in guessing the bit. This is done by choosing randomly $r \in_R Z_N$ and $s \in_R Z_N^*$, considering $\hat{w} = w \cdot g^r \cdot s^N$ and querying $\mathcal{O}(g, \hat{w})$ on several randomized $\hat{w}$'s.

Notice that if $c + r < N$ the oracle returns as output $c_1 + r_1 \bmod 2$, and since we know r_1 we can compute c_1. A majority vote on the result of all the queries will be the correct c_1 with very high probability.

In order to ensure that $c + r < N$, we somewhat "reduce" the size of c. We guess the top $\gamma = 1 - \log \epsilon$ bits of c, and zero them accordingly, i.e.

$$w'_d = g^{2^{n-\gamma}d} w$$

for all 2^γ choices of d (note that is is a polynomial, in n, number of choices).

Of course if we guessed incorrectly the actual top bits of w'_d will not be zeroed, however for one of our guesses they will be, and this guess will yield the correct answer.

Observe that, since we zeroed the leading γ bits of c, the sum $r + c$ can wrap around N only if the γ most significant bits of r are all 1. Thus the probability of $r + c > N$ is smaller that $2^{-\gamma} = \epsilon/2$. We can add this probability to the error probability of the oracle. Consequently the oracle is now correct with probability $\frac{1}{2} + \frac{\epsilon}{2}$. This simply implies that we need to increase the number of randomized queries accordingly.

Once c_1 is known, we zero it and we perform a shift to the right as before. We then repeat the process for the remaining bits. Since the correct d is still unknown, we obtain a (polynomially sized) set of candidate values for c. Notice that we cannot check, given w, which one of the c's is the correct one. However this still implies an algorithm to output c correctly with probability $1/poly$, which contradicts Definition 1. □

3.2   Simultaneous Security of Many Bits

It is not hard to show that $Class_g(\cdot)$ hides $O(\log n)$ bits simultaneously (this can be shown using standard techniques). In this section we show that by slightly strenghtening the computational assumption about computing residuosity class then we can increase the number of simultaneously secure bits, up to $O(n)$.

What we require is that $Class_g(\cdot)$ is hard to compute even when c is chosen at random from $[0..B]$ where B is a bound smaller than N. More formally:

Definition 2. We say that computing the function $Class_g(\cdot)$ is B-hard if, for every probabilistic polynomial time algorithm $\mathcal{A}$, there exists a negligible function $\mathsf{negl}()$ such that

$$\Pr \begin{bmatrix} p,q \leftarrow \mathcal{PRIMES}(n/2); & N = pq; \\ g \leftarrow Z_{N^2}^* \text{ s.t. } ord(g) > N; \\ c \leftarrow [0..B]; & z \leftarrow Z_N^*; & w = g^c z^N \bmod N^2; \\ \mathcal{A}(N, g, w) = c \end{bmatrix} = \mathsf{negl}(n)$$

Clearly in order for the $Class_g$ to be B-hard, it is necessary that the bound B be sufficiently large. If we had only a polynomial (in n) number of guesses, then the definition would be clearly false. Thus when we assume that $Class_g$ is B-hard we implicitly assume that $b = \log B = \omega(\log n)$.

Theorem 2. Let N be a random n-bit RSA modulus; $B = 2^b$. If the function $Class_g(\cdot)$ is B-hard (see Definition 2) then it has $n - b$ simultaneously hard-core bits.

3.3   Proof of Theorem 2

In order to prove Theorem 2 we first need to show that the bits in positions $1, 2, \ldots, n - b$ are individually secure. Then we prove simultaneous security.

INDIVIDUAL SECURITY. Let i be an integer $1 \leq i \leq n - b$ and assume that we are given an oracle $\mathcal{O}_i$ which on input N, g and $u \in_R Z_{N^2}^*$ computes correctly the ith-bit of $Class_g(u)$ with a probability (over u) of $1/2 + \epsilon(n)$, where again $\epsilon(n) > 1/poly(n)$.

In order to show that $Class_g(\cdot)$ is not B-hard, we will show how to build an algorithm $\mathcal{A}$ which uses $\mathcal{O}_i$ and given $w \in Z^*_{N^2}$ with $Class_g(w) < B$, computes $c = Class_g(w)$. Let $\gamma = 1 - \log \epsilon = O(\log n)$.

We split the proof in two parts: the first case has $1 \leq i < n - b - \gamma$. The second one deals with $n - b - \gamma \leq i \leq n - b$.

If $1 \leq i < n - b - \gamma$ the inversion algorithm works as follows. We are given $w \in Z^*_{N^2}$ where $w = g^c y^N \bmod N^2$ and we know that $c = Class_g(w) < B$. We compute c bit by bit; let c_i denote the i-th bit of c. To compute c_1 we square w, i times computing $w_i = w^{2^i} \bmod N^2$. This will place c_1 in the i-th position (with all zeroes to its right). Since the oracle may be correct only slightly more than half of the times, we need to randomize the query. Thus we choose $r \in_R Z_N$ and $s \in_R Z^*_N$ and finally query the oracle on $\hat{w} = w_i g^r s^N \bmod N^2$. Notice the following:

- Given the assumptions on B and i we know that $w_i = w^{2^i} = g^{2^i c} z^{2^i N}$ and $2^i c$ is not taken $\bmod N$ since it will not "wrap around".
- $Class_g(\hat{w}) = 2^i c + r \bmod N$. But since $2^i c$ has at least γ leading zeroes the probability (over r) that $2^i c + r$ wraps around is $\leq \epsilon/2$.
- Since c_1 has all zeroes to its right, there are no carries in the i-th position of the sum. Thus by subtracting r_i to the oracle's answer we get c_1 unless $2^i c + r$ wraps around or the oracle provides a wrong answer.

In conclusion we get the correct c_1 with probability $1/2 + \epsilon/2$, thus by repeating several (polynomially many) times the process and taking majority we get the correct c_1 with very high probability.

Once we get c_1, we "zero" it in the squared w_i by setting $w_i \leftarrow w_i g^{-c_1 2^i} \bmod N^2$. Then we perform a "shift to the right" using Lemma 1, setting $w_i \leftarrow w_i^z \bmod N^2$ where $z = 2^{-1} \bmod N$. At this point we have c_2 in the oracle position and we can repeat the randomized process to discover it. We iterate the above process to discover all the bits of c [3].

Since each bit is determined with very high probability, the value $c = c_b \ldots c_1$ will be correct with non-negligible probability.

If $n - b - \gamma < i < n - b$ the above procedure may fail since now $2^i c$ does not have γ leading zeroes anymore. We fix this problem by guessing the γ leading bits of c (i.e. $c_{b-\gamma}, \ldots, c_b$). This is only a polynomial number of guesses.

For each guess, we "zero" those bits (let α be the γ-bit integer corresponding to each guess and set $w \leftarrow wg^{-2^{b-\gamma}\alpha} \bmod N^2$). Now we are back in the situation we described above and we can run the inversion algorithm. This will give us a polynomial number of guesses for c and we output one of them randomly chosen which will be the correct one with non-negligible probability. Notice that we are not able to verify if the solution is the correct one, but in any case the algorithm violates our security assumption (see Definition 2.)

---

[3] We note that Lemma 1 is necessary to perform "shifts to the right" only for the bits in position $i = 1, \ldots, b$. For the other ones we can shift to the right by simply "undoing" the previous squaring operations.

SIMULTANEOUS SECURITY. Notice that in the above inversion algorithm, every time we query $\mathcal{O}_i$ with the value $\hat{w}$ we know all the bits in position $1, \ldots, i-1$ of $Class_g(\hat{w})$. Indeed these are the first $i-1$ bits of the randomizer r. Thus we can substitute the above oracle with the weaker one $\hat{\mathcal{O}}_i$ which expects $\hat{w}$ and the bits of $Class_g(\hat{w})$ in position $1, \ldots, i-1$. $\square$

3.4   Security Analysis

We note here that the class problem can be considered a weaker version of a composite discrete log problem. Let $d = \gcd(p-1, q-1)$ and let C_m denote the cyclic group of m elements, then for any t dividing λ we have

$$Z^*_{N^2} \simeq C_d \times C_{\lambda/t} \times C_{Nt}.$$

Let $g_2, g_1, g \in Z^*_{N^2}$ be the preimages, under such an isomorphism, of generators of $C_d, C_{\lambda/t}$ and C_{Nt} respectively. Thus we can represent any element of $Z^*_{N^2}$ uniquely as $g_2^{e_2} g_1^{e_1} g^e$, where $e_2 \in Z_d$, $e_1 \in Z_{\lambda/t}$ and $e \in Z_{Nt}$. For a given $g, g_1, g_2 \in Z^*_{N^2}$ the composite discrete logarithm problem we consider is to find these e, e_1, e_2 for any given $w \in Z^*_{N^2}$. For a given g, the class problem is to find just $e \bmod N$ for any given $w \in Z^*_{N^2}$.

Obviously if one can solve the composite discrete logarithm problem, one can solve the class problem; in particular

$$w \equiv g_2^{e_2} g_1^{e_1} g^e \equiv g_2^{e_2} g_1^{e_1} g^{lN+x} \equiv g^x \left(g_2^{k_2 e_2} g_1^{k_1 e_1} g^l \right)^N \equiv g^x y^N \bmod N^2$$

where $k_2 = N^{-1} \bmod d$, and $k_1 = N^{-1} \bmod \lambda/t$, where we note we can make sure $x \in \{0 \ldots N\}$ by a suitable choice of l, and we can force $y \in Z^*_N$ since $(y + kN)^N \equiv y^N \bmod N^2$.

However there is a very important distinction between the class problem and the discrete log problem. In the composite discrete logarithm problem, if we are given g, g_1, g_2, e, e_1, e_2 and w we can verify (in polynomial time) that we do indeed have the discrete logarithm of x. A fascinating and open question in the class problem, is to determine the complexity of an algorithm that verifies the class is correct given only g, $e \bmod N$ and w. Equation 1 shows that this is no harder than factoring, but nothing more is presently known.

Assuming that the function $Class_g$ is hard to compute even in the case that $c < B$ may seem a very strong requirement. It is in some way non-standard.

In order to partially justify it, we notice that not even a trivial exhaustive search algorithm (running in time $O(B)$) seems to work, since even if one is given a candidate c there is no way to verify that it is correct. Verification is equivalent to determining if one has an N'th residue modulo N^2, and this seems a hard problem.

Of course if one did have a verification algorithm that ran in time M then the trivial exhaustive search method would take time $O(MB)$ and there may well be a baby-step, giant-step extension of the method that took time $O(M\sqrt{B})$.

Without an efficient verification algorithm it seems hard to exploit the fact that $c < B$.

Of course because this is a new assumption we are not able to make any stronger claim on its security. Further research in this direction will either validate the security assumption or lead us to learn and discover new things about the residuosity class problem. Though we note that our main theorem still holds (because we can choose B to be large enough to prevent $O(\sqrt{B})$ attacks) even if there were an efficient verification algorithm.

4   A General Result

In this section we briefly show how to generalize the results from the previous section to any family of trapdoor functions with some well defined properties. We show two theorems: the first is a direct generalization of Theorem 2; the second theorem holds for the weaker case in which we do not know the order of the group on which the trapdoor function operates. In this case we are able to extract less hard-core bits.

Theorem 3. *Let M be an m-bit odd integer, and G a group with respect to the operation of multiplication. Let $f : Z_M \to G$ be a one-way, trapdoor isomorphic function (i.e. such that $f(a + b \bmod M) = f(a) \cdot f(b) \in G$). Then, under the assumption that f remains hard to invert when its input belongs to the closed interval $[0 \dots B]$, with $B = 2^b$, it follows that f has $m - b$ simultaneously hard bits.*

It is not hard to see that the techniques of the proof of Theorem 2 can be extended to the above case.

The above theorem assumes that M is exactly known. Let us now consider the case in which M is not known, but we have a very close upper bound on it. I.e. we know $\hat{M} > M$ such that $\frac{\hat{M}-M}{M} = \mathsf{negl}(m)$. Moreover we assume that f is computable on any integer input (but taken $\bmod M$), i.e. we assume that there is an efficient algorithm A that takes as input any integer x and returns as output $A(x) = f(x \bmod M)$.

Theorem 4. *Under the assumption that f remains hard to invert when its input belongs to the closed interval $[0 \dots B]$, with $B = 2^b < \sqrt{M}$, f has $m - 2b$ simultaneously hard bits.*

Proof. The proof follows the same outline of the proof of Theorem 2 except that in this case we are not able to perform "shifts to the right" as outlined in Lemma 1 since we do not know M exactly. Thus the proof succeeds only for the bits in location $b + 1, \dots, m - b$ Notice that this implies $b < m/2$, i.e. $B < \sqrt{M}$.

Again, we first show that each bit is individually secure. We then extend this to prove simultaneous hardness.

INDIVIDUAL SECURITY. Let i be an integer $b \leq i \leq n - b$ and assume that we are given an oracle $\mathcal{O}_i$ which on input M and $u \in_R G$ computes correctly

$(f^{-1}(u))_i$ with probability $1/2 + \epsilon(m)$ where $\epsilon(m) > 1/poly(m)$. As in the proof of theorem 2 prove the statement by providing an algorithm $\mathcal{A}$ which uses $\mathcal{O}_i$ and given $w \in G$ with $f^{-1}(w) < B$, computes $c = f^{-1}(w)$.

The inversion algorithm works almost as the one proposed in the proof of theorem 2. The main difference is that, this time we cannot use lemma 1 to perform shifts to the right. However, in order to let the proof go through, we adopt the following trick: once c_i is known we "zero" it in the original w by setting $w \leftarrow wf(-2^{i-1}c_i \bmod M)$. We then repeat the process with the other bits. The only differences with the above process when computing c_j are that:

- we need to square w only $i - j + 1$ times (actually by saving the result of the intermediate squarings before, this is not necessary).
- to zero c_j once we found it we need to set $w \leftarrow wf(-2^{j-1}c_j \bmod M)$

Since each bit is determined with very high probability, the value $c = c_b \ldots c_1$ will be correct with non-negligible probability.

The simultaneous security of the bits in position $b, b+1, \ldots, n-b$ easily follows, as described in the proof of theorem 2. Details will appear in the final version of this paper.

5   Applications to Secure Encryption

In this section we show how to construct a secure encryption scheme based on our results.

For concreteness let us focus on fixed parameters, based on today's computing powers. We can assume that $n = 1024$ is the size of the RSA modulus N and $m = 128$ (the size of a block cipher key) to be the size of the message M that has to be securely exchanged.

OUR SOLUTION. Using Paillier's $Class_g(\cdot)$ function with our proof methods, it is possible to securely hide the message M with a single invocation of the function. In order to encrypt 128 bits we need to set $n - b > 128$, which can be obtained for the maximum possible choice of $b - 896$ (i.e. the weakest possible assumption). In other words we need to assume that $Class_g$ is hard to invert when $c < N^{.875}$.

To encrypt M one sets $c - r_1|M$ where r_1 is a random string, chooses $y \in_R Z_N^*$ and sends $w = g^c y^N$. This results in two modular exponentiation for the encryption and one exponentiation to decrypt (computations are done mod N^2). The ciphertext size is $2n$.

RSA. In the case of plain RSA we can assume also that the RSA function hides only one bit per encryption (see [5]). In this scenario to securely encrypt (and also decrypt) the message we need 128 exponentiations mod N. The size of the ciphertext is $mn = 128$ Kbit. Encryption speed can be much improved by considering RSA with small public exponent. In any case our scheme is better for decryption speed and message blow-up.

BLUM-GOLDWASSER. Blum and Goldwasser in [3] show how to encrypt with the RSA/Rabin function and pay the $O(m/\log n)$ penalty only in encryption. The idea is to take a random seed r and apply the RSA/Rabin function m times to it and each time output the hard bit. Then one sends the final result r^{e^m} and the masked ciphertext $M \oplus B$ where B is the string of hard bits. It is sufficient to compute r from r^{e^m} to decrypt and this takes a single exponentiation. The size of the ciphertext is $n + m$.

Using the Rabin function this costs only 128 multiplications to encrypt and a single exponentiation to decrypt. We clearly lose compared to this scheme.

Remark: It is worth to notice that even if the proposed solution is less efficient, in practice, than the Blum-Goldwasser one, it remains asymptotically better. As a matter of fact, we need only $O(m/k)$ (where $k = \omega(\log n)$ is the number of simultaneously hard bits produced) invocations of the trapdoor function, while all previously proposed schemes require many more invocations (in general, the number of invocations, has order $O(m/\log n)$). Basically for longer messages we may "catch up" with the other schemes.

The observed slow down, solely depends on the fact that the function used is less efficient than RSA or Rabin. It would be nice to come up with more efficient trapdoor functions that also hides many bits.

6   Conclusions

In this paper we presented the bit security analysis of the encryption scheme proposed by Paillier at Eurocrypt'99 [10]. We prove that the scheme hides the least significant bit of the N-residuosity class. Also by slightly strenghtening the computational assumption about residuosity classes we can show that Paillier's encryption scheme hides up to $O(n)$ bits.

An interesting theoretical implication of our results is that we presented the first candidate trapdoor functions that hide many (up to $O(n)$) bits. No such object was known previously in the literature.

There are several problems left open by this research. Are there trapdoor functions that hide $\omega(\log n)$ bits and are comparable in efficiency to RSA/Rabin? In the case of RSA/Rabin can we come up with a "restricted input assumption" that will allow us to prove that they also hide $\omega(\log n)$ bits? Regarding our new assumptions: is it possible to devise an algorithm to compute $Class_g(\cdot) < B$ that depends on B?

References

1. W. Alexi, B. Chor, O. Goldreich and C. Schnorr. *RSA and Rabin Functions: Certain Parts are as Hard as the Whole.* SIAM J. Computing, 17(2):194–209, April 1988.
2. J.C. Benaloh. Verifiable Secret-Ballot Elections. Ph.D. Thesis, Yale University, 1988.

3. M. Blum and S. Goldwasser. An efficient probabilistic public-key encryption scheme which hides all partial information. *Proc. of Crypto '84*, LNCS vol. 196, pages 289-302
4. M. Blum and S. Micali. How to Generate Cryptographically Strong Sequences of Pseudo-Random Bits. *SIAM Journal on Computing*, Vol. 13, No. 4:850-864, 1984
5. R. Fischlin and C.P. Schnorr. Stronger Security Proofs for RSA and Rabin Bits. *J. of Cryptology*, 13(2):221–244, Spring 2000.
6. O. Goldreich and L. Levin A hard-core predicate for all one-way functions. *Proc. 21st ACM Symposium on Theory of Computing*, 1989
7. S. Goldwasser and S. Micali. Probabilistic Encryption. *JCSS*, 28(2):270–299, April 1984.
8. J. Hastad, A. W. Schrift and A. Shamir. The Discrete Logarithm Modulo a Composite Hides $O(n)$ Bits. *JCSS* Vol. 47, pages 376-404, 1993.
9. T. Okamoto and S. Uchiyama. A New Public-Key Cryptosystem as Secure as Factoring In *Advances in Cryptology - Eurocrypt '97*, LNCS vol. 1233, Springer, 1997, pages 308-318.
10. P. Paillier. Public-Key Cryptosystems Based on Composite Degree Residuosity Classes. In *Advances in Cryptology - Eurocrypt '99*, LNCS vol. 1592, Springer, 1997, pages 223-238.
11. S. Patel and G. S. Sundaram. An Efficient Discrete Log Pseudo Random Generator. In *Advances in Cryptology - CRYPTO '98*, LNCS vol. 1492, Springer, 1998, pages 304-315.
12. M. Rabin. Digital Signatures and Public Key Encryptions as Intractable as Factorization. MIT Technical Report no. 212, 1979
13. R. Rivest, A. Shamir and L. Adelman. A Method for Obtaining Digital Signature and Public Key Cryptosystems. *Comm. of ACM*, 21 (1978), pp. 120–126
14. A. Yao. *Theory and Applications of Trapdoor Functions*. IEEE FOCS, 1982.

Assumptions Related to Discrete Logarithms: Why Subtleties Make a Real Difference

Ahmad-Reza Sadeghi and Michael Steiner

Fachrichtung Informatik, Universität des Saarlandes, D-66123 Saarbrücken, Germany,
{sadeghi,steiner}@cs.uni-sb.de

Abstract. The security of many cryptographic constructions relies on assumptions related to Discrete Logarithms (DL), e.g., the Diffie-Hellman, Square Exponent, Inverse Exponent or Representation Problem assumptions. In the concrete formalizations of these assumptions one has some degrees of freedom offered by parameters such as computational model, the problem type (computational, decisional) or success probability of adversary. However, these parameters and their impact are often not properly considered or are simply overlooked in the existing literature.

In this paper we identify parameters relevant to cryptographic applications and describe a formal framework for defining DL-related assumptions. This enables us to precisely and systematically classify these assumptions.

In particular, we identify a parameter, termed granularity, which describes the underlying probability space in an assumption. Varying granularity we discover the following surprising result: We prove that two DL-related assumptions can be reduced to each other for medium granularity but we also show that they are provably not reducible with generic algorithms for high granularity. Further we show that reductions for medium granularity can achieve much better concrete security than equivalent high-granularity reductions.

Keywords: Complexity Theory, Cryptographic Assumptions, Generic Algorithms, Discrete Logarithms, Diffie-Hellman, Square Exponent, Inverse Exponent.

1 Introduction

Most modern cryptographic algorithms rely on assumptions on the computational difficulty of some particular number-theoretic problem. One well-known class of assumptions is related to the difficulty of computing discrete logarithms in cyclic groups [1]. In this class a number of variants exists. The most prominent ones besides Discrete Logarithm (DL) itself are the computational and decisional Diffie-Hellman (DH) assumptions [2, 3, 4] and their generalization [5, 6]. Less known assumptions are Matching Diffie-Hellman [7, 8], Square Exponent[1](SE) [10, 11], and the Inverse Exponent (IE) [12], an assumption also

---

[1] This problem is called Squaring Diffie-Hellman in [9]

B. Pfitzmann (Ed.): EUROCRYPT 2001, LNCS 2045, pp. 244–261, 2001.

implicitly required for the security of [13, 14]. Several papers have studied relations among these assumptions, e.g., [15, 16, 17, 18, 9].

In the concrete formalizations of these assumptions one has various degrees of freedom offered by parameters such as computational model, problem type (computational, decisional or matching) or success probability of adversary. However, such aspects are often not precisely considered in the literature and related consequences are simply overlooked. In this paper, we address these aspects by identifying the parameters relevant to cryptographic assumptions. Based on this, we present an understandable formal framework and a notation for defining DL-related assumptions. This enables us to precisely and systematically classify these assumptions.

Among the specified parameters, we focus on a parameter we call *granularity* of the probability space which underlies an assumption. Granularity defines what part of the underlying algebraic structure (i.e., algebraic group and generator) is part of the probability space and what is fixed in advance: For high granularity an assumption has to hold for all groups and generators; for medium granularity the choice of the generator is included in the probability space and for low granularity the probability is taken over both the choice of the group and the generator. Assumptions with lower granularity are weaker than those with higher granularity. Yet not all cryptographic settings can rely on the weaker variants: Only when the choice of the system parameters is guaranteed to be random one can rely on a low-granularity assumption. Consider an anonymous payment system where the bank chooses the system parameters. To base the security of such a system a-priori on a low-granularity assumption would not be appropriate. A cheating bank might try to choose a weak group with trapdoors (easy problem instances) [19] to violate the anonymity of the customer. An average-case low-granular assumption would not rule out that infinitely many weak groups exist even though the number of easy problem instances is asymptotically negligible. However, if we choose the system parameters of the payment system through a random yet verifiable process we can resort to a weaker assumption with lower granularity. Note that to our knowledge no paper on anonymous payment systems does address this issue properly. Granularity was also overlooked in different contexts, e.g., [3] ignores that low-granular assumptions are not known to be random self-reducible which leads to a wrong conclusion.

In this paper we show that varying granularity can lead to surprising results. We extend the results of [9] to the problem class IE, i.e., we prove statements on relations between IE, DH and SE for both computational and decisional variants in the setting of [9] which corresponds to the high-granular case. We then consider medium granularity (with other parameters unchanged) and show the impact: We prove that the decisional IE and SE assumptions are equivalent for medium granularity whereas this is provably not possible for their high-granular variants, at least not in the generic model [15]. We also show that reductions between computational IE, SE and DH can offer much better concrete security for medium granularity than their high-granular analogues.

2   Terminology

2.1   Algebraic Structures

The following terms are related to the algebraic structures underlying an assumption.

Group G: All considered assumptions are based on cyclic finite groups. For brevity, however, we will omit the "cyclic finite" in the sequel and refer to them simply as "groups". The order of a group is associated with a security parameter k which classifies the group according to the difficulty of certain problems (e.g., DL).

Group family $\mathcal{G}$: A set of groups with the "same" structure/nature. An example is the family of the groups used in DSS [20], i.e., unique subgroups of $\mathbb{Z}_p^*$ of order q with p and q prime, $|q| \approx 2k$ and $p = rq + 1$ for an integer r sufficiently large to make DL hard to compute in security parameter k. Other examples are non-singular elliptic curves or composite groups $\mathbb{Z}_n^*$ with n a product of two safe primes.

Generator g: In the DL settings, we also need a generator g which generates the group G, i.e., $\forall y \in G \ \exists x \in \mathbb{Z}_{|G|} : \ y = g^x$.

Structure instance SI: The structure underlying the particular problem. In our case this means a group G together with a non-empty tuple of generators g_i. As a convention we abbreviate g_1 to g if there is only a single generator associated with a given structure instance.

2.2   Problem Families

The following two definitions characterize a particular problem underlying an assumption.

Problem family $\mathcal{P}$: A family of abstract and supposedly difficult relations. Examples are Discrete Logarithm (DL), Diffie-Hellman (DH), or the Representation Problem (RP). Note that the problem family ignores underlying algebraic groups and how parameters are chosen. Further, note that in the definition of problem families we don't distinguish between decisional or computational variants of a problem.

Problem instance PI: A list of concrete parameters fully describing a particular instance of a problem family, i.e., a structure instance SI and a tuple $(priv, publ, sol)$ where $priv$ is the tuple of secret values used to instantiate that problem, $publ$ is the tuple of information publicly known on that problem and sol is the solution of that problem instance. This presentation achieves a certain uniformity of description and allows a generic definition of problem types. For convenience, we define PI^{SI}, $PI^{\mathcal{P}}$, PI^{publ}, PI^{priv} and PI^{sol} to be the projection of a problem instance PI to its structure instance, problem family and public, private and solution part, respectively. When not explicitly stated, we can assume that $priv$ consists always of elements from $\mathbb{Z}_{|G|}$ and $publ$ and sol consists of elements from G. Furthermore, the structure instance SI is assumed to be publicly known.

If we take the DH problem for integers modulo a prime p as an example, PI_{DH} is defined by the tuple $(((\mathbb{Z}_p^*, p), (g)), ((x, y), (g^x, g^y), (g^{xy})))$ with $PI_{DH}{}^{\mathcal{P}} := DH$, $PI_{DH}{}^{SI} := ((\mathbb{Z}_p^*, p), (g))$, $PI_{DH}{}^{priv} := (x, y)$, $PI_{DH}{}^{publ} := (g^x, g^y)$, $PI_{DH}{}^{sol} := g^{xy}$, respectively.

3   Parameters of DL-Based Assumptions

In formulating intractability assumptions for problems related to DL we identified the following orthogonal parameters which suffice to describe assumptions relevant to cryptographic applications.[2]

Note that the labels of the following sublists (e.g., "u" and "n" for the first parameter) are used later in Section 4 to identify values corresponding to a given parameter (e.g., "Computational capability of adversary" for above example).

1. **Computational capability of adversary:** Potential algorithms solving a problem have to be computationally limited for number-theoretic assumptions to be meaningful (otherwise we could never assume their nonexistence). Here, we only consider algorithms (called adversary in the following) with running times bounded by a polynomial. The adversary can be of

   u (Uniform complexity): There is a single probabilistic Turing machine (TM) $\mathcal{A}$ which for any given problem instance from the proper domain returns a (not necessarily correct) answer in expected polynomial time in the security parameter k.

   n (Non-uniform complexity): There is an (infinite) family of TMs $\{\mathcal{A}_i\}$ with description size and running time of $\mathcal{A}_i$ bounded by a polynomial in the security parameter k.

   To make the definition of the probability spaces more explicit we model probabilistic TMs always as deterministic machines with the random coins given as explicit input $\mathcal{C}$ chosen from the uniform distribution $\mathcal{U}$.

   Finally, a note on notation: In the case a machine $\mathcal{A}$ has access to some oracles $\mathcal{O}_1, \dots, \mathcal{O}_n$ we denote that as $\mathcal{A}^{\mathcal{O}_1, \dots, \mathcal{O}_n}$.

2. **"Algebraic knowledge":** A second parameter describing the adversary's computational capabilities relates to the adversary's knowledge on the group family. It can be one of the following:

   σ (Generic): This means that the adversary doesn't know anything about the structure (representation) of the underlying algebraic group. More precisely this means that all group elements are represented using a random bijective encoding function $\sigma(\cdot) : \mathbb{Z}_{|G|} \to G$ and group operations can only be performed via the addition and inversion oracles $\sigma_+(\sigma(x), \sigma(y))$ and $\sigma_-(x) \leftarrow \sigma_-(x)$ respectively, which the adversary receives as a black box [15, 22, 23].

---

[2] For this paper we slightly simplified the classification. Further parameters and values and more details can be found in the full paper [21].

If we use σ in the following we always mean the (not further specified) random encoding used for generic algorithms with a group G and generator g implicitly implied in the context. In particular, by $\mathcal{A}^\sigma$ we refer to a generic algorithm.

(marked by absence of σ) (Specific): In this case the adversary can also exploit special properties (e.g., the encoding) of the underlying group.

3. **Success probability:** The adversary's success probability in solving problem instances (for a given security parameter k and probability distribution $\mathcal{D}$) can either be

1 (Perfect): The algorithm $\mathcal{A}$ must solve all problem instances from $\mathcal{D}$.

$1 - 1/\mathbf{poly}(k)$ (Strong): The algorithm $\mathcal{A}$ must be successful with overwhelming probability, i.e., at most a negligible (in k) amount of instances in $\mathcal{D}$ can remain unsolved.

ϵ (Invariant): The algorithm $\mathcal{A}$ must answer at least a constant fraction ϵ of the queries from $\mathcal{D}$ successfully.

$1/\mathbf{poly}(k)$ (Weak): The algorithm $\mathcal{A}$ must be successful with at least a non-negligible amount of queries from $\mathcal{D}$.

An assumption requiring the inexistence of perfect adversaries corresponds to worst-case complexity, i.e., if the assumption holds then there are at least a few hard instances. However, what is a-priori required in most cases in cryptography is an assumption requiring even the inexistence of weak adversaries, i.e., if the assumption holds then most instances are hard.

4. **"Granularity of probability space":** Depending on what part of the structure instance is a-priori fixed (i.e., the assumption has to hold for all such parameters) or not (i.e., the parameters are part of the probability space underlying an assumption) we distinguish among following situations:

l (Low-granular): The group family (e.g., prime order subgroups of $\mathbb{Z}_p^*$) is fixed but not the specific structure instance (e.g., parameters p, q and generators g_i for the example group family given above).

m (Medium-granular): The group (e.g., p and q) but not the generators g_i are fixed.

h (High-granular): The group as well as the generators g_i are fixed.

5. **Problem family:** Following problem families are useful (and often used) for cryptographic applications. As mentioned in Section 2.2 we describe the problem family (or more precisely their problem instances) by an (abstract) structure instance SI $(G, g_1, g_2, \dots)$ and an (explicit) tuple $(priv, publ, sol)$:

DL (Discrete Logarithm): $PI_{DL} := (SI, ((x), (g^x), (x)))$.

DH (Diffie-Hellman): $PI_{DH} := (SI, ((x, y), (g^x, g^y), (g^{xy})))$.

GDH (Generalized Diffie-Hellman): $PI_{GDH} := (SI, ((x_i | 1 \leq i \leq n \wedge n \geq 2), (g^{\prod_{i \in I} x_i} | \forall I \subset \{1, \dots, n\}), (g^{\prod_{i=1}^n x_i})))$.

SE (Square-Exponent): $PI_{SE} := (SI, ((x), (g^x), (g^{x^2})))$.

IE (Inverse-Exponent): $PI_{IE} := (SI, ((x), (g^x), (g^{x^{-1}})))$.

Note that $priv(x)$ has to be an element of $\mathbb{Z}_{|G|}^*$ here, contrary to the other problem families mentioned where $priv$ contains elements of $\mathbb{Z}_{|G|}$.

RP (Representation Problem): $PI_{RP} := (SI, ((x_i | 1 \leq i \leq n \wedge n \geq 2), (\prod_{i=1}^n g_i^{x_i}), (x_i | 1 \leq i \leq n)))$.

6. **Problem type**: Each problem can be formulated in three variants.

   C (Computational): For a given problem instance PI the algorithm $\mathcal{A}$ succeeds if and only if it can solve PI, i.e., $\mathcal{A}(PI^{publ}, \dots) = PI^{sol}$.

   D (Decisional): For a given problem instance PI, a random problem instance $PI_{\mathcal{R}}$ and a random bit b the algorithm $\mathcal{A}$ succeeds if and only if it can decide whether a given solution matches the given problem instance, i.e., $\mathcal{A}(PI^{publ}, b * PI^{sol} + \bar{b} * PI_{\mathcal{R}}{}^{sol}), \dots) = b$.

   M (Matching): For two given problem instances PI_0 and PI_1, and a random bit b the algorithm $\mathcal{A}$ succeeds if and only if it can correctly associate the solutions to their corresponding problem instances, i.e., $\mathcal{A}(PI_0{}^{publ}, PI_1{}^{publ}, PI_b{}^{sol}, PI_{\bar{b}}{}^{sol}, \dots) = b$.

7. **Group family**: We distinguish between group families with the following generic properties. The factorization of the group order contains

   lprim large prime factors (at least one)

   nsprim no small prime factor

   prim only a single and large prime factor

4 Defining Assumptions

Using the parameters and corresponding values defined in the previous section we can define intractability assumptions in a compact and precise way. The used notation is composed out of the labels corresponding to the parameter values of a given assumption. This is best illustrated in following example:[3] The term

$$1/\mathsf{poly}(k)\text{-DDH}^\sigma(\text{c:u}; \text{g:h}; \text{f:prim})$$

denotes the decisional (D) Diffie-Hellman (DH) assumption in prime-order groups (f:prim) with weak success probability ($1/\mathsf{poly}(k)$), limited to generic algorithms (σ) of uniform complexity (c:u), and with high granularity (g:h).

The formal assumption statement automatically follows from the parameter values implied by an assumption term. For space reasons we restrict ourselves again to an example as explanation: To assume that above-mentioned assumption $1/\mathsf{poly}(k)\text{-DDH}^\sigma(\text{c:u}; \text{g:h}; \text{f:prim})$ holds informally means that there are no generic algorithms of uniform complexity which are asymptotically able to distinguish a non-negligible amount of DH tuples from random ones in prime-order subgroups where the probability space is defined according to high granularity. Formally this assumption is given below. To give the reader a better feel for the newly introduced parameter granularity we specify also the corresponding assumptions with medium and low granularity.

A few explanations to the statements: S_G, S_g and S_{PI} are the probabilistic algorithms selecting groups, generators and problem instances, respectively; ExpectRunTime gives a bound on the expected run time of the algorithm and **Prob**$[\mathcal{S} :: \mathcal{PS}]$ gives the probability of statement $\mathcal{S}$ with the probability taken over a probability space defined by $\mathcal{PS}$. Furthermore, remember that PI_{DH} is $(SI, ((x, y), (g^x, g^y), (g^{xy})))$, $PI_{DH}{}^{publ}$ is (g^x, g^y) and $PI_{DH}{}^{sol}$ is (g^{xy}).

---

[3] A more thorough treatment is omitted here due to space reasons and will appear in [21].

1. Assumption $1/\mathsf{poly}(k)$-DDH^σ(c:u; g:h; f:prim), i.e., with high granularity:

   $\forall p_1, p_2 > 0;\ \forall \mathcal{A}^\sigma \in \mathrm{TM};\ \exists k_0;\ \forall k > k_0;$
   $\forall G \leftarrow S_G(\text{"prime-order groups"}, 1^k);\ \forall g \leftarrow S_g(G);\ SI \leftarrow (G, g);$
   $\mathrm{ExpectRunTime}(\mathcal{A}^\sigma(\mathcal{C}, G, g, PI_{DH})) < k^{p_2};$
   $(|\mathbf{Prob}[\mathcal{A}^\sigma(\mathcal{C}, G, g, PI_{DH}{}^{publ}, b * PI_{DH}{}^{sol} + \bar{b} * PI_{\mathcal{R}}{}^{sol}) = b ::$
   $\quad b \xleftarrow{\mathcal{R}} \{0,1\};\ \mathcal{C} \xleftarrow{\mathcal{R}} \mathcal{U};$
   $\quad PI_{DH} \leftarrow S_{PI}(DH, SI);\ PI_{\mathcal{R}} \leftarrow S_{PI}(PI_{DH}{}^{\mathcal{P}}, PI_{DH}{}^{SI});$
   $] - 1/2\ |\cdot 2) < 1/k^{p_1}$

2. As above except now with medium granularity
   $(1/\mathsf{poly}(k)$-DDH^σ(c:u; g:m; f:prim)):

   $\forall p_1, p_2 > 0;\ \forall \mathcal{A}^\sigma \in \mathrm{TM};\ \exists k_0;\ \forall k > k_0;$
   $\forall G \leftarrow S_G(\text{"prime-order groups"}, 1^k);$
   $\mathrm{ExpectRunTime}(\mathcal{A}^\sigma(\mathcal{C}, G, g, PI_{DH})) < k^{p_2};$
   $(|\mathbf{Prob}[\mathcal{A}^\sigma(\mathcal{C}, G, g, PI_{DH}{}^{publ}, b * PI_{DH}{}^{sol} + \bar{b} * PI_{\mathcal{R}}{}^{sol}) = b ::$
   $\quad b \xleftarrow{\mathcal{R}} \{0,1\};\ \mathcal{C} \xleftarrow{\mathcal{R}} \mathcal{U};$
   $\quad g \leftarrow S_g(G);\ SI \leftarrow (G, g);$
   $\quad PI_{DH} \leftarrow S_{PI}(DH, SI);\ PI_{\mathcal{R}} \leftarrow S_{PI}(PI_{DH}{}^{\mathcal{P}}, PI_{DH}{}^{SI});$
   $] - 1/2\ |\cdot 2) < 1/k^{p_1}$

3. As above except now with low granularity
   $(1/\mathsf{poly}(k)$-DDH^σ(c:u; g:l; f:prim)):

   $\forall p_1, p_2 > 0;\ \forall \mathcal{A}^\sigma \in \mathrm{TM};\ \exists k_0;\ \forall k > k_0;$
   $\mathrm{ExpectRunTime}(\mathcal{A}^\sigma(\mathcal{C}, G, g, PI_{DH})) < k^{p_2};$
   $(|\mathbf{Prob}[\mathcal{A}^\sigma(\mathcal{C}, G, g, PI_{DH}{}^{publ}, b * PI_{DH}{}^{sol} + \bar{b} * PI_{\mathcal{R}}{}^{sol}) = b ::$
   $\quad b \xleftarrow{\mathcal{R}} \{0,1\};\ \mathcal{C} \xleftarrow{\mathcal{R}} \mathcal{U};$
   $\quad G \leftarrow S_G(\text{"prime-order groups"}, 1^k);\ g \leftarrow S_g(G);\ SI \leftarrow (G, g);$
   $\quad PI_{DH} \leftarrow S_{PI}(DH, SI);\ PI_{\mathcal{R}} \leftarrow S_{PI}(PI_{DH}{}^{\mathcal{P}}, PI_{DH}{}^{SI});$
   $] - 1/2\ |\cdot 2) < 1/k^{p_1}$

To express relations among assumptions we will use following notation:

$A \implies B$ means that if assumption A holds, so does assumption B, i.e., A (B) is a weaker (stronger) assumption than B (A). Vice-versa, it also means that if there is a (polynomially-bounded) algorithm $\mathcal{A}_B$ breaking assumption B then we can build another (polynomially-bounded) algorithm $\mathcal{A}_A^{\mathcal{A}_B}$ with (oracle) access to $\mathcal{A}_B$ which breaks assumption A.

$A \iff B$ means that $A \implies B$ and $B \implies A$, i.e., A and B are assumptions of the same (polynomial) complexity.

Furthermore, if we are referring to oracle-assumption, i.e., assumptions where we give adversaries access to auxiliary oracles, we indicate it by listing the oracles at the end of the list in the assumption term. For example, the assumption $1/\mathsf{poly}(k)$-DDH^σ(c:u; g:h; f:prim; $\mathcal{O}_{1\text{-}\mathrm{DSE}(\mathrm{c:u;\ g:h;\ f:prim})}$) corresponds to the first assumption statement given above except that now the adversary also gets access to an oracle breaking the assumption 1-DSE(c:u; g:h; f:prim). Finally, if we use $*$ for a particular parameter in an assumption term we mean the class of assumptions where this parameter is varied over all possible values.

5   The Impact of Granularity

It would go beyond the scope (and space) of this paper to discuss all previously identified parameters and we will focus only on granularity. Before stating the actual results, let us first briefly repeat the practical relevance of granularity as alluded in the introduction. Assumptions with lower granularity are weaker and are so more desirable in principle. However, which of the granularity variants is appropriate in cryptographic protocols depends on how and by whom the parameters are chosen. A priori we have to use a high-granular assumption. Yet in following situations we can resort to a weaker less granular assumption: The security requirements of the cryptographic system guarantee that it's in the best (and only) interest of the chooser of the system parameters to choose them properly; the system parameters are chosen by a mutually trusted third party; or the system parameters are chosen in a verifiable random process.[4] Furthermore, care has to be taken for DL-related high and medium granular assumptions in $\mathbb{Z}_p^*$ and its subgroups. Unless we further constrain the set of valid groups with (expensive) tests as outlined by [19], we require, for a given security parameter, considerably larger groups than for the low granular counterpart of the assumptions.

6   Computational DH, SE and IE

Maurer and Wolf [10] prove the equivalence between the computational SE and DH assumption in their uniform and high-granular variant for both perfect and invariant success probabilities.

   We briefly review their results, extend their results to medium granularity and prove similar relations between IE and DH.

6.1   CSE versus CDH

Theorem 1 ([10]).
   $\epsilon\text{-CSE}(c{:}u; g{:}h; f{:}*) \iff \epsilon\text{-CDH}(c{:}u; g{:}h; f{:}*)$ □

More concretely, they show the following: Let $0 < \alpha_1 < 1$, $0 < \alpha_2 < 1$ be arbitrary constants and let G be a cyclic group with known order $|G|$. Then

(a) given an oracle $\mathcal{O}_{CDH}$ which breaks $\epsilon\text{-CDH}(c{:}u; g{:}h; f{:}*)$ in G with success probability at least $\epsilon = \alpha_1$, there exists an algorithm $\mathcal{A}^{\mathcal{O}_{CDH}}$ that breaks $\epsilon\text{-CSE}(c{:}u; g{:}h; f{:}*)$ in G with success probability at least $\epsilon = \alpha_1$.
(b) given an oracle $\mathcal{O}_{CSE}$ which breaks $\epsilon\text{-CSE}(c{:}u; g{:}h; f{:}*)$ in G with success probability at least $\epsilon = \alpha_2$, there exists an algorithm $\mathcal{A}^{\mathcal{O}_{CSE}}$ that breaks $\epsilon\text{-CDH}(c{:}u; g{:}h; f{:}*)$ in G with success probability at least $\epsilon = \alpha_2{}^3$.

From these reductions the theorem immediately follows.

---

[4] This can be done either through a joint generation using random coins [24] or using heuristics such as the one used for DSS key generation [20].

Remark 1. Maurer and Wolf showed the reduction for invariant success probability. However, the results easily extend also to all other variants related to success probability, i.e., weak, strong and perfect.   o

Above relation also holds for medium granularity as we show next.

Theorem 2.

$1/\mathsf{poly}(k)\text{-}\mathrm{CSE}(\mathrm{c:u};\mathrm{g:m};\mathrm{f:}*) \iff 1/\mathsf{poly}(k)\text{-}\mathrm{CDH}(\mathrm{c:u};\mathrm{g:m};\mathrm{f:}*).$ □

Proof (sketch). The proof idea of Theorem 1 can also be applied in this case. The only thing we have to show is that the necessary randomization in the reduction steps can be extended to the medium granularity variants of CDH and CSE. This can be done using standard techniques and is shown in the full version of this paper [21]. The rest of the proof remains then the same as the proof of Theorem 1.   ∎

Remark 2. Reduction proofs of a certain granularity can in general be easily applied to the lower granularity variant of the involved assumptions. The necessary condition is only that all involved randomizations extend to the wider probability space associated with the lower granularity parameter.

Remark 3. In all the mentioned problem families the necessary random self-reducibility exists for medium granularity and above remark always holds, i.e., we can transform proofs from a high-granular variant to the corresponding medium-granular variant. However, it does not seem to extend to low-granular variants. This would require to randomize not only over the public part of the problem instance *PI* and the generator *g* but also over the groups *G* with the same associated security parameter k; this seems impossible to do in the general case and is easily overlooked and can lead to wrong conclusions, e.g., the random self-reducibility as stated in [3] doesn't hold as the assumptions are (implicitly) given in their low-granular form.   o

6.2   CDH versus CIE

In the following we prove that similar relations as above also exist for CIE. In the high-granular case following theorem holds. Here as well as in following results related to IE we will restrict ourselves to groups of prime order. The results also extend to more general groups but the general case is more involved[5] and omitted in this version of the paper for space reasons.

Theorem 3.

$1/\mathsf{poly}(k)\text{-}\mathrm{CDH}(\mathrm{c:u};\mathrm{g:h};\mathrm{f:prim}) \iff 1/\mathsf{poly}(k)\text{-}\mathrm{CIE}(\mathrm{c:u};\mathrm{g:h};\mathrm{f:prim})$ □

More concretely, we show the following: Let G be a cyclic group with known prime order. Then

---

[5] Due to the difference in input domains between IE and other assumptions we have to deal with the distribution of $\mathbb{Z}^*_{|G|}$ over $\mathbb{Z}_{|G|}$. This results, e.g., in the success probability being reduced by a factor of $\varphi(|G|)/|G|$.

(a) given an oracle $\mathcal{O}_{CDH}$ which breaks $*$-CDH(c:u; g:h; f:prim) in G with success probability at least $* = \alpha_1(k)$, there exists an algorithm $\mathcal{A}^{\mathcal{O}_{CDH}}$ that solves CIE(c:u; g:h; f:prim) in G with success probability at least $\alpha_1(k)^{O(\log|G|)}$.[6]

(b) given an oracle $\mathcal{O}_{CIE}$ which breaks $*$-CIE(c:u; g:h; f:prim) in G with success probability at least $* = \alpha_2(k)$, there exists an algorithm $\mathcal{A}^{\mathcal{O}_{CIE}}$ that solves CSE(c:u; g:h; f:prim) in G with success probability at least $\alpha_2(k)^3$.

(c) following (b), there exists also an algorithm $\mathcal{A}^{\mathcal{O}_{CIE}}$ that, with success probability at least $\alpha_2(k)^9$, breaks $1/\mathsf{poly}(k)$-CDH(c:u; g:h; f:prim) in G.

From these reductions and Remark 4 the theorem immediately follows. The complete proof can be found in [21].

Remark 4. For strong and perfect success probabilities, i.e., $\alpha_1(k)$ is either 1 or $1 - 1/\mathsf{poly}(k)$, the resulting success probability in case (a) can always be polynomially bounded because $O(\log|G|) = O(\mathsf{poly}(k))$ and there always exist constants c and c' such that for large enough k it holds that $(1 - 1/k^{c'})^{O(\mathsf{poly}(k))} \geq 1/k^c$. However, for the weak and invariant success probability, i.e., $\alpha_1(k)$ is either ϵ or $1/\mathsf{poly}(k)$, the resulting error cannot be bounded polynomially. This implies that above reduction in (a) does not work directly in the case where the oracle $\mathcal{O}_{CDH}$ is only of the weak or invariant success probability flavor! The success probability of $\mathcal{O}_{CDH}$ has first to be improved by self-correction [15] to strong success probability, a task expensive both in terms of oracle calls and group operations.   ∘

Next, we prove above equivalence also for medium granularity. Similar to Theorem 2 we could argue that due to the existence of a randomization the result immediately follows also for the medium granularity case. However, we will show below that the reduction can be performed much more efficiently in the medium granular case than in the case above; thereby we improve the concrete security considerably.

Theorem 4.
$$1/\mathsf{poly}(k)\text{-CSE}(\text{c:u; g:m; f:prim}) \iff 1/\mathsf{poly}(k)\text{-CIE}(\text{c:u; g:m; f:prim}).\quad \square$$
Proof. "$\Rightarrow$": We construct $\mathcal{A}^{\mathcal{O}_{CIE}}$ as follows: Assume we are given a CSE instance g^x with respect to generator g. We set $h := g^x$ and $t := x^{-1}$, pass $g^x(=h)$ and $g(= h^t)$ to $\mathcal{O}_{CIE}$. Assuming the oracle answered correctly we get the desired solution to the CSE problem: $h^{t^{-1}} = (g^x)^{(x^{-1})^{-1}} = g^{x^2}$.
"$\Leftarrow$": Conversely we can exploit the identity $((g^x)^{(x^{-1})^2} = (g^x)^{x^{-2}} = g^{xx^{-2}} = g^{x^{-1}}$ to construct $\mathcal{A}^{\mathcal{O}_{CSE}}$ solving CIE with a single call to $\mathcal{O}_{CSE}$.   ■

Remark 5. For each direction we need now only a single oracle call. If we take also into account that with a single oracle call $1/\mathsf{poly}(k)$-CSE(c:u; g:m; f:prim) can be reduced to $1/\mathsf{poly}(k)$-CDH(c:u; g:m; f:prim) we achieve a reduction from $1/\mathsf{poly}(k)$-CIE(c:u; g:m; f:prim) to $1/\mathsf{poly}(k)$-CDH(c:u; g:m; f:prim) while retaining the success probability of the oracle.

---

[6] The exponent $O(\log|G|)$ stems from a square and multiply used in the reduction.

Remark 6. Above observation also implies that, contrary to the high-granular (Remark 4) case, this reduction directly applies to the invariant and weak success probability variant of the assumptions, i.e., no self-correction is required.   ∘

In particular the Remark 5 is of high significance. The reduction we get in the medium-granular case is much more efficient than the corresponding reduction in the high-granular case: With a single instead of $\log(|G|)$ (very expensive) oracle calls and $O(\log(|G|))$ instead of $O(\log(|G|)^2)$ group operations we achieve a success probability which is higher by a power of $O(\log(|G|))$!

7   Decisional DH, SE and IE

7.1   Difficulty in the Generic Model

We state first a Lemma which plays an important role for later proofs in the context of generic algorithms:

Lemma 1 ([25, 15]). *Let $P(X_1, X_2, \cdots, X_n)$ be a non-zero polynomial in $\mathbb{Z}_{p^e}[X]$ of total degree $d \geq 0$ $(p \in \mathbb{P}; e \in \mathbb{N})$. Then the probability that $P(x_1, x_2, \cdots, x_n) \equiv 0$ is at most d/p for a random tuple $(x_1, x_2, \cdots, x_n) \in_\mathcal{R} \mathbb{Z}_{p^e}^n$.*   □

Using Lemma 1, Wolf [9] shows that there exists no generic algorithm that can solve DSE (and consequently also DDH) in polynomial time if the order of the multiplicative group is not divisible by small primes, as summarized in following theorem:

Theorem 5 ([9]).
$$true \implies 1/\mathsf{poly}(k)\text{-DSE}^\sigma(\text{c:u; g:h; f:nsprim})$$ □

Remark 7. More precisely, Wolf shows that the probability that any $\mathcal{A}^\sigma$ can correctly distinguish correct DSE inputs from incorrect ones is at most $\frac{(T+4)(T+3)}{2p'}$ where p' is the smallest prime factor of $|G|$ and T is an upper bound on the algorithm's runtime.

Remark 8. It might look surprising that $1/\mathsf{poly}(k)\text{-DSE}^\sigma(\text{c:u; g:h; f:nsprim})$ always holds, i.e., it's a fact, not an assumption. Of course, the crucial aspect is the rather restricted adversary model (the σ in the assumption statement) which limits adversaries to generic algorithms. However, note that this fact means that to break DSE one has to exploit deeper knowledge on the actual structure of the used algebraic groups. In particular, for appropriately chosen prime-order subgroups of $\mathbb{Z}_p^*$ and elliptic or hyper-elliptic curves no such exploitable knowledge could yet be found and all of currently known efficient and relevant algorithms in these groups are generic algorithms, e.g., Pohlig-Hellman [26] or Pollard-ρ [27]. Nevertheless, care has to be applied when proving systems secure in the generic model [28].   ∘

In the following theorem we show that also DIE cannot be solved by generic algorithms if the order of the multiplicative group is not divisible by small primes.

Theorem 6.
$$true \implies 1/\mathsf{poly}(k)\text{-}\mathrm{DIE}^\sigma(\text{c:u; g:h; f:nsprim})$$ $\Box$

The proof is similar to the proof of Theorem 5 and can be found in [21].

7.2   DSE versus DDH

Wolf [9] shows following two results on the relation of DSE and DDH: DSE can easily be reduced to DDH but the converse doesn't hold; in fact, Theorem 8 shows that a DSE oracle, even when perfect, is of no help in breaking DDH assumptions.

Theorem 7 ([9]).
$$1/\mathsf{poly}(k)\text{-}\mathrm{DSE}(\text{c:u; g:h; f:}*) \implies 1/\mathsf{poly}(k)\text{-}\mathrm{DDH}(\text{c:u; g:h; f:}*)$$ $\Box$

Remark 9. Following Remark 2, this result easily extends also to the medium-granular variant.   ○

Theorem 8 ([9]).
$$true \implies 1/\mathsf{poly}(k)\text{-}\mathrm{DDH}^\sigma(\text{c:u; g:h; f:nsprim}; \mathcal{O}_{1\text{-}\mathrm{DSE}(\text{c:u; g:h; f:nsprim})})$$ $\Box$

Remark 10. More precisely, Wolf shows that the probability that any $\mathcal{A}^{\sigma, \mathcal{O}_{DSE}}$ can correctly distinguish correct DDH inputs from incorrect ones is at most $\frac{(T+5)(T+4)}{2p'}$ where p' is the smallest prime factor of $|G|$ and T is an upper bound on the algorithm's runtime.   ○

7.3   DIE versus DDH

In the following we prove that similar relations also hold among DDH and DIE: We show a reduction from DIE to DDH and prove that a DIE oracle, even when perfect, is of no help in breaking DDH assumptions.

Theorem 9.
$$1/\mathsf{poly}(k)\text{-}\mathrm{DIE}(\text{c:u; g:h; f:prim}) \implies 1/\mathsf{poly}(k)\text{-}\mathrm{DDH}(\text{c:u; g:h; f:prim})$$ $\Box$

Theorem 10,
$$true \implies 1/\mathsf{poly}(k)\text{-}\mathrm{DDH}^\sigma(\text{c:u; g:h; f:nsprim}; \mathcal{O}_{1\text{-}\mathrm{DIE}(\text{c:u; g:h; f:nsprim})})$$ $\Box$

Both proofs follow similar strategies than the proofs of Theorem 7 and 8 and can be found in [21]. One twist is that the input domains between IE and DH are different and the DIE-oracle cannot answer correctly to the queries not from its domain. However, since this limits the use of a DIE-oracle in solving DDH even further, this does not affect the desired result.

7.4   DSE versus DIE

In the next theorem we prove that an oracle breaking 1-DSE(c:u; g:h; f:*) is of no help in breaking $1/\mathsf{poly}(k)\text{-}\mathrm{DIE}^\sigma(\text{c:u; g:h; f:}*)$.

Theorem 11.
$$true \implies 1/\text{poly}(k)\text{-DIE}^\sigma(\text{c:u; g:h; f:nsprim}; \mathcal{O}_{1\text{-DSE}(\text{c:u; g:h; f:nsprim})}) \qquad \square$$

Proof. Similar to the proofs of Theorem 6 and 10 we define a Lemma which associates the minimal generic complexity of solving DIE directly to the smallest prime factor of the order of the underlying group G. Theorem 11 immediately follows from Lemma 2 and Remark 11. ∎

Remark 11. In the classical formulation of decision problems the adversary gets, depending on the challenge b, either the correct element or a random element as input, i.e., in the case of DIE the adversary gets g^x together with $g^{x^{-1}}$ if $b = 0$ and g^c if $b = 1$. The formulation used in the Lemma considers a slightly different variant of the decisional problem type: We consider here an adversary which receives, in random order, both the correct and a random element and the adversary has to decide on the order of the elements, i.e., the adversary gets g^x and $(g^{x^{-1}}, g^c)$ for $b = 0$ and $(g^c, g^{x^{-1}})$ for $b = 1$.

   This formulation makes the proofs easier to understand. However, note that both variants can be shown equivalent. ○

Lemma 2. *Let G be a cyclic group and g a corresponding generator, let p' be the smallest prime factor of $n = |G|$. Let $\mathcal{O}_{DSE}$ be a given oracle solving DSE tuples in G and let $\mathcal{A}^{\sigma, \mathcal{O}_{DSE}}$ be any generic algorithm for groups G with maximum run time T and oracle access to $\mathcal{O}_{DSE}$. Further let x_0, x_1 be random elements of $\mathbb{Z}^*_{|G|}$, $b \in_\mathcal{R} \{0,1\}$ a randomly and uniformly chosen bit and $C \xleftarrow{\mathcal{R}} \mathcal{U}$. Then it always holds that*

$$\boldsymbol{Prob}[\mathcal{A}^\sigma(C, (G, g), g^{x_0}, g^{x_b^{-1}}, g^{x_{\bar{b}}^{-1}}) = b] \leq \frac{(T+4)(T+3)}{2p'}$$
$$\square$$

Proof. For given $\sigma(1)$, $\sigma(x)$, $\{\sigma(x^{-1}), \sigma(c)\}$, assume that the algorithm $\mathcal{A}^{\sigma, \mathcal{O}_{DSE}}$ computes at most $T_1 + 4$ (images of) distinct linear combinations P_i of the elements $1, x, x^{-1}, c$ with $P_i(1, x, x^{-1}, c) = a_{i1} + a_{i2}x + a_{i3}x^{-1} + a_{i4}c$ such that

$$\sigma(P_i(1, x, x^{-1}, c)) = \sigma(a_{i1} + a_{i2}x + a_{i3}x^{-1} + a_{i4}c),$$

where a_{ij} are constant coefficients. Furthermore, it is not a-priori known to $\mathcal{A}^{\sigma, \mathcal{O}_{DSE}}$ which of the (known) values in $\{a_{i3}, a_{i4}\}$ is the coefficient for x^{-1} and which one corresponds to c. Assume that $\mathcal{A}^{\sigma, \mathcal{O}_{DSE}}$ makes T_2 calls to $\mathcal{O}_{DSE}$. $\mathcal{A}^{\sigma, \mathcal{O}_{DSE}}$ may be able to distinguish the coefficient by obtaining information from either of the following events:

E_a: $\mathcal{A}^{\sigma, \mathcal{O}_{DSE}}$ finds a collision between two distinct linear equations (P_i, P_j) with $i \neq j$, i.e.,

$$\sigma(P_i(1, x, x^{-1}, c)) = \sigma(P_j(1, x, x^{-1}, c)) \Rightarrow P_i(1, x, x^{-1}, c) = P_j(1, x, x^{-1}, c)$$

E_b: $\mathcal{A}^{\sigma,\mathcal{O}_{DSE}}$ gets at least one positive answer from $\mathcal{O}_{DSE}$ for a non-trivial query with $i \neq j$, i.e.,

$$\sigma(P_i(1, x, x^{-1}, c)) = \sigma((P_j(1, x, x^{-1}, c)^2).$$

Let E be the union of the events E_a and E_b. Now we compute an upper bound for the probability that either of these events occurs.

Case E_a: Consider P_i and P_j as polynomials. There are $\binom{T_1+4}{2} = \frac{(T_1+4)(T_1+3)}{2}$ possible distinct pairs. The probability of a collision for two linear combinations P_i, P_j is the probability of (randomly) finding the root of polynomial $x(P_i - P_j) \equiv 0 \bmod p^e$ for any prime factor p of $|G|$ with $p^e \| |G|$. Due to Lemma 1 this probability is at most $2/p^e$ $(\leq 2/p')$, because the degree of $x(P_i - P_j)$ is at most two.[7] It follows that $\mathbf{Prob}[E_a] \leq \frac{(T_1+4)(T_1+3)}{2} \frac{2}{p'} = \frac{(T_1+4)(T_1+3)}{p'}$.

Case E_b: For $i \neq j$ it is not possible to derive a relation $P_i = P_j{}^2$ except that P_i and P_j are both constant polynomials ($\neq 0$), meaning that the polynomial $x^2(P_i - P_j{}^2) \equiv 0 \bmod p^e$ for $x \neq 0$. The total degree of the polynomial $x^2(P_i - P_j{}^2)$ is at most 4 and the probability for E_b is at most $\frac{4}{p'}T_2$.

In total we have

$$\mathbf{Prob}[E] \leq \mathbf{Prob}[E_a] + \mathbf{Prob}[E_b] = \frac{(T_1+4)(T_1+3)}{p'} + T_2\frac{4}{p'} \leq \frac{(T+4)(T+3)}{p'},$$

with $T_1 + T_2 \leq T$. The success probability of $\mathcal{A}^{\sigma,\mathcal{O}_{DSE}}$ therefore is:

$$\mathbf{Prob}[\mathcal{A}^{\sigma,\mathcal{O}_{DSE}}(..) = b] = \mathbf{Prob}[E] + \frac{1}{2}\mathbf{Prob}[\bar{E}] =$$

$$\mathbf{Prob}[E] + \frac{1 - \mathbf{Prob}[E]}{2} = \frac{1}{2} + \frac{\mathbf{Prob}[E]}{2} \leq \frac{1}{2} + \frac{(T+4)(T+3)}{2p'}.$$

∎

In sharp contrast to the above mentioned high granular case, we prove in the following theorem that these assumptions are equivalent for their medium granular version (other parameters remain unchanged).

Theorem 12.
   $1/\mathrm{poly}(k)$-DSE(c:u; g:m; f:prim) $\iff$ $1/\mathrm{poly}(k)$-DIE(c:u; g:m; f:prim).   □
Proof. "$\Leftarrow$": Assume we are given a DIE tuple $I_{DIE} = (g, g^x, g^z)$ where g^z is either $g^{x^{-1}}$ or a random element of group G. Set $h := g^z$ then $g = h^t$ and $g^x = h^{tx}$ for some (unknown) $t \in \mathbb{Z}_{|G|}^*$. After reordering the components we obtain the tuple (h, h^t, h^{xt}).

---

[7] Note that P_i, P_j are also functions of $x^{-1}, x \neq 0$ and thus one can virtually think of the polynomial $x(P_i - P_j)$ by multiplying both sides of the equation $P_i = P_j$ with x. Furthermore, uniformly distributed random values $\bmod n$ are also randomly and uniformly distributed $\bmod p^e$.

If $z = x^{-1}$ then $t = x$ and the tuple (h, h^t, h^{xt}) will have the form (h, h^t, h^{t^2}) which represents a DSE tuple and can be solved by the given DSE oracle. The probability distribution is correct, since h is a group generator and h^t is a random element of G.

If $z \neq x^{-1}$ then $t \neq x$ and h^{xt} is a random group element (x is a random element of $\mathbb{Z}_{|G|}$) and the elements h, h^t, h^{tx} are independent.

"$\Rightarrow$": Assume, we are given a DSE tuple (g, g^x, g^z) where g^z is either g^{x^2} or a random group element. Set $h := g^x$ then $g = h^t$ and $g^z = h^{tz}$ for some (unknown) $t \in \mathbb{Z}^*_{|G|}$. After reordering the components we obtain the tuple (h, h^t, h^{tz}).

If $z = x^2$ then we have[8] $x = t^{-1}$ and $z = t^{-2}$ meaning that the tuple (h, h^t, h^{xt}) has the form $(h, h^t, h^{t^{-1}})$ representing a DIE tuple. Its probability distribution is correct because h is a group generator, h^t is a random element of G and the last element $h^{t^{-1}}$ has the correct form.

If $z \neq x^2$ then h^{zt} is a random group element, since t is a random element of $\mathbb{Z}^*_{|G|}$, and further the elements h, h^t and h^{tz} are independent.   ∎

8   Conclusions

In this paper, we identify the parameters relevant to cryptographic assumptions. Based on this we present a framework and notation for defining assumptions related to Discrete Logarithms. Using this framework these assumptions can be precisely and systematically classified. Wider adoption of such a terminology would ease the study and comparison of results in the literature, e.g., the danger of ambiguity and mistakes in lengthly stated textual assumptions and theorems would be minimized. Furthermore, clearly stating and considering these parameters opens an avenue to generalize results regarding the relation of different assumptions and to get a better understanding of them. This is the focus of our ongoing research and is covered to a larger extent in the full version of the paper [21].

A parameter in defining assumptions previously ignored in the literature is granularity. We show (as summarized in Figure 1) that varying this parameter leads to surprising results: we prove that some DL-related assumptions are equivalent in one case (medium granular) and provably not equivalent, at least not in a generic sense, in another case (high granular). Furthermore, we also show that some reductions for medium granularity are much more efficient than their high-granular version leading to considerably improved concrete security, in particular as medium granularity results in weaker assumptions than high-granular ones. However, we note that medium- or low-granular assumption apply in cryptographic settings only when the choice of system parameters is guaranteed to be truly random.

In this paper we only scratched the topic of granularity and interesting open questions remain to be answered: While for both CDL and CDH it can be shown

---

[8] This is because $h^x = g^{x^2} = h^{tz}$ which implies $h^x = h^{tx^2}$, $x = tx^2$ and $t = x^{-1}$.

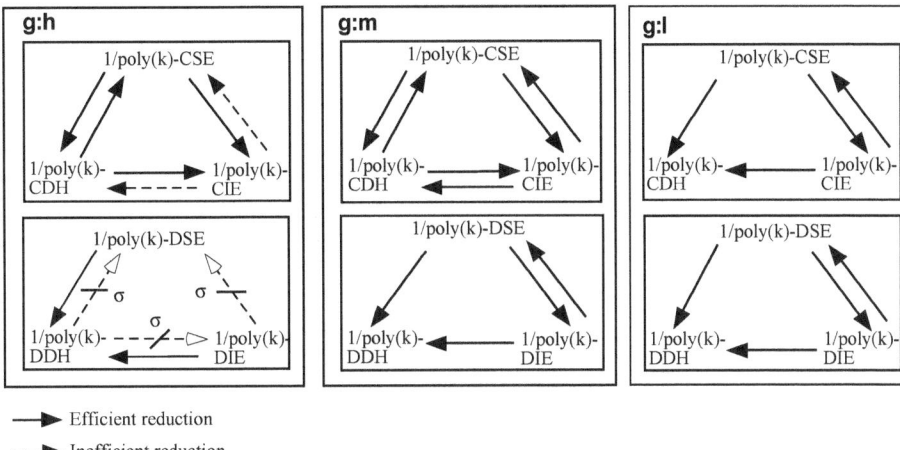

Efficient reduction
Inefficient reduction
Reduction impossible in generic model

Fig. 1. Summary of our results

that their high- and medium-granular assumptions are equivalent, this is not yet known for DDH (also briefly mentioned as an open problem in [29]). Only few relations can be shown for low-granular assumption as no random self-reducibility is yet known. However, achieving such "full" random self-reducibility seems very difficult in general (if not impossible) in number-theoretic settings [30] contrary to, e.g., lattice settings used in [31].

Acknowledgements

We thank Birgit Pfitzmann, Matthias Schunter, and the anonymous reviewers for their helpful comments.

References

[1] Kevin S. McCurley. The discrete logarithm problem. In Carl Pomerance, editor, *Cryptology and Computational Number Theory*, volume 42 of *Proceedings of Symposia in Applied Mathematics*, pages 49–74, Providence, 1990. American Mathematical Society.

[2] Whitfield Diffie and Martin Hellman. New directions in cryptography. *IEEE Transactions on Information Theory*, IT-22(6):644–654, November 1976.

[3] Dan Boneh. The Decision Diffie-Hellman problem. In *Third Algorithmic Number Theory Symposium*, number 1423 in Lecture Notes in Computer Science, pages 48–63. Springer-Verlag, Berlin Germany, 1998.

[4] Ronald Cramer and Victor Shoup. A practical public key cryptosystem provably secure against adaptive chosen ciphertext attack. In Hugo Krawczyk, editor, *Advances in Cryptology – CRYPTO '98*, number 1462 in Lecture Notes in Computer Science, pages 13–25. International Association for Cryptologic Research, Springer-Verlag, Berlin Germany, August 1998.

[5] Moni Naor and Omer Reingold. Number-theoretic constructions of efficient pseudo-random functions. In *38th Symposium on Foundations of Computer Science (FOCS)*, pages 458–467. IEEE Computer Society Press, 1997.

[6] Michael Steiner, Gene Tsudik, and Michael Waidner. Key agreement in dynamic peer groups. *IEEE Transactions on Parallel and Distributed Systems*, 11(8):769–780, August 2000.

[7] Yair Frankel, Yiannis Tsiounis, and Moti Yung. "indirect discourse proofs": Achieving fair off-line cash (FOLC). In K. Kim and T. Matsumoto, editors, *Advances in Cryptology – ASIACRYPT '96*, number 1163 in Lecture Notes in Computer Science, pages 286–300. Springer-Verlag, Berlin Germany, 1996.

[8] Helena Handschuh, Yiannis Tsiounis, and Moti Yung. Decision oracles are equivalent to matching oracles. In *International Workshop on Practice and Theory in Public Key Cryptography '99 (PKC '99)*, number 1560 in Lecture Notes in Computer Science, Kamakura, Japan, March 1999. Springer-Verlag, Berlin Germany.

[9] Stefan Wolf. *Information-theoretically and Computionally Secure Key Agreement in Cryptography*. PhD thesis, ETH Zürich, 1999.

[10] Ueli M. Maurer and Stefan Wolf. Diffie-Hellman oracles. In Koblitz [32], pages 268–282.

[11] Mike Burmester, Yvo Desmedt, and Jennifer Seberry. Equitable key escrow with limited time span (or, how to enforce time expiration cryptographically). In K. Ohta and D. Pei, editors, *Advances in Cryptology – ASIACRYPT '98*, number 1514 in Lecture Notes in Computer Science, pages 380–391. Springer-Verlag, Berlin Germany, 1998.

[12] Birgit Pfitzmann and Ahmad-Reza Sadeghi. Anonymous fingerprinting with direct non-repudiation. In Okamoto [33], pages 401–414.

[13] Jan Camenisch, Ueli Maurer, and Markus Stadler. Digital payment systems with passive anonymity-revoking trustees. In E. Bertino, H. Kurth, G. Martella, and E. Montolivo, editors, *Proceedings of the Fourth European Symposium on Research in Computer Security (ESORICS)*, number 1146 in Lecture Notes in Computer Science, pages 33–43, Rome, Italy, September 1996. Springer-Verlag, Berlin Germany.

[14] George Davida, Yair Frankel, Yiannis Tsiounis, and Moti Yung. Anonymity control in e-cash systems. In *Proceedings of the First Conference on Financial Cryptography (FC '97)*, number 1318 in Lecture Notes in Computer Science, pages 1–16, Anguilla, British West Indies, February 1997. International Financial Cryptography Association (IFCA), Springer-Verlag, Berlin Germany.

[15] Victor Shoup. Lower bounds for discrete logarithms and related problems. In Walter Fumy, editor, *Advances in Cryptology – EUROCRYPT '97*, number 1233 in Lecture Notes in Computer Science, pages 256–266. International Association for Cryptologic Research, Springer-Verlag, Berlin Germany, 1997.

[16] Ueli M. Maurer and Stefan Wolf. Diffie-Hellman, Decision Diffie-Hellman, and discrete logarithms. In *IEEE Symposium on Information Theory*, page 327, Cambridge, USA, August 1998.

[17] Ueli M. Maurer and Stefan Wolf. Lower bounds on generic algorithms in groups. In Kaisa Nyberg, editor, *Advances in Cryptology – EUROCRYPT '98*, number 1403 in Lecture Notes in Computer Science, pages 72–84. International Association for Cryptologic Research, Springer-Verlag, Berlin Germany, 1998.

[18] Eli Biham, Dan Boneh, and Omer Reingold. Breaking generalized Diffie-Hellman modulo a composite is no easier than factoring. *Information Processing Letters*,

70:83–87, 1999. Also appeares in Theory of Cryptography Library, Record 97-14, 1997.

[19] Daniel M. Gordon. Designing and detecting trapdoors for discrete log cryptosystems. In E.F. Brickell, editor, *Advances in Cryptology – CRYPTO '92*, volume 740 of *Lecture Notes in Computer Science*, pages 66–75. International Association for Cryptologic Research, Springer-Verlag, Berlin Germany, 1993.

[20] National Institute of Standards and Technology (NIST). The digital signature standard (DSS). FIPS PUB 186-2, January 2000.

[21] Ahmad-Reza Sadeghi and Michael Steiner. Assumptions related to discrete logarithms: Why subtleties make a real difference. Full version of paper, available from http://www.semper.org/sirene/lit/abstrA1.html.

[22] Dan Boneh and Richard J. Lipton. Algorithms for black box fields and their application to cryptography. In Koblitz [32], pages 283–297.

[23] V. I. Nechaev. Complexity of a determinate algorithm for the discrete logarithm. *Mathematical Notes*, 55(2):165–172, 1994. Translated from Matematicheskie Zametki, 55(2):91–101, 1994.

[24] Christian Cachin, Klaus Kursawe, and Victor Shoup. Random oracles in Constantinople: practical asynchronous Byzantine agreement using cryptography. In *Proceedings of the 19th Annual ACM Symposium on Principles of Distributed Computing*, Portland, Oregon, July 2000. ACM. Full version appeared as Cryptology ePrint Archive Report 2000/034 (2000/7/7).

[25] J. T. Schwartz. Fast probabilistic algorithms for verification of polynomial identities. *Journal of the ACM*, 27(4):701–717, 1980.

[26] S.C. Pohlig and M. E. Hellman. An improved algorithm for computing logarithms over GF(p) and its cryptographic significance. *IEEE Transactions on Information Theory*, 24:106–110, 1978.

[27] J. M. Pollard. Monte carlo methods for index computation mod p. *Mathematics of Computation*, 32:918–924, 1978.

[28] Marc Fischlin. A note on security proofs in the generic model. In Okamoto [33], pages 458–469.

[29] Victor Shoup. On formal models for secure key exchange. Research Report RZ 3120 (#93166), IBM Research, April 1999. A revised version 4, dated November 15, 1999, is available from http://www.shoup.net/papers/.

[30] Dan Boneh. Personal Communication, October 2000.

[31] Miklós Ajtai and Cynthia Dwork. A public-key cryptosystem with worst-case/average-case equivalence. In *29th Annual Symposium on Theory Of Computing (STOC)*, pages 284–293, El Paso, TX, USA, May 1997. ACM Press.

[32] Neal Koblitz, editor. *Advances in Cryptology – CRYPTO '96*, number 1109 in Lecture Notes in Computer Science. International Association for Cryptologic Research, Springer-Verlag, Berlin Germany, 1996.

[33] T. Okamoto, editor. *Advances in Cryptology – ASIACRYPT '2000*, number 1976 in Lecture Notes in Computer Science, Kyoto, Japan, 2000. International Association for Cryptologic Research, Springer-Verlag, Berlin Germany.

On Adaptive vs. Non-adaptive Security of Multiparty Protocols

Ran Canetti[1], Ivan Damgaard[2], Stefan Dziembowski[2],
Yuval Ishai[3,*], and Tal Malkin[4,**]

[1] IBM Watson
[2] Aarhus University, BRICS[‡]
[3] DIMACS and AT&T.
[4] AT&T Labs – Research.

Abstract. Security analysis of multiparty cryptographic protocols distinguishes between two types of adversarial settings: In the non-adaptive setting, the set of corrupted parties is chosen in advance, before the interaction begins. In the adaptive setting, the adversary chooses who to corrupt during the course of the computation. We study the relations between adaptive security (i.e., security in the adaptive setting) and non-adaptive security, according to two definitions and in several models of computation. While affirming some prevailing beliefs, we also obtain some unexpected results. Some highlights of our results are:

- According to the definition of Dodis-Micali-Rogaway (which is set in the information-theoretic model), adaptive and non-adaptive security are equivalent. This holds for both honest-but-curious and Byzantine adversaries, and for any number of parties.
- According to the definition of Canetti, for honest-but-curious adversaries, adaptive security is equivalent to non-adaptive security when the number of parties is logarithmic, and is strictly stronger than non-adaptive security when the number of parties is super-logarithmic. For Byzantine adversaries, adaptive security is strictly stronger than non-adaptive security, for any number of parties.

1 Introduction

Security analysis of cryptographic protocols is a delicate task. A first and crucial step towards meaninful analysis is coming up with an appropriate definition of security of the protocol problem at hand. Formulating good definitions is non-trivial: They should be compehensive and stringent enough to guarantee security against a variety of threats and adversarial behaviors. On the other hand, they should be as simple, workable, and as permissive as possible, so as to facilitate design and analysis of secure protocols, and to avoid unnessecary requirements.

---

[*] Work partially done while the author was at the Technion and while visiting IBM Watson
[**] Work partially done while the author was at MIT and while visiting IBM Watson
[‡] Basic Research in Computer Science, Center of the Danish National Research Foundation

B. Pfitzmann (Ed.): EUROCRYPT 2001, LNCS 2045, pp. 262–279, 2001.

Indeed, in contrast with the great advances in constructing cryptographic protocols for a large variety of protocol problems, formalizing definitions of security for crypographic protocol problems has been progressing more slowly. The first protocols appearing in the literature use only intuitive and ad-hoc notions of security, and rigorous security analysis was virtually non-existent. Eventually, several general definitions of security for cryptographic protocols have appeared in the literature. Most notable are the works of Goldwasser and Levin [GL90], Micali and Rogaway [MR91], Beaver [B91], Canetti [C00] and Dodis and Micali [DM00] (that concentrate on the task of *secure function evaluation* [Y82,Y86,GMW87]), and Pfitzmann and Waidner [PW94], Pfitzmann Schunter and Waidner [PSW00], and Canetti [C00a] (that discuss general reactive tasks). In particular, only recently do we have precise and detailed definitions that allow rigorous study of "folklore beliefs" regarding secure protocols.

This work initiates a comparative study of notions of security, according to different definitions. We concentrate on secure function evaluation, and in particular the following aspect. Adversarial behavior of a computational environment is usually modelled via a single algorithmic entity, the **adversary**, the capabilities of which represent the actual security threats. Specifically, in a network of communicating parties the adversary is typically allowed to control (or, **corrupt**) some of the parties. Here the following question arises: How are the corrupted parties chosen? One standard model assumes that the set of corrupted parties is fixed before the computation starts. This is the model of **non-adaptive** adversaries. Alternatively, the adversary may be allowed to corrupt parties during the course of the computation, when the identity of each corrupted party may be based on the information gathered so far. We call such adversaries **adaptive**.

Indeed, attackers in a computer network (hackers, viruses, insiders) may break into computers during the course of the computation, based on partial information that was already gathered. Thus the adaptive model seems to better represent realistic security threats, and so provide a better security guarantee. However, defining and proving security of protocols is considerably easier in the non-adaptive model. One quintessential example for the additional complexity of guaranteeing adaptive security is the case of using encryption to transform protocols that assume ideally secure channels into protocols that withstand adversaries who hear all the communication. In the non-adaptive model standard Chosen-Ciphertext-Attack secure encryption [DDN91,CS98,S99] (or even plain semantically secure encryption [GM84], if used appropriately) is sufficient. To obtain adaptively secure encryption, it seems that one needs to either trust data erasures [BH92], or use considerably more complex constructs [CFGN96,B97,DN00].

Clearly, adaptive security implies non-adaptive security, under any reasonable definition of security. However, is adaptive security really a stronger notion than non-adaptive security? Some initial results (indicating clear separation in some settings) are provided in [CFGN96]. On the other hand, it is a folklore belief that in an "information theoretic setting" adaptive and non-adaptive security should be equivalent. Providing more complete answers to this question, in several models of computation, is the focus of this work. While some of our results affirm common beliefs, other results are quite surprising, and may considerably simplify the design and analysis of protocols.

Models of computation. We study the additional power of adaptive adversaries in a number of standard adversary models, and according to two definitions (the definition of Dodis, Micali, and Rogaway [MR91,DM00], and that of Canetti [C00]). To develop the necessary terminology for presenting our results let us very shortly outline the structure of definitions of security of protocols. (The description below applies to both definitions. The [MR91,DM00] definition imposes some additional requirements, sketched in a later section.)

As mentioned above, both definitions concentrate on the task of Secure Function Evaluation. Here the parties wish to jointly evaluate a given function at a point whose value is the concatenation of the inputs of the parties. In a nutshell, protocols for secure function evaluation are protocols that "emulate" an *ideal process* where all parties privately hand their inputs to an imaginary trusted party who privately computes the desired results, hands them back to the parties, and vanishes. A bit more precisely, it is required that for any adversary $\mathcal{A}$, that interacts with parties running a secure protocol π and induces some global output distribution, there exists an "ideal-process" adversary $\mathcal{S}$, that manages to obtain essentially the same global output distribution *in the ideal process*. The global output contains the adversary's output (which may be assumed to be his entire view of the computation), together with the identities and outputs of the uncorrupted parties. (Adversary $\mathcal{S}$ is often called a simulator, since it typically operates by simulating a run of $\mathcal{A}$.) The following parameters of the adversarial models turn out to be significant for our study.

ADVERSARIAL ACTIVITY: The adversary may be either passive (where even corrupted parties follow the prescribed protocol, and only try to gather additional information), or active, where corrupted parties are allowed to arbitrarily deviate from their protocol. Passive (resp., active) adversaries are often called honest-but-curious (resp., Byzantine).

NUMBER OF PLAYERS: We distinguish between the case of a small number of players, where n, the number of players, is $O(\log k)$, and a large number of players, where n is $\omega(\log k)$. (Here k is the security parameter.)

COMPLEXITY OF ADVERSARIES: We consider three cases. Information-Theoretic (IT) security does not take into account any computational complexity considerations. That is, both adversaries $\mathcal{A}$ and $\mathcal{S}$ have unbounded resources regardless of each other's resources. Universal security allows $\mathcal{A}$ unbounded resources, but requires $\mathcal{S}$ to be efficient (i.e., expected polynomial) in the complexity of $\mathcal{A}$. Computational security restricts both $\mathcal{A}$ and $\mathcal{S}$ to expected polynomial time (in the security parameter). Note that universal security implies both IT security and computational security (all other parameters being equal). However, IT security and computational security are incomparable. See [C00] for more discussion on the differences between these notions of security and their meaning.

QUALITY OF EMULATION: We consider either perfect emulation (where the output distributions of the real-life computation and of the ideal process must be identically distributed), or **statistical emulation** (where the output distributions shuld be statistically indistinguishable), or **computational emulation** (where the output distributions shuld be computationally indistinguishable).

The rest of the Introduction overviews the state of affairs regarding the added power of adaptivity, as discovered by our investigation. We do not attempt here to explain "why" things are as they are. Such (inevitably subjective) explanations require more familiarity with the definitions and are postponed to the body of the paper.

Our results: Canetti's definition. This definition is stated for several models of computation. We concentrate by default on the *secure channels* model, where the communication channels are perfectly secret and *universal security* is required. The same results hold also for the *computational* setting, where the adversary sees all communication but is restricted to polynomial time. Finally, we also consider a weaker variant of this definition, not considered in [C00], where only IT security is required (and the communication channels are secure).

The most distinctive parameter here seems to be whether the adversary is active or passive. If the adversary is active (i.e., Byzantine) then adaptive security is strictly stronger than non-adaptive security, regardless of the values of all other parameters. We show this via a protocol for three parties, that is non-adaptively universally secure with perfect emulation, but adaptively *in*secure, even if the adversary is computationally bounded and we are satisfied with computational emulation. This is the first such example involving only a constant number of players, for *any* constant.

In the case of passive adversaries the situation is more involved. Out of the nine settings to be considered (IT, universal, or computational security, with perfect, statistical, or computational emulation), we show that for one – IT security and perfect emulation – adaptive and non-adaptive security are equivalent, for any number of players. In all other eight settings we show that, roughly speaking, adaptive security is equivalent to non-adaptive security when the number of players is small, and is strictly stronger when the number of players is large. We elaborate below.

For a large number of players, it follows from an example protocol shown in [CFGN96] that for statistical or computational emulation, adaptive security is strictly stronger than non-adaptive security. We show separation also for perfect emulation, where universal or computational security is required. We complete the picture by showing that for a small number of players, and perfect emulation, adaptive and non-adaptive security are equivalent. Equivalence holds even in the case of statistical or computational emulation, if n is $O(\log k / \log \log k)$. (Notice that there is a small gap between this equivalence result and the known separating example for $n \in \omega(\log k)$. To close this gap, we also show that if one relaxes slightly the demands to the complexity of simulators and allows them to be expected polynomial time *except with negligible probability*, then this gap can be closed: equivalence holds for all $n \in O(\log k)$. In many cases, this definition of "efficient simulation" seems to be as reasonable as the standard one.)

Equivalence of adaptive and non-adaptive security for the case of passive adversaries and a small number of players is very good news: Many protocol problems (for instance, those related to threshold cryptography) make most sense in a setting where the number of parties is *fixed*. In such cases, when concentrating on passive adversaries, adaptivity comes "for free", which significantly simplifies the construction and analysis of these protocols.

Our results: Dodis-Micali-Rogaway definition. This definition holds for the *secure channels* setting only. It is incomparable to the definition of [C00]: On the one hand, it makes a number of additional requirements. On the other hand, only IT security is required. Here, to our surprise, adaptive and non-adaptive security turn out to be equivalent, even for active adversaries, and regardless of the number of players.

Two properties of the Dodis-Micali-Rogaway definition are essential for our proof of equivalence to work. The first is that only IT security is required. The second property may be roughly sketched as follows. It is required that there exists a stage in the protocol execution where all the parties are "committed" to their contributed input values; this stage must occur strictly before the stage where the output values become known to the adversary. (In order to formally state this requirement one needs to make some additional technical restrictions, amounting to what is known in the jargon as "one-pass black-box simulation". See more details within.)

Organization. Section 2 presents our results relating to the definition of [C00]. Section 3 presents our results relating to the definition of Dodis-Micali-Rogaway.

2   Adaptivity vs. Non-adaptivity in the Definition of Canetti

This section describes our results relative to the [C00] definition of security. The main aspects of the definition that we will rely on were shortly described in the Introduction. A more detailed overview is deleted for lack of space and appears in the full version of this work [CDDIM01]. Section 2.1 shows a separating example for the case of active adversary, Section 2.2 describes separating examples for passive adversary and a large number of players, Section 2.3 proves the equivalence for passive adversaries and a small number of players, and Section 2.4 shows the equivalence for passive adversaries in the setting of IT security and perfect emulation.

2.1   Separation for Active Adversaries

This section shows that adaptive and non-adaptive security are not equivalent in the case of active adversaries, for all settings considered here: information-theoretic, universal, and computational security, with perfect, statistical, or computational emulation. This is proved by an example of a simple protocol for secure function evaluation which is non-adaptively secure, but adaptively insecure, in all above settings.

Our protocol involves three players D, R_1, R_2, where R_1, R_2 have no input, and D's input consists of two bits $s_1, s_2 \in \{0, 1\}$. The function f_{act} to be computed is the function that returns no output for D, s_1 for R_1, and s_2 for R_2. The *adversary structure* $\mathcal{B}$ (the collection of player subsets that can be corrupted) contains all subsets of $\{D, R_1\}$, namely the only restriction is that R_2 cannot be corrupted. The protocol π_{act} proceeds as follows.

1. D sends s_1 to R_1.
2. D sends s_2 to R_2.
3. Each R_i outputs the bit that was sent to it by D, and terminates. D outputs nothing and terminates.

Claim 1. *The protocol π_{act} non-adaptively, perfectly emulates f_{act} with universal security, against active adversary structure $\mathcal{B}$.*

Proof. Consider a non-adaptive real-life adversary $\mathcal{A}$ that corrupts D. The ideal-process simulator $\mathcal{S}$ proceed as follows. $\mathcal{S}$ corrupts D in the ideal model, and provides $\mathcal{A}$ with the inputs s_1, s_2 of D. $\mathcal{A}$ generates s'_1 to be sent to R_1 and s'_2 to be sent to R_2. $\mathcal{S}$ gives s'_1, s'_2 to the trusted party as D's input, outputs $\mathcal{A}$'s output, and terminates. It is easy to see that the global output generated by $\mathcal{S}$ in the ideal model is identical to the global output with the real-life $\mathcal{A}$.

The above simulator can be easily modified for the case that $\mathcal{A}$ breaks into both D and R_1 (here $\mathcal{S}$ may hand in to the trusted party $0, s'_2$ as the input of D, where s'_2 is the message prepared by $\mathcal{A}$ to be sent to R_2).

Finally, consider $\mathcal{A}$ that corrupts only R_1. The simulator $\mathcal{S}$ proceeds as follows. $\mathcal{S}$ corrupts R_1 in the ideal model, hands the empty input to the trusted party, and obtains the output s_1 in the ideal model. $\mathcal{S}$ then hands s_1 to $\mathcal{A}$ as the message that was sent from D to R_1, outputs $\mathcal{A}$'s output, and terminates. Again it is easy to see that the global output generated by $\mathcal{S}$ is identical to the global output with $\mathcal{A}$.

Claim 2. *The protocol π_{act} is adaptively insecure for evaluating the function f_{act}, with either universal, IT or computational security, against active adversary structure $\mathcal{B}$.*

Proof. We show an adaptive efficient real life adversary $\mathcal{A}$, such that there is no (even computationally unbounded) adaptive ideal-model adversary (simulator) $\mathcal{S}$ that can emulate the global view induced by $\mathcal{A}$ (even if the emulation is only required to be computational). Intuitively, the goal of our adversary is to ensure that whenever $s_1 = s_2$, R_2 will output 0, whereas we do not care what happens in other cases. $\mathcal{A}$ starts by corrupting R_1 and receiving s_1 in the first stage of the protocol. If $s_1 = 0$, $\mathcal{A}$ terminates. If $s_1 = 1$, $\mathcal{A}$ corrupts D and sends $s'_2 = 0$ to R_2 in the second stage of the protocol.

To prove that this $\mathcal{A}$ cannot be simulated in the ideal world, note that in the real world, $\mathcal{A}$ never corrupts D when D's input is $s_1 = s_2 = 0$, but always corrupts D when D's input is $s_1 = s_2 = 1$. In both these cases, R_2 always outputs 0. Now let $\mathcal{S}$ be an arbitrary unbounded adaptive ideal-process simulator.

(Below "overwhelming probability" refers to $1 - neg$ for some negligible function neg.) If, when interacting with $\mathcal{S}$ in the ideal model, whenever $s_1 = s_2 = 1$, R_2 outputs 0 with overwhelming probability, then it must be that with overwhelming probability, whenever $s_1 = s_2 = 1$, $\mathcal{S}$ corrupts D before D hands s_1, s_2 to the trusted party. However, in the ideal process, before the trusted party takes the inputs and computes the function, corrupting a party provides only its input, and no other information. Thus, in our case, before D is corrupted $\mathcal{S}$ cannot gain any information. It follows that $\mathcal{S}$ corrupts D before D hands s_1, s_2 to the trusted party with the same probability for any input s_1, s_2, and in particular when the input is $s_1 = s_2 = 0$. However in the real world, $\mathcal{A}$ never corrupts D in this case, and so the global views are significantly different.

Claim 1 and Claim 2 together imply that our example separates adaptive security from non-adaptive security for active adversaries in all settings considered. Thus we have:

Theorem 3. *For active adversaries, adaptive security is strictly stronger than non-adaptive security, under any notion of security, as long as there are at least three parties.*

Discussion. The essential difference between adaptive and non-adaptive security is well captured by the simplicity of the protocol used in our separating example, which at first look may seem like a very "harmless" protocol. Indeed, π_{act} is a straight-forward implementation of the function f_{act}, which just "mimics" the ideal-world computation, replacing the trusted party passing input from one party to the output of another party, by directly sending the message between the parties. For the non-adaptive setting, this intuition translates into a proof that any adversary $\mathcal{A}$ can be simulated by an adversary $\mathcal{S}$ in the ideal world. However, as we have shown, the protocol is susceptible to an attack by an adaptive adversary.

In the heart of this separation is the idea that some information in the protocol (the value of s_1 in our example) is revealed prematurely before the parties have "committed" to their inputs. Thus, an adaptive adversary may take advantage of that by choosing whether to corrupt a party (and which one) based on this information, and then changing the party's input to influence the global output of the execution.

On the other hand, as we will show, for a passive adversary and information theoretic security, non-adaptive security is equivalent to adaptive security. This may suggest the intuition that even for active adversaries, in the information-theoretic setting, adaptive and non-adaptive security may be equivalent for a subclass of protocols that excludes examples of the above nature; that is, for protocols where "no information is revealed before the parties have committed to their inputs". This is in fact the case for many existing protocols (cf., [BGW88,CDM98]), and furthermore, the definition of Dodis-Micali-Rogaway *requires* this condition. In Section 3 we indeed formalize and prove this intuition, showing equivalence for the definition of Dodis-Micali-Rogaway.

Finally, we remark that for *two* parties and active adversaries, the situation is more involved: In the IT setting, adaptive security is equivalent to non-adaptive security. In the universal and computational settings, we have a separating example showing that adaptive security is strictly stronger, assuming perfectly hiding bit-commitment exists (which holds under standard complexity assumptions). However, this example heavily relies on a technical requirement, called post-execution corruptibility (PEC), which is part of the definition of adaptive security, needed in order to guarantee secure composability of protocols (the technical meaning of the requirement is described along with the definition in [CDDIM01]). In contrast, the above three party separating example holds in all settings, regardless of whether the PEC requirement is imposed or not.[1]

2.2 Separation for Passive Adversaries and a Large Number of Players

In [CFGN96], Canetti et al. show an example protocol that separates adaptive and non-adaptive security for passive adversaries and a large number of players, when only statistical or computational emulation is required. This separation holds for universal, IT, and computational security. Very roughly, the protocol is based on sharing a secret among a large set of players, making the identity of the set very hard to guess for a non-adaptive adversary, but easy for an adaptive one. We refer the reader to [CFGN96] for details of the example.

To complete the picture, we show an example that, under standard complexity assumptions, separates adaptive and non-adaptive security even when perfect emulation is required, for the universal or computational security model. The example is only sketched here, and the complete proof and definitions of the standard primitives used, are deferred to the final version of the paper.

Our example relies on the existence of perfectly hiding bit commitment schemes and collision-intractable hash functions.[2] For n players, we will need to hash n commitments in a collision-intractable manner. Thus, the number of players required depends on the strength of the assumption: For n that is polynomial in the security parameter k, this is a standard assumption, whereas for $n = \omega(\log k)$ this requires a stronger assumption. For simplicity, we refer below to a large number of players, instead of making the explicit distinction based on the quality of computational assumption.

The protocol involves players $P_0, P_1, \ldots, P_n$, where the input of P_0 is a function h from a family of collision intractable hash functions, and a public key pk for a perfectly hiding bit commitment scheme. The input of each other P_i is a bit b_i. The output of each player is h, pk. The protocol proceeds as follows:

---

[1] The setting of two parties *without* PEC is only of interest if we are considering a 2-party protocol as a standalone application, without composing it with multi-player protocols. For this setting, we can prove equivalence of adaptive and non-adaptive security in the secure channels model or when the simulation is black box.

[2] This example is an extension of another example given in [CFGN96], which uses only bit commitment, and works only for black-box simulators.

1. P_0 sends h, pk to all other players.
2. Each P_i, $i \geq 1$ computes and broadcasts a commitment $c_i = commit(pk, b_i, r_i)$.
3. All players output h, pk.

We allow the adversary to corrupt P_0 and in addition any subset of size $n/2$ of the other players.

Then this protocol is non-adaptively universally secure, with perfect emulation (since the bit commitment used is perfectly hiding). However, the protocol is adaptively insecure (both universally and computationally): Consider an adversary $\mathcal{A}$ that first corrupts P_0, computes the hash function on all the commitments sent, and interprets it as a subset of $n/2$ players to which $\mathcal{A}$ subsequently breaks. It can then be shown that any simulator for $\mathcal{A}$ can be used to either break the commitment scheme, or find collisions in h.

We thus have the following theorem.

Theorem 4. *For passive adversaries and a large number of parties, adaptive security is strictly stronger than non-adaptive security, under all notions of security except IT with perfect emulation. This holds unconditionally for either statistical or computational emulation, and under the assumption that a perfectly hiding bit commitment scheme and a collision intractable hash function family exist, for perfect emulation.*

2.3 Equivalence for Passive Adversaries and a Small Number of Parties

This section proves that adaptive and non-adaptive security against a *passive* adversary are equivalent when the number of parties is small.

Before going into our results, we need to elaborate on a simplifying assumption we make in this section. As previously mentioned, the [C00] definition of adaptive security (as well as [B91,MR91], in different ways) include a special technical requirement, called post-execution corruptibility (PEC). This requirement is in general needed in order to guarantee secure composition of protocols in the adaptive setting (see [CDDIM01] for more technical details about PEC).

However, in the particular setting of this section, i.e. passive adversaries and a small number of players, it turns out that PEC is an "overkill" requirement for guaranteeing composability of protocols. Very informally, the argument for this is the following. Let π and ρ be protocols that are adaptively secure without the PEC property. These protocols are (of course) also non-adaptively secure. Since the non-adaptive definition of security is closed under (non-concurrent) composition [C00], it follows that the 'composed' protocol, $\pi \circ \rho$, is non-adaptively secure. By our result given below, the composed protocol is also adaptively secure (without PEC).

We conclude that in the setting of this section, PEC is not needed to guarantee adaptively secure composition, and therefore we discuss in this section only results that hold without assuming the PEC requirement.[3]

---

[3] If we were to assume the PEC requirement, we can in fact show a two-party protocol which is non-adaptively secure, but adaptively insecure (this is the same example

We first note that the general definition takes a simpler form in the passive case. In particular, in the passive case we may assume without loss of generality that the real-life adversary waits until the protocol terminates, and then starts to adaptively corrupt the parties; corrupting parties at an earlier stage is clearly of no advantage in the passive case. Similarly, the ideal-process adversary may be assumed to corrupt parties after the ideal function evaluation terminates. To further ease the exposition, we will make in the remainder of this section the following simplifying assumptions: (1) assume that the adversary is deterministic; (2) assume that the function computed by the protocol is deterministic; and (3) ignore auxiliary inputs. The results in this section generalize to hold without the above assumptions.

The Card Game. In attempting to prove equivalence between non-adaptive and adaptive security, it may be helpful to picture the following game. Let $\mathcal{B} \subseteq 2^{[n]}$ be a monotone adversary structure. The game involves two players, the *adversary* and the *simulator*, and n distinct cards. The two players are bound to different rules, as specified below.

Adversary. When the adversary plays, the faces of the n cards are picked from some (unknown) joint distribution $V = (V_1, \ldots, V_n)$ and are initially covered. The adversary proceeds by sequentially uncovering cards according to a fixed deterministic strategy; that is, the choice of the next card to be uncovered is determined by the contents of previously uncovered cards. Moreover, the index set of uncovered cards should always remain within the confines of the structure $\mathcal{B}$. After terminating, the adversary's output consists of the identity and the contents of all uncovered cards.

Simulator. The simulator plays in a different room. It is initially given n distinct *blank* cards, all of which are covered. Similarly to the adversary, it is allowed to gradually uncover cards, as long as the set of uncovered cards remains in $\mathcal{B}$. Its goal is to fill the blank uncovered cards with content, so that the final configuration (including the identity and contents of uncovered cards) is "similarly" distributed to the adversary's output. (The precise sense of this similarity requirement will depend on the specific security setting.) Note that unless the simulator has some form of access to the unknown distribution V, the game would not make much sense. Indeed, we grant the simulator the following type of restricted access to V. At each stage, when the set of uncovered cards is some $b \in \mathcal{B}$, the simulator may freely sample from some fixed distribution $\tilde{V}_b$ which is guaranteed to be "similar" to V_b, the restriction of V to b. (Again, the type of this similarity depends on the setting.) The $|\mathcal{B}|$ distributions $\tilde{V}_b$ may be arbitrarily (or adversarially) fixed, as long as they conform to the above similarity condition.

---

based on perfectly hiding bit commitment which was mentioned in the end of Section 2.1). Thus, strictly speaking, there is a separation in this setting under the [c00] definition. The results in other sections hold regardless of whether the PEC requirement is imposed or not.

Let us briefly explain the analogy between the above game and the question of non-adaptive versus adaptive security. Fix some n-party protocol π computing a deterministic function f, and suppose that π is non-adaptively secure against a passive $\mathcal{B}$-limited adversary. The n cards correspond to the n parties. The distribution V corresponds to the parties' joint view under an input x, which is a-priori unknown. Uncovering the i-th card by the adversary and learning V_i corresponds to corrupting the i-th party P_i in the real-life process and learning its entire view: its input, random input, communication messages, and output. Uncovering the i-th card by the simulator corresponds to corrupting P_i in the ideal-model process. Finally, each distribution $\tilde{V}_b$ from which the simulator can sample corresponds to a simulation of a non-adaptive adversary corrupting b, which exists under the assumption that π is non-adaptively secure. Note that the simulator can access $\tilde{V}_b$ only when all cards in b are uncovered; this reflects the fact that the non-adaptive simulation cannot proceed without learning the inputs and outputs of corrupted parties. The types of similarity between V_b and $\tilde{V}_b$ we will consider are *perfect*, *statistical*, and *computational*, corresponding to the type of non-adaptive emulation we assume. We will also consider the relation between the computational complexity of the adversary and that of the simulator, addressing the security variants in which the simulator is computationally bounded.

Remark. The above game models a secure channels setting, in which the adversary has no information before corrupting a party. To model open channels (or a "broadcast" channel), the distribution V should be augmented with an additional entry V_0, whose card is initially uncovered. The analysis that will follow can be easily adapted to deal with this more general setting.

Perfect Emulation. We first deal with perfect emulation, i.e., the case where $\tilde{V}_b = V_b$ for all $b \in \mathcal{B}$. In this setting, we show how to construct an adaptive simulator running in (expected) time polynomial in the time of the adversary and the size of the adversary structure. The construction from this section will allow us to prove equivalence of non-adaptive and adaptive security both in the information-theoretic case (see Section 2.4) and, when the adversary structure is small, in the universal case.

A black-box simulator. To prove equivalence between non-adaptive and adaptive security it suffices to show that for any adversary strategy $\mathcal{A}$ there exists a simulator strategy $\mathcal{S}$, such that under any distribution V the simulator wins. In fact, we will construct a single simulator $\mathcal{S}$ with a black-box access to $\mathcal{A}$, and later analyze it in various settings.

A CONVENTION FOR MEASURING THE RUNNING TIME OF BLACK-BOX SIMULATORS. In the following we view adaptive simulators as algorithms supplied with two types of oracles: *distribution oracles* $\tilde{V}_b$, implemented by a non-adaptive ideal-process adversary (to be referred to as a *non-adaptive simulator*), and an adaptive *adversary oracle* $\mathcal{A}$. In measuring the running time of a simulator, each oracle call will count as a single step. This convention is convenient for proving universal security: If the protocol has universal non-adaptive security and the black-box simulator $\mathcal{S}$ runs in expected time poly(k) then, after substituting appropriate implementations of the oracles, the expected running time of $\mathcal{S}$ is polynomial in k and the expected running time of $\mathcal{A}$.[4]

In the description and analysis of $\mathcal{S}$ we will use the following additional notation. By v_b, where v is an n-tuple (presumably an instance of V) and $b \subseteq [n]$ is a set, we denote the restriction of v to its b-entries. For notational convenience, we assume that the entries of a partial view v_b, obtained by restricting v or by directly sampling from $\tilde{V}_b$ or V_b, are labeled by their corresponding b-elements (so that b can be inferred from v_b). We write $v \xrightarrow{\text{A}} b$ if the joint card contents (view) v leads the adversary $\mathcal{A}$ to uncover (corrupt) the set b at some stage. For instance, $v \xrightarrow{\text{A}} \emptyset$ always holds. An important observation is that whether $v \xrightarrow{\text{A}} b$ holds depends only on v_b. This trivially follows from the fact that cards cannot be covered once uncovered. Hence, we will also use the notation $v' \xrightarrow{\text{A}} b$, where v' is a $|b|$-tuple representing a partial view.

In our description of the simulator we will adopt the simplified random variable notation from the game described above, but will revert to the original terminology of corrupting parties rather than uncovering cards.

Before describing our simulator $\mathcal{S}$, it is instructive to explain why a simpler simulation attempt fails. Consider a "straight line" simulator which proceeds as follows. It starts by corrupting $b = \emptyset$. At each iteration, it samples $\tilde{V}_b$ and runs the adversary on the produced view to find the first party outside b it would corrupt. The simulator corrupts this party, adds it to b, and proceeds to the next iteration (or terminates with the adversary's output if the adversary would terminate before corrupting a party outside b). This simulation approach fails for the following reason. When sampling $\tilde{V}_b$, the produced view is independent of the event which has lead the simulator to corrupt b. This makes it possible, for instance, that the simulator corrupts a set which cannot be corrupted at all in the real-life execution. The simulator $\mathcal{S}$, described next, will fix this problem by insisting that the view sampled from $\tilde{V}_b$ be consistent with the event that the adversary corrupts b.

Algorithm of $\mathcal{S}$:

1. Initialization:
   Let $b_0 = \emptyset$. The set b_i will contain the first i parties corrupted by the simulator.
2. For $i = 0, 1, 2, \ldots$ do:
   (a) Repeatedly sample $v' \xleftarrow{\text{R}} \tilde{V}_{b_i}$ (by invoking the non-adaptive simulator) until $v' \xrightarrow{\text{A}} b_i$ (i.e., the sampled partial view would lead $\mathcal{A}$ to corrupt b_i). Let v_i be the last sampled view. (Recall that v' includes the identities of parties in b_i.)

---

[4] Note that when the protocol has universal non-adaptive security, a distribution $\tilde{V}_b$ can be sampled in expected polynomial time from the view of an ideal-process adversary corrupting b.

(b) Invoke $\mathcal{A}$ on v_i to find the index p_{i+1} of the party which $\mathcal{A}$ is about to corrupt next (if any). If there is no such party (i.e., $\mathcal{A}$ terminates), output v_i. Otherwise, corrupt the p_{i+1}-th party, let $b_{i+1} = b_i \cup \{p_{i+1}\}$, and iterate to the next i.

The analysis of the simulator $\mathcal{S}$, appearing in [CDDIM01], shows that in the case of a perfect non-adaptive emulation ($\tilde{V}_b \stackrel{\mathrm{d}}{=} V_b$): (1) $\mathcal{S}$ perfectly emulates $\mathcal{A}$, and (2) the expected running time of $\mathcal{S}$ is linear in $|\mathcal{B}|$. We may thus conclude the following:

Theorem 5. *For function evaluation protocols with passive adversary, universal perfect security, and $n = O(\log k)$ parties, adaptive and non-adaptive security are equivalent.*

Imperfect Emulation. We next address the cases of statistical and computational security against a passive adversary. Suppose that we are given an imperfect (statistical or computational) non-adaptive simulator and attempt to construct an adaptive one. If we use exactly the same approach as before, some technical problems arise: with imperfect non-adaptive emulation, it is possible that a real life adversary $\mathcal{A}$ corrupts some set with a very small probability, whereas this set is *never* corrupted in emulated views. As a result, the loop in step (2a) of the algorithm of $\mathcal{S}$ will never terminate, and the expected time will be infinite. Consequently, it is also unclear whether $\mathcal{S}$ will produce a good output distribution when given access to imperfect non-adaptive simulation oracles $\tilde{V}_b$.

We start by showing that when the size of the adversary structure is polynomial, the simulator $\mathcal{S}$ will indeed produce a (statistically or computationally) good output distribution even when given access to (statistically or computationally) imperfect non-adaptive simulators. Moreover, it will turn out that when the adversary structure is polynomial, the expected running time of $\mathcal{S}$ is polynomial *except with negligible probability*. Later, we define a more sophisticated simulator $\mathcal{S}'$ which achieves strict expected-polynomial time simulation, at the expense of requiring a stronger assumption on the size of the adversary structure.

Specifically, these results can be summarized by the following theorem, whose proof appears in [CDDIM01].

Theorem 6. *For function evaluation protocols with passive adversary and $n = O(\log k / \log \log k)$ parties, adaptive and non-adaptive security are equivalent under any notion of security. Moreover, with a relaxed notion of efficiency allowing a negligible failure probability, the bound on the number of parties can be improved to $n = O(\log k)$.*

We remark that Theorem 6 is essentially tight in the following sense: when $n = \omega(\log k)$, adaptive security is separated from non-adaptive security even if the adaptive simulator is allowed to be computationally unbounded.

2.4 Equivalence for Passive Adversaries and IT Security

The analysis of the simulation from the previous section implies the following:

Theorem 7. *For function evaluation protocols with passive adversary and perfect information-theoretic security, adaptive and non-adaptive security are equivalent.*

Note that there is no dependence on the number of players in the above theorem.

3 Adaptivity vs. Non-adaptivity in the Definition of Dodis-Micali-Rogaway

3.1 Review of the Definition

For completeness, we start with a very short summary of the definition of secure multiparty computation by Micali and Rogaway, more specifically the version that appears in the paper by Dodis and Micali [DM00]. For additional details, please refer to [DM00].

We have n players, each player P_i starts with a value x_i as input and auxiliary input a_i. We set $a = (a_1, ... a_n); x = (x_1, ..., x_n)$.

To satisfy the definition, a protocol π must have a fixed *committal round CR*, the point at which inputs become uniquely defined, as follows: The *traffic* of a player consists of all messages he sends and receives. π must specify *input- and output functions* that map traffic to input- and output values for the function f computed. The *effective inputs* $\hat{x}_1^\pi, ..., \hat{x}_n^\pi$ are determined by applying the input functions to the traffic of each player up to and including CR. So these values are the ones that players "commit to" as their inputs. The *effective outputs* $\hat{y}_1^\pi, ..., \hat{y}_n^\pi$ are determined by applying the output functions to the entire traffic of each player.

For adversary $\mathcal{A}$ (taking random input and auxiliary input α), random variable $View(\mathcal{A}, \pi)$ is the view of $\mathcal{A}$ when attacking π. We define:

$$History(\mathcal{A}, \pi) = View(\mathcal{A}, \pi), \hat{x}^\pi, \hat{y}^\pi$$

The way $\mathcal{A}$ interacts with the protocol is as follows: In each round, $\mathcal{A}$ sees all messages from honest players in this round. He may then issue some number of corruption requests adaptively, and only then must he generate the messages to be sent to the remaining honest players.

The definition calls for existence of a simulator $\mathcal{S}$ which may depend on the protocol in question, but not the adversary. The goal of the simulator is to sample the distribution of $History(\mathcal{A}, \pi)$. To do so, it is allowed to interact with $\mathcal{A}$, but it is restricted to one-pass black-box simulation with no bound on the simulator's running time, i.e., $\mathcal{A}$ interacts with $\mathcal{S}$ in the same way it interacts with π, and $\mathcal{S}$ is not allowed to rewind $\mathcal{A}$. The simulator $\mathcal{S}$ gets an oracle O as help (where the oracle knows x, a):

- If P_j is corrupted before CR, the oracle sends x_j, a_j to $\mathcal{S}$.
- At CR, $\mathcal{S}$ applies the input functions to the view of $\mathcal{A}$ it generated so far to get effective inputs of corrupted players $\hat{x}_j^{\mathcal{S}}$. It sends these values to O. O computes the function choosing random input r and using as input the values it got from $\mathcal{S}$ for corrupted players and the real x_j's for honest players. The result is $\hat{y}^{\mathcal{S}} = (\hat{y}_1^{\mathcal{S}}, ..., \hat{y}_n^{\mathcal{S}})$. O sends the results for corrupted players back to $\mathcal{S}$.
- If P_j is corrupted in or after CR, O sends $x_j, a_j, \hat{y}_j$ to $\mathcal{S}$.

The random variable $View(\mathcal{A}, \mathcal{S})$ is the view of $\mathcal{A}$ when interacting with $\mathcal{S}$. The effective inputs $\hat{x}^{\mathcal{S}}$ are as defined above, i.e., if a P_j is corrupted before CR, then his effective input $\hat{x}_j^{\mathcal{S}}$ is determined by the input function on his traffic, else $\hat{x}_j = x_j$. The effective outputs $\hat{y}^{\mathcal{S}}$ are defined as what the oracle outputs, i.e. $\hat{y}^{\mathcal{S}} = f(\hat{x}^{\mathcal{S}}, r)$.

$$History(\mathcal{A}, \mathcal{S}) = View(\mathcal{A}, \mathcal{S}), \hat{x}^{\mathcal{S}}, \hat{y}^{\mathcal{S}}$$

We can now define that π computes f securely iff there exists a simulator $\mathcal{S}$ such that for every adversary $\mathcal{A}$, and every x, a, α,

$$History(\mathcal{A}, \mathcal{S}) \equiv History(\mathcal{A}, \pi)$$

i.e., the two variables have identical distributions.

At first sight it may seem strange that the definition does not explicitly require that players who are honest up to CR actually commit to their real inputs, or that players who are never corrupted really receive "correct" values. But this follows from the definition:

Lemma 1. *If π computes f securely, then the input- and output functions are such that if P_j remains honest up to CR, then $\hat{x}_j^\pi = x_j$. And if P_j is never corrupted, then $\hat{y}_j^\pi$ is the j'th component of $f(\hat{x}^\pi, r)$, for a random r.*

Proof. Consider an adversary $\mathcal{A}_j$ that never corrupts P_j. Then the first claim follows from $x_j = \hat{x}_j^{\mathcal{S}}$ and $History(\mathcal{A}_j, \mathcal{S}) \equiv History(\mathcal{A}_j, \pi)$. The second follows from $History(\mathcal{A}_j, \mathcal{S}) \equiv History(\mathcal{A}_j, \pi)$ and the fact that the correlation $\hat{y}_j^{\mathcal{S}} = f(\hat{x}^{\mathcal{S}}, r)_j$ between $\hat{x}^{\mathcal{S}}$ and $\hat{y}^{\mathcal{S}}$ always holds.

Note that this lemma continues to hold, even if we only assume static security.

3.2   Equivalence of Adaptive and Non-adaptive Security

It turns out to be convenient in the following to define the notion of a *partial history*, of an adversary $\mathcal{A}$ that either attacks π or interacts with a simulator. A partial history constrains the history up to a point at the start of, or inside round j for some j. That is, round $j - 1$ has been completed but round j has not. If $j \leq CR$, then such a partial history consists of a view of the adversary up to round j, and possibly including some part of round j. If $j > CR$, but the protocol is not finished, a partial history consists of a partial view of $\mathcal{A}$ as described before plus the effective inputs. Finally, if the protocol is finished at round j, the history is as defined earlier: complete view of $\mathcal{A}$ plus the effective inputs and outputs.

Note that if $\mathcal{S}$ is such that $History(\mathcal{A}, \pi) \equiv History(\mathcal{A}, \mathcal{S})$, then trivially it also holds that the partial histories of $\mathcal{A}, \pi$ and of $\mathcal{A}, \mathcal{S}$ ending at any point are identically distributed. Moreover, since $\mathcal{S}$ never rewinds, the value of the partial history of $\mathcal{A}, \mathcal{S}$ at some point in time will be fixed as soon as $\mathcal{S}$ has reached that point in the simulation.

We can then slightly extend the actions an adversary can take: a *halting adversary* $\mathcal{A}'$ is one that interacts with protocol or simulator in the normal way, but may at any point output a special halting symbol and then stop. In the simulation, if the simulator receives such a symbol, the simulation process also stops. The histories $History(\mathcal{A}', \pi), History(\mathcal{A}', \mathcal{S})$ are defined to be whatever the partial history is at the point when $\mathcal{A}$ stops.

Trivially protocol π is secure in the above definition if and only if, for any halting adversary $\mathcal{A}'$, $History(\mathcal{A}', \pi) \equiv History(\mathcal{A}', \mathcal{S})$. Note that this extension of the definition does not capture any new security properties, it is simply a "hack" that turns out to be convenient in the proof of the following theorem.

In the following we assume that there exists a static (non-adaptive) simulator $\mathcal{S}_0$ such that for every *static* adversary $\mathcal{A}_0$, and every x, a, α,

$$History(\mathcal{A}_0, \mathcal{S}_0) \equiv History(\mathcal{A}_0, \pi)$$

We want to make a general simulator $\mathcal{S}$ that shows that π in fact is secure against any adaptive adversary $\mathcal{A}$, in other words, we claim

Theorem 8. *Adaptive and non-adaptive security are equivalent under the Dodis-Micali-Rogaway definition.*

To this end, we construct a static adversary $\mathcal{A}_B$ (of the halting type), for every set B that it is possible for $\mathcal{A}$ to corrupt. $\mathcal{A}_B$ plays the following strategy, where we assume that $\mathcal{A}_B$ is given black-box access to (adaptive) adversary $\mathcal{A}$, running with some random and auxiliary inputs $r_\mathcal{A}$ and α[5]:

Algorithm of $\mathcal{A}_B$

1. Corrupt the set B initially. For each $P_j \in B$, initialize the *honest* algorithm for P_j, using as input x_j, a_j learnt from corrupting P_j (and fresh random input).
2. Start executing the protocol, initially letting the players in B play honestly, but keeping a record of their views. At the same time, start running $\mathcal{A}$.
3. Whenever $\mathcal{A}$ issues a corruption request for player P_j, we do the following: if $P_j \in B$, we provide $\mathcal{A}$ with x_j, a_j and all internal data of P_j. After this point, all messages for P_j are sent to $\mathcal{A}$, and we let $\mathcal{A}$ decide the actions of P_j from this point. If $P_j \notin B$, output a halt symbol and stop.

---

[5] We could also have given $r_\mathcal{A}, \alpha$ as input to $\mathcal{A}_B$, letting it simulate the algorithm of $\mathcal{A}$, but the set-up we use is more convenient in the following.

The idea in the following is to use the assumed ability (by S_0) to generate histories of $\mathcal{A}_B$ attacking π to generate histories of $\mathcal{A}$ attacking π. Note that in any round of π, the current history of $\mathcal{A}_B$ contains both the (so far honest) history of 0 or more players that $\mathcal{A}$ has not yet corrupted, plus the view so far of $\mathcal{A}$. So for any such (partial) history u of $\mathcal{A}_B$, we let $Aview(u)$ be the view of $\mathcal{A}$ that can be extracted from u in the natural way.

In particular, if u is a history of $\mathcal{A}_B$ that ends after the final round of the protocol, then $Aview(u)$ is a complete view of $\mathcal{A}$ where $\mathcal{A}$ corrupted only players in B, whereas if u ends before the protocol is complete, $Aview(u)$ ends in the some round where $\mathcal{A}$ requested to corrupt some player outside B.

We are now ready to describe how a simulator $\mathcal{S}$ can be constructed. The full algorithm and analysis of $\mathcal{S}$ are omitted for lack of space, and appear in [CDDIM01]. Here we give only the basic idea:

From the beginning, A has not corrupted any players. So we can create the start of a history by running $(A_\emptyset, S_0)$ (recall that $A_\emptyset$ runs A "in the background"). This will stop as soon as A corrupts the first player P_j. Say this happens in round i. Let v be the view of A we obtain from this. Recall that S_0 provides perfect emulation. This means that in real life when A attacks F, we would with the same probability obtain a history where up to round i, A obtains view v and all players including P_j have been honest.

Now, by construction of $A_{\{P_j\}}$ this same history up to round i can also be realized by $A_{\{P_j\}}$ attacking F: the only difference is that from the beginning $A_{\{P_j\}}$ and not the j'th player runs the honest algorithm of P_j. And again by assumption on S_0, the history can also be realized by $A_{\{P_j\}}$ interacting with S_0.

We can therefore (by exhaustive search over the random inputs) generate a random history of S_0 interacting with $A_{\{P_j\}}$, conditioned on the event that the view v for A is produced in the first i rounds (and moreover, this can be done without rewinding A). This process may be inefficient, but this is no problem since we consider IT-security here. Once we succeed, we let $(S_0, A_{\{P_j\}})$ continue to interact until they halt, i.e., we extend the history until the protocol is finished or A corrupts the next player (say $P_{j'}$). In the former case, we are done, and otherwise we continue in the same way with $A_{\{P_j, P_{j'}\}}$.

Once we finish the CR, the effective inputs will be determined, and we will get resulting outputs from the oracle. Note here that since we only consider one-pass blackbox simulation, we will never need to rewind back past the CR, which might otherwise create problems since then A could change its mind about the effective inputs. Thus also the one-pass black-box requirement is essential for the proof.

References

B91. D. Beaver, "Secure Multi-party Protocols and Zero-Knowledge Proof Systems Tolerating a Faulty Minority", J. Cryptology, Springer-Verlag, (1991) 4: 75-122.
B97. D. Beaver, "Plug and Play Encryption", CRYPTO 97.
BH92. D. Beaver and S. Haber, "Cryptographic Protocols Provably secure Against Dynamic Adversaries", *Eurocrypt,* 1992.

BGW88. M. Ben-Or, S. Goldwasser and A. Wigderson, "Completeness Theorems for Non-Cryptographic Fault-Tolerant Distributed Computation", *20th Symposium on Theory of Computing (STOC)*, ACM, 1988, pp. 1-10.

C00. R. Canetti, "Security and Composition of Multiparty Cryptographic Protocols", *Journal of Cryptology*, Vol. 13, No. 1, Winter 2000. On-line version at http://philby.ucsd.edu/cryptolib/1998/98-18.html.

C00a. R. Canetti, "A unified framework for analyzing security of Protocols", manuscript, 2000. Available at http://eprint.iacr.org/2000/067.

CDDIM01. R. Canetti, I. Damgaard, S. Dziembowski, Y. Ishai and T. Malkin, "On adaptive vs. non-adaptive security of multiparty protocols", http://eprint.iacr.org/2001.

CFGN96. R. Canetti, U. Feige, O. Goldreich and M. Naor, "Adaptively Secure Computation", *28th Symposium on Theory of Computing (STOC)*, ACM, 1996. Fuller version in MIT-LCS-TR #682, 1996.

CDM98. R.Cramer, I.Damgaard and U.Maurer: *General Secure Multiparty Computation from Any Linear Secret-Sharing Scheme*, EuroCrypt 2000.

CCD88. D. Chaum, C. Crepeau, and I. Damgaard. Multi-party Unconditionally Secure Protocols. In *Proc. 20th Annual Symp. on the Theory of Computing (STOC)*, pages 11–19, ACM, 1988.

CS98. R. Cramer and V. Shoup, "A paractical public-key cryptosystem provably secure against adaptive chosen ciphertext attack", *CRYPTO '98*, 1998.

DN00. I. Damgaard and J. Nielsen, "Improved non-committing encryption schemes based on a general complexity assumption", CRYPTO 2000.

DM00. Y. Dodis and S. Micali, "Parallel Reducibility for Information-Theoretically Secure Computation", CRYPTO 2000.

DDN91. D. Dolev, C. Dwork and M. Naor, "Non-malleable cryptography", SICOMP, to appear. Preliminary version in STOC 91.

GMW87. O. Goldreich, S. Micali and A. Wigderson, "How to Play any Mental Game", *19th Symposium on Theory of Computing (STOC)*, ACM, 1987, pp. 218-229.

GL90. S. Goldwasser, and L. Levin, "Fair Computation of General Functions in Presence of Immoral Majority", *CRYPTO '90, LNCS 537*, Springer-Verlag, 1990.

GM84. S. Goldwasser and S. Micali, "Probabilistic encryption", *JCSS*, Vol. 28, No 2, April 1984, pp. 270-299.

MR91. S. Micali and P. Rogaway, "Secure Computation", unpublished manuscript, 1992. Preliminary version in *CRYPTO '91, LNCS 576*, Springer-Verlag, 1991.

PW94. B. Pfitzmann and M.Waidner, "A General Framework for Formal Notions of Secure Systems", Hildesheimer Informatik-Berichte, ISSN 0941-3014, April 1994.

PSW00. B. Pfitzmann, M. Schunter and M.Waidner, "Secure Reactive Systems", IBM Technical report RZ 3206 (93252), May 2000.

S99. A. Sahai, "Non malleable, non-interactive zero knowlege and adaptive chosen ciphertext security", FOCS 99.

Y82. A. Yao, "Protocols for Secure Computation", In *Proc. 23rd Annual Symp. on Foundations of Computer Science (FOCS)*, pages 160–164. IEEE, 1982.

Y86. A. Yao, "How to generate and exchange secrets", In *Proc. 27th Annual Symp. on Foundations of Computer Science (FOCS)*, pages 162–167. IEEE, 1986.

Multiparty Computation
from Threshold Homomorphic Encryption

Ronald Cramer, Ivan Damgård, and Jesper B. Nielsen

BRICS[*] Department of Computer Science
University of Århus
Ny Munkegade
DK-8000 Arhus C, Denmark
{cramer,ivan,buus}@brics.dk

Abstract. We introduce a new approach to multiparty computation (MPC) basing it on homomorphic threshold crypto-systems. We show that given keys for any sufficiently efficient system of this type, general MPC protocols for n parties can be devised which are secure against an active adversary that corrupts any minority of the parties. The total number of bits broadcast is $O(nk|C|)$, where k is the security parameter and $|C|$ is the size of a (Boolean) circuit computing the function to be securely evaluated. An earlier proposal by Franklin and Haber with the same complexity was only secure for passive adversaries, while all earlier protocols with active security had complexity at least quadratic in n. We give two examples of threshold cryptosystems that can support our construction and lead to the claimed complexities.

1   Introduction

The problem of multiparty computation (MPC) dates back to the papers by Yao [Yao82] and Goldreich et al. [GMW87]. What was proved there was basically that a collection of n parties can efficiently compute the value of an n-input function, such that everyone learns the correct result, but no other new information. More precisely, these protocols can be proved secure against a polynomial time bounded adversary who can *corrupt* a set of less than $n/2$ parties initially, and then make them behave as he likes, we say that the adversary is *active*. Even so, the adversary should not be able to prevent the correct result from being computed and should learn nothing more than the result and the inputs of corrupted parties. Because the set of corrupted parties is fixed from the start, such an adversary is called *static* or non-adaptive.

There are several proposals on how to define formally the security of such protocols [MR91,Bea91,Can00], but common to them all is the idea that security means that the adversary's view can be *simulated* efficiently by a machine that has access to only those data that the adversary is *entitled* to know.

---

[*] Basic Research in Computer Science,
Centre of the Danish National Research Foundation.

B. Pfitzmann (Ed.): EUROCRYPT 2001, LNCS 2045, pp. 280–300, 2001.
© Springer-Verlag Berlin Heidelberg 2001

Proving correctness of a simulation in the case of [GMW87] requires a complexity assumption, such as existence of trapdoor one-way permutations. This is because the model of communication considered there is such that the adversary may see every message sent between parties, this is known as the *cryptographic model*. Later, *unconditionally* secure MPC protocols were proposed by Ben-Or et al. and Chaum et al.[BGW88,CCD88], in the model where *private* channels are assumed between every pair of parties. In this paper, however, we are only interested in the cryptographic model with an active and static adversary.

Over the years, several protocols have been proposed which, under specific computational assumptions, improve the efficiency of general MPC, see for instance [CDM00,GRR98]. Virtually all proposals have been based on some form of *verifiable secret sharing* (VSS), i.e., a protocol allowing a dealer to securely distribute a secret value s among the parties, where the dealer and/or some of the parties may be cheating. The basic paradigm is to ensure that all inputs and intermediate values in the computation are VSS'ed; this prevents the adversary from causing the protocol to terminate early or with incorrect results. In all these earlier protocols, the number of bits sent was $\Omega(n^2 k|C|)$, where n is the number of parties, k is a security parameter, and $|C|$ is the size of a circuit computing the function. Here, C may be a Boolean circuit, or an arithmetic circuit over a finite field, depending on the protocol.

In [FH96] Franklin and Haber propose a protocol for *passive* adversaries which achieves complexity $O(nk|C|)$. This protocol is not based on VSS (there is no need since the adversary is passive) but instead on a so called joint encryption scheme, where a ciphertext can only be decrypted with the help of all parties, but still the length of an encryption is independent of the number of parties.

2   Our Results

In this paper, we present a new approach to building multiparty computation protocols with active security, namely we start from any secure threshold encryption scheme with certain extra homomorphic properties. This allows us to avoid the need to VSS all values handled in the computation, and therefore leads to more efficient protocols, as detailed below.

The MPC protocols we construct here can be proved secure against an active and static adversary who corrupts any minority of the parties. Like the protocol of [FH96], our construction requires an initial phase where keys for the threshold cryptosystem are set up. This can be done by a trusted party, or by any suitable MPC. In particular, the techniques of Damgård and Koprowski [DK01] could be used to make this phase reasonably efficient for the example cryptosystems we present here (see below). We stress that unlike some earlier proposals for preprocessing in MPC, the complexity of this phase does not depend on the number or the size of computations to be done later. In the following we therefore focus on the complexity of the actual computation. In our protocol the computation can be done only by broadcasting a number of messages, no encryption is needed to set up private channels. The complexities we state are therefore simply the number of bits broadcast. This does not invalidate comparison with earlier protocols because first, the same measure was used in [FH96] and second, the earlier protocols with active security have complexity quadratic in n even if one only counts the bits broadcast. Our protocol has complexity $O(nk|C|)$ bits and requires $O(d)$ rounds, where d is the depth of C. To the best of our knowledge, this is the most efficient general MPC protocol proposed to date for active adversaries. Note that all complexities stated here and in the previous section are for computing deterministic functions. Probabilistic functions can be handled using standard techniques, see Section 8.1 for details.

Here, C is an arithmetic circuit over a ring R determined by the cryptosystem used, e.g., $R = \mathbf{Z}_N$ for an RSA modulus N, or $R = GF(2^k)$. While such circuits can simulate any Boolean circuit with a small constant factor overhead, this also opens the possibility of building an ad-hoc circuit over R for the desired function, possibly exploiting the fact that with a large R, we can manipulate many bits in one arithmetic operation.

The complexities given here assume existence of sufficiently efficient threshold cryptosystems. We give two examples of such systems with the right properties. One is based on Paillier's cryptosystem [Pai99] and Damgård and Jurik's generalisation thereof[DJ01], the other one is a variant of Franklin and Haber's cryptosystem [FH96], which is secure assuming that both the QR assumption and the DDH assumption are true (this is essentially the same assumption as the one made in [FH96]). While the first example is known (from [DJ01,FPS00]), the second is new and may be of independent interest.

Franklin and Haber in [FH96] left as an open problem to study the communication requirements for active adversaries. We can now say that under the same assumption as theirs, active security comes essentially for free.

2.1   Concurrent Related Work

In concurrent independent work, Jacobson and Juels[MJ00] present an idea for MPC somewhat related to ours, the *mix-and-match approach*. It too is based on threshold encryption (with extra algebraic properties, similar to, but different from the ones we use). Beyond this, the techniques are completely different. For Boolean circuits and in the random oracle model, they get the same message complexity as we obtain (without using random oracles). The round complexity is larger than ours (namely $O(n+d)$). Another difference is that mix-and-match is inherently limited to circuits where gates can be specified by constant size truth-tables, thus excluding arithmetic circuits over large rings. On the other hand, while mix-and-match can be based on the DDH assumption alone, it is not known if this is possible for our notion of threshold *homomorphic* encryption.

In [MH00], Hirt, Maurer and Przydatek show an MPC protocol designed for the private channels model. It can be transformed to our setting by implementing the channels using secure public-key encryption. This results in protocol that can be based on any secure public-key encryption scheme, with essentially the same communication complexity as ours, but with lower resilience, i.e. tolerating only less than $n/3$ active cheaters.

3   An Informal Description

In this section, we give an informal introduction to some main ideas. All the concepts introduced here will be treated more formally later in the paper. We will assume that from the start, the following scenario has been established: we have a semantically secure threshold public-key system given, i.e., there is a public encryption key pk known by all parties, while the matching private decryption key has been shared among the parties, such that each party holds a share of it.

The message space of the cryptosystem is assumed to be a ring R. In practice R might be $\mathbf{Z}_N$ for some RSA modulus N. For a plaintext $a \in R$, we let $\bar{a}$ denote an encryption of a. We then require certain homomorphic properties: from encryptions $\bar{a}, \bar{b}$, anyone can easily compute an encryption of $a + b$, which we denote $\bar{a} \boxplus \bar{b}$. We also require that from an encryption $\bar{a}$ and a constant $\alpha \in R$, it is easy to compute a random encryption of αa.

Finally we assume that three secure and efficient protocols are available:

Proving you know a plaintext If P_i has created an encryption $\bar{a}$, he can give a zero-knowledge proof of knowledge that he knows a.

Proving multiplications correct Assume that P_i is given an encryption $\bar{a}$, chooses a constant α, computes a random encryption $\overline{\alpha a}$ and broadcasts $\bar{\alpha}, \overline{\alpha a}$. He can then give a zero-knowledge proof that indeed $\overline{\alpha a}$ contains the product of the values contained in $\bar{\alpha}$ and $\bar{a}$.

Threshold decryption For the third protocol, we have common input pk and an encryption $\bar{a}$, in addition every party also uses his share of the private key as input. The protocol computes securely a as output for everyone.

We can then sketch how to perform securely a computation specified as a circuit doing additions and multiplications in R.

The MPC protocol would simply start by having each party publish encryptions of his input values and give zero-knowledge proofs that he knows these values and also, if we are simulating a Boolean circuit, that the values are 0 or 1. Then any operation involving addition or multiplication by constants can be performed with no interaction. This leaves only the following problem: Given encryptions $\bar{a}, \bar{b}$ (where it may be the case that no parties knows a nor b), compute securely an encryption of $c = ab$. This can be done by (a slightly more elaborate version of) the following protocol:

1. The parties generate an additive secret sharing of a:
    (a) Each party P_i chooses at random a value $d_i \in R$, broadcasts an encryption $\bar{d_i}$, and proves he knows d_i. Let d denote $\sum_{i=1}^{n} d_i$.
    (b) The parties use the third protocol to decrypt $\bar{a} \boxplus \bar{d_1} \boxplus ... \boxplus \bar{d_n}$.
    (c) Party P_1 sets $a_1 = (a + d) - d_1$, all other parties P_i set $a_i = -d_i$. Note that $a = \sum_{i=1}^{n} a_i$.
2. Each P_i broadcasts an encryption $\overline{a_i b}$, and invoke the second protocol with inputs $\bar{b}, \overline{a_i}$ and $\overline{a_i b}$.

3. Let H be the set of parties for which the previous step succeeded, and let C be the complement of H. The parties decrypt $\boxplus_{i \in C} \overline{a_i}$, learn $a_C = \sum_{i \in C} a_i$, and compute $\overline{a_C b}$. From this, and $\{\overline{a_i b} \mid i \in H\}$, all parties can compute an encryption $(\boxplus_{i \in H} \overline{a_i b}) \boxplus \overline{a_C b}$, which is indeed an encryption of ab.

At the final stage we know encryptions of the output values, which we can just decrypt. Intuitively this is secure if the encryption is secure because, other than the outputs, only random values and values already known to the adversary are ever decrypted. We will give proofs of this intuition in the following.

The above multiplication protocol is a more efficient version of a related idea from [FH96], where we have exploited the homomorphic properties to add protection against faults without loosing efficiency. Other papers have exploited homomorphic properties to construct efficient protocols for multiplication. In [GV87] Goldreich and Vainish used the homomorphic properties of the QR problem to construct a two-party protocol for multiplication over $GF(2)$ and in [KK91] Kurosawa and Kotera generalised it to $GF(L)$ for small L using the homomorphic properties of the Lth residuosity problem. Using also the Lth residuosity problem [Kur91] constructs an efficient ZKIP for multiplication over $GF(L)$.

4   Preliminaries and Notation

Let A be a probabilistic polynomial time (PPT) algorithm, which on input $x \in \{0,1\}^*$ and random bits $r \in \{0,1\}^*$ outputs a value $y \in \{0,1\}^*$. We write $y \leftarrow A(x)[r]$ to denote that y should be computed by running A on x and r. By $y \leftarrow A(x)$ we mean that y should be computed using uniformly random r and by $y \in A(x)$ we mean that y is among the values, that $A(x)$ outputs.

4.1   The MPC Model

We prove security in the MPC model from [Can00], with an open authenticated broadcast channel, against active non-adaptive adversaries corrupting any minority of the parties. We index the parties by $N = \{1, \dots, n\}$ and let $C \subset N$ denote the subset of corrupted parties, k is the security parameter, and x_i is the secret input of party P_i. We study functions giving a common output $y = f(x_1, \dots, x_n)$. We extend the model to handle this by saying that oracles for such functions broadcast y on the broadcast channel. This to assure that the ideal-model adversary learns the public output even though no party is corrupted. The technical report [CDN00] contains a more detailed description of the model. In the following we let secure mean secure against minority adversaries.

4.2 Σ-protocols

In this section, we look at two-party zero-knowledge protocols of a particular form. Assume we have a binary relation R consisting of pairs (x, w), where we think of x as a (public) instance of a problem and w as a witness, a solution to the instance. Assume also that we have a 3-move proof of knowledge for R: this protocol gets a string x as common input for prover and verifier, whereas the prover gets as private input w such that $(x, w) \in R$. Conversations in the protocol are of form (a, e, z), where the prover sends a, the verifier chooses e at random, the prover sends z, and the verifier accepts or rejects. There is a security parameter k, such that the length of both x and e are linear in k. We will only look at protocols where also the length of a and z are linear in k. Such a protocol is said to be a Σ-protocol if we have the following:

- The protocol is *complete*: if the prover gets as private input w such that $(x, w) \in R$, the verifier always accepts.
- The protocol is *special honest verifier zero-knowledge*: from a challenge value e, one can efficiently generate a conversation (a, e, z), with probability distribution equal to that of conversation between the honest prover and verifier where e occurs as challenge.
- A cheating prover can answer only one of the possible challenges: more precisely, from the common input x and any pair of accepting conversations $(a, e, z), (a, e', z')$ where $e \neq e'$, one can compute efficiently w such that $(x, w) \in R$.

It is easy to see that the definition of Σ-protocols is closed under parallel composition.

5 Threshold Homomorphic Encryption

In this section we formalise the notion of threshold homomorphic encryption.

Definition 1 (Threshold Encryption Scheme). *We call the tuple $(K, \mathrm{KD}, R, E, \mathrm{Decrypt})$ a* threshold encryption scheme *if the following holds.*

Key space *The* key space $K = \{K_k\}_{k \in \mathbf{N}}$ *is a family of finite sets of keys of the form $(pk, sk_1, \dots, sk_n)$. We call pk the* public key *and call sk_i the* private key share *of party i. There exists a PPT* key-generator K *which given k generates a random key $(pk, sk_1, \dots, sk_n) \leftarrow K(k)$ from K_k. By sk_C for $C \subset N$ we denote the family $\{sk_i\}_{i \in C}$.*

Key-generation *There exists a n-party protocol KD securely evaluating the key-generator K.*

Message Sampling *There exists a PPT algorithm R, which on input pk outputs a uniformly random element from a set R_{pk}. We write $m \leftarrow R_{pk}$.*

Encryption *There exists a PPT algorithm E, which on input pk and $m \in R_{pk}$ outputs an encryption $\overline{m} \leftarrow E_{pk}(m)$ of m. By C_{pk} we denote the set of possible encryptions for the public key pk.*

Decryption *There exists a secure protocol $\mathrm{Decrypt}$ which on common input $(\overline{M}, pk)$, and secret input sk_i for the honest party P_i, where sk_i is the secret* *key share of the public key pk and $\overline{M}$ is a set of encryptions of the messages $M \subset R_{pk}$, returns M as common output.*[1]

Threshold semantic security *Let A be any PPT algorithm, which on input 1^k, C such that $|C| < n/2$, public key pk, and corresponding private keys sk_C outputs two messages $m_0, m_1 \in R_{pk}$ and some arbitrary value $s \in \{0,1\}^*$. Let $X_i(k,C)$ denote the distribution of (s, c_i), where $(pk, sk_1, \dots, sk_n) \leftarrow K(k)$, $(m_0, m_1, s) \leftarrow A(1^k, C, pk, sk_C)$, and $c_i \leftarrow E_{pk}(m_i)$. Then $X_i = \{X_i(k,C)\}_{k \in N, C:|C|<n/2}$ for $i = 0,1$ are distribution ensembles over the index set $\{C \| |C| < n/2\}$ and we require that $X_0 \overset{c}{\approx} X_1$.*

A threshold homomorphic encryption scheme in addition has these properties:

Message ring For all public keys pk, the message space R_{pk} is a ring in which we can compute efficiently using the public key only. We denote the ring $(R_{pk}, \cdot_{pk}, +_{pk}, 0_{pk}, 1_{pk})$.

$+_{pk}$-homomorphic There exists a PPT algorithm, which given public key pk and encryptions $\overline{m}_1 \in E_{pk}(m_1)$ and $\overline{m}_2 \in E_{pk}(m_2)$ outputs a uniquely determined encryption $\overline{m} \in E_{pk}(m_1 +_{pk} m_2)$. We write $\overline{m} \leftarrow \overline{m}_1 \boxplus_{pk} \overline{m}_2$. Further more there exists a similar algorithm, $\boxminus_{pk}$, for subtraction.

Multiplication by constant There exists a PPT algorithm, which on input pk, $m_1 \in R_{pk}$ and $\overline{m}_2 \in E_{pk}(m_2)$ outputs a random encryption $\overline{m} \leftarrow E_{pk}(m_1 \cdot_{pk} m_2)$. We write $\overline{m} \leftarrow m_1 \boxdot_{pk} \overline{m}_2 \in E_{pk}(m_1 \cdot_{pk} m_2)$. We assume that we can also multiply a constant from the right.

Blindable There exists a PPT algorithm Blind, which on input $pk, \overline{m} \in E_{pk}(m)$ outputs an encryption $\overline{m}' \in E_{pk}(m)$ such that $\overline{m}'$ is distributed identically to $E_{pk}(m)[r]$, where r is chosen uniformly random.

Check of ciphertextness Given $y \in \{0,1\}^*$ and pk, where pk is a public key, it is easy to check whether $y \in C_{pk}$.[2]

Proof of plaintext knowledge Let $L_1 = \{(pk, y) | pk \text{ a public key} \wedge y \in C_{pk}\}$. There exists a Σ-protocol for the relation over $L_1 \times (\{0,1\}^*)^2$ given by $(pk, y) \sim (x, r) \Leftrightarrow x \in R_{pk} \wedge y = E_{pk}(x)[r]$.

Proof of correct multiplication Let $L_2 = \{(pk, x, y, z) | pk \text{ is a public key} \wedge x, y, z \in C_{pk}\}$. There exists a Σ-protocol for the relation over $L_2 \times (\{0,1\}^*)^3$ given by $(pk, x, y, z) \sim (d, r_1, r_2) \Leftrightarrow y = E_{pk}(d)[r_1] \wedge z = (d \boxdot_{pk} x)[r_2]$.

---

[1] We need that the Decrypt protocol is secure when executed in parallel. The MPC-model in [Can00] is however not security preserving under parallel composition, so we have to state this required property of the Decrypt protocol by simply letting the input be sets of ciphertexts.

[2] This check can be either directly or using a Σ-protocol: we will always use the test in a context, where a party publishes an encryption and then the recipients either check locally that $y \in C_{pk}$ or the publisher proves it using a Σ-protocol. In the following sections we adopt the terminology to the case, where the recipients can perform the test locally. Details for the case where a Σ-protocol is used are easy extractable.

6   Multiparty Σ-Protocols

In Section 7 we describe how to implement general multiparty computation from a threshold homomorphic encryption scheme, but as the first step towards this we show how one can generally and efficiently extend two-party Σ-protocols, as those for proof of plaintext knowledge and proof of correct multiplication in a threshold homomorphic encryption scheme, into secure multiparty protocols. We will need two essential tools in this section: the notion of *trapdoor commitments* and a multiparty protocol for generating a sufficiently random bit string. Our underlying purpose here is to allow a party to prove a claim using a Σ-protocol such that all other parties will be convinced and to do it much more efficiently than doing the original Σ-protocol independently with each of the other parties.

6.1   Generating (Almost) Random Common Challenges

First of all we want to be able to generate a common challenge for the Σ-protocols. Suppose first that $n \leq 16k$. Then we create a challenge by letting every party choose at random a $\lceil 2k/n \rceil$-bit string, and concatenate all these strings. This produces an m-bit challenge, where $2k \leq m \leq 16k$. We can assume without loss of generality that the basic Σ-protocol allows challenges of length m bits (if not, just repeat it in parallel a number of times). It is easy to see that with this construction, at least k bits of a challenge are chosen by honest parties and are therefore random, since a majority of parties are assumed to be honest. This is equivalent to doing a Σ-protocol where the challenge length is the number of bits chosen by honest parties. The cost of doing such a proof is $O(k)$ bits. If $n > 16k$, we will assume, as detailed later, that an initial preprocessing phase returns as public output a description of a random subset A of the parties, of size $4k$. It is easy to see that, except with probability exponentially small in k, A will contain at least k honest parties. We then generate a challenge by letting each party in A choose one bit at random, and then continue as above.

6.2   Trapdoor Commitments

A trapdoor commitment scheme can be described as follows: first a public key pk is chosen based on a security parameter k, by running a PPT *generator* G.

There is a fixed function *commit* that the committer C can use to compute a commitment c to s by choosing some random input r, computing $c = commit(s, r, pk)$, and broadcasting c. Opening takes place by broadcasting s, r; it can then be checked that $commit(s, r, pk)$ is the value C broadcasted originally.

We require the following:

(Perfect) Hiding For a pk correctly generated by G, uniform r, r' and any s, s', the distributions of $commit(s, r, pk)$ and $commit(s', r', pk)$ are identical.

(Computational) Binding For any C running in expected polynomial time (in k) the probability that C on input pk computes s, r, s', r' such that $commit(s, r, pk) = commit(s', r', pk)$ and $s \neq s'$ is negligible.

Trapdoor Property The algorithm for generating pk also outputs a string t, the trapdoor. There is an efficient algorithm which on input t, pk outputs a commitment c, and then on input any s produces r such that $c = commit(s, r, pk)$. The distribution of c is identical to that of commitments computed in the usual way.

In other words, the commitment scheme is binding if you know only pk, but given the trapdoor, you can cheat arbitrarily. Finally, we also assume that the length of a commitment to s is linear in the length of s.[3] Existence of commitments with all these properties follow in general merely from existence of Σ-protocols for hard relations, and this assumption in turn follows from the properties we already assume for the threshold cryptosystems. For concrete examples that would fit with the examples of threshold encryption we use, see [CD98].

6.3   Putting Things Together

In our global protocol, we assume that the initial preprocessing phase independently generates for each party P_i a public key k_i for the trapdoor commitment scheme and distributes it to all participating parties. We may assume in the following that the simulator for our global protocol knows the trapdoors t_i for (some of) these public keys. This is because it is sufficient to simulate in the hybrid model where parties have access to a trusted party that will output the k_i's on request. Since this trusted party gets no input from the parties, the simulator can imitate it by running G itself a number of times, learning the trapdoors, and showing the resulting k_i's to the adversary.

In our global protocol there are a number of *proof phases*. In each such phase, each party in some subset N'[4] is supposed to give a proof of knowledge: each P_i in the subset has broadcast an x_i and claims he knows w_i such that (x_i, w_i) is in some relation R_i which has an associated Σ-protocol. We then do the following:

1. Each P_i in N' computes the first message a_i in his proof and broadcasts $c_i = commit(a_i, r_i, k_i)$.[5]

---

[3] In principle any commitment scheme can be transformed to fulfil this. Assume that a scheme C has commitments of length k^c and consider the modified scheme C' which on security parameter k runs C on security parameter $k' = k^{1/c}$. This scheme is still a commitment scheme as $\delta(k')$ is still negligible and now the commitments has length k. However in the new scheme the basic cryptographic primitives providing the security is instantiated at a much lower key-size, and indeed such a reduction is only *weakly security preserving*[pac96]. The remaining reductions in this paper are all *polynomially security preserving* and for the security of the protocol to be polynomially preserved relative to the underlying computational assumptions the above reduction should be avoided.

[4] The subset N' is the subset of the parties that still participate, i.e. have not been excluded due to deviation from the protocol.

[5] The intuition behind the use of (independently generated instances) of perfectly hiding trapdoor commitments in the proofs of knowledge of e.g. a plaintext is to avoid malleability issues, and to ensure "independence of inputs" where necessary.

2. Make random challenge e according to the method described earlier.
3. Each P_i in N' computes the answer z_i to challenge e, and broadcasts a_i, r_i, z_i
4. Every party can check every proof given by verifying $c_i = commit(a_i, r_i, k_i)$ and that (a_i, e, z_i) is an accepting conversation.

It is clear that such a proof phase has communication complexity no larger than n times the complexity of a single Σ-protocol, i.e. $O(nk)$ bits. We denote the execution of the protocol by $(\mathcal{A}', N'') \leftarrow \Sigma(\mathcal{A}, x_{N'}, w_{H \cap N'}, k_N)$, where $\mathcal{A}$ is the state of the adversary before the execution, $x_{N'} = \{x_i\}_{i \in N'}$ are the instances that the parties N' are to prove that they know a witness to, $w_{H \cap N'} = \{w_i\}_{i \in H \cap N'}$ are witnesses for the instances x_i corresponding to honest P_i, $k_N = \{k_i\}_{i \in N}$ is the commitment keys for all the parties, $\mathcal{A}'$ is the state of the adversary after the execution, and $N'' \subset N'$ is the subset of the parties completing the proof correctly. The reason why the execution only depends on the witnesses $w_{H \cap N'}$ is that the corrupted parties are controlled by the adversary and their witnesses, if even well-defined, are included in the start-state $\mathcal{A}$ of the adversary.

Now let $t_H = \{t_i\}_{i \in H}$ be the commitment trapdoors for the honest parties. We describe a procedure $(\mathcal{A}', N'', w_{N'' \cap C}) \leftarrow S_\Sigma(\mathcal{A}, x_{N'}, t_H, k_N)$ that will be used as subroutine in the simulation of our global protocol. $S_\Sigma(\mathcal{A}, x_{N'}, k_N, t_H)$ will have the following properties:

- $S_\Sigma(\mathcal{A}, x_{N'}, k_N, t_H)$ runs in expected polynomial time and the part $(\mathcal{A}', N'')$ of the output is perfectly indistinguishable from the output of a real execution $\Sigma(\mathcal{A}, x_{N'}, w_{H \cap N'}, k_N)$ given the start state $\mathcal{A}$ of the adversary (which we assume includes $x_{N'}$ and k_N).
- Except with negligible probability $w_{N'' \cap C} = \{w_i\}_{i \in N'' \cap C}$ are valid witnesses to the instances x_i corresponding the corrupted parties completing the proofs correctly.

The algorithm of S_Σ is as follows:

1. For each P_i: if P_i is honest, use the trapdoor t_i for k_i to compute a commitment c_i that can be opened arbitrarily and show c_i to the adversary. If P_i is corrupt, receive c_i from the adversary.
2. Run the procedure for choosing the challenge, choosing random contributions on behalf of honest parties. Let e_0 be the challenge produced.
3. For each P_i do (where the adversary may choose the order in which parties are handled): *If P_i is honest,* run the honest verifier simulator to get an accepting conversation (a_i, e_0, z_i). Use the commitment trapdoor to compute r_i such that $c_i = commit(a_i, r_i)$ and show (a_i, r_i, z_i) to the adversary. *If P_i is corrupt,* receive (a_i, r_i, z_i) from the adversary.
   The current state $\mathcal{A}'$ of the adversary and the subset N'' of parties correctly completing the proof is copied to the output from this simulation subroutine. In addition, we now need to find witnesses for x_i from those corrupt P_i that sent a correct proof in the simulation. This is done as follows:
4. For each corrupt P_i that sent a correct proof in the view just produced, execute the following loop:

(a) Rewind the adversary to its state just before the challenge is produced.
(b) Run the procedure for generating the challenge using fresh random bits on behalf of the honest parties. This results in a new value e_1.
(c) Receive from the adversary proofs on behalf of corrupted parties and generate proofs on behalf of honest parties, w.r.t. e_1, using the same method as in Step 3. If the adversary has made a correct proof a'_i, r'_i, e', z'_i on behalf of P_i, exit the loop. Else go to Step 4a.

If $e_0 \neq e_1$ and $a_i = a'_i$ compute and output a witness for x_i, from the conversations (a_i, e_0, z_i), (a'_i, e_1, z'_i). Else output $c_i, a_i, r_i, a'_i, r'_i$ (this will be a break of the commitment scheme). Go on to next corrupt P_i.

It is clear by inspection and assumptions on the commitments and Σ-protocols that the part $(\mathcal{A}', N'')$ of the output is distributed correctly. For the running time, assume P_i is corrupt and let ϵ be the probability that the adversary outputs a correct a_i, r_i, z_i given some fixed but arbitrary value $View$ of the adversary's view up to the point just before e is generated. Observe that the contribution from the loop to the running time is ϵ times the expected number of times the loop is executed before terminating, which is $1/\epsilon$, so that to the total contribution is $O(1)$ times the time to do one iteration, which is certainly polynomial. As for the probability of computing correct witnesses, observe that we do not have to worry about cases where ϵ is negligible, say $\epsilon < 2^{-k/2}$, since in these cases $P_i \notin N''$ with overwhelming probability. On the other hand, assume $\epsilon \geq 2^{-k/2}$, let $\bar{e}$ denote the part of the challenge e chosen by honest parties, and let $pr()$ be the probability distribution on $\bar{e}$ given the view $View$ and given that the choice of $\bar{e}$ leads to the adversary generating a correct answer on behalf of P_i. Clearly, both $\bar{e}_0$ and $\bar{e}_1$ are distributed according to $pr()$. Now, the a priori distribution of $\bar{e}$ is uniform over at least 2^k values. This, and $\epsilon \geq 2^{-k/2}$ implies by elementary probability theory that $pr(\bar{e}) \leq 2^{-k/2}$ for any e, and so the probability that $\bar{e}_0 = \bar{e}_1$ is $\leq 2^{-k/2}$. We conclude that except with negligible probability, we will output either the required witnesses, or a commitment with two different valid openings. However, the latter case occurs with negligible probability. Indeed, if this was not the case, observe that since the simulator never uses the trapdoors of k_i for corrupt P_i, the simulator together with the adversary could break the binding property of the commitments. Formulating a reduction proving this formally is straightforward and is left to the reader.

7   General MPC
from Threshold Homomorphic Encryption

Assume that we have a threshold homomorphic encryption scheme as described in Section 5. In this section we describe the FuncEval$_f$ protocol which securely computes any polynomial time computable n-party function $y \leftarrow f(x_1, \dots, x_n)$ using a uniform polynomially sized family of arithmetic circuits over R_{pk}.

Our approach works for any reasonable encoding of f as an arithmetic circuit. This can allow for efficient encodings of arithmetic function if one can exploiting knowledge about the rings R_{pk} over which the function is evaluated. For simplicity we will however here assume that f is encoded using a circuit taking inputs from $\{0_{pk}, 1_{pk}\}$, using $+$, $-$, and $\cdot$ gates, and using the same circuit for a fixed security parameter. Since our encryption scheme is only $+$ and $-$-homomorphic we will be needing a protocol Mult for securely computing an encryption of $m_1 \cdot_{pk} m_2$ given encryptions of m_1 and m_2.

We assume that the parties has access to a trusted party Preprocess, which at the beginning of the protocol outputs a public value $(k_1, \ldots, k_n)$, where k_i is a random public commitment key for a trapdoor commitment scheme as described in Section 6.2. If $n > 16k$ then further more the trusted party returns a public description of a random $4k$-subset of the parties as described in Section 6.1[6]. As described in Section 6.3, we can then from the Σ-protocols of the threshold homomorphic encryption scheme for proof of plaintext knowledge and correct multiplication construct n-party versions, which we call POPK resp. POCM. The corresponding versions of our general simulation routine S_Σ for these protocols will be called S_{POPK} resp. S_{POCM}.

7.1   The Mult Protocol

Description. All honest parties P_i know public values $k_N = \{k_i\}_{i \in N}$, pk, and encryptions $\overline{a}, \overline{b} \in E_{pk}(a)$, for some possible unknown $a, b \in R_{pk}$, and private values sk_i. Further more a set N' of participating parties, those that have not been caught deviating from the protocol, is known to all parties. The corrupted parties are controlled by an adversary and the parties want to compute a common value $\overline{c} \in E_{pk}(ab)$ without anyone learning anything new about a, b, or $a \cdot b$.

Implementation.

1. First all participating parties additively secret share the value of a.
   (a) P_i, for $i \in N'$ chooses a value d_i uniformly at random in R_{pk}, computes an encryption $\overline{d}_i \leftarrow E(d_i)$, broadcasts it, and participates in POPK to check that each P_i knows r_i and d_i such that $\overline{d}_i = E_{pk}(d_i)[r_i]$.
   (b) Let N'' denote parties completing the proof in (a) and let $d = \sum_{i \in N''} d_i$. All parties compute $\overline{d} = \boxplus_{i \in N''} \overline{d}_i$ and $\overline{e} = \overline{a} \boxplus \overline{d}$.
   (c) The parties in N'' call Decrypt to compute the value $a + d$ from $\overline{e}$.
   (d) The party in N'' with smallest index sets $\overline{a}_i \leftarrow \overline{e} \boxminus \overline{d}_i$ and $a_i \leftarrow a + d - d_i$. The other parties in N'' set $\overline{a}_i \leftarrow \boxminus \overline{d}_i$ and $a_i \leftarrow -d_i$.
2. Each party P_i for $i \in N''$ computes $\overline{f}_i \leftarrow a_i \boxdot \overline{b}$, broadcasts $\overline{f}_i$, and participates in POCM to check that all $\overline{f}_i$ was computed correctly. Let X be the subset failing the proof and let $N''' = N'' \setminus X$.
3. The parties compute $\overline{a}_X = \boxplus_{i \in X} \overline{a}_i$ and decrypt it using Decrypt to obtain $a_X = \sum_{i \in X} a_i$.
4. All parties compute $\overline{c} \leftarrow (\boxplus_{i \in N'''} \overline{f}_i) \boxplus (a_X \boxdot \overline{b}) \in E_{pk}(ab)$.

---

[6] In the following we present the case where $n \leq 16k$.

Theorem 1. *There exists a simulator for the* Mult *protocol, which produces a view in the ideal-world that is computationally indistinguishable from the view produced by an execution of the* Mult *protocol in the real-world.*

Proof outline: We give an outline of the main ideas used in the simulation. The main obstacle is Step 1c, where $a + d$ should be handed to the adversary to simulate an oracle call to the Decrypt oracle. By choosing $\bar{d}_i$ correctly for honest parties and using S_{POPK} the simulator can learn d, but it cannot learn a. We handle this by letting one of the honest party P_s choose $\bar{d}_s$ as $\text{Blind}(E_{pk}(d'_s) \boxminus \bar{a})$ for uniformly random $d'_s \in R_{pk}$. Then all values are still correctly distributed, but now the simulator can compute $a + d$ as $(\sum_{i \in N'' \setminus \{s\}} d_i) + d'_s$. Observe that the simulator now cannot compute the value a_s which is necessary later. Doing the same computation on d'_s it can however compute $a'_s = a_s + a$. Now assume that the simulator has access to an oracle returning an encryption $\bar{c}'$ of $a \cdot b$. Then it simulates Step 2 by computing $a_s \boxdot \bar{b}$ as $\text{Blind}((a'_s \boxdot \bar{b}) \boxminus \bar{c}')$ and running S_{POCM}. Step 3 is simulated by giving $\sum_{i \in X} a_i$ to the adversary (this value the simulator can compute as $s \notin X$) and Step 4 is simulated by doing the correct computation as the necessary values are available.

By the properties of the knowledge extractors and Blind it follows that the view produced as above is statistically indistinguishable from the view of a real-world execution. Now instead of the oracle value $\bar{c}'$ use a random encryption of 0_{pk}. If this changes the view produced by the simulator except computationally negligible, then *as the simulator does not use the secret keys of any honest party* the simulator would be a distinguisher of encryptions of $a \cdot b$ and encryptions of 0_{pk} contradicting the semantic security of the encryption scheme. The technical report [CDN00] contains a detailed proof. □

As the Decrypt protocol is assumed to be secure under parallel composition and our multiparty zero-knowledge proofs have been proven to be secure, so are then trivially the Mult protocol.

7.2   The FuncEval$_f$ Protocol

Now assume that the description of a arithmetic circuit for evaluating f is given. The parties then evaluate f in the following manner. First the parties run the Preprocess and the KD oracles and obtain the keys $(k_1, \dots, k_n)$ for the trapdoor commitment scheme and $(pk, sk_1, \dots, sk_n)$ for the encryption scheme. The key sk_i is private to P_i. Each party P_i then does a bitwise encryption $\bar{x}_{i,j} \leftarrow E_{pk}(x_{i,j})$ of its input x_i, broadcasts the encryptions, and proves in zero-knowledge, that $\bar{x}_{i,j}$ does in fact contain either 0 or 1.[7] For those P_i failing the above proof the parties exclude them, take x_i to be $000 \dots 00$, and compute $\bar{x}_{i,j} \leftarrow E_{pk}(x_{i,j})[r]$ for some fixed agreed upon string $r \in \{0,1\}^{p(k)}$. In this way all parties get to know common legal encrypted circuit inputs for all parties. Then the circuit is evaluated. In each round all gates that are ready to be evaluated are evaluated

---

[7] We will be needing a Σ-protocol for doing this, but such protocol is easy to implement for the examples of threshold encryption that we give in Section 8.

in parallel. Addition and subtraction gates are evaluated locally using $\boxplus$ and $\boxminus$; and multiplication gates are evaluated using the Mult protocol. Finally the parties decrypt the output gates and output the reviled values.

Theorem 2. *The* FuncEval$_f$ *protocol as described above securely evaluates f in the presence of active non-adaptive adversary corrupting at most a minority of the parties.[8] The round complexity is in the order of the depth of the circuit for f and the communication complexity of the protocol is $O((nk+d)|f|)$ bits, where $|f|$ denotes the size of the circuit for evaluating f and d denotes the communication complexity of a decryption.*

Proof outline: Given a real-world adversary $\mathcal{A}$ we construct a simulator $\mathcal{S}(\mathcal{A})$ running in the ideal-world. The initial oracle calls are simulated by running the generators locally and giving the appropriate values to $\mathcal{A}$. The simulator saves $(k_1, \ldots, k_n, pk, \{sk_i\}_{i \in C})$ for later use, *but discards sk_i for all honest parties.* Assume for now that $\mathcal{S}$ has access to an oracle giving it the values $\overline{x}_{i,j}$ for all honest parties. The simulator then gives the $\overline{x}_{i,j}$ values to $\mathcal{A}$ and receive $\overline{x}_{i,j}$ for all corrupted parties from $\mathcal{A}$. Using the knowledge extractor the simulator then learns all $x_{i,j}$ for corrupted parties that completed the proof that $\overline{x}_{i,j}$ contains 0 or 1. It uses these values as input to the ideal evaluation of f and learns the output y. Now the gate evaluation is simulated using the simulator for the Mult protocol. The decryption (by oracle call) of output gates are simulated by just handing the correct value to $\mathcal{A}$. These values are known as the correct output y of the computation is known to the simulator. This simulation is by the properties of the knowledge extractors and the Mult simulator computationally indistinguishable from that of a real-world execution. We get rid of the oracle values $\overline{x}_{i,j}$ as we did in the proof of Theorem 1.

The round complexity follows by inspection. The gates that give rise to communication are the input, multiplication, and output gates. The communication used to handle these gates is in the order of n encryptions ($O(nk)$ bits), n zero-knowledge proofs ($O(nk)$ bits as we have assumed that the Σ-protocols have communication complexity $O(k)$) and 1 decryption ($O(d)$ bits by definition). The total communication complexity therefore is $O((nk+d)|f|)$ as claimed. The technical report [CDN00] contains a detailed proof. $\square$

The threshold homomorphic encryption schemes we present in Section 8 both have $d = O(kn)$. It follows that for deterministic f the FuncEval$_f$ protocol based on any of these schemes has communication complexity $O(nk|f|)$ bits.

8   Examples of Threshold Homomorphic Cryptosystems

In this section, we describe some concrete examples of threshold systems meeting our requirements, including Σ-protocols for proving knowledge of plaintexts, correctness of multiplications and validity of decryptions.

---

[8] Generally we can make the protocol secure against any corruption structure for which the threshold cryptosystem is secure and for which one can generate short random challenges containing at least k random bits as in Section 6.1.

Both our examples involve choosing as part of the public key a k-bit RSA modulus $N = pq$, where p, q are chosen such that $p = 2p' + 1, q = 2q' + 1$ for primes p', q' and both p and q have $k/2$ bits. For convenience in the proofs to follow, we will assume that the length of the challenges in all the proofs is $k/2-1$.

8.1   Basing It on Paillier's Cryptosystem

In [Pai99], Paillier proposes a probabilistic public-key cryptosystem where the public key is a k-bit RSA modulus N and an element $g \in Z_{N^2}^*$ of order divisible by N. The plaintext space for this system is Z_N. In [DJ01] the crypto-system is generalised to have plaintext space Z_{N^s} for any s smaller than the factors of N and there g has order divisible by N^s. To encrypt $a \in Z_{N^s}$, one chooses $r \in Z_{N^{s+1}}^*$ at random and computes the ciphertext as $\bar{a} = g^a r^{N^s} \mod N^{s+1}$. The private key is the factorisation of N, i.e., $\phi(N)$ or equivalent information. Under an appropriate complexity assumption given in [Pai99], this system is semantically secure, and it is trivially homomorphic over Z_{N^s} as we require here: we can set $\bar{a} \boxplus \bar{b} = \bar{a} \cdot \bar{b} \mod N^{s+1}$. Furthermore, from α and an encryption $\bar{a}$, a *random* encryption of αa can be obtained by multiplying $\bar{a}^\alpha \mod N^{s+1}$ by a random encryption of 0. In [DJ01] a threshold version of this system has been proposed, based on a variant of Shoup's [Sho00] technique for threshold RSA. A multiplication protocol was also given in [DJ01], though for a slightly different setting. We will not go into further details here, but note that using known techniques the multiplication protocol can be modified to meet our definition of a threshold homomorphic encryption scheme. The technical report [CDN00] contains more details.

Generalisations of FuncEval. Using standard techniques, the FuncEval-pro-tocol can be extended to handle probabilistic functions. In this section we de-scribe how this works when we instantiate using the Paillier cryptosystem. We show how random Z_{N^s}-gates (outputting a uniformly random element from Z_{N^s} unknown to all parties) and random 0/1-gates (outputting a uniformly random element from $\{0,1\}$ unknown to all parties) can be implemented securely in a constant number of rounds.

As a step-stone we also recall how to implement inversion gates and do un-bounded fan-in multiplication of invertible elements in a constant number of rounds (see [BB89]). Due to lack of space only the protocols are given, but they can all be proven secure using the techniques of the previous sections.

In the following, if $n \geq 16k$, let the random group denote the $4k$-subset A given in the preprocessing and if $n < 16k$ let the random group be all the parties. Assume that the parties in the random group are indexed $1, \ldots, r(n, k)$. Observe that $r(n, k) \in O(\min(n, k))$ and that except with negligible probability the random group contains a honest party.

Random Z_{N^s}-gates. All the parties in the random group pick a uniformly ran-dom element $a_i \in Z_{N^s}$, broadcast an encryption $\bar{a_i}$, and prove knowledge of a_i.

Then all the parties compute $\bar{a} = \boxplus_{i=1}^{r(n,k)} \overline{a_i}$ and a is the secret uniformly random $\mathbf{Z}_{N^s}$-element. The communication complexity is $O(r(n,k)k)$.

Inversion gates. Given $\bar{a}$, where a is invertible, the parties generates $\bar{b}$ using a random $\mathbf{Z}_{N^s}$-gate. Since b is invertible except with negligible probability, we assume in the following that it is. The parties compute $\overline{ab}$ and reveals ab (this is secure since ab is a uniformly random invertible element). The parties computes $(ab)^{-1}$ and $\overline{a^{-1}} = (ab)^{-1} \boxdot \bar{b}$. The communication complexity is $O(nk)$.

Constant-round unbounded fan-in multiplication of invertible elements. Given encryptions $\overline{x_1}, \dots, \overline{x_l}$ of invertible elements the parties generate secret random $\mathbf{Z}_{N^s}$-elements $\overline{y_0}, \dots, \overline{y_l}$, compute $\overline{y_0^{-1}}, \dots, \overline{y_l^{-1}}$, and compute and reveal $z_i = y_{i-1} x_i y_i^{-1}, i = 1, \dots, l$. Then they compute $\overline{\prod_{i=1}^{l} x_i} = \overline{y_0^{-1}} (\prod_{i=1}^{l} z_i) \overline{y_l}$. The communication complexity of this is $O(lnk)$.

Random 0/1-gates. Each party in the random group generates a random bit b_i, publishes $\overline{b_i}$, proves knowledge of $b_i \in \{0,1\}$, and all the parties compute $\bar{b} = \left[\left(\boxdot_{i=1}^{r(n,k)} (1 \boxminus 2 \boxdot \overline{b_i}) \right) \boxplus 1 \right] \boxdot 2^{-1}$. The communication complexity of this is $O(r(n,k)nk)$.

8.2 Basing It on QRA and DDH

In this section, we describe a cryptosystem which is a simplified variant of Franklin and Haber's system [FH96], a somewhat similar (but non-threshold) variant was suggested by one the authors of the present paper and appears in [FH96].

For this system, we choose an RSA modulus $N = pq$, where p, q are chosen such that $p = 2p' + 1, q = 2q' + 1$ for primes p', q'. We also choose a random generator g of $SQ(N)$, the subgroup of quadratic residues modulo N (which here has order $p'q'$). We finally choose x at random modulo $p'q'$ and let $h = g^x \bmod N$. The public key is now N, g, h while x is the secret key.

The plaintext space of this system is $\mathbf{Z}_2$. We set $\Delta = n!$ (recall that n is the number of parties). Then to encrypt a bit b, one chooses at random r modulo N^2 and a bit c and computes the ciphertext

$$((-1)^c g^r \bmod N, \ (-1)^b h^{4\Delta^2 r} \bmod N)$$

The purpose of choosing r modulo N^2 is to make sure that g^r will be close to uniform in the group generated by g even though the order of g is not public. It is clear that a ciphertext can be decrypted if one knows x. The purpose of having $h^{4\Delta^2 r}$ (and not h^r) in the ciphertext will be explained below.

The system clearly has the required homomorphic properties, we can set:

$$(\alpha, \beta) \boxplus (\gamma, \delta) = (\alpha\gamma \bmod N, \beta\delta \bmod N)$$

Finally, from an encryption (α, β) of a value a and a known b, one can obtain a *random* encryption of value $ba \bmod 2$ by first setting (γ, δ) to be a random encryption of 0 and then outputting $(\alpha^b \gamma \bmod N, \beta^b \delta \bmod N)$.

We now argue that under the Quadratic Residuosity Assumption (QRA) and the Decisional Diffie Hellman Assumption (DDH), the system is semantically secure. Recall that DDH says that the distributions $(g, h, g^r \bmod p, h^r \bmod p)$ and $(g, h, g^r \bmod p, h^s \bmod p)$ are indistinguishable, where g, h both generate the subgroup of order p' in $\mathbf{Z}_p^*$ and r, s are independent and random in $\mathbf{Z}_{p'}$. By the Chinese remainder theorem, this is easily seen to imply that also the distributions $(g, h, g^r \bmod N, h^r \bmod N)$ and $(g, h, g^r \bmod N, h^s \bmod N)$ are indistinguishable, where g, h both generate $SQ(N)$ and r, s are independent and random in $\mathbf{Z}_{p'q'}$. Omitting some tedious details, we can then conclude that the distributions

$$(g, h, (-1)^c g^r \bmod N, h^{4\Delta^2 r} \bmod N)$$
$$(g, h, (-1)^c g^r \bmod N, h^{4\Delta^2 s} \bmod N)$$
$$(g, h, (-1)^c g^r \bmod N, -h^{4\Delta^2 s} \bmod N)$$
$$(g, h, (-1)^c g^r \bmod N, -h^{4\Delta^2 r} \bmod N)$$

are indistinguishable, using (in that order) DDH, QRA and DDH.

Threshold Decryption. Shoup's method for threshold RSA [Sho00] can be directly applied here: he shows that if one secret-shares x among the parties using a polynomial computed modulo $p'q'$ and publishes some extra verification information, then the parties can jointly and securely raise an input number to the power $4\Delta^2 x$. This is clearly sufficient to decrypt a ciphertext as defined here: to decrypt the pair (a, b), compute $ba^{-4\Delta^2 x} \bmod N$. We do not describe the details here, as the protocol from [Sho00] can be used directly. We only note that decryption can be done by having each party broadcast a single message and prove by a Σ-protocol that it is correct. The communication complexity of this is $O(nk)$ bits. In the original protocol the random oracle model is used when parties prove that they behave correctly. However, the proofs can instead be done according to our method for multiparty Σ-protocols without loss of efficiency (Section 6). This also immediately implies a protocol that will decrypt several ciphertexts in parallel.

Proving You Know a Plaintext. We will need an efficient way for a party to prove in zero-knowledge that a pair (α, β) he created is a legal ciphertext, and that he knows the corresponding plaintext. A pair is valid if and only if α, β both have Jacobi symbol 1 (which can be checked easily) and if for some r we have $(g^2)^r = \alpha^2 \bmod N$ and $(h^{8\Delta^2})^r = \beta^2 \bmod N$. This last pair of statements can be proved non-interactively and efficiently by a standard equality of discrete log proof appearing in [Sho00]. Note that the squarings of α, β ensure that we are working in $SQ(N)$, which is necessary to ensure soundness.

This protocol has the standard 3-move form of a Σ-protocol. It proves that an r fitting with α, β *exists*. But it does not prove that the prover *knows* such an r (and hence knows the plaintext), unless we are willing to also assume the strong RSA assumption[9]. With this assumption, on the other hand, the equality of discrete log proof is indeed a proof of knowledge.

However, it is possible to do without this extra assumption: observe that if β was correctly constructed, then the prover knows a square root of β (namely $h^{2\Delta^2 r} \bmod N$) iff $b = 0$ and he knows a root of $-\beta$ otherwise. One way to exploit this observation is if we have a commitment scheme available that allows committing to elements in $\mathbf{Z}_N$. Then P_i can commit to his root α, and prove in zero-knowledge that he knows α and that $\alpha^4 = \beta^2 \bmod N$. This would be sufficient since it then follows that α^2 is β or $-\beta$.

Here is a commitment scheme (already well known) for which this can be done efficiently: choose a prime P, such that N divides $P - 1$ and choose elements G, H of order N modulo P, but where no party knows the discrete logarithm of H base G. This can all be set up initially (recall that we already assume that keys are set up once and for all). Then a commitment to α has form $(G^r \bmod P, G^\alpha H^r \bmod P)$, and is opened by revealing α, r. It is easy to see that this scheme is unconditionally binding, and is hiding under the DDH assumption (which we already assumed). Let $[\alpha]$ denote a commitment to α and let $[\alpha][\beta] \bmod P$ be the commitment you obtain in the natural way by component-wise multiplication modulo P. It is then clear that $[\alpha][\beta] \bmod P$ is a commitment to $\alpha + \beta \bmod N$.

It will be sufficient for our purposes to make a Σ-protocol that takes as input commitments $[\alpha], [\beta], [\gamma]$, shows that the prover knows α and shows that $\alpha\beta = \gamma \bmod N$. Here follows such a protocol:

1. Inputs are commitments $[\alpha], [\beta], [\gamma]$ where P_i claims that $\alpha\beta = \gamma \bmod N$. P_i chooses a random δ and makes commitments $[\delta], [\delta\beta]$.
2. The verifier send a random e.
3. P_i opens the commitment $[\alpha]^e[\delta] \bmod P$ to reveal a value e_1. P_i opens the commitment $[\beta]^{e_1}[\delta\beta]^{-1}[\gamma]^{-e} \bmod P$ to reveal 0.
4. The verifier accepts iff the commitments are correctly opened as required.

Using standard techniques, it is straightforward to show that this protocol is a Σ-protocol. The technical report [CDN00] contains more details.

Proving Multiplications Correct. Finally, we need to consider the scenario where party P_i has been given an encryption C_a of a, has chosen a constant b, and has published encryptions C_b, D, of values b, ba, and where D has been constructed by P_i as we described above. It follows from this construction that if $b = 1$, then $D = C_a \boxplus E$ where E is a random encryption of 0. Assuming $b = 1$, E can be easily reconstructed from D and C_a.

---

[9] That is, assume that it is hard to invert the RSA encryption function, even if the adversary is allowed to choose the public exponent

Now we want a Σ-protocol that P_i can use to prove that D contains the correct value. Observe that this is equivalent to the statement

$$((C_b \text{ encrypts } 0) \; AND \; (D \text{ encrypts } 0)) \; OR$$
$$((C_b \text{ encrypts } 1) \; AND \; (E \text{ encrypts } 0))$$

We have already seen how to prove by a Σ-protocol that an encryption (α, β) contains a value b, by proving that you know a square root of $(-1)^b \beta$. Now, standard techniques from [CDS94] can be applied to building a new Σ-protocol proving a monotone logical combination of statements such as we have here.

9 An Optimisation of the FuncEval Protocol

The following optimisation of the FuncEval-protocol was brought to our attention by an anonymous referee. The optimisation applies to the situation where at most $(\frac{1}{2} - c)n$ parties, for some $c > 0$, can be corrupted and n is larger than k. In that case we can use the random group for doing the entire computation. The decryption keys for the threshold cryptosystem are distributed only to the random group and all parties are given the public key. All parties then broadcast encryptions of their inputs as before. Then the parties in the random group do the actual computation and broadcast the result. The communication complexity of this is $O(k^2|C|)$ as the initial broadcast of inputs are dominated by the computation. This is better than $O(kn|C|)$ if $n > k$. The same optimisation applies to any MPC protocol by letting the parties secret share their input among the random group initially. This typically reduces a complexity of $O(k^d n^e |C|)$ to $O(k^{d+e}|C|)$. Finally k^{d+e} can be replaced by k to obtain a communication complexity of $O(k|C|)$ using the *weakly security preserving* reduction of Footnote 3. Note that the last part of the transformation has no practical value, it is a property of the security model allowing to sell security for cuts in complexity.

References

ACM88.   *Proceedings of the Twentieth Annual ACM STOC*, Chicago, Illinois, 2–4 May 1988.

BB89.   J. Bar-Ilan and D. Beaver. Non-cryptographic fault-tolerant computing in constant number of rounds of interaction. In *Proc. ACM PODC'89*, pages 201–209, 1989.

Bea91.   D. Beaver. Foundations of secure interactive computing. In Joan Feigenbaum, editor, *Advances in Cryptology - Crypto '91*, pages 377–391, Berlin, 1991. Springer-Verlag. LNCS Vol. 576.

BGW88.   Michael Ben-Or, Shafi Goldwasser, and Avi Wigderson. Completeness theorems for non-cryptographic fault-tolerant distributed computation (extended abstract). In ACM [ACM88], pages 1–10.

Can00.   Ran Canetti. Security and composition of multiparty cryptographic protocols. *Journal of Cryptology*, 13(1):143–202, winter 2000.

CCD88.   David Chaum, Claude Crépeau, and Ivan Damgård. Multiparty unconditionally secure protocols (extended abstract). In ACM [ACM88], pages
11–19.

CD98.   Ronald Cramer and Ivan Damgaard. Zero-knowledge proofs for finite field
arithmetic, or: Can zero-knowledge be for free. In Hugo Krawczyk, editor,
Advances in Cryptology - Crypto '98, pages 424–441, Berlin, 1998. Springer-
Verlag. LNCS Vol. 1462.

CDM00.   Ronald Cramer, Ivan Damgård, and Ueli Maurer. General secure multi-
party computation from any linear secret-sharing scheme. In Bart Preneel,
editor, *Advances in Cryptology - EuroCrypt 2000*, pages 316–334, Berlin,
2000. Springer-Verlag. LNCS Vol. 1807.

CDN00.   Ronald Cramer, Ivan B. Damgård, and Jesper B. Nielsen. Multiparty computation from threshold homomorphic encryption. Research Series RS-
00-14, BRICS, Department of Computer Science, University of Aarhus,
June 2000. Updated version available at Cryptology ePrint Archive, record
2000/064, http://eprint.iacr.org/.

CDS94.   R. Cramer, I. B. Damgård, and B. Schoenmakers. Proofs of partial knowledge and simplified design of witness hiding protocols. In Yvo Desmedt,
editor, *Advances in Cryptology - Crypto '94*, pages 174–187, Berlin, 1994.
Springer-Verlag. LNCS Vol. 839.

DJ01.   Ivan Damgård and Mads Jurik. A generalisation, a simplification and some
applications of paillier's probabilistic public-key system. In *Public Key
Cryptography, Fourth International Workshop on Practice and Theory in
Public Key Cryptography, PKC 2001, Proceedings*, 2001. LNCS. Obtainable
from http://www.daimi.au.dk/~ivan.

DK01.   Ivan Damgård and Maciej Koprowski. Practical threshold RSA signatures
without a trusted dealer. In *these proceedings*.

FH96.   Matthew Franklin and Stuart Haber. Joint encryption and message-efficient
secure computation. *Journal of Cryptology*, 9(4):217–232, Autumn 1996.

FPS00.   P. Fouque, G. Poupard, and J. Stern. Sharing decryption in the context of
voting or lotteries. In *Proceedings of Financial Crypto 2000*, 2000.

GMW87.   Oded Goldreich, Silvio Micali, and Avi Wigderson. How to play any mental
game or a completeness theorem for protocols with honest majority. In
Proceedings of the Nineteenth Annual ACM STOC, pages 218–229, New
York City, 25–27 May 1987.

GRR98.   R. Gennaro, M. Rabin, and T. Rabin. Simplified VSS and fast-track multi-
party computations with applications to threshold cryptography. In *Proc.
ACM PODC'98*, 1998.

GV87.   O. Goldreich and R. Vainish. How to solve any protocol problem - an
efficiency improvement. In Carl Pomerance, editor, *Advances in Cryptology
- Crypto '87*, pages 73–86, Berlin, 1987. Springer-Verlag. LNCS Vol. 293.

KK91.   Kaoru Kurosawa and Motoo Kotera. A multiparty protocol for modulo
operations. Technical Report SCIS91-3B, 1991.

Kur91.   Kaoru Kurosawa. Zero knowledge interactive proof system for modulo
operations. In *IEICE Trans.*, volume E74, pages 2124–2128, 1991.

MH00.   Bartosz Przydatek, Martin Hirt, and Ueli M. Maurer. Efficient secure multi-
party computation. In Tatsuaki Okamoto, editor, *Advances in Cryptology
- ASIACRYPT 2000*, pages 143–161, Berlin, 2000. Springer. LNCS Vol.
1976.

MJ00.   Ari Juels and Markus Jakobsson. Mix and match: Secure function evalua-
          tion via ciphertexts. In Tatsuaki Okamoto, editor, *Advances in Cryptology
          - ASIACRYPT 2000*, pages 162–177, Berlin, 2000. Springer. LNCS Vol.
          1976.
MR91.   S. Micali and P. Rogaway. Secure computation. In Joan Feigenbaum,
          editor, *Advances in Cryptology - Crypto '91*, pages 392–404, Berlin, 1991.
          Springer-Verlag. LNCS Vol. 576.
pac96.   *Pseudorandomness and Cryptographic Applications*. Princeton University
          Press, 1996.
Pai99.   P. Paillier. Public-key cryptosystems based on composite degree residue
          classes. In Jacques Stern, editor, *Advances in Cryptology - EuroCrypt '99*,
          pages 223–238, Berlin, 1999. Springer-Verlag. LNCS Vol. 1592.
Sho00.   Victor Shoup. Practical threshold signatures. In Bart Preneel, editor,
          Advances in Cryptology - EuroCrypt 2000, pages 207–220, Berlin, 2000.
          Springer-Verlag. LNCS Vol. 1807.
Yao82.   Andrew C. Yao. Protocols for secure computations (extended abstract).
          In *23rd Annual Symposium on Foundations of Computer Science*, pages
          160–164, Chicago, Illinois, 3–5 November 1982. IEEE.

On Perfect and Adaptive Security
in Exposure-Resilient Cryptography

Yevgeniy Dodis[1], Amit Sahai[2], and Adam Smith[3]

[1] Department of Computer Science, New York University,
251 Mercer St, New York, NY 10012, USA. dodis@cs.nyu.edu
[2] Department of Computer Science, Princeton University,
35 Olden St, Princeton, NJ 08540, USA. sahai@cs.princeton.edu
[3] Laboratory for Computer Science, Massachusetts Institute of Technology,
545 Main St, Cambridge, MA 02139, USA. asmith@theory.lcs.mit.edu

Abstract. We consider the question of *adaptive* security for two related cryptographic primitives: all-or-nothing transforms and exposure-
resilient functions. Both are concerned with retaining security when an
intruder learns some bits of a string which is supposed to be secret:
all-or-nothing transforms (AONT) protect their input even given partial
knowledge of the output; *exposure-resilient functions* (ERF) hide their
output even given partial exposure of their input. Both of these primitives can be defined in the perfect, statistical and computational settings and have a variety of applications in cryptography. In this paper,
we study how these notions fare against adaptive adversaries, who may
choose which positions of a secret string to observe on the fly.
In the *perfect* setting, we prove a new, strong lower bound on the con-
structibility of (perfect) AONT. This applies to both standard and adaptively secure AONT. In particular, to hide an input as short as $\log n$
bits, the adversary must see *no more than half* of the n-bit output. This
bound also provides a new impossibility result on the existence of (ramp)
secret-sharing schemes [6] and relates to a combinatorial problem of independent interest: finding "balanced" colorings of the hypercube.
In the statistical setting, we show that adaptivity adds strictly more
power to the adversary. We relate and reduce the construction of adaptive ERF's to that of *almost-perfect resilient functions* [19], for which
the adversary can actually *set* some of the input positions and still
learn nothing about the output. We give a probabilistic construction of
these functions which is essentially optimal and substantially improves
on previous constructions of [19,5]. As a result, we get nearly optimal
adaptively secure ERF's and AONT's. Finally, extending the statistical
construction we obtain optimal *computational* adaptive ERF's, "public-
value" AONT's and resilient functions.

1 Introduction

Recently, there has been an explosion of work [23,9,10,20,18,7,1,26,14] surround-
ing an intriguing notion introduced by Rivest called the *All-Or-Nothing Transform* (AONT) [23]. Roughly speaking, an AONT is a randomized mapping which

B. Pfitzmann (Ed.): EUROCRYPT 2001, LNCS 2045, pp. 301–324, 2001.

can be efficiently inverted if given the output in *full*, but which leaks *no* information about its input to an adversary even if the adversary obtains *almost all* the bits of the output. The AONT has been shown to have important cryptographic applications ranging from increasing the efficiency of block ciphers [20,18,7] to protecting against almost complete exposure of secret keys [10]. The first formalization and constructions for the AONT were given by Boyko [9] in the Random-Oracle model. However, recently Canetti et al. [10] were able to formalize and exhibit efficient constructions for the AONT in the standard computational model. They accomplished this goal by reducing the task of constructing AONT's to constructing a related primitive which they called an *Exposure-Resilient Function* (ERF) [10]. An ERF is a deterministic function whose output looks random to an adversary even if the adversary obtains *almost all* the bits of the input. A salient feature of the work of [10] is the fact that they were able to achieve good results for the computational (and most cryptographically applicable) versions of these notions by first focusing on the perfect and statistical forms of AONT's and ERF's.

1.1 Background

We first recall informally the definitions of the two main notions we examine in this paper. An ℓ-AONT [23,9,10] is an efficiently computable and *invertible* randomized transformation T, which transforms any string x into a pair of strings (y_s, y_p), respectively called the *secret* and the *public* part of T. While the invertability of T allows to reconstruct x from the *entire* $T(x) = (y_s, y_p)$, we require that any adversary learning all of y_p and all but ℓ bits of y_s obtains "no information" about x.

On the other hand, an ℓ-ERF [10] is an efficiently computable deterministic function f on strings such that even if an adversary learns all but ℓ bits of a *randomly chosen* input r, it still cannot distinguish the output $f(r)$ from a random string. As usual, we can define *perfect*, *statistical*, and *computational* versions of these notions. It is easy to see that in the perfect or statistical settings, the length of the output of an ℓ-ERF can be at most ℓ; whereas for perfect or statistical ℓ-AONT's, the length of the input is at most ℓ. To beat these trivial bounds, one must examine the computational forms of ERF's and AONT's. Indeed, if we are given a pseudorandom generator, it is easy to see that by applying the generator to the output of a perfect or statistical ERF, we can obtain ERF's with arbitrary (polynomial) output size.

Canetti et al. [10] showed that the following simple construction suffices to construct AONT's from ERF's. Given an ℓ-ERF f mapping $\{0,1\}^n$ to $\{0,1\}^k$, we construct an ℓ-AONT T transforming k bits to n bits of secret output and k bits of public output: $T(x) = \langle r, \ f(r) \oplus x \rangle$. Intuitively, if at least ℓ bits of r are missed, then $f(r)$ "looks" random. Hence $f(r) \oplus x$ also looks random, thus hiding all information about the input x.

APPLICATIONS. The All-Or-Nothing Transform and its variants have been applied to a variety of problems. In the perfect setting, it is a special case of a *ramp scheme* [6], useful for sharing secrets efficiently. Its statistical variant can be used to provide secure communication over the "wire-tap channel II", a partly public channel where the adversary can observe almost all the bits communicated (but the sender and the receiver do not know which) [22,3]. In the computational setting, it also has many uses. Rivest [23], and later Desai [14], use it to enhance the security of block ciphers against brute-force key search. Matyas et al. [20] propose to use AONT to increase the efficiency of block ciphers: rather than encrypt all blocks of the message, apply an AONT to the message and encrypt only one or very few blocks. The same idea is used in various forms by Jackobson et al. [18] and Blaze [7] to speed up remotely-keyed encryption. Similarly, it can be combined with authentication to yield a novel encryption technique [24,1]. Several other applications have been suggested by [9,26].

Another class of applications for (computational) AONT's was suggested by Canetti et al. [10]. They considered a situation where one of our most basic cryptographic assumptions breaks down — the secrecy of a key can become partially compromised (a problem called *partial key exposure*). [10] point out that most standard cryptographic definitions do not guarantee (and often violate) security once even a small portion of the key has been exposed. The AONT offers a solution to this problem. Namely, rather than store a secret key x, one stores $y = T(x)$ instead. Now the adversary gets no information about the secret key even if he manages to get all but ℓ bits of y. The problem of *gradual* key exposure is also raised by [10], where information about a (random) private key is slowly but steadily leaked to an adversary. In this situation, the private key can be "renewed" using an ERF to protect it against discovery by the adversary, while additionally providing forward security when the "current" key is totally compromised.

1.2 Adaptive Security

In many of the applications above, the question of adaptive security arises naturally. For example, in the problem of partial key exposure, it is natural to consider an adversary that is able to first gain access to some fraction of the bits of the secret, and then decides which bits to obtain next as a function of the bits the adversary has already seen.

PERFECT AONT'S AND ADAPTIVE SECURITY. In the definition of a *perfect* ℓ-AONT, we demand that any subset of all but ℓ bits of the output must be *completely independent* of the input x.[1] In this case, it is trivial to observe that there is no difference between adaptive and non-adaptive security. Hence, if we could construct good perfect AONT's, this would also solve the problem of constructing adaptively secure AONT's.

Consider ℓ-AONT's that transform k bits to n bits. [10] show how to construct perfect ℓ-AONT's where $\ell = n(\frac{1}{2} + \varepsilon)$ for any $\varepsilon > 0$ (at the expense of smaller $k = \Omega(n)$), but were unable to construct perfect AONT's with $\ell < n/2$ (i.e. perfect AONT's where the adversary could learn more than half of the output).

---

[1] In the perfect setting, public output is not needed (e.g., can be fixed a-priori).

PERFECT AONT'S — OUR CONTRIBUTION. In our work, we show that unfortunately this limitation is inherent. More precisely, whenever $n \leq 2^k$, the adversary *must* miss at least half of the output in order not to learn anything about the input. We prove this bound by translating the question of constructing perfect ℓ-AONT's to the question of finding *"ℓ-balanced" weighted colorings of the hypercube*, which is of independent combinatorial interest. Namely, we want to color and weight the nodes of the n-dimensional hypercube $\mathcal{H} = \{0, 1\}^n$ using $c = 2^k$ colors, such that every ℓ-dimensional subcube of $\mathcal{H}$ is "equi-colored" (i.e. has the same total weight for each of the c colors). We prove our result by non-trivially extending the beautiful lower bound argument of Friedman [15] (which only worked for unweighted colorings) to our setting. Our bound also gives a new bound on *ramp secret sharing schemes* [6]. In such schemes one divides the secret of size k into n schares such that there are two thresholds t and $(t - \ell)$ such that any t shares suffice to reconstruct the secret but no $(t - \ell)$ shares yield any information. To our knowledge, the best known bound for ramp schemes [8,17,21] was $\ell \geq k$. Our results imply a much stronger bound of $\ell \geq t/2$ (when each share is a bit; over larger alphabets of size q we get $\ell > t/q$).

Therefore, we show that despite their very attractive perfect security, perfect AONT's are of limited use in most situations, and do not offer a compelling way to achieve adaptive security.

STATISTICAL ERF'S AND ADAPTIVE SECURITY. The definition of a *perfect ℓ-ERF* (mapping n bits to k bits) states that the output, when considered jointly with any subset of $(n - \ell)$ bits of the input, must be truly uniform. In this case, clearly once again adaptive and non-adaptive security collapse into one notion. The definition of a (non-adaptive) *statistical ℓ-ERF*, however, allows for the the joint distribution above to be merely *close* to uniform. In this case, the non-adaptive statistical definition does *not* imply adaptive security, and in particular the construction given in [10] of statistical ERF's fails to achieve adaptive security.[2] Intuitively, it could be that a small subset of the input bits S_1 determines some non-trivial boolean relation of another small subset of the input bits S_2 with the output of the function (e.g., for a fixed value of the bits in S_1, one output bit might depend only on bits in S_2). In the adaptive setting, reading S_1 and then S_2 would break an ERF. In the non-adaptive setting, however, any *fixed* subset of the input bits is very unlikely to contain $S_1 \cup S_2$. (A similar discussion applies to AONT's.) In other words, statistical constructions of [10] were able to produce statistical ℓ-ERF's (and ℓ-AONT's) with nearly optimal $\ell = k + o(k)$, but failed to achieve adaptive security, while perfect ERF's achieve adaptive security, but are limitted to $\ell > n/2$ [15].

STATISTICAL ERF'S — OUR CONTRIBUTION. Thus, we seek to identify notions lying somewhere in between perfect and statistical (non-adaptive) ERF's that would allow us to construct adaptively secure ERF's (and AONT's), and yet achieve better parameters than those achievable by perfect ERF's (and AONT's). In this task, we make use of *resilient functions* (RF's). These were first defined

---

[2] For more details, see Section 2.2.

in the perfect setting by Vazirani [28] and first studied by Chor et al. [12] and independently by Bennett et al. [3]. An ℓ-RF is identical to an ℓ-ERF except that the adversary, instead of merely *observing* certain bits of the input, gets to *set* all but ℓ bits of the input.[3] Note that the notions of ERF and RF are the same when considered in the perfect setting. A statistical variant of resilient functions (no longer equivalent to ERF's) was first considered by Kurosawa et al. [19], who also gave explicit constructions of such functions (improved by [5]).

We show that the strong notion of statistical RF's introduced by Kurosawa et al. [19] suffices to construct adaptively secure ERF's (and AONT's). While the construction of Kurosawa et al. [19] already slightly beats the lower bound for perfect ERF's, it is very far from the trivial lower bound of $\ell > k$ (in fact, it is still limited to $\ell > n/2$). We present an efficient probabilistic construction of such "almost-perfect" RF's achieving optimal $\ell = k + o(k)$. While not fully deterministic, our construction has to be run only *once and for all*, after which the resulting efficient function is "good" with probability exponentially close to 1, and can be *deterministically* used in all the subsequent applications. As a result of this construction and its relation to adaptive ERF's and AONT's, we achieve essentially optimal security parameters for adaptive security by focusing on a stronger notion of almost-perfect RF's.

We also take the opportunity to study several variants of statistical RF's and (static/adaptive) ERF's, and give a complete classification of these notions, which may be of additional, independent interest.

COMPUTATIONAL SETTING. As we pointed out, [10] used their statistical (non-adaptive) constructions to get ERF's and AONT's in the computational setting. We show that the same techniques work with our adaptive definitions. Coupled with our statistical constructions, we get nearly optimal computational constructions as well.

LARGER ALPHABETS. To simplify the presentation and the discussion of the results in this paper, as well as to relate them more closely with the previous work, we restrict ourselves to discussing exposure-resilient primitives over the alphabet $\{0, 1\}$. However, all our notions and results can be easily generalized to larger alphabets.

1.3   Organization

In Section 2, we define the central objects of study in our paper, and review some of the relevant previous work of [10]. In Section 3 we study perfect AONT's, relate them to hypecube colorings and prove the strong lower bound on ℓ (showing the limitations of perfect AONT's). Finally, in Section 4 we study variants of statistical ERF's will allow us to achieve adaptive security. We show that "almost-rerfect" RF's of [19] achieve this goal, and exhibit a simple and almost optimal (probabilistic) construction of such functions. In particular, we show

---

[3] In much of the literature about resilient functions, such a function would be called an $(n - \ell)$-resilient function. We adopt our notation for consistency.

the existence of adaptively secure AONT's and ERF's with essentially optimal parameters.

2 Preliminaries

Let $\left\{\begin{smallmatrix}n\\\ell\end{smallmatrix}\right\}$ denote the set of size-ℓ subsets of $[n] = \{1 \ldots n\}$. For $L \in \left\{\begin{smallmatrix}n\\\ell\end{smallmatrix}\right\}$, $y \in \{0,1\}^n$, let $[y]_{\bar{L}}$ denote y restricted to its $(n-\ell)$ bits *not* in L. We say a function $\epsilon(n)$ is negligible (denoted by $\epsilon = negl(n)$) if for every constant c, $\epsilon(n) = O\left(\frac{1}{n^c}\right)$. We denote an algorithm $\mathcal{A}$ which has oracle access to some string y (i.e., can query individual bits of y) by $\mathcal{A}^y$.

2.1   Definitions for Non-adaptive Adversaries

For static adversaries, the definitions of AONT and ERF can be stated quite efficiently in terms of perfect, statistical or computational indistinguishability (see [16]). For consistency we have also provided a definition of RF (where adaptivity does not make sense, and hence the adversary can be seen as "static").

Note that for full generality, we follow the suggestion of [10] and allow the all-or-nothing transform to have two outputs: a *public* part which we assume the adversary always sees; and a *secret* part, of which the adversary misses ℓ bits.

Definition 1. *A polynomial-time randomized transformation* $T : \{0,1\}^k \to \{0,1\}^s \times \{0,1\}^p$ *is an* ℓ-AONT (all-or-nothing transform) *if*

1. T *is polynomial-time invertible, i.e. there exists efficient* I *such that for any* $x \in \{0,1\}^k$ *and any* $y = (y_1, y_2) \in T(x)$, *we have* $I(y) = x$. *We call* y_1 *is the* secret part *and* y_2, *the* public part *of* T.
2. *For any* $L \in \left\{\begin{smallmatrix}s\\\ell\end{smallmatrix}\right\}, x_0, x_1 \in \{0,1\}^k$: $\langle x_0, x_1, [T(x_0)]_{\bar{L}}\rangle \approx \langle x_0, x_1, [T(x_1)]_{\bar{L}}\rangle$[4]
   Here $\approx$ *can refer to perfect, statistical or computational indistinguishability.*

If $p = 0$, *the resulting* AONT *is called* secret-only.

Definition 2. *A polynomial time function* $f : \{0,1\}^n \to \{0,1\}^k$ *is an* ℓ-ERF (exposure-resilient function) *if for any* $L \in \left\{\begin{smallmatrix}n\\\ell\end{smallmatrix}\right\}$ *and for a randomly chosen* $r \in \{0,1\}^n$, $R \in \{0,1\}^k$, *we have:* $\langle [r]_{\bar{L}}, f(r)\rangle \approx \langle [r]_{\bar{L}}, R\rangle$.
Here $\approx$ *can refer to perfect, statistical or computational indistinguishability.*

Definition 3. *A polynomial time function* $f : \{0,1\}^n \to \{0,1\}^k$ *is* ℓ-RF (resilient function) *if for any* $L \in \left\{\begin{smallmatrix}n\\\ell\end{smallmatrix}\right\}$, *for any assignment* $w \in \{0,1\}^{n-\ell}$ *to the positions not in* L, *for a randomly chosen* $r \in \{0,1\}^n$ *subject to* $[r]_{\bar{L}} = w$ *and random* $R \in \{0,1\}^k$, *we have:* $\langle f(r) \mid [r]_{\bar{L}} = w\rangle \approx \langle R\rangle$.
Here $\approx$ *can refer to perfect, statistical or computational indistinguishability.*

---

[4] Notice, for $L \in \left\{\begin{smallmatrix}s\\\ell\end{smallmatrix}\right\}$ we have notationally that $[(y_1, y_2)]_{\bar{L}} = ([y_1]_{\bar{L}}, y_2)$.

As an obvious note, a ℓ-RF is also a static ℓ-ERF (as we shall see, this will *no longer hold* for adaptive ERF; see Lemma 5).

PERFECT PRIMITIVES. It is clear that perfect ERF are the same as perfect RF. Additionally, perfect AONT's are easy to construct from perfect ERF's. In particular one could use the simple one-time pad construction of [10]: $T(x) = \langle r, f(r) \oplus x \rangle$, where r is the secret part of the AONT. However, we observe that (ignoring the issue of efficiency) there is no need for the public part in the perfect AONT (i.e., we can fix it to any valid setting y_2 and consider the restriction of the AONT where the public part is always y_2). Setting $y_2 = \mathbf{0}$ in the one-time pad construction implies an AONT where we output a random r subject to $f(r) = x$. Thus, in the perfect setting the "inverse" of an ℓ-ERF *is* an ℓ-AONT, and we get:

Lemma 1. *(Ignoring issues of efficiency) A perfect ℓ-ERF $f : \{0,1\}^n \to \{0,1\}^k$ implies the existence of a perfect (secret-only) ℓ-AONT $T : \{0,1\}^k \to \{0,1\}^n$.*

While the reduction above does *not* work with statistical ERF (to produce statistical AONT), we will show that it works with a stronger notion of *almost-perfect* RF (to produce statistical AONT). See Lemma 7.

2.2   Definitions for Adaptive Adversaries

ADAPTIVELY SECURE AONT. In the ordinary AONT's the adversary has to "decide in advance" which $(s - \ell)$ bits of the (secret part of) the output it is going to observe. This is captured by requiring the security for all *fixed* sets L of cardinality ℓ. While interesting and non-trivial to achieve, in many applications (e.g. partial key exposure, secret sharing, protecting against exhaustive key search, etc.) the adversary potentially has the power to choose which bits to observe *adaptively*. For example, at the very least it is natural to assume that the adversary could decide which bits of the secret part to observe after it learns the public part. Unfortunately, the constructions of [10] do not even achieve this minimal adaptive security, invalidating their claim that "public part requires no protection and can be given away for free". More generally, the choice of which bit(s) to observe next may partially depend on which bits the adversary has already seen. Taken to the most extreme, we can allow the adaptive adversary to read the bits of the secret part "one-bit-at-a-time", as long as he misses at least ℓ of them.

Definition 4. *A polynomial time randomized transformation $T : \{0,1\}^k \to \{0,1\}^s \times \{0,1\}^p$ is a (perfect, statistical or computational) adaptive ℓ-AONT (adaptive all-or-nothing transform) if*

1. *T is efficiently invertible, i.e. there is a polynomial time machine I such that for any $x \in \{0,1\}^k$ and any $y = (y_1, y_2) \in T(x)$, we have $I(y) = x$.*
2. *For any adversary $\mathcal{A}$ who has oracle access to string $y = (y_s, y_p)$ and is required not to read at least ℓ bits of y_s, and for any $x_0, x_1 \in \{0,1\}^k$, we have: $\left| \Pr(\mathcal{A}^{T(x_0)}(x_0, x_1) = 1) - \Pr(\mathcal{A}^{T(x_1)}(x_0, x_1) = 1) \right| \leq \epsilon$, where*

- *In the perfect setting $\epsilon = 0$.*
- *In the statistical setting $\epsilon = negl(s + p)$.*
- *In the computational setting $\epsilon = negl(s + p)$ for any PPT $\mathcal{A}$.*

We stress that the adversary can base its queries on x_0, x_1, the public part of the output, as well as those parts of the secret output that it has seen so far. We also remark that in the perfect setting this definition is *equivalent* to that of an ordinary perfect ℓ-AONT. Thus, adaptivity does not help the adversary in the perfect setting (because the definition of a perfect AONT is by itself very strong!). In particular, good perfect AONT's are good adaptive AONT's. Unfortunately, we will later show that very good perfect AONT's do not exist.

ADAPTIVELY SECURE ERF. In the original definition of ERF [10], the adversary has to "decide in advance" which $(n - \ell)$ input bits it is going to observe. This is captured by requiring the security for all *fixed* sets L of cardinality ℓ. However, in many situations (e.g., the problem of gradual key exposure [10]), the adversary has more power. Namely, it can decide which $(n - \ell)$ bits of the secret to learn *adaptively* based on the information that it has learned so far. In the most extreme case, the adversary would decide which bits to observe "one-bit-at-a-time". Unfortunately, the definition and the construction of [10] do not satisfy this notion.

There is one more particularity of adaptive security for ERF's. Namely, in some applications (like the construction of AONT's using ERF's [10]) the adversary might observe some partial information about the secret output of the ERF, $f(r)$, *before it starts to compromise the input r*. Is it acceptable in this case that the adversary can learn more partial information about $f(r)$ than he already has? For example, assume we use $f(r)$ as a stream cipher and the adversary learns the first few bits of $f(r)$ before it chooses which $(n - \ell)$ bits of r to read. Ideally, we will not want the adversary to be able to learn some information about the remaining bits of $f(r)$ — the ones that would be used in the stream cipher in the future. Taken to the extreme, even if the adversary sees either the *entire* $f(r)$ (i.e., has complete information on $f(r)$), or a random R, and *only then* decides which $(n - \ell)$ bits of r to read, it cannot distinguish the above two cases.

As we argued, we believe that a good notion of adaptive ERF should satisfy both of the properties above, which leads us to the following notion.

Definition 5. *A polynomial time function $f : \{0,1\}^n \to \{0,1\}^k$ is a (perfect, statistical or computational) adaptive ℓ-ERF (adaptive exposure-resilient function) if for any adversary $\mathcal{A}$ who has access to a string r and is required not to read at least ℓ bits of r, when r is chosen at random from $\{0,1\}^n$ and R is chosen at random from $\{0,1\}^k$, we have: $|\Pr(\mathcal{A}^r(f(r)) = 1) - \Pr(\mathcal{A}^r(R) = 1)| \le \epsilon$, where*

- *In the perfect setting $\epsilon = 0$.*
- *In the statistical setting $\epsilon = negl(n)$.*
- *In the computational setting $\epsilon = negl(n)$ for any PPT $\mathcal{A}$.*

Notice that in the perfect setting this definition is equivalent to that of an ordinary (static) perfect ℓ-ERF, since for any L, the values $[r]_{\bar{L}}$ and $f(r)$ are uniform and independent. In the statistical setting, the notions are no longer equivalent: indeed, the original constructions of [10] fail dramatically under an adaptive attack. We briefly mention the reason. They used so-called randomness extractors in their construction of statistical ERF's (see [10] for the definitions). Such extractors use a small number of truly random bits d to extract all the randomness from any "reasonable" distribution X. However, it is crucial that this randomness d is chosen independently from and *after* the distribution X is specified. In their construction d was part of the input r, and reading upto $(n - \ell)$ of the remaining bits of r defined the distribution X that they extracted randomness from. Unfortunately, an adaptive adversary can first read d, and only then determine which other bits of r to read. This alters X depending on d, and the notion of an extractor does not work in such a scenario. In fact, tracing the particular extractors that they use, learning d first indeed allows an adaptive adversary to break the resulting static ERF.

Also notice that once we have good adaptive statistical ERF's, adaptive computational ERF's will be easy to construct in same same way as with regular ERF [10]: simply apply a good pseudorandom generator to the output of an adaptive statistical ERF. Finally, we notice that the generic one-time pad construction of [10] of AONT's from ERF's extends to the adaptive setting, as long as we use the strong adaptive definition of ERF given above. Namely, the challenge has to be given first, since the adversary for the AONT may choose which bits of the secret part r to read when having already read the entire public part — either $f(r) \oplus x_0$ or $f(r) \oplus x_1$ (for known x_0 and x_1!). Thus, we get

Lemma 2. *If $f : \{0,1\}^n \to \{0,1\}^k$ is an adaptive ℓ-ERF, then $T(x) = \langle r, x \oplus f(r) \rangle$ is an adaptive ℓ-AONT with secret part r and public part $x \oplus f(r)$.*

3   Lower Bound on Perfect AONT

In this section we study perfect AONT's. We show that there exists a strong limitation in constructing perfect AONT's: the adversary must miss at least half of the n-bit output, even if the input size k is as small as $\log n$. Recall that perfect AONT's are more general than perfect ERF's (Lemma 1), and thus our bound non-trivially generalizes the lower bound of Friedman [15] (see also another proof by [4]) on perfect ERF. As we will see, the proof will follow from the impossibility of certain weighted "balanced" colorings of an n-dimensional hypercube, which is of independent interest.

Theorem 1. *If $T : \{0,1\}^k \to \{0,1\}^n$ is a perfect (secret-only) ℓ-AONT, then*

$$\ell \geq 1 + n \cdot \frac{2^{k-1} - 1}{2^k - 1} = \frac{n}{2} + \left(1 - \frac{n}{2(2^k - 1)}\right) \tag{1}$$

In particular, for $n \leq 2^k$ we get $\ell > \frac{n}{2}$, so at least half of the output of T has to remain secret even if T exponentially expands its input! Moreover, the equality can be achieved only by AONT's constructed from ERF's via Lemma 1.

3.1   Balanced Colorings of the Hypercube

A *coloring* of the n-dimensional hypercube $\mathcal{H} = \{0,1\}^n$ with c colors is any map which associates a color from $\{1, \dots, c\}$ to each node in the graph. In a *weighted coloring*, each node y is also assigned a non-negative real weight $\chi(y)$. We will often call the nodes of weight 0 *uncolored*, despite them having an assigned nominal color. For each color i, we define the weight vector χ_i of this color by assigning $\chi_i(y) = \chi(y)$ if y has color i, and 0 otherwise. We notice that for any given $y \in \mathcal{H}$, $\chi_i(y) > 0$ for at most one color i, and also $\sum \chi_i = \chi$. A coloring where all the nodes are uncolored is called *empty*. Since we will never talk about such colorings, we will assume that $\sum_{y \in \mathcal{H}} \chi(y) = 1$. A *uniform* coloring has all the weights equal: $\chi(y) = 2^{-n}$ for all y.

An ℓ-*dimensional subcube* $\mathcal{H}_{L,a}$ of the hypercube is given by a set of ℓ "free" positions $L \in \{^n_\ell\}$ and an assignment $a \in \{0,1\}^{n-\ell}$ to the remaining positions, and contains the resulting 2^ℓ nodes of the hypercube consistent with a.

Definition 6. *We say a weighted coloring of the hypercube is ℓ-balanced if, within every subcube of dimension ℓ, each color has the same weight. That is, for each L and a, $\sum_{y \in \mathcal{H}_{L,a}} \chi_i(y)$ is the same for all colors i.*

Notice, ℓ-balanced coloring is also ℓ'-balanced for any $\ell' > \ell$, since an ℓ' dimensional subcube is the disjoint union of ℓ-dimensional ones. We study balanced colorings since they exactly capture the combinatorial properties of ℓ-AONT's and ℓ-ERF's. We get the following equivalences.

Lemma 3. *Ignoring efficiency, the following equivalences hold in the perfect setting:*

1. *ℓ-AONT's from k to n bits $\Longleftrightarrow$ weighted ℓ-balanced colorings of n-dimensional hypercube with 2^k colors.*
2. *ℓ-ERF's from n to k bits $\Longleftrightarrow$ uniform ℓ-balanced colorings of n-dimensional hypercube with 2^k colors.*

Proof Sketch. For the first equivalence, the color of node $y \in \mathcal{H}$ corresponds to the value if the inverse map $I(y)$, and its weight corresponds to $\Pr_{x,T}(T(x) = y)$. For the second equivalence, the color of node $y \in \mathcal{H}$ is simply $f(y)$.   □

Notice, the lemma above also gives more insight into why perfect AONT's are more general than perfect ERF's (and an alternative proof of Lemma 1). We now restate our lower bound on perfect AONT's in Theorem 1 in terms of weighted ℓ-balanced colorings of $\mathcal{H}$ with $c = 2^k$ colors (proving it for general c).

Theorem 2. *Any (non-empty) ℓ-balanced weighted coloring of the n-dimensional hypercube using c colors must have $\ell \geq \frac{n}{2} + \left(1 - \frac{n}{2(c-1)}\right)$. Moreover, equality can hold only if the coloring is uniform and no two adjacent nodes of positive weight have the same color.*

We believe that the theorem above is interesting in its own right. It says that once the number of colors is at least 3, it is impossible to find a c-coloring (even weighted!) of the hypercube such that all ℓ-dimensional subcubes are "equi-colored", unless ℓ is very large (linear in n).

3.2   Proof of the Lower Bound (Theorem 2)

In our proof of Theorem 2, we will consider the 2^n-dimensional vector space V consisting of real-valued (not boolean!) vectors with positions indexed by the strings in $\mathcal{H}$, and we will use facts about the Fourier decomposition of the hypercube.

FOURIER DECOMPOSITION OF THE HYPERCUBE.   Like the original proof of Friedman [15] for the case of uniform colorings, we use the adjacency matrix A of the hypercube. A is a $2^n \times 2^n$ dimensional 0-1 matrix, where the entry $A_{x,y} = 1$ iff x and y (both in $\{0,1\}^n$) differ in exactly one coordinate. Recall that a non-zero vector $\mathbf{v}$ is an *eigenvector* of the matrix A corresponding to an *eigenvalue* λ, if $A\mathbf{v} = \lambda\mathbf{v}$. Since A is symmetric, there is an orthonormal basis of $\mathbb{R}^{2^n}$ in which all 2^n vectors are eigenvectors of A. For two strings in x, z in $\{0,1\}^n$, let $x \cdot z$ denote their inner product modulo 2 and let $weight(z)$ be the number of positions of z which are equal to 1. Then:

Fact 1 *A has an orthonormal basis of eigenvectors $\{\mathbf{v}_z : z \in \{0,1\}^n\}$, where the eigenvalue of $\mathbf{v}_z$ is $\lambda_z = n - 2 \cdot weight(z)$, and the value of $\mathbf{v}_z$ at position y is $\mathbf{v}_z(y) = \frac{1}{\sqrt{2^n}} \cdot (-1)^{z \cdot y}$.*

We will use the notation $\langle \mathbf{u}, \mathbf{v} \rangle = \mathbf{u}^\top \mathbf{v} = \sum_i u_i v_i$ to denote the inner product of $\mathbf{u}$ and $\mathbf{v}$, and let $\|\mathbf{u}\|^2 = \langle \mathbf{u}, \mathbf{u} \rangle = \sum_i u_i^2$ denote the square of the Euclidean norm of $\mathbf{u}$. We then get the following useful fact, which follows as an easy exercise from Fact 1 (it is also a consequence of the Courant-Fischer inequality).

Fact 2 *Assume $\{\mathbf{v}_z : z \in \{0,1\}^n\}$ are the eigenvectors of A as above, and let $\mathbf{u}$ be a vector orthogonal to all the $\mathbf{v}_z$'s corresponding to z with $weight(z) < t$: $\langle \mathbf{u}, \mathbf{v}_z \rangle = 0$. Then we have:* $\mathbf{u}^\top A \mathbf{u} \leq (n - 2t) \cdot \|\mathbf{u}\|^2$. *In particular, for any* $\mathbf{u}$ *we have:* $\mathbf{u}^\top A \mathbf{u} \leq n \cdot \|\mathbf{u}\|^2$.

EXPLOITING BALANCEDNESS.   Consider a non-empty ℓ-balanced weighted coloring χ of the hypercube using c colors. Let χ_i be the characteristic weight vector corresponding to color i (i.e. $\chi_i(y)$ is the weight of y when y has color i and 0 otherwise). As we will show, the χ_i's have some nice properties which capture the balancedness of the coloring χ. In particular, we know that for any colors i and j and for any ℓ-dimensional subcube of $\mathcal{H}$, the sum of the components of χ_i and of χ_j are the same in this subcube. Hence, if we consider the difference $(\chi_i - \chi_j)$, we get that the sum of its coordinates over any ℓ-dimensional subcube is 0.

To exploit the latter property analytically, we consider the quantity $(\chi_i - \chi_j)^\top A(\chi_i - \chi_j)$, where A is the adjacency matrix of the n-dimensional hypercube. As suggested by Fact 2, we can bound this quantity by calculating the Fourier coefficients of $(\chi_i - \chi_j)$ corresponding to large eigenvalues. We get:

Lemma 4. *For any $i \neq j$, we have:* $(\chi_i - \chi_j)^\top A(\chi_i - \chi_j) \leq (2\ell - n - 2) \cdot \|\chi_i - \chi_j\|^2$.

We postpone the proof of this crucial lemma until the the end of this section, and now just use it to prove our theorem. First, note that the lemma above only gives us information on two colors. To simultaneously use the information from all pairs, we consider the *sum* over all pairs i, j, that is

$$\Delta \overset{\text{def}}{=} \sum_{i,j} (\chi_i - \chi_j)^\top A(\chi_i - \chi_j) \tag{2}$$

We will give upper and lower bounds for this quantity (Equation (3) and Equation (4), respectively), and use these bounds to prove our theorem. We first give the upper bound, based on Lemma 4.

Claim.

$$\Delta \le 2 \, (2\ell - n - 2) \, (c - 1) \cdot \sum_i \|\chi_i\|^2 \tag{3}$$

Proof. We can ignore the terms of Δ when $i = j$ since then $(\chi_i - \chi_j)$ is the 0 vector. Using Lemma 4 we get an upper bound:

$$\sum_{i,j} (\chi_i - \chi_j)^\top A(\chi_i - \chi_j) \le (2\ell - n - 2) \cdot \sum_{i \ne j} \|\chi_i - \chi_j\|^2$$

Now the vectors χ_i have disjoint supports (since each $y \in \mathcal{H}$ is assigned only one color), so we have $\|\chi_i - \chi_j\|^2 = \|\chi_i\|^2 + \|\chi_j\|^2$. Substituting into the equation above, we see that each $\|\chi_i\|^2$ appears $2(c-1)$ times (recall that c is the number of colors), which immediately gives the desired bound in Equation (3). □

Second, we can expand the definition of Δ to directly obtain a lower bound.

Claim.

$$\Delta \ge -2n \cdot \sum_i \|\chi_i\|^2 \tag{4}$$

Proof. Since A is symmetric we have $\chi_i^\top A\chi_j = \chi_j^\top A\chi_i$. Then:

$$\sum_{i,j} (\chi_i - \chi_j)^\top A(\chi_i - \chi_j) = \sum_{i,j} (\chi_i^\top A\chi_i + \chi_j^\top A\chi_j - 2\chi_i^\top A\chi_j)$$

$$= 2c \cdot \sum_i \chi_i^\top A\chi_i - 2 \cdot \sum_{i,j} \chi_i^\top A\chi_j$$

Let us try to bound this last expression. On the one hand, we know that $\chi_i^\top A\chi_i \ge 0$ since it is a product of matrices and vectors with non-negative entries. On the other hand, we can rewrite the last term as a product:

$$\sum_{i,j} \chi_i^\top A\chi_j = \left(\sum_i \chi_i \right)^\top A \left(\sum_i \chi_i \right)$$

This quantity, however, we can bound using the fact that the maximum eigenvalue of A is n (see Fact 2). We get:

$$\left(\sum_i \chi_i\right)^\top A \left(\sum_i \chi_i\right) \le n \cdot \left\|\sum_i \chi_i\right\|^2$$

Since the vectors χ_i have disjoint support (again, each node y is assigned a unique color), they are orthogonal and so $\|\sum_i \chi_i\|^2 = \sum_i \|\chi_i\|^2$. Combining these results, we get the desired lower bound:

$$\sum_{i,j} (\chi_i - \chi_j)^\top A(\chi_i - \chi_j) \ge 0 - 2n \cdot \sum_i \|\chi_i\|^2 = -2n \cdot \sum_i \|\chi_i\|^2 \qquad \square$$

Combining the lower and the upper bounds of Equation (3) and Equation (4), we notice that $\sum_i \|\chi_i\|^2 > 0$ and can be cancelled out (since the coloring χ is non-empty). This gives us $2(2\ell - n - 2)(c - 1) \ge -2n$, which exactly implies the needed bound on ℓ.

PROOF OF LEMMA 4. It remains to prove Lemma 4, i.e. $(\chi_i - \chi_j)^\top A(\chi_i - \chi_j) \le (2\ell - n - 2) \cdot \|\chi_i - \chi_j\|^2$. By Fact 2, it is sufficient show that all the Fourier coefficients of $(\chi_i - \chi_j)$ which correspond to eigenvalues $\lambda_z \ge 2\ell - n = n - 2(n - \ell)$ are 0. In other words, that $(\chi_i - \chi_j)$ is orthogonal to all the eigenvectors $\mathbf{v}_z$ whose eigenvalues are at least $(n - 2(n - \ell))$, i.e. $weight(z) \le n - \ell$. But recall that by the definition of balancedness, on any subcube of dimension at least ℓ, the components of $(\chi_i - \chi_j)$ sum to 0! On the other hand, the eigenvectors $\mathbf{v}_z$ are constants on very large-dimensional subcubes of $\mathcal{H}$ when λ_z is large (see Fact 1). These two facts turn out to be exactly what we need to in order to show that $\langle \mathbf{v}_z, \chi_i - \chi_j \rangle = 0$ whenever $\lambda_z \ge 2\ell - n$, and thus to prove Lemma 4.

Claim. For any $z \in \{0, 1\}^n$ with $weight(z) \le n - \ell$ (i.e. $\lambda_z \ge 2\ell - n$), we have: $\langle \mathbf{v}_z, \chi_i - \chi_j \rangle = 0$.

Proof. Pick any vector $z = (z_1, \dots, z_n) \in \{0, 1\}^n$ with $weight(z) \le n - \ell$, and let S be the support of z, i.e. $S = \{j : z_j = 1\}$. Note that $|S| \le n - \ell$. Also, recall that $\mathbf{v}_z(y) = \frac{1}{\sqrt{2^n}} \cdot (-1)^{z \cdot y}$ (see Fact 1). Now consider any assignment a to the variables of S. By letting the remaining variables take on all possible values, we get some subcube of the hypercube, call it $\mathcal{H}_a$.

One the one hand, note that $\mathbf{v}_z$ is constant (either $1/\sqrt{2^n}$ or $-1/\sqrt{2^n}$) on that subcube, since if y and y' differ only on positions *not* in S, we will have $z \cdot y = z \cdot y'$. Call this value C_a. On the other hand, since the coloring is ℓ-balanced and since $|S| \le n - \ell$, the subcube $\mathcal{H}_a$ has dimension at least ℓ and so we know that both colors i and j have equal weight on $\mathcal{H}_a$. Thus summing the values of $(\chi_i - \chi_j)$ over this subcube gives 0.

Using the above two observations, we show that $\langle \chi_i - \chi_j, \mathbf{v}_z \rangle = 0$ by rewriting the inner product as a sum over all assignments to the variables in S:

$$\langle \chi_i - \chi_j, \mathbf{v}_z \rangle = \sum_{y \in \mathcal{H}} \mathbf{v}_z(y)[\chi_i(y) - \chi_j(y)] = \sum_{a \in \{0,1\}^{|S|}} \left(\sum_{y \in \mathcal{H}_a} \mathbf{v}_z(y)[\chi_i(y) - \chi_j(y)] \right)$$

$$= \sum_a C_a \cdot \left(\sum_{y \in \mathcal{H}_a} \chi_i(y) - \sum_{y \in \mathcal{H}_a} \chi_j(y) \right) = \sum_a C_a \cdot 0 = 0 \qquad \square$$

EQUALITY CONDITIONS. We now determine the conditions on the colorings so that we can achieve equality in Theorem 2 (and also Theorem 1). Interestingly, such colorings are very structured, as we can see by tracing through our proof. Namely, consider the lower bound proved in Equation (4), i.e. that $\sum_{i,j}(\chi_i - \chi_j)^\top A(\chi_i - \chi_j) \leq -2n \sum_i \|\chi_i\|^2$. Going over the proof, we see that equality can occur only if two conditions occur.

On the one hand, we must have $\chi_i^\top A \chi_i = 0$ for all colors i. An easy calculation shows that $\chi_i^\top A \chi_i$ is 0 only when there is no edge of non-zero weight connecting two nodes of color i. Thus, this condition implies that the coloring is in fact a c-coloring in the traditional sense of complexity theory: no two adjacent nodes will have the same color. On the other hand, the inequality $(\sum_i \chi_i)^\top A(\sum_i \chi_i) \leq n \cdot \|\sum_i \chi_i\|^2$ must be tight. This can only hold if the vector $\chi = \sum_i \chi_i$ is parallel to $(1, 1, \ldots, 1)$ since that is the only eigenvector with the largest eigenvalue n. But this means that all the weights $\chi(y)$ are the same, i.e. that the coloring must be *uniform*.

We also remark that Chor et al. [12] showed (using the Hadamard code) that our bound is tight for $k \leq \log n$.

3.3 Extension to Larger Alphabets

Although the problem of constructing AONT's is usually stated in terms of bits, it is natural in many applications (e.g., secret-sharing) to consider larger alphabets, namely to consider $T : \{0, \ldots, q-1\} \rightarrow \{0, \ldots, q-1\}^n$. All the notions from the "binary" case naturally extend to general alphabets as well, and so does our lower bound. However, the lower bound we obtain is mostly interesting when the alphabet size q is relatively small compared to n. In particular, the threshold $n/2$, which is so crucial in the binary case (when we are trying to encode more than $\log n$ bits), becomes n/q (recall, q is the size of the alphabet). Significantly, this threshold becomes meaningless when $q > n$. This isn't surprising, since in this case we can use Shamir's secret sharing [25] (provided q is a prime power) and achieve $\ell = k$. We also remark that our bound is tight if $q^k \leq n$ and can be achieved similarly to the binary case by using the q-ary analog of the Hadamard code.

Theorem 3. *For any integer $q \geq 2$, let $T : \{0, \ldots, q-1\}^k \rightarrow \{0, \ldots, q-1\}^n$ be a perfect ℓ-AONT. Then*

$$\ell \geq \frac{n}{q} + \left(1 - \frac{q-1}{q} \cdot \frac{n}{q^k - 1} \right)$$

In particular, $\ell > n/q$ when $q^k > n$.

Similarly to the binary case, there is also a natural connection between ℓ-AONT's and weighted ℓ-balanced colorings of the "multi-grid" $\{0, \dots, q-1\}^n$ with $c = q^k$ colors. And again, the bound of Theorem 2 extends here as well and becomes $\ell \geq \frac{n}{q} + \left(1 - \frac{q-1}{q} \cdot \frac{n}{c-1}\right)$.

The proof techniques are essentially identical to those for the binary case. We now work with the graph $\{0, \dots, q-1\}^n$, which has an edge going between every pair of words that differ in a single position. We think of vertices in this graph as vectors in $\mathbb{Z}_q^n$. If ω is a primitive q-th root of unity in $\mathbb{C}$, then a orthonormal basis of eigenvectors of the adjacency matrix is given by the q^n-dimensional complex vectors $\mathbf{v}_z$ for $z \in \{0, \dots, q-1\}^n$, where $\mathbf{v}_z(y) = \frac{1}{\sqrt{q^n}} \cdot \omega^{z \cdot y}$ (here, $z \cdot y$ is the standard dot product modulo q). Constructing upper and lower bounds as above, we eventually get $(q\ell - n - q)(c-1) \sum_i \|\chi_i\|^2 \geq -n(q-1) \sum_i \|\chi_i\|^2$ which implies the desired inequality. Equality conditions are the same.

4   Adaptive Security in the Statistical Setting

We now address the question of adaptive security in the *statistical* setting. Indeed, we saw that both perfect ERF's and perfect AONT's have strong limitations. We also observed in Lemma 2 that we only need to concentrate on ERF's — we can use them to construct AONT's. Finally, we know that applying a regular pseudorandom generator to a good adaptively secure statistical ERF will result in a good adaptively secure *computational* ERF. This leaves with the need to construct adaptive statistical ERF's (recall that unfortunately, the construction of [10] for the static case is not adaptively secure). Hence, in this section we discuss only the statistical setting, and mainly resilient functions (except for Section 4.3; see below).

More specifically, in Section 4.1 we discuss several flavors of statistical resilient functions, and the relation among them, which should be of independent interest. In particular, we argue that the notion of almost-perfect resilient functions (APRF) [19] is the strongest one (in particular, stronger than adaptive ERF). In Section 4.2 we show how to construct APRF's. While seemingly only slightly weaker than perfect RF's, we show that we can achieve much smaller, optimal resilience for such functions: $\ell \approx k$ (compare with $\ell \geq n/2$ for perfect RF's). In particular, this will imply the existence of nearly optimal statistical RF's and adaptive statistical ERF's with the same parameters. Finally, in Section 4.3 we will show that APRF's can also be used to show the existence of optimal *secret-only* adaptive statistical AONT's (which improves the one-time pad construction from Lemma 2 and was not known even in the non-adaptive setting of [10]).

4.1   Adaptive ERF and Other Flavors of Resilient Functions

The definition presented in section 2 for adaptive security of an ERF is only one of several possible notions of adaptive security. Although it seems right for most applications involving resilience to *exposure*, one can imagine stronger attacks in which the security of resilient functions (RF), which tolerate even partly fixed inputs, would be desired. In this section we relate these various definitions, and reduce them to the stronger notion of an *almost-resilient* function [19], which are of independent combinatorial interest.

There are several parameters which one naturally wants to vary when considering "adaptive" security of an ERF, which is in its essence an extractor for producing good random bits from a partially compromised input.

1. Does the adversary get to see the challenge (output vs. a random string) before deciding how to "compromise" the input?
2. Does the adversary get to decide on input positions to "compromise" one at a time or all at once?
3. Does the adversary get to fix (rather than learn) some of the positions?

FLAVORS OF RESILIENT FUNCTIONS. To address the above questions, we lay out the following definitions. Unless stated otherwise, f denotes an efficient function $f : \{0,1\}^n \to \{0,1\}^k$, $L \in \{^n_\ell\}$, r is chosen uniformly from $\{0,1\}^n$, R is chosen uniformly from $\{0,1\}^k$. Finally, the adversary $\mathcal{A}$ is computationally unbounded, and has to obtain a non-negligible advantage in the corresponding experiment.

1. **(Weakly) Static ERF:** (This is the original notion of [10].)
   $r \in \{0,1\}^n$ is chosen at random. The adversary $\mathcal{A}$ specifies L and learns $w = [r]_{\bar{L}}$. $\mathcal{A}$ is then given the challenge Z which is either $f(r)$ or R. $\mathcal{A}$ must distinguish between these two cases.
2. **Strongly Static ERF:** (In this notion, the challenge is given first).
   $r \in \{0,1\}^n$ is chosen at random. The adversary $\mathcal{A}$ is then given the challenge Z which is either $f(r)$ or R. Based on Z, $\mathcal{A}$ specifies L, then learns $w = [r]_{\bar{L}}$, and has to distinguish between $Z = f(r)$ and $Z = R$.
3. **Weakly Adaptive ERF:** (This is a natural notion of adaptivity for ERF.)
   $r \in \{0,1\}^n$ is chosen at random. The adversary $\mathcal{A}$ learns up to $(n - \ell)$ bits of r, one at a time, basing each of his choices on what he has seen so far. $\mathcal{A}$ is then given the challenge Z which is either $f(r)$ or R, and has to distinguish between these two cases.
4. **(Strongly) Adaptive ERF:** (This is the notion defined in Section 2.)
   $r \in \{0,1\}^n$ is chosen at random. The adversary $\mathcal{A}$ is then given the challenge Z which is either $f(r)$ or R. Based on Z, $\mathcal{A}$ learns up to $(n - \ell)$ bits of r, one at a time, and has to distinguish between $Z = f(r)$ and $Z = R$.
5. **Statistical RF:** (This is the extension of resilient functions [12,3] to the statistical model, also defined in Section 2.)
   $\mathcal{A}$ chooses any set $L \in \{^n_\ell\}$ and any $w \in \{0,1\}^{n-\ell}$. $\mathcal{A}$ requests that $[r]_{\bar{L}}$ is set to w. The remaining ℓ bits of r in L are set at random. $\mathcal{A}$ is then given a challenge Z which is either $f(r)$ or R, and has to distinguish between these two cases. (Put another way, $\mathcal{A}$ loses if for any $L \in \{^n_\ell\}$ and any $w \in \{0,1\}^{n-\ell}$, the distribution induced by $f(r)$ when $[r]_{\bar{L}} = w$ and the other ℓ bits of r chosen at random, is statistically close to the uniform on $\{0,1\}^k$.)

6. **Almost-Perfect** RF (APRF): (This is the notion of [19].)
$\mathcal{A}$ chooses any set $L \in \binom{n}{\ell}$ and any $w \in \{0,1\}^{n-\ell}$. $\mathcal{A}$ requests that $[r]_{\bar{L}}$ is set to w. The remaining ℓ bits of r in L are set at random and $Z = f(r)$ is evaluated. $\mathcal{A}$ wins if there exists $y \in \{0,1\}^k$ such that $\Pr(Z = y)$ in this experiment does not lie within $2^{-k}(1 \pm \epsilon)$, where ϵ is negligible.[5]

Note that for each of the first five notions above, we can define the "error parameter" ϵ as the advantage of the adversary in the given experiment (for the sixth notion, ϵ is already explicit).

Let us begin by discussing the notion we started with — adaptive ERF. First, it might seem initially like the notion of weakly adaptive ERF is all that we need. Unfortunately, we have seen that to construct adaptive AONT's from ERF's via Lemma 2, we need strong adaptive ERF's. Second, the "algorithmic" adaptive behavior of the adversary is difficult to deal with, so it seems easier to deal with a more combinatorial notion. For example, one might hope that a statistical RF is by itself an adaptive ERF (notice, such RF is clearly a *static* ERF), and then concentrate on constructing statistical RF's. Unfortunately, this hope is false, as stated in the following lemma.

Lemma 5. *There are functions which are statistical* RF *but not statistical adaptive (or even strongly static!)* ERF.

Proof Sketch. Let n be the input size. Let f' be an statistical RF from $n' = \frac{n}{2}$ bits to $k' = \frac{n}{6}$ bits such that $\ell' = \frac{n}{4}$. Such functions exist, as we prove in Section 4.2.

Define f as follows: on an n-bit input string r, break r into two parts r_1 and r_2 both of length $\frac{n}{2}$. Apply f' to r_1 to get a string s of length $\frac{n}{6}$. Now divide s into $\frac{n}{6(\log n - 1)}$ blocks of size $\log \frac{n}{2}$, which can be interpreted as a random subset S from $\{1, \ldots, \frac{n}{2}\}$ with $\frac{n}{6(\log n - 1)}$ elements. Let $\bigoplus S$ be the parity of the bits in $[r_2]_S$. The output of f is the pair $\langle s, \bigoplus S \rangle$. Thus $k \approx \frac{n}{6}$.

Now let $\ell = n - \frac{n}{6(\log n - 1)}$. Clearly, an adversary who sees the challenge first, can (non-adaptively) read the bits $[r_2]_S$ and check the parity (giving him advantage at least $1/2$ over the random string). Thus, f is not an adaptively secure ERF. On the other hand, an adversary who can fix only $(n - \ell) \approx n/6 \log(n)$ input bits can still not learn anything about the output of f' and thus is unlikely to know the value of *all* the bits in S. Such an adversary will always have negligible advantage. Hence f is a statistical RF. □

Since the opposite direction (from adaptive ERF's to statistical RF's) is obviously false as well, we ask if some notion actually can *simultaneously* achieve both adaptive security for ERF, and statistical security for RF. Fortunately, it turns that by satisfying the stronger condition of an almost-perfect resilient function (APRF) [19], one obtains an adaptive ERF. Since APRF's will play such a crucial role in our study, we give a separate, more formal definition.

---

[5] Note that in [19] the error parameter was measured slightly differently: they define ϵ as the maximum absolute deviation. Our convention makes sense in the cryptographic setting since then the adversary's advantage at distinguishing $f(r)$ from random in any of the above experiments is comparable ϵ, as opposed to $\epsilon 2^k$.

Definition 7. *A polynomial time function* $f : \{0,1\}^n \rightarrow \{0,1\}^k$ *is* ℓ-APRF *(almost-perfect resilient function) if for any* $L \in \{\binom{n}{\ell}\}$, *for any assignment* $w \in \{0,1\}^{n-\ell}$ *to the positions not in* L, *for a randomly chosen* $r \in \{0,1\}^n$ *and for some negligible* $\epsilon = negl(n)$, *we have:*

$$\Pr(f(r) = y \mid [r]_{\bar{L}} = w) = (1 \pm \epsilon)2^{-k} \qquad (5)$$

While it is obvious that any APRF is a statistical RF (by summing over 2^k values of y), the fact that it is also an adaptive ERF is less clear (especially considering Lemma 5), and is shown below.

Theorem 4. *If* f *is an* APRF, *then* f *is a statistical adaptive* ERF.

Proof. By assumption, f is an ℓ-APRF with error ϵ: for every set $L \in \{\binom{n}{\ell}\}$ and every assignment w to the variables not in L, Equation (5) above holds when r is chosen at random. Now suppose that we have an adaptive adversary $\mathcal{A}$ who, given either $Z = f(r)$ or $Z = R$ and (limited) access to r, can distinguish between the two cases with advantage ϵ'. We will show that $\epsilon' \leq \epsilon$.

At first glance, this may appear trivial: It is tempting to attempt to prove it by conditioning on the adversary's view at the end of the experiment, and concluding that there must be *some* subset L and appropriate fixing w which always leads to a good chance of distinguishing. However, this argument fails since the adversary $\mathcal{A}$ may base his choice of L on the particular challenge he receives, and on the bits he considers.

So we use a more sophisticated argument, although based on a similar intuition. First, we can assume w.l.o.g. that the adversary $\mathcal{A}$ is deterministic, because there is some setting of his random coins conditioned on which he will distinguish with advantage at least ϵ', and so we may as well assume that he always uses those coins.

Following the intuition above, we consider the adversary's view at the end of the experiment, just before he outputs his answer. This view consists of two components: the input challenge Z and the $(n - \ell)$ observed bits $w = w_1, \dots, w_{n-\ell}$ (which equal $[r]_{\bar{L}}$ for some set L of size at least ℓ). Significantly, L need not be explicitly part of the view: since $\mathcal{A}$ is deterministic, L is a function of Z and w.

Denote by $\mathsf{View}_{\mathcal{A}}^{(Z)}$ the view of $\mathcal{A}$ on challenge Z. When $Z = R$, it is easy to evaluate the probability that $\mathcal{A}$ will get a given view. Since the values $r \in \{0,1\}^n$ and $R \in \{0,1\}^k$ are independent, we have

$$\Pr\left[\mathsf{View}_{\mathcal{A}}^{(R)} = (y, w)\right] = 2^{-(n-\ell+k)}$$

On the other hand, when $Z = f(r)$, we have to be careful. If L is the subset corresponding to $\mathcal{A}$'s choices on view (y, w), then we do indeed have:

$$\Pr\left[\mathsf{View}_{\mathcal{A}}^{(f(r))} = (y, w)\right] = \Pr\left[f(r) = y \wedge [r]_{\bar{L}} = w\right]$$

This last equality holds even though the choice of L may depend on y. Indeed, $\mathcal{A}$ is deterministic and so he will always choose the subset L when $[r]_{\bar{L}} = w$, regardless of the other values in r. *Thus, we can in some sense remove the adversary from the discussion entirely.* Now this last probability can be approximated by conditioning and using Equation (5):

$$\Pr\left[f(r) = y \wedge [r]_{\bar{L}} = w\right] = \Pr\left[f(r) = y \mid w = [r]_{\bar{L}}\right] \Pr\left[w = [r]_{\bar{L}}\right]$$
$$= (1 \pm \epsilon)2^{-k} \cdot 2^{-(n-\ell)}$$
$$= (1 \pm \epsilon)2^{-(n-\ell+k)}$$

We can now explicitly compute the adversary's probability of success in each of the two experiments we are comparing. Let $A(y, w) = 1$ if $\mathcal{A}$ accepts on view (y, w) and 0 otherwise. Then:

$$\epsilon' = \left|\Pr\left[\mathcal{A}^r(f(r)) = 1\right] - \Pr\left[\mathcal{A}^r(R) = 1\right]\right|$$
$$= \left|\sum_{y,w}\left(\Pr\left[\text{View}_{\mathcal{A}}^{(f(r))} = (y, w)\right] - \Pr\left[\text{View}_{\mathcal{A}}^{(R)} = (y, w)\right]\right) \cdot A(y, w)\right|$$
$$\leq \sum_{y,w}\left|(1 \pm \epsilon)2^{-(n-\ell+k)} - 2^{-(n-\ell+k)}\right| \leq \epsilon$$

Thus $\epsilon' \leq \epsilon$, and so f is a statistical adaptive ERF.   □

CLASSIFICATION OF RESILIENT FUNCTIONS. In fact, we can completely relate all the six notions of resilient functions that we introduced:

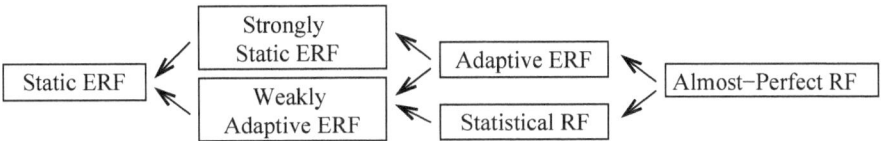

This diagram is complete: if there is no path from notion A to notion B, then there is a function which satisfies A but not B. We notice that except for the two proofs above, only one non-trivial proof is needed in order to complete the diagram: the separation between weakly adaptive ERF's and static ERF's (other implications and separations are easy exercises). However, this separation follows from the static construction of Canetti et al. [10], which, as we mentioned, need not yield a weakly adaptive ERF.

We also remark that while the diagram above is useful from a structural point of view, in the next section we show how to build APRF's — the strongest of the above notions — achieving $l \approx k$, which is nearly optimal even for *static* ERF's — the weakest of the above notions. Thus, all the above notions are almost the "same" in terms of the optimal parameters they achieve (which are also substantially better than those possible in the perfect setting).

4.2   Obtaining Nearly Optimal Almost-Resilient Functions

Given the discussion of the previous section, it is natural to try to construct good APRF's. These were first defined and studied by Kurosawa et al. [19]. Using techniques from coding theory, they construct[6] ℓ-APRF such that $\ell \geq \frac{n+k}{2} + 2\log\left(\frac{1}{\epsilon}\right)$. Although this beats the lower bound on perfect ERF of [15,4], it is very far from the trivial lower bound $\ell \geq k$, especially when $k = o(n)$. Thus, it is natural to ask whether this is a fundamental limitation on APRF's, or whether indeed one can approach this simplistic lower bound.

As a first step, we can show that if f is picked at random from all the functions from $\{0,1\}^n$ to $\{0,1\}^k$, it is very likely to be a good APRF (we omit the proof since we subsume it later). However, this result is of little practical value: storing such a function requires $k \cdot 2^n$ bits. Instead, we replace the random function with a function from a t-wise independent hash family [11] for t roughly on the order of n. Functions in some such families (e.g., the set of all degree $t-1$ polynomials over the field $GF(2^n)$) require as little as tn bits of storage, and are easy to evaluate.

Using tail-bounds for t-wise independent random variables, one can show that with very high probability we will obtain a good APRF:

Theorem 5. *Fix any n, ℓ and ϵ. Let $\mathcal{F}$ be a family of t-wise independent functions from n bits to k bits, where $t = n/\log n$ and*

$$k = \ell - 2\log\left(\frac{1}{\epsilon}\right) - O(\log n)$$

Then with probability at least $(1 - 2^{-n})$ a random function f sampled from $\mathcal{F}$ will be an ℓ-APRF (and hence adaptive ℓ-ERF and statistical ℓ-RF) with error ϵ.

Corollary 1. *For any $\ell = \omega(\log n)$, there exists an efficient statistical adaptive ℓ-ERF $f : \{0,1\}^n \to \{0,1\}^k$ with $k = \ell - o(\ell)$.*

The proof of Theorem 5 uses the following lemma, which is used (implicitly) in the constructions of deterministic extractors of [27]. Recall that a distribution X over $\{0,1\}^n$ has *min-entropy* m, if for all x, $\Pr(X = x) \leq 2^{-m}$.

Lemma 6. *Let $\mathcal{F}$ be a family of t-wise independent functions (for even $t \geq 8$) from n to k bits, let X be a distribution over $\{0,1\}^n$ of min-entropy m, and let $y \in \{0,1\}^k$. Assume for some $\alpha > 0$*

$$k \leq m - \left(2\log\frac{1}{\epsilon} + \log t + 2\alpha\right). \qquad (6)$$

Let f be chosen at random from $\mathcal{F}$ and x be chosen according to X. Then

$$\Pr_{f \in \mathcal{F}}\left(\left|\Pr_x(f(x) = y) - \frac{1}{2^k}\right| \geq \epsilon \cdot \frac{1}{2^k}\right) \leq 2^{-\alpha t} \qquad (7)$$

---

[6] This result *looks* (but is not) different from the one stated in [19] since we measure ϵ differently.

In other words, for any $y \in \{0,1\}^k$, if f is chosen from $\mathcal{F}$ then with overwhelming probability we have that the probability that $f(X) = y$ is $\frac{1}{2^k}(1 \pm \epsilon)$.

Theorem 5 follows trivially from this lemma. Indeed, set $\alpha = 3 \log n$, $t = n/\log n$. Notice that for any $L \in \{^n_\ell\}$ and any setting w of bits not in L, the random variable $X = \langle r \mid [r]_{\bar{L}} = w \rangle$ has min-entropy $m = \ell$. Then k given in Theorem 5 indeed satisfies Equation (6). Now we apply Lemma 6 and take the union bound in Equation (7) over all possible fixings of some $(n - \ell)$ input bits, and over all $y \in \{0,1\}^n$. Overall, there are at most $\binom{n}{\ell} 2^{n-\ell} 2^k \leq 2^{2n}$ terms in the union bound, and each is less than $2^{-\alpha t} = 2^{-3n}$, finishing the proof of Theorem 5.

For completeness, we give a simple proof of Lemma 6. We will make use of the following "tail inequality" for sums of t-wise independent random variables proven by Bellare and Rompel [2]. There they estimate $\Pr[\|Y - \mathsf{Exp}[Y]\| > A]$, where Y is a sum of t-wise independent variables. We will only be interested in $A = \epsilon \cdot \mathsf{Exp}[Y]$, where $\epsilon \leq 1$. In this case, tracing the proof of Lemma 2.3 (and Lemma A.5 that is used to prove it) of [2], we get the following:

Theorem 6 ([2]). *Let t be an even integer, and assume $Y_1, \ldots, Y_N$ are t-wise independent random variables in the interval $[0,1]$. Let $Y = Y_1 + \ldots + Y_N$, $\mu = \mathsf{Exp}[Y]$ and $\epsilon < 1$. Then*

$$\Pr(|Y - \mu| \geq \epsilon\mu) \leq C_t \cdot \left(\frac{t}{\epsilon^2 \mu}\right)^{t/2} \tag{8}$$

where the constant $C_t < 3$ and in fact $C_t < 1$ for $t \geq 8$.

Now we can prove Lemma 6:

Proof. Let p_x denote the probability that $X = x$, and let q denote the random variable (only over the choice of f) which equals to the probability (over the choice of x given f) that $f(x) = y$, i.e.

$$q = \sum_{x \in \{0,1\}^n} p_x \cdot I_{\{f(x)=y\}}$$

where $I_{\{f(x)=y\}}$ is an indicator variable which is 1 if $f(x) = y$ and 0 otherwise. Since for any x the value of $f(x)$ is uniform over $\{0,1\}^k$, we get that $\mathsf{Exp}_f[I_{\{f(x)=y\}}] = 2^{-k}$, and thus $\mathsf{Exp}_f[q] = 2^{-k}$. Notice also that the variables $I_{\{f(x)=y\}}$ are t-wise independent, since f is chosen at random from a family of t-wise independent functions. And finally notice that since X has min-entropy m, we have that all $p_x \leq 2^{-m}$.

Thus, if we let $Q_x = 2^m \cdot p_x \cdot I_{\{f(x)=y\}}$, and $Q = \sum_{x \in \{0,1\}^n} Q_x = 2^m q$, we get that the variables Q_x are t-wise independent, all reside in the interval $[0,1]$, and $\mathsf{Exp}[Q] = 2^m \mathsf{Exp}[q] = 2^{m-k}$. Now we can apply the tail inequality given in Theorem 6 and obtain:

$$\Pr_f \left[\left| q - \frac{1}{2^k} \right| \geq \epsilon \cdot \frac{1}{2^k} \right] = \Pr_f \left[|Q - 2^{m-k}| \geq \epsilon \cdot 2^{m-k} \right]$$

$$\leq \left(\frac{t}{\epsilon^2 \cdot 2^{m-k}} \right)^{t/2} = \left(\frac{1}{2^{m-k-2\log\frac{1}{\epsilon}-\log t}} \right)^{t/2}$$

$$\leq 2^{-\alpha t}$$

where the last inequality follows from Equation (6).   □

4.3   Adaptively Secure AONT

We already remarked that that the construction of optimal adaptive statistical ERF's implies the construction of adaptive computational ERF's. Combined with Lemma 2, we get optimal constructions of AONT's as well. We notice also that the public part of these AONT construction is k. In the statistical setting, where we achieved optimal $\ell = k+o(k)$, we could then combine the public and the secret part of the AONT to obtain a *secret-only* adaptive AONT with $\ell = 2k + o(k)$. One may wonder if there exist statistical secret-only AONT's with $\ell = k + o(k)$, which would be optimal as well. Using our construction of almost-perfect resilient functions, we give an affirmative answer to this question. Our construction is not efficient, but the existential result is interesting because it was not known even in the static setting.

Lemma 7. *Ignoring the issue of efficiency, there exist adaptive statistical secret-only ℓ-AONT $T : \{0,1\}^k \to \{0,1\}^n$ with $\ell = k + o(k)$.*

Proof. Recall, Lemma 1 used an inverse of a perfect RF (or ERF, which is the same) to construct perfect secret-only AONT. We now show that the same construction can be made to work in the statistical setting provided we use APRF rather than weaker statistical RF. In particular, let $f : \{0,1\}^n \to \{0,1\}^k$ be an ℓ-APRF. We know that we can achieve $\ell = k + o(k)$. We define $T(x)$ to be a random $r \in \{0,1\}^n$ such that $f(r) = x$. (This is well-defined since APRF's are surjective.)

Now take any distingusher $\mathcal{A}$, any $x \in \{0,1\}^k$ and any possible view of $\mathcal{A}$ having oracle access to $T(x) = r$. Since we can assume that $\mathcal{A}$ is deterministic, this view can be specified by the $(n-\ell)$ values w that $\mathcal{A}$ read from r (in particular, the subset L is also determined from w). Now, we use Bayes law to estimate $\Pr(\text{View}_{\mathcal{A}}^{(T(x))} = w)$. Notice, since $r = T(x)$ is a random preimage of x, we could assume that r was chosen at random from $\{0,1\}^n$, and use conditioning on $f(r) = x$. This gives us:

$$\Pr(\text{View}_{\mathcal{A}}^{(T(x))} = w) = \Pr(\text{View}_{\mathcal{A}}^{(r)} = w \mid f(r) = x) = \Pr([r]_{\bar{L}} = w \mid f(r) = x)$$

$$= \frac{\Pr(f(r) = x \mid [r]_{\bar{L}} = w) \cdot \Pr([r]_{\bar{L}} = w)}{\Pr(f(r) = x)}$$

$$= \frac{(1 \pm \epsilon) \cdot 2^{-k} \cdot 2^{\ell-n}}{(1 \pm \epsilon) \cdot 2^{-k}} = (1 \pm 2\epsilon) \cdot 2^{\ell-n}$$

Notice that this bound is independent on $\mathcal{A}$, x and w. Hence, for any x_0, x_1 and any adversary $\mathcal{A}$, $\text{View}_{\mathcal{A}}^{(T(x_0))}$ and $\text{View}_{\mathcal{A}}^{(T(x_1))}$ are within statistical distance 4ϵ from each other, implying that T is an adaptive statistical AONT. $\qquad\square$

Acknowledgments

It is our pleasure to thank Dan Spielman for help in simplifying the proof of Theorem 2, and Salil Vadhan for extremely helpful discussions and for showing us an early version of [27].

References

1. M. Bellare and A. Boldyreva. The Security of Chaffing and Winnowing. In *Proc. of Asiacrypt*, 2000.
2. M. Bellare, J. Rompel. Randomness-Efficient Oblivious Sampling. In *Proc. of 35th FOCS*, pp. 276–287, 1994.
3. C. Benett, G. Brassard, J. Robert. Privacy Amplification by public discussion. In *SIAM J. on Computing*, pp. 17(2):210–229, 1988.
4. J. Bierbrauer, K. Gopalakrishnan, D. Stinson. Orthogonal Arrays, resilient functions, error-correcting codes and linear programming bounds. In *SIAM J. of Discrete Math*, 9:424–452, 1996.
5. J. Bierbrauer, H. Schellwat. Almost Independent and Weakly Biased Arrys: Efficient Constructions and Cryptologic Applications. In *Proc. of CRYPTO*, pp. 531–543, 2000.
6. G. R. Blakley and C. Meadows. Security of Ramp Schemes. In *Proc. of CRYPTO*, pp. 242–268, 1984.
7. M. Blaze. High Bandwidth Encryption with low-bandwidth smartcards. In *Fast Software Encryption*, pp. 33–40, 1996.
8. C. Blundo, A. De Santis, U. Vaccaro. Efficient Sharing of Many Secrets. In *Proc. of STACS*, LNCS 665, pp. 692-703, 1993.
9. V. Boyko. On the Security Properties of the OAEP as an All-or-Nothing Transform. In *Proc. of Crypto*, pp. 503–518, 1999.
10. R. Canetti, Y. Dodis, S. Halevi, E. Kushilevitz and A. Sahai. Exposure-Resilient Functions and All-Or-Nothing Transforms. In *Proc. of EuroCrypt*, 2000.
11. L. Carter and M. Wegman. Universal classes of hash functions. *JCSS*, vol. 18, pp. 143–154, 1979.
12. B. Chor, J. Friedman, O. Goldreich, J. Håstad, S. Rudich, R. Smolensky. The Bit Extraction Problem or t-resilient Functions. In *Proc. of FOCS*, pp. 396–407, 1985.
13. Y. Dodis. Exposure-Resilient Cryptography. Ph.D. Thesis., MIT, 2000.
14. A. Desai. The security of All-or-Nothing Encryption: Protecting Against Exhaustive Key Search In *Proc. of CRYPTO*, pp. 359–375, 2000.
15. J. Friedman. On the Bit Extraction Problem In *Proc. of FOCS*, pp. 314–319, 1992.
16. O. Goldreich. Foundations of Cryptography (Fragments of a Book). URL: `http://www.wisdom.weizmann.ac.il/home/oded/public_html/frag.html`
17. W.-A. Jackson and K. Martin. A Combinatorial Interpretation of Ramp Schemes *Australasian Journal of Combinatorics*, **14** (1996), 51–60.

18. M. Jakobsson, J. Stern, M. Yung. Scramble All, Encrypt Small. In *Proc. of Fast Software Encryption*, pp. 95–111, 1999.

19. K. Kurosawa, T. Johansson and D. Stinson. Almost k-wise independent sample spaces and their cryptologic applications. Submitted to *J. of Cryptology*, preliminary version appeared in *Proc. of EuroCrypt*, pp. 409–421, 1997.

20. S. Matyas, M. Peyravian and A. Roginsky. Encryption of Long Blocks Using a Short-Block Encryption Procedure. Available at `http://grouper.ieee.org/groups/1363/P1363a/LongBlock.html`.

21. W. Ogata and K. Kurosawa. Some Basic Properties of General Nonperfect Secret Sharing Schemes. *Journal of Universal Computer Science*, Vol.**4**, No. 8 (1998), pp. 690–704.

22. L. H. Ozarow and A. D. Wyner. Wire-Tap Channel II In *Proc. of EUROCRYPT*, pp. 33–50, 1984.

23. R. Rivest. All-or-Nothing Encryption and the Package Transform. In *Fast Software Encryption, LNCS*, 1267:210–218, 1997.

24. R. Rivest. Chaffing and Winnowing: Confidentiality without Encryption. *CryptoBytes (RSA Laboratories)*, 4(1):12–17, 1998.

25. A. Shamir. How to share a secret. In *Communic. of the ACM*, 22:612-613, 1979.

26. S. U. Shin, K. H. Rhee. Hash functions and the MAC using all-or-nothing property. In *Proc. of Public Key Cryptography, LNCS*, 1560:263–275, 1999.

27. L. Trevisan and S. Vadhan. Extracting randomness from samplable distributions In *Proc. of FOCS*, 2000.

28. U. Vazirani. Towards a Strong Communication Complexity Theory or Generating Quasi-Random Sequences from Two Communicating Semi-Random Sources. In *Combinatorica*, 7(4):375–392, 1987.

Cryptanalysis of Reduced-Round MISTY

Ulrich Kühn

Dresdner Bank AG
Group Information Technology
RS Research
D-60301 Frankfurt
Germany
Ulrich.Kuehn@dresdner-bank.com

Abstract. The block ciphers MISTY1 and MISTY2 proposed by Matsui are based on the principle of provable security against differential and linear cryptanalysis. This paper presents attacks on reduced-round variants of both ciphers, without as well as with the key-dependent linear functions FL. The attacks employ collision-searching techniques and impossible differentials. KASUMI, a MISTY variant to be used in next generation cellular phones, can be attacked with the latter method faster than brute force when reduced to six sounds.

1 Introduction

The MISTY algorithms proposed by Matsui [8] are designed to be resistant against differential [3] and linear [7] cryptanalysis. One design criterion is that no single differential or linear characteristic with a usable probability does hold for the cipher. An additional feature is the use of key-dependent linear functions which were introduced to counter other than differential and linear attacks.

Previous attacks by Tanaka, Hisamatsu and Kaneko [12] on MISTY1 and by Sugita [10] on MISTY2 employ higher order differentials against 5-round variants without the linear FL functions. A cryptographic weakness of the round construction of MISTY2 was pointed out by Sakurai and Zheng [9].

In this paper we present attacks on reduced-round variants of MISTY1 and MISTY2, both without and with the key-dependent linear functions FL. The round function involves a huge amount of keying material, so it is one purpose of this paper to point out properties of the round function that allow to use divide-and-conquer techniques on the subkeys in order to improve basic attacks which make use of impossible differentials [2, 5] and collision-searching [1]; the latter technique is extended by using multiple permutations. Furthermore reduced-round KASUMI, a MISTY variant to be used in next generation cellular phones, is attacked with impossible differentials. Table 1 shows a summary of the attacks.

This paper is organised as follows. The MISTY algorithms are described in Section 2; properties of the key scheduling and the round function that are used here are explained in Section 3; the new attacks on MISTY1 resp. MISTY2 are described in Section 4 resp. 5. A comparison to KASUMI is made in Section 6. Conclusions are drawn in Section 7.

B. Pfitzmann (Ed.): EUROCRYPT 2001, LNCS 2045, pp. 325–339, 2001.

Table 1. Summary of attacks on MISTY variants.

Cipher	FL functions	Rounds	Complexity [data]	[time]	Comments
MISTY1	–	5	11×2^7	2^{17}	[12] (previously known)
	–	5	2^6	2^{38}	[10, 11] (previously known)
	–	6	2^{39}	2^{106}	impossible differential (new)
	–	6	2^{54}	2^{61}	impossible differential (new)
	✓	4	2^{23}	$2^{90.4}$	impossible differential (new)
	✓	4	2^{38}	2^{62}	impossible differential (new)
	✓	4	2^{20}	2^{89}	collision-search (new)
	✓	4	2^{28}	2^{76}	collision-search (new)
MISTY2	–	5	2^7	2^{39}	[10, 11] (previously known)
	✓	5	2^{23}	2^{90}	impossible differential (new)
	✓	5	2^{38}	2^{62}	impossible differential (new)
	✓	5	2^{20}	2^{89}	collision-search (new)
	✓	5	2^{28}	2^{76}	collision-search (new)
KASUMI	✓	6	2^{55}	2^{100}	impossible differential (new)

2   Description of MISTY

The MISTY algorithms [8] are symmetric block ciphers with a block size of 64 bits and a key size of 128 bits. There are two flavors called MISTY1 and MISTY2, which differ by their global structure (see Figure 1). MISTY1 is a Feistel network with additional key-dependent linear functions FL placed in the data path before every second round. MISTY2 has a different structure that allows parallel execution of round functions during encryption. The FL functions are applied in MISTY2 to both halves of the data before every fourth round and also in every second round just before XORing the right to the left half of the data. In both ciphers the linear functions are also used as an output transformation.

MISTY has a recursive structure, that is, the round function consists of a network with a smaller block size using the function FI that itself is again a smaller network; the structure of both the round function FO and the function FI is that of MISTY2. Figure 2 shows FO, FI and FL in a representation that is equivalent to the original description [8]. This equivalent description[1] is the result moving the mixing of the leftmost seven bits of each KI_{ij} in each FI (as given in the specification [8]) out of FI and to the end of its superstructure FO; this is possible because these key bits do not affect any S-box inside the instance of FI where they are inserted. Due to the recursive structure a huge amount of keying material is involved in each round, i.e. 112 bits for FO in

---
[1] For another equivalent description of MISTY's round function see [12].

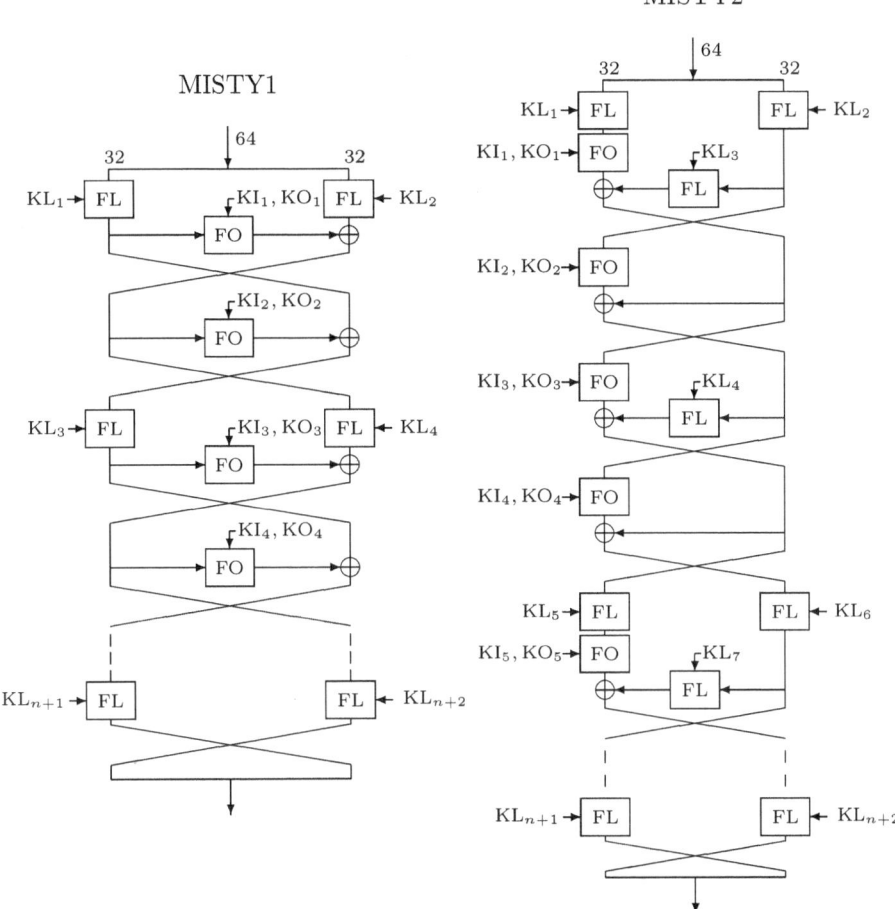

Fig. 1. Global structure of MISTY1 and MISTY2.

the original description; the equivalent description has a key size of 107 bits. Additional subkey bits are used if the round contains FL functions. The ciphers are proposed with 8 (MISTY1) resp. 12 (MISTY2) rounds.

The key scheduling takes as input a 128 bit key consisting of 16 bit values $K_1, \ldots, K_8$ and computes additional 16 bit values $K'_t = \mathrm{FI}_{K_{t+1}}(K_t)$, $1 \le t \le 8$ where $K_9 = K_1$. The subkeys of each round are (i is identified with $i - 8$ for $i > 8$):

Subkey	KO_{i1}	KO_{i2}	KO_{i3}	KO_{i4}	KI_{i1}	KI_{i2}	KI_{i3}	KL_i
Value	K_i	K_{i+2}	K_{i+7}	K_{i+4}	K'_{i+5}	K'_{i+1}	K'_{i+3}	$(K_{\frac{i+1}{2}}\|\|K'_{\frac{i+1}{2}+6})$ (odd i)
								$(K'_{\frac{i}{2}+2}\|\|K_{\frac{i}{2}+4})$ (even i)

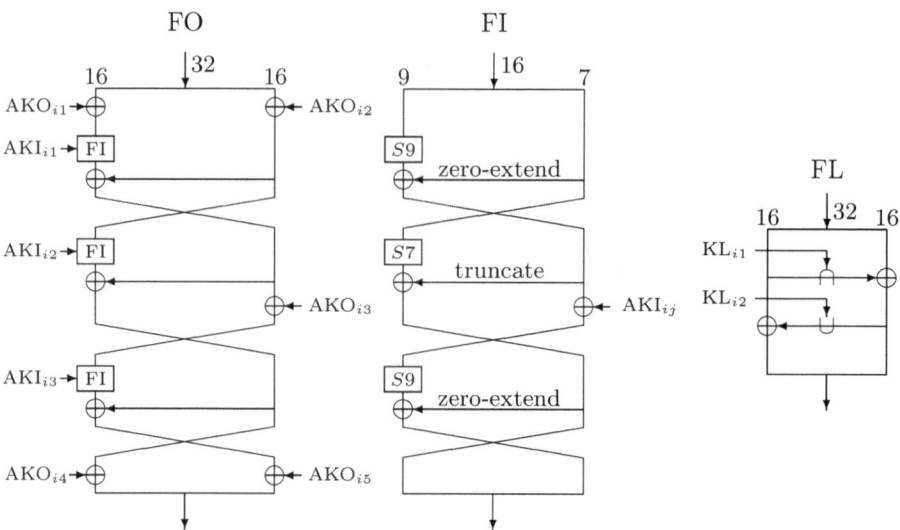

Fig. 2. The functions FO and FI in a form equivalent to the original specification which eliminates the left 7 bits of the key to FI. $S7$ and $S9$ are bijective 7×7 resp. 9×9 S-boxes; in FL the operators $\cap$ resp. $\cup$ denote the bitwise AND resp. OR.

Given $\text{KO}_i = (\text{KO}_{i1}, \ldots, \text{KO}_{i4})$, $\text{KI}_i = (\text{KI}_{i1}, \ldots, \text{KI}_{i3})$, then AKO_{ij} and AKI_{ij} of our equivalent description relate to the original subkeys as follows. Let $||$ denote the concatenation of bitstrings and $[x]_{i..j}$ the selection of the bits $i..j$ from x where bit 0 is the rightmost bit. Let KI'_{ij} denote the 16 bits $[\text{KI}_{ij}]_{15..9}||00||[\text{KI}_{ij}]_{15..9}$. Then the actual subkeys are

$$\begin{aligned}
\text{AKO}_{ik} &= \text{KO}_{ik}, \quad \text{with } 1 \le k \le 2 \\
\text{AKO}_{i3} &= \text{KO}_{i2} \oplus \text{KO}_{i3} \oplus \text{KI}'_{i1} \\
\text{AKO}_{i4} &= \text{KO}_{i2} \oplus \text{KO}_{i4} \oplus \text{KI}'_{i1} \oplus \text{KI}'_{i2} \\
\text{AKO}_{i5} &= \text{KO}_{i2} \oplus \text{KI}'_{i1} \oplus \text{KI}'_{i2} \oplus \text{KI}'_{i3} \\
\text{AKI}_{ik} &= [\text{KI}_{ik}]_{8..0}, \quad \text{with } 1 \le k \le 3
\end{aligned} \tag{1}$$

Notation. Throughout this paper all differences are taken as XOR of the appropriate values. Let L_i resp. R_i denote the left resp. right half of the input to round i, X_i the input to the round function FO, and Z_i its output; so L_1 resp. R_1 denotes the left resp. right half of the plaintext data. If round i uses FL in its data path (for example every odd round in MISTY1) let X_i resp. Y_i denote the left resp. right half of the data after the transformation through FL, and set $X_i = L_i$, $Y_i = R_i$ otherwise. For MISTY2 let $\tilde{Y}_i$ denote the possibly transformed value of Y_i that is XORed to Z_i to form the half of the round's output that becomes R_{i+1} after the swap.

3   Observations on the Key Scheduling and the Round Function

Key Scheduling. The key scheduling is designed such that every round is affected by all key bits. This causes major problems in terms of complexity when exhaustively guessing the subkey of one round with a distinguisher for the other rounds, but it also allows to recover the whole key with reasonable effort once a large part of one round subkey is known.

For example consider the first round's subkeys AKO_{11}, AKO_{12}, AKO_{13} and AKI_{11}, AKI_{12}, AKI_{13}. By equation 1 (and the key scheduling table in Section 2) the 16 bits of key words K_1 and K_3 are known. AKI_{12} resp. AKI_{13} provides a 9 bit condition for K_2 resp. K_4 and K_5. After guessing the 7 bits of KI'_{11} in AKO_{13} there is – knowing AKI_{11} – a 16 bit condition for K_6 and K_7; also the word K_8 is known from AKO_{13}. Using a factor of 2 for the 8 computations of FI in the key schedule the total complexity of exhaustive search is about $2 \cdot 2^{128-32} \cdot 2^{-9} \cdot 2^{-9} \cdot 2^{-32} = 2^{47}$ encryptions using two or three known plaintexts and corresponding ciphertexts.

Round Function in Differential and Collision-Searching Attacks. The subkeys AKO_{i4} and AKO_{i5} are invisible in our attacks as they introduce fixed constants after all non-linearities when FO is applied in forward direction. The following properties of FO allow divide-and-conquer techniques for the other 75 subkey bits at the cost of increased chosen plaintext or ciphertext requirements.

Property 1. In forward direction, consider FO in round i having an output XOR of the form (β, β) where β is a nonzero 16 bit value. Then the input and output XOR of the third instance of FI must be zero, so (AKO_{i3}, AKI_{i3}) does not influence the output XOR. The input XOR to FO must be (α_l, α_r) such that α_r cancels the output XOR of the first FI under key (AKO_{i1}, AKI_{i1}) when the input XOR is α_l from the given input values. The value of β is solely influenced by (AKO_{i2}, AKI_{i2}).

Property 2. In forward direction, consider inputs to FO in round i of the form (a_i, b) where the a_i are all different (thus forming a permutation in the notation of [1]) and b is a constant. Then the output of the second FI is a constant that depends on AKO_{i2} and AKI_{i2}; the input of the third FI is a permutation, namely the XOR of the output of the first FI and $b \oplus AKO_{i2} \oplus AKO_{i3}$. As long as $AKO_{i2} \oplus AKO_{i3}$ has the same value as for the unknown key, and AKO_{i1}, AKI_{i1} and AKI_{i3} are also correct, the output of FO is the same as for the correct subkey, up to XORing with a constant. So one can set $AKO_{i2} = 0$, $AKI_{i2} = 0$ in a first step, making sure that $AKO_{i2} \oplus AKO_{i3}$ has the correct value.

Directional Asymmetry. Due to the Feistel network, FO is used in MISTY1 in forward direction both for encrypting and decrypting data. But for MISTY2 this is not the case. In forward direction the output of the second FI does not affect the input of the third FI; this fact is inherently used in both properties explained above. In backward direction, the output of the first and second FI each affect the input of every subsequent FI, which makes analysis harder in this direction. This is the reason that for MISTY2 the attacks presented in this paper use the chosen ciphertext model of attack, as then FO in the first round can be used in forward direction.

4   Attacks on Reduced-Round MISTY1

In this section we present attacks on MISTY1; it is assumed that the final swap of MISTY is also present in the reduced variant. One attack finds the last round subkey of 6 rounds of MISTY1 without FL functions, other attacks find the last round subkey of MISTY1 reduced to 4 rounds with FL functions but without the final output transformation; these attacks break exactly half of the cipher.

4.1   Attacking MISTY1 without FL Functions

This attack is based on the generic 5-round impossible differential for Feistel networks with bijective round functions

$$(0, 0, \alpha_l, \alpha_r) \overset{5R}{\not\to} (0, 0, \alpha_l, \alpha_r), \quad (\alpha_l, \alpha_r) \neq (0, 0),$$

discovered by Knudsen [5]. The attack looks for differences $(\beta_l, \beta_r, \alpha_l, \alpha_r)$ after 6 rounds (including the final swap) and rules out all subkeys that can yield $(\alpha_l, \alpha_r) \to (\beta_l, \beta_r)$ from the given outputs, as that is impossible.

The basic attack uses a structure of 2^{32} chosen plaintexts $P_i = (x, y, a_i, b_i)$ with some fixed values x, y and (a_i, b_i) running through all 2^{32} values. After obtaining the corresponding ciphertexts (c_i, d_i, e_i, f_i) by encryption under the unknown key set up a list of values $w_i = (a_i, b_i) \oplus (e_i, f_i)$. For a pair i, j such that $w_i = w_j$ the input difference is $(0, 0, \alpha_l, \alpha_r)$ with $(\alpha_l, \alpha_r) = (a_i \oplus a_j, b_i \oplus b_j)$; the output difference after six rounds and the final swap is $(\beta_l, \beta_r, \alpha_l, \alpha_r)$ with $(\beta_l, \beta_r) = (c_i \oplus c_j, d_i \oplus d_j)$. Now check for all 75 bit subkeys $k = (AKO_{61}, AKI_{61}, \ldots, AKO_{63}, AKI_{63})$ if $FO_k((e_i, f_i)) \oplus FO_k((e_j, f_j)) = (\beta_l, \beta_r)$. Such a subkey is wrong while a correct guess never yields this difference.

About $\binom{2^{32}}{2} \cdot 2^{-32} \approx 2^{31}$ pairs $w_i = w_j$ are expected in a structure. A wrong key has a probability of about 2^{-32} to cause a given output XOR, so a fraction of $(1 - 2^{-32})^{2^{31}} = e^{-1/2}$ of the wrong subkeys are discarded. After repeating this basic step $75/\log_2(e^{1/2}) \approx 104$ times only the correct subkey is expected to survive.

This attack takes about $104 \cdot 2^{32} \approx 2^{39}$ chosen plaintexts. The time complexity is $2 \cdot 2^{31}$ computations of FO per guessed key and per structure, so the total complexity is about $\left(\sum_{i=0}^{103} (e^{-1/2})^i \right) \cdot 2^{75+32} \approx 2^{108.4}$ evaluations of FO which is equivalent to about 2^{106} encryptions of 6-round MISTY1 without FL functions; this is hardly a practical attack.

It is possible to reduce the amount of work at the cost of increased chosen plaintext requirements using Property 1 of FO (see Section 3). Using the above structure of plaintexts and their ciphertexts set up a list of (w_i, u_i) with $w_i = (a_i \oplus e_i, b_i \oplus f_i)$ and $u_i = c_i \oplus d_i$. Now only matches with $w_i = w_j$ and $u_i = u_j$ are of interest which yield $(c_i, d_i, e_i, f_i) \oplus (c_j, d_j, e_j, f_j) = (\beta, \beta, \alpha_l, \alpha_r)$. About $\binom{2^{32}}{2} \cdot 2^{-32} \cdot 2^{-16} \approx 2^{15}$ matches are expected with this form; these pairs are analysed. We determine subkeys that yield $(\alpha_l, \alpha_r) \to (\beta, \beta)$ via FO as follows (such a subkey cannot be the correct one). For each (AKO_{61}, AKI_{61}) we check if the first FI gives output XOR α_r from e_i, e_j. Then each guess of (AKO_{62}, AKI_{62}) is checked if it yields output XOR β by the second instance of FI. Each part results in about 2^9 candidates due to a 16 bit restriction.

Each structure is expected to discard about $2^{9+9} \cdot 2^{15} = 2^{33}$ 50 bit key candidates. Due to collisions a fraction of $1/e$ of the wrong keys is expected to remain after 2^{17} structures, but using in total $2^{17} \ln 2^{50} \approx 2^{17} \cdot 2^5$ structures, only the correct subkey remains. Thus about $2^{32} \cdot 2^{22} = 2^{54}$ chosen plaintexts with about $2^{15} \cdot 2^{22} = 2^{37}$ analysed pairs are needed. The time complexity of this part is $2 \cdot 2 \cdot 2^{25}$ evaluations of FI per analysed pair equivalent to about 2^{26} evaluations of FO. In total this is $2^{26} \cdot 2^{37} = 2^{63}$ evaluations of FO equivalent to about 2^{61} encryptions of 6-round MISTY1 without FL functions.

It remains to determine the 25 key bits (AKO_{63}, AKI_{63}) using the basic attack with $25/\log_2(e^{1/2}) \approx 35$ structures requiring $2^{38.2}$ chosen plaintexts which can be reused from previous structures. The time complexity of this second part is about $\left(\sum_{i=0}^{34} (e^{-1/2})^i \right) \cdot 2^{25+32} \approx 2^{58.4}$ evaluations of FO equivalent to about 2^{56} encryptions, which is much less than for the first part.

In total this attack needs about 2^{54} chosen plaintexts and time comparable to 2^{61} encryptions; about 2^{37} pairs are analysed.

4.2   Attacking MISTY1 with FL Functions

Here we show two attacks on 4-round MISTY1 where FL functions are present with the exception of the final output transformation. One attack uses an impossible differential, the other uses the collision-searching technique of Biham's attack on Ladder-DES [1]; in order to use Property 2 we extend this technique by employing multiple permutations.

Differential-Style Attack. The impossible differential used to attack MISTY1 without the FL functions does not work here. The problem occurs because FL changes nonzero differences.

Lemma 1. *The generic 5-round impossible differential for Feistel networks does not work for MISTY1 with the keyed linear functions FL.*

Proof. Assume that the differential starts at an odd-numbered round, i.e. a round where the FL functions are applied in, w.l.o.g. at round 1. The difference in the R_i is changed by FL for $i \in \{1, 3, 5\}$. For the impossible differential to work the differences in Y_3 and L_4 have to be the same, and thus the output XOR of the round function must be zero (which is impossible). But as the application of FL in general changes the differences, this cannot be assured. In the second case the differential starts at an even numbered round, i.e. a round where FL is not applied in; here the reasoning goes along the same lines. $\square$

The following 3-round impossible differential does work since FL cannot change zero differences. An input difference $(0, 0, \alpha_l, \alpha_r)$ to round 1 with some nonzero values α_l, α_r cannot yield an output difference of $(0, 0, \delta_l, \delta_r)$ before the swap in round 3 for a nonzero values δ_l, δ_r. After round 1 the difference is $(\beta_l, \beta_r, 0, 0)$ for some nonzero β_l, β_r as FL is bijective. Going backwards from round 3, the output difference of round 2 (before the swap) must have been $(\gamma_l, \gamma_r, 0, 0)$ with nonzero γ_l, γ_r which is only possible if $(\gamma_l, \gamma_r) = (\beta_l, \beta_r)$ and if FO causes a zero output difference which is impossible. Basically the same argument works when the differential starts at round 2, where the nonzero part of the difference is changed in round 3.

The attack works along similar lines as in Section 4.1 but uses structures of 2^{16} plaintexts $P_i = (x, y, a_i, b_i)$ where x, y are constant and the (a_i, b_i) all different. Let (c_i, d_i, e_i, f_i) denote the ciphertexts. For each structure all about 2^{31} pairs can be used which rule out a fraction of about $e^{-1/2}$ of the wrong keys. This attack requires about $75/\log_2(e^{1/2}) \approx 104$ structures (2^{23} chosen plaintexts) and about $\left(\sum_{i=0}^{74} (e^{-1/2})^i \right) \cdot 2^{75+16} \approx 2^{92.4}$ evaluations of FO comparable to $2^{90.4}$ encryptions.

We can improve this result by using Property 1. From the ciphertexts a list $u_i = c_i \oplus d_i$ is set up. So we can easily find those pairs which yield an output XOR $(\beta, \beta, \alpha_l, \alpha_r)$; their number is expected to be 2^{15} per structure. The analysis of the first part from the improved analysis in Section 4.1 can be used for finding $AKO_{41}, AKO_{42}, AKI_{41}$, and AKI_{42} requiring about $2^{17} \cdot 2^5 = 2^{22}$ structures (2^{38} chosen plaintexts) and $2^{22} \cdot 2^{16} \cdot 2^{26} = 2^{64}$ computations of FO comparable to 2^{62} encryptions. The second part for recovering AKO_{43} and AKI_{43} needs another $\left(\sum_{i=0}^{24} (e^{-1/2})^i \right) \cdot 2^{25+16} \approx 2^{42.4}$ computations of FO where the needed plaintexts/ciphertexts are reused. In total this attack needs 2^{38} chosen plaintexts and work of about 2^{62} encryptions.

Attack Using Collisions. Biham's attack on Ladder-DES [1] is also applicable to 4 round MISTY1 with FL functions, as these are bijective and thus cannot produce collisions. Consider a collection of chosen plaintexts of the form (x, y, a_i, b_i) with $i \in I$ for some index set I where x, y are constants and (a_i, b_i) different random values. Using the notation from [1] this property of the collection of $\{(a_i, b_i)\}_{i \in I}$ is called a *permutation*, that is, there can be no collision.

By the FL functions X_1 is a constant (x', y'), and Y_1 is a permutation, say $\{(a'_i, b'_i)\}_{i \in I}$. Z_1 is another fixed constant (x'', y'') derived from (x', y') by FO, so $L_2 = X_2$ is the permutation $\{(a'_i \oplus x'', b'_i \oplus y'')\}_{i \in I}$ while $R_2 = Y_2$ is constant. Then Z_2 is yet another permutation, and so is L_3. X_3 is still a permutation after the FL in round 3, as is Z_3, but $Z_3 \oplus Y_3$ behaves like a pseudo-random function.

The attack proceeds as follows. Prepare 2^{20} plaintexts $P_i = (x, y, a_i, b_i)$ and get their encryptions $C_i = (c_i, d_i, e_i, f_i)$ under the unknown key. For all guesses $k = (\text{AKO}_{41}, \text{AKI}_{41}, \ldots, \text{AKO}_{43}, \text{AKI}_{43})$ of the last round's FO 75 bit key decrypt the ciphertexts one round

$$w_i = \text{FO}_k((e_i, f_i)) \oplus (c_i, d_i).$$

If $w_i = w_j$ for some i, j then the key guess is wrong. The one-round decryption with a wrong key behaves like a pseudo-random function, so on average about 2^{16} decryptions are needed to eliminate a wrong guess; a correct guess never produces a collision. The attack needs 2^{20} chosen plaintexts and at most $2^{20} \cdot 2^{75} = 2^{95}$ evaluations of FO. But on average a wrong guess should be ruled out after about 2^{16} tries, so the workload is expected to be about $2^{16} \cdot 2^{75} = 2^{91}$ evaluations of FO equivalent to 2^{89} encryptions.

The probability of each wrong key guess to survive is the probability that all 2^{20} decrypted values are distinct. By the birthday paradox this probability is $\exp(-2^{20}(2^{20} - 1)/(2 \cdot 2^{32})) \approx \exp(-2^7) \approx 2^{-184}$, so for all keys the probability for a false guess to survive is 2^{-109}.

This attack can be improved using Property 2 at the cost of more chosen plaintexts. This version uses 2^{28} chosen plaintexts $P_i = (x, y, a_i, b_i)$ with constants x, y and all different (a_i, b_i). The ciphertexts $C_i = (c_i, d_i, e_i, f_i)$ are partitioned into sets B_t, $t \in \{0, \ldots, 2^{16} - 1\}$, such that $C_i \in B_t \Leftrightarrow f_i = t$. First, set $\text{AKO}_{42} = 0$, $\text{AKI}_{42} = 0$. For each guess $k = (\text{AKO}_{41}, \text{AKI}_{41}, k_{23}, \text{AKI}_{43})$ of 50 bits with k_{23} in the role of AKO_{43} and each B_t, $0 \le t \le 2^{16} - 1$ decrypt all $C_i \in B_t$ one round yielding $w_i^t = \text{FO}_{k_t}((e_i, f_i)) \oplus (c_i, d_i)$. If at one point $w_i^t = w_j^t$ then this key is discarded, and the procedure is started with the next guess. This takes at most $2^{50} \cdot 2^{28} = 2^{78}$ evaluations of FO comparable to 2^{76} encryptions to complete.

Once a correct k with $k_{23} = \text{AKO}_{42} \oplus \text{AKO}_{43}$ has been found the correct 25 bits $\text{AKO}_{42}, \text{AKI}_{42}$ with $\text{AKO}_{43} = k_{23} \oplus \text{AKO}_{42}$ have to be found. This time ciphertexts are used such that f_i varies. Here about 2^{20} ciphertexts from the collection of the 2^{28} should be sufficient to find the correct key. This requires work of at most $2^{20} \cdot 2^{25} = 2^{45}$ evaluations of FO equivalent to 2^{43} encryptions. The time and chosen plaintext requirements are dominated by the first part (2^{76} work and 2^{28} chosen plaintexts).

The first part uses several permutations, with the complication that the sum of the number of elements over all permutations is a constant. The probability of success can be estimated using methods from convexity theory [6]; we show that the case that all permutations are of equal size is the worst case. Let $m_t = |B_t|$ and $N = 2^{32}$. For each B_t a wrong key survives the test with probability $p_t = \exp(-\frac{m_t(m_t - 1)}{2N})$ with $0 \le |m_t| \le 2^{28}$ and $\sum_{t=0}^{2^{16}-1} |m_t| = 2^{28}$. The product of all p_t is the probability of failure to eliminate the wrong key.

Lemma 2. *The function* $p(m_0, \ldots, m_{2^{16}-1}) = \prod p_t$ *with* $m_i \in \{0, \ldots, M\}$, $\sum_{i=0}^{2^{16}-1} m_i = M > 0$ *has its maximum for* $m_0 = \cdots = m_{2^{16}-1} = M/2^{16}$.

Proof. Consider the function $f(m) = \exp(-\frac{m(m-1)}{2N})$; it is clear that $\ln(f(m))$ is a concave function for $0 \le m \le M$. It follows from [6, Prop. E.1] that $p(m_0, \ldots, m_{2^{16}-1})$ is Schur-concave and thus has its maximum when all m_i are equal, as claimed.   □

By Lemma 2 the maximum probability of each wrong key guess to survive is

$$\left(\exp(-\frac{2^{12}(2^{12}-1)}{2 \cdot 2^{32}})\right)^{2^{16}} \approx \left(\exp(-2^{-9})\right)^{2^{16}} = \exp(-2^7) \approx 2^{-184}.$$

It follows that also the probability is negligible that a single wrong key guess survives the first part. The probability that a wrong guess survives in the second part is, by the birthday paradox, about 2^{-184}, so for all 25 key bits this is about 2^{-159} which is also negligible.

5   Attacks on Reduced-Round MISTY2

While the attacks given in this section work for 5-round MISTY2 both with and without FL functions, the attacks on MISTY2 without FL functions have a much higher complexity than the one given in [10]; therefore we present here only the attacks on MISTY2 with FL functions; again we assume that the final swap but no output transformation is present.

Because of the asymmetry of the round function described in Section 3 it seems to help to attack MISTY2 in the chosen ciphertext model, as then the round function is used in the forward direction when testing a guessed value of a subkey.

Differential-Style Attack. This attack on 5-round MISTY2 makes use of the following impossible differential:

Proposition 1. *Given MISTY2 without FL, any input XOR $(\alpha_l, \alpha_r, 0, 0)$ with nonzero (α_l, α_r) to round i cannot yield a difference $(\delta_1, \delta_2, \delta_1, \delta_2)$ for any (δ_1, δ_2) in round $i + 3$. Conversely, a difference $(\delta_1, \delta_2, \delta_1, \delta_2)$, $(\delta_1, \delta_2) \ne (0, 0)$, in round $i + 3$ cannot decrypt to a difference $(\alpha_l, \alpha_r, 0, 0)$ before round i.*

For MISTY2 with FL functions this differential is also impossible provided that $\tilde{Y}_{i+3} = Y_{i+3}$, i.e. round $i + 3$ does not apply FL to the right half before it is XORed to the left half.

Proof. This differential uses the miss-in-the-middle approach (see [2]) where two differentials with probability 1 are concatenated such that a contradiction arises. The 2-round differential used here has input difference $(\alpha_l, \alpha_r, 0, 0)$ and output difference $(\beta_1, \beta_2, \beta_1, \beta_2)$ which happens with probability 1. The input difference of $(\alpha_l, \alpha_r, 0, 0)$ causes a nonzero input difference for the first FO, which then becomes output difference $(\beta_1, \beta_2) \ne (0, 0)$ as FO is bijective. The XOR with the right hand side zero difference does not change this. So at the beginning of round 2 the difference is $(0, 0, \beta_1, \beta_2)$ which FO cannot change. After round 2 the difference is $(\beta_1, \beta_2, \beta_1, \beta_2)$. The same reasoning works for the backwards direction, where an output difference $(\delta_1, \delta_2, \delta_1, \delta_2) \neq (0,0,0,0)$ decrypts always to $(\gamma_1, \gamma_2, 0, 0)$. Connecting two instances of this differential yields the contradiction.

When FL functions are present, the assumption on round $i + 3$ ensures that the output difference of FO in this round is zero. Application of FL in the first two rounds cannot yield a zero difference in the right half input to round $i + 2$, so the contradiction between rounds $i + 1$ and $i + 2$ still occurs.   □

In oder to use this impossible differential the condition of a missing FL function in the last round must be met. From the specification of MISTY2 it is clear that if a group of 4 rounds does not employ FL functions in the fourth round the round preceeding this group also does not use FL, so no additional key material has to be guessed besides the subkey for FO. This holds for example for rounds 2 to 6.

The attack works as follows. Set up a structure of 2^{16} ciphertexts $C_i = (e_i, f_i, e_i \oplus x, f_i \oplus y)$ where x, y are constants and (e_i, f_i) are different values. Get the plaintexts $P_i = (a_i, b_i, c_i, d_i)$ by decryption under the unknown key. Every pair of ciphertexts fulfills the ciphertext condition of the impossible differential. For each pair P_i, P_j any key k to the first round that encrypts P_i and P_j to a difference $(\alpha_1, \alpha_2, 0, 0)$ must be a wrong guess, while a correct guess never yields such a contradiction. There are about 2^{31} such pairs, so that a fraction of $(1 - 2^{-32})^{2^{31}} = e^{-1/2}$ of the wrong keys survives. Thus about $75/\log_2(e^{1/2}) \approx 104$ structures (about 2^{23} chosen ciphertexts) are required to eliminate all wrong keys. The work complexity is $\left(\sum_{i=0}^{103} (e^{-1/2})^i \right) \cdot 2^{75+16} \approx 2^{92.4}$ computations of FO roughly comparable to 2^{90} decryptions.

An improvement of the work factor can be reached using Property 1 in a similar way as for MISTY1 in sections 4.1 and 4.2. For the attack we use the same structures as above. From their decryptions $P_i = (a_i, b_i, c_i, d_i)$ we make a list $w_i = c_i \oplus d_i$. All matches $w_i = w_j$, $i \neq j$ yield a plaintext difference $P_i \oplus P_j = (\alpha_l, \alpha_r, \beta, \beta)$ for some value of β; these are the analysed pairs. With the input resp. output XOR (α_l, α_r) resp. (β, β) for FO in the first round we determine subkeys (AKO_{11}, AKI_{11}), (AKO_{12}, AKI_{12}) that yield this output difference from (a_i, b_i) and (a_j, b_j) as follows. For each (AKO_{11}, AKI_{11}) we check if the first FI gives output XOR α_r from a_i, a_j. Then each guess for (AKO_{12}, AKI_{12}) is checked if it yields output XOR β by the second FI. Each part is expected to result in about 2^9 candidates due to the 16 bit restriction. Each of the expected 2^{18} combinations is a wrong guess by the impossible differential.

In each structure there are about 2^{31} pairs, each of which has a chance of 2^{-16} to have a plaintext difference (β, β) in the right half. So about 2^{15} pairs are analysed, each of which excludes about 2^{18} not necessarily distinct subkey guesses. After about $2^{17} \cdot \ln(2^{50}) \approx 2^{17} \cdot 2^5$ structures (2^{38} chosen ciphertexts, 2^{37} analysed pairs) there is only a single remaining key expected. The time complexity per pair is $2 \cdot 2^{25}$ evaluations each for the first and the second FI, which is about 2^{26} evaluations of FO. In total this is about $2^{26} \cdot 2^{38} = 2^{64}$ evaluations of FO equivalent to about 2^{62} encryptions.

Determining the last 25 subkey bits AKO_{13} and AKI_{13} can be done with the basic attack and $25/\log_2(e^{1/2}) \approx 35$ structures with about $2^{21.2}$ chosen ciphertexts reused from previous structures. The work requirements are about $\left(\sum_{i=0}^{34}(e^{-1/2})^i\right) \cdot 2^{25+16} \approx 2^{42.4}$ evaluations of FO which is approximately 2^{40} encryptions, much less than for the first part.

In total about 2^{38} chosen ciphertexts and work of about 2^{62} encryptions is required to find the first round's 75 bit subkey; about 2^{37} pairs are analysed.

Attack Using Collisions. This attack on 5-round MISTY2 with FL functions but without the output transformation works with collision-searching; it is based on the following observation in the chosen ciphertext model.

Proposition 2. *Given four rounds of MISTY2 starting at round n such that $\tilde{Y}_{n+3} = Y_{n+3}$ holds, i.e. Y_{n+3} is not transformed via FL before the XOR. Assume that no output transformation with FL takes place. Given a set of ciphertexts $C_i = (e_i, f_i, x \oplus e_i, y \oplus f_i)$ where x, y are constant and $\{(e_i, f_i)\}$ form a permutation. After decryption the right half R_n is a permutation.*

Proof. Z_{n+3} is always the constant (x, y) and thus X_{n+3} as well as L_{n+3} is a constant, say (x', y'). On the other hand, R_{n+3} is the permutation $\{(e_i, f_i)\}$. After being XORed with (x', y') this becomes Z_{n+2}, so that also X_{n+2} and L_{n+2} are permutations while R_{n+2} is a constant. Z_{n+1} is a permutation which is the XOR of a constant and a permutation $\tilde{Y}_{n+1}$ which is L_{n+2} possibly transformed by an instance of FL. So X_{n+1} is a permutation. Now the claim follows.   □

The attack using Proposition 2 works for example on the five rounds of MISTY2 from round 2 to round 6. Both round 2 and round 6 do not apply any FL functions. An attack using 2^{89} work and 2^{20} chosen ciphertexts works straightforward as in section 4.2 with the same analysis, so the detailed description is omitted here.

In order to use the observation on reducing the amount key material to be guessed the attack uses 2^{28} chosen ciphertexts of the form $C_i = (e_i, f_i, x \oplus e_i, y \oplus f_i)$ where x, y are constants and $\{(e_i, f_i)\}$ form a permutation. Encryption under the unknown key yields plaintexts $P_i = (a_i, b_i, c_i, d_i)$ which we partition into 2^{16} sets B_t such that $P_i \in B_{b_i}$; thus all $P_i \in B_t$ for a given t have the same value $b_i = t$. First, set $AKO_{i1} = 0$, $AKI_{12} = 0$. For each 50 bit key guess $k = (AKO_{11}, AKI_{11}, k_{23}, AKI_{13})$ with k_{23} in the role of AKO_{13}, and for each B_t, $t \in \{0, \ldots, 2^{16} - 1\}$ encrypt all $P_i \in B_t$ one round yielding $w_i^t = FO_k((a_i, b_i)) \oplus (c_i, d_i)$. If we find $w_i^t = w_j^t$, $i \neq j$ then this key must be discarded, and the procedure is started with the next key guess. This takes at most $2^{50} \cdot 2^{28} = 2^{78}$ evaluations of FO comparable to 2^{76} encryptions to complete.

Once a correct k with $k_{23} = AKO_{12} \oplus AKO_{13}$ has been found, we have to find the correct 25 bits (AKO_{12}, AKI_{12}) and set $AKO_{13} = k_{23} \oplus AKO_{12}$. Here we use plaintexts P_i where both a_i and b_i vary; about 2^{20} plaintexts from the collection of 2^{28} plaintexts should be sufficient. This requires work of at most $2^{20} \cdot 2^{25} = 2^{45}$ evaluations of FO equivalent to 2^{43} encryptions. The time and chosen ciphertext requirements are dominated by the first part (2^{76} work, 2^{28} chosen ciphertexts).

The probability that a wrong key guess survives is bounded with the same arguments as in Section 4.2 by Lemma 2 and the birthday paradox, so the details are omitted here.

6   Comparison to KASUMI

The algorithm KASUMI [4] is a MISTY variant that is to be used in next-generation cellular phones. The global structure is a Feistel network with 8 rounds including the final permutation. Its round function consists of FO and FL, applied before resp. after FO in odd resp. even-numbered rounds. The FO function has the same structure as in MISTY with the subkeys KO_{ij}, $1 \leq j \leq 3$ being applied by XOR before FI but a lacking final XOR of KO_{i4} after all non-linearities; the FI function involves an additional fourth round, and the FL function uses left rotations by one bit before each XOR. The S-boxes $S7$ and $S9$ are bijective, but different from those of MISTY. Each round uses 128 key bits, 32 bits for FL and 96 bits for FO. These are derived by revolving through the key bytes and applying rotations and bitwise additions of constants.

The usage of the basic Feistel structure without FL functions in the data path makes KASUMI susceptible to an attack based on the same 5-round impossible differential as used in Section 4.1, but with the additional difficulty that FL is part of the round functions and FO uses more keying material. The differential can be used as both FO and FL are bijective. It should be noted that a property similar to Property 1 does also hold for KASUMI's FO when it is preceded by FL as it happens in odd-numbered rounds:

Property 3. Assume that the concatenation of FL and FO has a nonzero output XOR (δ, δ). Denote the input XOR to FL by (α_l, α_r) and its output XOR (the input to FO) by (β_l, β_r). The difference β_r is solely determined by the first round of FL, so is the right half of the data in the first round of FO. In order to have the given output XOR of FO the third round's output and input XOR must both be zero which means that (KO_{i3}, KI_{i3}) can be ignored. The output XOR (δ, δ) is determined by the second round of FO from the inputs with XOR β_r; additionally, β_r is canceled by the output XOR β_r of the FI in the first round of FO, coming from the left halfs of the inputs with XOR β_l.

The attack on rounds 2 to 7 of KASUMI including the last swap works as follows. In round 7 the function FL is applied before FO, so we can rely on Property 3. The attack uses the same structures as were used in Section 4.1 and looks for pairs with ciphertext XOR $(\delta, \delta, \alpha_l, \alpha_r)$ with the same methods. We expect about such 2^{15} pairs per structure which will be analysed. Let (c_i, d_i, e_i, f_i) and (c_j, d_j, e_j, f_j) be such a pair. In order to use Property K we first fix a guess of the first round subkey KL_{71} of FL in round 7, yielding f_i', f_j' with $\beta_r := f_i' \oplus f_j'$. Then we determine which guesses of $(KL_{72}, KO_{71}, KI_{71})$ yield the XOR β_r after the first FI. We expect about $2^{48}/2^{16} = 2^{32}$ guesses to fulfill this condition. Then, independently, we check which guess for $(\mathrm{KO}_{72}, \mathrm{KI}_{72})$ yields output XOR δ after the second FI from inputs f_i' and f_j'; here we expect about $2^{32}/2^{16} = 2^{16}$ guesses. Combinations of all these guesses are wrong subkeys and can be discarded. Their expected number is 2^{48} for each guess of KL_{71}, so each analysed pair is expected to discard about 2^{64} subkeys á 96 bits.

After about 2^{17} structures an expected number of $2^{96}/e$ distinct subkeys are discarded. In total we need about $2^{17} \ln(2^{96}) \approx 67 \cdot 2^{17} \approx 2^{23}$ structures with 2^{55} chosen plaintexts and about 2^{38} analysed pairs to single out the right subkey.

The work requirements for each pair and each guess of KL_{71} are $2 \cdot 2^{48} + 2 \cdot 2^{32} \approx 2^{49}$ computations of the second round of FL and FI. In total this is about 2^{103} computations of FL and FI roughly equivalent to 2^{100} encryptions. Although this is much faster than brute force it is hardly a practical attack because of the high data and work requirements.

7   Conclusion

For MISTY1 the use of keyed linear functions inhibits the attack using the 5-round impossible differential of Feistel networks with bijective round functions; for MISTY2 we cannot make this claim as we did not find an impossible differential longer than 4 rounds.

The attacks on MISTY2 suggest that this structure might be one round weaker than the Feistel structure, at least when the linear functions FL are present. The directional asymmetry of the MISTY2 structure used in FO with embedded 3-round FI suggests that this structure might be stronger in the backwards direction compared to the forward direction.

By adding a fourth round to FI – like done for KASUMI – its equivalent description of FO would not reduce the number of key bits, so the attacks would only need to guess 7 bits more for each FI. If FO had one more round the properties used to improve both the differential and collision-searching attacks would not hold, leaving only the basic forms of attack; but this would require more keying material.

Instead, the changes for KASUMI, i.e. adding a round to FI and employing the linear functions as part of the round function does not require more keying material and seems to make an analysis of the round function very demanding.

Acknowledgments

Thanks are due to Mitsuru Matsui for providing reference [10] and to the anonymous referees for helpful comments and the suggestion to apply the techniques to KASUMI. The author is grateful to Dr. Uwe Deppisch, Alfred Goll and the colleagues of the IT research department of Dresdner Bank for the encouraging work conditions that were very helpful in conducting this research.

References

[1] E. Biham. Cryptanalysis of Ladder-DES. In E. Biham, editor, *Fast Software Encryption: 4th International Workshop*, Volume 1267 of *Lecture Notes in Computer Science*, pages 134–138, Haifa, Israel, 20–22 Jan. 1997. Springer-Verlag.

[2] E. Biham, A. Biryukov, and A. Shamir. Miss in the middle attacks on IDEA and Khufu. In L. Knudsen, editor, *Fast Software Encryption, 6th international Workshop*, Volume 1636 of *Lecture Notes in Computer Science*, pages 124–138, Rome, Italy, 1999. Springer-Verlag.

[3] E. Biham and A. Shamir. *Differential Cryptanalysis of the Data Encryption Standard*. Springer Verlag, Berlin, 1993.

[4] ETSI/SAGE. Specification of the 3GPP Confidentiality and Integrity Algorithms – Document 2: KASUMI Specification, Version 1.0. 3G TS 35.202, December 23, 1999. `http://www.etsi.org/dvbandca/3GPP/3GPPconditions.html`.

[5] L. R. Knudsen. DEAL — A 128-bit block cipher. Technical Report 151, Department of Informatics, University of Bergen, Bergen, Norway, Feb. 1998.

[6] A. W. Marshal and I. Olkin. *Inequalities: Theory of Majorization and Its Applications*, volume 143 of *Mathematics in Science and Engineering*. Academic Press, New York, 1979.

[7] M. Matsui. Linear cryptanalysis method for DES cipher. In T. Helleseth, editor, *Advances in Cryptology - EuroCrypt '93*, Volume 765 of *Lecture Notes in Computer Science*, pages 386–397, Berlin, 1993. Springer-Verlag.

[8] M. Matsui. New block encryption algorithm MISTY. In E. Biham, editor, *Fast Software Encryption: 4th International Workshop*, Volume 1267 of *Lecture Notes in Computer Science*, pages 54–68, Haifa, Israel, 20–22 Jan. 1997. Springer-Verlag.

[9] K. Sakurai and Y. Zheng. On non-pseudorandomness from block ciphers with provable immunity against linear cryptanalysis. *IEICE Trans. Fundamentals*, E80-A(1):19–24, January 1997.

[10] M. Sugita. Higher order differential attack of block ciphers MISTY1,2. Technical Report ISEC98-4, Institute of Electronics, Information and Communication Engineers (IEICE), 1998.

[11] M. Sugita. Personal communication, January 2001.

[12] H. Tanaka, K. Hisamatsu, and T. Kaneko. Strength of MISTY1 without FL function for higher order differential attack. In M. Fossorier, H. Imai, S. Lin, and A. Poli, editors, *Proc. Applied algebra, algebraic algorithms, and error-correcting codes: 13th international symposium, AAECC-13*, Volume 1719 of *Lecture Notes in Computer Science*, pages 221–230, Hawaii, USA, 1999. Springer Verlag.

The Rectangle Attack –
Rectangling the Serpent*

Eli Biham[1,**], Orr Dunkelman[1,***], and Nathan Keller[2,†]

[1] Computer Science department,
Technion – Israel Institute of Technology,
Haifa 32000, Israel
[2] Mathematics department,
Technion – Israel Institute of Technology,
Haifa 32000, Israel

Abstract. Serpent is one of the 5 AES finalists. The best attack published so far analyzes up to 9 rounds. In this paper we present attacks on 7-round, 8-round, and 10-round variants of Serpent. We attack a 7-round variant with all key lengths, and 8- and 10-round variants with 256-bit keys. The 10-round attack on the 256-bit keys variants is the best published attack on the cipher. The attack enhances the amplified boomerang attack and uses better differentials. We also present the best 3-round, 4-round, 5-round and 6-round differential characteristics of Serpent.

1 Introduction

Serpent [1] is a block cipher which was suggested as a candidate for the Advanced Encryption Standard (AES) [8], and was selected to be among the five finalists.

In [4] a modified variant of Serpent in which the linear transformation was modified into a permutation was analyzed. The permutation allows one active S box to activate only one S box in the consecutive round, a property that cannot occur in Serpent. Thus, it is not surprising that this variant is much weaker than Serpent, and that it can be attacked with up to 35 rounds.

In [6] the 256-bit variant of Serpent up to 9 rounds is attacked using an amplified boomerang attack. The attack is based on building a 7-round distinguisher for Serpent, and using it for attacking up to 9 rounds. The distinguisher is built using the amplified boomerang technique. It uses a 4-round differential characteristic in rounds 1–4, and a 3-round characteristic in rounds 5–7.

In this paper we enhance the amplified boomerang attack, and present the best 3-round, 4-round, 5-round and 6-round differential characteristics of Serpent

---

* The work described in this paper has been supported by the European Commission through the IST Programme under Contract IST-1999-12324 and by the fund for the promotion of research at the Technion.
** biham@cs.technion.ac.il, http://www.cs.technion.ac.il/~biham/
*** orrd@cs.technion.ac.il, http://vipe.technion.ac.il/~orrd/me/
† nkeller@tx.technion.ac.il

B. Pfitzmann (Ed.): EUROCRYPT 2001, LNCS 2045, pp. 340–357, 2001.
© Springer-Verlag Berlin Heidelberg 2001

published so far. We use these characteristic to devise an attack on 7-round Serpent with all key lengths, and an attack on 8-round Serpent with 256-bit keys. We also use these results to develop the best known distinguisher for 8-round Serpent by presenting a new cryptanalytic tool — the rectangle attack. This tool is then used to attack 10-round 256-bit key Serpent.

The paper is organized as follows: In Section 2 we give the description of Serpent. In Section 3 we present a differential attack on 7-round Serpent, and a differential attack on 8-round 256-bit key Serpent. In Section 4 we present the *Rectangle Attack*, and in Section 5 we describe the 8-round distinguisher and implement the attack on 10-round 256-bit key Serpent. Section 6 summarizes the paper. In the appendices we describe new 3-round, 4-round, 5-round and 6-round differential characteristics, which are the best known so far.

2 A Description of Serpent

Serpent [1] is a block cipher with block size of 128 bits and 0–256 bit keys. It is an SP-network, consisting of alternating layers of key mixing, S boxes and linear transformation. Serpent has an equivalent bitsliced description, which makes it very efficient.

The key scheduling algorithm of serpent accepts 256-bit keys. Shorter keys are padded by 1 followed by as many 0's needed to have a total length of 256 bits. The key is then used to derive 33 subkeys of 128 bits.

We use the notations of [1]. Each intermediate value of the round i is denoted by $\hat{B}_i$ (which is a 128-bit value). The rounds are numbered from 0 to 31. Each $\hat{B}_i$ is composed of four 32-bit words X_0, X_1, X_2, X_3.

Serpent has 32 rounds, and a set of eight 4-bit to 4-bit S boxes. Each round function R_i ($i \in \{0, \ldots, 31\}$) uses a single S box 32 times in parallel. For example, R_0 uses S_0, 32 copies of which are applied in parallel. Thus, the first copy of S_0 takes bits 0 from X_0, X_1, X_2, X_3 and returns the output to the same bits (0). This is implemented as a boolean expression of the 4 registers.

The set of eight S-boxes is used four times. S_0 is used in round 0, S_1 is used in round 1, etc. After using S_7 in round 7 we use S_0 again in round 8, then S_1 in round 9, and so on. The last round is slightly different from the others: apply S_7 on $\hat{B}_{31} \oplus \hat{K}_{31}$, and XOR the result with $\hat{K}_{32}$ rather than applying the linear transformation.

The cipher may be formally described by the following equations:

$$\hat{B}_0 := P$$
$$\hat{B}_{i+1} := R_i(\hat{B}_i)$$
$$C := \hat{B}_{32}$$

where

$$R_i(X) = LT(\hat{S}_i(X \oplus \hat{K}_i)) \qquad i = 0, \ldots, 30$$
$$R_i(X) = \hat{S}_i(X \oplus \hat{K}_i) \oplus \hat{K}_{32} \quad i = 31$$

where $\hat{\mathcal{S}}_i$ is the application of the S-box $S_{i \bmod 8}$ thirty two times in parallel, and LT is the linear transformation.

The linear transformation is as follows: The 32 bits in each of the output words are linearly mixed by

$$X_0, X_1, X_2, X_3 := \hat{\mathcal{S}}_i(\hat{B}_i \oplus \hat{K}_i)$$
$$X_0 := X_0 <<< 13$$
$$X_2 := X_2 <<< 3$$
$$X_1 := X_1 \oplus X_0 \oplus X_2$$
$$X_3 := X_3 \oplus X_2 \oplus (X_0 << 3)$$
$$X_1 := X_1 <<< 1$$
$$X_3 := X_3 <<< 7$$
$$X_0 := X_0 \oplus X_1 \oplus X_3$$
$$X_2 := X_2 \oplus X_3 \oplus (X_1 << 7)$$
$$X_0 := X_0 <<< 5$$
$$X_2 := X_2 <<< 22$$
$$\hat{B}_{i+1} := X_0, X_1, X_2, X_3$$

where $<<<$ denotes rotation, and $<<$ denotes shift. In the last round, this linear transformation is replaced by an additional key mixing: $B_{32} := S_7(B_{31} \oplus K_{31}) \oplus K_{32}$.

3 Differential Attack on 7- and 8-Round Serpent

In this section we present attacks on 7-round and 8-round Serpent from round 4 to round 10 (or round 11 in the 8-round variant), i.e., encryption starts with S_4 and ends with S_2 (S_3 for the 8-round variant)[1]. In Appendix D a 6-round differential characteristic between round 4 and round 9 with probability 2^{-93} is presented. In the rest of this paper we keep the round numbers as in the corresponding rounds of Serpent, i.e., from round 4 to round 10, rather than from round 0 to round 6.

We adopt the representation of the differential characteristics using figures as in [5], but add more data to the figures. The figures describe data blocks by rectangles of 4 rows and 32 columns. The rows are the bitsliced 32-bit words, and each column is the input to a different S box. The upper line represents X_0, the lower line represents X_3, and the rightmost column represents the least significant bits of the words. A thin arrow represents a probability of 1/8 for the specific S box (given the input difference, the output difference is achieved with probability 1/8), and a fat arrow stands for probability 1/4. If there is a

---

[1] Attacks starting from other rounds do not necessarily have the same complexities since the S boxes used in the various rounds are different.

difference in a bit, the box related to it is filled. Example for our notation can be found in Figure 1, in which in the first S box (S box 0; related to bits 0) the input difference 1 causes an output difference 3 with probability 1/4, and in S box 30 input difference 3 causes an output difference 1 with probability 1/8.

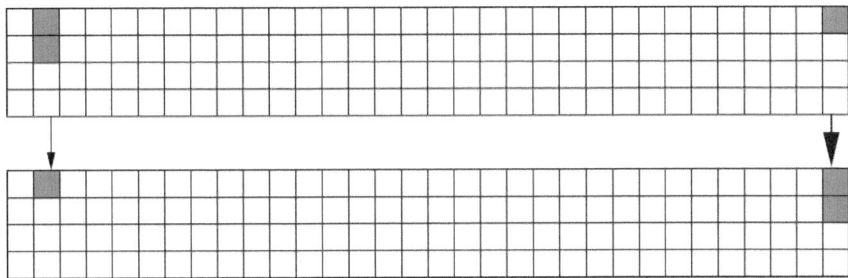

Fig. 1. Difference Representation Example

The attack uses 2^{14} characteristics with different input differences but the same output difference. The 2^{14} characteristics differ only in the first round, in which they have the same active S boxes with different input differences. All the characteristics have the same differences after the first round, and all have the same probability 2^{-93}. The input difference for one of the 6-round characteristics is presented in Figure 2, and the common output is presented in Figure 3 (the full characteristic is presented in Appendix D).

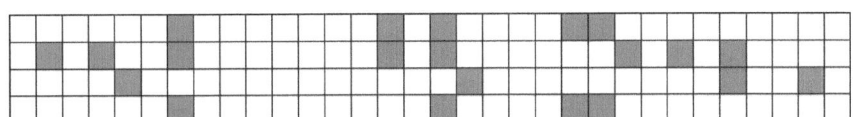

Fig. 2. The Input Difference of the 6-Round Differential Characteristic

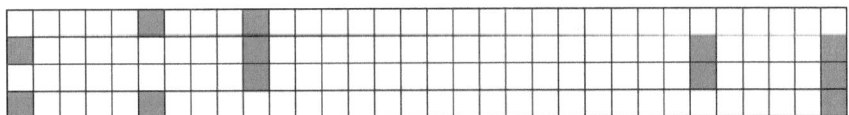

Fig. 3. The Output Difference of the 6-Round Differential Characteristic

The attack requires 2^{32} structures of 2^{52} chosen plaintexts each. In each structure all the inputs to the 19 inactive S boxes in the first round are fixed to some (random) value, while the 52 bits of input to the 13 active S boxes receive all the 2^{52} possible values. In these data structures there are $2^{32} \cdot 2^{51} = 2^{83}$

pairs for each possible characteristic. Each characteristic has probability 2^{-93}, therefore, we expect that about $2^{83} \cdot 2^{14} \cdot 2^{-93} = 2^4$ pairs satisfy one of the characteristics. We call these pairs *right pairs*. The number of possible pairs in each structure is $(2^{52})^2/2 = 2^{103}$, thus we have $2^{103} \cdot 2^{32} = 2^{135}$ pairs to consider in total.

Each pair satisfying one of the characteristics has 19 active S boxes in round 10, thus any pair with non-zero output difference in one of the remaining 13 S boxes can be automatically discarded. Thus, about $2^{103-52} = 2^{51}$ candidates for right pairs remain from each structure.

Moreover, in 3 S boxes only 4 output differences are possible if one of the characteristics is satisfied; in 6 S boxes only 6 output differences are possible; in 9 S boxes only 7 output differences are possible, and in the remaining S box eight output differences are possible. Discarding any pair with a wrong output difference using the above filter should keep only a fraction of $\frac{4}{16}^3 \cdot \frac{6}{16}^6 \cdot \frac{7}{16}^9 \cdot \frac{8}{16} \approx 2^{-26.22}$ of the pairs. Thus, only about $2^{51} \cdot 2^{-26.22} = 2^{24.78}$ pairs remain for each structure.

For each structure, we check whether the remaining pairs satisfy one of the 2^{14} possible plaintext differences (corresponding to the 2^{14} characteristics). As there are about 2^{52} possible input differences, only a fraction of about $2^{-52} \cdot 2^{14} = 2^{-38}$ of the pairs remain at this stage. Thus, the expected number of remaining pairs in all the 2^{32} structures is $2^{24.78} \cdot 2^{-38} \cdot 2^{32} = 2^{18.78}$.

For each remaining pair we compute a list of possible whitening subkeys of the 19 active S boxes in the last round. For each active S box, each pair suggests at most 4 values for the subkey of that S box. Thus, counting on m S boxes results in at most $2^{18.78} \cdot 4^m = 2^{18.78+2m}$ hits. The average number of hits (for a wrong value) is $2^{18.78+2m}/2^{4m}$, which is smaller than 1 for $m \geq 10$. On the other hand, the correct subkey is counted for each right pair, i.e., about 16 times, and thus it can be easily identified when $m \geq 10$. Then, we count on the remaining 9 S boxes and take the only value suggested more than two or three times. Note that even if we got more than one possible subkey after counting on 10 S boxes, only one of them is expected to remain after this stage. In total we retrieve 76 subkey bits using at most $2^{38.78}$ one round encryptions and 2^{40} 4-bit counters. We can retrieve 52 additional bits by analyzing the first round as well.

After we retrieve 128 bits of subkey material we can easily find a 128-bit key using linear equations. For 192- and 256-bit keys we can take another set of characteristics. The new set includes the original characteristics used in the attack rotated one bit to the left, i.e., if we have a difference in the least significant bit of X_0 in the original characteristics, we have a difference in the second bit (bit 1) of X_0 in the new set. There is an additional set, in which the rotation is by two bits. (Note that rotation by 3 bits does not make good characteristics). This way we obtain additional 36 subkey bits from round 4 (as out of the 52 bits in the input to the 13 active S boxes there are 16 common bits). This phase of the attack is much simpler, as we already know the common 16 subkey bits, and can easily discard wrong pairs. We also get 32 additional bits from round 10, thus obtaining additional 68 bits (36 from round 4, and 32 from round 10). For

192-bit keys, this information is sufficient to recover the key. For 256-bit keys we can use other differentials with probability 2^{-94} (which are just equivalent to the differential we have used with slight modifications in the last round of the characteristic) using similar techniques and retrieve the remaining unknown bits.

We conclude that the attack requires $2^{52} \cdot 2^{32} = 2^{84}$ plaintexts for 128-bit keys, and twice as much for 192-bit and 256-bit keys. The time complexity of the attack is 2^{85} memory accesses. The memory requirements are 2^{40} 4-bit counters and 2^{52} cells for a hash table.

In order to reduce the time of analysis we perform the algorithm in the following way:

1. For each structure:
   (a) Insert all the ciphertexts into a hash table according the 52 ciphertext's bits of the inactive S boxes in the last round.
   (b) For each entry with collision (a pair of ciphertext with equal 52-bit values) check whether the plaintexts' difference (in round 4) is one of the 2^{14} characteristics' input difference.
   (c) If a pair passes the above test, check whether the difference (in the 76 bits) can be caused by the output difference of the characteristics.
   (d) If a pair passes also the above test, we add 1 to the counter related to the 40 bits of the subkey (as there are $4m$ subkey bits, and for $m = 10$ we get the best results).
2. Collect all the (few) subkeys whose counter has at least 10 hits. With a high probability the correct subkey is in this list (and it is the only one in it).
3. For each pair suggesting a value in the list, we complete the subkey of the other 9 S boxes in round 10, and the 13 S boxes from round 4. As we should have only right pairs (with very few additional wrong pairs), and as the right pairs agree on the rest of the subkey, we can identify the right subkey by intersecting the sets proposed by the various pairs.

For each structure 2^{52} memory accesses are performed for the hashing. In the hash table about $1/e$ of the entries are empty, and $1/e$ of the entries contain only one plaintext (and no pairs need to be analyzed). Counting on all the possibilities for the number of plaintexts in each entry of the hash table we conclude that 2^{51} pairs from each structure need to be analyzed. Most of them are discarded by the first filter, and about 2^{13} pairs remain for the second filtering and counting. Therefore, we can estimate the work for each structure as the work needed to hash all plaintexts and then to look at the hash table afterwards, and to perform the search whenever there are more than two plaintexts in one hash entry. The number of pairs we expect to check is 2^{51} and most of them can be discarded almost immediately. We perform about 2^{53} memory accesses for each structure, and the amount of work needed for the whole attack is equivalent to about $2^{33} \cdot 2^{52} = 2^{85}$ memory accesses.

3.1   8-Round 256-Bit Key Serpent

One can easily extend our attack to 8 rounds for the 256-bit key variant by guessing the subkey of round 11. For each possible value of the subkey of round 11 we decrypt the last round and use the attack from the previous subsection. This way, there is no need to make the extra work of completing the key by retrieving other subkeys. The data complexity remains the same 2^{84}, and the time complexity is $2^{128} \cdot 2^{85} = 2^{213}$ memory accesses with 2^{40} counters.

4   The Rectangle Attack

4.1   Amplified Boomerang Attack

The main idea of the amplified boomerang attack [6] is to use two short differential characteristics instead of one long characteristic. Therefore, this technique is very useful when we have good short differential characteristics and very bad long ones.

Let a cipher $E : \{0,1\}^n \times \{0,1\}^k \rightarrow \{0,1\}^n$ be composed of two encryption functions E_0 and E_1. Thus, $E = E_1 \circ E_0$. We assume that a good differential is not known for E, but for E_0 we have a differential characteristic $\alpha \rightarrow \beta$ with probability p, and for E_1 we have a differential characteristic $\gamma \rightarrow \delta$ with probability q, where $pq \gg 2^{-n/2}$.

The basic attack is based on building quartets of plaintexts (x, y, z, w) which satisfy several differential conditions. Assume that $x \oplus y = \alpha$ and $z \oplus w = \alpha$. Each pair has probability p to satisfy the characteristic $\alpha \rightarrow \beta$ in E_0. We denote by x', y', z', w' the encrypted values of x, y, z, w under E_0, respectively $(x' = E_0(x), \ldots, w' = E_0(w))$. We are interested in the cases where $x' \oplus y' = \beta$, $z \oplus w' = \beta$ and $x' \oplus z' = \gamma$, as in these cases $y' \oplus w' = (x' \oplus \beta) \oplus (z' \oplus \beta) = \gamma$ as well. We receive two pairs for E_1 each with input difference γ. When encrypting those x', y', z', w' by E_1, in some of the cases the input difference γ becomes δ, and we look for the cases where both differences become $x'' \oplus z'' = \delta$ and $y'' \oplus z'' = \delta$ after E_1. A quartet satisfying all these differential requirements is called a *right quartet*. An outline of such a quartet is shown in Figure 4.

The question which rises is what is the fraction of the right quartets among all the quartets. If we have m pairs with difference α, a fraction of about p of them satisfies the characteristic for E_0. Thus, we have about mp pairs with output difference β in the input to E_1, giving about $(mp)^2/2$ quartets consisting of two such pairs. Assuming that the intermediate encryption values distribute uniformly over all possible values, then with probability 2^{-n} we get x' and z' such that $x' \oplus z' = \gamma$, but once this occurs we automatically get another pair with input difference γ (the pairs are (x', z') and (y', w')). Note that x' and w' have also a probability 2^{-n} to have a difference $x' \oplus z' = \gamma$, thus, given two pairs (x', y') and (z', w') we have two ways to use them as a quartet, with probability 2^{-n+1}. Therefore, we have $(mp)^2/2 \cdot 2^{-n+1}$ quartets which might satisfy our requirements. Each of the pairs satisfies the second characteristic for E_1 with

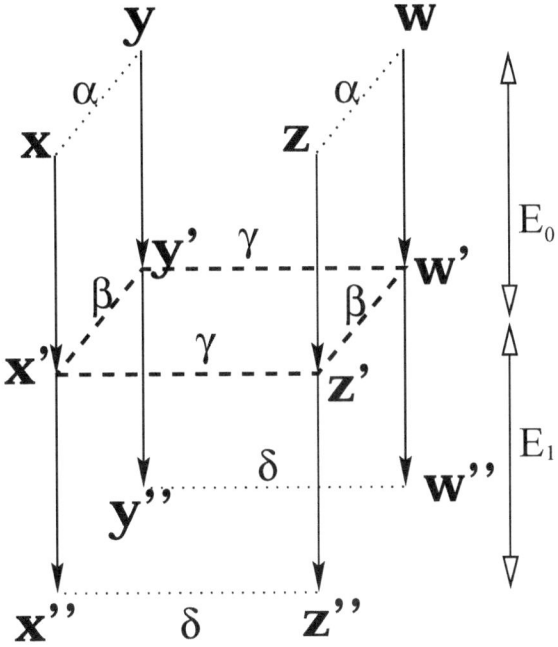

Fig. 4. Right Quartet for the Rectangle Attack

probability q. Thus, starting with m pairs $(x, y),(z, w)$, the expected number of right quartets is:

$$\binom{mp}{2} \cdot 2^{-n+1} \cdot q^2 \cdot 2^{-n} = m^2 \cdot 2^{-n} \cdot (pq)^2.$$

Therefore, the distinguisher counts quartets (x, y, z, w) of plaintexts which satisfy that $x'' \oplus z'' = y'' \oplus w'' = \delta$.

For a random permutation (or for a random value of α and δ) the expected number of quartets is $m^2 \cdot 2^{-2n}$, as there are m^2 possible quartets (there are $m^2/2$ pairs of pairs, and each pair of pairs can create two quartets e.g., $(x, y), (z, w)$ and $(x, y), (w, z)$). For each pair (x, z) or (y, w) the probability of having a specific difference in the output is 2^{-n}. Therefore, if $pq > 2^{-n/2}$, we would count more quartets than random noise. This way when m is sufficiently large we can have a distinguisher which distinguishes between E and a random cipher.

4.2   Rectangling the Boomerang

The first improvement was suggested in [6], in which it was observed that instead of requiring a specific γ, we can count on all possible γ' values for which $\gamma' \to \delta$ by E_1.

Therefore, the probability $\Pr^2(\gamma \to \delta) = q^2$ for the pairs (x', z') and (y', w') to have output difference δ is replaced by the probability $\sum_{any\ \gamma'} \Pr^2(\gamma' \to \delta)$, and we have about

$$\binom{m \cdot \Pr(\alpha \to \beta)}{2} \cdot 2^{-n+1} \cdot \sum_{any\ \gamma'} \Pr^2(\gamma' \to \delta)$$

quartets satisfying the rectangle conditions. As a result, we might prefer the difference $x'' \oplus z''$ to be some value δ which has many lower probability characteristics instead of an optimal δ with one characteristic with the highest probability.

Our second improvement is quite similar. Instead of discarding pairs with wrong β value, we sort the pairs into piles according to the output difference (β) of E_0. For each possible pile we perform the original attack. For each pile we have probability $\sum_{any\ \gamma'} \Pr^2(\gamma' \to \delta)$ to have a quartet at the end. The number of pairs in each pile β' is

$$\binom{m \cdot \Pr(\alpha \to \beta')}{2}$$

Thus, we have about

$$\sum_{any\ \beta'} \binom{m \cdot \Pr(\alpha \to \beta')}{2} \cdot 2^{-n+1} \cdot \sum_{any\ \gamma'} \Pr^2(\gamma' \to \delta) =$$

$$m^2 \cdot 2^{-n} \cdot \sum_{any\ \beta'} \Pr^2(\alpha \to \beta') \cdot \sum_{any\ \gamma'} \Pr^2(\gamma' \to \delta)$$

quartets for the second step of the attack.

Our third improvement is based on the first two. We can take into consideration more quartets. Assume that for the first pair the difference α causes some difference a, and for the second pair $\alpha \to b$. Then, we can count also characteristics for which $\gamma \to \delta$ and $\gamma \oplus a \oplus b \to \delta$. This way the number of quartets is

$$m^2 \cdot 2^{-n} \cdot \sum_{a,b} \left[\Pr(\alpha \to a) \Pr(\beta \to b) \cdot \sum_{\gamma} \Pr(\gamma \to \delta) \Pr(\gamma \oplus a \oplus b \to \delta) \right]$$

Note that this improvement counts all the quartets with plaintext difference α and ciphertext difference δ. However, it is very hard to do the exact calculation.

5 Attacking 10-Round Serpent

In Section 4 we presented a method to build a distinguisher for a function $E = E_1 \circ E_0$. We now present a method to use the distinguisher to find subkey material.

We attack a 10-round 256-bit key Serpent (round 0 to round 9) using an 8-round rectangle distinguisher. In this distinguishing attack E_0 is rounds 1–4 of Serpent, and E_1 is rounds 5–8. The basic differential characteristic $(\alpha \rightarrow \beta)$ used in rounds 1–4 is also the best known 4-round differential characteristic of Serpent. This characteristic and the basic differential characteristic used in rounds 5–8 are presented in Appendix B. α and δ are presented in Figure 5 and Figure 6, respectively.

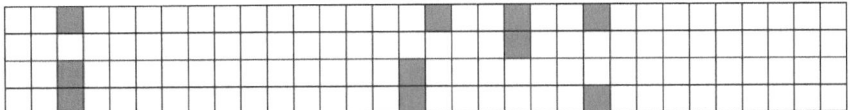

Fig. 5. The Input Difference α of the Rectangle Attack

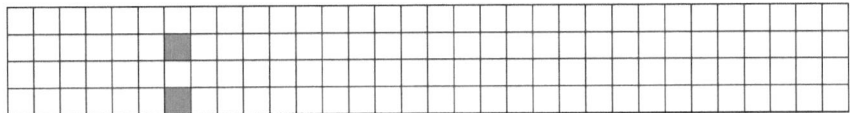

Fig. 6. The Output Difference δ of the Rectangle Attack

The first differential has probability of 2^{-29}. Using the second improvement and counting on all possible output differences of round 4, we receive $\sum_{\beta'} \Pr^2[\alpha \rightarrow \beta'] = 2^{-50.8}$. The second differential has probability of 2^{-47}. Using the first improvement and counting on a large set of characteristics (all are very similar to the basic one, and have the same last two rounds) we receive $\sum_{\gamma'} \Pr^2[\gamma' \rightarrow \delta] = 2^{-69.8}$. These probabilities were computed by a computer program which scanned characteristics similar to those presented in Appendix B.

For $m = 2^{125.8}$ pairs with the difference α of Figure 5 about $2 \cdot (2^{125.8} \cdot 2^{-25.4})^2/2 = 2^{200.8}$ quartets can be formed after the fourth round. The probability to get a specific γ is about 2^{-128}, thus the expected number of quartets with a given γ is about $2^{72.8}$ quartets. As $\sum_{\gamma'} \Pr^2[\gamma' \rightarrow \delta] = 2^{-69.8}$, the number of right quartets is 8.

To attack 10-round Serpent (rounds 0–9) we use a similar technique to the one used in [6]. We request $2^{62.8}$ structures of 2^{64} plaintexts each. The structures are chosen so that each structure varies over all the possible inputs to the active S boxes in round 1, while the input of the rest of the S boxes is kept fixed (this is done by checking which S boxes in round 0 affect the active bits in round 1, and trying all the inputs to these S boxes). Using this procedure for choosing the plaintexts we get $2^{125.8}$ pairs with difference α after round 0.

We keep all the plaintexts and their corresponding ciphertexts in a large table (whose size is $2^{126.8} \cdot 2 \cdot 16 = 2^{131.8}$ bytes of memory), and keep 2^{84} 4-bit counters, where each counter corresponds to one of the possible values of the 84 bits of the subkeys we search for (64 bits entering 16 S boxes in the first round, and 20 bits entering 5 S boxes in the last round). In order to count the number of quartets with the given α and δ we perform the following algorithm:

1. Initialize the counter's array with 0's.
2. For each 64-bit subkey value in round 0, for each 20-bit value subkey value in round 9, and for each plaintext x:
   - Partially encrypt x through round 0 in the 16 S boxes and denote the value we get by x_1.
   - Calculate $x_1 \oplus \alpha$ and denote this value by y_1.
   - Partially decrypt y_1 through round 0 in the 16 S boxes, and find the corresponding plaintext, which we denote by y (this plaintext y exists in our data, due to the way we choose the structures). The value of the plaintext bits of y related to the other 16 S boxes is the same as of x.
   - Let x'' and y'' be the corresponding ciphertexts of x and y respectively. then,
   - Partially decrypt through the 5 active S boxes x'' and y'', denote the value you get by x_9'' and y_9'', respectively.
   - Partially encrypt $x_9'' \oplus \delta$ and $y_9'' \oplus \delta$ and check whether the corresponding ciphertexts exist in our data. If these ciphertexts exist, we check their corresponding plaintexts, whether under the guessed 64-bit subkey of round 0 we get a difference α. If so, we increase the corresponding counter by 1.
3. Run over all counters, and print the corresponding indices whose counter is greater than or equal to 7.

The inner loop is performed at most $2^{84} \cdot 2^{126.8}$ times, and includes at most 4 times encrypting 16 S boxes (equivalent to two rounds of Serpent) and 4 times decrypting 5 S boxes (equivalent to 5/8 rounds of Serpent). Thus, the time complexity of the attack is at most $2^{84} \cdot 2^{125.8} \cdot 2\frac{5}{8}/10 \approx 2^{208.4}$ 10-round Serpent encryptions. The time complexity can be reduced by half by building in advance an equivalent table in which each entry i contains $S_1(S_1^{-1}(i) \oplus \delta)$ and use it in the last round, and similarly computing a table with $S_0^{-1}(S_0(i) \oplus \alpha)$ for the first round.

5.1   Reducing Time Requirements

One can also use the technique of hash tables presented in [5] to reduce the time complexity to 2^{205} memory accesses, in exchange for increasing the memory complexity to 2^{196} bytes of RAM.

6   Summary

In this paper we presented the best published attack on 10-round 256-bit key Serpent. The attack requires $2^{126.8}$ chosen plaintexts, $2^{207.4}$ time and $2^{131.8}$ bytes of RAM. A variant of the attack requires 2^{205} time but 2^{196} bytes of RAM.

We presented a differential attack on 7-round Serpent, which works for all key sizes, with data complexity of 2^{84} chosen plaintexts, time complexity of 2^{85} memory accesses and 2^{52} memory (blocks of 128-bit). We presented an attack on 8-round 256-bit key Serpent requiring 2^{84} chosen plaintexts, 2^{213} time and 2^{84} memory (blocks of 128-bit). We summarize these results in Table 1.

We also presented the best known 3-round, 4-round, 5-round and 6-round differential characteristics of Serpent, whose probabilities are 2^{-15}, 2^{-29}, 2^{-60} and 2^{-93}, respectively. In Table 2 we summarize these characteristics and the best previously published characteristics.

Table 1. Summary of Differential Attacks on Serpent with Reduced Numbers of Rounds

Rounds	Key Size	Complexity			Source
		Data	Time	Memory	
6	all	2^{83}	2^{90}	2^{40}	[5] - Section 3.2
	all	2^{71}	2^{103}	2^{75}	[5] - Section 3.3
	192 & 256	2^{41}	2^{163}	2^{45}	[5] - Section 3.4
7	256	2^{122}	2^{248}	2^{126}	[5] - Section 3.5
	all	2^{84}	2^{85} MA	2^{52}	This paper
8	192 & 256	2^{128}	2^{163}	2^{133}	[5] - Section 4.2
	192 & 256	2^{110}	2^{175}	2^{115}	[5] - Section 5.3
	256	2^{84}	2^{213} MA	2^{84}	This paper
9	256	2^{110}	2^{252}	2^{212} bytes	[5] - Section 5.4
10	256	$2^{126.8}$	$2^{207.4}$	$2^{131.8}$ bytes	This paper
	256	$2^{126.8}$	2^{205}	2^{196} bytes	This paper

MA - Memory Accesses
Memory unit is one block, unless written otherwise

Table 2. Summary of the Differential Characteristics of Serpent

Number of Rounds	Paper	Starting from	Number of Active S boxes	Probability
3	[5]	S_5	7	2^{-16}
	This paper*	S_2	7	2^{-15}
4	[5]	S_1	14	2^{-31}
	[10]	S_6	14	2^{-34}
	This paper	S_1	13	2^{-29}
5	[5]	S_1	38	2^{-80}
	[10]	S_5	24	2^{-61}
	This paper	S_5	25	2^{-60}
6	[10]	S_1	41	2^{-97}
	This paper	S_4	38	2^{-93}

* This is also the upper bound presented in this paper.

Acknowledgment

The authors would like to thank Prof. R. Adler of the Industrial Engineering Department at the Technion for the fruitful discussions related to probability theory. The authors would also like to thank R. Anderson for his remarks concerning this paper.

References

1. R. Anderson, E. Biham, L.R. Knudsen, *Serpent: A Proposal for the Advanced Encryption Standard*, NIST AES Proposal, 1998.
2. E. Biham, *A Note on Comparing the AES Candidates*, Second AES Candidate Conference, 1999.
3. E. Biham, A. Shamir, *Differential Cryptanalysis of the Data Encryption Standard*, Springer-Verlag, 1993.
4. O. Dunkelman, *An Analysis of Serpent-p and Serpent-p-ns*, presented at the rump session of the Second AES Candidate Conference, 1999. Available on-line at *http://vipe.technion.ac.il/~orrd/crypt/*.
5. T. Kohno, J. Kelsey, B. Schneier, *Preliminary Cryptanalysis of Reduced-Round Serpent*, Third AES Candidate Conference, 2000.
6. J. Kelsey, T. Kohno, B. Schneier, *Amplified Boomerang Attacks Against Reduced-Round MARS and Serpent*, proceedings of Fast Software Encryption 2000, to appear.
7. L.R. Knudsen, *Truncated and Higher Order Differentials*, proceedings of Fast Software Encryption 2, Springer-Verlag, LNCS 1008, pp. 196–211, 1995.
8. NIST, *A Request for Candidate Algorithm Nominations for the AES*, available on-line at *http://www.nist.gov/aes/*.
9. D. Wagner, *The Boomerang Attack*, proceedings of Fast Software Encryption 1999, Springer Verlag, LNCS 1636, pp. 156–170, 1999.
10. X.Y. Wang, L.C.K. Hui, C.F. Chong, W.W. Tsang, H.W. Chan, *The Differential Cryptanalysis of an AES Finalist - Serpent*, Technical Report TR-2000-04. Available on-line at: *http://www.csis.hku.hk/research/techreps/*.

A   A 3-Round Differential Characteristic

Our 3-round differential characteristic is based on the one found in [5], where a 3-round differential characteristic with 7 active S boxes and probability 2^{-16} is presented. The characteristic is based on 4 active S boxes in the first round, 1 in the second round and 2 in the last round. The problem in finding characteristics is not finding the first round's input and the last round's output of the S boxes, as they can be chosen to have maximal probability. The problem is to have a minimal number of active S boxes, which is related to the output of the first round (which passes the linear transformation), the second round, and the input for the last round (as this determines the number of active S boxes in the last round).

We start by selecting the differences of the second round in a similar way to [5]. We observe that if we use the second round of the characteristic having S_3

instead of S_6 and having probability $1/8$ in S_3, we can ensure that all active S boxes in rounds 2 and 4 of the cipher (which are the first and third rounds of the characteristic, respectively) have probability $1/4$, thus having a total probability of 2^{-15}.

The 3-round differential characteristic with probability 2^{-15} that we get is as follows: In round 2 (or 10 or 18 or any other round having S_2) the following characteristic holds with probability 2^{-8}:

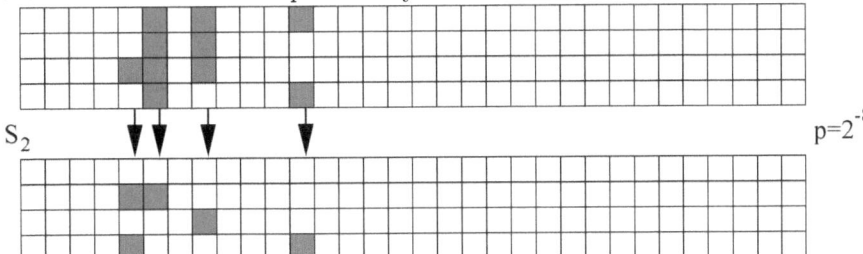

After the linear transformation and the application of S_3 we get the following differential characteristic with probability 2^{-3}:

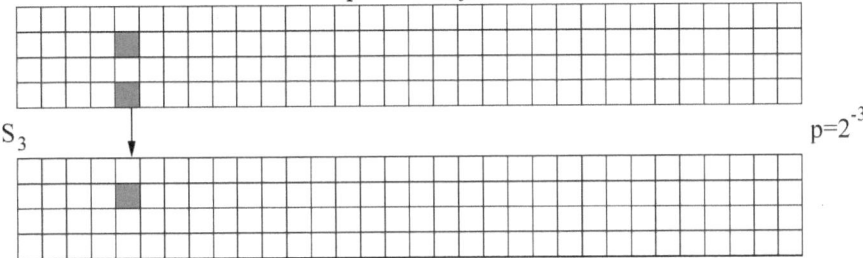

After the linear transformation and the application of S_4 we get the following differential characteristic with probability 2^{-4}:

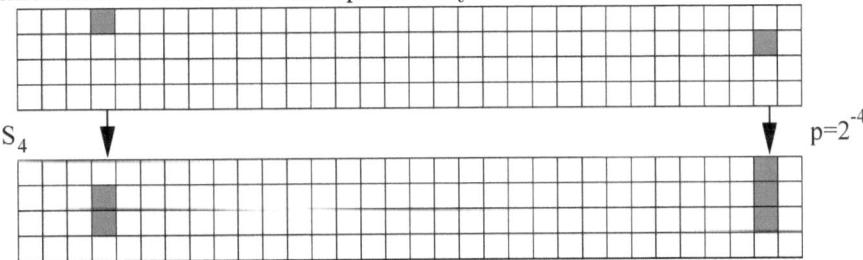

During the search for the best characteristic, we exhaustively checked all possible 3-round characteristics with 7 active S boxes and found this to be the best possible characteristic. As all 3-round characteristics have at least 7 active S boxes, and with 8 active S boxes the probability of the characteristic is at most 2^{-16}, this proves that this is the best 3-round differential characteristic of Serpent.

B   A 4-Round Differential Characteristics

B.1   A 4-Round Characteristic for Rounds 1–4

One option for achieving a minimal number of S boxes (13 according to [1]) is to have in the second round's S box S_2 $5 \rightarrow 4$ and $4 \rightarrow A_x$, and in the third round to have an active S box S_3 with $4 \rightarrow A_x$. Of course we would like to maximize the probabilities of these entries.

Checking the S boxes for such instances we found out that the best characteristic is when the first round of the 4-round characteristic is set at rounds using S_1. We receive the following 4-round differential characteristic with probability 2^{-29}:

In round 1 (or any other round having S_1) the following characteristic holds with probability 2^{-11}:

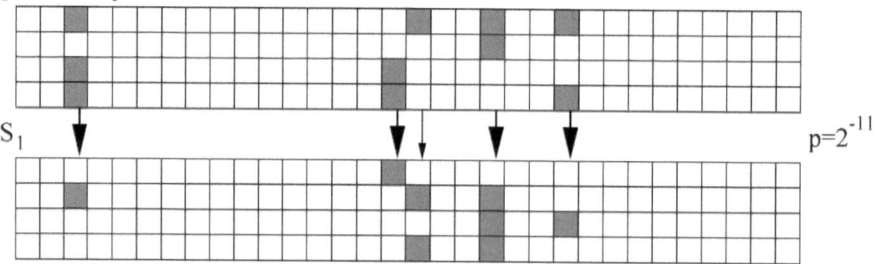

After the linear transformation and the application of S_2 we get the following differential characteristic with probability 2^{-5}:

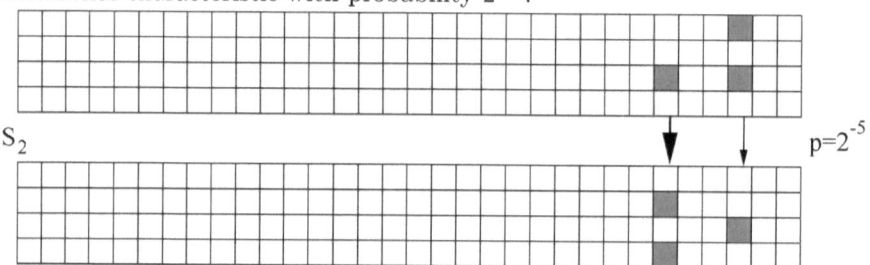

After the linear transformation and the application of S_3 we get the following differential characteristic with probability 2^{-3}:

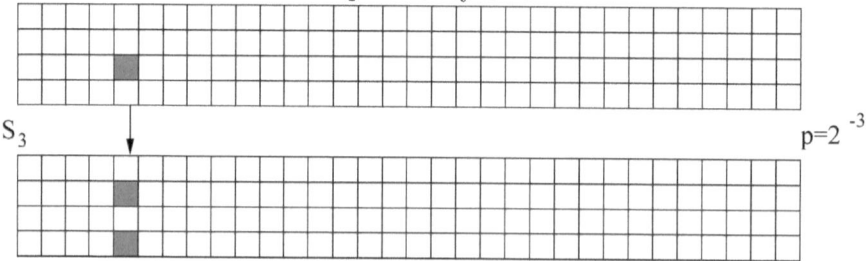

After the linear transformation and the application of S_4 we get the following differential characteristic with probability 2^{-10}:

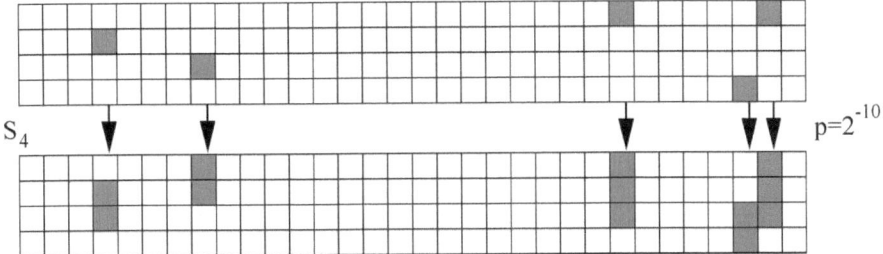

B.2   A 4-Round Characteristics for Rounds 5–8

This second 4-round differential characteristic is used along with the previous one in the attack of Section 5, and has probability of 2^{-47}. We use the basic characteristic described in [10], where a 5-round differential characteristic with probability 2^{-61} is described. As we need a characteristic of round 5–8, we remove the last round and get a 4-round characteristic with probability 2^{-48}. As part of our efforts to find higher probability differential characteristics for the amplified boomerang attack, we try a technique found very useful in previous attempts: we add another active S box in the first round. This might seem a bad thing (as this reduces the probability) but we found out that in exchange we get 3 more entries with probability $1/4$ instead of $1/8$. Thus, our characteristic has probability of 2^{-47}.

In round 5 (or any other round having S_5) the following characteristic holds with probability 2^{-24}:

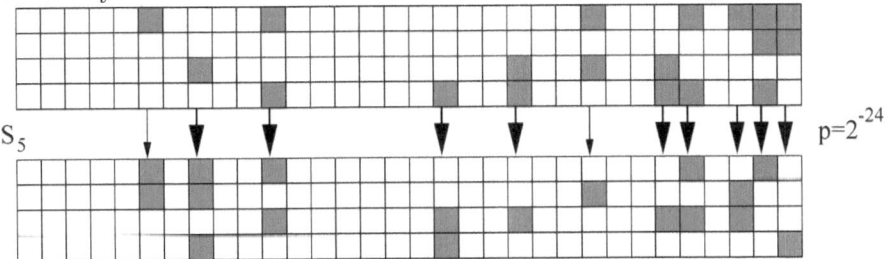

After the linear transformation and the application of S_6 we get the following differential characteristic with probability 2^{-16}:

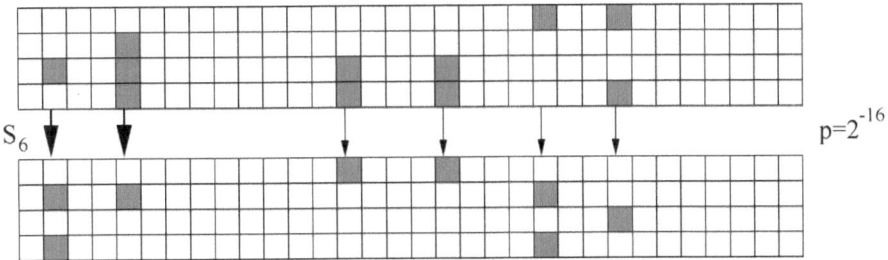

After the linear transformation and the application of S_7 we get the following differential characteristic with probability 2^{-5}:

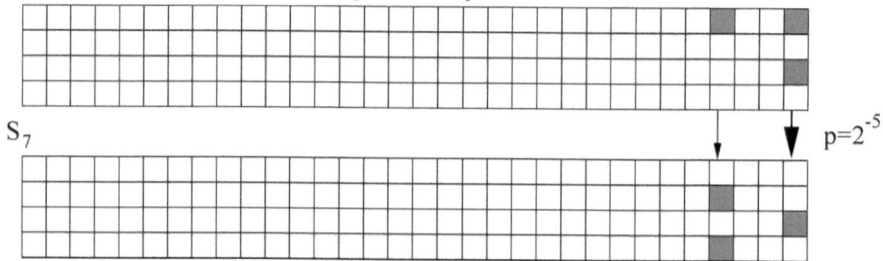

After the linear transformation and the application of S_0 we get the following differential characteristic with probability 2^{-2}:

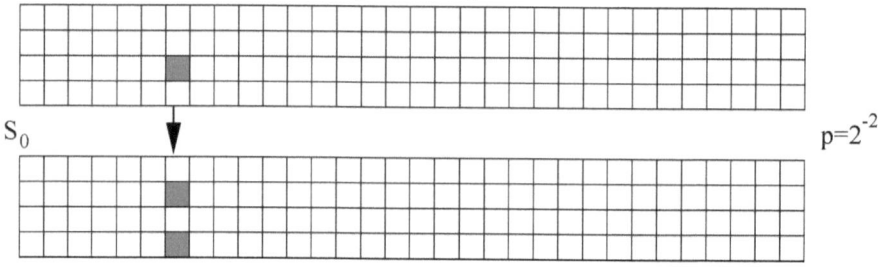

Note that the last two rounds are the same as in [10].

C   A 5-Round Differential Characteristic

As stated in Appendix B, we took a 5-round characteristic from [10], truncated it and improved it to have 4-round characteristic. By adding the last round from [10] back to the characteristic we get a 5-round characteristic with probability 2^{-60}.

Thus, we add after the 4th round of the characteristic from Appendix B.2 the following round, which apply S_1, and has probability of 2^{-13}:

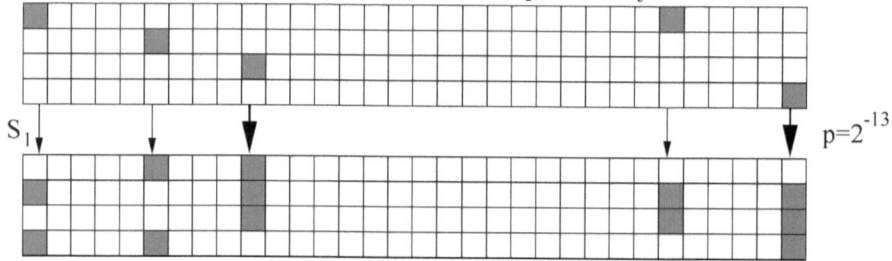

We have found another 5-round differential characteristic with probability 2^{-60}, and three more characteristics with probability 2^{-61} which are quite similar to the this one.

D A 6-Round Differential Characteristic

In order to get the best 6-round characteristic we can, we add a round before the 5-round characteristic from Appendix C and alter the first two rounds of it.

Thus, the 6-round characteristic starts in a round using S_4, and has probability 2^{-93}.

In round 4 (or any other round having S_4) the following characteristic holds with probability 2^{-28}:

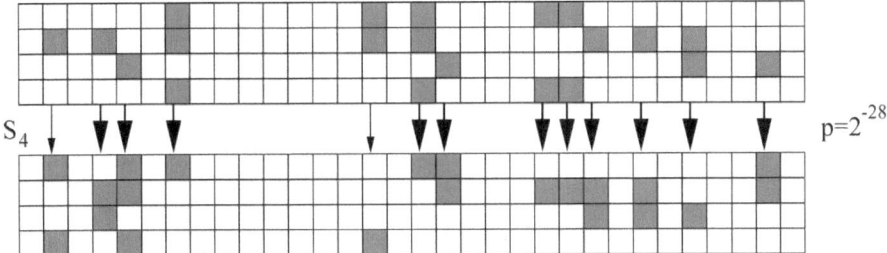

After the linear transformation and the application of S_5 we get the following differential characteristic with probability 2^{-29}:

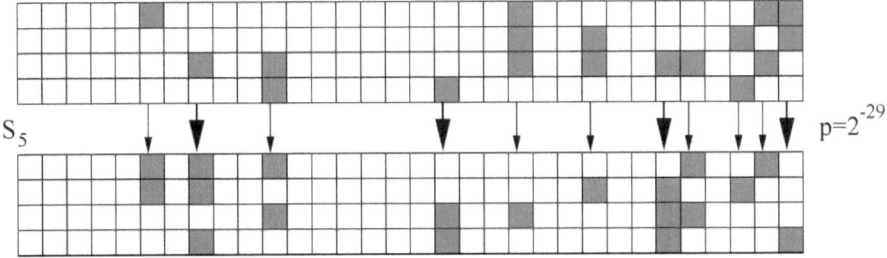

After the linear transformation and the application of S_6 we get the following differential characteristic with probability 2^{-16}:

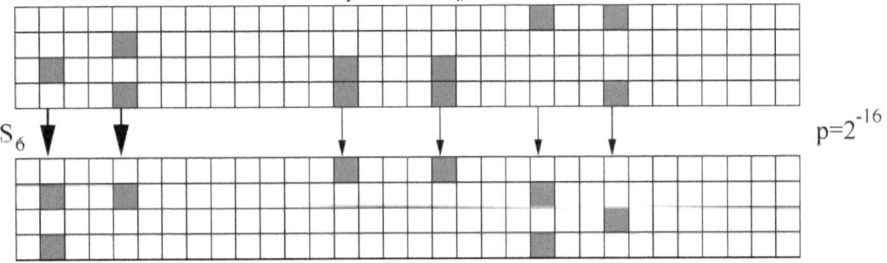

After this round the characteristic is the same as described for S_7, S_0 and S_1 in Appendices B.2 and C.

We observe that there are 2^{14} 6-round characteristics with the same last 5-rounds (only the input difference of the first round changes). This follows from the fact that in 2 S boxes in the first round we have 8 possible input differences with the same probability for the given output differences and in 8 S boxes we have two possibilities.

We also observed that by rotating all the characteristics one bit to the left (or two) the characteristics remain valid with the same probability (rotation by three or more bits does not work).

Efficient Amplification of the Security of Weak Pseudo-random Function Generators

Steven Myers

Department of Computer Science
University of Toronto
Toronto, Ontario, Canada
myers@cs.toronto.edu

Abstract. We show that given a PRFG (pseudo-random function generator) G which is $\frac{1}{c}$-partially secure, the construction $g_1(x \oplus r_1) \oplus \cdots \oplus g_{\log^2 n}(x \oplus r_{\log^2 n})$ produces a strongly secure PRFG, where $g_i \in G$ and r_i are strings of random bits. Thus we present the first "natural" construction of a (totally secure) PRFG from a partially secure PRFG. Using results of Luby and Rackoff, this result also demonstrates how to "naturally" construct a PRPG from partially secure PRPG.

1 Introduction

Cryptographers have noted that the Data Encryption Standard (DES) is effectively the composition of 16 insecure permutation generators. Because DES has withstood much cryptanalysis it is often both considered to be secure (given its small key size) and conjectured to be a Pseudo-Random Permutation Generator(PRPG). This construction has led some cryptographers to attempt to provide evidence that supports the apparent observation that the composition of permutation generators can amplify security.

Following this line of research, Luby and Rackoff [8] defined the notion of a partially secure PRPG to be a permutation generator which produces permutations that cannot be efficiently distinguished from random permutations by small circuits with a probability better than $\frac{1}{c}$, for some constant $c > 1$. They proved that the composition of a constant number of partially secure PRPGs results in a partially secure PRPG with stronger security then any of its constituent components. Unfortunately, Luby and Rackoff's result did not permit the construction of a PRPG from a partially secure PRPG.

It was known that a partially secure PRPG implied a totally secure PRPG. The construction used the following chain of results. It is possible to construct a weak one-way function from a partially secure PRPG; then, using [13,7], construct a one-way function; then, using Yao's XOR Lemma [4], construct a Pseudo-random number generator (PRNG); then, using [2], construct a PRFG; and then finally, using [9,12], construct a PRPG. However, this construction is obviously neither "natural" nor efficient.

In this paper we give a natural, efficient and parallelizable construction for generating a Pseudo-Random Function Generator(PRFG) from a partially se-

B. Pfitzmann (Ed.): EUROCRYPT 2001, LNCS 2045, pp. 358–372, 2001.
© Springer-Verlag Berlin Heidelberg 2001

cure PRFG. Our proof follows from the ideas of Luby and Rackoff [8]. Further, since partially secure PRPG are a special case of partially secure PRFG, we can use a partially secure PRPG to construct a PRFG. Then, using a previous result by Luby and Rackoff [9], or more recent work by Naor and Reingold [12], we can "naturally" and efficiently construct a PRPG from the PRFG. If $F = \{F^n\}$ is a "partially secure" pseudo-random function generator, then our construction is as follows:

$$f_1^n(x \oplus r_1) \oplus \cdots \oplus f_m^n(x \oplus r_m),$$

where the f_i^n's are randomly chosen from F^n, and the r_i's are randomly chosen from $\{0,1\}^n$. The key for this new generator consists of all the keys for the functions (f_i's), and all of the strings of random bits (r_i's).

Our construction is similar to an XOR product, and in this light, our proof might be considered an XOR lemma for PRFG. Further support for this this view is found in the fact that our proof closely follows that of Levin's in [7].

Given the relatively few number of proofs showing security amplification in an unrestricted adversarial model, we think this result will be of interest to those researchers interested in security amplification.

Further, we believe that this result can be viewed as one step in the long journey to developing a good theory for the development of block-ciphers. Currently, block-ciphers are developed primarily using heuristics, with little theory to guide the development of their underlying architecture. Thus, while there are no natural examples of partially secure PRFG that the author is aware of, should cipher-designers develop efficient function generators which they have reason to believe are partially secure, then they can use the construction suggested in this paper, and have good reason to believe that the resulting cipher has stronger security properties than its constituents.

For the purposes of example only, suppose block-cipher designers had reason to believe that an 8-round version of DES was a "partially" secure PRFG[1]. Then designers could have some faith that the suggested construction could be used amplify the security of this "partially" secure generator. Further, the parallelizability of the construction might allow designers to make certain time/space trade-offs. For example, the designers might trade-off the time required for more rounds of DES, with the circuit size required to implement the above construction with a version of DES with fewer rounds.

1.1 Related Work

There are very few results in cryptography which demonstrate the amplification of security in a general, non-restrictive adversarial model. The first such result was Yao's XOR Lemma [13], which now has several proofs ([7,5,3]). All of these results apply to the security amplification of weak one-way functions and predicates. In a domain closer to that of PRFG, Luby and Rackoff [8] give a direct product lemma for PRPG where the direct product is taken via the composition of weak PRPG. Unfortunately, their proof falls short of demonstrating that the

---

[1] We use the quotes around "partially" as DES in not an asymptotic notion

direct product of a sufficient number of weak PRPG yields a a strongly secure PRPG. The reason for this is explained in further detail in the sequel. A direct product theorem for PRFG is given by Myers [11], where the direct product is based on the composition and exclusive-or of PRFG. Unfortunately, this result also fails to achieve a strongly secure PRFG for reasons similar to those of [8]. Further complicating the matters with the result in [11] is the fact that the size of the constructed generator is super-polynomial after $\omega(\log n)$ applications of the direct product.

Therefore, our result presents the first efficient and natural direct product theorem achieving strongly secure PRFG from weakly secure PRFG in a general adversarial model.

Since Luby and Rackoff proposed their partial security model in [8], cryptographers have developed other models where it is possible to demonstrate some manner of security amplification. Kilian and Rogaway [6] propose a model where component permutation generators are replaced with completely random permutation generators. Constructions using the generators are then analyzed, and their security compared to that of a *random* permutation generator. Note that in this model, since the permutation generators are random, attacks can only be performed on the construction, and not the underlying component generators. Kilian and Rogaway call such attacks *generic*, as they do not make use of the underlying structure of the permutation generator.

As previously alluded to, under this model Kilian and Rogaway [6] have shown that the DESX construction increases the effective key length of DES. Also under the same model, Aiello et al. [1] have shown that the composition of multiple random permutation generators results in a permutation generator which is more secure than a random generator.

2   Notation, Definitions & the Model

Below we introduce some notation and terminology which will be used in the paper.

Notation 1. *For $\mu, \nu \in \{0,1\}^*$, let $\mu \bullet \nu$ denote their concatenation.*

Notation 2. *Let $\mathcal{F}^{l \to p}$ denote the set of all functions $f : \{0,1\}^l \to \{0,1\}^p$, and let $\mathcal{F}^n$ be the set $\mathcal{F}^{n \to n}$.*

Notation 3. *For $\alpha, \beta \in \{0,1\}^n$, let $\alpha \oplus \beta$ denote the bit-by-bit exclusive-or of α and β. For $f, g \in \mathcal{F}^n$, let $(f \oplus g)(\alpha)$ denote $f(\alpha) \oplus g(\alpha)$.*

Notation 4. *For any set A, let $x \in A$ be the action of uniformly at random choosing an element x from A. For any distribution $\mathcal{D}$, let $x \in \mathcal{D}$ be the action of randomly choosing an element according to $\mathcal{D}$.*

It will be clear from context when $\in$ is used to refer to an element in a set, and when it refers to choosing from a distribution.

Definition 1. *Let* $\mathcal{D}_1, \mathcal{D}_2, ...$ *be a sequence of distributions, and let e represent a series of events* $e_1, e_2, ...$ *such that for all i,* e_i *is an event of* $\mathcal{D}_i$. *We say that e occurs with* significant *probability if for some constant* $c > 0$ *and for infinitely many n the* $\mathrm{Pr}_{\mathcal{D}_n}(e_n) \geq \frac{1}{n^c}$. *We say that an event e occurs with* negligible *probability if, for all constants* $c > 0$ *and for all sufficiently large n,* $\mathrm{Pr}_{\mathcal{D}_n}(e_n) < \frac{1}{n^c}$.

2.1   Circuits

In the definition of each cryptographic primitive there exists the notion of an adversary. Abstractly, its purpose is to break an effect that a primitive is trying to achieve. Resource bounds are imposed on the adversaries, so that they model the computational power "real world" adversaries might feasibly have access to. There are two standard computational models which are used to define resource bounded adversaries: uniform and non-uniform. In this paper we will consider only non-uniform adversaries.

A non-uniform adversary is a sequence of circuits $(C_1, C_2, ...)$, where circuit C_i is used on inputs of size i. We wish to model efficient computation on the part of the adversary, so we assume that the size of each circuit C_i is bounded by $p(i)$, for some polynomial p. The size of a circuit is defined to be the number of gates, and the number of connections between gates in the circuit. For simplicity we assume we have gates for all 16 binary and 4 unary functions.

In order to model the adversaries of certain primitives, we allow the circuits to have access to an oracle. This is modeled by defining oracle gates to be gates of unit size which compute a specified function. The gates are otherwise treated like normal gates. An oracle function will normally be considered an input to the circuit.

We stress that the description of the circuit family need not be efficiently computable, even though each circuit is of small size relative to the size of its input.

Definition 2. *Let C be a circuit whose outputs are in the range* $\{0, 1\}$. *Then we say C is a* decision *circuit. Let x be an input to C. Then we say C* accepts *x if* $C(x) = 1$, *and we say that C* rejects *x if* $C(x) = 0$.

Definition 3. *We say a circuit C is* probabilistic, *if it requires as input a sequence of random bits.*

Notation 5. *Let* $\mathcal{D}$ *be a distribution over the inputs of a decision circuit C. Then we use as a shorthand* $\mathrm{Pr}_{d \in \mathcal{D}}(C(d))$ *to represent* $\mathrm{Pr}_{d \in \mathcal{D}}[C(d) = 1]$.

Definition 4. *Let* $\mathcal{D}$ *be a distribution over the inputs of a decision circuit C. We say that C accepts a fraction* $\mathrm{Pr}_{d \in \mathcal{D}}(C(d))$ *of its inputs, and rejects a fraction* $1 - \mathrm{Pr}_{d \in \mathcal{D}}(C(d))$ *of its inputs.*

Notation 6. *We write C^f to represent a circuit C that has oracle gates which compute the function f in unit time. We wish to consider these gates as "input" to the circuit, and therefore if f is of the form $\{0,1\}^n \to \{0,1\}^{m(n)}$, for a polynomial m, then we say that f is part of C's input and it has size n.*

Notation 7. *Let C be a circuit with access to the oracle function f. Then let Q_C denote the number of oracle gates in C (Note: Q is short for query).*

In the remainder of the paper we shall assume that all circuits are standardized in the following manner: no circuit will ever repeat oracle queries, and all circuits C_n in a circuit family $\{C_n\}$ will perform exactly $m(n)$ queries, for some polynomial m (ie. $Q_{C_n} = m(n)$). Any polynomial sized family of circuits can easily be modified to satisfy the above two requirements.

2.2   Function Generators

Definition 5. *We call $G : \{0,1\}^\kappa \times \{0,1\}^n \to \{0,1\}^m$ a function generator. We say that $k \in \{0,1\}^\kappa$ is a key of G, and we write $G(k, \cdot)$ as $g_k(\cdot)$, and say that key k chooses the function g_k. Let $g \in G$ represent the act of uniformly at random choosing a key k from $\{0,1\}^\kappa$, and then using the key k to choose the function g_k.*

Let m and ℓ be polynomials, and let $\mathcal{N} \subseteq \mathbb{N}$ be an infinitely large set. For each $n \in \mathcal{N}$, let $G^n : \{0,1\}^{\ell(n)} \times \{0,1\}^n \to \{0,1\}^{m(n)}$ be a function generator. We call $G = \{G^n | n \in \mathcal{N}\}$ a function generator ensemble.

In an abuse of notation, we will often refer to both specific function generators and function generator ensembles as function generators. We hope it will be clear from the context which term is actually being referred to.

Definition 6 (ϵ-Distinguishing Adversary). *Let $\epsilon : \mathbb{N} \to [0,1]$, and let $\mathcal{D}^1 = \{\mathcal{D}_i^1 | i \in \mathbb{Z}^+\}$ and $\mathcal{D}^2 = \{\mathcal{D}_i^2 | i \in \mathbb{Z}^+\}$ be two sequence of distributions over oracle gates, where $\mathcal{D}_i^j$ is a distribution over oracle gates of input size i, for $j \in \{1,2\}$. If $\{C_n\}$ is an adversary with access to oracle gates, then we say it is capable of ϵ distinguishing $\mathcal{D}^1$ from $\mathcal{D}^2$ if, for some polynomial p and infinitely many n:*

$$\left| \Pr_{d_1 \in \mathcal{D}^1} \left[C_n^{d_1} = 1 \right] - \Pr_{d_2 \in \mathcal{D}^2} \left[C_n^{d_2} = 1 \right] \right| \geq \epsilon(n) + \frac{1}{p(n)}.$$

Definition 7 (Pseudo-Random Function Generator Ensembles). *Let m and ℓ be polynomials. For each n let $G^n : \{0,1\}^{\ell(n)} \times \{0,1\}^n \to \{0,1\}^{m(n)}$ be a function generator, computable in time bounded by a polynomial in n. Define $G = \{G^n | n \in \mathbb{N}\}$ to be the function generator ensemble. Define $\mathcal{F} = \{\mathcal{F}^{n \to m(n)} | n \in \mathbb{N}\}$.*

We say that G is $(1 - \epsilon(n))$ secure if there exists no adversary $\{C_n\}$, bound in size to be polynomial in n, which can ϵ distinguish G from $\mathcal{F}$.

We say that G is a pseudo-random function generator (PRFG) if it is 1 secure.

Definition 8. *If G is a 1-secure generator, we say it is* strongly secure. *If G is $\frac{1}{p(n)}$ secure, for some polynomial p, then we say that it is* partially secure. *If G is not partially secure, then we say it is* insecure.

2.3   Previously Known Lemmas

Below is a well known form of the Chernoff bound. For a proof of this result refer to [10] or any standard book on probabilistic computation.

Lemma 1 (Chernoff Bound). *Let $x_1, .., x_{n^t}$ be i.i.d.r.v. which take the values 0 or 1 with probabilities q or $p = 1 - q$ respectively. Let $X_{n^t} = \frac{1}{n^t} \sum_{i=1}^{n^t} x_i$. Then for any k and l, there exists a t such that:*

$$\Pr\left[|X_{n^t} - p| \geq \frac{1}{n^k}\right] \leq \frac{1}{2^{n^l}}.$$

The following lemma is a generalization of standard derandomization proofs in the non-uniform computation model. Before stating the lemma, we give the following intuition of its statement. Let $\mathcal{D}_1$ and $\mathcal{D}_2$ be two distribution over oracle functions, and P be a predicate with a domain over functions. Then if C is a probabilistic circuit such that $C^{\mathcal{D}_1}$ approximates $P(\mathcal{D}_1)$ and $C^{\mathcal{D}_2}$ approximates $P(\mathcal{D}_2)$, then there exists a derandomized version of C which approximates both $P(\mathcal{D}_1)$ and $P(\mathcal{D}_2)$.

Lemma 2 (Derandomization Lemma). *Let $C^w(r)$ be a probabilistic oracle-circuit, where w is an oracle function, and r is a string of random input bits. Let D^1 and D^2 be two distributions over $\mathcal{F}^n$, and let R be the distribution over C's random bits. Let $P : \mathcal{F}^n \times \mathbb{R} \to \{0, 1\}$ be a predicate. Then, If*

$$\Pr_{w \in D^1, r \in R}[P(w, C^w(r)) = 1] \geq 1 - p \quad and \quad \Pr_{w \in D^2, r \in R}[P(w, C^w(r)) = 1] \geq 1 - p,$$

then there exists an $\tilde{r} \in R$ such that $\Pr_{w \in D^i}[P(w, C^w(\tilde{r})) = 1] \geq 1 - 2p$, for $i \in \{1, 2\}$.

Proof. This result is a generalization of standard derandomization techniques for non uniform circuits. The details are left to the full version of the paper. □

3   Result

We will show that there is a "natural" construction which constructs strongly secure PRFGs from $1 - \delta$ secure PRFGs. The construction we present uses function generators that generate functions of the form $f : \{0, 1\}^n \to \{0, 1\}^n$, this is done to simplify the presentation. The result can easily be modified to generate functions of the form $f : \{0, 1\}^n \to \{0, 1\}^{m(n)}$, for any polynomial m. The construction is based on the operator generator described below.

Let f_1 and f_2 be two functions such that $f_i : \{0, 1\}^n \to \{0, 1\}^n$, for $i \in \{1, 2\}$. For each $r_1, r_2 \in \{0, 1\}^n$ we define the operator $\diamondsuit_{r_1 \bullet r_2}^n$, which acts on the functions f_1 and f_2 and produces a function of type $\{0,1\}^n \to \{0,1\}^n$ as defined below:

$$(f_1 \Diamond^n_{r_1 \bullet r_2} f_2)(x) = f_1(x \oplus r_1) \oplus f_2(x \oplus r_2).$$

We define the $\Diamond$ operator generator (read Diamond) as $\Diamond = \{\Diamond^n_{r_1,r_2} | n \in \mathbb{N} \wedge r_1, r_2 \in \{0,1\}^n\}$.

Before describing the construction, we will formally describe how to combine two function generators using the $\Diamond$ operator generator.

Definition 9. *Let* $G = \{G^n : \{0,1\}^{\ell(n)} \times \{0,1\}^n \to \{0,1\}^n | n \in \mathbb{N}\}$ *be a function generator ensemble. Let* $H = \{H^n : \{0,1\}^{\kappa(n)} \times \{0,1\}^n \to \{0,1\}^n | n \in \mathbb{N}\}$ *be a function generator ensemble. Let* $\Diamond$ *be the operator generator defined previously. Then let* $F = \{F^n : \{0,1\}^{\ell(n)+\kappa(n)+2 \cdot n} \times \{0,1\}^n \to \{0,1\}^n | n \in \mathbb{N}\}$ *be the function generator defined by* $F^n(k_1 \bullet k_2 \bullet k_3 \bullet k_4, x) = \left(g^n_{k_1} \Diamond^n_{k_3 \bullet k_4} h^n_{k_2}\right)(x)$, *where* $|k_1| = \ell(n)$, $|k_2| = \kappa(n)$ *and* $|k_3| = |k_4| = n$. *This is written in shorthand as* $F = G \Diamond H$.

Similarly, if $g : \{0,1\}^n \to \{0,1\}^n$, *then we write* $g \Diamond H$ *as short-hand for the function generator defined by* $F^n(k_2 \bullet k_3 \bullet k_4, x) = \left(g \Diamond^n_{k_3 \bullet k_4} h^n_{k_2}\right)(x)$, *where* $|k_2| = \kappa(n)$ *and* $|k_3| = |k_4| = n$.

3.1   The Construction

Let p be a polynomial. We construct the generator F from the generator G as follows:

$$F = \underbrace{G \Diamond \cdots \Diamond G}_{p(n)}.$$

Note that in order to compute a random function $f^n \in F$ it is sufficient to compute

$$\left(g_1(x \oplus r_1) \oplus \cdots g_{p(n)}(x \oplus r_n)\right),$$

where $g_i \in G$ and $r_i \in \{0,1\}^n$.

Observe that the key for F includes $p(n)$ keys for G and $p(n)$ random strings. The random strings are necessary for the security amplification, and a counter example to our security amplification claims can easily be constructed if they are omitted. For further discussion on this construction and several other plausible candidates see [11].

In order to prove the security of the construction we use the Diamond Isolation Lemma (the name for this lemma comes from the stylistically similar Isolation Lemma used by Levin [7] in proving Yao's XOR Lemma [13]) stated below. Intuitively, the lemma shows that the function generator which results from the combination of two partially secure function generators by the $\Diamond$ operator generator is more secure than either of the two constituent generators. The majority of the work in this paper goes towards proving this lemma correct.

Lemma 3 (Diamond Isolation Lemma). *There exists a fixed polynomial p (which is retrievable form the proof of this lemma) such that the following hold. Let $\epsilon, \delta : \mathbb{Z} \to [0,1]$ be functions. Let H and G be function generators, where*

$c_G(n)$ and $c_H(n)$ are polynomials which bound from above the size of the circuits which compute the function generators respectively.

Hypothesis: *There exists a family of decision-circuits $\{C_n\}$, where for each n the circuit C_n is of size bounded above by the polynomial $s_C(n)$, and for some $c > 0$ and infinitely many n:*

$$\left| \Pr_{g \in G^n \Diamond H^n} (C_n^g) - \Pr_{f \in \mathcal{F}_n} (C_n^f) \right| \geq \epsilon(n)\delta(n) + \frac{1}{n^c}.$$

Conclusion: *For infinitely many n there exists either a decision-circuit Υ_n of size $p(n^c \cdot c_G(n))s_C(n)$ for which:*

$$\left| \Pr_{h \in H^n} (\Upsilon_n^h) - \Pr_{f \in \mathcal{F}_n} (\Upsilon_n^f) \right| \geq \epsilon(n) + \frac{1}{n^{3c}};$$

or a decision-circuit Ξ_n of size $\leq (2Q_{C_n}c_H(n) + s_C(n))$, where $Q_{\Xi_n} = Q_{C_n}$, and for which:

$$\left| \Pr_{g \in G^n} (\Xi_n^g) - \Pr_{f \in \mathcal{F}_n} (\Xi_n^f) \right| \geq \delta(n) + \frac{1}{n^{2c}}.$$

Luby and Rackoff prove a similar lemma in [8]. It shows that the composition of two partially secure PRPGs results in a generator which is more secure than either of the constituents. Excluding the fact that their lemma is restricted to permutation generators instead of function generators, our lemma is stronger in two senses. First, the security requirement in the hypothesis is strictly weaker (ie. the improvement in security from combining the two generators is stronger in our result). Second, the size of the distinguishing circuit for G is only additively larger than the distinguishing circuit for G$\Diamond$H. In the Luby and Rackoff construction, the distinguishing circuits for G and H are both multiplicatively larger than the circuit which distinguishes G$\circ$H. It is this second fact that permits us to achieve PRFGs in our construction. Furthermore, this proof is simpler than that of Luby and Rackoff. Their proof contains a corollary which corresponds to Corollary 5 in our proof. However, unlike Corollary 5, their corollary is only proven true with respect to the computational security of G$\circ$H. This restriction is necessary for their construction, but increases the difficulty of the proof. We now prove that our construction produces a PRFG from a $1 - \epsilon$ secure PRFG.

Theorem 1 (Diamond Composition Theorem). *Let $0 < \epsilon < 1$ be a constant. Let G be a $1 - \epsilon$ secure PRFG. Then for each $p \in \Omega(\log^2 n) \cap \left(\cup_{i=1}^\infty \mathcal{O}(n^i)\right)$ the generator $\mathsf{F} = \underbrace{\mathsf{G}\Diamond \cdots \Diamond \mathsf{G}}_{p(n)}$ is a secure PRFG.*

Proof. (sketch) The intuition for this argument is as follows. We assume that F is not secure, and thus there is a family of distinguishing circuits for F. We apply the Isolation Lemma to the generator F. The result is either that the generator G is not $1 - \epsilon$ secure as claimed, or we have a family of distinguishing circuits (slightly larger than the original circuit family) for a generator smaller than F. We apply the Isolation Lemma inductively to this smaller generator until we are only left with an $\epsilon + \frac{1}{n^c}$ family of distinguishing circuits for the generator G, which contradicts its assumed $1 - \epsilon$ security. The theorem follows. The full details are left to the full version of the paper.   □

Before presenting a proof of the Diamond Isolation Lemma (Lemma 3), we first present two important technical lemmas. A complete proof of the Diamond Isolation Lemma follows.

3.2   Two Technical Lemmas

The first lemma and corollary demonstrate that the acceptance probability of an oracle-decision-circuit is the same whether the circuit is given an oracle chosen uniformly at random from the set of all functions; or given an oracle chosen uniformly at random from the set of all functions combined with any specific function using the $\diamond$ operator generator.

Lemma 4. *Given any decision-circuit C, for each $h \in \mathcal{F}^n$ and for each $r_1, r_2 \in \{0,1\}^n$:*

$$\Pr_{\phi \in h \diamond^n_{r_1 \bullet r_2} \mathcal{F}^n} (C^\phi) = \Pr_{f \in \mathcal{F}^n} (C^f).$$

Proof. First observe that for each $r_2 \in \{0,1\}^n$ the distribution $\{h'(x \oplus r_2) | h' \in \mathcal{F}^n\} = \mathcal{F}^n$. Then let $g(x) = h(x \oplus r_1)$, and observe that the distribution $g \oplus \mathcal{F}^n = \mathcal{F}^n$, proving the result.

Corollary 1. *Given any decision-circuit C, for each $h \in \mathcal{F}^n$: $\Pr_{\phi \in h \diamond \mathcal{F}^n}(C^\phi) = \Pr_{f \in \mathcal{F}^n}(C^f)$.*

The next lemma demonstrates that the probability of acceptance by a *polynomial sized* oracle-decision-circuit is "almost" the same whether given access to an oracle chosen uniformly at random from the set of all functions; or given an oracle chosen randomly from the set of functions specified by combining, via the $\diamond$ operator generator, any distribution of functions with "almost" any specific function.

Lemma 5. *Let $\{C_n\}$ be a polynomial sized family of decision-circuits. Then for every constant c, for sufficiently large n, for each $s \in \mathcal{F}^n$, for all but $\frac{1}{2^{n/4}}$ of the $w \in \mathcal{F}^n$:*

$$\left| \Pr_{g \in s \diamond w} (C_n^g) - \Pr_{f \in \mathcal{F}^n} (C_n^f) \right| < \frac{1}{n^c}.$$

Proof. (sketch) Below we outline the high-level ideas behind the proof of the lemma. We leave the detailed proof for the full version of the paper.

In the remainder of this proof sketch, when we say a value has a good approximation, we imply it approximates the value to within a $\frac{1}{\text{poly}(n)}$-additive factor, where poly(n) can be any polynomial. Further, when we say an approximation is good it is implicit that we mean that it is good with very high probability (greater than $(1 - \frac{1}{2^{cn}})$ for some $c > 0$).

We define an experiment that has a random variable that is a good approximation to both $\Pr_{g \in s \Diamond w}(C_n^g)$ and $\Pr_{f \in \mathcal{F}^n}(C_n^f)$. A direct result is that for most $w \in \mathcal{F}^n$ the value $|\Pr_{g \in s \Diamond w}(C_n^g) - \Pr_{f \in \mathcal{F}^n}(C_n^f)|$ is small, and the result follows.

The major work involved in proving this lemma involves showing that the random variable in the experiment approximates both of the aforementioned values.

We define an experiment in which we draw uniformly at random a function $w \in \mathcal{F}^n$ and a set of $p(n)$ keys from $\{0, 1\}^{2n}$ for the $\Diamond$ operator, $\{(k_i^1 \bullet k_i^2)\}$, where p is a polynomial. We define the random variable:

$$\frac{1}{p(n)} \sum_{i=1}^{p(n)} C_n^{(s \Diamond_{k_i^1 \bullet k_i^2} w)} \tag{1}$$

It is clear, by the Chernoff bound, that p can be chosen so that (1) is a good approximation of $\Pr_{g \in s \Diamond w}(C_n^g)$.

In order to demonstrate that (1) also approximates $\Pr_{f \in \mathcal{F}^n}(C_n^f)$, we show that it is a good approximation of a second random variable, which itself closely approximates $\Pr_{f \in \mathcal{F}^n}(C_n^f)$.

We define a second experiment as choosing uniformly at random $q(n)$ functions from $\mathcal{F}^n$, where q is a polynomial. We define the second random variable as:

$$\frac{1}{q(n)} \sum_{i=1}^{q(n)} C_n^{f_i}. \tag{2}$$

By the Chernoff bound, for an appropriate q, the random variable (2) is a good approximation for $\Pr_{f \in \mathcal{F}^n}(C_n^f)$. Therefore, it suffices to show that the random variable (1) is a good approximation for (2).

We show that (1) and (2) are good approximations of each other by defining a third experiment in which both random variables can be calculated. In this experiment, with very high probability the random variables are equal, and therefore they are good approximations of each other.

In the third experiment we draw uniformly at random a polynomial (in n) number of random strings, $\{r_i\}$, from $\{0, 1\}^n$ and a polynomial number of keys from $\{0, 1\}^{2n}$ for the $\Diamond$ operator, $\{k_i^1 \bullet k_i^2\}$.

Observe that the random variable (2) can be calculated in this experiment: any call to an oracle-gate during the computation of C^{f_i} can be answered with a random bit-string r_j. (Recall C is of a special form: it never makes the same oracle query twice.)

Unfortunately, it's not as easy to calculate (1) in the third experiment. As w was chosen at random in the first experiment, for any i we can calculate the value $C^{(s \Diamond_{k_i^1 \bullet k_i^2} w)}$ by replacing the outputs of the oracle gates with random bit-strings. Unfortunately, the calculation of (1) requires the evaluation of $C^{(s \Diamond_{k_i^1 \bullet k_i^2} w)}$ for a polynomial number of values of i. These evaluations are not independent, and therefore the scheme used to calculate (2) is not a valid method for computing (1). The problem is that during the evaluations of $C^{(s \Diamond_{k_a^1 \bullet k_a^2} w)}$ and $C^{(s \Diamond_{k_b^1 \bullet k_b^2} w)}$

the respective queries x and y could be made to oracle gates, where $x \oplus k_a^2 = y \oplus k_b^2$, and in such cases the outputs of the gates are dependent on each other. Fortunately, we can show that the probability of such an event occurring is negligible and that this is the only case in which we cannot replace the output of the oracle gates with random strings to simulate the calculation of (1). Therefore, with high probability, the values of (1) and (2) are equal in third experiment. The lemma follows. $\square$

3.3   Proof of the Isolation Lemma

Assume that there exists a polynomial-sized decision-circuit family $\{C_n\}$ which for some constant $c > 0$ and infinitely many n, $\left|\Pr_{g \in G^n \diamond H^n}(C_n^g) - \Pr_{f \in \mathcal{F}^n}(C_n^f)\right| \geq \epsilon(n)\delta(n) + \frac{1}{n^c}$. WLOG we assume that $\Pr_{g \in G^n \diamond H^n}(C_n^g) - \Pr_{f \in \mathcal{F}^n}(C_n^f) \geq \epsilon(n)\delta(n) + \frac{1}{n^c}$, as otherwise we can simply flip the output bit of C_n.

Lemma 6. *For $i > 0$ and for each n let*

$$K_n(i) = \Pr_{f \in \mathcal{F}^n}(C_n^f) + \frac{1}{n^i} \quad \text{and let } S^n(i) = \left\{ w \in \mathcal{F}^n \,\middle|\, \Pr_{g \in G^n \diamond w}(C_n^g) \geq K_n(i) \right\}.$$

Then for all i,j: $\Pr_{w \in \mathcal{F}^n}(w \in S^n(i)) \leq \frac{1}{n^j}$, for sufficiently large n.

Proof. Suppose for contradiction that there exists an i and j such that for infinitely many n $\Pr_{w \in \mathcal{F}^n}(w \in S^n(i)) \geq \frac{1}{n^j}$. We will show this contradicts Lemma 5. We first note that since $\Pr_{\phi \in G^n \diamond S^n(i)}(C_n^\phi) \geq \Pr_{f \in \mathcal{F}^n}(C_n^f) + \frac{1}{n^i}$, then by an averaging argument we can fix a $g \in G^n$ such that $\Pr_{h \in g \diamond S^n(i)}(C_n^h) \geq \Pr_{f \in \mathcal{F}^n}(C_n^f) + \frac{1}{n^i}$. Then using the first moment method we note that given g, there must be a fraction $\frac{1}{n^{2i}}$ of $w \in S^n(i)$ which have the "good" property that $\Pr_{\psi \in g \diamond w}(C_n^\psi) \geq \Pr_{f \in \mathcal{F}^n}(C_n^f) + \frac{1}{n^{2i}}$. Since $S^n(i)$ is also a "significant" $(\frac{1}{n^j})$-fraction of $\mathcal{F}^n$, the probability that a random w has the "good" property is $\frac{1}{n^{2i+j}}$, and this contradicts Lemma 5. $\square$

Lemma 7. *Either there exists a family of decision-circuits $\{\Xi_n\}$, where for each n the circuit Ξ_n is of size $\leq Q_{C_n} 2 c_H(n) + s_C(n)$; $Q_{\Xi_n} = Q_{C_n}$; and for infinitely many n:*

$$\left| \Pr_{g \in G^n}(\Xi_n^g) - \Pr_{f \in \mathcal{F}^n}(\Xi_n^f) \right| \geq \delta(n) + \frac{1}{n^{2c}};$$

or for all sufficiently large n and all $h^n \in H^n$:

$$\left| \Pr_{g \in G \diamond h^n}(C_n^g) - \Pr_{f \in \mathcal{F}^n}(C_n^f) \right| < \delta(n) + \frac{1}{n^{2c}}.$$

Proof. Suppose it is the case that for infinitely many n there exists an $h^n \in H^n$ such that $\left|\Pr_{g \in G \diamond h^n}(C_n^g) - \Pr_{f \in \mathcal{F}^n}(C_n^f)\right| \geq \delta(n) + \frac{1}{n^{2c}}$. For each such n we create a decision circuit Ξ_n, where $\Xi_n^w = C_n^{(w \diamond h^n)}$. We observe that:

$$\left| \Pr_{\psi \in G^n}(\Xi_n^\psi) - \Pr_{f \in \mathcal{F}^n}(\Xi_n^f) \right| = \left| \Pr_{\psi \in G^n \diamond h^n}(C_n^\psi) - \Pr_{f \in \mathcal{F}^n \diamond h^n}(C_n^f) \right|$$

$$= |\Pr_{\psi \in \mathsf{G}^n \Diamond h^n}(C_n^{\psi}) - \Pr_{f \in \mathcal{F}^n}(C_n^f)| \qquad \text{(Corollary 1)}$$

$$\geq \delta(n) + \frac{1}{n^{2c}}$$

It is easy to see that C_n can be modified, in a straightforward manner, by adding $Q_{C_n}(C_{\mathsf{H}}(n) + 10n)$ gates and wires to compute $\varXi_n$, while still using Q_{C_n} oracle gates. For simplicity of presentation in this paper we have assumed that $7n \leq C_{\mathsf{H}}(n)$, giving us a circuit of size $\leq s_C(n) + Q_{C_n}(2C_{\mathsf{H}}(n))$.   □

Main Argument. We now present the main argument for proving the Diamond Isolation Lemma. WLOG, we assume that

$$\Pr_{g \in \mathsf{G}^n \Diamond \mathsf{H}^n}(C_n^g) - \Pr_{f \in \mathcal{F}^n}(C_n^f) \geq \epsilon(n)\delta(n) + \frac{1}{n^c}, \qquad (3)$$

if this is not the case flip the output bit of C_n.

We assume that there exists no family of circuits $\{\varXi_n\}$, where each circuit $\varXi_n$ is of size $c_{\mathsf{H}}(n) + s_C(n)$, such that for infinitely many n:

$$\left|\Pr_{g \in \mathsf{H}^n}(\varXi_n^g) - \Pr_{f \in \mathcal{F}^n}(\varXi_n^f)\right| \geq \delta(n) + \frac{1}{n^{2c}}.$$

From the above assumption and Lemma 7, we know that for all sufficiently large n and all $h^n \in \mathsf{H}^n$:

$$\left|\Pr_{\psi \in \mathsf{G} \Diamond h^n}(C_n^{\psi}) - \Pr_{f \in \mathcal{F}^n}(C_n^f)\right| < \delta(n) + \frac{1}{n^{2c}}. \qquad (4)$$

We now outline the argument. By (3), C_n accepts a fraction of $\mathsf{G}^n \Diamond \mathsf{H}^n$ which is "significantly larger" than $\epsilon(n)\delta(n) + \Pr_{C_n}(\mathcal{F}^n)$. However, by (4), for each $h \in \mathsf{H}^n$ not much more than a $\delta(n) + \Pr_{C_n}(\mathcal{F}^n)$ fraction of the functions in $\mathsf{G}^n \Diamond h$ are accepted by C_n. As $\Pr_{\phi \in \mathsf{G} \Diamond h}(C_n^{\phi})$ is the expected value of $\Pr_{\phi \in \mathsf{G} \Diamond h}(C_n^{\phi})$ over the distribution H^n, it must be the case that $\Pr_{\phi \in \mathsf{G} \Diamond h}(C_n^{\phi})$ is "significantly larger" than $\Pr_{f \in \mathcal{F}^n}(C_n^f)$ for at least an $\epsilon(n)$ fraction of the $h \in \mathsf{H}^n$. Given a function ω our distinguishing circuit will approximate $\Pr_{\psi \in \mathsf{G} \Diamond \omega}(C_n^{\psi})$ and accept if it is "significantly larger" than $\Pr_{f \in \mathcal{F}^n}(C_n^f)$. By the above argument this will accept an $\epsilon(n)$ fraction of the functions in H^n and, by Lemma 6, the same circuit will accept almost no random functions in $\mathcal{F}^n$. We now give the details of the proof outlined above.

Since we cannot compute $\Pr_{\phi \in \mathsf{G}^n \Diamond \omega}(C_n^{\phi})$ in polynomial time, we approximate it with the probabilistic circuit A_n:

$$A_n^w = \frac{1}{n^b} \sum_{i=1}^{n^b} C_n^{(g_i \Diamond_{k_i^1 \bullet k_i^2}^n w)},$$

where $g_1, ..., g_{n^b} \in \mathsf{G}^n$ and $k_1^1, k_1^2, .., k_{n^b}^1, k_{n^b}^2 \in \{0, 1\}^n$ are randomly chosen. Let $\kappa(n)$ be the length of the key of H^n, and set (with foresight) $\alpha > 1$ so that

$n^\alpha > \kappa(n)$. Using the Chernoff Bound, b is chosen large enough such that:

$$\Pr_{w \in \mathcal{F}^n}\left[\left|A_n^w - \Pr_{\phi \in G^n \Diamond w}(C_n^\phi)\right| \geq \frac{1}{n^{3c}}\right] \leq \frac{1}{2n^{2\alpha}},$$

and

$$\Pr_{h \in \mathsf{H}^n}\left[\left|A_n^h - \Pr_{\phi \in G^n \Diamond h^n}(C_n^\phi)\right| \geq \frac{1}{n^{3c}}\right] \leq \frac{1}{2n^{2\alpha}}.$$

Since we want a deterministic circuit we derandomize A_n, by Lemma 2, to get the circuit B_n, such that for all but $\frac{1}{2n^\alpha}$ of the $w \in \mathcal{F}^n$:

$$\left|B_n^w - \Pr_{\phi \in G^n \Diamond w}(C_n^\phi)\right| < \frac{1}{n^{3c}}, \tag{5}$$

and for all of the $h \in \mathsf{H}^n$:

$$\left|B_n^h - \Pr_{\phi \in G^n \Diamond h}(C_n^\phi)\right| < \frac{1}{n^{3c}}, \tag{6}$$

since for each $k \in \{0,1\}^{\kappa(n)}$ the probability of picking h_k^n from H^n is at least $\frac{1}{2^{\kappa(n)}} > \frac{1}{2n^\alpha}$.

Claim.

$$\Pr_{h \in \mathsf{H}^n}[B_n^h \geq \Pr_{f \in \mathcal{F}^n}(C_n^f) + \frac{1}{n^{2c}}] \geq \epsilon(n) + \frac{1}{n^{2c}}.$$

Proof. Assume for contradiction that $\Pr_{h \in \mathsf{H}^n}[B_n^h \geq \Pr_{f \in \mathcal{F}^n}(C_n^f) + \frac{1}{n^{2c}}] < \epsilon(n) + \frac{1}{n^{2c}}$. Let $\mathcal{K}^n \subseteq \mathsf{H}^n$ be the set of functions $h \in \mathsf{H}^n$, for which $B_n^h \geq \Pr_{f \in \mathcal{F}^n}(C_n^f) + \frac{1}{n^{2c}}$, and let $\overline{\mathcal{K}^n}$ be its complement.

$$\Pr_{\phi \in G \Diamond \mathsf{H}}(C_n^\phi) - \Pr_{f \in \mathcal{F}^n}(C_n^f) = \sum_{h \in \mathcal{K}^n}\left(\left(\Pr_{\phi \in G \Diamond h}(C_n^\phi) - \Pr_{f \in \mathcal{F}^n}(C_n^f)\right)\Pr_{\psi \in \mathsf{H}^n}[\psi = g]\right) +$$

$$\sum_{h \in \overline{\mathcal{K}^n}}\left(\left(\Pr_{\phi \in G \Diamond h}(C_n^\phi) - \Pr_{f \in \mathcal{F}^n}(C_n^f)\right)\Pr_{\psi \in \mathsf{H}^n}[\psi = h]\right)$$

$$\leq \sum_{h \in \mathcal{K}^n}\left(\left(\Pr_{\phi \in G \Diamond h}(C_n^\phi) - \Pr_{f \in \mathcal{F}^n}(C_n^f)\right)\Pr_{\psi \in \mathsf{H}^n}[\psi = h]\right) +$$

$$\sum_{h \in \overline{\mathcal{K}^n}}\left(\left(\left(B_n^h - \Pr_{f \in \mathcal{F}^n}(C_n^f)\right) + \frac{1}{n^{2c}}\right)\Pr_{\psi \in \mathsf{H}^n}[\psi = h]\right)$$

$$\leq \sum_{h \in \mathcal{K}^n}\left(\left(\Pr_{\phi \in G \Diamond h}(C_n^\phi) - \Pr_{f \in \mathcal{F}^n}(C_n^f)\right)\Pr_{\psi \in \mathsf{H}^n}[\psi = h]\right) +$$

$$\left(1 - \epsilon(n) - \frac{1}{n^{2c}}\right)\frac{1}{n^{2c}} \tag{7}$$

$$\leq \left(\epsilon(n) + \frac{1}{n^{2c}}\right)\left(\delta(n) + \frac{1}{n^{2c}}\right) + \frac{1}{n^{2c}} \tag{8}$$

$$< \epsilon(n)\delta(n) + \frac{1}{n^c}. \qquad \text{(contradiction)} \tag{9}$$

Equation (7) follows from two facts. First, by assumption, the probability that a random $h \in H^n$ is in $\overline{\mathcal{K}^n}$ is $1 - \epsilon(n) - \frac{1}{n^{2c}}$. Second, for each $h \in \overline{\mathcal{K}^n}$, $B_n^h - \Pr_{f \in \mathcal{F}^n}(C_n^f) < \frac{1}{n^{2c}}$. Equation (8) follows from two facts. First, by assumption, $\Pr_{h \in H^n}[h \in \mathcal{K}^n] < \epsilon(n) + \frac{1}{n^{2c}}$. Second, by (4), for each $h \in H^n$, $\Pr_{\phi \in G \diamond h}(C_n^\phi) - \Pr_{f \in \mathcal{F}^n}(C_n^f) < \delta(n) + \frac{1}{n^{2c}}$. Equation (9) contradicts the fact that $\Pr_{\phi \in G \diamond H}(C_n^\phi) - \Pr_{f \in \mathcal{F}^n}(C_n^f) \geq \epsilon(n)\delta(n) + \frac{1}{n^c}$. $\qquad \Box$

We create the decision circuit Υ_n^w which accepts w iff $B_n^w \geq \Pr_{f \in \mathcal{F}^n}(C_n^f) + \frac{1}{n^{2c}}$.

$$\Pr_{h \in H^n}(\Upsilon_n^h) - \Pr_{f \in \mathcal{F}^n}(\Upsilon_n^f) \geq \epsilon(n) + \frac{1}{n^{2c}} - \Pr_{f \in \mathcal{F}^n}(\Upsilon_n^f)$$

$$\geq \epsilon(n) + \frac{1}{n^{2c}} - \frac{1}{2n^\alpha} -$$

$$\Pr_{w \in \mathcal{F}^n}\left[\Pr_{g \in G^n \diamond w}(C_n^g) \geq \Pr_{f \in \mathcal{F}^n}(C_n^f) + \frac{1}{n^{2c}} - \frac{1}{n^{3c}}\right] \quad (10)$$

$$\geq \epsilon(n) + \frac{1}{n^{2c}} - \frac{1}{n^{3c}} - \frac{1}{2n^\alpha} \quad (11)$$

$$\geq \epsilon(n) + \frac{1}{n^{3c}} \qquad \text{(For sufficiently large } n\text{)}$$

Equation (10) follows as B_n^ω approximates $\Pr_{g \in G^n \diamond \omega}(C_n^g)$ to within a factor of $\frac{1}{n^{3c}}$ for all but $\frac{1}{2n^\alpha}$ of the $\omega \in \mathcal{F}^n$. Equation (11) follows by a direct application of Lemma 6.

By performing the straightforward construction of Υ_n, we see that there does exist a fixed polynomial p mentioned in the statement of the lemma for which the size of Υ_n is bound by $p(n^c \cdot c_G(n))s_C(n)$. $\qquad \Box$

4   Discussion and Further Research

We have presented a relatively simple and efficient construction for transforming a partially secure PRFG into a strongly secure PRFG. We believe this construction could possibly be used to guide the development of block-ciphers in the future. However, as described in the introduction, the construction may be useful only in outer layers of the cipher, after a certain minimal amount of security has been achieved by other means – possibly by the time proven method of using composition.

Further, as one of the anonymous referees pointed out, it appears possible to show in the Kilian Rogaway model [6] that the construction can be used to increase the effective key-length of a block-cipher. This would appear to give further evidence of the benefit of using the construction in practice. Further, since the construction is parallelizable it may be preferable to 3-DES for extending the key-lengths of DES. However, since the resulting generator is a function generator and not a permutation generator, there will be systems and applications where this is an infeasible approach.

Acknowledgments

The author would like to thank Charles Rackoff for suggesting the problem and for many valuable discussions and suggestions.

References

1. W. Aiello, M. Bellare, G. Di Crescenzo, and R. Vekatesan. Security amplification by composition: The case of doubly-iterated, ideal ciphers. In H. Krawczyk, editor, *Advances in Cryptology - Crypto 98*, volume 1462 of *LNCS*, pages 390–407. Springer-Verlag, 1998.
2. O. Goldreich, S. Goldwasser, and S. Micali. How to construct random functions. *Journal of the ACM*, 33(4):792–807, 1986.
3. O. Goldreich, N. Nisan, and A. Wigderson. On yao's xor-lemma. http://theory.lcs.mit.edu/~oded/, 1995.
4. J. Hastad, R. Impagliazzo, L.A. Levin, and M. Luby. Construction of pseudorandom generator from any one-way function. *Accepted to the SIAM Journal of Computing*, 28(4):1364–1396, 1998.
5. R. Impagliazzo. Hard core distributions for somewhat hard problems. http://www-cse.ucsd.edu/~russell/, 1994.
6. J. Kilian and P. Rogaway. How to protect DES against exhaustive key search. In N. Koblitz, editor, *Advances in Cryptology – Crypto 96*, volume 1109 of *LNCS*, pages 252–267. Springer-Verlag, 1996.
7. L.A. Levin. One-way functions and pseudorandom generators. *Combinatorica*, 7(4):357–363, 1987.
8. M. Luby and C. Rackoff. Pseudo-random permutation generators and cryptographic composition. In *Proceedings of the 18th Annual Symposium on Theory of Computing*, pages 353–363. ACM, 1986.
9. M. Luby and C. Rackoff. How to construct pseudorandom permutations from pseudorandom functions. *SIAM Journal on Computing*, 17:373–386, 1988.
10. Rajeev Motwani and Prabhakar Raghavan. *Randomized Algorithms*. Cambridge University Press, 1995.
11. Steven Myers. *On the Development of Pseudo-Random Function Generators and Block-Ciphers using the XOR and Composition Operators*. M.Sc. Thesis. University of Toronto, Canada, 1999.
12. Moni Naor and Omer Reingold. On the construction of pseudo-random permutations: Luby-Rackoff revisited. *Journal of Cryptology*, 12(1):29–66, 1999.
13. Andrew Yao. Theory and applications of trapdoor functions (extended abstract). In *Proceedings of the 23rd Symposium on Foundations of Computer Science*, pages 80–91. IEEE, 1982.

Min-round Resettable Zero-Knowledge
in the Public-Key Model

Silvio Micali and Leonid Reyzin

Laboratory for Computer Science
Massachusetts Institute of Technology
Cambridge, MA 02139
reyzin@theory.lcs.mit.edu
http://theory.lcs.mit.edu/~reyzin

Abstract. In STOC 2000, Canetti, Goldreich, Goldwasser, and Micali put forward the strongest notion of zero-knowledge to date, *resettable zero-knowledge* (RZK) and implemented it in constant rounds in a new model, where the verifier simply has a public key registered before any interaction with the prover.

To achieve ultimate round efficiency, we advocate a slightly stronger model. Informally, we show that, as long as the honest verifier does not use a given public key more than a fixed-polynomial number of times, there exist 3-round (which we prove optimal) RZK protocols for all of NP.

1 Introduction

THE NOTION OF RESETTABLE ZERO-KNOWLEDGE. A zero-knowledge (ZK) proof [GMR89], is a proof that conveys nothing but the verity of a given statement. As put forward by Canetti, Goldreich, Goldwasser, and Micali [CGGM00], *resettable zero-knowledge* (RZK) is the strongest form of zero-knowledge known to date. In essence, an RZK proof is a proof that remains ZK even if a polynomial-time verifier can force the prover to execute the proof multiple times with the same coin tosses. More specifically,

- *The verifier can reset the prover.* In each execution, the verifier can choose whether the prover should execute the protocol with a new random tape or with one of the tapes previously used.
- *The verifier can arbitrarily interleave executions.* The verifier can always start (in particular, in a recursive way) new executions in the middle of old ones, and resume the old ones whenever it wants.
- *The prover is oblivious.* As far as it is concerned, the prover is always executing a single instance of the protocol.

Resettable ZK is a strengthening of Dwork, Naor and Sahai's [DNS98] notion of concurrent ZK (CZK). In essence, in a CZK protocol, a malicious verifier acts

D. Pfitzmann (Ed.): EUROCRYPT 2001, LNCS 2045, pp. 373–393, 2001.
© Springer-Verlag Berlin Heidelberg 2001

as in an RZK protocol, except that it lacks the power of resetting the prover's random tapes.

CONSTRUCTING RZK PROTOCOLS. Perhaps surprisingly, it is possible to implement such a strong notion: RZK proofs for NP-complete languages are constructed in [CGGM00] under standard complexity assumptions. Their construction is concretely obtained by properly modifying the CZK protocol of Richardson and Kilian [RK99]. Because this underlying CZK protocol is not constant-round, neither is the resulting RZK protocol. (The minimum known number of rounds for implementing the protocol of [RK99] is polylogarithmic in the security parameter, as shown by Kilian and Petrank [KP00].)

Unfortunately, it may not be possible to obtain a constant-round RZK protocol: at least in the black-box model, Canetti, Kilian, Petrank and Rosen [CKPR01] recently proved that no constant-round protocol exists even for CZK. However, [CGGM00] also put forward an appealingly simple model, which we call the *bare public-key* (BPK) model, and provide a 5-round[1] RZK argument for any NP language in this model.

Let us now quickly recall what their model is.

THE BARE PUBLIC-KEY MODEL. An interactive proof system in the BPK model simply assumes that the verifier V has a public key, PK, that is registered before any interaction with the prover begins. No special protocol needs to be run to publish PK, and no authority needs to check any property of PK. It suffices for PK to be a string known to the prover, and chosen by the verifier prior to any interaction with the prover.

The BPK model is very simple. In fact, it is a *weaker* version of the frequently used public-key infrastructure (PKI) model, which underlies any public-key cryptosystem or digital signature scheme. In the PKI case, a secure association between a key and its owner is crucial, while in the BPK case no such association is required. The single security requirement of the BPK model is that a bounded number of keys (chosen beforehand) are "attributable" to a given user.[2]

We have recently pointed out in [MR01] that the BPK model has four distinct notions of soundness, depending on the power enjoyed by a malicious prover P^*: informally,

1. *one-time soundness*, arising when P^* is allowed a single interaction with V per theorem statement;
2. *sequential soundness*, arising when P^* is allowed multiple but sequential interactions with V;

---

[1] Their paper actually presents two related constructions: (1) a 4-round protocol with an additional 3-round preprocessing stage with a trusted third party, and (2) an 8-round protocol without such preprocessing. Their constructions can be easily modified to yield the 5-round protocol attributed above.

[2] Indeed, having a prover P work with an incorrect public key for a verifier V does not affect soundness nor resettable zero-knowledgeness; at most, it may affect completeness. (Working with an incorrect key may only occur when an active adversary is present— in which case, strictly speaking, completeness does not even apply: this fragile property only holds when all are honest.)

3. *concurrent soundness*, arising when P^* is allowed multiple interleaved interactions with the same V; and
4. *resettable soundness*, arising when P^* is allowed to reset V with the same random tape *and* interact with it concurrently.

As we have already said, the BPK model permits constant-round RZK protocols for all of NP. Indeed, the CGGM protocol is 5-round and sequentially sound, and we have recently constructed a 4-round one that also is sequentially sound [MR01]. To achieve ultimate round efficiency, we advocate strengthening the BPK model a bit.

THE UPPERBOUNDED PUBLIC-KEY MODEL. How many public keys can a verifier establish before it interacts with the prover? Clearly, no more than a polynomial (in the security parameter) number of keys. Though innocent-looking, this bound is a source of great power for the BPK model: it allows for the existence of constant-round black-box RZK, which is impossible in the standard model.

How many times can a verifier use a given public key? Of course, at most a polynomial (in the security parameter) number of times. Perhaps surprisingly, we show that if such an innocent-looking polynomial upperbound U is made explicit *a priori*, then we can further increase the round efficiency of RZK protocols.

In our *upperbounded public-key* (UPK) model, the honest verifier is allowed to fix a polynomial upperbound, U, on the number of times a public key will be used; keep track, via a counter, of how many times the key has been used; and refuse to participate once the counter has reached the upperbound.

Let us now make the following remarks about the UPK model:

- In the RZK setting, the "strong party" is the verifier (who controls quite dramatically the prover's executions). Such a strong party, therefore, should have no problems in keeping a counter in order to save precious rounds of interaction.
- The UPK model does not assume that the prover knows the current value of the verifier's counter. (Guaranteeing the accuracy of such knowledge would de facto require public keys that "change over time.")
- While our RZK protocol satisfies interesting efficiency constraints with respect to U, we believe that these should be considered properties of our specific protocol rather than requirements of the UPK model.
  (For instance, our public key length is independent of U, while the secret key length and each execution of the protocol depend on U only logarithmically. Only the verifier's key-generation phase depends linearly on U —a dependency that hopefully will be improved by subsequent protocols.)
- The UPK model is somewhat similar to the one originally envisaged in [GMR88] for secure digital signatures, where the signer posts an explicit upperbound on the number of signatures he will produce relative to a given public key, and keeps track of the number of signatures produced so far. (The purpose and use of our upperbound, however, are totally different.)
- While sufficient for constant-round implementation of the stronger RZK notion, the UPK model is perhaps simpler than those of [DS98] (which uses

"timing assumptions") and that of [Dam00] (which uses trusted third parties to choose some system parameters) for efficient implementation of CZK.

3-ROUND RZK IN THE UPK MODEL. Because the powerful RZK notion seems to require substantial interaction, it is important to establish how many rounds a reasonable model can save. As we have already said, the BPK model can reduce the number of rounds to four [MR01]. We show that the UPK model can do even better, reducing the number of rounds to the minimum and, at the same time, increasing soundness:

Main Theorem: *In the UPK model there exist 3-round concurrently sound RZK arguments for any language in NP, assuming collision-resistant hashing and the subexponential hardness of discrete logarithm and integer factorization.*[3]

ROUND-OPTIMALITY OF THE UPK MODEL. Our result is optimal (in either the UPK or the BPK model), at least for black-box RZK. This fact is evident from the following argument. Assume that a 2-round RZK (or even just ZK!) protocol (P, V) existed, in the BPK or the UPK model, for a language $L \notin$ BPP. Then one could construct from it a 3-round ZK protocol (P', V') by adding an initial round in which the verifier sends its public key PK to the prover.[4] Protocol (P', V') would thus contradict the result of [GK96], which states that no 3-round, black-box ZK proofs or arguments exist for non-trivial languages.

NECESSITY OF THE UPK MODEL. In the cited [MR01], we also show that it is impossible in the BPK model to achieve 3-round ZK with concurrent soundness. Thus, to achieve 3-round RZK, one needs either to come up with a protocol that is sequentially (but not concurrently) sound, or to enhance the model in some reasonable fashion. The former approach seems quite elusive, and whether such a protocol exists remains an open problem. Our solution is an example of the latter approach.

2   Resettable Zero-Knowledge in the UPK Model

In this section, we define RZK in the UPK model. Let us refer the reader to the original exposition of [CGGM00] for motivation and intuition of RZK, which we do not provide here due to space constraints. Here we focus on:

---

[3]   We can replace the integer factorization assumption with the more general assumption that subexponentially secure dense public-key cryptosystems [DDP00] and subexponentially secure certified trapdoor permutations [FLS99] exist. Or we can replace both the DL and the factorization assumptions with the assumption that decision Diffie-Hellman is subexponentially hard.

[4]   Note that the so constructed (P', V') will not be RZK (else, being 4-round, it would contradict the recent lowerbound of [CKPR01]—and indeed even the older lowerbound of [KPR98]). However, it will still be ZK. To see this, observe that the old black-box simulator, designed to handle very powerful resetting malicious verifier (who can choose from among multiple public keys in the public file) can be also used with the weaker standard verifier (who simply uses only a single public key transmitted in the first message).

- RZK *arguments* (rather than proofs). That is, we assume that the prover is polynomial-time and we let soundness hold in a computational (rather than probabilistic) sense. Our protocol in Section 4 and the public key protocol of [CGGM00] are RZK arguments.
- *Black-box* zero-knowledgeness. That is, we demand that there exist a single simulator that works for all malicious verifiers V^* (given oracle access to V^*). This is a stronger notion, and is indeed the one we satisfy in Section 4.

The Players

Let

- A *public file F* be a polynomial-size collection of records (id, PK_{id}), where id is a string identifying a verifier, and PK_{id} is its (alleged) public key.
- A *prover P (for a language L)* be an interactive deterministic polynomial-time TM that is given as inputs (1) a security parameter 1^n, (2) a n-bit string $x \in L$, (3) an auxiliary input y, (4) a public file F, (5) a verifier identity id, and (6) a random tape ω.
  For simplicity of exposition, one can view P as a *non-interactive* TM that is given, as an additional input, the entire history of the messages already exchanged in the interaction, and outputs the next message. Fixing all inputs, this view allows one to think of $P(1^n, x, y, F, id, \omega)$ as a simple deterministic oracle, which is helpful in defining the notion of RZK below.
- A *U-bounded (honest) verifier V*, for a positive polynomial U, be an interactive polynomial-time TM that, on first input a security parameter 1^n, works in $U(n) + 1$ stages, with the ability of keeping state information. In the first *key generation* stage, on input a security parameter 1^n, V outputs a public key PK and remembers the corresponding secret key SK. In subsequent $U(n)$ *verification* stages, on input an n-bit string x, V performs an interactive protocol with a prover.
- An *(s,t)-resetting verifier V^**, for any two positive polynomials t and s, be a TM that runs in two stages so that, on first input 1^n,
  1. In stage 1, V^* receives $s(n)$ values $x_1, \ldots, x_{s(n)} \subset L$ of length n each, and outputs an arbitrary public file F and a list of $s(n)$ identities $id_1, \ldots, id_{s(n)}$.
  2. In stage 2, V^* starts in the final configuration of stage 1, is given oracle access to $s(n)^3$ provers, and then outputs whatever it desires (in particular, it can output its "view" of the interactions, which includes its random string).
  3. The total number of steps of V^* in both stages is at most $t(n)$.
- A *black-box simulator M* be a polynomial-time machine that is given oracle access to V^*. By this we mean that it can run V^* multiple times, each time picking V^*'s inputs, random tape and (because V^* makes oracle queries itself) the answers to all of V^*'s queries. M is also given $s(n)$ values $x_1, \ldots, x_{s(n)} \in L$ as input.

The Definitions

To define RZK in the UPK model, we must define (1) completeness, (2) soundness and (3) resettable zero-knowledgeness proper. For lack of space, we omit a formal discussion of completeness in the UPK model. (This property is the usual one for interactive proofs, except that it has to hold only for the first $U(n)$ interactions, and to assume that $\mathcal{P}$ gets the correct public key for $\mathcal{V}$.) For the same reason, we omit a formal discussion of concurrent soundness, the type of soundness actually enjoyed by our protocol and informally specified in our introduction. (The reader is referred to [MR01] for formal details.) The third notion is the same as in [CGGM00]. Nonetheless, we find it useful to recall it below.

Definition 1. $(\mathcal{P}, \mathcal{V})$ *is* black-box resettable zero-knowledge for an NP-language L *if there exists a simulator M such that for every pair of positive polynomials (s, t), for every (s, t)-resetting verifier $\mathcal{V}^*$, for every $x_1, \ldots, x_{s(n)} \in L$ and their corresponding NP-witnesses $y_1, \ldots, y_{s(n)}$, the following probability distributions are indistinguishable (in time polynomial in n):*

1. *The output of $\mathcal{V}^*$ obtained after choosing $\omega_1, \ldots, \omega_{s(n)}$ uniformly at random, running the first stage of $\mathcal{V}^*$ to obtain F, and then letting $\mathcal{V}^*$ interact in its second stage with the following $s(n)^3$ instances of $\mathcal{P}$: $\mathcal{P}(x_i, y_i, F, id_k, \omega_j)$ for $1 \le i, j, k \le s(n)$.*
2. *The output of M with input $x_1, \ldots, x_{s(n)}$ interacting with $\mathcal{V}^*$.*

3   Tools

Let us quickly recall the notation, the definitions and the constructions that we utilize in our protocol.

3.1   Probabilistic Notation

(The following is taken verbatim from [BDMP91] and [GMR88].) If $A(\cdot)$ is an algorithm, then for any input x, the notation "$A(x)$" refers to the probability space that assigns to the string σ the probability that A, on input x, outputs σ. If S is a probability space, then "$x \xleftarrow{R} S$" denotes the algorithm which assigns to x an element randomly selected according to S. If F is a finite set, then the notation "$x \xleftarrow{R} F$" denotes the algorithm that chooses x uniformly from F.

If p is a predicate, the notation $\mathrm{PROB}[x \xleftarrow{R} S; y \xleftarrow{R} T; \cdots : p(x, y, \cdots)]$ denotes the probability that $p(x, y, \cdots)$ will be true after the ordered execution of the algorithms $x \xleftarrow{R} S; y \xleftarrow{R} T; \cdots$. The notation $[x \xleftarrow{R} S; y \xleftarrow{R} T; \cdots : (x, y, \cdots)]$ denotes the probability space over $\{(x, y, \cdots)\}$ generated by the ordered execution of the algorithms $x \xleftarrow{R} S, y \xleftarrow{R} T, \cdots$.

3.2   Trapdoor Commitment Schemes

In this section we present trapdoor commitment schemes that are secure against subexponentially strong adversaries (satisfying an additional key-verification property).[5]

Informally, a trapdoor commitment scheme consists of a quintuple of algorithms. Algorithm TCGen generates a pair of matching public and secret keys. Algorithm TCCom takes two inputs, a value v to be committed to and a public key, and outputs a pair, (c, d), of commitment and decommitment values. Algorithm TCVer takes the public key and c, v, d and checks whether c was indeed a commitment to v.

What makes the commitment *computationally binding* is that without knowledge of the secret key, it is computationally hard to come up with a single commitment c and two different decommitments d_1 and d_2 for two different values v_1 and v_2 such that TCVer would accept both c, v_1, d_1 and c, v_2, d_2. What makes it *perfectly secret* is that the value c yields no information about the value v. Moreover, this has to hold even if the public key is chosen adversarially. Thus, there has to be an algorithm TCKeyVer that takes a public key as input and verifies whether the resulting commitment scheme is indeed perfectly secret. (More generally, TCKeyVer can be an interactive protocol between the committer and the key generator, rather than an algorithm; however, for our application, the more restricted view suffices).

Perfect secrecy ensures that, information-theoretically, any commitment c can be decommitted arbitrarily: for any given commitment c to a value v_1, and any value v_2, there exists d_2 such that TCVer accepts c, v_2, d_2 and the public key (indeed, if for some v_2 such d_2 did not exist, then c would leak information about the actual committed value v_1). The trapdoor property makes this assurance computational: knowing the secret key enables one to decommit arbitrarily through the use of the TCFake algorithm.

Definition 2. *A* Trapdoor Commitment Scheme *(TC) is a quintuple of probabilistic polynomial-time algorithms* TCGen, TCCom, TCVer, TCKeyVer *and* TCFake, *such that*

1. *Completeness.* $\forall n, \forall v,$

$$\text{PROB}[(TCPK, TCSK) \xleftarrow{R} \text{TCGen}(1^n) \, ; \, (c, d) \xleftarrow{R} \text{TCCom}(TCPK, v) \, :$$
$$\text{TCKeyVer}(TCPK, 1^n) = \text{TCVer}(TCPK, c, v, d) = \text{YES}] = 1$$

2. *Computational Soundness.* $\exists\, \alpha > 0$ *such that for all sufficiently large n and for all 2^{n^α}-gate adversaries* ADV

$$\text{PROB}[\, (TCPK, TCSK) \xleftarrow{R} \text{TCGen}(1^n) \, ;$$
$$(c, v_1, v_2, d_1, d_2) \xleftarrow{R} \text{ADV}(1^n, TCPK) \, :$$
$$\text{TCVer}(TCPK, c, v_1, d_1) = \text{YES and}$$
$$\text{TCVer}(TCPK, c, v_2, d_2) = \text{YES and } v_1 \neq v_2] < 2^{-n^\alpha}$$

We call α the soundness constant.

---

[5] We follow a similar discussion in [CGGM00] almost verbatim.

3. Perfect Secrecy. $\forall$ $TCPK$ such that $\text{TCKeyVer}(TCPK, 1^n) = \text{YES}$ and $\forall v_1, v_2$ of equal length, the following two probability distributions are identical:

$$[(c_1, d_1) \overset{R}{\leftarrow} \text{TCCom}(TCPK, v_1) : c_1] \quad \text{and}$$
$$[(c_2, d_2) \overset{R}{\leftarrow} \text{TCCom}(TCPK, v_2) : c_2]$$

4. Trapdoorness. $\forall$ $(TCPK, TCSK) \in \{\text{TCGen}(1^n)\}$, $\forall v_1, v_2$ of equal length the following two probability distributions are identical:

$$[\, (c, d_1) \overset{R}{\leftarrow} \text{TCCom}(TCPK, v_1)\,;$$
$$d_2' \overset{R}{\leftarrow} \text{TCFake}(TCPK, TCSK, c, v_1, d_1, v_2) : (c, d_2')\,] \quad \text{and}$$
$$[\, (c, d_2) \overset{R}{\leftarrow} \text{TCCom}(TCPK, v_2) : (c, d_2)\,]$$

(In particular, the above states that faked commitments are correct: indeed, $d_2' \overset{R}{\leftarrow} \text{TCFake}(TCPK, TCSK, c, v_1, d_1, v_2)$ implies that $\text{TCVer}(TCPK, c, v_2, d_2') = \text{YES}$)

In this paper, we will also require that the relation $(TCPK, TCSK)$ be polynomial-time; this is easy to satisfy by simply including the random string used in key generation into the secret key.

Such commitment schemes can be constructed, in particular, based on a subexponentially strong variant of the Discrete Logarithm assumption. We refer the reader to [BCC88] (where, in Section 6.1.2, it is called a DL-based "chameleon blob") for the construction.

3.3 Hash-Based Commitment Schemes

We also have a need of non-trapdoor, non-interactive, computationally-binding commitment schemes (which, unlike trapdoor commitments, need not be secure against subexponentially strong adversaries). Because of the absence of the trapdoor requirement, these simpler commitment schemes can be implemented more efficiently if one replaces perfect secrecy by the essentially equally powerful property of *statistical secrecy* (i.e., even with infinite time one can get only a statistically negligible advantage in distinguishing the commitments of any two different values). In particular [DPP97,HM96] show how to commit to any value by just one evaluation of a collision-free hash function $H : \{0,1\}^* \rightarrow \{0,1\}^k$. To differentiate trapdoor commitments from these simpler ones, we shall call them *hash-based commitments*.

Though the trapdoor property does not hold, we still insist that, given any commitment and any value, it is possible in time 2^k to decommit to that value.

Definition 3. *A Hash-Based Commitment Scheme (HC) is a pair of probabilistic polynomial-time algorithms* HCCom, HCVer, *along with the algorithm* HCFake *that runs in time 2^kpoly when its first input is k and poly is some polynomial in the size of its input, such that*

1. Completeness. $\forall k$, $\forall v$,

$$\text{PROB}[(c,d) \xleftarrow{R} \text{HCCom}(1^k, v) : \text{HCVer}(1^k, c, v, d) = \text{YES}] = 1$$

2. Computational Soundness. *For all probabilistic polynomial-time machines* ADV, *and all sufficiently large k,*

$$\text{PROB}[(c, v_1, v_2, d_1, d_2) \xleftarrow{R} \text{ADV}(1^k) :$$
$$v_1 \neq v_2 \text{ and } \text{HCVer}(1^k, c, v_1, d_1) = \text{YES} = \text{HCVer}(1^k, c, v_2, d_2)]$$

is negligible in k.

3. Statistical Secrecy. $\forall v_1, v_2$ *of equal length, the statistical difference between the following two probability distribution is negligible in k:*

$$[(c_1, d_1) \xleftarrow{R} \text{HCCom}(1^k, v_1) : c_1] \text{ and } [(c_2, d_2) \xleftarrow{R} \text{HCCom}(1^k, v_2) : c_2]$$

4. Breakability. $\forall v_1, v_2$ *of equal length, the statistical difference between the following two probability distribution is negligible in k:*
$[(c, d_1) \xleftarrow{R} \text{HCCom}(1^k, v_1) ; d_2' \xleftarrow{R} \text{HCFake}(1^k, c, v_1, d_1, v_2) : (c, d_2')]$ *and*
$[(c, d_2) \xleftarrow{R} \text{HCCom}(1^k, v_2) : (c, d_2)]$

We refer the reader to [DPP97,HM96] for the constructions of such schemes, which are based on the assumption that collisions-resistant hash functions exist.

3.4 Non-interactive Zero-Knowledge Proofs of Knowledge

Non-interactive zero-knowledge (NIZK) proofs for any language $L \in \text{NP}$ were put forward and exemplified in [BFM88,BDMP91]. Ordinary ZK proofs rely on interaction. NIZK proofs replace interaction with a random *shared string*, σ, that enters the view of the verifier that a simulator must reproduce. Whenever the security parameter is 1^n, σ's length is $\text{NI}\sigma\text{Len}(n)$, where $\text{NI}\sigma\text{Len}$ is a fixed, positive polynomial.

Let us quickly recall their definition, modified for polynomial-time provers and security against subexponentially strong adversaries.

Definition 4. *Let non-interactive prover* NIP *and non-interactive verifier* NIV *be two probabilistic polynomial-time algorithms, and let* $\text{NI}\sigma\text{Len}$ *be a positive polynomial. We say that* (NIP, NIV) *is a NIZK argument system for an NP-language L if*

1. Completeness. $\forall\ x \in L$ *of length n, σ of length $\text{NI}\sigma\text{Len}(n)$, and NP-witness y for x,*

$$\text{PROB}[\Pi \xleftarrow{R} \text{NIP}(\sigma, x, y) : \text{NIV}(\sigma, x, \Pi) = \text{YES}] = 1.$$

2. Soundness. $\forall\ x \notin L$ *of length n,*

$$\text{PROB}[\sigma \xleftarrow{R} \{0,1\}^{\text{NI}\sigma\text{Len}(n)} : \exists\ \Pi \text{ s. t. } \text{NIV}(\sigma, x, \Pi) = \text{YES}]$$

is negligible in n.

3. Zero-Knowledgeness. $\exists\ \alpha > 0$ and a probabilistic polynomial-time simulator NIS such that, $\forall$ sufficiently large n, $\forall$ x of length n and NP-witness y for x, the following two distributions are indistinguishable by any 2^{n^α}-gate adversary:

$$[(\sigma', \Pi') \overset{R}{\leftarrow} \text{NIS}(x) : (\sigma', \Pi')] \quad \text{and}$$
$$[\sigma \overset{R}{\leftarrow} \{0,1\}^{\text{NI}\sigma\text{Len}(n)} ; \Pi \overset{R}{\leftarrow} \text{NIP}(\sigma, x, y) : (\sigma, \Pi)]$$

We call α the zero-knowledgeness constant.

In [DP92], De Santis and Persiano propose to add a *proof of knowledge* property to NIZK. Let $R \subseteq \{0,1\}^* \times \{0,1\}^*$ be a polynomial-time relation (i.e., given a pair of strings (x, y), it is possible to check in time polynomial in $|x|$ whether $(x, y) \in R$). L be the NP language corresponding to R ($L = \{x : \exists\ y \text{ s.t. } (x, y) \in R\}$). Let (NIP, NIV) be a NIZK proof system for L. An *extractor* is a probabilistic polynomial-time TM that runs in two stages: in stage one, on input 1^n, it outputs a string σ of length $\text{NI}\sigma\text{Len}(n)$ (and saves any information it wants to use in stage two); in stage two, on input x of length n and a proof Π for x relative to shared string σ, it tries to find a witness y for x.

Definition 5. *An NIZK argument* (NIP, NIV) *is a NIZKPK if there exists an extractor* $\text{NIExt} = (\text{NIExt}_1, \text{NIExt}_2)$ *such that, for all probabilistic polynomial-time malicious provers* NIP^*, *for all constants* $a > 0$, *for all sufficiently large* n *and for all* x,

$$\text{PROB}\ [(\sigma, state) = \text{NIExt}_1(1^n) ; \Pi = \text{NIP}^*(\sigma, x) ;$$
$$y = \text{NIExt}_2(state, x, \Pi) : (x, y) \in R] \geq \quad p_{n,x}(1 - n^{-a}),$$

where $p_{n,x} = \text{PROB}[\sigma \overset{R}{\leftarrow} \{0,1\}^n; \Pi = \text{NIP}^*(\sigma, x) : \text{NIV}(\sigma, x, \Pi) = 1]$.

The authors of [DP92] show that NIZKPKs exist for all polynomial-time relations under the RSA assumption. Furthermore, the results of [DDP00] (combined with those of [FLS99]) show the same under more general assumptions: that dense public-key cryptosystems and certified trapdoor permutations exist. They also present constructions secure under the specific assumptions of factoring Blum integers or decision Diffie-Hellman. Because we need NIZKPKs to be secure against subexponentially strong adversaries, we need subexponentially strong versions of these assumptions. We refer the reader to these papers for details.

3.5   Additional Basic Tools

We also use two basic and commonly used tools, whose definitions are recalled in the appendix. The first is a Merkle tree [Mer89], which can be constructed based on a collision-resistant hash function. The second is a subexponentially-strong pseudorandom function [GGM86], i.e., one that is secure against adversaries of size 2^{n^α} (such α is called the *pseudorandomness constant*). It can be constructed based on subexponentially strong one-way functions [HILL99].

4   Our Construction

WHY THE OBVIOUS SOLUTION DOES NOT WORK.   Before we begin, let us demonstrate that our goal cannot be more easily achieved by the following simpler construction.

Let $cmax = U(n)$ be the upperbound on the number of uses of the verifier's public key (i.e., the max value for the verifier's counter). Take a four-round ZK protocol, and have the verifier post $cmax$ independently generated first-round messages in its public key. Then execution number c simply uses first-round message number c appearing in the public key, and performs the remaining three rounds of the protocol as before.

The above construction does not work, because the prover does not know the real value c of the verifier's counter. This enables a malicious verifier to choose the value of c after it sees the prover's first message. Thus, if such a verifier resets the prover while varying c, it will typically gain knowledge. (Typically, in a 4-round ZK protocol, the verifier commits to a question without revealing it, the prover sends a first message, the verifier asks the question, and the prover answers it. However, if the prover were to answer two different questions relative to the same first message, then zero-knowledgeness disappears. Now, in the above construction, varying c enables the verifier to ask different questions.)

HIGH-LEVEL DESCRIPTION.   As in the CGGM protocol, we use the NP-complete language of graph 3-colorability and the parallel repetition of the protocol of [GMW91] as our starting point. Thus, in the first round, $\mathcal{P}$ commits to a number of random recolorings of a graph G, in the second round $\mathcal{V}$ requests to reveal the colors of one edge for each committed recoloring, and in the third round $\mathcal{P}$ opens the relevant commitments.

To allow the RZK simulator to work, our protocol uses trapdoor commitment schemes as in many prior ZK protocols (e.g., the RZK one of [CGGM00], the CZK one of [DNS98], and the ZK one of [FS89]). That is, $\mathcal{V}$'s public key contains a key for a trapdoor commitment scheme, and $\mathcal{P}$'s first-round commitments with respect to that public key. If the simulator knows the trapdoor, then it can open the commitments any way it needs in the the third round.

To ensure that the simulator knows the trapdoor, the CGGM protocol uses a three-round proof-of-knowledge subprotocol, with $\mathcal{V}$ proving to $\mathcal{P}$ knowledge of the trapdoor. This requires $\mathcal{V}$ to send two messages to $\mathcal{P}$. Because we have a *total* of only three rounds, we cannot use such a subprotocol—in three rounds $\mathcal{V}$ only sends one message to $\mathcal{P}$. We therefore use *non-interactive* ZK proofs of knowledge. This, of course, requires $\mathcal{P}$ and $\mathcal{V}$ to agree on a shared random string σ.

It is because of the string σ that we cannot use the BPK model directly, and have to strengthen it with a counter. Let $cmax = U(n)$ be the bound on the number of times public key is used. During key generation, $\mathcal{V}$ generates $cmax$ random strings $\sigma_1, \ldots, \sigma_{cmax}$, and commits to each one of them using hash-based commitments (to make the public key length independent of $cmax$, the resulting commitments are then put into a Merkle tree). In its first message,

$\mathcal{P}$ sends a fresh random string $\sigma_\mathcal{P}$, and in its message $\mathcal{V}$ decommits σ_c, where c is the current counter value and provides the NIZKPK proof with respect to $\sigma = \sigma_\mathcal{P} \oplus \sigma_c$.

The RZK simulator, after seeing the value of σ_c can rewind the verifier and choose $\sigma_\mathcal{P}$ so that $\sigma = \sigma_\mathcal{P} \oplus \sigma_c$ allows it to extract the trapdoor from the NIZKPK proof. Of course, there is nothing to prevent a malicious verifier $\mathcal{V}^*$ from choosing a value of c after seeing $\sigma_\mathcal{P}$; but because the number of choices for $\mathcal{V}^*$ is only polynomial, the simulator has an inverse polynomial probability of guessing c correctly.

One question still remains unresolved: how to ensure that a malicious verifier $\mathcal{V}^*$ does not ask $\mathcal{P}$ multiple different queries for the same recoloring of the graph? If $\mathcal{V}^*$ resets $\mathcal{P}$, then it will get the same committed recolorings in the first round; if it can then ask a different set of queries, then it gain a lot of information about the coloring of the graph (eventually even recovering the entire coloring). To prevent this, the CGGM protocol makes the verifier commit to its queries before it receives any information from $\mathcal{P}$. Our protocol, however, cannot afford to do that, because we only have three rounds. Instead, during key generation the verifier commits (using hash-based commitments) to a seed $PRFKey$ for a pseudorandom function PRF, and adds the commitment to the public key. The verifier's queries are then computed using $PRF(PRFKey, \cdot)$ applied to the relevant information received from $\mathcal{P}$ in the first round and the counter value c. To prove to $\mathcal{P}$ that they are indeed computed correctly, the verifier has to include in its NIZKPK proofs of knowledge of $PRFKey$ that leads to such queries and knowledge of the decommitment to $PRFKey$.

A FEW MORE TECHNICAL DETAILS. In our protocol, just like in the CGGM protocol all probabilistic choices of the prover are generated as a pseudorandom function of the input. (This is indeed the first step towards resettability, as it reduces the advantages of resetting the prover with the same random tape.) Because the prover makes no probabilistic choices in its second step, we do not need to include the verifier's message in the input to the pseudorandom function.

To ensure soundness and avoid problems with malleability of $\mathcal{V}$'s commitments, we use complexity leveraging in a way similar to the CGGM protocol. That is, and we shall use two polynomially-related security parameters: n for all the components except the hash-based commitment scheme HC, and $k = n^\epsilon$ for HC.

This will ensure that any algorithm that is endowed with a subroutine for breaking HC commitments, but is polynomial-time otherwise, is still unable (simply by virtue of its running time) of breaking any other of our components. This property will be used in our proof of soundness.

We actually choose the constant ϵ in a particular way. Namely, we shall use a trapdoor commitment scheme TC with soundness constant α_1, an NIZKPK system (NIP, NIV) (for a relation to be specified later) with zero-knowledgeness constant α_2, and a pseudorandom function PRF with pseudorandomness constant α_3, and set $\epsilon < \min(\alpha_1, \alpha_2, \alpha_3)$.

THE FULL DESCRIPTION. The complete details of $\mathcal{P}$ and $\mathcal{V}$ are given below.

Key Generation Algorithm for $\mathcal{V}$

System Parameter:
  A polynomial U
Security Parameter:
  1^n
Procedure:
  1. Let $cmax = U(n)$.
  2. Generate random strings $\sigma_1, \ldots, \sigma_{cmax}$ of length NIσLen(n) each. (Note: to save secret key length, the strings σ_c can be generated using a pseudorandom function of c, whose short seed can be made part of the secret key).
  3. Let $k = n^\epsilon$.
  4. Commit to each σ_c using $(\sigma Com_c, \sigma Decom_c) \overset{R}{\leftarrow}$ HCCom$(1^k, \sigma_c)$.
  5. Combine the values σCom_c into a single Merkle tree with root R. (Note: If the values σ_c's are generated via a PRF to save on secret key length then also the values σCom_c, the resulting Merkle tree, etc. can be computed efficiently in space logarithmic in $cmax$.)
  6. Generate a random string $PRFKey$ of length n.
  7. Commit to the $PRFKey$ using
    $(PRFKeyCom, PRFKeyDecom) \overset{R}{\leftarrow}$ HCCom$(1^n, PRFKey)$.
  8. Generate keys for the trapdoor commitment scheme:
    $(TCPK, TCSK) \overset{R}{\leftarrow}$ TCGen(1^n).
Output:
  $PK = (R, PRFKeyCom, TCPK)$
  $SK = (\{(\sigma_c, \sigma Decom_c)\}_{c=1}^{cmax}, (PRFKey, PRFKeyDecom), TCSK)$.

Protocol $(\mathcal{P}, \mathcal{V})$

Public File:
  A collection F of records (id, PK_{id}), where PK_{id} is allegedly the output of the Key Generation Algorithm above
Common Inputs:
  A graph $G = (V, E)$, and a security parameter 1^n

$\mathcal{P}$ **Private Input:**
  A valid coloring of G, $col : V \to \{0, 1, 2\}$; $\mathcal{V}$'s id and the file F; a random string ω
$\mathcal{V}$ **Private Input:**
  A secret key SK, a counter value c, and a bound $cmax$.
$\mathcal{P}$ **Step One :**
  1. Using the random string ω as a seed for PRF, generate a sufficiently long "random" string from the input to be used in the remaining computation.

2. Find PK_{id} in F; let $PK_{id} = (R, PRFKeyCom, TCPK)$
   (if more than one PK_{id} exist in F, use the alphabetically first one).
3. Verify $TCPK$ by invoking TCKeyVer($1^n, TCPK$).
4. Let $\sigma_{\mathcal{P}}$ be a random string of length NIσLen(n).
5. Commit to random recolorings of the G as follows.
   Let $\pi_1, \ldots, \pi_n$ be random permutations on $\{0, 1, 2\}$.
   For all i $(1 \leq i \leq n)$ and $v \in V$, commit to $\pi_i(col(v))$ by computing
   $(cCom_{i,v}, cDecom_{i,v}) \xleftarrow{R} \text{TCCom}(TCPK, \pi_i(col(v)))$.
6. If all the verifications hold, send $\sigma_{\mathcal{P}}$ and $\{cCom_{i,v}\}_{1 \leq i \leq n, v \in V}$ to $\mathcal{V}$.

$\mathcal{V}$ **Step One:**
1. Increment c and check that it is no greater than $cmax$.
2. For each j $(1 \leq j \leq n)$, compute a challenge edge $e_j \in E$ by
   applying PRF to the counter value c, j and the
   commitments received from $\mathcal{P}$:
   $e_j = \text{PRF}(PRFKey, c \circ j \circ \{cCom_{i,v}\}_{1 \leq i \leq n, v \in V})$
3. Let $\sigma = \sigma_{\mathcal{P}} \oplus \sigma_c$. Compute a NIZKPK proof Π using NIP on σ
   and the following statement:
   "$\exists$ key K for PRF that generated the challenge edges $\{e_j\}_{1 \leq j \leq n}$;
   $\exists$ decommitment D s. t. HCVer($1^n, PRFKeyCom, K, D$) = YES;
   $\exists$ secret key S corresponding to the public key $TCPK$."
   (Note: this can computed efficiently because $\mathcal{V}$ knows witnesses
   $PRFKey$ for K, $PRFKeyDecom$ for D, and $TCSK$ for S).
4. Send c, σ_c, σCom_c together with its authenticating path in
   the Merkle tree, $\sigma Decom_c$, Π and $\{e_j\}_{1 \leq j \leq n}$ to $\mathcal{P}$.

$\mathcal{P}$ **Step Two:**
1. Verify the authenticating path of σCom_c in the Merkle tree
2. Verify that HCVer($1^k, \sigma_c, \sigma Com_c, \sigma Decom_c$) = YES.
3. Let $\sigma = \sigma_{\mathcal{P}} \oplus \sigma_c$. Verify Π using NIV.
4. If all the verifications hold, for each $e_j = (v_j^0, v_j^1)$ and $b \in \{0, 1\}$,
   send $c_j^b = \pi_j(col(v_j^b))$ and $cDecom_{j, v_j^b}$ to $\mathcal{V}$.

$\mathcal{V}$ **Step Two:**
1. Verify that, for all j $(1 \leq j \leq n)$, and for all $b \in \{0, 1\}$
   TCVer($TCPK, cCom_{j, v_j^b}, c_j^b, cDecom_{j, v_j^b}$) = YES.
2. Verify that for all j $(1 \leq j \leq n)$, $c_j^0 \neq c_j^1$.
3. If all the verifications hold, accept. Else reject.

Theorem 1. $(\mathcal{P}, \mathcal{V})$ *is a 3-round RZK protocol in the UPK model.*

As usual, completeness is easily verified. We address soundness in Section 4.1
and resettable zero-knowledgeness in Section 4.2.

4.1   Computational Soundness

Suppose G is a graph that is not 3-colorable, and $\mathcal{P}^*$ is a circuit of size $t < 2^k$ that
can make $\mathcal{V}$ accept $(G, 1^n)$ with probability $p > 1/2^k$. Then, we shall construct

a small circuit A that receives $TCPK$ as input, and, using $\mathcal{P}^*$, will output two trapdoor decommitments for the same TC commitment. The size of A will be $\mathrm{poly}(n) \cdot t \cdot 2^k / \mathrm{poly}(p)$. Thus, A will violate the soundness of TC, because its size is less (for a sufficiently large n) than $2^{n^{\alpha_1}}$ allowed by the soundness property of TC (recall in fact that $k = n^\epsilon$ and $\epsilon < \alpha_1$).

A is constructed as follows. It receives as input a public key $TCPK$ for TC generated by $\mathrm{TCGen}(1^n)$. A then generates PK as if it were the public key of the specified honest verifier $\mathcal{V}$, using the $\mathcal{V}$'s key generation procedure with the exception of step 7, for which it simply uses $TCPK$. Note that A knows all the components of corresponding secret key of $\mathcal{V}$, with the exception of $TCSK$. A selects an identity id and creates a file F to contain the single record (id, PK) (or embeds it into a larger such file containing other identities and public keys, but honestly generated).

A will now run $\mathcal{P}^*$ multiple times with inputs F and id (G and 1^n are already known to $\mathcal{P}^*$), each time with the same random tape. Thus, each time, $\mathcal{P}^*$ will send the same set of strings σ_P and $\{cCom_{i,v}\}_{1 \le i \le n, v \in V}$. Our goal, each time, is to allow A to respond with a different random set of challenges $\{e'_j\}_{1 \le j \le n}$. Then, after an expected number of tries that is inversely polynomial in p, there will exist a recoloring i and a node v such that $cCom_{i,v}$ has been opened by $\mathcal{P}^*$ in two different ways. That is, there will be a "break" of the commitment scheme TC.

Therefore, all there remains to be shown is *how* A can ask a different random set of challenges, despite the fact that it has committed to $\mathcal{V}$'s $PRFKey$ in PK. Recall that honest $\mathcal{V}$ executes the protocol at most $cmax$ time, and that the current value of $\mathcal{V}$'s counter will be known to P^*. If P^* has such an overall success probability p of proving G 3-colorable, then there exists a value of $\mathcal{V}$'s counter for which the success probability of $\mathcal{P}^*$ is at least p. Let c be such a value. Because of A's non-uniformity, we assume A "knows" c.

To issue a set of (different) random challenges in response to the same first message of $\mathcal{P}^*$, A uses the NIZKPK simulator NIS as follows. First, A selects a set of random challenges $\{e'_j\}_{1 \le j \le n}$. Second, it invokes NIS to obtain a "good looking proof" σ' and Π' for the following statement Σ:

$\Sigma =$ "$\exists$ key K for PRF that generated the challenge edges $\{e'_j\}_{1 \le j \le n}$;
$\qquad \exists$ decommitment D s. t. $\mathrm{HCVer}(1^n, PRFKeyCom, K, D) = \mathrm{YES}$;
$\qquad \exists$ secret key S corresponding to the public key $TCPK$."

(Note that Σ is potentially false, because it may be the case that no such K exists at all; we address this below.) Third, A sets $\tau = \sigma' \oplus \sigma_P$. Fourth, A comes up with a decommitment $\tau Decom$ that decommits σCom_c (the commitment to the c-th shared string computed during key generation) to τ rather than the originally committed σ_c. This can be done by implementing HCFake by means of a (sub)circuit of size $\mathrm{poly}(k)2^k$. Fifth, A sends $\tau, \sigma Com_c$ together with its authenticating path in the Merkle tree (A knows that path from key generation), $\tau Decom$, Π' and $\{e'_j\}_{1 \le j \le n}$ to $\mathcal{P}^*$.

Thus, all that's left to show is that $\mathcal{P}^*$ will behave the same way as it would for the true verifier $\mathcal{V}$, even though it received random, rather than pseudorandom, challenges, together with a faked decommitment and a simulated proof of a potentially false statement Σ. This is done by a properly constructed hybrid argument that relies on the zero-knowledgness of (NIP, NIV), the pseudorandomness of PRF and the statistical secrecy and breakability of HC.

First, note that random $\{e'_j\}_{1 \leq j \leq n}$ cannot be distinguished from pseudorandomly generated $\{e'_j\}_{1 \leq j \leq n}$ (wihout knowledge of $PRFKey$): otherwise, we'd violate the pseudorandomness of PRF. Moreover, this holds even in the presence of $PRFKeyCom$, because $PRFKeyCom$ is statistically secret, and thus reveals a negligible amount of information about $PRFKey$. It follows that the tuple $(PRFKeyCom, \{e'_j\}_{1 \leq j \leq n}, \sigma', \Pi')$ cannot be distinguished from the tuple $(PRFKeyCom, \{e_j\}_{1 \leq j \leq n}, \sigma'', \Pi'')$, where the challenge edges $\{e_j\}_{1 \leq j \leq n}$ are produced by the true PRF with the true committed-to $PRFKey$, and σ'', Π'' are produced by NIS. This, in turn, by zero-knowledgeness is indistinguishable from $(PRFKeyCom, \{e_j\}_{1 \leq j \leq n}, \sigma, \Pi)$, with the pseudorandomly generated $\{e_j\}_{1 \leq j \leq n}$, a truly random σ and Π honestly generated by NIP. By a hybrid argument, therefore, the tuple $(PRFKey, \{e_j\}_{1 \leq j \leq n}, \sigma, \Pi)$ is indistinguishable from the tuple $(PRFKey, \{e'_j\}_{1 \leq j \leq n}, \tau, \Pi')$. Of course, if we replace σ by the pair $(\sigma_P, \tau = \sigma \oplus \sigma_P)$ and σ' by the pair $(\sigma_P, \sigma_c = \sigma \oplus \sigma_P)$, the statement still holds. Moreover, it holds in the presence of σCom_c, because the commitment to σ_c is statistically secret (and thus is almost equally as likely to be a commitment to τ). The authenticating path of σCom_c in the Merkle tree is just a (randomized) function of σCom_c and root R of the tree, and thus does not affect indistinguishability. Finally, note that this indistinguishability holds with respect to any distinguishing circuit of size $2^k \text{poly}(n)$, because the zero-knowledgeness and pseudorandomness constants α_2 and α_3 are greater than ϵ. Therefore, indisntinguishability holds even in the presence of the decommitment $\tau Decom$ or $\sigma Decom_c$, because this decommitment can be computed by such a circuit from σCom_c using HCFake.

4.2   Resettable Zero-Knowledgeness (Sketch)

Let $\mathcal{V}^*$ be an (s, t)-resetting verifier. We will show how to construct the simulator M as required by Definition 1. Due to lack of space in this extended abstract, below we present only the essential points of our construction.

As alredy proven in [CGGM00], resettability is such a strong capability of a malicious verifier, that it has nothing else to gain by interleaving its executions with the honest prover. Thus, we can assume that our $\mathcal{V}^*$ executes with $\mathcal{P}$ only sequentially.

Recall that $\mathcal{V}^*$ runs in two stages. Then M operates as follows. First, M runs the first stage of $\mathcal{V}^*$ to obtain a public file F. Then, for every record (id, PK_{id}) in F, M remembers some information (whose meaning will be explained later on):

1. M remembers whether PK_{id} is "broken" or not
2. If PK_{id} is broken, M also remembers the value of $TCSK$
3. If PK_{id} is not broken, M also remembers a list of tuples $(c, \sigma_c, \sigma Com_c)$

Initially, every PK_{id} in F is marked as not broken, and the list of pairs for each record is empty.

Whenever V^* starts a new session for an id that is not broken and whose list of pairs is empty, M computes the "first prover message" as follows: it commits to arbitrary color values for graph G, and then selects σ_P at random. (Of course, if V^* dictates that M's random tape and inputs be equal to those in a prior interaction, M has no choice but to use the same first message as in that interaction.) When V^* responds with the verifier message, M takes $(c, \sigma_c, \sigma Com_c)$ from this message and adds it to the list of tuples maintained for PK_{id}. M then rewinds V^* to the beginning of V^*'s second stage.

Whenever V^* starts a new session for an id that is not broken but whose list of pairs is non-empty, M randomly chooses a tuple $(c', \sigma_{c'}, \sigma Com_{c'})$ from the list of tuples for PK_{id}. M then uses the extractor of the non-interactive ZK proof of knowledge, NIExt$_1$, to obtain a shared string σ, and sets $\sigma_P = \sigma \oplus \sigma_{c'}$. M then commits to arbitrary color values for graph G and sends the commitment and σ_P as the "first prover message" to V^*. When V^* responds with the verifier message, M compares the counter value c included in this response to the value c' from the pair chosen above.

1. If $c = c'$, then it must be the case that $\sigma_c = \sigma_{c'}$. (Otherwise, if the commitment $\sigma Com_{c'}$ previously stored by M is equal to the commitment σCom_c included in V^*'s response, σ_c and $\sigma_{c'}$ have been easily found, so as to violate the soundness of HC; and if the $\sigma Com_c \neq \sigma Com'_c$, then a collision has been easily found in the Merkle tree). Thus, the string Π, also included in the response of V^*, is an NIZK proof of knowledge with respect to the string σ output by NIExt$_1$. Therefore, M can use NIExt$_2$ to extract a witness $TCSK$ for the secret key of the commitment scheme. In this case, PK_{id} is marked as broken and M remembers $TCSK$.
2. If $c \neq c'$, then M has learned a potentially new tuple $(c, \sigma_c, \sigma Com_c)$, which it remembers as its list of pairs for PK_{id}.

M then rewinds V^* to the beginning of V^*'s second stage.

Whenever V^* starts a new session for an id that is broken, M can always simulate $\mathcal{P}$'s behavior because M knows the trapdoor to the commitment scheme. Thus, it can commit to arbitrary color values in its first message, and then decommit in its second message so that they look like a valid response to V^*'s challenge edges.

The expected running time of M is polynomial, because the expected number of rewinds before M breaks a given PK_{id} is polynomial in $cmax$ and inverse polynomial in the frequency with which V^* uses id.

It remains to show that V^* cannot ask for two different sets of challenge edges for the same first message of M (if it could, then, unless M knows the correct 3-coloring of the graph, it maybe unable to faithfully simulate the decommitments). However, if V^* has a non-negligible probability of doing so, then one can build a machine ADV to violate the soundness of HC in polynomial time with non-negligible probability, as follows.

ADV guesses, at random, for what instance of $\mathcal{P}$ the machine $\mathcal{V}^*$ will first give two different sets of challenges on the same first message. A also guesses, at random, the counter values c_1 and c_2 that $\mathcal{V}^*$ will use in these two cases. A then attempts to find out σ_{c_1} and σ_{c_2} by using the same technique as M. A then runs the second stage of $\mathcal{V}^*$ two more times: once to extract a witness K for $PRFKey$ and its decommitment D in the first case, and the other to extract a witness K' for $PRFKey$ and its decommitment D' in the second case (this witness extraction is done the same way as M). $K \neq K'$ and D and D' are valid decommitments, which violates soundness of HC.

Acknowledgements

We would like to thank Amit Sahai, Anna Lysyanskaya and the anonymous referees for helpful comments. The second author was supported, in part, by a National Science Foundation Graduate Research Fellowship and a grant from the NTT corporation.

References

BCC88.   G. Brassard, D. Chaum, and C. Crépeau. Minimum disclosure proofs of knowledge. *Journal of Computer and System Sciences*, 37(2):156–189, 1988.

BDMP91.   M. Blum, A. De Santis, S. Micali, and G. Persiano. Noninteractive zero-knowledge. *SIAM Journal on Computing*, 20(6):1084–1118, December 1991.

BFM88.   M. Blum, P. Feldman, and S. Micali. Non-interactive zero-knowledge and its applications (extended abstract). In *Proceedings of the Twentieth Annual ACM Symposium on Theory of Computing*, pages 103–112, 1988.

Bra89.   G. Brassard, editor. *Advances in Cryptology—CRYPTO '89*, volume 435 of *Lecture Notes in Computer Science*. Springer-Verlag, 1990.

CGGM00.   R. Canetti, O. Goldreich, S. Goldwasser, and S. Micali. Resettable zero-knowledge. In *Proceedings of the 32nd Annual ACM Symposium on Theory of Computing*, 2000. Updated version available at the Cryptology ePrint Archive, record 1999/022, http://eprint.iacr.org/.

CKPR01.   R. Canetti, J. Kilian, E. Petrank, and A. Rosen. Black-box concurrent zero-knowledge requires $\tilde{\Omega}(\log n)$ rounds. In *Proceedings of the Thirty-Second Annual ACM Symposium on Theory of Computing*, 6–8 July 2001.

Dam00.   I. Damgård. Efficient concurrent zero-knowledge in the auxiliary string model. In Bart Preneel, ed., *Advances in Cryptology—EUROCRYPT 2000*, volume 1807 of *Lecture Notes in Computer Science*, Springer-Verlag, 2000.

DDP00.   A. De Santis, G. Di Crescenzo, and G. Persiano. Necessary and sufficient assumptions for non-interactive zero-knowledge proofs of knowledge for all np relations. In U. Montanari, J. D. P. Rolim, and E. Welzl, editors, *Automata Languages and Programming: 27th International Colloquim (ICALP 2000)*, volume 1853 of *Lecture Notes in Computer Science*, pages 451–462. Springer-Verlag, July 9–15 2000.

DNS98.   C. Dwork, M. Naor, and A. Sahai. Concurrent zero knowledge. In *30th Annual ACM Symposium on Theory of Computing*, 1998.

DP92.   A. De Santis and G. Persiano. Zero-knowledge proofs of knowledge without interaction. In *33rd Annual Symposium on Foundations of Computer Science*, 1992.

DPP97.   I. B. Damgård, T. P. Pedersen, and B. Pfitzmann. On the existence of statistically hiding bit commitment schemes and fail-stop signatures. *Journal of Cryptology*, 10(3):163–194, Summer 1997.

DS98.   C. Dwork and A. Sahai. Concurrent zero-knowledge: Reducing the need for timing constraints. In H. Krawczyk, ed., *Advances in Cryptology— CRYPTO '98*, volume 1462 of *Lecture Notes in Computer Science*, 1998.

FLS99.   U. Feige, D. Lapidot, and A. Shamir. Multiple non-interactive zero knowledge proofs under general assumptions. *SIAM Journal on Computing*, 29(1):1–28, 1999.

FS89.   U. Feige and A. Shamir. Zero knowledge proofs of knowledge in two rounds. In Brassard [Bra89], pages 526–545.

GGM86.   O. Goldreich, S. Goldwasser, and S. Micali. How to construct random functions. *Journal of the ACM*, 33(4):792–807, October 1986.

GK96.   O. Goldreich and H. Krawczyk. On the composition of zero-knowledge proof systems. *SIAM Journal on Computing*, 25(1):169–192, February 1996.

GMR88.   S. Goldwasser, S. Micali, and R. L. Rivest. A digital signature scheme secure against adaptive chosen-message attacks. *SIAM Journal on Computing*, 17(2):281–308, April 1988.

GMR89.   S. Goldwasser, S. Micali, and C. Rackoff. The knowledge complexity of interactive proof systems. *SIAM Journal on Computing*, 18:186–208, 1989.

GMW91.   O. Goldreich, S. Micali, and A. Wigderson. Proofs that yield nothing but their validity or all languages in NP have zero-knowledge proof systems. *Journal of the ACM*, 38(1):691–729, 1991.

HILL99.   J. Håstad, R. Impagliazzo, L.A. Levin, and M. Luby. Construction of pseudorandom generator from any one-way function. *SIAM Journal on Computing*, 28(4):1364–1396, 1999.

HM96.   S. Halevi and S. Micali. Practical and provably-secure commitment schemes from collision-free hashing. In Neal Koblitz, editor, *Advances in Cryptology—CRYPTO '96*, volume 1109 of *Lecture Notes in Computer Science*, pages 201–215. Springer-Verlag, 18–22 August 1996.

KP00.   J. Kilian and E. Petrank. Concurrent zero-knowledge in polylogarithmic rounds. Technical Report 2000/013, Cryptology ePrint Archive, `http://eprint.iacr.org`, 2000.

KPR98.   J. Kilian, E. Petrank, and C. Rackoff. Lower bounds for zero-knowledge on the Internet. In *39th Annual Symposium on Foundations of Computer Science*, pages 484–492, Los Alamitos, California, November 1998. IEEE.

Mer89.   R. C. Merkle. A certified digital signature. In Brassard [Bra89], pages 218–238.

Mic.   Silvio Micali. CS proofs. SIAM Journal on Computing, to appear.

MR01.   S. Micali and L. Reyzin. Soundness in the public-key model. Unpublished manuscript, 2001.

NR97.   Moni Naor and Omer Reingold. Number-theoretic constructions of efficient pseudo-random functions. In *38th Annual Symposium on Foundations of Computer Science*, pages 458–467, Miami Beach, Florida, 20–22 October 1997. IEEE.

RK99.   R. Richardson and J. Kilian. On the concurrent composition of zero-knowledge proofs. In Jacques Stern, editor, *Advances in Cryptology—EUROCRYPT '99*, volume 1592 of *Lecture Notes in Computer Science*, pages 415–431. Springer-Verlag, 2–6 May 1999.

A   Merkle Trees

The description below is almost verbatim from [Mic].

Recall that a binary tree is a tree in which every node has at most two children, hereafter called the *0-child* and the *1-child*. A *Merkle tree* [Mer89] with security parameter n is a binary tree whose nodes store values, some of which are computed by means of a collision-free hash function $H : \{0,1\}^* \to \{0,1\}^n$ in a special manner. A leaf node can store any value, but each internal node should store a value that is the one-way hash of the concatenation of the values in its children. That is, if an internal node has a 0-child storing the value u and a 1-child storing a value v, then it stores the value $H(u \circ v)$. Thus, because H produces n-bit outputs, each internal node of a Merkle tree, including the root, stores an n-bit value. Except for the root value, each value stored in a node of a Merkle tree is said to be a *0-value*, if it is stored in a node that is the 0-child of its parent, a *1-value* otherwise.

The crucial property of a Merkle tree is that, unless one succeeds in finding a collision for H, *it is computationally hard to change any value in the tree (and, in particular, a value stored in a leaf node) without also changing the root value.* This property allows a party A to commit to L values, $v_1, \ldots, v_L$ (for simplicity assume that L is a power of 2 and let $d = \log L$), by means of a single n-bit value. That is, A stores value v_i in the i-th leaf of a full binary tree of depth d, and uses a collision-free hash function H to build a Merkle tree, thereby obtaining an n-bit value, R, stored in the root. This root value R "implicitly defines" what the L original values were. Assume in fact that, as some point in time, A gives R, but not the original values, to another party B. Then, whenever, at a later point in time, A wants to "prove" to B what the value of, say, v_i was, A may just reveal all L original values to B, so that B can recompute the Merkle tree and the verify that the newly computed root-value indeed equals R. More interestingly, A may "prove" what v_i was by revealing just $d + 1$ (that is, just $1 + \log L$) values: v_i together with its *authenticating path*, that is, the values stored in the siblings of the nodes along the path from leaf i (included) to the root (excluded), $w_1, \ldots, w_d$. Party B verifies the received alleged leaf-value v_i and the received alleged authenticating path $w_1, \ldots, w_d$ as follows. She sets $u_1 = v_i$ and, letting $i_1, \ldots, i_d$ be the binary expansion of i, computes the values $u_2, \ldots, u_d$ as follows: if $i_j = 0$, she sets $u_{j+1} = H(w_j \circ u_j)$; else, she sets $u_{j+1} = H(u_j \circ w_j)$. Finally, B checks whether the computed n-bit value u_d equals R.

B   Pseudorandom Functions

A pseudorandom function family, introduced by Goldreich, Goldwasser and Micali [GGM86] is a keyed family of efficiently computable functions, such that a function picked at random from the family is indistinguishable (via oracle access) from a truly random function with the same domain and range. More formally, let $\mathrm{PRF}(\cdot, \cdot) : \{0,1\}^n \times \{0,1\}^* \rightarrow \{0,1\}^n$ be an efficiently computable function. Our definition below is quite standard, except that it requires security against subexponentially strong adversaries.

Definition 6. *We say that* PRF *is a* pseudorandom function *if* $\exists\ \alpha > 0$ *such that for all sufficiently large n and all 2^{n^α}-gate adversaries* ADV, *the following difference is negligible in n:*

$$\mathrm{PROB}[PRFKey \stackrel{R}{\leftarrow} \{0,1\}^n : \mathrm{ADV}^{\mathrm{PRF}(PRFKey,\cdot)} = 1] -$$
$$\mathrm{PROB}[F \stackrel{R}{\leftarrow} (\{0,1\}^n)^{\{0,1\}^n \times \{0,1\}^*} : \mathrm{ADV}^{F(\cdot)} = 1]$$

We call α the pseudorandomness constant.

Pseudorandom functions can be constructed based on a variety of assumption. We refer the reader to [GGM86,NR97] (and references therein) for details.

Structural Cryptanalysis of SASAS

Alex Biryukov and Adi Shamir

Computer Science department
The Weizmann Institute
Rehovot 76100, Israel.

Abstract. In this paper we consider the security of block ciphers which contain alternate layers of invertible S-boxes and affine mappings (there are many popular cryptosystems which use this structure, including the winner of the AES competition, Rijndael). We show that a five layer scheme with 128 bit plaintexts and 8 bit S-boxes is surprisingly weak even when all the S-boxes and affine mappings are key dependent (and thus completely unknown to the attacker). We tested the attack with an actual implementation, which required just 2^{16} chosen plaintexts and a few seconds on a single PC to find the 2^{17} bits of information in all the unknown elements of the scheme.

Keywords: Cryptanalysis, Structural cryptanalysis, block ciphers, substitution permutation networks, substitution affine networks, Rijndael.

1   Introduction

Structural cryptanalysis is the branch of cryptology which studies the security of cryptosystems described by generic block diagrams. It analyses the syntactic interaction between the various blocks, but ignores their semantic definition as particular functions. Typical examples include meet in the middle attacks on double encryptions, the study of various chaining structures, and the properties of Feistel structures with a small number of rounds.

Structural attacks are often weaker than actual attacks on given cryptosystems, since they cannot exploit particular weaknesses (such as bad differential properties or weak avalanche effects) of concrete functions. The flip side of this is that they are applicable to large classes of cryptosystems, including those in which some of the internal functions are unknown or key dependent. Structural attacks often lead to deeper theoretical understanding of fundamental constructions, and thus they are very useful in establishing general design rules for strong cryptosystems.

The class of block ciphers considered in this paper are product ciphers which use alternate layers of invertible S-boxes and affine mappings. This structure is a generalization of substitution/permutation networks (in which the affine mapping is just a bit permutation), and a special case of Shannon's encryption paradigm which mixes complex local operations (called confusion) with simple global operations (called diffusion). There are many examples of substitution/affine ciphers in the literature, including Rijndael [4] which was recently

B. Pfitzmann (Ed.): EUROCRYPT 2001, LNCS 2045, pp. 394–405, 2001.

selected as the winner of the Advanced Encryption Standard (AES) competition. Rijndael is likely to become one of the most important block ciphers in the next 20-30 years, and thus there is a great interest in understanding its security properties.

The best non-structural attack on Rijndael (and its predecessor Square [3]) is based on the *square attack* which exploits the knowledge of the S-box, the simplicity of the key schedule and the relatively slow avalanche of the sparse affine mapping (which linearly mixes bytes only along the rows and columns of some matrix and adds a subkey to the result). It can break versions with six S-box layers and six affine layers (a seventh layer can be added if the attacker is willing to guess its 128 bit subkey in a nonpractical attack).

In our structural attacks we do not know anything about the S-boxes, the affine mappings, or the key schedule, since they can all be defined in a complex key-dependent way. In particular, we have to assume that the avalanche is complete after a single layer of an unknown dense affine mapping, and that any attempt to guess even a small fraction of the key would require a nonpractical amount of time. Consequently, we cannot use the square attack (even though we are influenced by some of its underlying ideas) and we have to consider a somewhat smaller number of layers.

In this paper we describe surprisingly efficient structural attacks on substitution/affine structures with five to seven layers. The main scheme we attack is the five layer scheme $S_3 A_2 S_2 A_1 S_1$ (see Figure 1) in which each S layer contains k invertible S-boxes which map m bits to m bits, and each A layer contains an invertible affine mapping of vectors of $n = km$ bits over $GF(2)$:

$$A_i(x) = L_i x \oplus B_i$$

The only information available to the attacker is the fact that the block cipher has this general structure, and the values of k and m. Since all the S-boxes and affine mappings are assumed to be different and secret, the effective key length of this five layer scheme is [1]:

$$\log(2^m!)^{3 \cdot \frac{n}{m}} + 2\log(0.29 \cdot 2^{n^2}) \approx 3 \cdot 2^m (m - 1.44) \cdot \frac{n}{m} + 2n^2.$$

The new attack is applicable to any choice of m and n, but to simplify the analysis we concentrate on the Rijndael-like parameters of $m = 8$ bit S-boxes and $n = 128$ bit plaintexts. The effective key length of this version is about $3 \cdot 2^{12} \cdot 6.56 + 2^{15} \approx 113,000 \approx 2^{17}$ bits, and thus exhaustive search or meet in the middle attacks are completely impractical. Our attack requires only 2^{16}

---

[1] The probability that m randomly chosen linear equations in m unknowns are linearly independent over $GF(2)$ is:

$$\left(\frac{2^m - 1}{2^m}\right)\left(\frac{2^m - 2}{2^m}\right)\left(\frac{2^m - 2^2}{2^m}\right) \cdots \left(\frac{2^m - 2^{m-1}}{2^m}\right) = \prod_{l=1}^{m}\left(1 - \frac{1}{2^l}\right) > 0.288788.$$

(1)

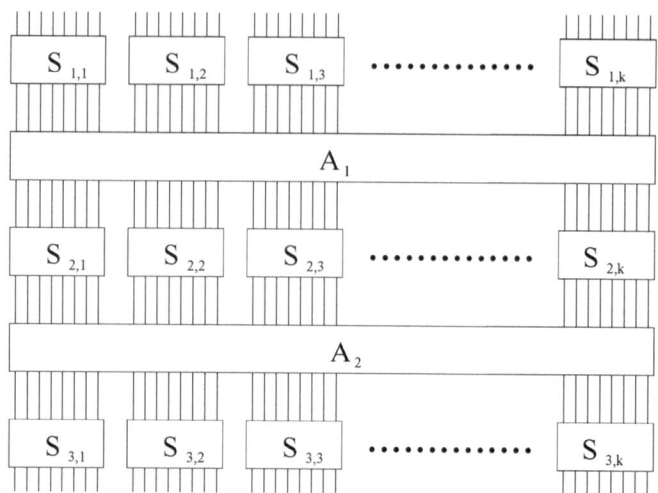

Fig. 1. Five-layer scheme.

chosen plaintexts and 2^{28} time to find all the unknown elements. This is quite close to the information bound since the 2^{16} given ciphertexts contain at most 2^{23} bits of information about the 2^{17} key bits.

It is important to note that not all the information about the S-boxes and the affine mappings can be extracted from the scheme, since there are many equivalent keys which yield the same mapping from plaintexts to ciphertexts. For example, we can change the order of the various S-boxes in a single layer and compensate for it by changing the definition of the adjacent affine mapping. In a similar way, we can move the additive constants in the affine mappings into the definition of the adjacent S-boxes. Our attack finds an equivalent representation of all the elements in the scheme which makes it possible to encrypt and decrypt arbitrary texts, but it may be different from the original definition of these elements.

A related structural attack on a five layer substitution/affine structure was recently published by Biham [2]. He attacked the slightly different structure $A_3S_2A_2S_1A_1$ (with two S-box layers and three affine layers) which was proposed by Patarin as a new algebraic public key cryptosystem called 2R. However, in Patarin's scheme the S-boxes are implemented by multivariate quadratic polynomials, which are non-bijective due to design constraints. The starting point of Biham's attack is the existence of random collisions created by such S-boxes, and its time and data complexities were forced by the birthday paradox to be at least 2^{60}. Biham's attack is thus inapplicable to substitution/affine structures with invertible operations which have no collisions, and has higher complexity than our attack.

2   The Multiset Attack

2.1   Multiset Properties

In this section we develop a calculus of multiset properties, which makes it possible to characterize intermediate values deep in the encryption structure even though nothing is known about the actual functions in it. Each multiset can be represented as a list of (value, multiplicity) pairs (e.g., the multiset $\{1, 1, 1, 2, 2, 2, 2, 7\}$ can also be represented as $(1, 3), (2, 4), (7, 1)$). The size of the multiset is the sum of all its multiplicities (8 in this example). We now define several multiset properties:

Definition 1 *A multiset M of m-bit values has property C (constant) if it contains an arbitrary number of repetitions of a single value.*

Definition 2 *A multiset M of m-bit values has property P (permutation) if it contains exactly once each one of the 2^m possible values.*

Definition 3 *A multiset M of m-bit values has property E (even) if each value occurs an even number of times (including no occurrences at all).*

Definition 4 *A multiset M of m-bit values has property B (balanced) if the XOR of all the values (taken with their multiplicities) is the zero vector 0^m.*

Definition 5 *A multiset M of m-bit values has property D (dual) if it has either property P or property E.*

We will consider now the issue of how the multiset properties defined above are transformed by various mappings. In general if a bijective function is applied to a multiset we get a new multiset with possibly new values, but the same collection of multiplicities. If a non-bijective function is applied to a multiset, then the multiplicities of several distinct input values that are mapped to a common output value are added. The following observations are easy to prove:

Lemma 1   *1. Any multiset with either property E or property P (when $m > 1$) also has property B.*

2. The E and C properties are preserved by arbitrary functions over m-bit values.

3. The P property is preserved by arbitrary bijective functions over m-bit values.

4. The B property is preserved by an arbitrary linear mapping from m bits to n bits when $m > 1$. It is preserved by arbitrary affine mappings when the size of the multiset is even.

Let us consider now blocks of larger size $n = k \cdot m$ with mixed multiset properties. For example, we denote by $C^{i-1}PC^{k-i}$ a multiset with the property that when we decompose each n bit value into k consecutive blocks of m contiguous bits, $k - 1$ of the blocks contain (possibly different) constants across the multiset, and the i-th block contains exactly once each one of the 2^m possible m-bit values.

Similarly, we denote by D^k a multiset that decomposes into k multisets each one of which has property D. This decomposition should be understood not as a cross product of k multisets but as a collection of k projections of n bit to m bit values. Note that this decomposition operation is usually nonreversible, since we lose the order in which the values in the various blocks are combined. For example the multiset decomposition

$$\{0, 1, 2, 3\}\{1, 1, 2, 2\}\{1, 1, 1, 1\}$$

(which has the multiset property PEC for $m = 2$) can be derived from several different multisets such as $\{(011), (111), (221), (321)\}$ or $\{(021), (121), (211), (311)\}$.

Let us consider now how these extended multiset properties are transformed by layers of S-boxes and affine mappings:

Lemma 2   1. Property $C^{i-1}PC^{k-i}$ is preserved by a layer of arbitrary S-boxes provided that the i-th S-box is bijective.
   2. Property D^k is transformed into property D^k by a layer of bijective S-boxes.
   3. Property D^k is transformed into B^k by an arbitrary linear mapping on n bits, and by an arbitrary affine mapping when the size of the multiset is even.
   4. Property $C^{i-1}PC^{k-i}$ is transformed into property D^k by an arbitrary affine mapping when the size of the multiset is even.

Proof
The only non-trivial claims are 3 and 4. Let us show why claim 3 holds. Denote by

$$y_j = \sum_{i=1}^{n} d_{ji} x_i$$

a bit y_j at the output of the linear mapping. Property B holds since for each j, the sum (mod 2) of y_j bits over the 2^m elements of the multiset is zero:

$$\sum_{s=1}^{2^m} y_j^s = \sum_{s=1}^{2^m} \sum_{i=1}^{n} d_{ji} x_i^s = \sum_{i=1}^{n} d_{ji} \sum_{s=1}^{2^m} x_i^s = 0.$$

The last expression is zero since by Lemma 1, claim 1, both P and E (and thus D) imply the B-property. The result remains true even when we replace the linear mapping by an affine mapping if we XOR the additive constant an even number of times.

Let us now show why claim 4 holds. Any affine mapping over $GF(2)$ can be divided into k distinct n to m-bit projections. Since $(k-1)m$ of the input bits are constant, we will be interested only in restrictions of these affine mappings to new affine mappings that map the i-th block of m bits (the one which has the P property) into some other m-bit block in the output:

$$y = A_{ij}(x) = L_{ij} \cdot x \oplus B_j, \ j = 1, \ldots k.$$

Here L_{ij} is an arbitrary $m \times m$ (not necessarily invertible) binary matrix and $B_j \in \{0, 1\}^m$. We can again ignore B_j since it is XOR'ed an even number of times. If L_{ij} is invertible over $GF(2)$, then $L_{ij} \cdot x$ is a 1-1 transform and thus $L_{ij} \cdot x$ gets all the 2^m possible values when x ranges over all the 2^m possible inputs, so it has property P.

Thus we are left with the case of non-invertible L_{ij}. Suppose that

$$rank(L_{ij}) = r < m.$$

The kernel is defined as the set of solutions of the homogeneous linear equation $L_{ij} \cdot x = 0$. Let x_0 be some solution of the non-homogeneous equation $L_{ij} \cdot x = y$. Then all the solutions of the non-homogeneous equation have the form $x_0 \oplus v_0$, where v_0 is any vector from the kernel. The size of the kernel is 2^{m-r}, and thus each y has either no preimages or exactly 2^{m-r} preimages. Since $r < m$ by assumption, 2^{m-r} is even, and thus the multiset of m-bit results has property E. Consequently each block of m bits of the output has either property P or property E, and thus the n bit output has property D^k, as claimed.

2.2 Recovering Layers S_1 and S_3.

The first phase of the attack finds the two outermost layers S_1 and S_3, in order to "peel them off" and attack the inner layers.

Consider a multiset of chosen plaintexts with property $C^{i-1}PC^{k-i}$. The key observations behind the attack are:

1. The given multiset is transformed by layer S_1 into a multiset with property $C^{i-1}PC^{k-i}$ by Lemma 2, claim 1.
2. The multiset $C^{i-1}PC^{k-i}$ is transformed by the affine mapping A_1 into a multiset with property D^k by Lemma 2, claim 4.
3. The multiset property D^k is preserved by layer S_2, and thus the output multiset is also D^k, by Lemma 2, claim 2.
4. The multiset property D^k is not necessarily preserved by the affine mapping A_2, but the weaker property B^k is preserved.
5. We can now express the fact that the collection of inputs to each S-box in S_3 satisfies property B by a homogeneous linear equation. We will operate with m-bit quantities at once as if working over $GF(2^m)$ (XOR and ADD are the same in this field). Variable z_i represents the m-bit input to the S-box which produces i as an output (i.e., the variables describe S^{-1}, which is well defined since S is invertible), and we use 2^m separate variables for each S-box in S_3. When we are given a collection of actual ciphertexts, we can use their m-bit projections as indices to the variables, and equate the XOR of the indexed variables to 0^m. Different collections of chosen plaintexts are likely to generate linear equations with different random looking subsets of variables (in which repetitions are cancelled in pairs). When sufficiently many linear equations are obtained we can solve the system by Gaussian elimination in order to recover all the S-boxes in S_3 in parallel.

Unfortunately, we cannot get a system of equations with a full rank of 2^m. Consider the truth table of the inverted S-box as a $2^m \times m$-bit matrix. Since the

S-box is bijective, the columns of this matrix are m linearly independent 2^m-bit vectors. Any linear combination of the S-box input bits (which are outputs of the inverted S-box) is also a possible solution, and thus the solution space must have a dimension of at least m. Moreover, since all our equations are XOR's of an even number (2^m) of variables, the bit complement of any solution is also a solution. Since the system of linear equations has a kernel of dimension at least $m+1$, there are at most $2^m - m - 1$ linearly independent equations in our system. When we tested this issue in an actual implementation of the attack for $m = 8$, we always got a linear system of rank 247 in 256 variables, as expected from the formula.

Fortunately, this rank deficiency is not a problem in our attack. When we pick any one of the non-zero solutions, we do not get the "true" S^{-1}, but $A(S^{-1})$, where A is an arbitrary invertible affine mapping over m-bits. By taking the inverse we obtain $S(A^{-1})$. This is the best we can hope for at this phase, since the arbitrarily chosen A^{-1} can be compensated for when we find $A(A_2) = A'_2$ instead of the "true" affine transform A_2, and thus the various solutions are simply equivalent keys which represent the same plaintext/ciphertext mapping.

A single collection of 2^m chosen plaintexts gives rise to one linear equation in the 2^m unknowns in each one of the k S-boxes in layer S_3. To get 2^m equations, we can use 2^{2m} (2^{16}) chosen plaintexts of the form (A, u, B, v, C), in which we place the P structures u and v at any two block locations, and choose A, B, C as arbitrary constants. For each fixed value of u, we get a single equation by varying v through all the possible 2^m values. However, we can get an additional equation by fixing v and varying u through all the 2^m possible values. Since we get $2 \cdot 2^m$ equations in 2^m unknowns, we can reduce the number of chosen plaintexts to $\frac{3}{4} \cdot 2^{2m}$ by eliminating the $\frac{1}{4}$ of the plaintexts in which u and v are simultaneously chosen in the top half of their range. The matrix of these (u, v) values has a missing top-right quarter, and we get half the equations we need from the full rows and half the equations we need from the full columns of this "L" shaped matrix.

Solving each system of linear equations by Gaussian elimination requires 2^{3m} steps, and thus we need $k2^{3m}$ steps to find all the S-boxes in S_3. For the Rijndael-like choice of parameters $n = 128$, $m = 8$ and $k = 16$, we get a very modest time complexity of 2^{28}.

To find the other external layer S_1, we can use the same attack in the reverse direction. However, the resultant attack requires both chosen plaintexts and chosen ciphertexts. In Section 3 we describe a slightly more complicated attack which requires only chosen plaintexts in all its phases.

2.3   Attacking the Inner Layers ASA

The second phase of the attack finds the middle three layers. We are left with a structure $A'_2 S_2 A'_1$ – two (possibly modified) affine layers and an S-box layer in the middle. In order to recover the affine layers we use Biham's low rank detection technique from [2]. Consider an arbitrary pair of known plaintexts P_1 and P_2 with difference $P_1 \oplus P_2$. With probability $k/2^m$, after A'_1 there will be no difference at the input to one of the k S-boxes in S_2. Thus there will also be no difference at the output of this S-box. Consider now the set of pairs $P_1 \oplus C_i$, $P_2 \oplus C_i$ for many randomly chosen n-bit constants C_i. Any pair in this set still has this property, and thus the set of all the obtained output differences after A_2' will have a rank of at most $n - m$, which is highly unusual for random n dimensional vectors. Consequently, we can confirm the desired property of the original pair P_1 and P_2 by applying this low rank test with about n modifiers C_i.

We want to generate and test pairs with zero input differences at each one of the k S-boxes. We choose a pool of t random vectors P_j and another pool of n modifiers C_i, and encrypt all the nt combinations $P_j \oplus C_i$. We have about $t^2/2$ possible pairs of P_j's, each one of them has a probability of $k/2^m$ to have the desired property at one of the S-boxes, and we need about $k \cdot log(k)$ random successes to cover all the k S-boxes. The critical value of t thus satisfies $t^2/2 \cdot k/2^m = k \cdot log(k)$ and thus $t = \sqrt{2^{m+1}log(k)}$. For $n = 128$ $m = 8$ and $k = 16$ we get $t = 2^{5.5}$, and thus the total number of chosen plaintexts we need is $nt = 2^{12.5}$, which is much smaller than the number we used in the first phase of the attack.

Now we use linear algebra in order to find the structure of A_2'. Consider the representation of A_2' as a set of n vectors $V_0, V_1, \ldots V_{n-1}$, $V_i \in \{0,1\}^n$, where A_2' transforms an arbitrary binary vector $b = b_0, b_1, \ldots b_{n-1}$ by producing the linear combination:

$$A_2'(b) = \bigoplus_{i=0}^{n-1} b_i V_i.$$

(we can ignore the affine constants viewing them as part of the S-box). From the data pool we extract information about k different linear subspaces of dimension $n - m \ (= 120)$. Then we calculate the intersection of any $k - 1 (= 15)$ of them. This intersection is an m-dimensional linear subspace which is generated by all the possible outputs from one of the S-boxes in layer S_2, after it is expanded from 8 bits to 128 bits by A_2'. We perform this operation for each S-box and by this we find a linear mapping A_2^* which is equivalent to the original choice. The complexity of this phase is that of Gaussian elimination on a set of $O(n - m)$ equations.

After finding and discarding A_2', we are left with the two layer structure $S_2 A_1'$. If we need to perform only decryption, we can recover this combined mapping by writing formal expressions for each bit, and then solving the linear equations with $k2^m \ (2^{12})$ variables. If we also need to perform encryption this trick will not work, since the formal expressions will be huge. However, we can just repeat our attack in the reverse direction by using chosen ciphertexts and recover A_1^*. After that we can find the remaining layer S_1 with about 2^m known plaintexts. Again we will find not the real S-box layer S_2 but the equivalent one which corresponds to the modified A_1^*, A_2^* that we have found in earlier phases.

Comment: for one of the mappings we need to know the order of the subspaces: we can assume arbitrary order of subspaces in A_2 together with arbitrary order of S-boxes in S_2, however at this point the order of subspaces in A_1 is no longer arbitrary. If after finding A_2 we mount the same attack on $S_2 A_1$ from the ciphertext direction, we can recover A_1' together with the correct ordering information.

The complete attack uses about 2^{2m} chosen plaintexts (2^{16}) and about $k2^{3m}$ ($16 \cdot 2^{24} = 2^{28}$) steps. We tested the attack with an actual implementation, and it always ended successfully after a few seconds of computation on a single PC. The attack remains practical even if we increase the size of the plaintexts from 128 to 1024 bits and replace the 8-bit S-boxes by 16-bit S-boxes, since with these parameters the attack requires 2^{32} chosen plaintexts and $64 \cdot 2^{3 \cdot 16} = 2^{54}$ time.

3   A Chosen Plaintext Attack on $ASAS$

In this section we show how to use a pure chosen plaintext attack, and avoid the less realistic chosen plaintext and chosen ciphertext attack. The modified attack has the same time and data complexities as the original attack.

After the first phase of the original attack we are left with a $A_2'S_2A_1S_1$ structure, since we can recover only one of the two external S-box layers. Since the inputs go through the additional S-box layer S_1, we can no longer argue that for any C_i, $P_1 \oplus C_i$ and $P_2 \oplus C_i$ will have a zero difference at the input to some S-box in S_2 whenever P_1 and P_2 have this property. We thus have to use a more structured set of modifiers which can be nonzero only at the inputs to the S-boxes in which P_1 and P_2 are identical.

For the sake of simplicity, we consider in this section only the standard parameters. We use 2^{16} chosen plaintexts with the multiset property PPC^{k-2} (the two P's could be placed anywhere, and we could reuse the chosen plaintexts from the first phase of the attack). There are 2^{15} different ways to choose a pair of values from the first P. For each such pair (a_1, a_2), we generate a group of 2^8 pairs of extensions of the form $(a_1, b_0, c, d, \ldots)$ and $(a_2, b_0, c, d, \ldots)$ where b_0 is any common element from the second P, and $c, d, \ldots$ are the constants from C^{k-2}. We claim that all these 2^8 pairs will have the same difference at the output of S_1, since the first S-box gets a fixed pair of values and the other S-boxes get identical inputs in each pair. We can now apply the low rank test since we have sufficiently many choices of (a_1, a_2) to get a zero difference at the input to each S-box in S_2 with high probability, and for any such (a_1, a_2) we have sufficiently many pairs with the same difference in order to reliably test the rank of the output vectors. Once we discover the partition of the output space into 16 different linear subspaces of dimension 120, we can again find the intersection of any 15 of them in order to find the 8 dimensional subspace generated by the outputs of each one of the 16 S-boxes. We fix A_2' by choosing any set of 8 arbitrary spanning vectors in each one of the 16 subspaces, and this is the best we can possibly do in order to characterize A_2' due to the existence of equivalent keys.

One possible problem with this compact collection of plaintexts is that the attack may fail for certain degenerate choices of affine mappings. For example, if both A_1 and A_2 are the identity mapping, the insufficiently mixed intermediate values always lead to very low output ranks. However, the attack was always successful when tested with randomly chosen affine mappings.

After peeling off the computed A'_2, we are now left with a $S'_2 A_1 S_1$ structure, which is different from the $A'_2 S_2 A'_1$ structure we faced in the original attack. We have already discovered in the previous part of the attack many groups of 256 pairs of plaintexts, where in each group we know that the XOR of each pair of inputs to any particular S-box in S'_2 is the same constant. We do not know the value of this constant, but we can express this property as a chain of homogeneous linear equations in terms of the values of the inverse S-box, which are indexed by the known outputs from the $S'_2 A_1 S_1$ structure. A typical example of the equations generated from one group is

$$S^{-1}(1) \oplus S^{-1}(72) = S^{-1}(255) \oplus S^{-1}(13) = S^{-1}(167) \oplus S^{-1}(217) = \ldots$$

If we need additional equations, we simply use another one of the 2^{15} possible groups of pairs, which yields a different chain of equations (with a different unknown constant). Note that these sparse linear equations are completely different from the dense equations we got in the first phase of the attack, which expressed the B property by equating the XOR's of various random looking subsets of 256 variables to 0^m.

We are finally left with a simple $A'_1 S_1$ structure. It can be attacked in a variety of ways, which are left as an exersise for the reader.

Comments:

- The attack works in exactly the same way if the affine mappings are over finite fields with even characteristic. In particular, it can be applied to Rijndael-like schemes in which the affine transforms are over $GF(2^8)$.
- The attack can be extended to the case where S_2 contains arbitrary random (not necessarily bijective) S-boxes with a small penalty in the number of chosen plaintexts. Direct application of our attack will not work, since the P property at the input to some S-box in layer S_2 may not result in a balanced output after S_2 if this particular S-box is non-bijective. In order to overcome this dificulty we can work with double-sized $2m$-bit S-boxes at layer S_1. We consider a projection mapping $PT1$ from $2m$ to m bits (in the affine mapping A_1) which necessarily has a non-zero kernel (and thus always has the E property which is preserved even by non-bijective S-boxes, and not the P property which is not preserved by non-bijective S-boxes). The attack works in exactly the same way with the exception that we pay a factor of 2^m in data and in the process of equation preparation (now each equation is the XOR of 2^{2m} variables instead of 2^m). The total complexity of the attack becomes 2^{3m} chosen plaintexts and $k2^{3m}$ steps.
- We can attack the scheme even if a sparse linear mapping (a bit permutation or a mapping that mixes small sets of bits like the Serpent [1] mappings) is added to the the input. The attack works as long as we can guess columns of the linear mapping that correspond to the inputs of one particular S-box in S_1. If we add an initial bit permutation with the standard parameters, we can guess which 8 plaintext bits enter this S-box, and construct the

$C^{i-1}PC^{k-i}$ structure we need to get each linear equation with just this knowledge. Note that to generate the P property we can choose these 8 bits in an unordered way, and to generate the other C^{k-1} property we don't care about the destination of the other bits under the bit permutation, and thus the number of cases we have to consider is at most $\binom{128}{8} \approx 2^{40}$. By increasing the time complexity of the attack by this number, we get a (barely practical) attack on this six layer scheme. By symmetry, we can also attack the scheme in which the additional bit permutation layer is added at the end, and with a somewhat higher complexity we can attack the seven layer scheme in which we add unknown bit permutations both at the beginning and at the end of the scheme. It is an open problem whether we can attack with reasonable complexity six layer schemes with a general affine mapping added either at the beggining or at the end.

– We can attack the scheme even if the S-boxes have inputs of different sizes which are unknown to the attacker, since this information will be revealed by rank analysis.
– We can attack modified schemes which have various types of feedback connections between the S-boxes in the first and last rounds (see Figure 2 for one example). The idea is that we still have some control over multisets in such construction: We can cause the rightmost S-box to run through all the possible inputs (if the XORed feedback is a constant) and thus can force multisets to have the $C^{k-1}P$ property after S_1 even when the indicated feedback connections are added. The extraction of the S-boxes in the last layer S_3 has to be carried out sequentially from right to left, in order to take into account the effect of the feedbacks at the bottom.
– The attack stops working if S_3 contains non-bijective S-boxes. One can estimate the sizes of the equivalence (collision) classes of the outputs of the particular S-box. However even writing the linear equations does not seem possible: If we get the same output value twice in our structure, we cannot tell which variables should be used as the input of the S-box in each case.

Acknowledgements

We thank David Wagner for a very useful early exchange of ideas, and Anton Mityagin for implementing and testing the attack described in this paper.

References

1. R. Anderson, E. Biham, L. Knudsen, *Serpent: A Proposal for the AES*, 1st AES Conference, 1998.
2. E. Biham, *Cryptanalysis of Patarin's 2-Round Public Key System with S-boxes (2R)*, proceedings of EUROCRYPT'2000, LNCS 1807, pp.408–416, Springer-Verlag, 2000.
3. J. Daemen, L. Knudsen, V. Rijmen, *The Block Cipher Square*, proceedings of FSE'97, LNCS 1267, pp.147–165, Springer-Verlag, 1997.
4. V. Rijmen, J. Daemen, *AES Proposal: Rijndael*, 1st AES Conference, 1998.

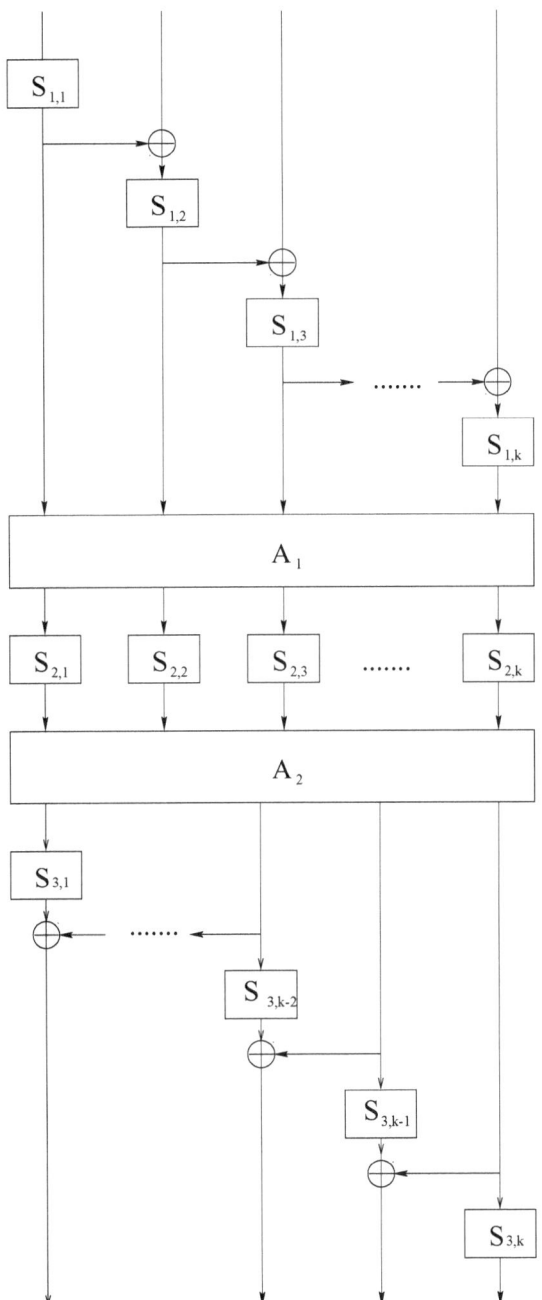

Fig. 2. Modified scheme with S-box feedbacks.

Hyper-bent Functions

Amr M. Youssef[1] and Guang Gong[2]

[1] Center for Applied Cryptographic Research
Department of Combinatorics & Optimization
University of Waterloo, Waterloo, Ontario N2L 3G1, CANADA
a2youssef@cacr.math.uwaterloo.ca
[2] Center for Applied Cryptographic Research
Department of Electrical and Computer Engineering
University of Waterloo, Waterloo, Ontario N2L 3G1, CANADA
ggong@cacr.math.uwaterloo.ca

Abstract. Bent functions have maximal minimum distance to the set of affine functions. In other words, they achieve the maximal minimum distance to all the coordinate functions of affine monomials. In this paper we introduce a new class of bent functions which we call hyper-bent functions. Functions within this class achieve the maximal minimum distance to all the coordinate functions of *all* bijective monomials. We provide an explicit construction for such functions. We also extend our results to vectorial hyper-bent functions.

Key words. Boolean functions, bent functions, hyper-bent functions, nonlinearity.

1 Introduction

Nonlinearity [19] is a crucial requirement for the Substitution boxes (S-boxes) in secure block ciphers. The success of linear cryptanalysis [16] depends on how well the S-boxes functions can be approximated by an affine function. Highly nonlinear functions provide good resistant towards linear cryptanalysis. On the other hand, even such functions can be attacked by higher order differential cryptanalysis [13] [25] if they have a low algebraic degree. From the view point of polynomials, Jakobsen and Knudsen [13] introduced the interpolation attack on block ciphers. This attack is useful for attacking ciphers using simple algebraic functions as S-boxes. In [13] Jakobsen extended this cryptanalysis method to attack block ciphers with probabilistic nonlinear relation of low degree. The complexity of both attacks depends on the degree of the polynomial approximation and/or on the number of terms in the polynomial approximation expression. Along the same line of research, Gong and Golomb [11] introduced a new criterion for the S-box design. By showing that many block ciphers can be viewed as a non linear feedback shift register with input, Gong and Golomb proposed that S-boxes should not be approximated by a bijective monomial. The reason is that, for $gcd(c, 2^n - 1) = 1$, the trace functions $Tr(\zeta_j x^c)$ and $Tr(\lambda x), x \in GF(2^n)$, are both m-sequences with the same linear span.

B. Pfitzmann (Ed.): EUROCRYPT 2001, LNCS 2045, pp. 406–419, 2001.
© Springer-Verlag Berlin Heidelberg 2001

For boolean functions with even number of input variables, bent functions achieve the maximal minimum distance to the set of affine functions. In other words, they achieve the maximal minimum distance to all the coordinate functions of affine monomials (I.e., functions in the form $Tr(\lambda x) + e$). However, this doesn't guarantee that such bent functions cannot be approximated by the coordinate functions of bijective monomials (I.e., functions in the form $Tr(\lambda x^c) + e, gcd(c, 2^n - 1) = 1$). For example, 120 bent functions out of the 896 bent functions with 4 input variables, have a minimum Hamming distance distance of 2 from the coordinate functions of the monomial x^7 and their complements.

A natural question is whether there exists a class of functions that have the same distance to all the coordinate functions of all bijective monomials. In this paper we give an affirmative answer to this question and provide an explicit construction method for such functions. Functions obtained by our construction also achieve the maximum algebraic normal form degree. We also extend our results to vectorial boolean functions.

We conclude this section by the notation and concepts which will be used throughout the paper. For the theory of shift register sequences and finite fields, the reader is referred to [9], [15].

- $\mathbb{K} = GF(2^{2n})$.
- $\mathbb{E} = GF(2^n)$.
- $\mathbb{F} = GF(2)$.
- α a primitive element of $\mathbb{K}$.
- $Tr_M^N(x)$, $M|N$, represents the trace function from $\mathbb{F}_{2^N}$ to $\mathbb{F}_{2^M}$, i.e., $Tr_M^N(x) = x + x^q + \cdots + x^{q^{l-1}}$ where $q = 2^M$ and $l = N/M$. If $M = 1$ and the context is clear, we write it as $Tr(x)$.
- $\underline{a} = \{a_i\}$, a binary sequence with period $s|2^{2n} - 1$. Sometimes, we also use a vector of dimension s to represent a sequence with period s. I.e., we also write $\underline{a} = (a_0, a_1, \cdots, a_{s-1})$.
- $Per(\underline{b})$, the least period of a sequence $\underline{b}$.
- $wt(s)$: the number of 1's in one period of the sequence s or the number of 1's in the set of images of the function $s(x) : GF(2^m) \to GF(2)$. This is the so called *the Hamming weight* of s whether s is a periodic binary sequence or a function from $GF(2^m)$ to $GF(2)$.

2   Preliminaries

There exists a 1-1 correspondence among the set of binary sequences with period $m|2^N - 1$, the set of polynomial functions from $GF(2^N)$ to $GF(2)$ and the set of Boolean functions in N variables through the trace representation of sequences. However, these connections are scattered in the literature. In this section, we will put this 1-1 correspondence together.

B. 1-1 Correspondence Among Periodic Sequences, Polynomial Functions and Boolean Functions

Let

- $\mathcal{S}$ be the set of all binary sequences with period $r|2^N - 1$,
- $\mathcal{F}$, the set of all (polynomial) functions from $GF(2^N)$ to $GF(2)$, and
- $\mathcal{B}$ the set of all Boolean functions in N variables.

There is a 1-1 correspondence among these three sets:

$$\mathcal{S} \longleftrightarrow \mathcal{F} \longleftrightarrow \mathcal{B}$$

which we will explain as follows.

B1. 1-1 Correspondence Between $\mathcal{S}$ and $\mathcal{F}$

Without loss of generality, assume that $f(0) = 0$. Any non-zero function $f(x) \in \mathcal{F}$ can be represented as

$$f(x) = \sum_{i=1}^{s} Tr_1^{m_{t_i}}(\beta_i x^{t_i}), \beta_i \in GF(2^{m_{t_i}})^*, \tag{1}$$

where t_i is a coset leader of a cyclotomic coset modulo $2^N - 1$, and $m_{t_i}|N$ is the size of the cyclotomic coset containing t_i. For any sequence $\underline{a} = \{a_i\} \in \mathcal{S}$, there exists $f(x) \in \mathcal{F}$ such that

$$a_i = f(\alpha^i), i = 0, 1, \cdots,$$

where α is a primitive element of $\mathbb{K}$. $f(x)$ is called *the trace representation* of $\underline{a}$. ($\underline{a}$ is also referred to as an *s*-term sequence.) If $f(x)$ is any function from $\mathbb{K}$ to $\mathbb{F}$, by evaluating $f(\alpha^i)$, we get a sequence over $\mathbb{F}$ with period dividing $2^N - 1$. Thus

$$\delta : \underline{a} \leftrightarrow f(x) \tag{2}$$

is a one-to-one correspondence between $\mathcal{F}$ and $\mathcal{S}$ through the trace representation in (1). We say that $f(x)$ is the *trace representation* of $\underline{a}$ and $\underline{a}$ is the *evaluation* of $f(x)$ at α. In this paper, we also use the notation $\underline{a} \leftrightarrow f(x)$ to represent the fact that $f(x)$ is the trace representation of $\underline{a}$. The set consisting of the exponents that appear in the trace terms of $f(x)$ is said to be the *null spectrum set* of $f(x)$ or $\underline{a}$.

If $s = 1$, i.e.,

$$a_i = Tr_1^N(\beta \alpha^i), i = 0, 1, \cdots, \beta \in \mathbb{K}^*,$$

then $\underline{a}$ is an m-sequence over $\mathbb{F}$ of period $2^N - 1$ of degree N. (For a detailed treatment of the trace representation of sequences, see [17].)

B2. 1-1 Correspondence between $\mathcal{F}$ and $\mathcal{B}$

Let $\{\alpha_1, \cdots \alpha_N\}$ be a basis of $\mathbb{K}/\mathbb{F}$ and let α be a primitive element of K. For $x \in \mathbb{K}$ we can represent x as

$$x = x_0\alpha_0 + x_1\alpha_1 + \cdots + x_{N-1}\alpha_{N-1}, x_i \in \mathbb{F}.$$

Thus we have

$$f(x) = f(\sum_{i=0}^{N-1} x_i\alpha_i) = g(x_0, \cdots, x_{N-1}),$$

i.e.,

$$\sigma : f(x) \to g(x_0, \cdots, x_{N-1}) \tag{3}$$

is a bijective map from $\mathcal{F}$ to $\mathcal{B}$. On the other hand, from the Lagrange interpolation [15], for a given Boolean function $g(x_0, \cdots, x_{N-1}) \in \mathcal{B}$, we can determine its polynomial representation $f(x)$ as follows

$$f(x) = \sum_{i=0}^{2^N-1} d_i x^i, \tag{4}$$

where

$$d_i = \sum_{x \in \mathbb{K}^*} (g(x_0, \cdots, x_{N-1}) - g(0, \cdots, 0))x^{-i},$$

where $x = \sum_{i=0}^{N-1} x_i\alpha_i$. Thus (4) gives a bijective map from $\mathcal{B}$ to $\mathcal{F}$ which is the inverse of (3). The correspondence among $\mathcal{S}$, $\mathcal{F}$ and $\mathcal{B}$ is shown in Figure 1.

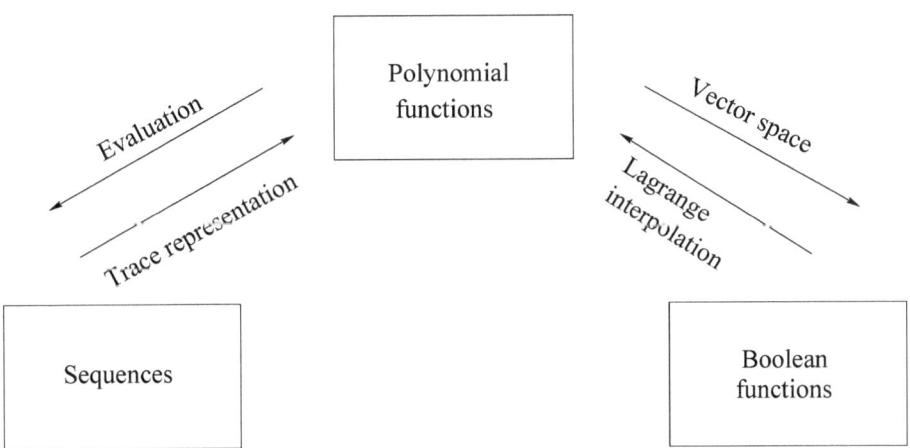

Fig. 1. Correspondence among $\mathcal{S}$, $\mathcal{F}$ and $\mathcal{B}$

From the above diagram, we have a 1-1 correspondence between $\mathcal{S}$, the set of all binary sequences with period dividing $2^N - 1$, and $\mathcal{B}$, the set of Boolean functions in N variables.

3   Extended Transform Domain Analysis for Boolean Functions

The Hadamard transform of $f : \mathbb{E} \to \mathbb{F}$ is defined by [1]

$$\hat{f}(\lambda) = \sum_{x \in \mathbb{E}} (-1)^{f(x) + Tr(\lambda x)}, \lambda \in \mathbb{E}. \tag{5}$$

The Hadamard transform spectrum of f exhibits the nonlinearity of f. More precisely, the nonlinearity of f is given by

$$NL(f) = 2^{n-1} - \frac{1}{2} \max_{\lambda \in \mathbb{E}} |\hat{f}(\lambda)|.$$

I.e., the absolute value of $\hat{f}(\lambda)$ reflects the difference between agreements and disagreements of $f(x)$ and the linear function $Tr(\lambda x)$. Only Bent functions [23] have a constant spectrum of their Hadamard transform. Gong and Golomb [11] showed that many block ciphers can be viewed as a non linear feedback shift register with input. In the analysis of shift register sequences [9], all m-sequences are equivalent under the decimation operation on elements in a sequence. The same idea can be used to approximate boolean functions, i.e., we can use monomial functions instead of linear functions to approximate Boolean functions. In other words, for $gcd(c, n-1) = 1$, the trace functions $Tr(\zeta_j x^c)$ and $Tr(\lambda x)$ have the same linear span. From the view point of m-sequences, both of the sequences $\{Tr(\zeta \alpha^{ic})\}_{i \geq 0}$ and $\{Tr(\lambda \alpha^i)\}_{i \geq 0}$ are m-sequences of period $2^n - 1$. The former can be obtained from the later by decimation c. Gong and Golomb [11] introduced the concept of extended Hadamard transform (*EHT*) for a function from $\mathbb{E}$ to $\mathbb{F}$. The extended Hadamard transform is defined as follows.

Definition 1. *Let $f(x)$ be a function from $\mathbb{E}$ to $\mathbb{F}$. Let*

$$\hat{f}(\lambda, c) = \sum_{x \in \mathbb{K}} (-1)^{f(x) + Tr(\lambda x^c)} \tag{6}$$

where $\lambda \in \mathbb{E}$ and c is a coset leader modulo $2^n - 1$ co-prime to $2^n - 1$. Then we call $\hat{f}(\lambda, c)$ an extended Hadamard transform of the function f.

Notice that the Hadamard transform of f, defined by (5), is $\hat{f}(\lambda, 1)$. The numerical results in [11] show that, for all the coordinate functions $f_i, i = 1, \cdots, 32$ of the DES s-boxes, the distribution of $\hat{f}_i(\lambda, c)$ in λ is invariant for all c.

Thus a new generalized nonlinearity measure can be defined as

$$NLG(f) = 2^{n-1} - \frac{1}{2} \max_{\substack{\lambda \in \mathbb{E}, \\ c : gcd(c, 2^n - 1) = 1}} |\hat{f}(\lambda, c)|.$$

This leads to a new criterion for the design of Boolean functions used in conventional cryptosystems. The *EHT* of Boolean functions should not have any large component.

Throughout the rest of the paper, we consider functions with even number of input variables.

4   Construction for Hyper-bent Functions

In this section we introduce a new class of functions which have a constant *EHT* spectrum. A binary function $f : \mathbb{K} \to \mathbb{F}$ is said to be hyper-bent if and only if the *EHT* of f,

$$\widehat{f}(\lambda, c) = \sum_{x \in \mathbb{K}} (-1)^{Tr(\lambda x^c) + f(x)} = \pm 2^n$$

for all $\lambda \in \mathbb{K}$ and for all c, $gcd(c, 2^{2n} - 1) = 1$. Clearly a hyper-bent function must be bent.

Let $\underline{b} = \{b_j\}_{j \geq 0}$ be a binary sequence with period $2^n + 1$. In the following, first we will give the criterion such that $g(x) \leftrightarrow \underline{b}, g(0) = 0$, is hyper-bent and count the number of such functions. Then we will show that all such hyper-bent functions obtained from our construction achieve maximal algebraic degree.

Theorem 1. *With the above notation, then $g(x)$ is hyper-bent if and only if $wt(\underline{b}) = 2^{n-1}$, i.e., $\underline{b}$ is balanced (Note that a sequence is said to be balanced if the disparity between the number of 1's and the number of 0's in one period is not to exceed 1).*

In order to prove Theorem 1 we need the following Lemmas. Let $r = 2^n - 1, d = 2^n + 1$. Write $u_i = Tr(\zeta \alpha^{ci}), i = 0, 1, \cdots$. Thus $\underline{u} = \{u_i\}$ is an m-sequence of period $2^{2n} - 1$. Let $\underline{v} = \underline{u} + \underline{b}$. Then $\underline{v}$ can be written into a (d, r)-interleaved sequence [10], i.e., $\underline{v}$ can be arranged into the following array

$$\begin{bmatrix} v_0 & v_1 & \cdots & v_{d-1} \\ v_d & v_{d+1} & \cdots & v_{2d-1} \\ \vdots & \vdots & \vdots & \vdots \\ v_{d(r-1)} & v_{d(r-1)+1} & \cdots & v_{rd-1} \end{bmatrix} = (\underline{v}_0, \underline{v}_1, \cdots \underline{v}_{d-1}),$$

where $\underline{v}_j$'s are columns of the matrix.

Lemma 1. $\underline{v}_j = \underline{u}_j + \underline{b}_j$ where $\underline{u}_j$'s are columns of the matrix

$$\begin{bmatrix} u_0 & u_1 & \cdots & u_{d-1} \\ u_d & u_{d+1} & \cdots & u_{2d-1} \\ \vdots & \vdots & \vdots & \vdots \\ u_{d(r-1)} & u_{d(r-1)+1} & \cdots & u_{rd-1} \end{bmatrix},$$

and $\underline{b}_j = (b_j, b_j, \cdots, b_j)$ is a constant sequence.

Proof. The result follows by noting that $Per(\underline{u}) = dr$ and $Per(\underline{b}) = d$.

□

Lemma 2. *With the notation in Lemma 1, we have*

$$wt(\underline{v}) = \sum_{j=0}^{d-1} wt(\underline{u}_j + \underline{b}_j).$$

Proof. Note that

$$wt(\underline{v}) = \sum_{j=0}^{d-1} wt(\underline{v}_j).$$

Applying Lemma 1, the result follows immediately.

□

Lemma 3. *Let* $wt(\underline{b}) = t$ *and*

$$\widehat{g}(\lambda, c) = \sum_{x \in \mathbb{K}} (-1)^{Tr(\lambda x^c) + g(x)}.$$

Then $\widehat{g}(\lambda, c) \in \{2t, 2(t - 2^n)\}, \forall \lambda \in \mathbb{K}^*$ *and* $\widehat{g}(0, c) = (2^n - 1)(d - 2t) + 1.$

Proof.

$$\widehat{g}(0, c) = \sum_{x \in K} (-1)^{g(x)} = 1 + \sum_{x \in K^*} (-1)^{g(x)} = 1 + r \sum_{k=0}^{d-1} (-1)^{b_k}$$
$$= 1 + r(d - 2wt(\underline{b})).$$

For $\lambda \neq 0$,

$$\widehat{g}(\lambda, c) = \sum_{x \in \mathbb{K}} (-1)^{Tr(\lambda x^c) + g(x)} = 1 + \sum_{i=0}^{2^{2n}-1} (-1)^{u_i + b_i} \qquad (7)$$
$$= 1 - wt(\underline{v}) + (q - 1 - wt(\underline{v})) = 2^{2n} - 2wt(\underline{v}).$$

So, we only need to determine the value of the Hamming weight of $\underline{v}$. From Lemma 2 we have

$$wt(\underline{v}) = \sum_{j=0}^{d-1} wt(\underline{u}_j + \underline{b}_j).$$

Note that $\{u_j\}$ is a binary m-sequence of period $2^{2n} - 1$. So one of the component sequences $\underline{u}_j$ is a zero sequence and the rest of 2^n sequences are shifts of the binary m-sequence $\{Tr(\alpha^{icd})\}$ with period r. Without loss of generality, we can suppose that that $\underline{u}_0 = (0, 0, \cdots 0)$ is the zero sequence. Therefore $wt(\underline{u}_0) = 0$ and $wt(\underline{u}_j) = 2^{n-1}, 1 \leq j < 2^n$. We have the following two cases
Case 1. $b_0 = 0$.

$$wt(\underline{v}) = \sum_{j>0, b_j=0} wt(u_j) + \sum_{j>0, b_j=1} wt(u_j + 1)$$
$$= (d - t - 1)2^{n-1} + t(2^{n-1} - 1)$$
$$= d2^{n-1} - t - 2^{n-1}.$$

Thus in this case, we have

$$\widehat{g}(\lambda, c) = 1 + (2^{2n} - 1) - 2(d2^{n-1} - t - 2^{n-1}) = 2^{2n} - d2^n + 2t + 2^n = 2t.$$

Case 2. $b_0 = 1$

$$wt(\underline{\mathbf{v}}) = 2^n - 1 + \sum_{j>0, b_j=0} wt(u_j) + \sum_{j>0, b_j=1} wt(u_j + 1)$$

$$= 2^n - 1 + (d-t)2^{n-1} + (2^{n-1}-1)(t-1))$$

$$= 2^{n-1} + d2^{n-1} - t.$$

Substituting into (7) we get

$$\widehat{g}(\lambda, c) = 2^{2n} - 2(2^{n-1} + d2^{n-1} - t) = 2^{2n} - 2^n - d2^n + 2t = 2(t - 2^n).$$

Thus $\widehat{g}(\lambda, c) \in \{2t, 2(t - 2^n)\}, \forall \lambda \in \mathbb{K}^*$.

$\square$

Proof of Theorem 1. If $g(x)$ is hyper-bent, then $|\widehat{g}(\lambda, c)| = 2^n$. From Lemma 3, we have $\widehat{g}(\lambda, c) = 2t$ or $2(t - 2^n)$. Thus we have $2t = 2^n \Leftrightarrow t = 2^{n-1}$ or $2(2^n - t) = 2^n \Leftrightarrow t = 2^{n-1}$. Thus if $g(x)$ is hyper-bent then $wt(\underline{\mathbf{b}}) = 2^{n-1}$. Conversely, if $wt(\underline{\mathbf{b}}) = 2^{n-1} = t$, then according to Lemma 3 we have

$$\widehat{g}(\lambda, c) = 2t = 2^n,$$

or

$$\widehat{g}(\lambda, c) = 2(t - 2^n) = 2(2^{n-1} - 2^n) = -2^n,$$

which implies that

$$|\widehat{g}(\lambda, c)| = 2^n, \forall \lambda \in \mathbb{K}^*.$$

and

$$\widehat{g}(0, c) = (2^n - 1)(2^n + (d - 2t)) + 1 = (2^n - 1)(2^n + 1 - 2^n) + 1 = 2^n.$$

Thus $g(x)$ is bent.
$\square$

In the following theorem, we will count how many hyper-bent functions can be obtained using this construction and show these functions achieve maximal algebraic degree.

Note. Let $f(x) = \sum_{i=0}^{2^m-1} c_i x^i$ be a function from $GF(2^m)$ to $GF(2)$. The algebraic degree of $f(x)$ is defined by

$$AD_f = max\{AD_{x^i} | i : c_i \neq 0\} \text{ where } AD_{x^i} = wt(i).$$

So the algebraic degree of $f(x)$ is equal to the algebraic degree of a boolean form of $f(x)$.

Theorem 2. *Let*

$$S(d) = \{\underline{\mathbf{b}} = \{b_i\}_{i \geq 0} | b_i \in \mathbb{F} \text{ and } Per(\underline{\mathbf{b}}) = d\},$$

and

$$\Gamma(d) = \{h(x) : \mathbb{K} \rightarrow \mathbb{F} | h(x) \leftrightarrow \underline{\mathbf{b}} \in S(d)\}.$$

Then there are $\binom{2^n + 1}{2^{n-1}}$ *hyper-bent functions in* $\Gamma(d)$ *and each of such functions has algebraic degree* n, *which is the maximal algebraic degree that bent functions in* $2n$ *variables can achieve.*

In order to prove Theorem 2 we need the following lemma.

Lemma 4. *Let* $0 < a < 2^n + 1$, $a \equiv 1 \mod 2$. *Then*

$$wt(a(2^n - 1)) = n.$$

Proof. If $a = 2^n$, then $wt(2^n(2^n - 1)) = wt(2^n - 1) = n$. If $0 < a < 2^n$, then we can write

$$a(2^n - 1) = a2^n - a = 2^n(a - 1) + (2^n - a).$$

Since $(2^n - a) < 2^n$ and $\geq 2^n 2^n(a - 1) \leq 2^{2n}$ then

$$wt(a(2^n - 1)) = wt(2^n(a - 1)) + wt(2^n - a). \tag{8}$$

Since $a \equiv 1 \mod 2$, we can write

$$a = 1 + 2^{i_1} + 2^{i_2} + \cdots + 2^{i_k},$$

where $0 < i_1 < i_2 < \cdots < i_k < n$. Thus

$$a - 1 = 2^{i_1} + \cdots + 2^{i_k} \Rightarrow wt(2^n(a - 1)) = k. \tag{9}$$

We also have

$$2^n - a = 2^n - 1 - (a - 1) = 1 + 2 + \cdots + 2^n - 1 - (2^{i_1} + \cdots + 2^{i_k}).$$

Therefore

$$wt(2^n - a) = n - k. \tag{10}$$

Substituting (9) and (10) into (8), we obtain that $wt(a(2^n - 1)) = k + n - k = n$.

□

Proof of Theorem 2. According to Theorem 1, $h(x)$ is hyper-bent if and only if $wt(\underline{b}) = 2^{n-1}$. I.e., there are 2^{n-1} 1's in $\{b_0, b_1, \cdots, b_{d-1}\}$) where $Per(\underline{b})|d$. In the following, we first show $Per(\underline{b}) = d$ provided $wt(\underline{b}) = 2^{n-1}$ and $Per(\underline{b})|d$. I.e., if $wt(\underline{b}) = 2^{n-1}$ and $Per(\underline{b})|d$, then $\underline{b} \in S(d)$. Let $d = kPer(\underline{b})$. Then $wt(\underline{b})k = 2^{n-1} \Longrightarrow k|2^c$. Since $gcd(d, 2) = 1$ and $k|d$, this forces $k = 1$. Therefore $Per(\underline{b}) = d$. Hence $\underline{b} \in S(d)$. Note that the number of sequences with weight 2^{n-1} in $S(d)$ is $\binom{2^n + 1}{2^{n-1}}$ which is just the number of bent functions in $\Gamma(d)$.

It can also be shown that for each $h(x) \in \Gamma(d)$, there exists $f : \mathbb{K} \to \mathbb{F}$ such that $h(x) = f(x^r)$ and the evaluation of $f(x)$ has period dr. Therefore, the exponent of any trace term in $h(x)$ has r as a divisor, i.e., it can be expressed as rs where $1 \leq s \leq 2^n$ and s is a coset leader. Applying Lemma 4, we have $wt(rs) = n$. Thus all exponents in $h(x)$ have weight n. Therefore $h(x)$ has algebraic degree n which is maximal since bent functions in $2n$ variables has algebraic degree $\leq n$ [21].

Remark 1. It is easy to show that the complement of the functions obtained from our constructions are also hyper-bent. Thus the total number of hyper-bent functions (with $2n$ input bits) obtained from our construction is $2 \binom{2^n + 1}{2^{n-1}}$.

Remark 2. After this paper was accepted for publications, Claude Carlet pointed out that this class of hyper-bent functions corresponds to the class of bent functions (family PS/ap) introduced by Dillon in his dissertation. In [6] Dillon constructed this class using the partial spread difference sets.

Remark 3. The special case where $g(x)$ in Theorem 1 is given by $Tr(\alpha x^{2^n-1})$ and $\alpha \in \mathbb{K}$ such that $\underline{\mathbf{b}} \leftrightarrow Tr(\alpha x^{2^n-1})$ is balanced, is equivalent to the difference set construction in [7]. In [7] Dillon showed that the difference sets constructed using $Tr(\alpha x^{2^n-1})$ is inequivalent to the Maiorana-McFarland construction [18]. Using a similar approach, it is easy to show that the construction in Theorem 1 is inequivalent to the Maiorana-McFarland construction.

5   Construction of Balanced Functions with Large NLG

By randomly complementing 2^{n-1} zeros of a hyper-bent functions we obtain a balanced functions with $NLG \geq 2^{2n-1} - 2^n$. Using this construction procedure, for $2n = 8$, we were able to obtain balanced functions with $NLG = 116$. (Note that the best known nonlinearity for balanced functions with 8 input variables is 116 [8], [22]).

Example 1. The $(17, 15)$ interleaved sequence corresponding the a balanced function $f : GF(2^8) \rightarrow GF(2)$ with $NLG = 116$ is shown below. The ones in the complemented positions is surrounded by brackets.

$$
\begin{bmatrix}
1\,1 & 0 & 0\,1\,1\,0\,1\,1\,0\,0\,0\,1 & 0 & 0\,1\,0 \\
1\,1 & (1) & 0\,1\,1\,0\,1\,1\,0\,0\,0\,1 & 0 & 0\,1\,0 \\
1\,1 & (1) & 0\,1\,1\,0\,1\,1\,0\,0\,0\,1 & (1) & 0\,1\,0 \\
1\,1 & 0 & 0\,1\,1\,0\,1\,1\,0\,0\,0\,1 & 0 & 0\,1\,0 \\
1\,1 & 0 & 0\,1\,1\,0\,1\,1\,0\,0\,0\,1 & 0 & 0\,1\,0 \\
1\,1 & 0 & 0\,1\,1\,0\,1\,1\,0\,0\,0\,1 & 0 & 0\,1\,0 \\
1\,1 & 0 & 0\,1\,1\,0\,1\,1\,0\,0\,0\,1 & 0 & 0\,1\,0 \\
1\,1 & (1) & 0\,1\,1\,0\,1\,1\,0\,0\,0\,1 & 0 & 0\,1\,0 \\
1\,1 & (1) & 0\,1\,1\,0\,1\,1\,0\,0\,0\,1 & 0 & 0\,1\,0 \\
1\,1 & 0 & 0\,1\,1\,0\,1\,1\,0\,0\,0\,1 & 0 & 0\,1\,0 \\
1\,1 & 0 & 0\,1\,1\,0\,1\,1\,0\,0\,0\,1 & 0 & 0\,1\,0 \\
1\,1 & (1) & 0\,1\,1\,0\,1\,1\,0\,0\,0\,1 & 0 & 0\,1\,0 \\
1\,1 & (1) & 0\,1\,1\,0\,1\,1\,0\,0\,0\,1 & 0 & 0\,1\,0 \\
1\,1 & (1) & 0\,1\,1\,0\,1\,1\,0\,0\,0\,1 & 0 & 0\,1\,0 \\
1\,1 & 0 & 0\,1\,1\,0\,1\,1\,0\,0\,0\,1 & 0 & 0\,1\,0
\end{bmatrix}
$$

6   Construction for Vectorial Hyper-bent Functions

Let $\{\eta_0, \eta_1, \cdots, \eta_{m-1}\}$ be a basis of $GF(2^m)$ over $GF(2)$. Then we can write $h : GF(2^{2n}) \rightarrow GF(2^m)$ as

$$
h(x) = \sum_{j=0}^{m-1} h_j(x)\eta_j.
$$

We call $h(x)$ *VHB* if and only if every nonzero linear combination of its output coordinates is a hyper-bent function, i.e., $h(x)$ is a *VHB* function if and only for any non-zero m-tuple $(c_0, c_1, \cdots, c_{m-1}) \in \mathbb{F}^m$, the function $\sum_{i=0}^{m-1} c_i h_i(x)$ from $\mathbb{K}$ to $\mathbb{F}$ is hyper-bent. Clearly, *VHB* functions are a sub-class of perfect nonlinear functions [21], [24]. It is known that such functions exist only for $m \le n$ [21]. In this section we will apply the results in Section 4 to present a new construction for vectorial hyper-bent (*VHB*) functions with the maximum possible number of outputs and maximal algebraic degree.

Keep r and d as defined before. Let $\gamma = \alpha^d$ and $\mathbb{E} = GF(2^n)$. Then γ is a primitive element of $\mathbb{E}$. Let $\{\beta_0, \cdots, \beta_{n-1}\}$ be the dual basis of $\{1, \gamma, \cdots, \gamma^{n-1}\}$ of $\mathbb{E}$ over $\mathbb{F}$ and $\pi : \mathbb{E} \to \mathbb{E}$ be a permutation. We define

$$
b_{i,j}^{\pi} = \begin{cases} Tr(\beta_j \pi(\gamma^i)), & 0 \le i \le 2^n - 2, \\ Tr(\beta_j \pi(0), & i = 2^n - 1, \\ 0, & i = 2^n, \end{cases} \tag{11}
$$

Let $\underline{\mathbf{b}}_j^{\pi} = \{b_{i,j}^{\pi}\}_{i \ge 0}$ where $b_{i,j}^{\pi} = b_{s,j}^{\pi}$ for $i = kd + s$ with $0 \le s < d$. From the construction above, it is easy to see that $Per(\underline{\mathbf{b}}_j^{\pi}) = d$.

Theorem 3. *With the notation above, let Π be the set consisting of all permutations of $\mathbb{E}$ and*

$$
P = \{h(x) : \mathbb{K} \to \mathbb{E} | h(x) = \sum_{j=0}^{n-1} h_j(x) \gamma^j, h_j(x) \leftrightarrow \underline{\mathbf{b}}_j^{\pi}, \pi \in \Pi\},
$$

Then any function $h \in P$ is a VHB function with maximum algebraic degree. Moreover, we have

$$
|P| = 2^n!.
$$

Proof. We can write $\pi(x) = \sum_{j=0}^{n-1} Tr(\beta_j \pi(x)) \gamma^j$. Since π is a permutation of $\mathbb{E}$, then $wt(Tr(\beta_j \pi(x)) = 2^{n-1}$ for $0 \le j < n$. From (11), $\underline{\mathbf{b}}_j^{\pi}$ is obtained from the evaluation of $Tr(\beta_j \pi(x))$ by lengthening it by one zero bit. Thus $wt(\underline{\mathbf{b}}_j^{\pi}) = 2^{n-1}$ for each $j : 0 \le j < n$. For any nonzero m-tuple $(c_0, c_1, \cdots, c_{m-1}) \in \mathbb{F}^m$, the evaluation of $D(x) = \sum_{i=0}^{m-1} c_i h_i(x)$ can be obtained from the evaluation of $C(x) = \sum_{i=0}^{m-1} c_i Tr(\beta_i \pi(x))$ by lengthening it by one zero bit. Note that

$$
C(x) = Tr \left(\sum_{i=0}^{m-1} c_i \beta_i \pi(x) \right) = Tr(\theta \pi(x))
$$

where $\theta = \sum_{i=0}^{m-1} c_i \beta_i \ne 0, \theta \in \mathbb{E}$. Thus $wt(C(x)) = wt(Tr(\theta \pi(x))) = 2^{n-1}$. Therefore the evaluation of $D(x)$ has weight 2^{n-1}. According to Theorem 4, $D(x)$ is hyper-bent. Thus $h(x)$ is a *VHB* function. The proof regarding the algebraic degree is identical to that in the proof of Theorem 2. Since the number of permutations of E is $2^n!$, then $|P| = 2^n!$.

□

Remark 4. Note that we can insert one zero bit in equation (11) at any place other than $i = 2^n$ and we still have $wt(\mathbf{b}_j) = 2^{n-1}$. Hence we have $(2^n+1)!/2$ such constructions, which corresponds to the number of ordered partitions of $2^n + 1$ elements of which 2 elements are the same and the rest of the $2^n - 1$ elements are all different. It is easy to show that the complement of the functions obtained from our constructions are also *VHB* functions. Thus the total number of *VHB* functions (with $2n$ input bits) obtained from our construction is

$$(2^n + 1)!.$$

Remark 5. Note that the notion of hyper-bent functions investigated in this paper is different from the one used in [5]. In fact, the class of functions considered in [5] are those Boolean functions on $(GF(2))^m$ (m even) such that, for a given even integer k ($2 \leq k \leq m - 2$), any of the Boolean functions on $(GF(2))^{m-k}$ obtained by fixing k coordinates of the variable is bent.

7   Conclusions

Boolean functions used in block cipher design should have a large Hamming distance to functions with simple algebraic description. In this paper, we presented a method to construct bent functions which achieve the maximal minimum distance to the set of all bijective monomials. Functions obtained from our construction achieve the maximum algebraic degree. These functions can be modified to achieve the balance property while maintaining large distance to bijective monomials. We also presented a method to construct vectorial bent functions for which every non zero linear combination of its coordinate functions satisfy the above property. These functions also achieve both the largest degree and the largest number of output bits. It should also be noted that while Rijndael (the NIST's Selection for the AES [14]) S-boxes are constructed by the monomial $x^{-1} = x^{254}$ over $GF(2^8)$, an affine transformation over $GF(2)$ is applied to the output of these S-boxes and hence the equivalent S-boxes will not have a simple algebraic description when looked at as a polynomial over $GF(2^8)$.

Acknowledgment

After this paper was accepted for publications, we realized that the term "hyper-bent" has been used by Claude Carlet to refer to a different class of Boolean functions. We would like to thank Claude Carlet for encouraging us to keep the "hyper-bent" term and for attracting our attention to references [6] and [7] .

References

1. R.E. Blahut, *Theory and Practice of Error Control Codes*, Addison-Wesley Publishing Company, 1983.

2. C. Carlet, A construction of bent functions, *Proc. of Third Conference of Finite Fields and Applications*, Glasgow, London Mathematical Society, Lecture Series 233, Cambridge University Press, 1996, pp. 47-58.

3. C. Carlet, Two new classes of bent functions, *Advances in Cryptology-EuroCrypt'85*, Lecture Notes in Computer Science, No. 765, Springer-Verlag, 1994, pp. 77-101.

4. C. Carlet , P. Charpin and V. Zinoviev, *Codes, bent functions and permutations suitable for DES-like cryptosystems*, Designs, Codes and Cryptography. vol.15, no.2; Nov. 1998; pp.125-156.

5. C. Carlet, Hyper-bent functions, *PRAGOCRYPT'96*, Czech Technical University Publishing House, Prague, pp. 145-155, 1996.

6. J. F. Dillon, *Elementary Hadamard Difference sets*, Ph.D. Dissertation, University of Maryland, 1974.

7. J. F. Dillon, *Elementary Hadamard Difference sets*, in Proc. Sixth S-E Conf. Comb. Graph Theory and Comp., 237-249, F. Hoffman et al. (Eds), Winnipeg Utilitas Math (1975) .

8. H. Dobbertin, Construction of bent functions and balanced Boolean functions with high nonlinearity, *Proceedings of Fast Software Encryption, Second International Workshop*, Springer-Verlag, 1995, pp. 61-74.

9. S.W. Golomb,*Shift Register Sequences*, Aegean Park Press. Laguna Hills, California. 1982.

10. G. Gong, Theory and applications of q-ary interleaved sequences, *IEEE Trans. on Inform. Theory*, vol. 41, No. 2, 1995, pp. 400-411.

11. G. Gong and S. W. Golomb, *Transform Domain Analysis of DES*, IEEE transactions on Information Theory. Vol. 45. no. 6. pp. 2065-2073. September, 1999.

12. T. Jakobsen and L. Knudsen, *The Interpolation Attack on Block Ciphers, LNCS 1267*, Fast Software Encryption. pp. 28-40. 1997.

13. T. Jakobsen, *Cryptanalysis of Block Ciphers with Probabilistic Non-linear Relations of Low Degree*, Proceedings of Crypto'99. LNCS 1462. pp. 213-222. 1999.

14. J. Daemen and V. Rijmen, *AES Proposal: Rijndael*, http://csrc.nist.gov/encryption/aes/rijndael/

15. R. Lidl and H. Niederreiter, *Finite Fields*, Encyclopedia of Mathematics and its Applications, Volume 20, Addison-Wesley, 1983.

16. M. Matsui, *Linear Cryptanalysis method for DES cipher* Advances in Cryptology, Proceedings of Eurocrypt'93, LNCS 765, pp. 386-397, Springer-Verlag, 1994.

17. R. J. McEliece, *Finite Fields For Computer Scientists and Engineers* , Kluwer Academic Publishers, Dordrecht, 1987.

18. R.L. McFarland, *A family of Noncyclic Difference Sets* , Journal of Comb. Th. (Series A) 15, pp. 1-10, 1973.

19. W. Meier and O. Staffelbach, *Nonlinearity criteria for cryptographic functions*, Proceedings of EUROCRYPT '89, Springer-Verlag, Berlin, Germany, 1990 pp. 549-62.

20. K. Nyberg, S-boxes and round functions with controllable linearity and differential uniformity, *Proceedings of Fast Software Encryption, Second International Workshop*, Springer-Verlag, Berlin, Germany, 1995, pp.111-130.

21. K. Nyberg, Perfect nonlinear S-boxes, *Proceedings of EUROCRYPT '91*, Springer-Verlag, Berlin, Germany, 1991, pp.378-86.

22. P. Sarkar and S. Maitra, Nonlinearity Bounds and Constructions of Resilient Boolean Functions, *Proceedings of CRYPTO '2000*, Springer-Verlag, Berlin, Germany, LNCS 1880, pp. 515-532.

23. O.S. Rothaus , On bent functions, *J. Combinatorial Theory*, vol. 20(A), 1976, pp.300-305.

24. T. Satoh , T. Iwata and K. Kurosawa, On cryptographically secure vectorial Boolean functions, *Proceedings of ASIACRYPT'99*, Springer-Verlag, Berlin, Germany, 1999, pp. 20-28.

25. T. Iwata and K. Kurosawa, Probabilistic higher order differential cryptanalysis and higher order bent functions. *Proceedings of ASIACRYPT'99*, Springer-Verlag, Berlin, Germany, 1999, pp. 62-74.

New Method for Upper Bounding
the Maximum Average Linear Hull Probability
for SPNs

Liam Keliher[1], Henk Meijer[1], and Stafford Tavares[2]

[1] Department of Computing and Information Science
Queen's University at Kingston, Ontario, Canada, K7L 3N6
{keliher,henk}@cs.queensu.ca
[2] Department of Electrical and Computer Engineering
Queen's University at Kingston, Ontario, Canada, K7L 3N6
tavares@ee.queensu.ca

Abstract. We present a new algorithm for upper bounding the maximum average linear hull probability for SPNs, a value required to determine provable security against linear cryptanalysis. The best previous result (Hong et al. [9]) applies only when the linear transformation branch number ($\mathcal{B}$) is M or $(M + 1)$ (maximal case), where M is the number of s-boxes per round. In contrast, our upper bound can be computed for any value of $\mathcal{B}$. Moreover, the new upper bound is a function of the number of rounds (other upper bounds known to the authors are not). When $\mathcal{B} = M$, our upper bound is consistently superior to [9]. When $\mathcal{B} = (M + 1)$, our upper bound does not appear to improve on [9]. On application to Rijndael (128-bit block size, 10 rounds), we obtain the upper bound $UB = 2^{-75}$, corresponding to a lower bound on the data complexity of $\frac{8}{UB} = 2^{78}$ (for 96.7% success rate). Note that this does not demonstrate the existence of a such an attack, but is, to our knowledge, the first such lower bound.

Keywords: substitution-permutation networks, linear cryptanalysis, maximum average linear hull probability, provable security

1 Introduction

The substitution-permutation network (SPN) [6] is a fundamental block cipher architecture designed to be a practical implementation of Shannon's principles of *confusion* and *diffusion* [15], through the use of substitution and linear transformation (LT), respectively. There has been a recent increase in interest in SPNs, in part because their simplicity lends itself to analysis, and, from an implementation viewpoint, because they tend to be highly parallelizable. This interest will no doubt be spurred on by the recent adoption of Rijndael (a straightforward SPN) as the U.S. Government Advanced Encryption Standard (AES)[5].

The two most powerful cryptanalytic attacks on block ciphers are generally considered to be linear cryptanalysis (LC) [11] and differential cryptanalysis

B. Pfitzmann (Ed.): EUROCRYPT 2001, LNCS 2045, pp. 420–436, 2001.
© Springer-Verlag Berlin Heidelberg 2001

(DC) [2]. There exists a strong duality between these two attacks which allows certain results related to one of the attacks to be translated into the corresponding results for the other attack [1][12]. This duality applies to the work of this paper; for this reason we will limit our focus to LC.

In carrying out LC, an attacker typically computes a vector called the *best linear characteristic*, for which the associated *linear characteristic probability* (LCP) is maximal. This LCP allows the attacker to estimate the number of chosen plaintexts required to mount a successful attack. In [14], Nyberg showed that the use of linear characteristics underestimates the success of LC. In order to guarantee *provable security*, a block cipher designer needs to consider *approximate linear hulls* instead of linear characteristics, and the *maximum average linear hull probability* instead of the LCP of the best linear characteristic.

In this paper we present a new method for computing an upper bound on the maximum average linear hull probability for SPNs. The best previous result is that of Hong et al. [9], which applies only to SPNs with highly diffusive LTs. In contrast, our method can be applied to an SPN with any LT (computation time may vary). Moreover, the upper bound we compute is a function of the number of rounds of the SPN; all other upper bounds known to the authors do not depend on the number of rounds. When the diffusiveness of the LT is one less than maximum (the relevant definition is given in Section 5.3), our upper bound is consistently superior to that of [9]. For LTs with maximum diffusiveness, our upper bound does not appear to improve on [9].

Application of our method to Rijndael (128-bit block size, 10 rounds), which involved extensive computation, yielded the upper bound $UB = 2^{-75}$, for a corresponding lower bound on the data complexity of LC of $\frac{8}{UB} = 2^{78}$ (for 96.7% success rate—see Section 3). Note that this does not demonstrate the existence of a such an attack, but is, to our knowledge, the first such lower bound.

Conventions

In what follows, $\{0,1\}^d$ denotes the set of all d-bit vectors, which we view as row vectors. For a vector or matrix $\mathbf{w}$, $\mathbf{w}'$ denotes the transpose of $\mathbf{w}$. We adopt the convention that numbering of the bits of a binary vector proceeds from left to right, beginning at 1. The Hamming weight of a vector $\mathbf{x}$ is written $wt(\mathbf{x})$. If $\mathbf{Z}$ is a random variable (r.v.), $E[\mathbf{Z}]$ denotes the expected value of $\mathbf{Z}$. And we use $\#\mathcal{A}$ to indicate the number of elements in the set $\mathcal{A}$.

2   Substitution-Permutation Networks

A block cipher is a bijective mapping from N bits to N bits (N is called the *block size*) parameterized by a bitstring called a *key*, denoted $\mathbf{k}$. Common block sizes are 64 and 128 bits. The input to a block cipher is called a *plaintext*, and the output is called a *ciphertext*.

An SPN encrypts a plaintext through a series of R simpler encryption steps called *rounds*. The input to round r ($1 \leq r \leq R$) is first bitwise XOR'd with an N-bit *subkey*, denoted $\mathbf{k}^r$, which is typically derived from the key, $\mathbf{k}$, via a separate *key-scheduling algorithm*. The *substitution stage* then partitions the resulting vector into M subblocks of size n ($N = Mn$), which become the inputs to a row of bijective $n \times n$ *substitution boxes* (*s-boxes*)—bijective mappings from $\{0,1\}^n$ to $\{0,1\}^n$. Finally, the *permutation stage* applies an invertible LT to the output of the s-boxes (classically, a bitwise permutation). Often the permutation stage is omitted from the last round. A final subkey, $\mathbf{k}^{R+1}$, is XOR'd with the output of round R to form the ciphertext. Figure 1 depicts an example SPN with $N = 16$, $M = n = 4$, and $R = 3$.

We assume the most general situation for the key, namely, that $\mathbf{k}$ is an *independent key* [1], a concatenation of $(R + 1)$ independent subkeys—symbolically, $\mathbf{k} = \langle \mathbf{k}^1, \mathbf{k}^2, \ldots, \mathbf{k}^{R+1} \rangle$. We use $\mathcal{K}$ to denote the set of all independent keys.

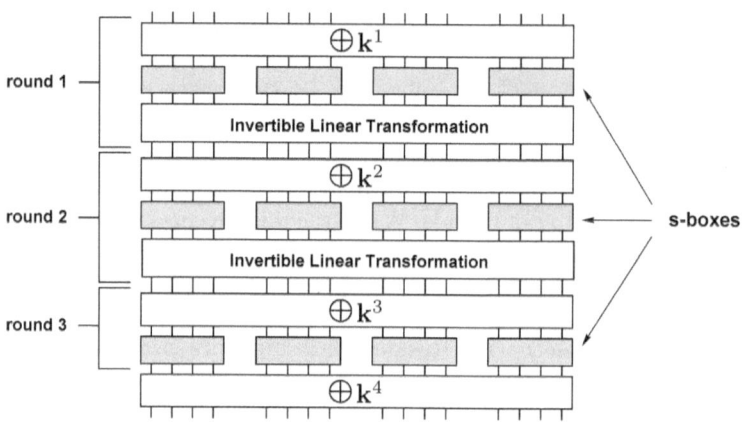

Fig. 1. SPN with $N = 16$, $M = n = 4$, $R = 3$

3   Linear Cryptanalysis

Linear cryptanalysis (LC) was introduced by Matsui in 1993 [11]. The more powerful version is known as Algorithm 2 (Algorithm 1 extracts only a single subkey bit). As applied to SPNs, Algorithm 2 can be used to extract the first subkey, $\mathbf{k}^1$. Once $\mathbf{k}^1$ is known, the first round can be stripped off, and LC can be reapplied to obtain $\mathbf{k}^2$, and so on.

Let $\mathbf{P}$, $\mathbf{C}$, and $\mathbf{X}$ be r.v.'s representing the plaintext, ciphertext, and intermediate input to round 2, respectively. The attacker attempts to identify the best correlation between the parity of a subset of the bits of $\mathbf{X}$ and the parity of a subset of the bits of $\mathbf{C}$. Symbolically, the attacker wants *masks* $\mathbf{a}, \mathbf{b} \in \{0,1\}^N \setminus \mathbf{0}$ which maximize the following *linear probability*:

$$LP_{\mathbf{k}}(\mathbf{a} \to \mathbf{b}) \overset{\text{def}}{=} (2 \cdot \text{Prob}\,\{\mathbf{a} \bullet \mathbf{X} = \mathbf{b} \bullet \mathbf{C}\} - 1)^2 , \qquad (1)$$

for a fixed key, $\mathbf{k}$ (the symbol $\bullet$ denotes the inner product over GF(2)). Note that $LP_{\mathbf{k}}(\mathbf{a} \to \mathbf{b}) \in [0,1]$. Given $\mathbf{a}$ and $\mathbf{b}$, the attack proceeds as in Figure 2.

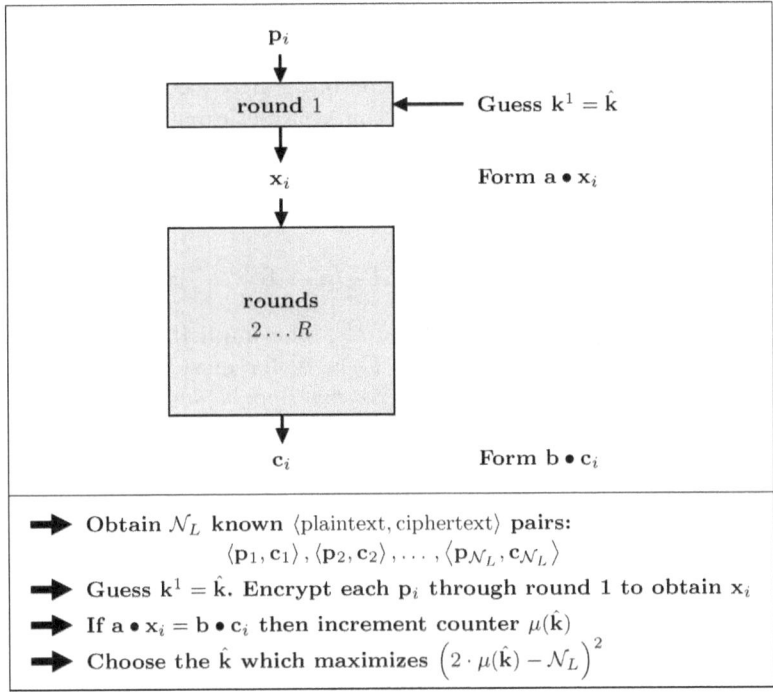

Obtain $\mathcal{N}_L$ known $\langle\text{plaintext}, \text{ciphertext}\rangle$ **pairs:**

$$\langle \mathbf{p}_1, \mathbf{c}_1 \rangle, \langle \mathbf{p}_2, \mathbf{c}_2 \rangle, \dots, \langle \mathbf{p}_{\mathcal{N}_L}, \mathbf{c}_{\mathcal{N}_L} \rangle$$

Guess $\mathbf{k}^1 = \hat{\mathbf{k}}$**. Encrypt each** $\mathbf{p}_i$ **through round 1 to obtain** $\mathbf{x}_i$

If $\mathbf{a} \bullet \mathbf{x}_i = \mathbf{b} \bullet \mathbf{c}_i$ **then increment counter** $\mu(\hat{\mathbf{k}})$

Choose the $\hat{\mathbf{k}}$ **which maximizes** $\left(2 \cdot \mu(\hat{\mathbf{k}}) - \mathcal{N}_L\right)^2$

Fig. 2. Summary of linear cryptanalysis (Algorithm 2)

The probability that Algorithm 2 will determine the correct value of $\mathbf{k}^1$ increases as the number of known $\langle\text{plaintext}, \text{ciphertext}\rangle$ pairs, $\mathcal{N}_L$, is increased. The value $\mathcal{N}_L$ is called the *data complexity* of the attack—this is what the attacker wants to minimize. Given an assumption about the behavior of round-1 output [11], Matsui shows that if $\mathcal{N}_L = \frac{c}{LP_{\mathbf{k}}(\mathbf{a} \to \mathbf{b})}$, then Algorithm 2 has the success rates in the following table, for various values of the constant, c.

c	2	4	8	16
Success rate	48.6%	78.5%	96.7%	99.9%

3.1 Notational Generalization

In describing Algorithm 2, we have discussed input and output masks ($\mathbf{a}$ and $\mathbf{b}$, respectively) and the associated linear probability for rounds $2 \dots R$ of an R-round SPN. It is useful to consider these and other related concepts as applying to any $T \geq 2$ consecutive rounds of an SPN. Hereafter, unless specified otherwise, terms such as "first round" and "last round" are relative to the T rounds under consideration. For Algorithm 2, then, $T = R - 1$, and the "first round," or "round 1," is actually round 2 of the SPN. And for simplicity, we will always assume that the LT is absent from round T (this does not affect LC).

4   Linear Characteristics

For fixed $\mathbf{a}, \mathbf{b} \in \{0,1\}^N$, direct computation of $LP_{\mathbf{k}}(\mathbf{a} \rightarrow \mathbf{b})$ is generally infeasible, first since it requires encrypting all N-bit vectors through rounds $1 \ldots T$, and second because $LP_{\mathbf{k}}(\mathbf{a} \rightarrow \mathbf{b})$ depends on the unknown key, $\mathbf{k}$. The latter is usually handled by working not with $LP_{\mathbf{k}}(\mathbf{a} \rightarrow \mathbf{b})$, but with the *average* (expected) value of $LP_{\mathbf{k}}(\mathbf{a} \rightarrow \mathbf{b})$ over all independent keys $\mathbf{k} \in \mathcal{K}$, denoted $E_T[\mathbf{a}, \mathbf{b}]$:

$$E_T[\mathbf{a}, \mathbf{b}] \overset{\text{def}}{=} E[LP_{\mathbf{K}}(\mathbf{a} \rightarrow \mathbf{b})] \tag{2}$$

($\mathbf{K}$ is an r.v. uniformly distributed over $\mathcal{K}$). The implicit assumption is that $LP_{\mathbf{k}}(\mathbf{a} \rightarrow \mathbf{b})$ is approximately equal to $E_T[\mathbf{a}, \mathbf{b}]$ for most values of $\mathbf{k}$ (Harpes et al. refer to this as the *Hypothesis of fixed-key equivalence* [7]). The data complexity of Algorithm 2 for masks $\mathbf{a}$ and $\mathbf{b}$ is now taken to be $\mathcal{N}_L = \frac{c}{E_T[\mathbf{a}, \mathbf{b}]}$. The problem of computational complexity is usually treated by approximating $E_T[\mathbf{a}, \mathbf{b}]$ through the use of *linear characteristics* (or simply *characteristics*).

4.1   One-Round and Multi-round Linear Characteristics

Note that the linear probability in (1) can be defined for any binary mapping—in particular, for a bijective $n \times n$ s-box, S. Let $\boldsymbol{\alpha}, \boldsymbol{\beta} \in \{0,1\}^n$, and let $\mathbf{X}$ be an r.v. uniformly distributed over $\{0,1\}^n$. Define

$$LP^S(\boldsymbol{\alpha} \rightarrow \boldsymbol{\beta}) \overset{\text{def}}{=} (2 \cdot \text{Prob}\{\boldsymbol{\alpha} \bullet \mathbf{X} = \boldsymbol{\beta} \bullet S(\mathbf{X})\} - 1)^2 \tag{3}$$

$$q \overset{\text{def}}{=} \max_{S \in \text{SPN}} \max_{\boldsymbol{\alpha}, \boldsymbol{\beta} \in \{0,1\}^n \setminus 0} LP^S(\boldsymbol{\alpha} \rightarrow \boldsymbol{\beta}) . \tag{4}$$

A *one-round* characteristic for round t, $1 \le t \le T$, is a pair $\Omega^t = \langle \mathbf{a}^t, \mathbf{b}^t \rangle$ in which $\mathbf{a}^t$ and $\mathbf{b}^t$ are input and output masks, respectively, for round t, *excluding the permutation stage*. The *linear characteristic probability* of Ω^t, denoted $LCP^t(\Omega^t)$ or $LCP^t(\mathbf{a}^t \rightarrow \mathbf{b}^t)$, is simply the linear probability obtained by viewing round t (minus the permutation stage) as an $N \times N$ s-box:

$$LCP^t(\Omega^t) \overset{\text{def}}{=} (2 \cdot \text{Prob}\{\mathbf{a}^t \bullet \mathbf{X} = \mathbf{b}^t \bullet S^t(\mathbf{X} \oplus \mathbf{k}^t)\} - 1)^2 , \tag{5}$$

where $S^t(\cdot)$ denotes application of the s-boxes of round t, and $\mathbf{X}$ is an r.v. uniformly distributed over $\{0,1\}^N$. (Note: It can be shown that $LCP^t(\Omega^t)$ is independent of the (unknown) subkey $\mathbf{k}^t$, and therefore the operation $\oplus \mathbf{k}^t$ can be removed from (5).) Let the M s-boxes of round t be enumerated from left to right as $S_1^t, S_2^t, \ldots, S_M^t$. Note that $\mathbf{a}^t$ and $\mathbf{b}^t$ determine input and output masks for each s-box in round t; let the masks for S_i^t be denoted $\boldsymbol{\alpha}_i^t$ and $\boldsymbol{\beta}_i^t$, respectively. Then by Matsui's Piling-up Lemma [11],

$$LCP^t(\Omega^t) = \prod_{i=1}^{M} LP^{S_i^t}(\boldsymbol{\alpha}_i^t \rightarrow \boldsymbol{\beta}_i^t) . \tag{6}$$

Definition 1. *Let* $\mathbf{L}$ *denote the N-bit LT of the SPN represented as a binary* $N \times N$ *matrix, i.e., if* $\mathbf{x}, \mathbf{y} \in \{0,1\}^N$ *are the input and output, respectively, for the LT, then* $\mathbf{y} = (\mathbf{L}\mathbf{x}')'$.

Lemma 1 ([3]). *If* $\mathbf{b} \in \{0,1\}^N$ *and* $\mathbf{a} = (\mathbf{L}'\mathbf{b}')'$, *then* $\mathbf{a} \bullet \mathbf{x} = \mathbf{b} \bullet \mathbf{y}$ *for all N-bit inputs to the LT,* $\mathbf{x}$, *and corresponding outputs,* $\mathbf{y}$ *(i.e., if* $\mathbf{b}$ *is an output mask for the LT, then* $\mathbf{a} = (\mathbf{L}'\mathbf{b}')'$ *is the (unique) corresponding input mask).*

Now given one-round characteristics for each of rounds $1 \ldots T$, $\Omega^1 = \langle \mathbf{a}^1, \mathbf{b}^1 \rangle$, $\Omega^2 = \langle \mathbf{a}^2, \mathbf{b}^2 \rangle$, ..., $\Omega^T = \langle \mathbf{a}^T, \mathbf{b}^T \rangle$, these can be *concatenated* to form a single T-round characteristic if $\mathbf{a}^{t+1}$ and $\mathbf{b}^t$ are corresponding output and input masks for the LT, respectively, for $1 \leq t \leq (T-1)$ (see Lemma 1). The resulting T-round characteristic is the tuple $\Omega = \langle \mathbf{a}^1, \mathbf{a}^2, \ldots, \mathbf{a}^T, \mathbf{b}^T \rangle$. The linear characteristic probability of Ω is again given by Matsui's Piling-up Lemma:

$$LCP(\Omega) = \prod_{t=1}^{T} LCP^t(\Omega^t) . \tag{7}$$

4.2   Choosing the Best Characteristic

In carrying out LC, the attacker typically runs an algorithm to find the T-round characteristic, Ω, for which $LCP(\Omega)$ is maximal; such a characteristic (not necessarily unique) is called the *best characteristic* [12]. If $\Omega = \langle \mathbf{a}^1, \mathbf{a}^2, \ldots, \mathbf{a}^T, \mathbf{b}^T \rangle$, and if the input and output masks used in Algorithm 2 are taken to be $\mathbf{a} = \mathbf{a}^1$ and $\mathbf{b} = \mathbf{b}^T$, respectively, then $E_T[\mathbf{a}, \mathbf{b}]$ (used to determine $\mathcal{N}_L = \frac{c}{E_T[\mathbf{a},\mathbf{b}]}$) is approximated by

$$E_T[\mathbf{a}, \mathbf{b}] \approx LCP(\Omega) . \tag{8}$$

5   Provable Security against Linear Cryptanalysis

The approximation in (8) has been widely used to evaluate the security of block ciphers against LC [8]. Knudsen calls a block cipher *practically secure* if the data complexity determined by this method is prohibitive [10]. However, in 1994 Nyberg demonstrated that this approach underestimates the success of LC [14]. We state Nyberg's results in the context of SPNs.

5.1   Approximate Linear Hulls

Definition 2 (Nyberg). *Given nonzero N-bit masks* $\mathbf{a}, \mathbf{b}$, *the approximate linear hull,* $ALH(\mathbf{a}, \mathbf{b})$, *is the set of all T-round characteristics, for the T rounds under consideration, having* $\mathbf{a}$ *as the input mask for round 1 and* $\mathbf{b}$ *as the output mask for round T, i.e., all characteristics of the form* $\Omega = \langle \mathbf{a}, \mathbf{a}^2, \mathbf{a}^3, \ldots, \mathbf{a}^T, \mathbf{b} \rangle$.

Remark: Recall that any characteristic $\Omega \in \text{ALH}(\mathbf{a}, \mathbf{b})$ determines an input and an output mask for each s-box in rounds $1 \ldots T$. If this yields at least one s-box for which the input mask is zero and the output mask is nonzero, or vice versa, the linear probability associated with that s-box will be 0 (see (3)) and therefore $LCP(\Omega) = 0$ by (6) and (7). We exclude such characteristics from consideration.

Definition 3. *For* $\mathbf{a}, \mathbf{b} \in \{0, 1\}^N \setminus \mathbf{0}$, *let* $\text{ALH}(\mathbf{a}, \mathbf{b})^*$ *consist of the elements* $\Omega \in \text{ALH}(\mathbf{a}, \mathbf{b})$ *such that for each s-box in rounds* $1 \ldots T$, *the input and output masks determined by* Ω *for that s-box are either both zero or both nonzero.*

Theorem 1 (Nyberg). *Let* $\mathbf{a}$ *and* $\mathbf{b}$ *be fixed nonzero N-bit input and output masks, respectively, for T rounds of an SPN. Then*

$$E_T[\mathbf{a}, \mathbf{b}] = \sum_{\Omega \in \text{ALH}(\mathbf{a}, \mathbf{b})^*} LCP(\Omega). \tag{9}$$

It follows immediately from Theorem 1 that (8) does not hold in general, since $E_T[\mathbf{a}, \mathbf{b}]$ is shown to be equal to the sum of terms $LCP(\Omega)$ over a (large) set of characteristics. Therefore, on average, the linear characteristic probability of the best characteristic will be strictly *less than* $E_T[\mathbf{a}, \mathbf{b}]$. An important implication of this is that the attacker will overestimate the number of $\langle$plaintext, ciphertext$\rangle$ pairs required for a given success rate. Indeed, Harpes et al. [7] comment that Matsui observed that his attacks performed better than expected.

5.2   Maximum Average Linear Hull Probability

An SPN is considered to be provably secure against LC if the *maximum average linear hull probability* (MALHP), $\max_{\mathbf{a}, \mathbf{b} \in \{0,1\}^N \setminus \mathbf{0}} E_T[\mathbf{a}, \mathbf{b}]$, is sufficiently small that the resulting data complexity is prohibitive for any conceivable attacker. Note that this must hold for $T = R - 1$, because Algorithm 2 as presented attacks the first round. Since variations of LC can be used to attack the first and last rounds of an SPN simultaneously, it may also be important that the data complexity remain prohibitive for $T = R - 2$.

5.3   Best Previous Result

Since evaluation of the MALHP appears to be infeasible in general, researchers have adopted the approach of upper bounding this value. If such an upper bound is sufficiently small, provable security can be claimed. Hong et al. [9] give the best previously known result for the SPN architecture, stated in Theorem 2 below. First we need the following concepts.

Definition 4 ([1]). *Any T-round characteristic,* Ω, *determines an input and an output mask for each s-box in rounds* $1 \ldots T$. *Those s-boxes having nonzero input and output masks are called* active.

Definition 5. *Let* $\Omega \in \mathrm{ALH}(\mathbf{a}, \mathbf{b})^*$, *and let* $\mathbf{v}$ *be one of the masks in* Ω. *Then* $\mathbf{v}$ *is either an input or an output mask for the substitution stage of some round of the SPN. By the definition of* $\mathrm{ALH}(\mathbf{a}, \mathbf{b})^*$ *(Definition 3), the active s-boxes in this round can be determined from* $\mathbf{v}$ *(without knowing the corresponding output/input mask). We define* $\gamma_{\mathbf{v}}$ *to be the M-bit vector which encodes this pattern of active s-boxes:* $\gamma_{\mathbf{v}} = \gamma_1 \gamma_2 \ldots \gamma_M$, *where* $\gamma_i = 1$ *if the ith s-box is active, and* $\gamma_i = 0$ *otherwise, for* $1 \leq i \leq M$.

Definition 6 ([4]). *The* branch number *of the LT, denoted* $\mathcal{B}$, *is the minimum number of active s-boxes in any two consecutive rounds. It can be given by*

$$\mathcal{B} = \min \left\{ wt(\gamma_{\mathbf{v}}) + wt(\gamma_{\mathbf{w}}) : \ \mathbf{w} \in \{0,1\}^N \setminus \mathbf{0} \ \text{and} \ \mathbf{v} = (\mathbf{L}'\mathbf{w}')' \right\} .$$

It is not hard to see that $2 \leq \mathcal{B} \leq (M+1)$.

Theorem 2 (Hong et al.). *If* $\mathcal{B} = (M+1)$, *then* $\max_{\mathbf{a},\mathbf{b} \in \{0,1\}^N \setminus \mathbf{0}} E_T[\mathbf{a}, \mathbf{b}] \leq q^M$, *and if* $\mathcal{B} = M$, *then* $\max_{\mathbf{a},\mathbf{b} \in \{0,1\}^N \setminus \mathbf{0}} E_T[\mathbf{a}, \mathbf{b}] \leq q^{M-1}$, *where q is defined as in* (4).

6   New Upper Bound
for Maximum Average Linear Hull Probability

In this section we present a new method for upper bounding the maximum average linear hull probability. Our main results are Theorem 3 and Theorem 4. The upper bound we compute depends on:

   (a) q, the maximum linear probability over all SPN s-boxes (see (4))
   (b) T, the number of rounds being approximated by Algorithm 2
   (c) the structure of the SPN LT (via the $W[\ ]$ table in Definition 7 below)

6.1   Definition and Technical Lemmas

Definition 7. *Let* $\gamma, \hat{\gamma} \in \{0,1\}^M$. *Then*

$$W[\gamma, \hat{\gamma}] \stackrel{\mathrm{def}}{=} \# \left\{ \mathbf{y} \subset \{0,1\}^N : \gamma_{\mathbf{x}} = \gamma, \gamma_{\mathbf{y}} = \hat{\gamma}, \ \text{where} \ \mathbf{x} = (\mathbf{L}'\mathbf{y}')' \right\} .$$

Remark: Informally, the value $W[\gamma, \hat{\gamma}]$ represents the number of ways the LT can "connect" a pattern of active s-boxes in one round (γ) to a pattern of active s-boxes in the next round ($\hat{\gamma}$).

Lemma 2. *Let* Ω *be a one-round or T-round characteristic that makes A s-boxes active. Then* $LCP(\Omega) \leq q^A$.

Proof. Follows directly from (4), (6), and (7).

Lemma 3. *Let* $1 \leq t \leq T$, *and* $\mathbf{a}, \mathbf{b}^t \in \{0,1\}^N$. *Then*

$$\sum_{\mathbf{x} \in \{0,1\}^N} E_T[\mathbf{a}, \mathbf{x}] = \sum_{\mathbf{x} \in \{0,1\}^N} LCP^t(\mathbf{x} \rightarrow \mathbf{b}^t) = 1 .$$

Proof. The second sum equals 1 by application of Parseval's Theorem[13] to round t. To see that the first sum is equal to 1, apply Parseval's Theorem to the decryption function for rounds $1 \ldots T$ (masked by $\mathbf{a}$), and take the expected value over the set of independent keys with uniform distribution.

Lemma 4. *Let $T \geq 2$, and let $\mathbf{a}, \mathbf{b} \in \{0,1\}^N \setminus \mathbf{0}$. For any $\mathbf{x} \in \{0,1\}^N$ viewed as an input mask for the LT, let $\mathbf{y}$ denote the unique corresponding output mask (via the relationship given in Lemma 1). Then*

$$E_T[\mathbf{a}, \mathbf{b}] = \sum_{\mathbf{x} \in \{0,1\}^N \setminus \mathbf{0}} E_{T-1}[\mathbf{a}, \mathbf{x}] \cdot LCP^T(\mathbf{y} \rightarrow \mathbf{b}) .$$

Proof. Follows immediately from (7) and (9).

Lemma 5. *Let $m \geq 2$, and suppose $\{c_i\}_{i=1}^m$, $\{d_i\}_{i=1}^m$ are sequences of nonnegative values. Let $\{\dot{c}_i\}_{i=1}^m$, $\left\{\dot{d}_i\right\}_{i=1}^m$ be the sequences obtained by sorting $\{c_i\}$ and $\{d_i\}$, respectively, in nonincreasing order. Then $\sum_{i=1}^m c_i d_i \leq \sum_{i=1}^m \dot{c}_i \dot{d}_i$.*

Proof. See Appendix A.

Lemma 6. *Suppose $\{\dot{c}_i\}_{i=1}^m$, $\{\ddot{c}_i\}_{i=1}^m$, and $\left\{\dot{d}_i\right\}_{i=1}^m$ are sequences of nonnegative values, with $\left\{\dot{d}_i\right\}$ sorted in nonincreasing order. Suppose there exists $\tilde{m}$, $1 \leq \tilde{m} \leq m$, such that*

   (a) $\ddot{c}_i \geq \dot{c}_i$, for $1 \leq i \leq \tilde{m}$
   (b) $\ddot{c}_i \leq \dot{c}_i$, for $(\tilde{m}+1) \leq i \leq m$
   (c) $\sum_{i=1}^m \dot{c}_i \leq \sum_{i=1}^m \ddot{c}_i$

Then $\sum_{i=1}^m \dot{c}_i \dot{d}_i \leq \sum_{i=1}^m \ddot{c}_i \dot{d}_i$.

Proof. See Appendix A.

6.2   Derivation of New Upper Bound

Our approach is to compute an upper bound for each nonzero pattern of active s-boxes in round 1 and round T ($T \geq 2$); that is, we compute $UB_T[\gamma, \hat{\gamma}]$, for $\gamma, \hat{\gamma} \in \{0,1\}^M \setminus \mathbf{0}$, such that the following holds:

UB Property for T.   For all $\mathbf{a}, \mathbf{b} \in \{0,1\}^N \setminus \mathbf{0}$, $E_T[\mathbf{a}, \mathbf{b}] \leq UB_T[\gamma_{\mathbf{a}}, \gamma_{\mathbf{b}}]$.

If the *UB Property for T* holds, then an upper bound for the MALHP is given by $\max_{\gamma, \hat{\gamma} \in \{0,1\}^M \setminus \mathbf{0}} UB_T[\gamma, \hat{\gamma}]$. We first handle the case $T = 2$ in Theorem 3, and then use a recursive technique for $T \geq 3$ in Theorem 4.

Theorem 3. *Let $\gamma, \hat{\gamma} \in \{0,1\}^M \setminus \mathbf{0}$, $f = wt(\gamma)$, $\ell = wt(\hat{\gamma})$, and $W = W[\gamma, \hat{\gamma}]$. If*

$$UB_2[\gamma, \hat{\gamma}] \stackrel{\text{def}}{=} \begin{cases} \min\left\{q^f, q^\ell\right\} & \text{if } \max\left\{q^f, q^\ell\right\} \cdot W > 1 \\ q^{f+\ell} \cdot W & \text{if } \max\left\{q^f, q^\ell\right\} \cdot W \leq 1 \end{cases} \tag{10}$$

then the UB Property for 2 holds.

Proof. Let $\gamma, \hat{\gamma} \in \{0,1\}^M \setminus \mathbf{0}$ be fixed, and let $\mathbf{a}, \mathbf{b} \in \{0,1\}^N \setminus \mathbf{0}$ such that $\gamma_{\mathbf{a}} = \gamma$ and $\gamma_{\mathbf{b}} = \hat{\gamma}$. We want to show that $E_2[\mathbf{a}, \mathbf{b}] \leq UB_2[\gamma, \hat{\gamma}]$. There are $W = W[\gamma, \hat{\gamma}]$ ways that the LT can "connect" the f active s-boxes in round 1 to the ℓ active s-boxes in round 2. Let $\mathbf{x}_1, \mathbf{x}_2, \cdots, \mathbf{x}_W$ be the corresponding input masks for the LT, and let $\mathbf{y}_1, \mathbf{y}_2, \cdots, \mathbf{y}_W$ be the respective output masks (so $\gamma_{\mathbf{x}_i} = \gamma$ and $\gamma_{\mathbf{y}_i} = \hat{\gamma}$). Let $c_i = LCP^1(\mathbf{a} \rightarrow \mathbf{x}_i)$ and $d_i = LCP^2(\mathbf{y}_i \rightarrow \mathbf{b})$, for $1 \leq i \leq W$.

From Lemma 4 we have $E_2[\mathbf{a}, \mathbf{b}] = \sum_{i=1}^{W} c_i d_i$. We know that $0 \leq c_i \leq q^f$, $0 \leq d_i \leq q^\ell$ (by Lemma 2) and $\sum_{i=1}^{W} c_i \leq 1$, $\sum_{i=1}^{W} d_i \leq 1$ (by Lemma 3). Without loss of generality, assume that $f \geq \ell$, so $\min\{q^f, q^\ell\} = q^f$ and $\max\{q^f, q^\ell\} = q^\ell$ (since $0 \leq q \leq 1$). Note that q^f always upper bounds $E_2[\mathbf{a}, \mathbf{b}]$, since $E_2[\mathbf{a}, \mathbf{b}] = \sum_{i=1}^{W} c_i d_i \leq q^f \sum_{i=1}^{W} d_i \leq q^f$; we use this upper bound in the first case of (10). On the other hand, $q^{f+\ell} \cdot W$ also upper bounds $E_2[\mathbf{a}, \mathbf{b}]$, since $\sum_{i=1}^{W} c_i d_i \leq \sum_{i=1}^{W} q^f q^\ell = q^{f+\ell} \cdot W$. If $q^\ell \cdot W = 1$, the two upper bounds are identical. If $q^\ell \cdot W < 1$, then $q^{f+\ell} \cdot W < q^f$, so we use $q^{f+\ell} \cdot W$ as the upper bound in the second case of (10).

Theorem 4. *Let $T \geq 3$. Assume that values $UB_{T-1}[\gamma, \hat{\gamma}]$ have been computed for all $\gamma, \hat{\gamma} \in \{0,1\}^M \setminus \mathbf{0}$ such that the UB Property for $(T-1)$ holds. Let values $UB_T[\gamma, \hat{\gamma}]$ be computed using the algorithm in Figure 3. Then the UB Property for T holds.*

Proof. Throughout this proof, "Line X" refers to the Xth line in Figure 3. Let $\mathbf{a}, \mathbf{b} \in \{0,1\}^N \setminus \mathbf{0}$. It suffices to show that if $\gamma = \gamma_{\mathbf{a}}$ in Line 1 and $\hat{\gamma} = \gamma_{\mathbf{b}}$ in Line 2, then the value $UB_T[\gamma, \hat{\gamma}]$ computed in Figure 3 satisfies $E_T[\mathbf{a}, \mathbf{b}] \leq UB_T[\gamma, \hat{\gamma}]$. Enumerate the nonzero output masks for the LT as $\mathbf{y}_1, \mathbf{y}_2, \ldots, \mathbf{y}_{2^N-1}$, and let the corresponding input masks be given by $\mathbf{x}_1, \mathbf{x}_2, \ldots, \mathbf{x}_{2^N-1}$, respectively. From Lemma 4 we have $E_T[\mathbf{a}, \mathbf{b}] = \sum_{i=1}^{2^N-1} E_{T-1}[\mathbf{a}, \mathbf{x}_i] \cdot LCP^T(\mathbf{y}_i \rightarrow \mathbf{b})$. If $\gamma_{\mathbf{y}_i} \neq \gamma_{\mathbf{b}} (= \hat{\gamma})$, then $LCP^T(\mathbf{y}_i \rightarrow \mathbf{b}) = 0$ (by the Piling-up Lemma), so these $\mathbf{y}_i$ can be removed from consideration, leaving $\bar{\mathbf{y}}_1, \bar{\mathbf{y}}_2, \ldots, \bar{\mathbf{y}}_L$, and corresponding input masks, $\bar{\mathbf{x}}_1, \bar{\mathbf{x}}_2, \ldots, \bar{\mathbf{x}}_L$, respectively.

Let $c_i = E_{T-1}[\mathbf{a}, \bar{\mathbf{x}}_i]$ and $d_i = LCP^T(\bar{\mathbf{y}}_i \rightarrow \mathbf{b})$, for $1 \leq i \leq L$. Then $E_T[\mathbf{a}, \mathbf{b}] = \sum_{i=1}^{L} c_i d_i$. Let $\ell = wt(\hat{\gamma})$ (Line 3), and let $u_i = UB_{T-1}[\gamma, \gamma_{\bar{\mathbf{x}}_i}]$, for $1 \leq i \leq L$. Then $0 \leq c_i \leq u_i$, $0 \leq d_i \leq q^\ell$ (the latter by Lemma 2), and $\sum c_i \leq 1$, $\sum d_i \leq 1$ (by Lemma 3). It follows immediately that $E_T[\mathbf{a}, \mathbf{b}] \leq q^\ell \cdot \sum_{i=1}^{L} d_i \leq q^\ell$. We use this upper bound in **Case I** (Lines 19, 20).

Now note that some of the terms in $\{u_i\}$ are identical, since if $1 \leq i < j \leq L$ and $\gamma_{\bar{\mathbf{x}}_i} = \gamma_{\bar{\mathbf{x}}_j}$, then $u_i = u_j$. We use this to define an equivalence relation on $\{\bar{\mathbf{x}}_i\}$: $\bar{\mathbf{x}}_i \equiv \bar{\mathbf{x}}_j$ iff $\gamma_{\bar{\mathbf{x}}_i} = \gamma_{\bar{\mathbf{x}}_j}$. It can be seen that the number of elements in the equivalence class of $\bar{\mathbf{x}}_i$ is $W[\gamma_{\bar{\mathbf{x}}_i}, \hat{\gamma}]$.

Select indices $j_1, j_2, \ldots, j_H$ such that $\{\bar{\mathbf{x}}_{j_h}\}_{h=1}^{H}$ consists of one representative from each equivalence class. Let $\gamma_h = \gamma_{\bar{\mathbf{x}}_{j_h}}$, $U_h = u_{j_h} = UB_{T-1}[\gamma, \gamma_h]$, and $W_h = W[\gamma_h, \hat{\gamma}]$, for $1 \leq h \leq H$. Without loss of generality, assume that the indices are ordered such that $U_1 \geq U_2 \geq \cdots \geq U_H$. It is an important observation that the values γ_h, U_h, and W_h are the same as those defined in Lines 5, 7, and 8. The following four facts are straightforward.

1.   For each $\gamma \in \{0,1\}^M \setminus \mathbf{0}$
2.   For each $\hat{\gamma} \in \{0,1\}^M \setminus \mathbf{0}$
3. $\ell \leftarrow wt(\hat{\gamma})$
4. $\Gamma \leftarrow \{\xi \in \{0,1\}^M \setminus \mathbf{0} : W[\xi, \hat{\gamma}] \neq 0\}$
5.   Order the elements of Γ as $\gamma_1, \gamma_2, \ldots, \gamma_H$ such that
6. $UB_{T-1}[\gamma, \gamma_1] \geq UB_{T-1}[\gamma, \gamma_2] \geq \cdots \geq UB_{T-1}[\gamma, \gamma_H]$

7. $U_h \leftarrow UB_{T-1}[\gamma, \gamma_h]$, for $1 \leq h \leq H$
8. $W_h \leftarrow W[\gamma_h, \hat{\gamma}]$, for $1 \leq h \leq H$
9. $S_u \leftarrow \sum_{h=1}^H U_h W_h$
10. $S_q \leftarrow q^\ell \cdot \sum_{h=1}^H W_h$

11. $H_u \leftarrow H$
12.   If $S_u > 1$ then
13. $H_u \leftarrow \min\left\{G : 1 \leq G \leq H, \ \sum_{h=1}^G U_h W_h > 1\right\}$
14. $\delta_u \leftarrow 1 - \sum_{h=1}^{H_u - 1} U_h W_h$

15. $H_q \leftarrow H$
16.   If $S_q > 1$ then
17. $H_q \leftarrow \min\left\{G : 1 \leq G \leq H, \ q^\ell \cdot \sum_{h=1}^G W_h > 1\right\}$
18. $\delta_q \leftarrow 1 - q^\ell \cdot \sum_{h=1}^{H_q - 1} W_h$

19. **(Case I)**   If $(S_q \leq 1 < S_u)$ or $(1 < S_u, S_q$ and $H_u < H_q)$ then
20. $UB_T[\gamma, \hat{\gamma}] \leftarrow q^\ell$

21. **(Case II)** Else if $(S_u, S_q \leq 1)$ then
22. $UB_T[\gamma, \hat{\gamma}] \leftarrow q^\ell S_u$

23. **(Case III)** Else if $(S_u \leq 1 < S_q)$ or $(1 < S_u, S_q$ and $H_u > H_q)$ then
24. $UB_T[\gamma, \hat{\gamma}] \leftarrow \left(q^\ell \cdot \sum_{h=1}^{H_q - 1} U_h W_h\right) + U_{H_q} \cdot \delta_q$

25. **(Case IV)** Else if $\left(1 < S_u, S_q$ and $H_u = H_q \stackrel{\text{def}}{=} \tilde{H}\right)$ then
26. $UB_T[\gamma, \hat{\gamma}] \leftarrow \left(q^\ell \cdot \sum_{h=1}^{\tilde{H} - 1} U_h W_h\right) + \min\left\{U_{\tilde{H}} \cdot \delta_q, \ q^\ell \cdot \delta_u\right\}$

Fig. 3. Algorithm to compute $UB_T[\ ]$ for $T \geq 3$

Fact 1 $\sum_{h=1}^H W_h = L$.

Fact 2 $\sum_{i=1}^L u_i = \sum_{h=1}^H U_h W_h = S_u$ (S_u is defined in Line 9).

Fact 3 $q^\ell L = q^\ell \cdot \sum_{h=1}^H W_h = S_q$ (S_q is defined in Line 10).

Using Fact 2, we get the upper bound $E_T[\mathbf{a}, \mathbf{b}] = \sum_{i=1}^L c_i d_i \leq q^\ell \cdot \sum_{i=1}^L u_i = q^\ell S_u$. If $S_u \leq 1$, this upper bound is no larger than that of **Case I**; if $S_u < 1$, it is strictly smaller. This is the upper bound we use in **Case II** (Lines 21, 22).

The proofs of **Cases I** and **II** have parallels to the proofs of the two cases in Theorem 3. For **Cases III** and **IV**, however, we require additional techniques, since the terms which upper bound the c_i (namely, the u_i) are not, in general, all the same (in the proof of Theorem 3, all the c_i are upper bounded by q^f). The intuition for what follows is this: Since $\sum_{i=1}^{L} d_i \le 1$, it is not necessary to replace all the d_i by the value q^ℓ if the consequence is that $\sum_{i=1}^{L} q^\ell > 1$. Instead, certain of the d_i are replaced by q^ℓ and the rest by 0, so that the resulting summation is 1 (a residue term may be required). To ensure an upper bound, it is necessary that the q^ℓ terms be multiplied by the *largest* of the u_i terms. This is the reason for the sorting in Lines 5–6. (A "cutoff" of the u_i terms at the value 1 is also applied.)

Sort $\{c_i\}$, $\{d_i\}$, and $\{u_i\}$ in nonincreasing order to obtain the sequences $\{\dot{c}_i\}$, $\{\dot{d}_i\}$, and $\{\dot{u}_i\}$, respectively. Clearly $\dot{c}_i \le \dot{u}_i$, for $1 \le i \le L$. Applying Lemma 5 we have $\sum_{i=1}^{L} c_i d_i \le \sum_{i=1}^{L} \dot{c}_i \dot{d}_i$. If $S_u = \sum_{i=1}^{L} \dot{u}_i \le 1$, let $\ddot{c}_i = \dot{u}_i$, for $1 \le i \le L$. If $S_u > 1$, let L_u $(1 \le L_u \le L)$ be minimum such that $\sum_{i=1}^{L_u} \dot{u}_i > 1$, and let $\{\ddot{c}_i\}$ consist of the first L terms of $\quad \dot{u}_1, \dot{u}_2, \dots, \dot{u}_{L_u-1}, \left(1 - \sum_{i=1}^{L_u-1} \dot{u}_i\right), 0, 0, 0, \dots$.

If $S_q = q^\ell L \le 1$, let $\ddot{d}_i = q^\ell$ for $1 \le i \le L$. Otherwise, if $S_q > 1$, let $L_q = \left\lfloor \frac{1}{q^\ell} \right\rfloor$, and let $\{\ddot{d}_i\}$ consist of the first L terms of $\quad \underbrace{q^\ell, \dots, q^\ell}_{L_q \text{ terms}}, \left(1 - q^\ell L_q\right), 0, 0, 0, \dots$.

Then $\{\dot{c}_i\}$, $\{\ddot{c}_i\}$, and $\{\dot{d}_i\}$ satisfy the conditions on the identically named sequences in the statement of Lemma 6, so $\sum_{i=1}^{L} \dot{c}_i \dot{d}_i \le \sum_{i=1}^{L} \ddot{c}_i \dot{d}_i$. Also, $\{\dot{d}_i\}$, $\{\ddot{d}_i\}$, and $\{\ddot{c}_i\}$ satisfy the conditions on the three sequences in the statement of Lemma 6 (in that order), and therefore $\sum_{i=1}^{L} \ddot{c}_i \dot{d}_i \le \sum_{i=1}^{L} \ddot{c}_i \ddot{d}_i$. Combining, we get $E_T[\mathbf{a}, \mathbf{b}] \le \sum_{i=1}^{L} \ddot{c}_i \ddot{d}_i$, so it remains to show that $\sum_{i=1}^{L} \ddot{c}_i \ddot{d}_i \le UB_T[\gamma, \hat{\gamma}]$. Define the partial sums $P_0 = 0$ and $P_h = \sum_{j=1}^{h} W_j$, for $1 \le h \le H$ (so $P_H = L$).

Case III (Lines 23, 24) If either condition in Line 23 holds, then
  (a) $\ddot{c}_i = U_h$ and $\ddot{d}_i = q^\ell$, for $(P_{h-1} + 1) \le i \le P_h$, $1 \le h \le (H_q - 1)$
  (b) $\ddot{c}_i = U_{H_q}$, for $(P_{H_q-1} + 1) \le i \le P_{H_q}$
  (c) $\sum_{i=(P_{H_q-1}+1)}^{P_{H_q}} \ddot{d}_i = \delta_q$
  (d) $\ddot{d}_i = 0$, for $i \ge (P_{H_q} + 1)$
It follows that $\sum_{i=1}^{L} \ddot{c}_i \ddot{d}_i = \left(q^\ell \cdot \sum_{h=1}^{H_q-1} U_h W_h\right) + U_{H_q} \cdot \delta_q$, which is the upper bound used in **Case III**.

Case IV (Lines 25, 26) If the condition in Line 25 holds, then (using the definition of $\tilde{H}$ in Line 25)
  (a) $\ddot{c}_i = U_h$ and $\ddot{d}_i = q^\ell$, for $(P_{h-1} + 1) \le i \le P_h$, $1 \le h \le (\tilde{H} - 1)$
  (b) $\sum_{i=(P_{\tilde{H}-1}+1)}^{P_{\tilde{H}}} \ddot{c}_i = \delta_u$ and $\sum_{i=(P_{\tilde{H}-1}+1)}^{P_{\tilde{H}}} \ddot{d}_i = \delta_q$
  (c) $\ddot{c}_i = \ddot{d}_i = 0$, for $i \ge (P_{\tilde{H}} + 1)$

Let $Y = q^\ell \cdot \sum_{h=1}^{\tilde{H}-1} U_h W_h$. For $(P_{\tilde{H}-1} + 1) \leq i \leq P_{\tilde{H}}$, replacing $\ddot{c}_i$ by its upper bound $U_{\tilde{H}}$ gives $\sum_{i=1}^{L} \ddot{c}_i \ddot{d}_i \leq Y + U_{\tilde{H}} \cdot \delta_q$. For i in the same range, replacing $\ddot{d}_i$ by its upper bound q^ℓ gives $\sum_{i=1}^{L} \ddot{c}_i \ddot{d}_i \leq Y + q^\ell \cdot \delta_u$. Combining the above, we get $\sum_{i=1}^{L} \ddot{c}_i \ddot{d}_i \leq Y + \min \left\{ U_{\tilde{H}} \cdot \delta_q, \ q^\ell \cdot \delta_u \right\}$, the right-hand side of which is the upper bound used in **Case IV**.

7   Application of New Upper Bound to Rijndael

To test our new upper bound, we generated random invertible LTs for SPNs with various parameters. We found that for LTs with branch number $\mathcal{B} = M$, our upper bound was consistently superior to that of Hong et al. [9]. We give the results for one such LT in Appendix B. For LTs with $\mathcal{B} = (M + 1)$, our upper bound did not appear to improve on that of [9].

However, the bulk of our analysis we reserved for Rijndael with the following parameters: $N = 128$, $R = 10$, $M = 16$, $n = 8$, $q = 2^{-6}$. Note that the result of [9] does not apply to Rijndael, since for Rijndael, $\mathcal{B} = 5 < M = 16$) [5]. Tailoring our algorithm to any particular SPN involves computation of the values in $W[\ ]$ (Definition 7), which for Rijndael is a $2^{16} \times 2^{16}$ table. The Rijndael LT is depicted in Figure 4.

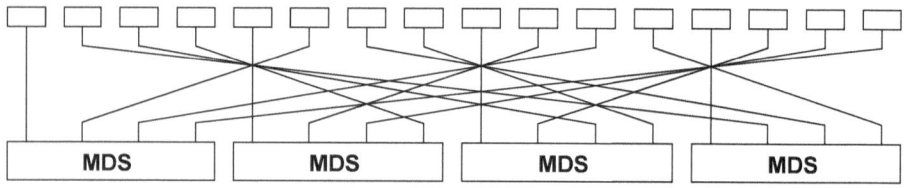

Fig. 4. Rijndael linear transformation

The 128-bit input block can be viewed as an array of 16 bytes. These bytes are first shuffled according to the figure, and then consecutive 4-byte sequences are fed into copies of the same highly diffusive 32-bit LT (based on maximum-distance-separable (MDS) codes). We first computed the $2^4 \times 2^4$ $W[\ ]$ table for the MDS LT, denoted $W_{\text{MDS}}[\ ]$, by transforming all 2^{32} output masks (see Definition 7). Given $\gamma \in \{0,1\}^{16}$ representing a pattern of active s-boxes for the Rijndael LT *input*, a corresponding 4-bit input pattern is determined for each copy of the MDS LT simply by tracing through the byte "shuffle": denote these $\gamma_1, \gamma_2, \gamma_3, \gamma_4 \in \{0,1\}^4$, from left to right, respectively. Then given $\hat{\gamma} \in \{0,1\}^{16}$ representing a pattern of active s-boxes for the Rijndael LT *output*, partition $\hat{\gamma}$ into consecutive 4-bit sequences representing output patterns for the MDS LT, denoted $\hat{\gamma}_1, \hat{\gamma}_2, \hat{\gamma}_3, \hat{\gamma}_4 \in \{0,1\}^4$. Then $W[\gamma, \hat{\gamma}] = \prod_{i=1}^{4} W_{\text{MDS}}[\gamma_i, \hat{\gamma}_i]$.

Since $W[\ ]$ turns out to be quite sparse (roughly 80,000,000 of the 2^{32} entries are nonzero, around 2%), we precompute it, and store the nonzero entries. By doing this first for the $W_{\mathrm{MDS}}[\,]$ table, computation of $W[\,]$ becomes fairly fast. Computing the upper bound in the case $T = 2$ using Theorem 4 is easy. The main work involves executing the algorithm in Figure 3 for $T = 3 \ldots 10$. Lines 3–26 are executed $(2^{16} - 1)^2$ times for each value of T ($3 \leq T \leq 10$), a total of $\approx 2^{35}$ iterations. Once the values $\gamma_1, \gamma_2, \ldots, \gamma_H$ in Line 5 are known, the time complexity of Lines 7–26 is $O(H)$. Since the values in Γ in Line 4 can be precomputed and stored during generation of $W[\,]$, the sorting specified in Lines 5–6 is the most expensive ($O(H \log H)$). The average value for H is 1191, although individual values vary widely.

For a fixed value of γ, computing $UB_T[\gamma, \hat{\gamma}]$ for all $\hat{\gamma} \in \{0,1\}^M \setminus \mathbf{0}$ and all T ($2 \leq T \leq 10$) takes approximately 40 minutes on a Sun Ultra 5, for a total running time in the range of 44,000 hours on that platform. We completed the computation by distributing it over roughly 60 CPUs for several weeks.

Our results for Rijndael are given in Figure 5. For $7 \leq T \leq 10$, the upper bound value is 2^{-75}, giving a corresponding lower bound on the data complexity of LC of 2^{78}, for a 96.7% success rate (see Section 3). Note that for Algorithm 2 as described in Section 3, $T = R - 1 = 9$.

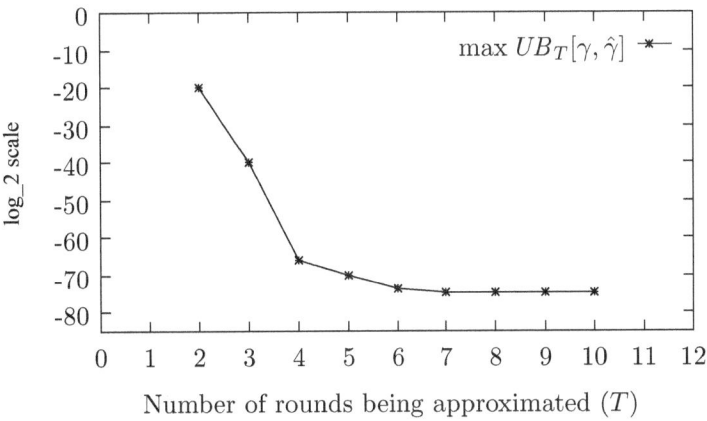

Fig. 5. New upper bound applied to Rijndael

8   Conclusion

We have presented a new method for computing an upper bound on the maximum average linear hull probability for SPNs. Our method has the advantage that it can be computed for an SPN with any LT layer, whereas the best previous result (Hong et al. [9]) applies only to SPNs with highly diffusive LTs, i.e., those having branch number $\mathcal{B} = M$ or $\mathcal{B} = (M + 1)$, where M is the number of s-boxes per round. In addition, our upper bound is a function of the number of rounds being approximated; other known upper bounds do not vary with the number of rounds. When applied to an SPN whose LT has branch number $\mathcal{B} = (M + 1)$ (the maximal case), our upper bound does not appear to improve on that of [9]. For SPNs whose LTs have branch number $\mathcal{B} = M$, our upper bound is consistently superior to that of [9].

A significant part of our work involved application of our method to Rijndael (with $N = 128$ and $R = 10$). This yielded the upper bound $UB = 2^{-75}$, for a corresponding lower bound on the data complexity of LC of $\frac{8}{UB} = 2^{78}$ (for a 96.7% success rate). Note that this does not demonstrate the existence of a such an attack, but is, to our knowledge, the first such lower bound.

Acknowledgments

The authors are grateful to the reviewers for comments which significantly improved the presentation of this paper. For help in obtaining the required computing resources, we are grateful to: the High Performance Computing Virtual Laboratory (Canada), Tom Bradshaw, Randy Ellis, Howard Heys, Alex MacPherson, Andrew Pollard, Carolyn Small, and Amr Youssef.

References

1. E. Biham, *On Matsui's linear cryptanalysis*, Advances in Cryptology—EUROCRYPT'94, Springer-Verlag, pp. 341–355, 1995.
2. E. Biham and A. Shamir, *Differential cryptanalysis of DES-like cryptosystems*, Journal of Cryptology, Vol. 4, No. 1, pp. 3–72, 1991.
3. J. Daemen, R. Govaerts, and J. Vandewalle, *Correlation matrices*, Fast Software Encryption : Second International Workshop, Springer-Verlag, pp. 275–285, 1995.
4. J. Daemen, L. Knudsen, and V. Rijmen, *The block cipher* SQUARE, Fast Software Encryption—FSE'97, Springer-Verlag, pp. 149–165, 1997.
5. J. Daemen and V. Rijmen, *AES proposal: Rijndael*,
   http://csrc.nist.gov/encryption/aes/round2/AESAlgs/Rijndael/Rijndael.pdf, 1999.
6. H. Feistel, *Cryptography and computer privacy*, Scientific American, Vol. 228, No. 5, pp. 15–23, May 1973.
7. C. Harpes, G. Kramer, and J. Massey, *A generalization of linear cryptanalysis and the applicability of Matsui's piling-up lemma*, Advances in Cryptology—EUROCRYPT'95, Springer-Verlag, pp. 24–38, 1995.
8. H.M. Heys and S.E. Tavares, *Substitution-permutation networks resistant to differential and linear cryptanalysis*, Journal of Cryptology, Vol. 9, No. 1, pp. 1–19, 1996.
9. S. Hong, S. Lee, J. Lim, J. Sung, and D. Cheon, *Provable security against differential and linear cryptanalysis for the SPN structure*, Fast Software Encryption (FSE 2000), Proceedings to be published by Springer-Verlag.
10. L.R. Knudsen, *Practically secure Feistel ciphers*, Fast Software Encryption, Springer-Verlag, pp. 211–221, 1994.
11. M. Matsui, *Linear cryptanalysis method for DES cipher*, Advances in Cryptology—EUROCRYPT'93, Springer-Verlag, pp. 386–397, 1994.
12. M. Matsui, *On correlation between the order of s-boxes and the strength of DES*, Advances in Cryptology—EUROCRYPT'94, Springer-Verlag, pp. 366–375, 1995.

13. W. Meier and O. Staffelbach, *Nonlinearity criteria for cryptographic functions*, Advances in Cryptology—EUROCRYPT'89, Springer-Verlag, pp. 549–562, 1990.
14. K. Nyberg, *Linear approximation of block ciphers*, Advances in Cryptology—EUROCRYPT'94, Springer-Verlag, pp. 439–444, 1995.
15. C.E. Shannon, *Communication theory of secrecy systems*, Bell System Technical Journal, Vol. 28, no. 4, pp. 656–715, 1949.

Appendix A

Proof (Lemma 5). Without loss of generality, assume that $\{d_i\}$ is already sorted in nonincreasing order, so $\dot{d}_i = d_i$. If $m = 2$ and $\{c_i\}$ is not sorted, i.e., if $c_1 < c_2$, then $\dot{c}_1 = c_2$ and $\dot{c}_2 = c_1$, so

$$\sum_{i=1}^{2} c_i d_i \leq \sum_{i=1}^{2} \dot{c}_i \dot{d}_i \iff c_1 d_1 + c_2 d_2 \leq c_2 \dot{d}_1 + c_1 \dot{d}_2$$

$$\iff (c_2 - c_1) d_2 \leq (c_2 - c_1) d_1$$

$$\iff d_2 \leq d_1 ,$$

which is true since $\{d_i\}$ was assumed to be sorted. Let $m \geq 3$ and assume the lemma holds for $m - 1$. Let s be the index of a minimal element in $\{c_i\}$, and let $\{\hat{c}_i\}_{i=1}^{m}$ be the sequence obtained by exchanging c_s and c_m in $\{c_i\}$. Then $\dot{c}_m = \hat{c}_m$, and therefore sorting $\{\hat{c}_i\}_{i=1}^{m-1}$ in nonincreasing order gives $\{\dot{c}_i\}_{i=1}^{m-1}$. By an argument similar to that of the $m = 2$ case, we have $\sum_{i=1}^{m} c_i d_i \leq \sum_{i=1}^{m} \hat{c}_i d_i$. Applying the induction hypothesis to the first $m-1$ terms of $\{\hat{c}_i\}$ and $\{d_i\}$ gives $\sum_{i=1}^{m-1} \hat{c}_i \dot{d}_i \leq \sum_{i=1}^{m-1} \dot{c}_i \dot{d}_i$. Combining these facts, we get

$$\sum_{i=1}^{m} c_i d_i \leq \sum_{i=1}^{m} \hat{c}_i d_i = \sum_{i=1}^{m-1} \hat{c}_i \dot{d}_i + \hat{c}_m \dot{d}_m \leq \sum_{i=1}^{m-1} \dot{c}_i \dot{d}_i + \dot{c}_m \dot{d}_m = \sum_{i=1}^{m} \dot{c}_i \dot{d}_i .$$

Proof (Lemma 6). Let

$$\dot{A} = \sum_{i=1}^{\tilde{m}} \dot{c}_i \qquad \dot{B} = \sum_{i=\tilde{m}+1}^{m} \dot{c}_i \qquad C = \sum_{i=1}^{m} \dot{c}_i$$

$$\ddot{A} = \sum_{i=1}^{\tilde{m}} \ddot{c}_i \qquad \ddot{B} = \sum_{i=\tilde{m}+1}^{m} \ddot{c}_i \qquad \ddot{C} = \sum_{i=1}^{m} \ddot{c}_i$$

By assumption, $\dot{A} \leq \ddot{A}$, $\dot{B} \geq \ddot{B}$, and $\dot{C} \leq \ddot{C}$. Let $\Delta A = \ddot{A} - \dot{A} \geq 0$ and $\Delta B = \dot{B} - \ddot{B} \geq 0$. Note that $\Delta A - \Delta B = \ddot{C} - \dot{C} \geq 0$. We have

$$\sum_{i=1}^{\tilde{m}} \ddot{c}_i \dot{d}_i \geq \sum_{i=1}^{\tilde{m}} \dot{c}_i \dot{d}_i + \Delta A \cdot \dot{d}_{\tilde{m}} \tag{11}$$

$$\sum_{i=\tilde{m}+1}^{m} \ddot{c}_i \dot{d}_i \geq \sum_{i=\tilde{m}+1}^{m} \dot{c}_i \dot{d}_i - \Delta B \cdot \dot{d}_{\tilde{m}+1} \tag{12}$$

Adding (11) and (12), we get

$$\sum_{i=1}^{m} \ddot{c}_i \dot{d}_i \geq \sum_{i=1}^{m} \dot{c}_i \dot{d}_i + \Delta A \cdot \dot{d}_{\tilde{m}} - \Delta B \cdot \dot{d}_{\tilde{m}+1}$$

$$\geq \sum_{i=1}^{m} \dot{c}_i \dot{d}_i + \Delta A \cdot \dot{d}_{\tilde{m}+1} - \Delta B \cdot \dot{d}_{\tilde{m}+1}$$

$$= \sum_{i=1}^{m} \dot{c}_i \dot{d}_i + (\Delta A - \Delta B) \cdot \dot{d}_{\tilde{m}+1}$$

$$\geq \sum_{i=1}^{m} \dot{c}_i \dot{d}_i .$$

Appendix B

Some of the LTs which we randomly generated were for SPNs with parameters $N = 24$, $M = 3$, and $n = 8$. For one example of such an LT for which $\mathcal{B} = M = 3$, we plot our upper bound against that of Hong et al. [9] in Figure 6, using a $\log_2$ scale on the y-axis. We also plot the value q^M (the upper bound of [9] for $\mathcal{B} = (M + 1) = 4$) for comparison purposes. On the x-axis we use *minimum nonlinearity*, $NL_{\min}$; for $n = 8$, the relationship between $NL_{\min}$ and q is given by $q = \left(1 - \frac{NL_{\min}}{128}\right)^2$. For this particular LT, it happened that our upper bound settled on a fixed value for $T = 2$, and did not decrease with an increasing number of rounds—this is the value we plot for each $NL_{\min}$.

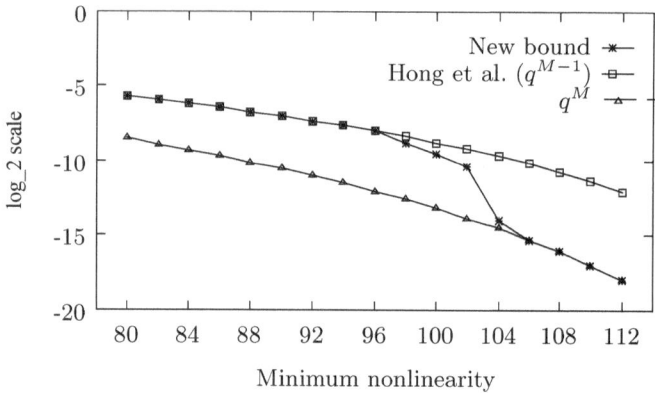

Fig. 6. Comparison of new upper bound with that of Hong et al. [9]

Lower Bounds
for Multicast Message Authentication

Dan Boneh[1,*], Glenn Durfee[1,**], and Matt Franklin[2]

[1] Computer Science Department, Stanford University,
Stanford CA 94305-9045, USA
{dabo,gdurf}@cs.stanford.edu
[2] Department of Computer Science, University of California,
Davis CA 95616-8562
franklin@cs.ucdavis.edu

Abstract. Message integrity from one sender to one receiver is typically
achieved by having the two parties share a secret key to compute a Message Authentication Code (MAC). We consider the "multicast MAC",
which is a natural generalization to multiple receivers. We prove that
one cannot build a short and efficient collusion resistant multicast MAC
without a new advance in digital signature design.

1 Introduction

We study the problem of message integrity in the context of a single source multicast. Consider a TV station, such as the Disney channel. The TV station is
broadcasting to n receivers. Each receiver would like to ensure that the broadcasts are indeed coming from the Disney channel rather than from a malicious
third party (who might be transmitting offensive material).

One natural approach would be to employ digital signatures. Suppose the
transmitter has a secret signing key and each of the receivers has the corresponding public key. To provide message integrity the transmitter signs every
message she broadcasts. No coalition of receivers can forge a message/signature
pair that will fool another receiver. Although signatures provide multicast message integrity they are fundamentally an overkill solution for this problem. First,
signatures are somewhat expensive to compute. Second, digital signatures provide non-repudiation: Any receiver can use the signature to prove to a third
party that the message came from the transmitter. However, non-repudiation is
unnecessary for message integrity.

Message integrity between two parties is usually done by sharing a secret
key k between the sender and receiver. When sending a message M the sender
computes a keyed hash function $\mathrm{MAC} = H_k(M)$ and transmits the MAC along
with the message. MACs are much faster than digital signatures, and do not
provide non-repudiation. We seek a generalization of MACs for the multicast

---

* Supported by NSF and the Packard Foundation.
** Supported by a Microsoft Graduate Research Fellowship.

B. Pfitzmann (Ed.): EUROCRYPT 2001, LNCS 2045, pp. 437–452, 2001.

setting. This would be a distribution of keys to sender and receivers, and a method for tagging messages by the sender that would be convincing to every receiver. We call this primitive a "multicast MAC" (MMAC).

One simple approach for a MMAC might be to share a global secret key k between the transmitter and all n receivers. The transmitter appends $H_k(M)$ to every transmitted message M. Each receiver can then verify the MAC sent by the transmitter. This is insecure since any receiver can forge messages that will fool any other receiver.

Another simple approach is secure but inefficient. The transmitter shares a distinct secret key k_i with each of the n receiver $u_1, \ldots, u_n$. When sending a message M the transmitter computes MMAC $= H_{k_1}(M) \| \ldots \| H_{k_n}(M)$ and transmits (M, MMAC). Each receiver u_i verifies the MMAC by using the entry that corresponds to the key k_i. This construction is secure, in the sense that no coalition of users can create a message/MMAC pair that will fool a user outside the coalition (since they do not have the outsider's MAC key). Unfortunately, the length of the MMAC is linear in the number of receivers. Hence, this construction is not very practical, even though it avoids non-repudiation.

Since none of the above solutions is perfect, it is tempting to try to build a MMAC that is as short as a signature (i.e., length independent of the number of receivers), but much more efficient. We give lower bounds that suggest that this might be a difficult task. Our main results show that if one could build practical (i.e. short) MMACs, then they could be converted into new efficient digital signature schemes. Consequently, it is unlikely that practical MMACs could be constructed without an unexpected advance in digital signature design.

We can relax our security requirement by saying that a MMAC is κ-secure if no coalition of size less than κ can fool another receiver. In Section 5 we generalize our lower bound and show that if one could build a κ-secure MMAC whose length is less than $\log_2 \sum_{i=0}^{\kappa} \binom{n}{i}$ then it could be converted into an efficient signature scheme. For small values of κ this lower bound is approximated by $O(\kappa \log n)$. This lower bound matches an upper bound construction based on pseudorandom functions due to Canetti et al. [1]. Hence our results show that for small values of κ the Canetti et al. construction is optimal.

Our results demonstrate the importance of recent constructions for practical multicast authentication [5,17,10,11,7,12]. Some of these constructions achieve great efficiency (well beyond what is implied by our bounds) by making use of additional assumptions, such as weak time synchronization between sender and receivers [10,11]. We emphasize that our lower bounds for MMACs suggest difficulty only for constructions that use the standard model for MACs, as described in the next section.

A fundamental result of theoretical cryptography is that a digital signature scheme can be derived from any one way function [9,13,2]. Since the existence of a multicast MAC implies the existence of a one way function, that would seem to imply a reduction of the form that we claim. However, this construction is far too inefficient to be considered for any practical purposes. In contrast, our results are achieved through direct reductions from multicast MACs to public key signature schemes. Our reductions are efficient, in the sense that the derived signature schemes have almost the same level of security as the underlying MMAC schemes.

1.1   Related Work

Previous work on multicast authentication followed two tracks: (1) the computational model, based on pseudorandom functions and hash functions, and (2) the information theoretic model, providing unconditional security. Constructions in the information theoretic model provide very strong security guarantees. This strong security comes at a price: The secret key can only be used for a small number of messages. MMACs built in the computational model are not as strong, since their security depends on a complexity assumption. However, computational MMACs can be used to authenticate many messages using relatively short keys. All of the results in this paper are set in the computational model.

In the computational model, Canetti et al. [1] construct a κ-secure MMAC by concatenating many pseudorandom functions whose output is a single bit. This construction does not provide non-repudiation. As mentioned above, our results show that this clever construction is optimal. We note that the security model in [1] is slightly different from our security model. They require that a coalition should not be able to create a forgery that can fool a specific receiver. In some cases a coalition might be content if a broadcast of a forged message fools *any* receiver. Hence, in our model, a forgery is considered successful if it fools *any* receiver outside the coalition. Adapting the construction of Canetti et al. to this stronger security model adds a factor of $\ln n$ to the length of their MMAC. The result is a MMAC of length $4e(\kappa + 1)\ln n \ln 1/\epsilon$ where n is the number of receivers, ϵ is the failure probability, and $e = 2.718$. For small values of κ, and a fixed ϵ, our lower bound of $O(\kappa \log n)$ asymptotically matches their upper bound.

In the information theoretic model, Multicast MACs were introduced by Desmedt, Frankel, and Yung [3] (see also Simmons [16] for the somewhat related notion of authentication codes with arbitration). They gave two constructions for κ-secure MMACs. Kurosawa and Obana [8] derived elegant lower bounds on the probability of success in impersonation and substitution attacks. They showed that the DFY construction is optimal. Safavi-Naini and Wang [14,15] show how to construct information theoretic MMACs using cover free set systems. Their constructions are similar to the ones given in [1]. Cover free set systems were also used by Fujii, et al. [4].

We briefly review the use of signatures as an alternative to MMACs for multicast authentication. There are two difficulties in using signatures for multicast MACs: (1) in streaming audio and video applications one cannot afford to buffer the entire message prior to signing it, and (2) multicast transmissions suffer from packet loss (multicast does not provide packet loss recovery), so one needs signature schemes for an unreliable transmission channel. Problem (1) is often solved by combining standard signatures with fast one time signatures [5,12]. Problem (2) is solved by introducing various types of redundancy during signature generation [17,12,10,7].

We note that the constructions in [10,11] provide short multicast message authentication without non-repudiation. The authentication tags in these constructions are shorter than our lower bounds predict since they rely on some weak timing synchorinization between sender and receivers. Our lower bounds suggest that one must resort to such assumptions to obtain practical multicast authentication without non-repudiation.

2   Definitions

We begin by giving precise definitions for MMACs secure against existential and selective forgeries. To reduce the number of definitions in the section we only consider the strongest adversaries, namely adversaries capable of adaptive chosen message attacks. For completeness, we briefly recall definitions of security for signatures schemes.

2.1   Multicast MACs

A Multicast MAC, or MMAC, is specified by three randomized algorithms (key-gen, mac-gen, mac-ver).

key-gen: takes a security parameter s and a number of receivers n and returns keys $\mathbf{sk}, \mathbf{rk}_1, \ldots, \mathbf{rk}_n \in \{0,1\}^*$. We call $\mathbf{sk}$ the *sender key* and $\mathbf{rk}_i$ the *ith receiver key*.

mac-gen: takes as input a message $M \in \{0,1\}^*$ and a key $K \in \{0,1\}^*$ and returns a tag $T = \mathsf{mac\text{-}gen}(M, K) \in \{0,1\}^\tau$ for some fixed tag length τ bits.

mac-ver: takes as input a message $M \in \{0,1\}^*$, a tag $T \in \{0,1\}^\tau$, and a key $K \in \{0,1\}^*$, and returns a bit: $\mathsf{mac\text{-}ver}(M, T, K) \in \{\text{'yes', 'no'}\}$.

These algorithms are subject to the constraint that for all $(\mathbf{sk}, \mathbf{rk}_1, \ldots, \mathbf{rk}_n)$ produced by key-gen(s, n) we have that

$$\forall M \in \{0,1\}^*, \ \forall i \in \{1, \ldots, n\} : \quad \mathsf{mac\text{-}ver}(M, \mathsf{mac\text{-}gen}(M, \mathbf{sk}), \mathbf{rk}_i) = \text{'yes'}$$

In other words, tags created by mac-gen using the correct sender key verify correctly for all receivers. Each of these algorithms must run in time polynomial in n, s, and the size of the message input.

MMAC security against selective forgery. A MMAC (key-gen, mac-gen, mac-ver) is said to be (t, ϵ, q)-*secure against selective forgery under an adaptive chosen message attack* if every t-time probabilistic algorithm A wins the game below with probability at most ϵ. We model the game as a communication between a challenger and the forging algorithm A. See Figure 1. We assume that the system parameters n and s are fixed ahead of time.

Step 1: The forging algorithm A starts the game by sending the challenger a target message $M \in \{0,1\}^*$. The forger's goal is to forge a MMAC for this message M. The forger also sends a subset $I \subseteq \{1, \ldots, n\}$. The subset I should be viewed as the set of receivers colluding to fool some other receiver.

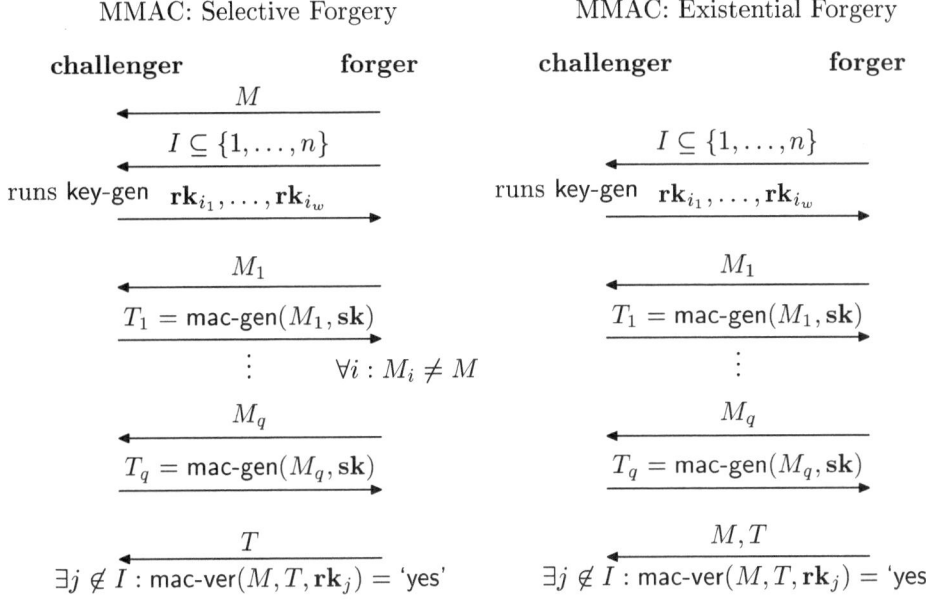

Fig. 1. The games used to define two security notions for a MMAC.

Step 2: The challenger runs algorithm key-gen(s, n) and obtains the MMAC keys $(\mathbf{sk}, \mathbf{rk}_1, \ldots, \mathbf{rk}_n)$. The challenger sends the subset $\{\mathbf{rk}_i\}_{i \in I}$ to A.

Step 3: Algorithm A then mounts a chosen message attack by sending queries $M_1, \ldots, M_q$ to the challenger, where $M_i \neq M$ for all $i = 1, \ldots, q$. The challenger responds with $T_i = \mathsf{mac\text{-}gen}(M_i, \mathbf{sk})$ for $i = 1, \ldots, q$. Note that these queries may be issued adaptively. That is, the adversary A might wait for a response T_i before issuing request M_{i+1}.

Step 4: Finally, A outputs a candidate MMAC, T, for the target message M.

We say that A wins this game if T verifies as a valid tag for M for some receiver j outside of I. More precisely, we say that A wins the game if

$$\exists j \notin I \quad \text{s.t.} \quad \mathsf{mac\text{-}ver}(M, T, \mathbf{rk}_j) = \text{'yes'}.$$

The probability that A wins this game is taken over the random coin flips of the algorithms key-gen, mac-gen, mac-ver, and the random coin flips of A.

The definition above assumes the adversary commits to the set of corrupt users I at the beginning of the game. One can also consider a stronger definition where the adversary is dynamic: the adversary adaptively chooses which users to corrupt during the game. Since our lower bounds already apply when the adversary is restricted to the static settings, the same lower bounds apply in the dynamic settings. Therefore, throughout the paper we only consider static adversaries.

MMAC security against existential forgery. A MMAC (key-gen, mac-gen, mac-ver) is said to be (t, ϵ, q)-*secure against existential forgery under an adaptive chosen message attack* if every t-time probabilistic algorithm A wins the following modified game with probability less than ϵ. The game is identical to the above, except that A does not commit to the message M in Step 1. Instead, the target message M is output by A in the last step (Step 4), at the same time as the candidate tag T. Note that we must have $M \neq M_i$ for all i. See Figure 1.

2.2 Signature Schemes

Our goal is to establish a relation between MMACs and digital signatures. We therefore briefly review two notions of security for digital signatures: security against selective forgery, and security against existential forgery [6]. We review both notions under a chosen message attack.

A signature scheme is specified by three probabilistic algorithms (skey-gen, sig-gen, sig-ver).

skey-gen: takes a security parameter s and returns keys $K_{sec}, K_{pub} \in \{0,1\}^*$. We call K_{sec} the *secret key* and K_{pub} the *public key*.
sig-gen: takes as input a message $M \in \{0,1\}^*$ and a key $K \in \{0,1\}^*$ and returns a signature $S = \text{sig-gen}(M, K) \in \{0,1\}^*$.
sig-ver: takes as input a message $M \in \{0,1\}^*$, a candidate signature $S \in \{0,1\}^*$, and a key $K \in \{0,1\}^*$, and returns a bit: $\text{sig-ver}(M, S, K) \in \{\text{'yes', 'no'}\}$.

These algorithms are subject to the constraint that for all pairs (K_{sec}, K_{pub}) produced by skey-gen(s), we have that

$$\forall M \in \{0,1\}^* : \quad \text{sig-ver}(M, \text{sig-gen}(M, K_{sec}), K_{pub}) = \text{'yes'}$$

Each of these algorithms must run in time polynomial in n, s, and the size of the input.

Signature security against selective and existential forgery. A signature scheme (skey-gen, sig-gen, sig-ver) is said to be (t, ϵ, q)-*secure against selective forgery under an adaptive chosen message attack* if every t-time probabilistic algorithm B wins the game below with probability at most ϵ. See Figure 2. We assume the security parameter s has already been fixed.

Step 1: The forging algorithm B outputs a target message $M \in \{0,1\}^*$.
Step 2: The challenger runs algorithm skey-gen(s) and obtains the keys (K_{sec}, K_{pub}). The challenger sends K_{pub} to B.
Step 3: Algorithm B then mounts a chosen message attack by querying the challenger with messages $M_1, \ldots, M_q \in \{0,1\}^*$, where $M_i \neq M$ for all $i = 1, \ldots, q$. The challenger responds with $S_i = \text{sig-gen}(M_i, K_{sec})$. Note that these queries may be issued adaptively.
Step 4: Finally, B outputs a candidate signature S for the target message M.

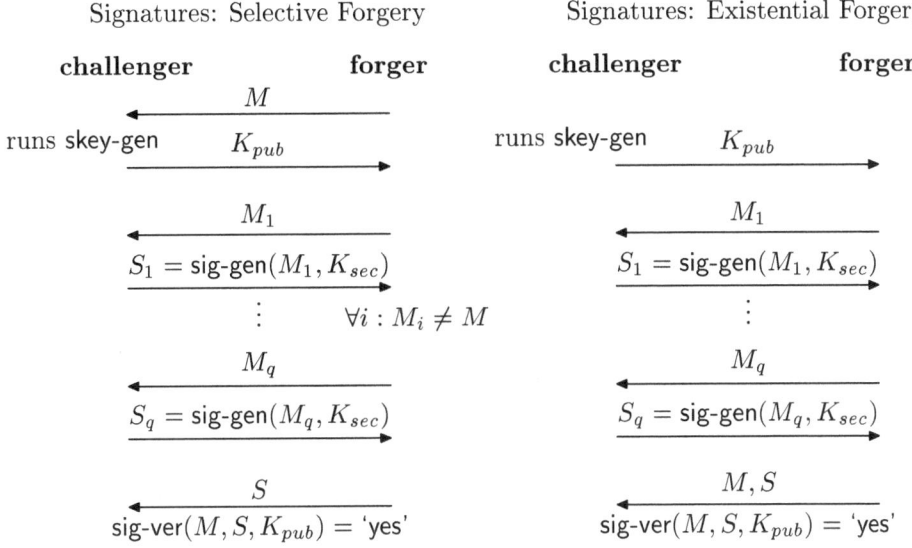

Fig. 2. Signature Scheme Security.

We say that B wins this game if S verifies as a valid signature on M. More precisely, we say that B wins this game if $\mathsf{sig\text{-}ver}(M, S, K_{\mathrm{pub}}) = $ 'yes'.

Similarly, a signature scheme is said to be (t, ϵ, q)-*secure against existential forgery under an adaptive chosen message attack* if every t-time probabilistic algorithm B wins a modified game with probability less than ϵ. The game is identical to the above, except that the target message M is output by B in the last step (Step 4), at the same time as the candidate signature S. See Figure 2.

3   Equivalence of MMAC and Signing for Selective Forgery

One can easily show that for each notion of security defined above, every (t, ϵ, q)-secure signature scheme is also a (t, ϵ, q)-secure multicast authentication code. Our goal in the next two sections is to show an approximate converse: any short MMAC gives rise to a signature scheme with an almost equal level of security. We begin by showing that a MMAC secure against selective forgery gives rise to a signature scheme secure against selective forgery. In the next section, we show a similar result for existential forgery.

The derived signature scheme: Given a MMAC (key-gen, mac-gen, mac-ver) we define the derived signature scheme (skey-gen, sig-gen, sig-ver) as follows:

skey-gen(k, n)   1. Run key-gen(k, n) to get $(\mathbf{sk}, \mathbf{rk}_1, \ldots, \mathbf{rk}_n)$.
2. Pick a random subset $I = \{i_1, \ldots, i_w\} \subseteq \{1, \ldots, n\}$.
3. Output $K_{\text{sec}} = \mathbf{sk}$ and $K_{\text{pub}} = (\mathbf{rk}_{i_1}, \ldots, \mathbf{rk}_{i_w})$.

sig-gen(M, K_{sec})   Output $T = \text{mac-gen}(M, K_{\text{sec}})$.

sig-ver(M, S, K_{pub})   Write $K_{\text{pub}} = (\mathbf{rk}_{i_1}, \ldots, \mathbf{rk}_{i_w})$. Output 'yes' if and only if for all $j = 1, \ldots, w$, mac-ver($M, S, \mathbf{rk}_{i_j}$) = 'yes'.

The following theorem shows that the derived signature scheme has nearly identical security properties as the MMAC.

Theorem 1. *Suppose the MMAC* (key-gen, mac-gen, mac-ver) *is* (t, ϵ, q)-*secure against selective forgery under an adaptive chosen message attack, and suppose the length of the output of* mac-gen($M, \mathbf{sk}$) *is bounded above by* $\tau = n - m$ *for all* M *and* $\mathbf{sk}$. *Then the derived signature scheme* (skey-gen, sig-gen, sig-ver) *is* $(t, \epsilon + \frac{1}{2^m}, q)$-*secure against selective forgery under an adaptive chosen message attack.*

Note that taking $m = 80$ already results in a sufficiently secure signature scheme. Hence, whenever the MMAC length is slightly shorter than the number of receivers, n, the MMAC is easily converted into a secure signature scheme.

Proof. Suppose we have a forger B that produces successful selective forgeries for the derived signature scheme (skey-gen, sig-gen, sig-ver). We build a forger A for the MMAC (key-gen, mac-gen, mac-ver). The proof will follow by contradiction. Recall that we model security as the probability of winning a game against a certain challenger. We describe how the algorithm A interacts with the challenger in this game, using B as a subroutine. See Figure 3.

Step 1: The algorithm A runs B to obtain the selected message M, which it forwards to the challenger as the message intended for its own selective forgery.

Step 2: Algorithm A chooses a random subset $I = \{i_1, \ldots, i_w\} \subseteq \{1, \ldots, n\}$ and sends this to the challenger. The challenger responds with $(\mathbf{rk}_{i_1}, \ldots, \mathbf{rk}_{i_w})$ for some $(\mathbf{sk}, \mathbf{rk}_1, \ldots, \mathbf{rk}_n)$ generated randomly by key-gen.

Step 3: The algorithm A sets $K_{\text{pub}} = (\mathbf{rk}_{i_1}, \ldots, \mathbf{rk}_{i_w})$ and sends K_{pub} to B. The distribution on K_{pub} is identical to the distribution on keys generated by skey-gen.

Step 4: Algorithm A now continues the execution of B, forwarding each query M_i to the challenger, and passing along each response T_i back to B. Note that T_i is a valid signature on M_i as defined by the derived signature scheme.

Step 5: After at most q queries, B outputs a signature forgery S for M. The algorithm A outputs S as its candidate MMAC forgery for M.

We show that A wins the selective forgery game for MMACs with probability at least ϵ. That is, S is a MMAC forgery with probability at least ϵ. The proof is based on the concept of a "bad pair". Let M' be a message in $\{0, 1\}^*$ and let

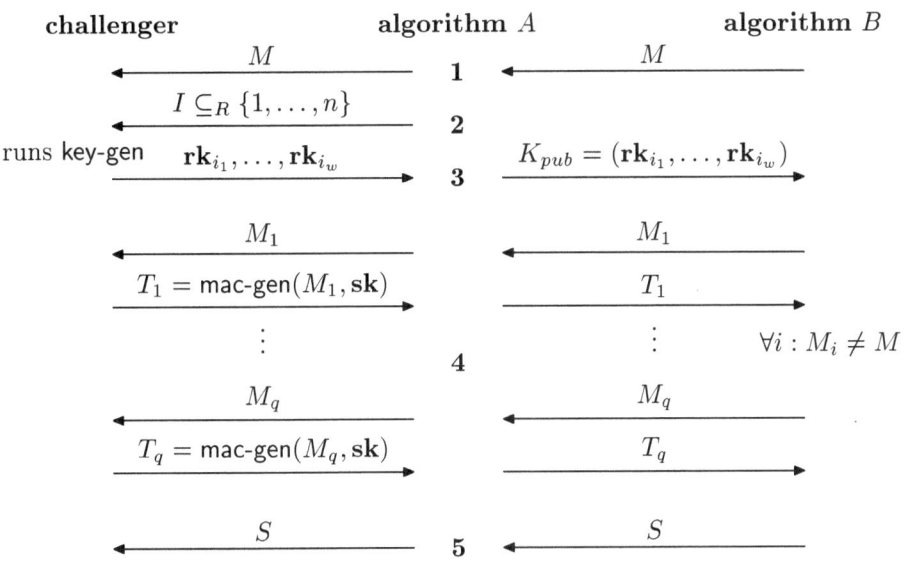

Fig. 3. MMAC forger A uses signature forger B to forge a MMAC.

I' be a coalition $I' \subseteq \{1, \ldots, n\}$. We say that the pair (M', I') is *bad* if there is some tag $T \in \{0, 1\}^{n-m}$ satisfying:

$$\forall i \in I' : \mathsf{mac\text{-}ver}(M', T, \mathbf{rk}_i) = \text{'yes'} \quad \text{and} \quad \forall j \notin I' : \mathsf{mac\text{-}ver}(M', T, \mathbf{rk}_j) = \text{'no'}.$$

In other words, (M', I') is bad if I' is precisely the subset of receiver keys for which some tag T verifies as a valid tag for M'. The following lemma shows that for a fixed message M there are few pairs (M, I) that are bad.

Lemma 1. *For any message M:*

$$\Pr[(M, I) \text{ is bad}] \leq \frac{1}{2^m}.$$

where the probability is over the choice of a random coalition $I \subseteq \{1, \ldots, n\}$.

Proof. For each tag $T \in \{0, 1\}^{n-m}$, let I_T be the set of receivers i for which $\mathsf{mac\text{-}ver}(M, T, \mathbf{rk}_i) = \text{'yes'}$. By definition, the pair (M, I_T) is bad. Notice that the collection

$$\{(M, I_T) \mid T \in \{0, 1\}^{n-m}\}$$

completely describes all bad pairs containing M in the first coordinate. Since there are only 2^{n-m} possible values for T, this set is of size at most 2^{n-m}. Since I is chosen independently of M, it follows that

$$\Pr_{I \subseteq \{1, \ldots, n\}}[(M, I) \text{ is bad}] \leq \frac{2^{n-m}}{2^n} = \frac{1}{2^m},$$

establishing the lemma. ∎

We are now ready to complete the proof Theorem 1.

Proof of Theorem 1.   We will establish the contrapositive. Suppose there is a forger B for the derived signature scheme (skey-gen, sig-gen, sig-ver) that runs in time t, makes q queries, and produces a successful selective forgery with probability at least $\epsilon + \frac{1}{2^m}$. We show the algorithm A described in Figure 3 wins the selective forgery game for the MMAC (key-gen, mac-gen, mac-ver) with probability at least ϵ.

We say that event $\mathcal{A}$ occurs when the pair (M, I) is **not** bad where M is the message chosen in Step 1, and I is the random set chosen in Step 2. We say the event $\mathcal{B}$ occurs when the algorithm B wins the signature forgery game by outputting a forgery S on M in the derived signature scheme. By assumption we know that $\Pr[\mathcal{B}] > \epsilon + \frac{1}{2^m}$. Now, when both events $\mathcal{A}$ and $\mathcal{B}$ occur, we deduce the following:

(1) Since S is a signature forgery for M we have that
$$\forall i \in I : \quad \mathsf{mac\text{-}ver}(M, S, \mathbf{rk}_i) = \text{'yes'};$$

(2) Since (M, I) is not bad, the set of users for which S is a valid MMAC cannot be I. Hence, by (1),
$$\exists j \notin I : \quad \mathsf{mac\text{-}ver}(M, S, \mathbf{rk}_j) = \text{'yes'}.$$

But the second condition is precisely what is needed for A to win the selective forgery game against the MMAC. Since by Lemma 1 we have that $\Pr[\neg \mathcal{A}] \le \frac{1}{2^m}$ we obtain the following:

$$\Pr[A \text{ wins MMAC forgery game}] \ge \Pr[\mathcal{B} \wedge \mathcal{A}] \ge \Pr[\mathcal{B}] - \Pr[\neg \mathcal{A}]$$
$$\ge \left(\epsilon + \frac{1}{2^m} \right) - \frac{1}{2^m} = \epsilon.$$

This probability is taken over the random coin flips of the challenger and of the algorithms A and B. Thus, the theorem follows.   ∎

4   Equivalence of MMAC and Signing for Existential Forgery

Next, we show that an existentially secure MMAC gives rise to an existentially secure signature scheme. The resulting bounds are a bit weaker than for selective forgery. Let (key-gen, mac-gen, mac-ver) be a MMAC, and let H be a collision-resistant hash function from $\{0, 1\}^*$ to $\{0, 1\}^h$. Define the derived signature scheme (skey-gen, sig-gen, sig-ver) as follows:

skey-gen(k, n) 1. Run key-gen(k, n) to get $(\mathbf{sk}, \mathbf{rk}_1, \ldots, \mathbf{rk}_n)$.
　　　　　　　　　　　2. Pick random subset $I = \{i_1, \ldots, i_w\} \subseteq \{1, \ldots, n\}$.
　　　　　　　　　　　3. Output $K_{\text{sec}} = \mathbf{sk}$ and $K_{\text{pub}} = (\mathbf{rk}_{i_1}, \ldots, \mathbf{rk}_{i_w})$.

sig-gen(M, K_{sec}) Output $T = \mathsf{mac\text{-}gen}(H(M), K_{\text{sec}})$.

sig-ver(M, S, K_{pub}) Write $K_{\text{pub}} = (\mathbf{rk}_{i_1}, \ldots, \mathbf{rk}_{i_w})$. Output 'yes' if and only if for all $j = 1, \ldots, w$, $\mathsf{mac\text{-}ver}(H(M), S, \mathbf{rk}_{i_j}) = \text{'yes'}$.

Suppose the MMAC (key-gen, mac-gen, mac-ver) is (t, ϵ, q)-secure against existential forgery under an adaptive chosen message attack, and suppose the length of the output of mac-gen(M) is bounded by $\tau = n - m$ for all M. Furthermore let H be chosen from a family of collision-resistant hash function. specifically, suppose no t-time algorithm can find $M_1 \neq M_2$ such that $H(M_1) = H(M_2)$ with success probability greater than some small ϵ_H. We show in the following theorem that the derived signature scheme retains nearly identical security properties.

Theorem 2. *The derived signature scheme* (skey-gen, sig-gen, sig-ver) *is* $(t, \epsilon + \frac{1}{2^{m-h}} + \epsilon_H, q)$-*secure against existential forgery under an adaptive chosen message attack.*

For example, suppose (key-gen, mac-gen, mac-ver) is (t, ϵ, q)-secure against existential forgery. Let H be the hash function SHA-1 : $\{0,1\}^* \rightarrow \{0,1\}^{160}$, with security $\epsilon_H \approx \frac{1}{2^{80}}$. Then taking $m = 240$ results in a sufficiently secure signature scheme. Hence, as soon as the MMAC length is slightly less than the number of receivers, n, we obtain an existentially secure signature scheme.

Proof of Theorem 2.   We will establish the contrapositive. Suppose we have a forger B that produces successful existential forgeries for the derived signature scheme (skey-gen, sig-gen, sig-ver). We build a MMAC forger A for (key-gen, mac-gen, mac-ver). Recall that we model security as the probability of winning a game against a certain challenger. We describe how the algorithm A interacts with the challenger in this game, using B as a subroutine.

Step 1: The algorithm A chooses a random subset $I = \{i_1, \ldots, i_w\} \subseteq \{1, \ldots, n\}$ and sends this to the challenger, which responds with $(\mathbf{rk}_{i_1}, \ldots, \mathbf{rk}_{i_w})$ for some $(\mathbf{sk}, \mathbf{rk}_1, \ldots, \mathbf{rk}_n)$ generated randomly by key-gen.
Step 2: Algorithm A sets $K_{\text{pub}} = (\mathbf{rk}_{i_1}, \ldots, \mathbf{rk}_{i_w})$ and sends K_{pub} to B.
Step 3: For each query M_i made by B, algorithm A sends the query $H(M_i)$ to the challenger. Algorithm A then passes the response T_i back to B.
Step 4: After at most q queries, B outputs a message M and a candidate signature forgery S for M. If $H(M_i) = H(M)$ for some $i \in \{1, \ldots, q\}$, the algorithm A aborts the forgery attempt, as a collision in H has been found. Otherwise, the algorithm A outputs the pair $(H(M), S)$ as its candidate MMAC forgery.

We claim that A wins the existential forgery game for MMACs with probability at least ϵ. The proof uses the following concept: we say that a subset of users $I' \subseteq \{1, \ldots, n\}$ is *bad* if there is some $H_m \in \{0,1\}^h$ and some tag $T \in \{0,1\}^{n-m}$ such that

$$\forall i \in I' : \mathsf{mac\text{-}ver}(H_m, T, \mathbf{rk}_i) = \text{'yes'}, \quad \text{and}$$
$$\forall j \notin I' : \mathsf{mac\text{-}ver}(H_m, T, \mathbf{rk}_j) = \text{'no'}.$$

That is, I' is bad when I' is precisely the subset of receiver keys for which some tag T verifies as a valid tag for some H_m in the range of the hash function H.

Lemma 2. *When I is a random subset of $\{1, \ldots, n\}$ we have that:*

$$\Pr[I \text{ is bad}] \leq \frac{1}{2^{m-h}}.$$

Proof. We use the bound of Lemma 1 on the probability that a *pair* (H_m, I') is bad, for any $H_m \in \{0,1\}^h$. We obtain the following:

$$\Pr_{I \subseteq \{1,\ldots,n\}}[I \text{ is bad}] = \Pr_{I \subseteq \{1,\ldots,n\}}[\exists H_m \in \{0,1\}^h \text{ s.t. } (H_m, I) \text{ is bad}]$$

$$\leq \sum_{H_m \in \{0,1\}^h} \Pr_{I \subseteq \{1,\ldots,n\}}[(H_m, I) \text{ is bad}] \leq 2^h \left(\frac{1}{2^m}\right) = \frac{1}{2^{m-h}},$$

as desired.   ∎

We can now complete the proof Theorem 2. Suppose there is a forger B for the derived signature scheme (skey-gen, sig-gen, sig-ver) that runs in time t, makes q queries, and produces a successful existential forgery with probability at least $\epsilon + \frac{1}{2^{m-h}} + \epsilon_H$. We claim algorithm A described above wins the existential forgery game for the MMAC (key-gen, mac-gen, mac-ver) with probability at least ϵ.

We say the event $\mathcal{A}$ occurs when the set I chosen in Step 1 of algorithm A is **not** bad. We say the event $\mathcal{B}$ occurs when the algorithm A does **not** abort in Step 4. Finally, we say the event $\mathcal{C}$ occurs when the algorithm B wins the existential forgery game by outputting a forgery S on M in the derived signature scheme. By assumption we know that $\Pr[\mathcal{C}] \geq \epsilon + \frac{1}{2^{m-h}} + \epsilon_H$.

Now, when events $\mathcal{A}$, $\mathcal{B}$, and $\mathcal{C}$ hold, we deduce the following:

(1) $\forall i \in I$: mac-ver$(H(M), S, \text{rk}_i) = $ 'yes'   (S is a signature forgery for M),
(2) $\forall i \in I$: $H(M) \neq H(M_i)$ (A does not abort),
(3) $\exists j \notin I$: mac-ver$(H(M), S, \text{rk}_j) = $ 'yes'   (by (1) and the fact that
                                                          I is not bad).

But the second and third conditions are precisely what is needed for A to win the existential forgery game against a MMAC. So, by Lemma 2 and the fact that H is collision-resistant:

$$\Pr[A \text{ wins MMAC forgery game}] \geq \Pr[\mathcal{C} \wedge \mathcal{A} \wedge \mathcal{B}]$$

$$\geq \Pr[\mathcal{C}] - \Pr[\neg\mathcal{A}] - \Pr[\neg\mathcal{B}]$$

$$\geq \left(\epsilon + \frac{1}{2^{m-h}} + \epsilon_H\right) - \frac{1}{2^{m-h}} - \epsilon_H = \epsilon.$$

This probability is taken over the random coin flips of the challenger and of the algorithms A and B. Thus, the theorem follows.   ∎

Note that the construction of the signature scheme above made use of a collision resistant hash function. The proof can be easily modified to only use one way universal hashing (OWUHF). Since OWUHF's can be constructed from one-way functions, there is no need to rely on collision resistance.

5   Coalitions of Limited Size

A MMAC (key-gen, mac-gen, mac-ver) is said to be (t, ϵ, q, κ)-*secure against selective forgery under an adaptive chosen message attack* if every t-time probabilistic algorithm A wins the game in Section 2 (depicted in Fig. 1) with probability less than ϵ, where the coalition I is subject to the constraint $|I| \leq \kappa$. Similarly, (t, ϵ, q, κ)-security against existential forgery is defined as (t, ϵ, q)-security against existential forgery where the coalition size $|I|$ is limited by κ. Note that for $\kappa = n$, these notions are exactly the same as those defined in Section 2; when $\kappa < n$, the security requirements are strictly weaker.

We show in this section that a (t, ϵ, q, κ)-secure MMAC with output length less than

$$\log_2 \sum_{i=0}^{\kappa} \binom{n}{i}$$

gives rise to a signature scheme of nearly equivalent security.

Let (key-gen, mac-gen, mac-ver) be a MMAC that is (t, ϵ, q, κ)-secure against selective forgery under an adaptive chosen message attack. Define the derived signature scheme (skey-gen, sig-gen, sig-ver) as in Section 3, with the modification that skey-gen(s, n) picks a random subset $I \subseteq \{1, \ldots, n\}$ subject to the constraint $|I| \leq \kappa$.

Suppose the length of the output of mac-gen(M) is bounded by

$$\tau := \left\lceil \log \sum_{i=0}^{\kappa} \binom{n}{i} \right\rceil - m$$

for all M. Then we show:

Theorem 3. *The derived signature scheme* (skey-gen, sig-gen, sig-ver) *is* $(t, \epsilon + \frac{1}{2^m}, q)$-*secure against selective forgery under an adaptive chosen message attack.*

The proof follows that of Theorem 1. Because of the restriction on the size of the coalition I, the following alternative to Lemma 1 is required.

Lemma 3. *For any fixed message M,*

$$\Pr[(M, I) \text{ is bad}] \leq \frac{1}{2^m}.$$

where the probability is over the choice of a random coalition $I \subseteq \{1, \ldots, n\}$ satisfying $|I| < \kappa$.

Proof. For each tag $T \in \{0, 1\}^{\tau}$, there is exactly one set I_T containing precisely those receivers i for which mac-ver$(M, T, \mathbf{rk}_i) = $ 'yes'. By definition, the pair (M, I_T) is bad. The collection

$$\{(M, I_T) \mid T \in \{0, 1\}^{\tau}\}$$

completely describes all bad pairs containing M in the first coordinate. Since there are only 2^τ possible values for T, this set is of size at most

$$2^\tau = 2^{-m} \sum_{i=0}^{\kappa} \binom{n}{i}.$$

Since I is chosen independently of M, it follows that

$$\Pr_{\substack{I \subseteq \{1,\ldots,n\} \\ |I| \leq \kappa}} [(M, I) \text{ is bad}] \leq \frac{2^{-m} \sum_{i=0}^{\kappa} \binom{n}{i}}{\sum_{i=0}^{\kappa} \binom{n}{i}} = \frac{1}{2^m},$$

establishing the lemma.   ∎

With this lemma in place, Theorem 3 follows just as Theorem 1.

An analogous theorem may be shown for security against existential forgery. Let (key-gen, mac-gen, mac-ver) be a MMAC that is (t, ϵ, q, κ)-secure against existential forgery under an adaptive chosen message attack. Define the derived signature scheme (skey-gen, sig-gen, sig-ver) as in Section 3, with the modification that skey-gen(k, n) picks a random subset $I \subseteq \{1, \ldots, n\}$ subject to the constraint $|I| \leq \kappa$.

Suppose the MMAC (key-gen, mac-gen, mac-ver) is (t, ϵ, q)-secure against existential forgery under an adaptive chosen message attack, and suppose the length of the output of mac-gen(M) is bounded by $\tau = (\log \sum_{i=0}^{\kappa} \binom{n}{i}) - m$ for all M. Furthermore assume that H is a collision-resistant hash function with security parameter ϵ_H. Then one can show

Theorem 4. *The derived signature scheme* (skey-gen, sig-gen, sig-ver) *is* $(t, \epsilon + \frac{1}{2^{m-h}} + \epsilon_H, q)$-*secure against selective forgery under an adaptive chosen message attack.*

The proof is similar to the proof of Theorem 2 with the appropriate modification to Lemma 2.

6   Conclusions

We gave precise definitions for Multicast MACs (MMACs) secure against selective and existential forgeries. Our main results show that a short collision-resistant multicast MAC can be easily converted into a signature scheme. This shows a gap between the cryptographic resources needed for two party MACs (where signatures are not needed) and the resources needed for Multicast MACs. Our bounds justify the recent effort into designing signature schemes for a multicast environment [5,12,10,7,12]. Such schemes require minimal buffering on the sender's side and resist packet loss. We also note the constructions of [10,11] that provide a short MMAC without non-repudiation by using some weak timing assumptions.

For small values of κ, our lower bound for κ-secure MMACs asymptotically matches the upper bound construction of Canetti et al. [1]. Hence, the Canetti et al. construction has optimal length (up to a small constant factor) for a MMAC that is based purely on pseudorandom functions.

References

1. R. Canetti, J. Garay, G. Itkis, D. Micciancio, M. Naor, and B. Pinkas, "Multicast Security: A taxonomy and some efficient constructions", in IEEE INFOCOM '99, vol. 2, pp. 708–716, 1999.
2. A. De Santis and M. Yung, "On the design of provably-secure cryptographic hash functions", in Proc. of Eurocrypt '90, LNCS 473, pp. 412–431, 1990.
3. Y. Desmedt, Y. Frankel, and M. Yung, "Multi-receiver/Multi-sender network security: efficient authenticated multicast/feedback", in IEEE INFOCOM '92, pp. 2045–2054, 1992.
4. F. Fujii, W. Kachen, and K. Kurosawa, "Combinatorial bounds and design of broadcast authentication", in IEICE Trans., vol. E79-A, no. 4, pp. 502–506, 1996.
5. R. Gennaro and P. Rohatgi, "How to sign digital streams", in Proc. of Crypto '97, 1997.
6. S. Goldwasser, S. Micali, and R. Rivest, "A digital signature scheme secure against adaptive chosen-message attacks", SIAM Journal of Computing, vol. 17, pp. 281–308, 1988.
7. P. Golle, N. Modadugo, "Streamed authentication in the presence of random packet loss", in Proc. of 8th Annual Internet Society Symposium on Network and Distributed System Security (NDSS '01), San Diego, 2001.
8. K. Kurosawa, S. Obana, "Characterization of (k, n) multi-receiver authentication", in Information Security and Privacy, ACISP '97, LNCS 1270, pp. 205–215, 1997.
9. M. Naor and M. Yung, "Universal one-way hash functions and their cryptographic applications", in Proc. of 21st Annual ACM Symposium on Theory of Computing, pp. 33–43, 1989.
10. A. Perrig, R. Canetti, D. Tygar, D. Song, "Efficient Authentication and Signature of Multicast Streams over Lossy Channels", in Proc. of 2000 IEEE Symposium on Security and Privacy, Oakland, 2000.
11. A. Perrig, R. Canetti, D. Song, D. Tygar, "Efficient and Secure Source Authentication for Multicast", in Proc. of 8th Annual Internet Society Symposium on Network and Distributed System Security (NDSS '01), San Diego, 2001.
12. P. Rohatgi, "A compact and fast hybrid signature scheme for multicast packet authentication", in Proc. of 6th ACM conference on Computer and Communication Security, 1999.
13. J. Rompel, "One-way functions are necessary and sufficient for secure signatures", in Proc. of 22nd Annual ACM Symposium on Theory of Computing, pp. 387–394, 1990.
14. R. Safavi-Naini, H. Wang, "Multireceiver authentication codes: models, bounds, constructions and extensions", Information and Computation, vol. 151, no. 1/2, pp. 148–172, 1999.
15. R. Safavi-Naini, H. Wang, "New results on multireceiver authentication codes", in Proc. of Eurocrypt '98, LNCS 1403, pp. 527–541, 1998.

16. G. Simmons, "A cartesian product construction for unconditionally secure authentication codes that permit arbitration", *J. Cryptology*, vol. 2, no. 2, pp. 77–104, 1990.
17. C. K. Wong, S. S. Lam, "Digital signatures for flows and multicasts", IEEE ICNP '98. Also, University of Texas at Austin, Computer Science Technical report TR 98-15.

Analysis of Key-Exchange Protocols and Their Use for Building Secure Channels[*]

Ran Canetti[1] and Hugo Krawczyk[2],[**]

[1] IBM T.J. Watson Research Center, Yorktown Heights, New York 10598.
canetti@watson.ibm.com
[2] EE Department, Technion, Haifa, Israel.
hugo@ee.technion.ac.il

Abstract. We present a formalism for the analysis of key-exchange protocols that combines previous definitional approaches and results in a definition of security that enjoys some important analytical benefits: (i) any key-exchange protocol that satisfies the security definition can be composed with symmetric encryption and authentication functions to provide provably secure communication channels (as defined here); and (ii) the definition allows for simple modular proofs of security: one can design and prove security of key-exchange protocols in an idealized model where the communication links are perfectly authenticated, and then translate them using general tools to obtain security in the realistic setting of adversary-controlled links.

We exemplify the usability of our results by applying them to obtain the proof of two classes of key-exchange protocols, Diffie-Hellman and key-transport, authenticated via symmetric or asymmetric techniques.

1 Introduction

Key-exchange protocols (KE, for short) are mechanisms by which two parties that communicate over an adversarially-controlled network can generate a common secret key. KE protocols are essential for enabling the use of shared-key cryptography to protect transmitted data over insecure networks. As such they are a central piece for building secure communications (a.k.a "secure channels"), and are among the most commonly used cryptographic protocols (contemporary examples include SSL, IPSec, SSH, among others).

The design and analysis of secure KE protocols has proved to be a non-trivial task, with a large body of work written on the topic, including [15,30,10,7,16,5,6], [26,2,34] and many more. In fact, even today, after two decades of research, some important issues remain without satisfactory treatment. One such issue is how to guarantee the adequacy of KE protocols for their most basic application: the generation of shared keys for implementing secure channels. Providing this

---

[*] This proceedings version is a condensed high-level outline of the results in this work; for a complete self-contained treatment the reader is referred to [13].
[**] Supported by Irwin and Bethea Green & Detroit Chapter Career Development Chair.

B. Pfitzmann (Ed.): EUROCRYPT 2001, LNCS 2045, pp. 453–474, 2001.
© Springer-Verlag Berlin Heidelberg 2001

guarantee (with minimal requirements from KE protocols) is the main focus and objective of this work. The other central goal of the paper is in simplifying the usability of the resultant security definitions via a modular approach to the design and analysis of KE protocols. We exemplify this approach with a proof of security for two important classes of KE protocols.

This paper adopts a methodology for the analysis of KE protocols that results from the combination of two previous works in this area: Bellare and Rogaway [5] and Bellare, Canetti and Krawczyk [2]. A main ingredient in the formalization of [5] is the use of the indistinguishability approach of [20] to defining security: roughly speaking, a key-exchange protocol is called secure if under the allowed adversarial actions it is infeasible for the attacker to distinguish the value of a key generated by the protocol from an independent random value. Here we follow this exact same approach but replace the adversarial model of [5] with an adversarial model derived from [2]. This combination allows to achieve the above two main objectives. We elaborate on these main aspects of our work.

First, the formalization of [2] captures not only the specific needs of KE protocols but rather develops a more general model for the analysis of security protocols. This allows formulating and proving the statement that KE protocols proven secure according to our definition (we call these protocols SK-secure) can be used in standard ways to provide "secure channels". More specifically, consider the common security practice by which pairs of parties establish a "secure channel" by first exchanging a *session key* using a KE protocol and then using this key to encrypt and authenticate the transmitted data under symmetric cryptographic functions. We prove that if in this setting one uses an SK-secure KE protocol together with secure MAC and encryption functions combined appropriately then the resultant channel provides both authentication and secrecy (in a sense that we define precisely) to the transmitted data. While this property of ensuring secure channels is indeed an obvious requirement from a secure KE protocol it turns out that formalizing and proving this property is non-trivial. In fact, there are "seemingly secure" key exchange protocols that do not necessarily guarantee this (e.g. those that use the session key during the exchange itself), as well as proposed definitions of secure key-exchange that do not suffice to guarantee this either (e.g., the definitions in [5,8,9,2]). Moreover, although several works have addressed this issue (see Section 1.1), to the best of our knowledge the notion of secure channels was never formalized in the context of KE protocols, let alone demonstrating that some definition of KE protocols suffices for this basic task. Indeed, one of the contributions of this work is a formalization of the secure channels task. While this formalization is not intended to provide general composability properties for arbitrary cryptographic settings, it arguably provides sufficient security guarantee for the central task of protecting the integrity and authenticity of communications over adversarially-controlled links.

Second, the approach of [2] allows for a substantial simplification in designing KE protocols and proving their security. This approach postulates a two-step methodology by which protocols can first be designed and analyzed in a much simplified adversarial setting where the communication links are assumed to be ideally authenticated (i.e., the attacker is not allowed to insert or change information transmitted over the communication links between parties). Then, in a second step, these protocols are "automatically" transformed into secure protocols in the realistic scenario of fully adversary-controlled communications by applying a protocol translation tool (or "compiler") called an *authenticator*. Fortunately, simple and efficient realizations of authenticators based on different cryptographic functions exist [2] thus making it a useful and practical design and analysis tool. (We stress that our framework does not *mandate* this methodology; i.e., it is possible of course to prove security of a KE protocol directly in the fully adversarial model.)

We use this approach to prove the security of two important classes of key-exchange protocols: Diffie-Hellman and key-transport protocols. All one needs to do is to simply prove the security of these protocols in the ideal authenticated-links model and then, thanks to the above modular approach, one obtains versions of these protocols that are secure in a realistic adversary-controlled network. The "authenticated" versions of the protocols depend on the authenticators in use. These can be based either on symmetric or asymmetric cryptographic techniques (depending on the trust model) and result in natural and practical KE protocols. The security guarantees that result from these proofs are substantial as they capture many of the security concerns in real communications settings including the asynchronous nature of contemporary networks, the run of multiple simultaneous sessions, resistance to man-in-the-middle and known-key attacks, maintaining security of sessions even when other sessions are compromised, and providing "perfect forward secrecy", i.e., protection of past sessions in case of the compromise of long-term keying material.

1.1 Related Work

Since its introduction in the seminal work of Diffie and Hellman [15] the notion of a key-exchange protocol has been the subject of many works (see [28] for an extensive bibliography). Here we mention some of the works that are more directly related to the present work.

Among the early works on this subject we note [30,10,7,16] as being instrumental in pointing out to the many subtleties involved in the analysis of KE protocols. The first complexity-theoretic treatment of the notion of security for KE protocols is due to Bellare and Rogaway [5] who formalize the security of KE protocols in the realistic setting of concurrent sessions running in an adversary-controlled network. As said above, [5] apply the indistinguishability definitional approach that we follow here as well. While [5] focused on the shared-key model of authentication, other works [8,9,4] extended the techniques to the public-key setting. One important contribution of [4] is in noting and fixing a shortcoming in the original definition of [5]; this fix, that we adopt here, is fundamental for proving our results about secure channels.

Bellare, Canetti, and Krawczyk [2] present a model for studying general session-oriented security protocols that we adopt and extend here. They also introduce the "authenticator" techniques that allow for greatly simplifying the analysis of protocols and that we use as a basic tool in our work. In addition, [2] proposes a definition of security of KE protocols rooted in the simulatability (or "ideal third party") approach used to define security of multiparty computation [19,29,1,11]. While this definitional approach is intuitively appealing the actual KE security definition of [2] comes short of the expectations. On one hand, it seems over-restrictive, in the sense that it rules out protocols that seem to provide sufficient security (and as demonstrated here can be safely used to obtain secure channels); on the other, it is not clear whether their definition suffices to prove composition theorems even in the restricted sense of secure channels as dealt with in this paper.

More recently, Shoup [34] presents a framework for the definition of security of KE protocols that follows the basic simulatability approach as in [2] but introduces significant modifications in order to overcome some of the shortcomings of the KE definition in [2] as well as to seek composability with other cryptographic applications. In particular, [34] states as a motivational goal the construction of "secure sessions" (similar to our secure channels), and it informally claims the sufficiency of its definitions to achieve this goal. A more rigorous and complete elaboration of that work will be needed to assess the correctness of these claims. In addition, [34] differs from our work in several other interesting aspects. In order to keep this introduction short, we provide a more extensive comparison with [34] in Appendix A.

A promising general approach for the analysis of reactive protocols and their concurrent composition has been developed by Pfitzmann, Schunter and Waidner [32,31,33] and Canetti [12]. This approach, that follows the simulatability tradition, can be applied to the task of key exchange to obtain a definition of KE protocols that guarantees secure concurrent composition with any set of protocols that make use of the generated keys. See more details in [14].

A subjective discussion. The above works follow two main distinct approaches to defining security of KE protocols: *simulation-based* and *indistinguishability-based*. The former is more intuitively appealing (due to its modeling of security via an ideally-trusted third party), and also appears to be more amenable to demonstrating general composability properties of protocols. On the other hand, the complexity of the resulting definitions, once all details are filled in, is considerable and makes for definitions that are relatively complex to work with. In contrast, the indistinguishability-based approach yields definitions that are simpler to state and easier to work with, however their adequacy for modeling the task at hand seems less clear at first glance. The results in this paper indicate the suitability of the indistinguishability-based approach in the context of KE protocols — if the goal is the application of KE protocols to the specific task of secure channels as defined here. By following this approach we gain the benefit of simpler analysis and easier-to-write proofs of security. At the same time, our work borrows from the simulation-based approach the modularity of building proofs via the intermediate ideally-authenticated links model, thus enjoying the "best of both worlds".

Organization.   Due to lack of space, the presentation here is kept at an informal level, and omits some important pieces. A complete and rigorous treatment, including model details and proofs, appears in [13]. Section 2 presents an overview of the protocol and adversary models used throughout this work. The definition of SK-security for KE protocols is presented in Section 3. Section 4 states the security of sample protocols. Section 5 demonstrates the suitability of our notion of security to realizing secure channels.

2   Protocol and Adversary Models: An Overview

In order to to define what is meant by the security of a key-exchange (KE) protocol we first need to establish a formalism for the most basic notions: what is meant by a protocol in general and a key-exchange protocol in particular, what are sessions, and what is an 'attacker' against such protocols. Here we use a formalism based on the approach of [2], where a general framework for studying the security of session-based multi-party protocols over insecure channels is introduced. We extend and refine this formalism to better fit the needs of practical KE protocols.

In order to motivate and make the formalism easier to understand, *we start by providing a high-level overview of our model.* The precise technical description appears in [13]. After introducing the protocol and adversary models we proceed to define the security of KE protocols in Section 3.

2.1   Protocols, Sessions and Key-Exchange

Message-driven protocols. We consider a set of parties (probabilistic polynomial-time machines), which we usually denote by $P_1, \ldots, P_n$, interconnected by point-to-point links over which messages can be exchanged. Protocols are collections of interactive procedures, run concurrently by these parties, that specify a particular processing of incoming messages and the generation of outgoing messages. Protocols are initially triggered at a party by an external "call" and later by the arrival of messages. Upon each of these events, and according to the protocol specification, the protocol processes information and may generate and transmit a message and/or wait for the next message to arrive. We call these message-driven protocols. (We note the asynchronous nature of protocols defined in this way which reflects the prevalent form of communication in today's networks.)

Sessions and protocol output. Protocols can trigger the initiation of subprotocols (i.e. interactive subroutines) or other protocols, and several copies of such protocols may be simultaneously run by each party. We call each copy of a protocol run at a party a session. Technically, a session is an interactive subroutine executed inside a party. Each session is identified by the party that runs it, the parties with whom the session communicates and by a session-identifier. These identifiers are used in practice to bind transmitted messages to their corresponding sessions. Each invocation of a protocol (or session) at a given party creates a local state for that session during execution, and produces local outputs by that party. This output can be a quantity (e.g a session key) returned to the calling program, or it can be the recording of a protocol event (such as a successful or failed termination). These local outputs serve to represent the "history" of a protocol and are important to formalize security. When a session ends its run we call it complete and assume that its local state is erased.

Key-exchange protocols.   Key-exchange (KE) protocols are message-driven protocols (as defined above) where the communication takes place between pairs of parties and which return, upon completion, a secret key called a session key. More specifically, the input to a KE protocol within each party P_i is of the form $(P_i, P_j, s, role)$, where P_j is the identity of another party, s is a session id, and role can be either initiator or responder. A session within P_i and a session within P_j are called matching if their inputs are of the form $(P_i, P_j, s, \text{initiator})$ and $(P_j, P_i, s, \text{responder})$. The inputs are chosen by a "higher layer" protocol that "calls" the KE protocol. We require the calling protocol to make sure that the session id's of no two KE sessions in which the party participates are identical. Furthermore, we leave it to the calling protocol to make sure that two parties that wish to exchange a key will activate matching sessions. Note that this may require some communication before the actual KE sessions are activated.[1]

Upon activation, the partners P_i and P_j of two matching sessions exchange messages (the initiator goes first), and eventually generate local outputs that include the name of the partners of the session, the session identifier, and the value of the computed session key. A key establishment event is recorded only when the exchange is completed (this signals, in particular, that the exchanged key can be used by the protocol that called the KE session). We note that a session can be completed at one partner but not necessarily at the other.

After describing these 'mechanics" of a KE protocol we need to define what is meant by a "secure" KE protocol. This is the subject of Section 3 and it is based on the adversarial model that we introduce next.

2.2   The Unauthenticated-Links Adversarial Model (UM)

In order to talk about the security of a protocol we need to define the adversarial setting that determines the capabilities and possible actions of the attacker. We want these capabilities to be as generic as possible (as opposed to, say, merely representing a list of possible attacks) while not posing unrealistic requirements. We follow the general adversarial formalism of [2] but specialize and extend it

---

[1]   Indeed, in practice protocols for setting up a secure session typically exchange some messages before the actual cryptographic key-exchange starts. The IKE protocol of the IPSEC standard is a good example [23].

here for the case of KE protocols. Using the terminology of [2] we call this model the Unauthenticated Links Model (UM).

Basic attacker capabilities. We consider a probabilistic polynomial-time (PPT) attacker that has full control of the communications links: it can listen to all the transmitted information, decide what messages will reach their destination and when, change these messages at will or inject its own generated messages. The formalism represents this ability of the attacker by letting the attacker be the one in charge of passing messages from one party to another. The attacker also controls the scheduling of all protocol events including the initiation of protocols and message delivery.

Obtaining secret information. In addition to these basic adversarial capabilities (given "for free" to the attacker), we let the attacker obtain secret information stored in the parties memories via explicit attacks. We consider all the secret information stored at a party as potentially vulnerable to break-ins or other forms of leakage. However, when defining security of a protocol it is important to guarantee that the leakage of some form of secret information has the least possible effect on the security of other secrets. For example, we will want to guarantee that the leakage of information specific to one session (such as the leakage of a session key or ephemeral state information) will have no effects on the security of other sessions, or that even the leakage of crucial long-term secrets (such as private keys) that are used across multiple sessions will not necessarily compromise secret information from all past sessions. In order to be able to differentiate between various vulnerabilities and to be able to guarantee as much security as possible in the event of information exposures, we classify attacks into three categories depending on the type of information accessed by the adversary:

Session-state reveal. The attacker provides the name of a party and a session identifier of a *yet incomplete* session at that party and receives the internal state of that session (since we see sessions as procedures running inside a party then the internal state of a session is well defined). An important point here is what information is included in the local state of a session. We leave this to be specified by each KE protocol. Therefore, our definition of security is parameterized by the type and amount of information revealed in this attack. For instance, the information revealed in this way may be the exponent x used by a party to compute a value g^x in a Diffie-Hellman key-exchange protocol, or the random bits used to encrypt a quantity under a probabilistic encryption scheme during a session. Typically, the revealed information will include all the local state of the session and its subroutines, except for the local state of the subroutines that directly access the long-term secret information, e.g. the local signature/decryption key of a public-key cryptosystem, or the long-term shared key.

Session-key query. The attacker provides a party's name and a session identifier of a *completed* session at that party and receives the value of the key generated by the named session This attack provides the formal modeling for leakage of information on specific session keys that may result from events such as break-ins, cryptanalysis, careless disposal of keys, etc. It will also serve, indirectly, to ensure that the unavoidable leakage of information produced by the use of session keys in a security application (e.g., information leaked on a key by its use as an encryption key) will not help in deriving further information on this and other keys.

Party corruption. The attacker can decide at any point to corrupt a party, in which case the attacker learns *all* the internal memory of that party including long-term secrets (such as private keys or master shared keys used across different sessions) and session-specific information contained in the party's memory (such as internal state of incomplete sessions and session-keys corresponding to completed sessions). Since by learning its long term secrets the attacker can impersonate a party in all all its actions then a party is considered completely controlled by the attacker from the time of corruption and can, in particular, depart arbitrarily from the protocol specifications.

Terminology: if a session is subject to any of the above three attacks (i.e. a session-state reveal, a session-key query or the corruption of the party holding the session) then the session is called locally exposed. If a session or its matching session is locally exposed then we call the session exposed.

Session expiration. One important additional element in our security model is the notion of session expiration. This takes the form of a protocol action that when activated causes the erasure of the named session key (and any related session state) from that party's memory. We allow a session to be expired at one party without necessarily expiring the matching session. The effect of this action in our security model is that the value of an expired session key cannot be found via any of the above attacks if these attacks are performed after the session expired. This has two important consequences: it allows us to model the common (and good) security practice of limiting the life-time of individual session keys and it allows for a simple modeling of the notion of perfect forward secrecy (see Section 3.2). We note that in order for a session to be locally exposed (as defined above) the attack against the session must happen *before* the session expires.

Bootstrapping the security of key-exchange protocols. Key-exchange protocols, as other cryptographic applications, require the bootstrapping of security (especially for authentication) via some assumed-secure means. Examples include the secure generation of parties' private keys, the installation of public keys of other parties, or the installation of shared "master" keys. Here too we follow the approach of [2] where the bootstrapping of the authentication functions is abstracted into an initialization function that is run prior to the initiation of any key-exchange protocol and that produces in a secure way (i.e. without adversarial participation) the required (long-term) information. By abstracting out this initial phase we allow for the combination of different protocols with different initialization functions: in particular, it allows our analysis of protocols (such as Diffie-Hellman) to be applicable under the two prevalent settings of authentication: symmetric and a-symmetric authentication. Two points to note are (1) the specification of the initialization function is part of the definition of each KE protocol; and (2) secret information generated by this function at a given party can be discovered by the attacker only upon corruption of that party. We stress that while this abstraction adds to the simplicity and applicability of our analysis techniques, the bootstrapping of security in actual protocols is an element that must be carefully analyzed (e.g., the interaction with a CA in the case of public-key based protocols). Integrating these explicit elements into the model can be done either directly as done in [34], or in a more modular way via appropriate protocol composition.

2.3   The AM, Protocol Emulation and Authenticators

A central ingredient in our analyses is the methodology introduced in [2] by which one can design and analyze a protocol under the highly-simplifying assumption that the attacker cannot change information transmitted between parties, and then transform these protocols and their security assurance to the realistic UM where the adversary has full control of the communication links. The main components in the formalization of this methodology are shortly described here (see [2,13] for complete details).

First, an adversarial model called authenticated-links model (denoted AM) is defined in a way that is identical to the UM with one fundamental difference: *the attacker is restricted to only deliver messages truly generated by the parties without any change or addition to them.* Then, the notion of "emulation" is introduced in order to capture the equivalence of functionality between protocols in different adversarial models, in particular between the UM and AM. Roughly speaking, a protocol π' emulates protocol π in the UM if for any adversary that interacts with π' in the UM there exists an adversary that interacts with π in the AM such that the two interactions "look the same" to an outside observer. Finally, special algorithms called authenticators are developed with the property that on input the description of a protocol π the authenticator outputs the description of a protocol π' such that π' emulates protocol π in the UM. That is, authenticators act as an automatic "compiler" that translate protocols in the AM into equivalent (or "as secure as") protocols in the UM.

In order to simplify the construction of authenticators, [2] offers the following methodology. First consider a very simple one-flow protocol in the AM, called MT, whose sole functionality is to transmit a single message from sender to recipient. Now build a restricted-type authenticator, called MT-authenticator, required to provide emulation for this particular MT protocol only. Finally, to any such MT-authenticator λ one associates an algorithm (or compiler) C_λ that translates any input protocol π into another protocol π' as follows: to each of the messages defined in protocol π apply the MT-authenticator λ. It is proven in [2] that C_λ is an authenticator (i.e., the resultant protocol π' emulates π in the UM). Particular realizations of MT-authenticators are presented in [2] based on different type of cryptographic functions (e.g., digital signatures, public-key encryption, MAC, etc.)

3   Session-Key Security

After having defined the basic formal model for key-exchange protocols and adversarial capabilities, we proceed to define *what is meant for a key-exchange protocol to be secure*. While the previous section was largely based on the work of [2], our definition of security closely follows the definitional approach of [5]. The resultant notion of security, that we call *session-key security* (or *SK-security*), focuses on ensuring the security of individual session-keys as long as the session-key value is not obtained by the attacker via an explicit key exposure (i.e. as long as the session is *unexposed* – see the terminology in the previous section). We want to capture the idea that the attacker "does not learn anything about the value of the key" from interacting with the key-exchange protocol and attacking other sessions and parties. As it is standard in the semantic-security approach this is formalized via the infeasibility to distinguish between the real value of the key and an independent random value.

We stress that this formulation of SK-security is very careful about tuning the definition to offer enough strength as required for the use of key-exchange protocols to realize secure channels (Section 5), as well as being realistic enough to avoid over-kill requirements which would prevent us from proving the security of very useful protocols (Section 4). We further discuss these aspects after the presentation of the definition.

3.1   Definition of SK-Security

We first present the definition for the UM. The formalization in the AM is analogous. We start by defining an "experiment" where the attacker $\mathcal{U}$ chooses a session in which to be "tested" about information it learned on the session-key; specifically, we will ask the attacker to differentiate the real value of the chosen session key from a random value. (Note that this experiment is an artifact of the definition of security, and not an integral part of the actual key-exchange protocols and adversarial intervention.)

For the sake of this experiment we extend the usual capabilities of the adversary, $\mathcal{U}$, in the UM by allowing it to perform a *test-session query*. That is, in addition to the regular actions of $\mathcal{U}$ against a key-exchange protocol π, we let $\mathcal{U}$ to choose, at any time during its run, a *test-session* among the sessions that are completed, unexpired and unexposed at the time. Let κ be the value of the corresponding session-key. We toss a coin b, $b \overset{R}{\leftarrow} \{0,1\}$. If $b = 0$ we provide $\mathcal{U}$ with the value κ. Otherwise we provide $\mathcal{U}$ with a value r randomly chosen from the probability distribution of keys generated by protocol π. The attacker $\mathcal{U}$ is now allowed to continue with the regular actions of a UM-adversary but *is not allowed to expose the test-session* (namely, it is not allowed session-state reveals, session-key queries, or partner's corruption on the test-session or its matching

session.[2]) At the end of its run, $\mathcal{U}$ outputs a bit b' (as its guess for b).
We will refer to an attacker that is allowed test-session queries as a KE-adversary.

Definition 1. *A* KE *protocol* π *is called* SK-secure *if the following properties hold for any* KE-*adversary* $\mathcal{U}$ *in the* UM.

1. *Protocol* π *satisfies the property that if two* uncorrupted *parties complete matching sessions then they both output the same key; and*
2. *the probability that* $\mathcal{U}$ *guesses correctly the bit* b *(i.e., outputs* $b' = b$*) is no more than* $1/2$ *plus a negligible fraction in the security parameter.*

If the above properties are satisfied for all KE-*adversaries in the* AM *then we say that* π *is* SK-secure *in the* AM.

The first condition is a "consistency" requirement for sessions completed by two *uncorrupted* parties. We have no requirement on the session-key value of a session where one of the partners was corrupted before the session completed – in fact, most KE protocols allow a corrupted party to strongly influence the exchanged key. The second condition is the "core property" for SK-security. We note that the term 'negligible' refers, as customary, to any function (in the security parameter) that diminishes asymptotically faster than any polynomial fraction. (This formulation allows, if so desired, to quantify security via a concrete security treatment. In this case one quantifies the attacker's power via specific bounds on computation time, number of corruptions, etc., while its advantage is bounded through a specific parameter ε.)

Discussion. We highlight three aspects of Definition 1.

- The attacker can keep running and attacking the protocol even after receiving the response (either real or random) to its test-session query. This ability (which represents a substantial strengthening of security relative to [5], see also [4]) is *essential* for proving the main property of SK-security shown in this paper, namely its guarantee of security when used to generate secure channels as described in Section 5.
- The attacker is not allowed to corrupt partners to the test-session or issue any other exposure command against that session while unexpired. This reflects the fact that there is no way to guarantee the secure use of a session-key that was exposed via an attacker's break-in (or cryptanalysis). In particular, this restriction is instrumental for proving the security of specific important protocols (e.g., Diffie-Hellman key exchange) as done in Section 4.
- The above restriction on the attacker by which it cannot corrupt a partner to the test-session is lifted as soon as the session expires at that partner. In this case the attacker should remain unable to distinguish between the real

---

[2] We stress, however, that the attacker *is allowed* to corrupt a partner to the test-session as soon as the test-session (or its matching session) expires at that party. See the discussion below. This may be the case even if the other partner has not yet expired the matching session or not even completed it.

value of the key from a random value. This is the basis to the guarantee of "perfect forward secrecy" provided by our definition and further discussed in Section 3.2.

We stress that in spite of its "compact" formulation Definition 1 is very powerful and can be shown to ensure many specific properties that are required from a good key-exchange protocol (see, for example, chapter 12 of [28]). Some of these properties include the guarantee that session-keys belong to the right probability distribution of keys (except if one of the partners is corrupted at time of exchange), the "authenticity" of the exchange (namely, a correct and consistent binding between keys and parties' identities), resistance to man-in-the-middle attacks (for protocols proven SK-secure in the UM), resistance to known-key attacks, forward secrecy, and more. However, we note that all these properties (which are sometimes listed as a replacement to a formal definition of security) in combination do not suffice to guarantee the most important aspect of key-exchange security that SK-security enjoys: namely, the composition of the key-exchange protocols with cryptographic functions to enable secure channels (e.g., the original definition of security in [5] does satisfy the above list of properties but is insufficient to guarantee secure channels).

We finally remark that Definition 1 makes security requirements from a KE protocol only in case that the protocol completes KE-sessions. No guarantee is made that KE-sessions will ever return, or that they will not be aborted, i.e., that the corresponding session key will not be null. (In fact, a KE protocol where all KE-sessions "hang" and never return satisfies the definition.) One can add an explicit termination requirement for sessions in which the parties are uncorrupted and all messages are correctly delivered by the attacker. For simplicity, we choose to leave the analysis of the termination properties of protocols out of the scope of the definition of security.

3.2   Forward Secrecy

Informally, the notion of "perfect forward secrecy" (PFS) [22,16] is stated as the property that "compromise of long-term keys does not compromise past session keys". In terms of our formalism this means that even if a party is corrupted (in which case all its stored secrets – short-term and long-term – become known to the attacker) then nothing is learned about sessions within that party that were previously unexposed and *expired* before the party corruption happened.

The provision that *expired* session-keys remain indistinguishable from random values even if a partner to that session is corrupted guarantees the perfect forward secrecy of SK-secure protocols. Put in other words, when proving a protocol to be SK-secure using Definition 1 one automatically gets a proof that that protocol guarantees PFS.

On the other hand, while PFS is a very important security property it is not required for all application scenarios, e.g., when only authentication is required, or when short-term secrecy suffices. Indeed, it is common to find in practice protocols that do not provide PFS and still are not considered insecure. One such typical case are "key-transport protocols" in which public key encryption is used to communicate a session-key from one party to another. (In this case, even if session-keys are erased from memory when no longer required, the corruption of a party may allow an attacker to compute, via the discovered long-term private keys, all the past session-keys.) Due to the importance of such protocols (they are commonly used in, e.g., SSL), and given that achieving PFS usually has a non-negligible computational cost, we define a notion of "SK-security without PFS" by simply disallowing the protocol's action of key expiration. That is, under this modified model, *session-keys never expire*. This results in a weaker notion of security since now by virtue of Definition 1 the attacker is *never allowed* to corrupt a partner to the test-session (or in other words, this weaker definition of security does not guarantee the security of a session-key for which one of the partners is ever corrupted).

Definition 2. *We say that a* KE *protocol satisfies* SK-security without PFS *if it enjoys SK-security relative to any* KE-*adversary in the* UM *that is* not *allowed to expire keys. (Similarly, if the above holds for any such adversaries in the* AM *then we say that π is* SK-secure without PFS *in the* AM.*)*

4   SK-Secure Protocols

This section demonstrates the usability of our definition of SK-security for proving the security of some simple and important key-exchange protocols. One is the original Diffie-Hellman protocol, the other is a simple "key transport" protocol based on public-key encryption. We first show that these protocols are secure in the simpler authenticated-links model (AM). Then, using the methodology from [2] we can apply to these protocols a variety of (symmetric or asymmetric) authentication techniques to obtain key-exchange protocols that are secure in the realistic UM model. Namely, applying any MT-authenticator (see Section 2.3) to the messages of the AM-protocol results in a secure KE protocol in the UM The next Theorem (proven in [13]) states that this methodology does work for our purposes.

Theorem 1. *Let π be a SK-secure key-exchange protocol in the* AM *with* PFS *(resp., without* PFS*) and let λ be an authenticator. Then* $\pi' = C_\lambda(\pi)$ *is a SK-secure key-exchange protocol in the* UM *with* PFS *(resp., without* PFS*).*

For lack of space we only describe here the protocol based on Diffie-Hellman exchange. The key-transport protocol based on public-key encryption is presented and analyzed in [13].

4.1   Two-Move Diffie-Hellman

We demonstrate that under the Decisional Diffie-Hellman (DDH) assumption the 'classic' two-move Diffie-Hellman key-exchange protocol designed to work

against eavesdroppers-only is SK-secure in the AM. We denote this protocol by
2DH and describe it in Figure 1. Here and in the sequel all exponentiations are
modulo the defined prime p.

Protocol 2DH

Common information: Primes $p, q,\ q/p{-}1$, and g of order q in Z_p^*.

Step 1: The initiator, P_i, on input (P_i, P_j, s), chooses $x \xleftarrow{\text{R}} Z_q$ and sends
$(P_i, s, \alpha = g^x)$ to P_j.

Step 2: Upon receipt of (P_i, s, α) the responder, P_j, chooses $y \xleftarrow{\text{R}} Z_q$, sends
$(P_j, s, \beta = g^y)$ to P_i, *erases* y, and outputs the session key $\gamma = \alpha^y$ under
session-id s.

Step 3: Upon receipt of (P_j, s, β), party P_i computes $\gamma' = \beta^x$, *erases* x,
and outputs the session key γ' under session-id s.

Fig. 1. The two-move Diffie-Hellman protocol in the AM

Theorem 2. *Assuming the Decisional Diffie-Hellman (DDH) assumption, protocol* 2DH *is SK-secure in the* AM.

Using Theorem 1 we can apply any authenticator to this protocol to obtain
a secure Diffie-Hellman exchange against realistic UM attackers. For illustration,
a particular instance of such a SK-secure protocol in the UM, using digital signatures for authentication, is shown in Section 4.2. Other flavors of authenticated
DH protocols can be derived in a similar way by using other authenticators (e.g.
based on public key encryption or on pre-shared keys [2]).

4.2   SK-Secure Diffie-Hellman Protocol in the UM

Here we apply the signature-based authenticator of [2] to the protocol 2DH from
Figure 1 to obtain a Diffie-Hellman key-exchange that is SK-secure in the UM.
We present the resultant protocol in Figure 2 (it is very similar to a protocol
specified in [24]). Its SK-security follows from Theorems 1 and 2.

Remarks on protocol SIG-DH. The protocol is the result of applying the
signature-based authenticator of [2] to the 2-pass Diffie-Hellman protocol of Figure 1 where the values α and β (the DH exponentials) serve as the challenges
required by the signature-based authenticator. This assumes (as specified in
protocol 2DH) that these exponentials are chosen afresh for each new exchange
(otherwise each party can add an explicit nonce to the messages which is also
included under the signature). We note that the identity of the destination party
included under the signatures is part of the specification of the signature-based
authenticator of [2] and is fundamental for the security of the protocol.

The description of SIG-DH in Figure 2 assumes, as formalized in our model,
that the value s of the session-id is provided to the parties. In practice, one

---

Protocol SIG-DH

Initial information: Primes p, q, $q/p–1$, and g of order q in Z_p^*. Each player has a private key for a signature algorithm SIG, and all have the public verification keys of the other players.

Step 1: The initiator, P_i, on input (P_i, P_j, s), chooses $x \stackrel{R}{\leftarrow} Z_q$ and sends $(P_i, s, \alpha = g^x)$ to P_j.

Step 2: Upon receipt of (P_i, s, α) the responder, P_j, chooses $y \stackrel{R}{\leftarrow} Z_q$, and sends to P_i the message $(P_j, s, \beta = g^y)$ together with its signature $\text{SIG}_j(P_j, s, \beta, \alpha, P_i)$; it also computes the session key $\gamma = \alpha^y$ and *erases* y.

Step 3: Upon receipt of (P_j, s, β) and P_j's signature, party P_i verifies the signature and the correctness of the values included in the signature (such as players identities, session id, the value of exponentials, etc.). If the verification succeeds then P_i sends to P_j the message $(P_i, s, \text{SIG}_j(P_i, s, \alpha, \beta, P_j))$, computes $\gamma' = \beta^x$, *erases x*, and outputs the session key γ' under session-id s.

Step 4: Upon receipt of the triple (P_i, s, sig), P_j verifies P_i's signature sig and the values it includes. If the check succeeds it outputs the session key γ under session-id s.

Fig. 2. Diffie-Hellman protocol in the UM: authentication via signatures.

---

usually generates the session identifier s as a pair (s_1, s_2) where s_1 is a value chosen by P_i and different (with very high probability) from all other such values chosen by P_i in his other sessions with P_j. Similarly, s_2 is chosen by P_j with an analogous uniqueness property. These values s_1, s_2 can be exchanged by the parties as a prologue to the above protocol (this may be the case of protocols that implement such a prologue to exchange some other system information and to negotiate exchange parameters; see for example [23]). Alternatively, s_1 can be included by P_i in the first message of SIG-DH, and s_2 be included by P_j in the second message. In any case, it is important that these values be included under the parties' signatures.

5   Applications to Secure Channels

It is common practice to protect end-to-end communications by letting the end parties exchange a secret session key and then use this key to authenticate and encrypt the transmitted data under symmetric cryptographic functions. In order for a key-exchange protocol to be considered secure it needs to guarantee that the above strategy for securing data works correctly, namely, that by using a shared key provided by the KE protocol one achieves sound authentication and secrecy. As it is customary, we will refer to a link between a pair of parties that achieves these properties as a **secure channel**. While secure channels may have different formalizations, here we restrict our treatment to the above setting of securing communications using symmetric cryptography with a key derived from a key-exchange protocol. *We prove that an SK-secure key-exchange protocol, together with a secure MAC and symmetric encryption appropriately combined, suffices for realizing such secure channels.*

For lack of space, this extended abstract contains only our treatment of the task of authenticating the communication. Full treatment, including the task of providing secrecy (both in the AM and in the UM) appears in [13].

A Template Protocol: Network Channel. We start by formalizing a "template protocol" that captures a generic session-oriented KE-based protocol for secure channels between pairs of parties in a multi-party setting with parties $P_1, \ldots, P_n$. This template protocol, called NetChan, applies to the unauthenticated-links model UM as well as to the authenticated-links model AM. Later we will see specific implementations of this template protocol where the generic 'send' and 'receive' primitives defined there are instantiated with actual functions (e.g., for providing authentication and/or encryption). We will also define what it means for such an implementation to be "secure".

A (session-based) network channel protocol, NetChan(π, snd, rcv), is defined on the basis of a KE protocol π, and two generic functions snd and rcv. (A more general treatment can be obtained by considering these functions as interactive protocols but we leave this more general approach beyond the scope of the present paper.) Both snd and rcv are probabilistic functions that take as arguments a session-key (we denote this key as a subscript to the function) and a message m. The functions may also depend on other session information such as a session-id and partner identifiers. The output of snd is a single value m', while the output of rcv is a pair (v, ok) where ok is a bit and v an arbitrary value. (The bit ok will be used to return a verification value, e.g. the result of verifying an authentication tag.) On the basis of such functions we define NetChan(π, snd, rcv) in Figure 3.

Network Authentication. On the basis of the above formalism, we treat the case of network channels that provide authentication of information over adversary-controlled channels. Namely, we are interested in a NetChan protocol that runs in the unauthenticated-links model UM and yet provides authenticity of transmitted messages. This implementation of NetChan (which we call NetAut) will be aimed at capturing the practice by which communicating parties use a key-exchange protocol to establish a shared session key, and use that key to authenticate (via a message authentication function, MAC) the information exchanged during that session. Namely, if P_i and P_j share a matching session s and P_i wants to send a message m to P_j during that session then P_i transmits m together with $\text{MAC}_\kappa(m)$ where κ is the corresponding session key. Thus, in this case we will instantiate the snd and rcv functions of NetChan with a MAC function as follows.

Protocol NetAut. Let π be a KE protocol and let f be a MAC function. Protocol NetAut(π, f) is protocol NetChan(π, snd, rcv) as defined in Figure 3, where functions snd and rcv are defined as:

 – On input m, $\text{snd}_\kappa(m)$ produces output $m' = (m, t) = (m, f_k(m))$.

Protocol NetChan(π, snd, rcv)

NetChan(π, snd, rcv) is initialized with the same initialization function I of the KE protocol π. It can then be invoked within a party P_i under the following activations:

1. establish-session($P_i, P_j, s, role$): this triggers a KE-session under π within P_i with partner P_j, session-id s and $role \in \{\text{initiator, responder}\}$. If the KE-session completes P_i records in its local output "`established session` s `with` P_j" and stores the generated session key.

2. expire-session(P_i, P_j, s): P_i marks session (P_i, P_j, s) (if it exists at P_i) as expired and the session key is erased. P_i records in its local output "`session` s `with` P_j `is expired`".

3. send(P_i, P_j, s, m): P_i checks that session (P_i, P_j, s) has been completed and not expired, if so it computes $m' = \text{snd}_\kappa(m)$, using the corresponding session key κ, sends (P_i, s, m') to P_j, and records "`sent message` m `to` P_j `within session` s" in the local output.

4. On incoming message (P_j, s, m'), P_i checks that the session (P_i, P_j, s) has been completed and not expired, if so it computes $(m, \text{ok}) = \text{rcv}_\kappa(m')$ under the corresponding session key κ. If ok $= 1$ then P_i records "`received message` m `from` P_j `within session` s." If ok $= 0$ then no further action is taken.

Fig. 3. A generic network channels protocol

- On input m', $\text{rcv}_\kappa(m')$ outputs (v, ok) as follows. If m' is of the form (m, t), and the pair (m, t) passes the verification function of f under key κ, then ok $= 1$ and $v = m$. Otherwise, ok $= 0$ and $v = null$.

In order to simplify and shorten presentation we assume that no two send activations within a session contain the same message. One can easily implement this assumption by specifying that the sender concatenates to the message a unique message id. In the cases where we care about preventing replay of messages by the attacker (as it is usually the case when providing message authentication) then message id's need to be specified in a way that the receiver can check their uniqueness (in this case sender and receiver maintain a shared state).

Our goal is to show that if the key-exchange protocol π is SK-secure and the MAC function f is secure (against chosen-message attacks) then the resultant network channels protocol NetAut(π, f) provides authenticated transmission of information. This requirement can be formulated under the property that "any message recorded by P_i as received from P_j has been necessarily recorded as sent by P_j, except if the pertinent session is exposed". We will actually strengthen this requirement and ask that a network channels protocol provides authentication if it *emulates* (i.e. imitates) the transmission of messages in the *ideally* authenticated-links model AM. Formally, we do so using the notion of protocol emulation and the formalization (see Section 2.3) of the message transmission protocol (MT) in the AM as done in [2]. Recall that MT is a simple protocol that defines the transmission of individual messages in the AM. Here we extend the

basic definition of MT to a *session-based* message transmission protocol called
SMT. By proving that the network channels protocol NetAut emulates SMT in the
UM we get the assurance that transmitting messages over unauthenticated-links
using NetAut is as secure as transmitting them in the presence of an attacker
that is not allowed to change or inject messages over the communication links.

The SMT protocol. We extend protocol MT from [2] to fit our session-based
setting in which transmitted messages are grouped into different sessions. We call
the extended protocol a **session-based message transmission protocol** (SMT), and
define it in Figure 4. Note the structural similarity between SMT and NetChan –
the differences are that no actual key-exchange is run in SMT, and the functions
snd and rcv are instantiated to simple "identity functions".

Protocol SMT

Protocol SMT can be invoked within a party P_i under the following activations:

1. establish-session(P_i, P_j, s): in this case P_i records in its local output
   "`established session s with` P_j".
2. expire-session(P_i, P_j, s): in this case P_i records in its local output
   "`session s with` P_j `is expired`".
3. send(P_i, P_j, s, m): in this case P_i checks that session (P_i, P_j, s) has been
   established and not expired, if so it sends message m to P_j together
   with the session-id s (i.e., the values m and s are sent over the ideally-
   authenticated link between P_i and P_j); P_i records in its local output
   "`sent message` m `to` P_j `within session s`".
4. On incoming message (m, s) received over its link from P_j, P_i checks
   that session (P_i, P_j, s) is established and not expired, if so it records in
   the local output "`received message` m `from` P_j `within session s`".

Fig. 4. SMT: The session-based MT protocol in the AM.

Protocol SMT represents a perfectly authenticated exchange of messages. An
implementation of protocol NetChan is said to be a **secure network authentication
protocol** if it *emulates* (see Section 2.3) protocol SMT in the UM.

Definition 3. *Protocol* NetChan(π, snd, rcv) *is called a* secure network authentication protocol *if it emulates protocol* SMT *in the* UM.

The following theorem is proven in [13]:

Theorem 3. *If π is a SK-secure key-exchange protocol in the* UM *and* snd, rcv
*are based as described above on a MAC function f that is secure against chosen
message attacks, then protocol* NetAut(π, snd, rcv) *is a secure network authentication protocol.*

Network Encryption and Secure Channels Protocols. For lack of space,
we omit from this extended abstract two basic components in our work (the

complete treatment appears in [13]). One is the formalization of a network encryption protocol and its security, the other is the definition of a secure channels protocol. These formalizations are based, as in the case of the network authentication protocol, on the above generic network channels template. In the case of the network encryption protocol, security (in the sense of secrecy) is formulated following the indistinguishability approach. Secure channels are then defined as network channel protocols that are simultaneously secure network authentication and secure network encryption protocols. Implementations of such secure protocols are presented using SK-secure key-exchange protocols and secure encryption and authentication functions. One particularly interesting aspect of our work is highlighted by recent results in [25] where it is demonstrated that the specific ordering of encryption and authentication as applied here is instrumental for achieving secure channels (if one assumes the standard strength, i.e. against chosen-plaintext attacks, of the encryption function). As it turns out other common orderings of these functions do not guarantee secure channels in this case (even if the KE protocol in use is secure).

References

1. D. Beaver, "Secure Multi-party Protocols and Zero-Knowledge Proof Systems Tolerating a Faulty Minority", J. Cryptology (1991) 4: 75-122.
2. M. Bellare, R. Canetti and H. Krawczyk, "A modular approach to the design and analysis of authentication and key-exchange protocols", *30th STOC*, 1998.
3. M. Bellare, A. Desai, D. Pointcheval, and P. Rogaway, "Relations Among Notions of Security for Public-Key Encryption Schemes", *Advances in Cryptology - CRYPTO'98 Proceedings*, Lecture Notes in Computer Science Vol. 1462, H. Krawczyk, ed., Springer-Verlag, 1998, pp. 26–45.
4. M. Bellare, E. Petrank, C. Rackoff and P. Rogaway, "Authenticated key exchange in the public key model," manuscript 1995–96.
5. M. Bellare and P. Rogaway, "Entity authentication and key distribution", *Advances in Cryptology, - CRYPTO'93*, Lecture Notes in Computer Science Vol. 773, D. Stinson ed, Springer-Verlag, 1994, pp. 232-249.
6. M. Bellare and P. Rogaway, "Provably secure session key distribution– the three party case," *Annual Symposium on the Theory of Computing (STOC)*, 1995.
7. R. Bird, I. Gopal, A. Herzberg, P. Janson, S. Kutten, R. Molva and M. Yung, "Systematic design of two-party authentication protocols," *IEEE Journal on Selected Areas in Communications* (special issue on Secure Communications), 11(5):679–693, June 1993. (Preliminary version: Crypto'91.)
8. S. Blake-Wilson, D. Johnson and A. Menezes, "Key exchange protocols and their security analysis," *Proceedings of the sixth IMA International Conference on Cryptography and Coding*, 1997.
9. S. Blake-Wilson and A. Menezes, "Entity authentication and key transport protocols employing asymmetric techniques", *Security Protocols Workshop*, 1997.
10. M. Burrows, M. Abadi and R. Needham, "A logic for authentication," DEC Systems Research Center Technical Report 39, February 1990. Earlier versions in *Proceedings of the Second Conference on Theoretical Aspects of Reasoning about Knowledge*, 1988, and *Proceedings of the Twelfth ACM Symposium on Operating Systems Principles*, 1989.

11. R. Canetti, "Security and Composition of Multiparty Cryptographic Protocols", *Journal of Cryptology*, Vol. 13, No. 1, 2000.
12. R. Canetti, "A unified framework for analyzing security of Protocols", manuscript, 2000. Available at http://eprint.iacr.org/2000/067.
13. R. Canetti and H. Krawczyk, "Analysis of Key-Exchange Protocols and Their Use for Building Secure Channels (Full Version)", http://eprint.iacr.org/2001.
14. R. Canetti and H. Krawczyk, "Proving secure composition of key-exchange protocols with any application", in preparation.
15. W. Diffie and M. Hellman, "New directions in cryptography," *IEEE Trans. Info. Theory* IT-22, November 1976, pp. 644–654.
16. W. Diffie, P. van Oorschot and M. Wiener, "Authentication and authenticated key exchanges", *Designs, Codes and Cryptography*, 2, 1992, pp. 107–125.
17. O. Goldreich, *"Foundations of Cryptography (Fragments of a book)"*, Weizmann Inst. of Science, 1995. (Available at http://philby.ucsd.edu/cryptolib.html)
18. O. Goldreich, S. Goldwasser and S. Micali, "How to construct random functions," *Journal of the ACM*, Vol. 33, No. 4, 210–217, (1986).
19. S. Goldwasser, and L. Levin, "Fair Computation of General Functions in Presence of Immoral Majority", *CRYPTO '90, LNCS 537*, Springer-Verlag, 1990.
20. S. Goldwasser and S. Micali, Probabilistic encryption, *JCSS*, Vol. 28, No 2, April 1984, pp. 270-299.
21. S. Goldwasser, S. Micali and C. Rackoff, "The Knowledge Complexity of Interactive Proof Systems", *SIAM Journal on Comput.*, Vol. 18, No. 1, 1989, pp. 186-208.
22. C.G. Günther, "An identity-based key-exchange protocol", *Advances in Cryptology - EUROCRYPT'89*, Lecture Notes in Computer Science Vol. 434, Springer-Verlag, 1990, pp. 29-37.
23. D. Harkins and D. Carrel, ed., "The Internet Key Exchange (IKE)", *RFC 2409*, November 1998.
24. ISO/IEC IS 9798-3, "Entity authentication mechanisms — Part 3: Entity authentication using asymmetric techniques", 1993.
25. H. Krawczyk, "The order of encryption and authentication for protecting communications (Or: how secure is SSL?)", manuscript.
26. H. Krawczyk, "SKEME: A Versatile Secure Key Exchange Mechanism for Internet,", *Proceedings of the 1996 Internet Society Symposium on Network and Distributed System Security*, Feb. 1996, pp. 114-127.
27. P. Lincoln, J. Mitchell, M. Mitchell, A. Schedrov, "A Probabilistic Poly-time Framework for Protocol Analysis", *5th ACM Conf. on Computer and System Security*, 1998.
28. A. Menezes, P. Van Oorschot and S. Vanstone, "Handbook of Applied Cryptography," CRC Press, 1996.
29. S. Micali and P. Rogaway, "Secure Computation", unpublished manuscript, 1992. Preliminary version in *CRYPTO 91*.
30. R. Needham and M. Schroeder, "Using encryption for authentication in large networks of computers," *Communications of the ACM*, Vol. 21, No. 12, December 1978, pp. 993–999.
31. B. Pfitzmann, M. Schunter and M. Waidner, "Secure Reactive Systems", IBM Research Report RZ 3206 (#93252), IBM Research, Zurich, May 2000.
32. B. Pfitzmann and M. Waidner, "A General Framework for Formal Notions of 'Secure' System", Hildesheimer Informatik-Berichte 11/94 Institut für Informatik, Universität Hildesheim, April 1994.

33. B. Pfitzmann and M. Waidner, "A model for asynchronous reactive systems and its application to secure message transmission", IBM Research Report RZ 3304 (#93350), IBM Research, Zurich, December 2000.
34. V. Shoup, "On Formal Models for Secure Key Exchange", Theory of Cryptography Library, 1999. Available at: http://philby.ucsd.edu/cryptolib/1999/99-12.html.

A   A Comparison with [34]

Section 1.1 mentioned the work by Shoup [34] on definitions and analysis of KE protocols. This appendix further expands on some of the differences between that work and ours.

We start with a short summary of the relevant parts of [34]. Shoup's definitions are based on the simulatability approach of [2] with some significant modifications. Three levels of security are presented: *Static security* (i.e., security against adversaries that corrupt parties only at the onset of the computation), *adaptive security* (where the adversary obtains only the long-term information of a newly corrupted party) and *strongly adaptive security* where the adversary obtains all the private information of corrupted parties. (Oddly, strongly adaptive security does not imply adaptive security.) In addition, two definitions based on the indistinguishability approach of Bellare and Rogaway [5] are presented. The first is aimed at capturing security without perfect forward secrecy (PFS), and is shown to be equivalent to the static variant of the simulation-based definition. The second is aimed at capturing security with PFS, and is claimed to be equivalent to the adaptive variant of the simulation-based definition. Sufficiency of the definitions to constructing secure-channel protocols is informally argued, but is not proved nor rigorously claimed.

While the first variant of the indistinguishability-based definition is roughly equivalent to the non-PFS variant presented here (modulo the general differences mentioned below), the second variant is strictly weaker than our PFS formulation of SK-security. Specifically, the definition in [34] accepts as secure protocols that do not erase sensitive ephemeral data (e.g. protocol DHKE-1 in [34]), while the definition here treats these protocols as insecure.

There are several other technical and methodological differences between the two works that we mention next. (a) A major methodological difference is our use of the authenticated-links model and authenticators as a simplifying analysis tool. While our formalization of security does not mandate the use of this methodology we carefully build our definitions to accomodate the use of this tool. (b) Shoup allows the adversary a more general attack than session-key query, namely an *application attack* that reveals an arbitrary function of the key. Our modeling does not define this explicit attack as it turns out to be unnecessary for guaranteeing secure channels. (c) Here we consider an additional adversarial behavior that is not treated in [34]. Specifically, we protect against adversaries that obtain the internal state of corrupted sessions (even without fully corrupting the corresponding parties) by requiring that such exposure will not compromise other protocol sessions run by the same parties. This protection is not guaranteed by some protocols suggested in [34] (e.g., protocol DHKE). (d) The treatment of the interaction with the certificate authority (CA). In [34] the interaction with the CA is an integral part of every KE protocol, whereas here this interaction with the CA is treated as a separate protocol. We make this choice for further modularity and ease of proof. Yet, as we already remarked in Section 2.2, the CA protocol needs to be taken into consideration with any full specification and analysis of actual KE protocols. (e) The treatment of the session-id's. In [34] the session-id's are artificially given to the parties by the model which results, in our view, in a more cumbersome formalization of the security conditions. In contrast, here we adopt a more natural approach where the session-id's are generated by the calling protocol and security is guaranteed only when these session-id's satisfy some minimal (and easy to implement) conditions. In particular, this formalism can be satisfied by letting the parties jointly generate the session-id (as is common in practice).

Overall, we believe that the approaches in this work and in [34] are not "mutually exclusive" and both can be useful depending on a particular setting or even taste. However, for [34] to be truly useful, and for a full comparison and assessment to be possible, many of the missing definition and proof details in that work will need to be completed. Especially, rigorous proofs of protocols and a definition of secure channels is needed to assess the sufficiency of these protocols for providing the basic secure-channels functionality.

Efficient Password-Authenticated Key Exchange Using Human-Memorable Passwords

Jonathan Katz[1], Rafail Ostrovsky[2], and Moti Yung[3]

[1] Telcordia Technologies and
Department of Computer Science, Columbia University.
jkatz@cs.columbia.edu
[2] Telcordia Technologies, Inc., 445 South Street, Morristown, NJ 07960.
rafail@research.telcordia.com
[3] CertCo, Inc.
moti@cs.columbia.edu

Abstract. There has been much interest in password-authenticated key-exchange protocols which remain secure even when users choose passwords from a very small space of possible passwords (say, a dictionary of English words). Under this assumption, one must be careful to design protocols which cannot be broken using *off-line dictionary attacks* in which an adversary enumerates all possible passwords in an attempt to determine the correct one. Many heuristic protocols have been proposed to solve this important problem. Only recently have formal validations of security (namely, proofs in the idealized random oracle and ideal cipher models) been given for specific constructions [3,10,22].

Very recently, a construction based on general assumptions, secure in the standard model with human-memorable passwords, has been proposed by Goldreich and Lindell [17]. Their protocol requires no public parameters; unfortunately, it requires techniques from general multi-party computation which make it impractical. Thus, [17] only proves that solutions are possible "in principal". The main question left open by their work was finding an efficient solution to this fundamental problem.

We show an efficient, 3-round, password-authenticated key exchange protocol with human-memorable passwords which is provably secure under the Decisional Diffie-Hellman assumption, yet requires only (roughly) 8 times more computation than "standard" Diffie-Hellman key exchange [14] (which provides no authentication at all). We assume public parameters available to all parties. We stress that we work in the standard model only, and do not require a "random oracle" assumption.

1 Introduction

1.1 Background

Protocols which allow for mutual authentication of two parties and for generating a cryptographically-strong shared key between them (*authenticated key*

B. Pfitzmann (Ed.): EUROCRYPT 2001, LNCS 2045, pp. 475–494, 2001.

exchange) underly most interactions taking place on the Internet. The importance of this primitive has been realized for some time by the security community (see [11] for exhaustive references), followed by an increasing recognition that precise definitions and formalization were needed. The first formal treatments [4,6,2,20,9,28,11] were in a model in which participants already share some cryptographically-strong information: either a secret key which can be used for encryption/authentication of messages, or a public key which can be used for encryption/signing of messages. The setting arising most often in practice — in which (human) users are only capable of storing "human-memorable" passwords (*password-authenticated key exchange*) — remains much less studied, though many heuristic protocols exist. Indeed, only recently have formal definitions of security for this setting appeared [3,10,22,17].

The problem (in the standard model; i.e., without random oracles) is difficult precisely because it requires "bootstrapping" from a weak shared secret to a strong one. In fact, it is not even *a priori* clear that a solution is possible. Completeness results for multi-party computation [18] do not directly apply here due to the strong adversarial model considered (see Section 2). In particular, the adversary may ask for concurrent (arbitrarily-interleaved) executions of the protocol, may modify messages or even prevent their delivery, may impersonate participants in the protocol and act as a "man-in-the-middle", and may corrupt *all* protocol participants. Nevertheless, in a very recent paper, Goldreich and Lindell [17] have shown that in principle, this problem is solvable based on any trapdoor permutation (leaving open the question of whether a practical solution is possible). We show, perhaps somewhat surprisingly, the existence of an *efficient* solution for human-memorable passwords under the Decisional Diffie-Hellman assumption.

1.2   The Adversarial Model

The setting is as follows (a formal discussion appears in Section 2): two parties within a larger network who share a *weak* (low-entropy) password wish to authenticate each other and generate a strong session key for protecting their subsequent communication. An adversary controls all communication in the network. Thus, messages may be tampered with, delivered out-of-order, or not delivered at all; the adversary may also ask for arbitrarily-interleaved executions of the protocol. Finally, the adversary may corrupt selected instances (see below) of the participants and obtain the session keys generated by successful executions of the protocol. The adversary succeeds if he can cause a participant to compute a session key which the adversary can then distinguish from random.

Since the space of possible passwords is small, an adversary who has monitored a conversation may enumerate all possible passwords and try to match the recorded conversation to each one. As an example, any challenge-response protocol in which one party sends challenge N and the other responds with $f(\text{password}, N)$ is trivially susceptible to this attack, regardless of f (note that such an attack is *not possible* by a poly-time adversary, for appropriate choice of f, when the parties share a *high-entropy* password). Additionally, the fact that the adversary can corrupt instances and determine the actual session key means that the protocol must ensure consistency between the recorded conversation and these session keys, even while not revealing any information about the password. These complications make this problem much harder than the case in which participants already share a strong key at the outset of the protocol.

What does security mean in a model which is inherently insecure? Indeed, since passwords are chosen from a small space, an adversary can always try each possibility one at a time in an impersonation (on-line) attack. Thus, we say a protocol is secure (informally) if this exhaustive guessing is the best an adversary can do. For a real-world adversary, such on-line attacks are the hardest to mount, and they are also the easiest to detect. It is very realistic to assume that the number of on-line attacks an adversary is allowed is severely limited, while other attacks (eavesdropping, off-line password guessing) are not.

1.3   Previous Work

The problem of off-line attacks in password-authenticated protocols was first noted by Bellovin and Merritt [7], followed by a flurry of work in the security community providing additional solutions with heuristic arguments for their security (see [11] for exhaustive references). More recently, two formal models for password-authenticated key exchange have been proposed: one by Bellare, Pointcheval, and Rogaway [3], based on [4,6] with extensions suggested by [21]; and a second by Boyko, MacKenzie, and Patel [10], following [2] with extensions given in [28]. While both models have their advantages, we choose to work in the first model and review the appropriate definitions in Section 2.

These models all assume that two parties wishing to communicate share only a human-memorable password; in particular, they do not assume a public-key infrastructure (PKI) which allows participants to generate and share public keys. Definitions for security in this setting have also been proposed [20,9,28] and, in fact, the first protocols resistant to off-line dictionary attacks were given in this model. However, the requirement of a secure PKI is a strong one, and we wish to avoid it.

Only recently have formal validations of security for specific protocols appeared [3,10,22]. However, these validations are not proofs in the standard model; [3] relies on ideal ciphers, while [10,22] rely on random oracles. More recently, Goldreich and Lindell [17] have shown a protocol based on general assumptions which is secure in the standard model. Interestingly, in contrast to the present work, their protocol does *not* require public parameters. Unfortunately, their construction requires a non-constant number of rounds and also requires techniques from generic multi-party computation [18]. Thus, their scheme serves as a general plausibility result (a terminology coined in [16]), but is much too inefficient for practical use. Finally, as pointed out by the authors themselves, the solution of [17] does not allow for concurrent executions of the protocol between parties using the same password.

1.4   Our Contribution

Security validation via proofs in the random oracle and ideal cipher models are useful, as they lend a measure of confidence to protocols whose security would otherwise be only heuristic. On the other hand, proofs of security in these models do not necessarily translate to real-world security [12], so it is important to have proofs under standard cryptographic assumptions. We prove the security of our construction using only the Decisional Diffie-Hellman (DDH) assumption.

Efficiency is especially important in this setting, where security concerns are motivated by very practical considerations (human users' inability to remember long secrets). We stress that our scheme, though provably secure, is very practical even when compared to heuristically-secure protocols such as [3,10] or the original Diffie-Hellman protocol [14] (which does not provide *any* authentication). Our protocol requires only three rounds and has communication and computational complexity only (roughly) 5-8 times greater than the above solutions. Furthermore, we are able to construct our scheme without making stronger assumptions (the DDH assumption is used in [14,3,10]).

Although our solution relies on public-key techniques (in fact, this is necessary [20]) we emphasize that our protocol is not a "public-key solution" (as in [2,20,9]). In particular, we do not require *any* participant to have a public key, but instead rely on one set of common parameters shared by everyone in the system. This avoids problems associated with public key infrastructures (such as revocation, centralized trust, key management issues, etc.), and also allows new servers and clients to join the network *at any time* during execution of the protocol without requiring access to an on-line, centralized (trusted) authority (in fact, they do not even need to inform anyone else of their presence). Furthermore, no participants know the "secret key" associated with the public parameters. This eliminates the risk that compromise of a participant will compromise the security of the entire system.

The construction given here is secure under both the notion of basic security and the stronger notion of "forward security" (in the weak corruption model). In this initial version we concentrate on basic security only, and leave the topic of forward security for the final version.

2   Model and Definitions

The reader is assumed to be familiar with the model of [3], which is the model in which we prove security of our protocol. For completeness, we review the main points of their definition here, and refer the reader to [3] for more details.

PRINCIPALS, PASSWORDS, AND INITIALIZATION. We have a fixed set of protocol participants (*principals*) each of which is either a client $C \in$ Client or a server $S \in$ Server (Client and Server are disjoint). We let User $\stackrel{\text{def}}{=}$ Client $\cup$ Server. Each $C \in$ Client has a password pw_C. Each $S \in$ Server has a vector $PW_S = \langle pw_C \rangle_{C \in \text{Client}}$ which contains the passwords of each of the clients (we assume that all clients share passwords with all servers). Recall that pw_C is what client C remembers to log in; therefore, it is assumed to be chosen from a relatively small space of possible passwords.

Before the protocol is run, an initialization phase occurs during which public parameters are set and passwords are chosen for each client. We assume that passwords for each client are chosen independently and uniformly[1] at random from the set $\{1, \ldots, N\}$, where N is a constant, independent of the security parameter.

EXECUTION OF THE PROTOCOL. In the real world, protocol P determines how principals behave in response to signals (input) from their environment. Each principal is able to execute the protocol multiple times with different partners; this is modeled by allowing each principal an unlimited number of *instances* in which to execute the protocol (see [6]). We denote instance i of user U as Π_U^i. A given instance is used only once. The adversary is assumed to have complete control over all communication in the network. Thus, the adversary's interaction with the principals is modeled via access to oracles whose inputs may range over $U \in$ User and $i \in \mathbb{N}$; this allows the adversary to "interact with" different instances. Global state is maintained throughout the entire execution for each instance with which the adversary interacts (this global state is not directly visible to the adversary); the global state for an instance may be updated by an oracle during an oracle call, and the oracle's output may depend upon this state. The oracle types, as defined in [3], are:

- Send(U, i, M) — This sends message M to instance Π_U^i. The oracle runs this instance as in a real execution, maintaining state as appropriate. The output of Π_U^i is given to the adversary in addition to other information; see [3].
- Execute(C, i, S, j) — This oracle executes the protocol between instances Π_C^i and Π_S^j, where $C \in$ Client and $S \in$ Server, and outputs a transcript of this execution. This transcript includes everything an adversary would see when eavesdropping on a real-world execution of the protocol, as well as other information; see [3].
- Reveal(U, i) — This outputs the session key sk_U^i (stored as part of the global state) of instance Π_U^i.
- Test(U, i) — This query is allowed only once, at any time during the adversary's execution. A random bit b is generated; if $b - 1$ the adversary is given sk_U^i, and if $b = 0$ the adversary is given a random session key.

ADVANTAGE OF THE ADVERSARY. Event Succ occurs (adversary $\mathcal{A}$ *succeeds*) if she asks a single Test query, outputs a bit b', and $b' = b$ (where b is the bit chosen by the Test oracle). The advantage of $\mathcal{A}$ in attacking protocol P, is defined as as $\mathsf{Adv}_{P,\mathcal{A}}^{\mathrm{ake}} \stackrel{\mathrm{def}}{=} 2\Pr[\mathsf{Succ}] - 1$. If the adversary were unrestricted, success would be trivial (since the adversary could submit a Reveal query for the same instance submitted to the Test oracle). Clearly, some restrictions must be imposed. Before describing these, we formalize the idea of *partnering*. Intuitively, instances Π_U^i

---

[1] This is for ease of presentation only, as our analysis can be extended easily to handle arbitrary distributions, including users with inter-dependent passwords.

and $\Pi^j_{U'}$ are partnered if they have jointly run protocol P. Formally, we define a *session-id* (*sid*) for each instance, and say that two instances are partnered if they hold the same *sid* (which is not null). Here, we define the *sid* as the concatenation of all messages sent and received by an instance (i.e., a transcript of the execution). The following restriction may now be imposed on an adversary whose Test query is (U, i): that a Reveal query may not be called on (U, i) or on (U', j), where $\Pi^j_{U'}$ is partnered with Π^i_U. Furthermore, instance Π^i_U must have completed execution, and therefore have a non-null session key defined.

A poly-time adversary will be able to break any protocol by attempting to impersonate a user and trying all passwords one-by-one (the size of the password space is independent of the security parameter — indeed, this is what distinguishes the problem from that of [4,6]). So, we say that a given protocol is *secure* when this kind of attack is the best an adversary can do. More formally, let q_{send} be the number of calls the adversary makes to the Send oracle. A protocol is secure if, when passwords are chosen from a dictionary of size N, the adversary's advantage in attacking the protocol is bounded by

$$\mathcal{O}(q_{\text{send}}/N) + \varepsilon(k),$$

for some negligible function $\varepsilon(\cdot)$. The first term represents the fact that the adversary can (essentially) do no better than guess a password during each call to the Send oracle[2]. In particular, even polynomially-many calls to the Execute oracle (i.e., passive observations of valid executions) and the Reveal oracle (i.e., compromise of short-term session keys) are of no help to an adversary; only on-line impersonation attacks (which are harder to mount and easier to detect) give the adversary a non-negligible advantage.

Concrete security is particularly important in this setting since the adversary's advantage is non-negligible (assuming Send queries are made). We quantify an adversary's maximum advantage as a function of the adversary's running time t and the number of queries made to the Send, Execute, and Reveal oracles ($q_{\text{send}}, q_{\text{execute}},$ and q_{reveal} respectively).

3   A Provably Secure Protocol for Password-AKE

3.1   Building Blocks

Our protocol and proof of security rely on a number of building blocks. First, our protocol uses the Cramer-Shoup cryptosystem [13] which is secure under adaptive chosen-ciphertext attack. Actually, we require an extension of the Cramer-Shoup cryptosystem, which remains secure under adaptive chosen-ciphertext attack. Our extension defines two "types" of encryption: client-encryption and

---

[2] A tighter definition of security would require that the adversary's advantage be bounded by $q_{\text{send}}/rN + \varepsilon(k)$, where r is the minimum number of messages an adversary needs to send in order to cause (completion of the protocol and) a non-null session key to be defined. An analysis of our proof proof indicates that the security of our construction is indeed tight in this respect.

server-encryption. Details appear in Appendix B. We will also need a one-time signature scheme [15] secure against existential forgery [19]. Finally, our proof of security relies on the Decisional Diffie-Hellman (DDH) assumption [14,8] (note that the security of the Cramer-Shoup cryptosystem requires the DDH assumption already). We review these components in Appendix A, and also explicitly quantify their (in)security which is necessary for an explicit analysis of the adversary's maximum advantage in attacking the key exchange protocol.

Chosen-ciphertext-secure encryption has been used previously in the context of secure key exchange [2,20,9,28]. However, as pointed out above, our protocol differs from these works in that it does *not* require the assumption of a public-key infrastructure, and no participant holds a secret key or publishes a public key. Indeed, "decryption" is never performed during execution of our protocol.

3.2   The Protocol

A high-level description of the protocol is given in Figure 1. Let p, q be primes such that $q|p-1$, and let $\mathcal{G}$ be a subgroup of $\mathbb{Z}_p^*$ of order q in which the DDH assumption holds. During the initialization phase, generators $g_1, g_2, h, c, d \in \mathcal{G}$ and a function $\mathcal{H}$ from a family of universal one-way hash functions [23] (which can be based on any one-way function [26]) are chosen at random and published. Note that this public information is *not* an added assumption[3]; "standard" Diffie-Hellman key exchange [14] typically assumes that parties use a fixed generator g (although this is not necessary), and [3,10] seem to require a public generator g for their proofs of security. However, we *do* require that no one know the discrete logarithms of any of the generators with respect to any other, and thus we need either a trusted party who generates the public information or else a source of randomness which can be used to publicly derive the information.

As part of the initialization phase, passwords are chosen randomly for each client. We assume that all passwords lie in (or can be mapped to) $\mathbb{Z}_q$. For typical values of $|q|$, this will be a valid assumption for human-memorable passwords.

Execution of the protocol is as follows (see Figure 1): When client C wants to connect to server S, the client first runs the key generation algorithm for the one-time signature scheme, giving VK and SK. Then, the client computes a client-encryption (see Appendix B) of $g_1^{pw_C}$. This, along with the client's name, is sent to the server as the first message. The server chooses random elements x_2, y_2, z_2, w_2 from $\mathbb{Z}_q$, computes α' using the first message, and forms $g_1^{x_2} g_2^{y_2} h^{z_2} (cd^{\alpha'})^{w_2}$. The server then computes a server-encryption (see Appendix B) of $g_1^{pw_C}$. This is sent back to the client as the second message. The client selects random elements x_1, y_1, z_1, w_1 from $\mathbb{Z}_q$, computes β' using the second message, and forms $K = g_1^{x_1} g_2^{y_1} h^{z_1} (cd^{\beta'})^{w_1}$. Finally, β' and K are signed using the signing key which was generated in the first step. The *sid* is defined as the transcript of the entire conversation. A formal description of the protocol appears in Appendix C.

---

[3] The protocols of [22,17], however, do not require any public information.

Public information: $p, q, g_1, g_2, h, c, d, \mathcal{H}$

Client   Server

$(\mathsf{VK}, \mathsf{SK}) \leftarrow \mathsf{SigGen}(1^k)$

$r_1 \leftarrow \mathbb{Z}_q$

$A = g_1^{r_1}; B = g_2^{r_1}$

$C = h^{r_1} g_1^{pw_C}$

$\alpha = \mathcal{H}(Client \,|\mathsf{VK}|A|B|C)$

$D = (cd^\alpha)^{r_1}$ $\underline{\quad Client \mid \mathsf{VK} \mid A \mid B \mid C \mid D \quad}\rightarrow$

$\qquad\qquad\qquad\qquad\qquad\qquad x_2, y_2, z_2, w_2, r_2 \leftarrow \mathbb{Z}_q$

$\qquad\qquad\qquad\qquad\qquad\qquad \alpha' = \mathcal{H}(Client \,|\mathsf{VK}|A|B|C)$

$\qquad\qquad\qquad\qquad\qquad\qquad E = g_1^{x_2} g_2^{y_2} h^{z_2} (cd^{\alpha'})^{w_2}$

$\qquad\qquad\qquad\qquad\qquad\qquad F = g_1^{r_2}; G = g_2^{r_2}$

$\qquad\qquad\qquad\qquad\qquad\qquad I = h^{r_2} g_1^{pw_C}$

$\qquad\qquad\qquad\qquad\qquad\qquad \beta = \mathcal{H}(Server \,|E|F|G|I)$

$\underline{\quad Server \mid E \mid F \mid G \mid I \mid J \quad}\leftarrow$ $J = (cd^\beta)^{r_2}$

$x_1, y_1, z_1, w_1 \leftarrow \mathbb{Z}_q$

$\beta' = \mathcal{H}(Server \,|E|F|G|I)$

$K = g_1^{x_1} g_2^{y_1} h^{z_1} (cd^{\beta'})^{w_1}$

$\mathsf{Sig} = \mathsf{Sign}_{\mathsf{SK}}(\beta' \mid K)$ $\underline{\qquad\quad K \mid \mathsf{Sig} \qquad\quad}\rightarrow$

$\qquad\qquad\qquad\qquad\qquad\qquad \text{if } \mathsf{Verify}_{\mathsf{VK}}((\beta \mid K), \mathsf{Sig}) = 1$

$\qquad\qquad\qquad\qquad\qquad\qquad C' = C/g_1^{pw_C}$

$I' = I/g_1^{pw_C}$ $\qquad\qquad\qquad sk_S = K^{r_2} A^{x_2} B^{y_2} (C')^{z_2} D^{w_2}$

$sk_C = E^{r_1} F^{x_1} G^{y_1} (I')^{z_1} J^{w_1}$ $\qquad\qquad \text{else } sk_S \leftarrow \mathcal{G}$

Fig. 1. The protocol for password-AKE. See text for details.

The protocol description in Figure 1 omits many implementation details which are important for the proof of security to hold. Most important is for both client and server to perform a "validity check" on the messages they receive. In particular, each side should check that the values they receive are actually in the group $\mathcal{G}$ and are not the identity (in other words, it is required to check that the group elements indeed have order q). Note that such validity checks are required even for chosen-ciphertext security of the underlying Cramer-Shoup cryptosystem.

CORRECTNESS. In an honest execution of the protocol, C and S calculate identical session keys. To see this, first note that $\alpha = \alpha'$ and $\beta = \beta'$ in an honest execution. Then:

$$sk_C = (g_1^{x_2} g_2^{y_2} h^{z_2} (cd^\alpha)^{w_2})^{r_1} g_1^{r_2 x_1} g_2^{r_2 y_1} h^{r_2 z_1} (cd^\beta)^{r_2 w_1}$$

and

$$sk_S = (g_1^{x_1} g_2^{y_1} h^{z_1} (cd^\beta)^{w_1})^{r_2} g_1^{r_1 x_2} g_2^{r_1 y_2} h^{r_1 z_2} (cd^\alpha)^{r_1 w_2},$$

and one can verify that these are equal.

MUTUAL AUTHENTICATION. We note that the protocol as presented above achieves key exchange only, and not mutual authentication. However, we can trivially add mutual authentication by adding a fourth message to the protocol. Details will appear in the final version.

3.3 Practical Considerations

In practice, a collision resistant hash function (say, SHA-1) can be used instead of a universal one-way hash function. This has the advantage of increased efficiency, at the expense of requiring a (possibly) stronger assumption for security.

Efficient one-time signatures [15] can be based on (presumed) one-way functions like SHA-1 or DES. In particular, one-time signatures are much more efficient than signature schemes which are secure against adaptive (polynomially-many) chosen message attacks.

Client computation can be reduced (which is important when the client is smartcard-based) as follows: instead of using a one-time signature scheme where fresh keys need to be generated each time a connection is made, a signing key/verification key can be generated once (upon initialization) and used for the lifetime of the client. Particularly suited for such applications are "on-the-fly" signature schemes such as [27,24,25]. This initialization step may be done by a host computer (with the keys then downloaded to the smartcard) or this step may be done off-line before the first connection is made. The proof of security given in Section 4 still holds. The disadvantage is that this signature scheme is now required to be secure against existential forgeries even when polynomially-many messages are signed (and not just a single message). In some cases, however, this tradeoff may be acceptable.

Finally, note that we may store $g_1^{pw_C}$ at the server instead of pw_C and thereby avoid computing the exponentiation each time the protocol is executed.

4 Security of the Protocol

We concentrate here on the basic security of the protocol, and leave the corresponding results about forward security to the full paper. The following theorem indicates that the protocol is secure, since all lower order terms are negligible in k (see Appendix A for definitions of the lower order terms).

Theorem 1. *Let P be the protocol of Figure 1, where passwords are chosen from a dictionary of size N, and let $k = |q|$ be the security parameter. Let A be an adversary which runs in time t and asks $q_{execute}$, q_{send}, and q_{reveal} queries to the respective oracles. Then:*

$$\mathsf{Adv}_{P,\mathcal{A}}^{\mathrm{ake}} < \frac{q_{\mathrm{send}}}{2N} + 2q_{\mathrm{send}}\varepsilon_{\mathrm{sig}}(k,t) + 2\varepsilon_{\mathrm{ddh}}(k,t) + 2q_{\mathrm{send}}\varepsilon_{\mathrm{cs}}(k,t,q_{\mathrm{send}}/2)$$
$$+ \frac{min\{2q_{\mathrm{reveal}}, q_{\mathrm{send}}\}}{q} + \frac{2\,min\{q_{\mathrm{reveal}}, q_{\mathrm{execute}}\}}{q^2}.$$

It will be helpful to develop some intuition and notation before presentation of the full proof. First, note that the Execute oracle cannot help the adversary. The reason is that Diffie-Hellman key exchange [14] forms the "heart" of this protocol, and this is secure under a passive attack.

Next, consider active "impersonation attacks" by the adversary. The protocol has three flows. When an adversary tries to impersonate a client (in an attempt to determine the eventual session key of a server), the adversary must send the first and third messages; when the adversary wants to impersonate a server (in an attempt to determine the eventual session key of a client), the adversary must "prompt" the client to generate the first message and must then send the second message. Consider an adversary impersonating a client, and let the first message (which comes from the adversary) be $\langle Client|\mathsf{VK}|A|B|C|D\rangle$. We say this message is *valid* if:

$$\log_{g_1} A = \log_{g_2} B = \log_h(C/g_1^{pw_C}) = \log_{cd^{\alpha'}} D, \tag{1}$$

where $\alpha' = \mathcal{H}(Client, \mathsf{VK}, A, B, C)$, and pw_C is the password for *Client*. We define *valid* analogously for the second message of an adversary impersonating a server (note that here the password which determines validity depends upon the name of the client to which the adversary sends the message). We do not define any notion of validity for the third message. The following fact is central to our proof:

Fact 1 *When an invalid message is sent to an instance, the session key computed by that instance is* information-theoretically *independent of all messages sent and received by that instance. This holds for both clients and servers.*

Proof. Consider the case of an adversary interacting with a server, with the first message as above. Let $\theta_1 \stackrel{\mathrm{def}}{=} \log_{g_1} g_2; \theta_2 \stackrel{\mathrm{def}}{=} \log_{g_1} h$; and $\theta_3 \stackrel{\mathrm{def}}{=} \log_{g_1}(cd^{\alpha'})$. Consider the random values x_2, y_2, z_2, w_2 (see Figure 1) used by the server instance during its execution. Element E of the second message constrains these values as follows:

$$\log_{g_1} E = x_2 + y_2\theta_1 + z_2\theta_2 + w_2\theta_3. \tag{2}$$

The session key is calculated as K^{r_2} multiplied by $sk_S' = A^{x_2}B^{y_2}(C/g_1^{pw_C})^{z_2}D^{w_2}$. But we have:

$$\log_{g_1} sk_S' = x_2\log_{g_1} A + y_2\theta_1\log_{g_2} B + z_2\theta_2\log_h(C/g_1^{pw_C}) + w_2\theta_3\log_{cd^{\alpha'}} D. (3)$$

When equation (1) does not hold (i.e., the message is invalid), equations (2) and (3) are linearly independent and $sk_S' \in_R \mathcal{G}$ is information-theoretically independent of the transcript of the execution. A similar argument holds for the case of an adversary interacting with a client. ∎

Let Π_U^i be an instance to which the adversary has sent an invalid message. Fact 1 implies that the adversary has advantage 0 in distinguishing the session key generated by this instance from a random session key. Thus, an adversary's (non-zero) advantage can come about only by sending a valid message to an instance.

We call a message sent by an adversary *previously-used* if the message was previously output by a client or server running the protocol (that is, the adversary has simply "copied" and re-used the message), and is *new* otherwise. The following lemma bounds the adversary's probability of coming up with a new, valid first or second message:

Lemma 1. *An adversary's probability of sending, at any point during the protocol, a first or second message which is both new and valid is bounded by* $\mathcal{O}(q_{\text{send}}/N) + \varepsilon(k)$, *for some negligible function* $\varepsilon(\cdot)$.

This lemma essentially follows from the chosen-ciphertext security (and hence non-malleability) of extended Cramer-Shoup encryption (see [13] and Appendix B). Detail appear in the full proof, below. The lemma reflects the fact that the adversary can (trivially) "guess" the appropriate password[4] each time he sends a first or second message.

The only remaining point to argue is that previously-used messages cannot significantly help the adversary. First note that if an adversary re-uses a first message, the adversary will (with high probability) not be able to compute a valid signature to include with the third message. If an adversary re-uses a second message, the full proof indicates that without knowing the randomness used to generate that message, the adversary will gain only negligible advantage.

Proof (of Theorem 1). We refer to the formal specification of the protocol as it appears in Appendix C. The number of clients and servers is polynomial in the security parameter, and this number is fixed in advance[5] and public.

We imagine a simulator who controls all oracles to which the adversary has access. The simulator runs the protocol initialization as described in Appendix C, Figure 2, including selecting passwords for each client[6]. The simulator answers the adversary's oracle queries as defined in Appendix C, Figures 3 and 4. The adversary succeeds if it can guess the bit h that the simulator uses during the Test query (see Section 1 for additional details).

We define a sequence of transformations $P_1, \ldots$ to the original protocol P_0, and bound the effect each transformation has on the adversary's advantage.

---

[4] The lemma assumes that passwords are chosen uniformly at random from the password space, but can be appropriately modified to handle arbitrary distributions.

[5] As mentioned in Section 1.4, clients and servers can in fact be dynamically added to the protocol *during execution* at the request of the adversary (and even with passwords chosen by the adversary, when forward security is considered). For simplicity, we focus on the static case.

[6] For simplicity we assume that users choose passwords independently and with uniform distribution. The analysis can easily be modified to accommodate arbitrary distributions.

Then, we bound the adversary's advantage in the final (transformed) protocol; this gives an explicit bound on the adversary's advantage in the original protocol.

Consider the verification keys output by the Send_0 oracle during the course of the protocol. We may restrict ourselves to the case where the adversary is unable to forge a new message/signature pair for any of these keys during the course of the protocol. This can change the adversary's success probability (as a simple hybrid argument shows) by at most $q_{\mathsf{send}_0} \varepsilon_{\mathsf{sig}}(k, t) \leq q_{\mathsf{send}} \varepsilon_{\mathsf{sig}}(k, t)$.

In protocol P_1, calls to the $\mathsf{Execute}$ oracle are answered as before, except that C and I are chosen at random from $\mathcal{G}$. The following bounds the effect on the adversary's advantage:

Lemma 2. *The adversary's success probability in P_1 differs by at most $\varepsilon_{\mathsf{ddh}}(k, t)$ from its advantage in P_0.*

Proof. The simulator uses the adversary as a black box to distinguish Diffie-Hellman quadruples from random quadruples. Given quadruple (g, h, s, t) and group $\mathcal{G}$, it runs the initialization as follows:

$$a, b, \ell \leftarrow \mathbb{Z}_q$$
$$g_1 = g; g_2 = g^a; c = g^b, d = g^\ell$$
$$\mathcal{H} \leftarrow \mathsf{UOWH}$$

Publish parameters $(q, g_1, g_2, h, c, d, \mathcal{H})$ and group $\mathcal{G}$
$$\langle pw_C \rangle_{C \in \mathsf{Client}} \leftarrow \{1, \ldots, N\}$$

By a random self-reducibility property [28,1], the simulator can generate s_T, t_T (for $T = 1, \ldots$) such that, if (g, h, s, t) is a Diffie-Hellman quadruple, so is (g, h, s_T, t_T); on the other hand, if (g, h, s, t) is a random quadruple, then (g, h, s_T, t_T) is distributed among random quadruples with g and h fixed. The T-th call to $\mathsf{Execute}$ is answered as:

$\mathsf{Execute}(\textit{Client}, i, \textit{Server}, j)$ —
$$(\mathsf{VK}, \mathsf{SK}) \xleftarrow{R} \mathsf{SigGen}(1^k) \qquad x_1, x_2, y_1, y_2, z_1, z_2, w_1, w_2 \xleftarrow{R} \mathbb{Z}_q$$
$$A = s_{2T}; \; B = s_{2T}^a; \; C = t_{2T} \cdot g_1^{pw_C} \qquad \alpha = \mathcal{H}(\textit{Client} \mid \mathsf{VK} | A | B | C)$$
$$D = s_{2T}^{b+a\ell} \qquad msg\text{-}out_1 \longleftarrow \langle \textit{Client} \mid \mathsf{VK} \mid A \mid B \mid C \mid D \rangle$$
$$E = g_1^{x_1} g_2^{x_1} h^{z_1} (cd^\alpha)^{w_1} \qquad F = s_{2T+1}; \; G = s_{2t+1}^a; \; I = t_{2T+1} \cdot g_1^{pw_C}$$
$$\beta = \mathcal{H}(\textit{Server} | E | F | G | I)$$
$$J = s_{2T+1}^{b+\beta\ell} \qquad msg\text{-}out_2 \longleftarrow \langle \textit{Server} \mid E \mid F \mid G \mid I \mid J \rangle$$
$$K = g_1^{x_2} g_2^{y_2} h^{z_2} (cd^\beta)^{w_2} \qquad msg\text{-}out_3 \longleftarrow \langle K \mid \mathsf{Sign}_{\mathsf{SK}}(\beta | K) \rangle$$
$$sk_S^j \leftarrow sk_C^i \leftarrow A^{x_1} B^{y_1} (C \cdot g_1^{-pw_C})^{z_1} D^{w_1} F^{x_2} G^{y_2} (I \cdot g_1^{-pw_C})^{z_2} J^{w_2}$$
$$sid_S^j \leftarrow sid_C^i \leftarrow \langle msg\text{-}out_1 \mid msg\text{-}out_2 \mid msg\text{-}out_3 \rangle$$
return $\langle msg\text{-}out_1, msg\text{-}out_2, msg\text{-}out_3 \rangle$

If (g, h, s, t) is a Diffie-Hellman quadruple, this is an exact simulation of P_0; on the other hand, if it is a random quadruple, this is an exact simulation of P_1. ∎

In protocol P_2, calls to $\mathsf{Execute}$ are answered as before except that the session key is chosen randomly from $\mathcal{G}$. The adversary's view (and thus its success probability) is within statistical distance $\min\{q_{\mathsf{reveal}}, q_{\mathsf{execute}}\}/q^2$ from the adversary's view in protocol P_1. Indeed, Fact 1 shows that the session key is independent of the transcript of the execution seen by the adversary whenever $msg\text{-}out_1$ or $msg\text{-}out_2$ are not valid (for the appropriate password). But when C and I are chosen randomly, the probability that both $msg\text{-}out_1$ and $msg\text{-}out_2$ are valid is exactly $1/q^2$.

In protocol P_3, the public parameters are generated by choosing g_1 and g_2 randomly from $\mathcal{G}$, then choosing x_1, x_2, y_1, y_2, and z randomly from $\mathbb{Z}_q$ and setting $c = g_1^{x_1} g_2^{x_2}$, $d = g_1^{y_1} g_2^{y_2}$, and $h = g_1^z$; $\mathcal{H}$ is chosen as before. Furthermore, the Send_3 oracle is changed as follows: the simulator first checks whether *first-msg-in* (which was the message sent to the Send_1 oracle for the same instance) is previously-used (see above). If so, the current query to the Send_3 oracle is answered normally. Otherwise, let *first-msg-in* $= \langle Client|\mathsf{VK}|A|B|C|D\rangle$. The simulator computes $\alpha = \mathcal{H}(Client|\mathsf{VK}|A|B|C)$ and checks whether $A^{x_1+y_1\alpha} B^{x_2+y_2\alpha} = D$ and $g_1^{pw_C} A^z = C$. If so, *first-msg-in* is said to *appear valid*, and the query is answered normally. If not, *first-msg-in* is said to *appear non-valid*, and the query is answered normally except that the session key is chosen randomly from $\mathcal{G}$.

Calls to $\mathsf{Send}_2(Client, i, msg\text{-}in)$ are answered in similar fashion. If *msg-in* is previously-used, the query is answered normally. Otherwise, let $msg\text{-}in = \langle Server|E|F|G|I|J\rangle$. The simulator computes $\beta=\mathcal{H}(Server|E|F|G|I)$ and checks whether $F^{x_1+y_1\beta} G^{x_2+y_2\beta} = J$ and $g_1^{pw_C} F^z = I$. If so, *msg-in* is said to *appear valid*, and the query is answered normally. If not, *msg-in* is said to *appear non-valid*, and the query is answered normally but the session key for instance Π_C^i is chosen randomly from $\mathcal{G}$.

The adversary's view of this protocol is exactly equivalent to its view of protocol P_2. When *first-msg-in* or *msg-in* appear non-valid, they are in fact not valid for password pw_C, and Fact 1 shows that the resulting session key is independent of the adversary's view. On the other hand, a message which appears valid may in fact be invalid, but since the query is answered normally the adversary's view is not affected.

In protocol P_4, the definition of the adversary's success is changed:

- If, during the course of answering a Send_3 oracle query, *first-msg-in* is new and appears valid, the session key is set to the special value ∇. If the adversary ever asks a Reveal query for this instance, the simulator halts immediately and the adversary succeeds.
- If, during the course of answering a Send_2 oracle query, *msg-in* is new and appears valid, the session key is set to the special value ∇. If the adversary ever asks a Reveal query for this instance, the simulator halts immediately and the adversary succeeds.
- Otherwise, the adversary succeeds, as before, by guessing the bit b.

This can only increase the advantage of the adversary.

In protocol P_5, calculation of the session key by the Send_3 oracle is modified. First, every time K is computed by the simulator when answering a call to the Send_2 oracle, the simulator stores K along with its associated values of x, y, z, w. When a call is made to the Send_3 oracle with $msg\text{-}in = \langle K|\mathsf{Sig}\rangle$, there are four possibilities:

- *first-msg-in* is new and appears valid. In this case the session key is set to ∇ and the simulator behaves as in P_4 (above).
- *first-msg-in* is new and appears non-valid. In this case, the simulator chooses the session key randomly (as in P_3, P_4).
- *first-msg-in* is previously-used and $\mathsf{Verify}_{\mathsf{VK}}((\beta|K), \mathsf{Sig}) = 0$. In this case, the simulator chooses the session key randomly (as in $P_0, \ldots, P_4$).
- *first-msg-in* is previously-used and $\mathsf{Verify}_{\mathsf{VK}}((\beta|K), \mathsf{Sig}) = 1$. It must be the case that $\langle K, \mathsf{Sig} \rangle$ was previously output by the Send_2 oracle (since we assume the adversary has not forged any new message/signature pairs). The simulator therefore knows values x', y', z', w' such that $K = g_1^{x'} g_2^{y'} h^{z'} (cd^\beta)^{w'}$. Let *first-msg-out* $= \langle Server|E|F|G|I|J \rangle$ and $I^* = I \cdot g_1^{-pw_C}$. The simulator calculates the session key as:

$$sk_S^i \longleftarrow A^x B^y (C^*)^z D^w F^{x'} G^{y'} (I^*)^{z'} J^{w'}.$$

The adversary's view is exactly equivalent to the adversary's view in P_4 (since K^r does equal $F^{x'} G^{y'} (I^*)^{z'} J^{w'}$ when *first-msg-out* is a valid message; it is valid since it was generated by the simulator who knows the appropriate password).

In protocol P_6 we change oracle Send_1 so that component I is chosen at random from $\mathcal{G}$. This cannot change the adversary's success probability by more than $q_{\mathsf{send}_1} \varepsilon_{\mathsf{cs}}(k, t, q_{\mathsf{send}_2} + q_{\mathsf{send}_3})$. If it did, the simulator could break extended-CS encryption under a chosen ciphertext attack as follows: parameters for extended-CS encryption become the public parameters for the protocol. During the course of the protocol, the simulator may determine whether a new message appears valid by submitting it to the decryption oracle and checking whether the returned plaintext is equal to the appropriate password. When calls to the Send_1 oracle are made, the simulator submits the appropriate password as the plaintext along with the server name, the value α, and a request for a server-encryption (see Appendix B). In return, the simulator is given $\langle Server|E|F|G|I|J \rangle$ (which may be an encryption of either the appropriate password or a random group element) along with x, y, z, w such that $E = g_1^x g_2^y h^z (cd^\alpha)^w$. A simple hybrid argument bounds the change in the adversary's success probability.

In protocol P_7, the Send_2 oracle is changed so that whenever *msg-in* was previously-used the session key is chosen at random from $\mathcal{G}$. To ensure consistency, the $\mathsf{Send}_3(Server, i, *)$ oracle is changed as follows: if sid_S^i matches sid_C^j for some other instance Π_C^j, then sk_S^i is set equal to sk_C^j. The statistical difference between the adversary's view in this protocol and the previous one is bounded by $\min\{q_{\mathsf{reveal}}, q_{\mathsf{send}_2}\}/q$. Indeed, Fact 1 shows that the views are equivalent when *msg-in* is invalid. Furthermore, the probability that *msg-in* is valid for the appropriate password is $1/q$ (since I was chosen at random).

In protocol P_8, the Send_0 oracle is changed so that component C is chosen randomly from $\mathcal{G}$. Following a similar analysis to that of protocol P_6, this cannot change the adversary's success probability by more than $q_{\mathsf{send}_0} \varepsilon_{\mathsf{cs}}(k, t, q_{\mathsf{send}_2} + q_{\mathsf{send}_3})$. Finally, in protocol P_9 the Send_3 oracle is changed so that a random session key is chosen when *first-msg-in* is previously-used. Following a similar analysis to that of protocol P_7, the statistical difference between the adversary's view in this protocol and the previous protocol is bounded by $\min\{q_{\text{reveal}}, q_{\text{send}_3}\}/q$.

Consider the adversary's advantage in protocol P_9. The adversary's view is entirely independent of the passwords chosen by the simulator unless the adversary manages to submit a new *msg-in* which appears valid at some point during execution of the protocol; i.e., succeeds in guessing the password. The adversary's probability of guessing the password, however, is precisely $(q_{\text{send}_2} + q_{\text{send}_3})/N$ (this assumes that passwords are selected uniformly; an analogous calculation can be done when this is not the case). The adversary's advantage in protocol P_9 is thus bounded by $q_{\text{send}}/2N$ (note that the adversary must ask a q_{send_0} query for a q_{send_2} query to be meaningful, and similarly must ask a q_{send_1} query for a q_{send_3} query to be meaningful). The adversary's advantage in the original protocol is therefore bounded by the expression in Theorem 1. ∎

Acknowledgments

Thanks to Yehuda Lindell, Philip MacKenzie, and Steven Myers for many helpful discussions on the topic of password-authenticated key exchange.

References

1. M. Bellare, A. Boldyreva, and S. Micali. Public-Key Encryption in a Multi-User Setting: Security Proofs and Improvements. Eurocrypt 2000.
2. M. Bellare, R. Canetti, and H. Krawczyk. A Modular Approach to the Design and Analysis of Authentication and Key Exchange Protocols. STOC '98.
3. M. Bellare, D. Pointcheval, and P. Rogaway. Authenticated Key Exchange Secure Against Dictionary Attacks. Eurocrypt 2000.
4. M. Bellare and P. Rogaway. Entity Authentication and Key Distribution. Crypto '93.
5. M. Bellare and P. Rogaway. Random Oracles are Practical: A Paradigm for Designing Efficient Protocols. ACM CCCS '93.
6. M. Bellare and P. Rogaway. Provably Secure Session Key Distribution: the Three Party Case. STOC '95.
7. S. Bellovin and M. Merritt. Encrypted Key Exchange: Password-Based Protocols Secure against Dictionary Attacks. IEEE Symposium on Security and Privacy, 1992.
8. D. Boneh. The Decision Diffie-Hellman Problem. Proceedings of the Third Algorithmic Number Theory Symposium, 1998.
9. M. Boyarsky. Public-Key Cryptography and Password Protocols: The Multi-User Case. ACM CCCS '99.
10. V. Boyko, P. MacKenzie, and S. Patel. Provably Secure Password-Authenticated Key Exchange Using Diffie-Hellman. Eurocrypt 2000.
11. V. Boyko. On All-or-Nothing Transforms and Password-Authenticated Key Exchange Protocols. PhD Thesis, MIT, Department of Electrical Engineering and Computer Science, Cambridge, MA, 2000.
12. R. Canetti, O. Goldreich, and S. Halevi. The Random Oracle Methodology, Revisited. STOC '98.

13. R. Cramer and V. Shoup. A Practical Public Key Cryptosystem Provably Secure Against Chosen Ciphertext Attack. Crypto '98.
14. W. Diffie and M. Hellman. New Directions in Cryptography. *IEEE Trans. Info. Theory*, 22(6): 644–654, 1976.
15. S. Even, O. Goldreich, and S. Micali. On-Line/Off-Line Digital Signatures. Crypto '89.
16. O. Goldreich. On the Foundations of Modern Cryptography. Crypto '97.
17. O. Goldreich and Y. Lindell. Personal Communication and Crypto 2000 Rump Session. Session-Key Generation using Human Passwords Only. Available at http://eprint.iacr.org/2000/057.
18. O. Goldreich, S. Micali, and A. Wigderson. How to Play Any Mental Game, or a Completeness Theorem for Protocols with an Honest Majority. STOC '87.
19. S. Goldwasser, R. Rivest, and S. Micali. A Digital Signature Scheme Secure Against Adaptive Chosen Message Attacks. SIAM J. Comp. 17(2): 281–308, 1988.
20. S. Halevi and H. Krawczyk. Public-Key Cryptography and Password Protocols. ACM Transactions on Information and System Security, 2(3): 230–268, 1999.
21. S. Lucks. Open Key Exchange: How to Defeat Dictionary Attacks Without Encrypting Public Keys. Proceedings of the Workshop on Security Protocols, 1997.
22. P. MacKenzie, S. Patel, and R. Swaminathan. Password-Authenticated Key Exchange Based on RSA. Asiacrypt 2000.
23. M. Naor and M. Yung. Universal One-Way Hash Functions and Their Cryptographic Applications. STOC '89.
24. G. Poupard and J. Stern. Security Analysis of a Practical "on the fly" Authentication and Signature Generation. Eurocrypt '98.
25. G. Poupard and J. Stern. On the Fly Signatures Based on Factoring. ACM CCCS '99.
26. J. Rompel. One-Way Functions are Necessary and Sufficient for Secure Signatures. STOC '90
27. C.-P. Schnorr. Efficient Signature Generation by Smartcards. J. Crypto. 4(3): 161–174 (1991).
28. V. Shoup. On Formal Models for Secure Key Exchange. Available at http://philby.ucsd.edu/cryptolib.

A   Building Blocks

DECISIONAL DIFFIE-HELLMAN (DDH) ASSUMPTION (see [8]). For concreteness, we let $\mathcal{G}$ be a subgroup of $\mathbb{Z}_p^*$ of order q where p, q are prime, $q|p-1$, and $|q| = k$, the security parameter. Let g be a generator of $\mathcal{G}$. The DDH assumption states that it is infeasible for an adversary to distinguish between the following distributions:

$$\{x, y, z \leftarrow \mathbb{Z}_q : (g^x, g^y, g^{xz}, g^{yz})\} \text{ and } \{x, y, z, w \leftarrow \mathbb{Z}_q : (g^x, g^y, g^z, g^w)\}.$$

More precisely, choose at random one of the above distributions and give adversary $\mathcal{A}$ an element chosen from this distribution. The adversary succeeds by guessing which distribution was chosen; the advantage is defined as usual. Let $\varepsilon_{\mathrm{ddh}}(k, t)$ be the maximum advantage of any adversary which runs in time t. The DDH assumption is that for $t = \mathrm{poly}(k)$, the advantage $\varepsilon_{\mathrm{ddh}}(k, t)$ is negligible.

ONE-TIME DIGITAL SIGNATURES (see [19,15]). Let $\mathsf{SigGen}(1^k)$ be a probabilistic algorithm generating a public verification key/private signing key $(\mathsf{VK}, \mathsf{SK})$. Signing message M is denoted by $\mathsf{Sig} \leftarrow \mathsf{Sign}_{\mathsf{SK}}(M)$, and verification is denoted by $b = \mathsf{Verify}_{\mathsf{PK}}(M, \mathsf{Sig})$ (the signature is correct if $b = 1$). Consider the following experiment: $\mathsf{SigGen}(1^k)$ is run to generate $(\mathsf{VK}, \mathsf{SK})$. Message M is chosen, and the signature $\mathsf{Sig} \leftarrow \mathsf{Sign}_{\mathsf{SK}}(M)$ is computed. Adversary $\mathcal{A}$ is given $(\mathsf{PK}, M, \mathsf{Sig})$ and outputs a pair (M', Sig') which is not equal to the message/signature pair it was given. The adversary's advantage is defined as the probability that $\mathsf{Verify}_{\mathsf{PK}}(M', \mathsf{Sig}') = 1$. Let $\varepsilon_{\mathrm{sig}}(k, t)$ be the maximum possible advantage of any adversary which runs in time t. The assumption is that for $t = \mathrm{poly}(k)$, this value is negligible. Note that a signature scheme meeting this requirement can be constructed [15,26] given any one way function[7].

EXTENDED CRAMER-SHOUP ENCRYPTION ([13]). The Cramer-Shoup cryptosystem is an encryption scheme secure under adaptive chosen ciphertext attack (see [13] for formal definitions). We extend their cryptosystem, as discussed in Appendix B; our extension remains secure under adaptive chosen ciphertext attack. The extension gives two "types" of encryption algorithms: a client-encryption algorithm and a server-encryption algorithm, both using the identical public parameters.

Consider the following experiment: $\mathsf{ExtCSGen}(1^k)$ is run to generate public key/private key pair (pk, sk). Adversary $\mathcal{A}$ is given pk and is also given access to a decryption oracle which, given ciphertext C, returns the corresponding plaintext P (or $\perp$ if the ciphertext is invalid). The adversary outputs a plaintext x, and may request either a client-encryption or a server-encryption of x. A random bit b is chosen; if $b = 0$ the adversary is given a random encryption (of the type requested) of x, while if $b = 1$ the adversary is given a random encryption (of the type requested) of a random element. The adversary may continue to submit queries to the decryption oracle, but cannot ask for decryption of the challenge ciphertext. The adversary succeeds by guessing b; the advantage is defined as usual. Let $\varepsilon_{\mathrm{cs}}(k, t, d)$ be the maximum possible advantage of any adversary which runs in time t and asks at most d decryption oracle queries. In [13] (see also Appendix B) it is proved that for $t, d - \mathrm{poly}(k)$, the advantage $\varepsilon_{\mathrm{cs}}(k, t, d)$ is negligible (under the DDH assumption). A concrete security bound can be found in [1].

B   Extended Cramer-Shoup Encryption

We consider here an extension of the Cramer-Shoup encryption scheme [13] which is chosen-ciphertext secure. No new techniques are used, and the proof of security for the modified scheme is exactly the same as for the original with the exception of a few details which one must be careful to get right.

Public parameters are generators $g_1, g_2, h = g_1^z, c = g_1^{x_1} g_2^{x_2}, d = g_1^{y_1} g_2^{y_2} \in \mathcal{G}$ along with a universal one-way hash function $\mathcal{H}$. Ciphertexts are of the form:

---

[7] The DDH assumption implies that $f(x) = g^x$ is a one-way function.

$\langle A|B|C|D|E|F \rangle$. Decryption is done as in [13]: first, $\alpha = \mathcal{H}(A, B, C, D, E)$ is computed, and the following condition is checked:

$$C^{x_1 + y_1 \alpha} D^{x_2 + y_2 \alpha} \overset{?}{=} F.$$

If it fails output $\perp$. Otherwise, output the plaintext E/C^z.

The essential difference lies in the definition of the encryption oracle. The adversary submits a plaintext m but also submits additional information, and the encryption oracle returns some side information in addition to the ciphertext. More precisely, the adversary includes a bit $b \in \{0, 1\}$, which determines whether the plaintext is encrypted via *client-encryption* or *server-encryption*. For the case of client-encryption, the adversary also includes $Client \in$ Client. For the case of server-encryption, the adversary includes $Server \in$ Server and a value $\alpha \in \mathbb{Z}_q$. The encryption oracle sets $m' = m$ with probability $1/2$ and chooses m' randomly from $\mathcal{G}$ otherwise. Encryption is then carried out as follows:

Client-encryption(m', *Client*)
  (VK, SK) $\leftarrow$ SigGen(1^k)
  $A = Client; B = $ VK
  $r \leftarrow \mathbb{Z}_q$
  $C = g_1^r; D = g_2^r; E = h^r m'$
  $\alpha = H(A, B, C, D, E)$
  $F = (cd^\alpha)^r$
  return($\langle A, B, C, D, E, F \rangle$, SK)

Server-encryption(m', *Server*, α)
  $x, y, z, w, r \leftarrow \mathbb{Z}_q$
  $A = Server; B = g_1^x g_2^y h^z (cd^\alpha)^w$
  $C = g_1^r; D = g_2^r; E = h^r m'$
  $\beta = H(A, B, C, D, E)$
  $F = (cd^\beta)^r$
  return($\langle A, B, C, D, E, F \rangle, x, y, z, w$)

Theorem 2. *The encryption scheme outlined above is secure (in the sense of indistinguishability) under an adaptive chosen ciphertext attack.*

Sketch of Proof (Informal)   The proof of security exactly follows [13], and it can be easily verified that the additional information given to the adversary does not improve her advantage. One point requiring careful consideration is the adversary's probability of finding a collision in $\mathcal{H}$. If $\mathcal{H}$ is collision resistant (a stronger assumption than being universal one-way), there is nothing left to prove. If $\mathcal{H}$ is universal one-way, however, it can first be noted that VK or B could be selected by a simulator before $\mathcal{H}$ is given to it (if the simulator prepares the public key such that it knows $\log_{g_1} g_2$ it can produce a representation of B for any value α given to it by the adversary). But, we must also deal with the fact that the adversary gets to choose A (and the bit b which determines whether client-encryption or server-encryption is used) *after* seeing $\mathcal{H}$. However, since the set User is fixed in advance, and (at worst) of size polynomial in the security parameter, the simulator can "guess" the adversary's choices in advance (before being given $\mathcal{H}$) and this will only affect the simulator's probability of finding a collision by a polynomial factor (details omitted). ∎

C Formal Specification of the Protocol

Initialize(1^k) —
    Select p, q prime with $|p| = k$ and $q|p-1$; this defines group $\mathcal{G}$
    Choose random generators $g_1, g_2, h, c, d \leftarrow \mathcal{G}$
    $\mathcal{H} \leftarrow$ UOWHF
    Publish parameters $(q, p, g_1, g_2, h, c, d, H)$
    $\langle pw_C \rangle_{C \in \mathsf{Client}} \leftarrow \{1, \ldots, N\}$

Fig. 2. Specification of protocol initialization.

Execute($Client, i, Server, j$) —
    $(\mathsf{VK}, \mathsf{SK}) \overset{R}{\leftarrow} \mathsf{SigGen}(1^k)$ $x_1, x_2, y_1, y_2, z_1, z_2, w_1, w_2, r_1, r_2 \overset{R}{\leftarrow} \mathbb{Z}_q$
    $A = g_1^{r_1}$; $B = g_2^{r_1}$; $C = h^{r_1} g_1^{pw_C}$ $\alpha = \mathcal{H}(Client \,|\, \mathsf{VK}|A|B|C)$
    $D = (cd^\alpha)^{r_1}$ $msg\text{-}out_1 \longleftarrow \langle Client \,|\, \mathsf{VK} \,|\, A \,|\, B \,|\, C \,|\, D \rangle$
    $E = g_1^{x_1} g_2^{x_1} h^{z_1} (cd^\alpha)^{w_1}$ $F = g_1^{r_2}$; $G = g_2^{r_2}$; $I = h^{r_2} g_1^{pw_C}$
    $\beta = \mathcal{H}(Server \,|\, E|F|G|I)$
    $J = (cd^\beta)^{r_2}$ $msg\text{-}out_2 \longleftarrow \langle Server \,|\, E \,|\, F \,|\, G \,|\, I \,|\, J \rangle$
    $K = g_1^{x_2} g_2^{y_2} h^{z_2} (cd^\beta)^{w_2}$ $msg\text{-}out_3 \longleftarrow \langle K \,|\, \mathsf{Sign}_{\mathsf{SK}}(\beta|K) \rangle$
    $sk_S^j \leftarrow sk_C^i \leftarrow A^{x_1} B^{y_1} (C \cdot g_1^{-pw_C})^{z_1} D^{w_1} F^{x_2} G^{y_2} (I \cdot g_1^{-pw_C})^{z_2} J^{w_2}$
    $sid_S^j \leftarrow sid_C^i \leftarrow \langle msg\text{-}out_1 \,|\, msg\text{-}out_2 \,|\, msg\text{-}out_3 \rangle$
    return $\langle msg\text{-}out_1, msg\text{-}out_2, msg\text{-}out_3 \rangle$

Reveal($User, i$) —
    return sk_U^i

Test($User, i$) —
    $b \overset{R}{\leftarrow} \{0, 1\}, sk \leftarrow \mathcal{G}$
    if $b = 0$ return sk else return sk_U^i

Fig. 3. Specification of the Execute, Reveal, and Test oracles to which the adversary has access. Note that $q, g_1, g_2, h, c, d, \mathcal{H}$ are public, and $\mathcal{G}$ is the underlying group. Subscript S refers to the server, and C to the client.

$\mathsf{Send}_0(\mathit{Client}, i, \mathit{Server})$ —

$\quad (\mathsf{VK}, \mathsf{SK}) \xleftarrow{R} \mathsf{SigGen}(1^k) \qquad r \xleftarrow{R} \mathbb{Z}_q$

$\quad A = g_1^r; \; B = g_2^r; \; C = h^r g_1^{pw_C} \qquad \alpha = \mathcal{H}(\mathit{Client}|\mathsf{VK}|A|B|C)$

$\quad \mathit{msg\text{-}out} \longleftarrow \langle \mathit{Client} \mid \mathsf{VK} \mid A \mid B \mid C \mid (cd^\alpha)^r \rangle$

$\quad \mathit{state}_C^i \longleftarrow \langle \mathsf{SK}, r, \mathit{msg\text{-}out} \rangle$

$\quad \text{return } \mathit{msg\text{-}out}$

$\mathsf{Send}_1(\mathit{Server}, i, \langle \mathit{Client}, \mathsf{VK}, A, B, C, D \rangle)$ —

$\quad x, y, z, w, r \xleftarrow{R} \mathbb{Z}_q \qquad \alpha = \mathcal{H}(\mathit{Client}|\mathsf{VK}|A|B|C) \qquad E = g_1^x g_2^y h^z (cd^\alpha)^w$

$\quad F = g_1^r; \; G = g_2^r; \; I = h^r g_1^{pw_C} \qquad \beta = \mathcal{H}(\mathit{Server} \mid E|F|G|I)$

$\quad \mathit{msg\text{-}out} \longleftarrow \langle \mathit{Server} \mid E \mid F \mid G \mid I \mid (cd^\beta)^r \rangle$

$\quad \mathit{state}_S^i \longleftarrow \langle \mathit{msg\text{-}in}, x, y, z, w, r, \beta, \mathit{msg\text{-}out} \rangle$

$\quad \text{return } \mathit{msg\text{-}out}$

$\mathsf{Send}_2(\mathit{Client}, i, \langle \mathit{Server}, E, F, G, I, J \rangle)$ —

$\quad \langle \mathsf{SK}, r, \mathit{first\text{-}msg\text{-}out} \rangle \longleftarrow \mathit{state}_C^i \qquad \beta = \mathcal{H}(\mathit{Server}|E|F|G|I)$

$\quad x, y, z, w \xleftarrow{R} \mathbb{Z}_q \qquad K = g_1^x g_2^y h^z (cd^\beta)^w$

$\quad \mathit{msg\text{-}out} \longleftarrow \langle K \mid \mathsf{Sign}_{\mathsf{SK}}(\beta|K) \rangle \qquad \mathit{sid}_C^i \longleftarrow \langle \mathit{first\text{-}msg\text{-}out} \mid \mathit{msg\text{-}in} \mid \mathit{msg\text{-}out} \rangle$

$\quad I^* = I \cdot g_1^{-pw_C} \qquad \mathit{sk}_C^i \longleftarrow E^r F^x G^y (I^*)^z J^w$

$\quad \text{return } \mathit{msg\text{-}out}$

$\mathsf{Send}_3(\mathit{Server}, i, \langle K, \mathsf{Sig} \rangle)$ —

$\quad \langle \mathit{first\text{-}msg\text{-}in}, x, y, z, w, r, \beta, \mathit{first\text{-}msg\text{-}out} \rangle \longleftarrow \mathit{state}_S^i$

$\quad \langle \mathsf{VK}, A, B, C, D \rangle \longleftarrow \mathit{first\text{-}msg\text{-}in}$

$\quad \mathit{sid}_S^i \longleftarrow \langle \mathit{first\text{-}msg\text{-}in} \mid \mathit{first\text{-}msg\text{-}out} \mid \mathit{msg\text{-}in} \rangle$

$\quad \text{if } \mathsf{Verify}_{\mathsf{VK}}((\beta|K), \mathsf{Sig}) = 1$

$\quad\quad C^* = C \cdot g_1^{pw_C} \qquad \mathit{sk}_S^i \longleftarrow A^x B^y (C^*)^z D^w K^r$

$\quad \text{else}$

$\quad\quad \mathit{sk}_S^i \xleftarrow{R} \mathcal{G}$

$\quad \text{return } \varepsilon$

Fig. 4. Specification of the Send oracles to which the adversary has access. Note that $q, g_1, g_2, h, c, d, \mathcal{H}$ are public, and $\mathcal{G}$ is the underlying group. Subscript S refers to the server, and C to the client. The third argument to the Send oracles is denoted $\mathit{msg\text{-}in}$.

Identification Protocols
Secure against Reset Attacks

Mihir Bellare[1], Marc Fischlin[2], Shafi Goldwasser[3], and Silvio Micali[3]

[1] Dept. of Computer Science & Engineering, University of California at San Diego,
9500 Gilman Drive, La Jolla, California 92093, USA.
mihir@cs.ucsd.edu.
www-cse.ucsd.edu/users/mihir.
[2] Dept. of Mathematics (AG 7.2), Johann Wolfgang Goethe-University,
Postfach 111932, 60054 Frankfurt/Main, Germany.
marc@mi.informatik.uni-frankfurt.de
www.mi.informatik.uni-frankfurt.de
[3] MIT Laboratory for Computer Science,
545 Technology Square, Cambridge MA 02139, USA.

Abstract. We provide identification protocols that are secure even when the adversary can reset the internal state and/or randomization source of the user identifying itself, and when executed in an asynchronous environment like the Internet that gives the adversary concurrent access to instances of the user. These protocols are suitable for use by devices (like smartcards) which when under adversary control may not be able to reliably maintain their internal state between invocations.

1 Introduction

An identification protocol enables one entity to identify itself to another as the legitimate owner of some key. This problem has been considered in a variety of settings. Here we are interested in an asymmetric setting. The entity identifying itself is typically called the prover, while the entity to which the prover is identifying itself is called the verifier. The prover holds a secret key sk whose corresponding public key pk is assumed to be held by the verifier.

The adversary's goal is to impersonate the prover, meaning to get the verifier to accept it as the owner of the public key pk. Towards this goal, it is allowed various types of attacks on the prover. In the model of smartcard based identification considered by [11], the adversary may play the role of verifier and interact with the prover, trying to learn something about sk, before making its impersonation attempt. In the model of "Internet" based identification considered by [6,1,5], the adversary is allowed to interact concurrently with many different prover "instances" as well as with the verifier. Formal notions of security corresponding to these settings have been provided in the works in question, and there are many protocol solutions for them in the literature.

In this work we consider a novel attack capability for the adversary. We allow it, while interacting with the prover, to reset the prover's internal state. That

B. Pfitzmann (Ed.): EUROCRYPT 2001, LNCS 2045, pp. 495–511, 2001.
© Springer-Verlag Berlin Heidelberg 2001

is, it can "backup" the prover, maintaining the prover's coins, and continue its interaction with the prover. In order to allow the adversary to get the maximum possible benefit from this new capability, we also allow it to have concurrent access to different prover instances. Thus, it can interact with different prover instances and reset each of them at will towards its goal of impersonating the prover. The question of the security of identification protocols under reset attacks was raised by Canetti, Goldreich, Goldwasser and Micali [8], who considered the same issue in the context of zero-knowledge proofs.

1.1   The Power of Reset Attacks

AN EXAMPLE. Let us illustrate the power of reset attacks with an example. A popular paradigm for smartcard based identification is to use a proof of knowledge [11]. The prover's public key is an instance of a hard NP language L, and the secret key is a witness to the membership of the public key in L. The protocol enables the prover to prove that it "knows" sk. A protocol that is a proof of knowledge for a hard problem, and also has an appropriate zero-knowledge type property such as being witness hiding [12], is a secure identification protocol in the smartcard model [11].

A simple instance is the zero-knowledge proof of quadratic residuosity of [15]. The prover's public key consists of a composite integer N and a quadratic residue $u \in Z_N^*$. The corresponding secret key is a square root $s \in Z_N^*$ of u. The prover proves that it "knows" a square root of u, as follows. It begins the protocol by picking a random $r \in Z_N^*$ and sending $y = r^2 \bmod N$ to the verifier. The latter responds with a random challenge bit c. The prover replies with $a = rs^c \bmod N$, meaning it returns r if $c = 0$ and $rs \bmod N$ if $c = 1$. The verifier checks that $a^2 \equiv yu^c \bmod N$. (This atomic protocol has an error probability of $1/2$, which can be lowered by sequential repetition. The Fiat-Shamir protocol [13] can be viewed as a parallelized variant of this protocol.)

Now suppose the adversary is able to mount reset attacks on the prover. It can run the prover to get y, feed it challenge 0, and get back $a = r$. Now, it backs the prover up to the step just after it returned y, and feeds it challenge 1 to get answer $a' = rs$. From a and a' it is easily able to extract the prover's secret key s. Thus, this protocol is not secure under reset attacks.

Generalizing from the example, we see that in fact, all proof of knowledge based identification protocols can broken in the same way. Indeed, in a proof of knowledge, the prover is defined to "know a secret" exactly when this secret can be extracted by a polynomial time algorithm (the "extractor") which has oracle access to the prover and is allowed to reset the latter [11,4]. An attacker allowed a reset attack can simply run the extractor, with the same result, namely it gets the secret. So the bulk of efficient smartcard based identification protocols in the literature are insecure under reset attacks.

MOUNTING RESET ATTACKS. Resetting or restoring the computational state of a device is particularly simple in the case the device consists of a smartcard which the enemy can capture and experiment with. If the card is manufactured with secure hardware, the enemy may not be able to read its secret content, but it could disconnect its battery so as to restore the card's secret internal content to some initial state, and then re-insert the battery and use it with that state a number of times. If the smart card implements a proof of knowledge prover for ID purposes, then such an active enemy may impersonate the prover later on.

Other scenarios in which such an attack can be realized is if an enemy is able to force a crash on the device executing the prover algorithm, in order to force it to resume computation after the crash in an older "computational state", thereby forcing it to essentially reset itself.

CAN WE USE RESETTABLE ZERO-KNOWLEDGE? Zero-knowledge proofs of membership secure under reset attack do exist [8], but for reasons similar to those illustrated above, are not proofs of knowledge. Accordingly, they cannot be used for identification under a proof of knowledge paradigm. One of the solution paradigms we illustrate later however will show how proofs of membership, rather than proofs of knowledge, can be used for identification.

1.2   Notions of Security

Towards the goal of proving identification protocols secure against reset attacks, we first discuss the notions of security we define and use.

We distinguish between two types of resettable attacks CR1 (Concurrent-Reset-1) and CR2 (Concurrent-Reset-2). In a CR1 attack, Vicky (the adversary) may interact concurrently, in the role of verifier, with many instances of the prover Alice, resetting Alice to initial conditions and interleaving executions, hoping to learn enough to be able to impersonate Alice in a future time. Later, Vicky will try to impersonate Alice, trying to identify herself as Alice to Bob (the verifier).

In a CR2 attack, Vicky, *while* trying to impersonate Alice (i.e attempting to identify herself as Alice to Bob the verifier), may interact concurrently, in the role of verifier, with many instances of the prover Alice, resetting Alice to initial conditions and interleaving executions. Clearly, a CR1 attack is a special case of a CR2 attack.

A definition of what it means for Vicky to win in the CR1 setting is straightforward: Vicky wins if she can make the verifier Bob accept. In the CR2 setting Vicky can make the verifier accept by simply being the woman-in-the-middle, passing messages back and forth between Bob and Alice. The definitional issues are now much more complex because the woman-in-the-middle "attack" is not really an attack and the definition must take this into account. We address these issues based on definitional ideas from [6,5], specifically by assigning session-ids to each completed execution of an ID protocol, which the prover must generate and the verifier accept at the completion of the execution. For reasons of brevity we do not discuss the CR2 setting much in this abstract, and refer the reader to the full version of this paper [3].

We clarify that the novel feature of our work is the consideration of reset attacks for identification. However our settings are defined in such a way that the traditional concurrent attacks as considered by [6,10] and others are incorporated, so that security against these attacks is achieved by our protocols.

1.3   Four Paradigms for Identification Secure Aagainst Reset Attack

As we explained above, the standard proof of knowledge based paradigm fails to provide identification in the resettable setting. In that light, it may not be clear how to even prove the existence of a solution to the problem. Perhaps surprisingly however, not only can the existence of solutions be proven under the minimal assumption of a one-way function, but even simple and efficient solutions can be designed.

This is done in part by returning to some earlier paradigms. Zero-knowledge proofs of knowledge and identification are so strongly linked in contemporary cryptography that it is sometimes forgotten that these in fact replaced earlier identification techniques largely due to the efficiency gains they brought. In considering a new adversarial setting it is thus natural to first return to older paradigms and see whether they can be "lifted" to the resettable setting. We propose in particular signature and encryption based solutions for resettable identification and prove them secure in both the CR1 and the CR2 settings. We then present a general method for transforming identification protocols secure in a concurrent but non-reset setting to ones secure in a reset setting. Finally we return to the zero-knowledge ideas and provide a new paradigm based on zero-knowledge proofs of membership as opposed to proofs of knowledge.

SIGNATURE BASED IDENTIFICATION. The basic idea of the signature based paradigm is for Alice convinces Bob that she is Alice, by being "able to" sign random documents of Bob's choice. This is known (folklore) to yield a secure identification scheme in the serial non-reset setting of [11] as long as the signature scheme is secure in the sense of [16]. It is also known to be secure in the concurrent non-reset setting [1]. But it fails in general to be secure in the resettable setting because an adversary can obtain signatures of different messages under the same prover coins. What we show is that the paradigm yields secure solutions in the resettable setting if certain special kinds of signature schemes are used. (The signing algorithm should be deterministic and stateless.) In the CR1 setting the basic protocol using such signature schemes suffices. The CR2 setting is more complex and we need to modify the protocol to include "challenges" sent by the prover. Since signature schemes with the desired properties exist (and even efficient ones exist) we obtain resettable identification schemes proven secure under minimal assumptions for both the CR1 and the CR2 settings, and also obtain some efficient specific protocols.

ENCRYPTION BASED IDENTIFICATION. In the encryption based paradigm, Alice convinces Bob she is Alice, by being "able to" decrypt ciphertexts which Bob created. While the basic idea goes back to symmetric authentication techniques of the seventies, modern treatments of this paradigm appeared more recently in [9,1,10] but did not consider reset attacks. We show that under an appropriate condition on the encryption scheme —namely that it be secure against chosen-ciphertext attacks— a resettable identification protocol can be obtained. As before the simple solution for the CR1 setting needs to be modified before it will work in the CR2 setting.

TRANSFORMING STANDARD PROTOCOLS. Although Fiat-Shamir like identification protocols are not secure in the context of reset attacks, with our third paradigm we show how to turn practical identification schemes into secure ones in the CR1 and CR2 settings. The solution relies on the techniques introduced in [8] and utilizes pseudorandom functions and trapdoor commitments. It applies to most of the popular identification schemes, like Fiat-Shamir [13], Okamoto-Schnorr [20,18] or Okamoto-Guillou-Quisquater [17,18].

ZK PROOF OF MEMBERSHIP BASED IDENTIFICATION. In the zero-knowledge proofs of membership paradigm, Alice convinces Bob she is Alice, by being "able to" prove membership in a hard language L, rather than by proving she has a witness for language L. She does so by employing a resettable zero-knowledge proof of language membership for L as defined in [8] . Both Alice and Bob will need to have a public-key to enable the protocol. Alice's public-key defines who she is, and Bob's public-key enables him to verify her identity in a secure way. We adopt the general protocol for membership in NP languages of [8] for the purpose of identification. The identification protocols are constant round. What makes this work is the fact that the protocol for language membership $(x \in L)$ being zero-knowledge implies "learning nothing" about x in a very strong sense — a verifier cannot subsequently convince anyone else that $x \in L$ with non-negligible probability. We note that while we can make this approach work using resettable zero-knowledge proofs, it does not seem to work using resettable witness indistinguishable proofs for ID protocols.

PERSPECTIVE. Various parts of the literature have motivated the study of zero-knowledge protocols secure against strong attacks such as concurrent or reset in part by the perceived need for such tools for the purpose of applications such as identification in similar attack settings. While the tools might be sufficient for identification, they are not necessary. Our results demonstrate that identification is much easier than zero-knowledge and the latter is usually an overkill for the former.

2   Definitions

If $A(\cdot, \cdot, \ldots)$ is a randomized algorithm then $y \leftarrow A(x_1, x_2, \ldots; R)$ means y is assigned the unique output of the algorithm on inputs $x_1, x_2, \ldots$ and coins R, while $y \leftarrow A(x_1, x_2, \ldots)$ is shorthand for first picking R at random (from the set of all strings of some appropriate length) and then setting $y \leftarrow A(x_1, x_2, \ldots; R)$. If $x_1, x_2, \ldots$ are strings then $x_1 \| x_2 \| \cdots$ denotes an encoding under which the constituent strings are uniquely recoverable. It is assumed any string x can be uniquely parsed as an encoding of some sequence of strings. The empty string is denoted ε.

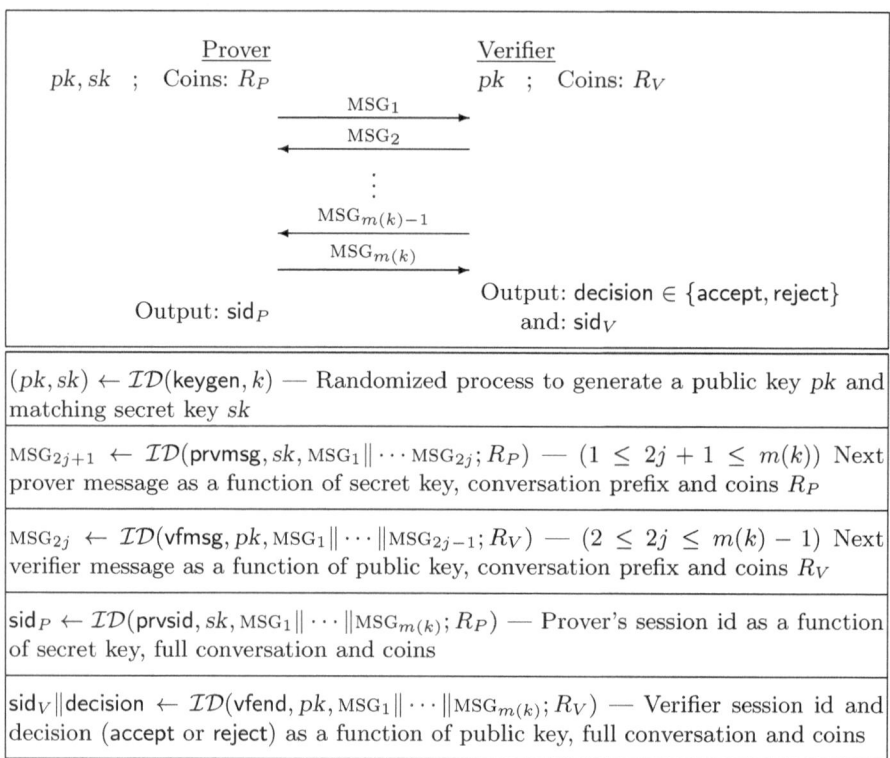

$(pk, sk) \leftarrow \mathcal{ID}(\mathsf{keygen}, k)$ — Randomized process to generate a public key pk and matching secret key sk

$\mathrm{MSG}_{2j+1} \leftarrow \mathcal{ID}(\mathsf{prvmsg}, sk, \mathrm{MSG}_1 \| \cdots \mathrm{MSG}_{2j}; R_P)$ — $(1 \le 2j + 1 \le m(k))$ Next prover message as a function of secret key, conversation prefix and coins R_P

$\mathrm{MSG}_{2j} \leftarrow \mathcal{ID}(\mathsf{vfmsg}, pk, \mathrm{MSG}_1 \| \cdots \| \mathrm{MSG}_{2j-1}; R_V)$ — $(2 \le 2j \le m(k) - 1)$ Next verifier message as a function of public key, conversation prefix and coins R_V

$\mathsf{sid}_P \leftarrow \mathcal{ID}(\mathsf{prvsid}, sk, \mathrm{MSG}_1 \| \cdots \| \mathrm{MSG}_{m(k)}; R_P)$ — Prover's session id as a function of secret key, full conversation and coins

$\mathsf{sid}_V \| \mathsf{decision} \leftarrow \mathcal{ID}(\mathsf{vfend}, pk, \mathrm{MSG}_1 \| \cdots \| \mathrm{MSG}_{m(k)}; R_V)$ — Verifier session id and decision (accept or reject) as a function of public key, full conversation and coins

Fig. 1. The prover sends the first and last messages in an $m(k)$-move identification protocol at the end of which the verifier outputs a decision and each party optionally outputs a session id. The *protocol description* function $\mathcal{ID}$ specifies all processes associated to the protocol.

An identification protocol proceeds as depicted in Figure 1. The prover has a secret key sk whose matching public key pk is held by the verifier. (In practice the prover might provide its public key, and the certificate of this public key, as part of the protocol, but this is better slipped under the rug in the model.) Each party computes its next message as a function of its keys, coins and the current conversation prefix. The number of moves $m(k)$ is odd so that the first and last moves belong to the prover. (An identification protocol is initiated by the prover who at the very least must provide a request to be identified.) At the end of the protocol the verifier outputs a decision to either accept or reject. Each party may also output a session id. (Sessions ids are relevant in the CR2 setting but can be ignored for the CR1 setting.) A particular protocol is described by a (single) *protocol description* function $\mathcal{ID}$ which specifies how all associated processes — key generation, message computation, session id or decision computation— are implemented. (We say that $\mathcal{ID}$ is for the CR1 setting if $\mathsf{sid}_P = \mathsf{sid}_V = \varepsilon$, meaning no session ids are generated.) The second part of Figure 1 shows how it works: the first argument to $\mathcal{ID}$ is a keyword —one of keygen, prvmsg, vfmsg, prvsid, vfend— which invokes the subroutine responsible for that function on the other arguments.

Naturally, a correct execution of the protocol (meaning one in the absence of an adversary) should lead the verifier to accept. To formalize this "completeness" requirement we consider an *adversary-free execution* of the protocol $\mathcal{ID}$ which proceeds as described in the following experiment:

$(pk, sk) \leftarrow \mathcal{ID}(\mathsf{keygen}, k)$; Choose tapes R_P, R_V at random
$\mathrm{MSG}_1 \leftarrow \mathcal{ID}(\mathsf{prvmsg}, sk, \varepsilon; R_P)$
For $j = 1$ to $\lfloor m(k)/2 \rfloor$ do
    $\mathrm{MSG}_{2j} \leftarrow \mathcal{ID}(\mathsf{vfmsg}, pk, \mathrm{MSG}_1 \| \cdots \| \mathrm{MSG}_{2j-1}; R_V)$
    $\mathrm{MSG}_{2j+1} \leftarrow \mathcal{ID}(\mathsf{prvmsg}, sk, \mathrm{MSG}_1 \| \cdots \| \mathrm{MSG}_{2j}; R_P)$
EndFor
$\mathsf{sid}_P \leftarrow \mathcal{ID}(\mathsf{prvsid}, sk, \mathrm{MSG}_1 \| \cdots \| \mathrm{MSG}_{m(k)}; R_P)$
$\mathsf{sid}_V \| \mathsf{decision} \leftarrow \mathcal{ID}(\mathsf{vfend}, pk, \mathrm{MSG}_1 \| \cdots \| \mathrm{MSG}_{m(k)}; R_V)$

The *completeness condition* is that, in the above experiment, the probability that $\mathsf{sid}_P = \mathsf{sid}_V$ and $\mathsf{decision} = \mathsf{accept}$ is 1. (The probability is over the coin tosses of $\mathcal{ID}(\mathsf{keygen}, k)$ and the random choices of R_P, R_V.) As always, the requirement can be relaxed to only ask for a probability close to one.

Fix an identification protocol description function $\mathcal{ID}$ and an adversary I. Associated to them is **Experiment**$_{\mathcal{ID},I}^{\mathsf{id\text{-}cr1}}(k)$, depicted in Figure 2, which is used to define the security of $\mathcal{ID}$ in the CR1 setting. (In this context it is understood that $\mathcal{ID}$ is for the CR1 setting, meaning does not produce session ids.) The experiment gives the adversary appropriate access to prover instance oracles $\mathsf{Prover}^1, \mathsf{Prover}^2, \ldots$ and a single verifier oracle, let it query these subject to certain restrictions imposed by the experiment, and then determine whether it "wins". The interface to the prover instance oracles and the verifier oracle (which, in the experiment, are implicit, never appearing by name) is via oracle queries; the experiment enumerates the types of queries and shows how answers are provided to them.

The experiment begins with some initializations which include choosing of the keys. Then the adversary is invoked on input the public key. A WakeNewProver query activates a new prover instance Prover^p by picking a random tape R_p for it. (A random tape for a prover instance is chosen exactly once and all messages of this prover instance are then computed with respect to this tape. The tape of a specific prover instance cannot be changed, or "reset", once chosen.) A $\mathsf{Send}(\mathsf{prvmsg}, i, x)$ query —viewed as sent to prover instance Prover^i— results in the adversary being returned the next prover message computed as $\mathcal{ID}(\mathsf{prvmsg}, sk, x; R_i)$. (It is assumed that $x = \mathrm{MSG}_1 \| \cdots \| \mathrm{MSG}_{2j}$ is a *valid conversation prefix*, meaning contains an even number of messages $2j < m(k)$, else the query is not valid.) Resetting is captured by allowing arbitrary (valid) conversation prefixes to be queried. (For example the adversary might try $\mathrm{MSG}_1 \| \mathrm{MSG}_2$ for many different values of MSG_2, corresponding to successively resetting the prover instance to the point where it receives the second protocol move.) Concurrency is captured by the fact that any activated prover instances can be queried.

Experiment$_{\mathcal{ID},I}^{\text{id-cr1}}(k)$ — Execution of protocol $\mathcal{ID}$ with adversary I and security parameter k in the CR1 setting

Initialization:

(1) $(pk, sk) \leftarrow \mathcal{ID}(\text{keygen}, k)$ // Pick keys via randomized key generation algorithm //

(2) Choose tape R_V for verifier at random ; $C_V \leftarrow 0$ // Coins and message counter for verifier //

(3) $p \leftarrow 0$ // Number of active prover instances //

Execute adversary I on input pk and reply to its oracle queries as follows:

- When I makes query WakeNewProver // Activate a new prover instance //

  (1) $p \leftarrow p + 1$; Pick a tape R_p at random ; **Return** p

- When I makes query $\text{Send}(\text{prvmsg}, i, \text{MSG}_1 \| \cdots \| \text{MSG}_{2j})$ with $0 \leq 2j < m(k)$ and $1 \leq i \leq p$

  (1) **If** $C_V \neq 0$ **then Return** $\perp$ // Interaction with prover instance allowed only before interaction with verifier begins //

  (2) $\text{MSG}_{2j+1} \leftarrow \mathcal{ID}(\text{prvmsg}, sk, \text{MSG}_1 \| \cdots \| \text{MSG}_{2j}; R_i)$

  (3) **Return** MSG_{2j+1}

- When I makes query $\text{Send}(\text{vfmsg}, \text{MSG}_1 \| \cdots \| \text{MSG}_{2j-1})$ with $1 \leq 2j-1 \leq m(k)$

  (1) $C_V \leftarrow C_V + 2$

  (2) **If** $2j < C_V$ **then Return** $\perp$ // Not allowed to reset the verifier //

  (3) **If** $2j-1 < m(k)-1$ **then** $\text{MSG}_{2j} \leftarrow$
      $\mathcal{ID}(\text{vfmsg}, pk, \text{MSG}_1 \| \cdots \| \text{MSG}_{2j-1}; R_V)$; **Return** MSG_{2j}

  (4) **If** $2j-1 = m(k)$ **then** decision $\leftarrow \mathcal{ID}(\text{vfend}, pk, \text{MSG}_1 \| \cdots \| \text{MSG}_{2j}; R_V)$

  (5) **Return** decision

Did I win? When I has terminated set $\text{WIN}_I = \text{true}$ if decision = accept.

Fig. 2. Experiment describing execution of identification protocol $\mathcal{ID}$ with adversary I and security parameter k in the CR1 setting.

A $\text{Send}(\text{vfmsg}, x)$ query is used to invoke the verifier on a conversation prefix x and results in the adversary being returned either the next verifier message computed as $\mathcal{ID}(\text{vfmsg}, pk, x; R_V)$ —this when the verifier still has a move to make— or the decision computed as $\mathcal{ID}(\text{vfend}, pk, x; R_V)$ —this when x corresponds to a full conversation. (Here R_V was chosen at random in the experiment initialization step. It is assumed that $x = \text{MSG}_1 \| \cdots \| \text{MSG}_{2j-1}$ is a valid conversation prefix, meaning contains an odd number of messages $1 \leq 2j - 1 \leq m(k)$, else the query is not valid.) Unlike a prover instance, resetting the (single) verifier instance is not allowed. (Our signature and encryption based protocols are actually secure even if verifier resets are allowed, but since the practical need to consider this attack is not apparent, the definition excludes it.) This is enforced explicitly in the experiments via the verifier message counter C_V.

In the CR1 setting, the adversary's actions are divided into two phases. In the first phase it interacts with the prover instances, not being allowed to interact with the verifier; in the second phase it is denied access to the prover instances and tries to convince the verifier to accept. $\mathbf{Experiment}_{\mathcal{ID},I}^{\text{id-cr1}}(k)$ enforces this by returning $\perp$ in reply to a $\mathsf{Send}(\mathsf{prvmsg}, i, x)$ unless $C_V = 0$.

The adversary wins if it makes the verifier instance accept. The parameter WIN_I is set accordingly in $\mathbf{Experiment}_{\mathcal{ID},I}^{\text{id-cr1}}(k)$. The definition of the protocol is responsible for ensuring that both parties reject a received conversation prefix if it is inconsistent with their coins. It is also assumed that the adversary never repeats an oracle query. We can now provide definitions of security for protocol $\mathcal{ID}$.

Definition 1. [Security of an ID protocol in the CR1 setting] *Let $\mathcal{ID}$ be an identification protocol description for the* CR1 *setting. Let I be an adversary (called an impersonator in this context) and let k be the security parameter. The advantage of impersonator I is*

$$\mathbf{Adv}_{\mathcal{ID},I}^{\text{id-cr1}}(k) = \Pr[\,\text{WIN}_I = \mathsf{true}\,]$$

where the probability is with respect to $\mathbf{Experiment}_{\mathcal{ID},I}^{\text{id-cr1}}(k)$. *Protocol $\mathcal{ID}$ is said to be polynomially-secure in the* CR1 *setting if* $\mathbf{Adv}_{\mathcal{ID}}^{\text{id-cr1}}(\cdot)$ *is negligible for any impersonator I of time-complexity polynomial in k.* ∎

We adopt the convention that the *time-complexity* $t(k)$ of an adversary I is the execution time of the entire experiment $\mathbf{Experiment}_{\mathcal{ID},I}^{\text{id-cr1}}(k)$, including the time taken for initialization, computation of replies to adversary oracle queries, and computation of WIN_I. We also define the *query-complexity* $q(k)$ of I as the number of $\mathsf{Send}(\mathsf{prvmsg}, \cdot, \cdot)$ queries made by I in $\mathbf{Experiment}_{\mathcal{ID},I}^{\text{id-cr1}}(k)$. It is always the case that $q(k) \leq t(k)$ so an adversary of polynomial time-complexity has polynomial query-complexity. These definitions and conventions can be ignored if polynomial-security is the only concern, but simplify concrete security considerations to which we will pay some attention later.

A definition of security for the CR2 setting can be found in [3].

3   CR1-Secure Identification Protocols

Four paradigms are illustrated: signature based, encryption based, identification based, and zero-knowledge based.

3.1   A Signature Based Protocol

We assume knowledge of background in digital signatures as summarized in [3].

SIGNATURE BASED IDENTIFICATION. A natural identification protocol is for the verifier to issue a random challenge CH_V and the prover respond with a signature of CH_V computed under its secret key sk. (Prefix the protocol with an initial start move by the prover to request start of an identification process, and

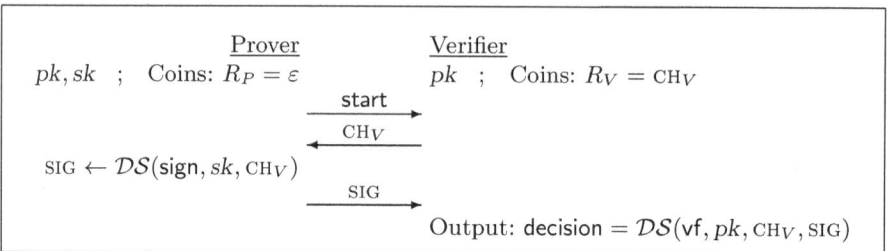

Fig. 3. Reset-secure identification protocol $\mathcal{ID}$ for the CR1 setting based on a deterministic, stateless digital signature scheme $\mathcal{DS}$.

you have a three move protocol.) This simple protocol can be proven secure in the serial, non-resettable (ie. standard smartcard) setting of [11] as long as the signature scheme meets the notion of security of [16] . (This result seems to be folklore.) The same protocol has also been proven to provide authentication in the concurrent, non-resettable (ie. standard network) setting [1]. (The intuition in both cases is that the only thing an adversary can do with a prover oracle is feed it challenge strings and obtain their signatures, and if the scheme is secure against chosen-message attack this will not help the adversary forge a signature of a challenge issued by the verifier unless it guesses the latter, and the probability of the last event can be made small by using a long enough challenge.) This protocol is thus a natural candidate for identification in the resettable setting.

However this protocol does not always provide security in the resettable setting. The intuition described above breaks down because resetting allows an adversary to obtain the signatures of different messages under the same set of coins. (It can activate a prover instance and then query it repeatedly with different challenges, thereby obtaining their signatures with respect to a fixed set of coin tosses.) As explained in [3], this is not covered by the usual notion of a chosen-message attack used to define security of signature schemes in [16]. And indeed, for many signature schemes it is possible to forge the signature of a new message if one is able to obtain the signatures of several messages under one set of coins. Similarly, if the signing algorithm is stateful, resetting allows an adversary to make the prover release several signatures computed using one value of the **state** variable —effectively, the prover does not get a chance to update its state is it expects to— again leading to the possibility of forgery on a scheme secure in the standard sense.

The solution is simple: restrict the signature scheme to be stateless and deterministic. In [3] we explain how signatures schemes can be imbued with these attributes so that stateless, deterministic signature schemes are available.

PROTOCOL AND SECURITY. Let $\mathcal{DS}$ be a deterministic, stateless signature scheme. Figure 3 illustrates the flows of the associated identification protocol $\mathcal{ID}$. A parameter of the protocol is the length $vcl(k)$ of the verifier's random challenge. The prover is deterministic and has random tape ε while the verifier's random tape is CH_V. Refer to Definition 1 and [3] for the meanings of terms used in the theorem below, and to [3] for the proof.

Theorem 1. [Concrete security of the signature based ID scheme in the CR1 setting] *Let $\mathcal{DS}$ be a deterministic, stateless signature scheme, let $vcl(\cdot)$ be a polynomially-bounded function, and let $\mathcal{ID}$ be the associated identification scheme as per Figure 3. If I is an adversary of time-complexity $t(\cdot)$ and query-complexity $q(\cdot)$ attacking $\mathcal{ID}$ in the CR1 setting then there exists a forger F attacking $\mathcal{DS}$ such that*

$$\mathbf{Adv}_{\mathcal{ID},I}^{\text{id-cr1}}(k) \leq \mathbf{Adv}_{\mathcal{DS},F}^{\text{ds}}(k) + \frac{q(k)}{2^{vcl(k)}} \ . \tag{1}$$

Furthermore F has time-complexity $t(k)$ and makes at most $q(k)$ signing queries in its chosen-message attack on $\mathcal{DS}$. ∎

This immediately implies the following:

Corollary 1. [Polynomial-security of the signature based ID scheme in the CR1 setting] *Let $\mathcal{DS}$ be a deterministic, stateless signature scheme, let $vcl(k) = k$, and let $\mathcal{ID}$ be the associated identification scheme as per Figure 3. If $\mathcal{DS}$ is polynomially-secure then $\mathcal{ID}$ is polynomially-secure in the CR1 setting.* ∎

We show in [3] that this implies:

Corollary 2. [Existence of an ID scheme polynomially-secure in the CR1 setting] *Assume there exists a one-way function. Then there exists an identification scheme that is polynomially-secure in the CR1 setting.*

3.2 An Encryption Based Protocol

ENCRYPTION BASED IDENTIFICATION. The idea is simple: the prover proves its identity by proving its ability to decrypt a ciphertext sent by the verifier. This basic idea goes back to early work in entity authentication where the encryption was usually symmetric (ie. private-key based). These early protocols however had no supporting definitions or analysis. The first "modern" treatment is that of [9] who considered the paradigm with regard to providing deniable authentication and identified non-malleability under chosen-ciphertext attack —equivalently, indistinguishability under chosen-ciphertext attack [2,9]— as the security property required of the encryption scheme. Results of [1,10,9] imply that the protocol is a secure identification scheme in the concurrent non-reset setting, but reset attacks have not been considered before.

PROTOCOL AND SECURITY. Let $\mathcal{AE}$ be an asymmetric encryption scheme polynomially-secure against chosen-ciphertext attack. Figure 4 illustrates the flows of the associated identification protocol $\mathcal{ID}$. A parameter of this protocol is the length $vcl(k)$ of the verifier's random challenge. The verifier sends the prover a ciphertext formed by encrypting a random challenge, and the prover identifies itself by correctly decrypting this to send the verifier back the challenge. The prover is deterministic, having random tape ε. We make the coins R_e used by the encryption algorithm explicit, so that the verifier's random tape consists of the challenge —a random string of length $vcl(k)$ where vcl is a parameter of the

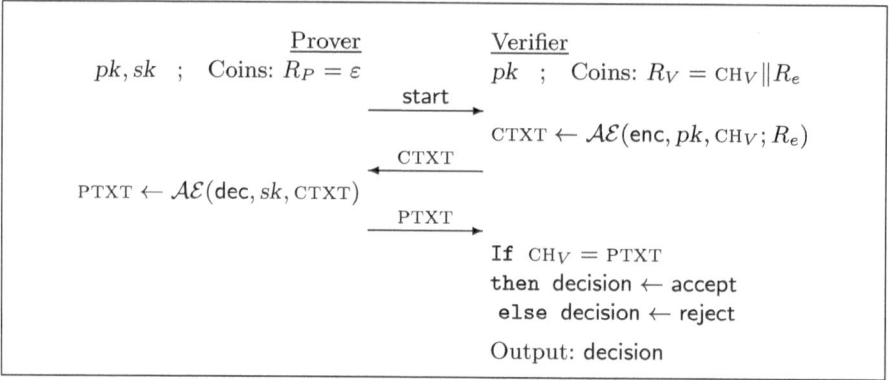

Fig. 4. Reset-secure identification protocol $\mathcal{ID}$ for the CR1 setting based on a chosen-ciphertext attack secure asymmetric encryption scheme $\mathcal{AE}$.

protocol— and coins sufficient for one invocation of the encryption algorithm. Refer to Definition 1 and [3] for the meanings of terms used in the theorem below, and to [3] for the proof.

Theorem 2. **[Concrete security of the encryption based ID scheme in the CR1 setting]** *Let $\mathcal{AE}$ be an asymmetric encryption scheme, let $vcl(\cdot)$ a polynomially-bounded function, and let $\mathcal{ID}$ be the associated identification scheme as per Figure 4. If I is an adversary of time-complexity $t(\cdot)$ and query-complexity $q(\cdot)$ attacking $\mathcal{ID}$ in the CR1 setting then there exists an eavesdropper E attacking $\mathcal{AE}$ such that*

$$\mathbf{Adv}_{\mathcal{ID},I}^{\text{id-cr1}}(k) \leq \mathbf{Adv}_{\mathcal{AE},E}^{\text{lr-cca}}(k) + \frac{2q(k)+2}{2^{vcl(k)}} \ . \tag{2}$$

Furthermore E has time-complexity $t(k)$, makes one query to its lr-encryption oracle, and at most $q(k)$ queries to its decryption oracle. ∎

This immediately implies the following:

Corollary 3. **[Polynomial-security of the encryption based ID scheme in the CR1 setting]** *Let $\mathcal{AE}$ be an asymmetric encryption scheme, let $vcl(k) = k$, and let $\mathcal{ID}$ be the associated identification scheme as per Figure 4. If $\mathcal{AE}$ is polynomially-secure against chosen-ciphertext attack then $\mathcal{ID}$ is polynomially-secure in the CR1 setting.* ∎

3.3   An Identification Based Protocol

IDENTIFICATION BASED PROTOCOL. As discussed in the introduction, proof of knowledge based identification protocols of the Fiat-Shamir type cannot be secure against reset attacks. In this section, however, we present a general transformation of such identification schemes into secure ones in the CR1 setting. We start with identification schemes that consists of three moves, an initial commitment COM of the prover, a random value CH$_V$, the challenge, of the verifier and a conclusive response RESP from the prover. We call a protocol obeying this structure a canonical identification scheme.

Loosely speaking, we will assume that the underlying canonical identical scheme $\mathcal{CID}$ is secure against non-resetting attacks in the CR1 model, i.e., against attacks where the adversary merely runs concurrent sessions with the prover without resets before engaging in a verification. In addition to the Fiat-Shamir system [13], most of the well-known practical identification schemes also achieve this security level, for example Ong-Schnorr [19,21] for some system parameters, Okamoto-Guillou-Quisquater [17,18] and Okamoto-Schnorr [20,18]. Nonetheless, there are also protocols which are only known to be secure against sequential attacks (e.g. [22]).

To avoid confusion with the derived scheme $\mathcal{ID}$, instead of writing Send(prvmsg, . . .) and Send(vfmsg, . . .), we denote the algorithms generating the commitment, challenge and response message for the CID-protocol $\mathcal{CID}$ by $\mathcal{CID}(\text{cmt}, . . .)$, $\mathcal{CID}(\text{chall}, . . .)$, and $\mathcal{CID}(\text{resp}, . . .)$, respectively, and the verification step by $\mathcal{CID}(\text{vf}, . . .)$. We also write $\mathbf{Adv}_{\mathcal{CID}, I_{\text{CID}}}^{\text{id-nr-cr1}}(k)$ for the probability that an impersonator I_{CID} succeeds in an attack on scheme $\mathcal{CID}$ in the non-resetting CR1 setting.

PROTOCOL AND SECURITY. Our solution originates from the work of [8] about resettable zero-knowledge. In order to ensure that the adversary does not gain any advantage from resetting the prover, we insert a new first round into the CID-identification protocol in which the verifier non-interactively commits to his challenge CH$_V$. The parameters for this commitment scheme become part of the public key. This keeps the adversary from resetting the prover to the challenge-message and completing the protocol with different challenges.

In addition, we let the prover determine the random values in his identification by applying a pseudorandom function to the verifier's initial commitment. Now, if the adversary resets the prover (with the same random tape) to the outset of the protocol and commits to a different challenge then the prover uses virtually independent randomness for this execution, although having the same random tape. On the other hand, using pseudorandom values instead of truly random coins does not weaken the original identification protocol noticeably. Essentially, this prunes the CR1 adversary into a non-resetting one concerning executions with the prover.

In order to handle the intrusion try we use use a special, so-called trapdoor commitment scheme $\mathcal{TDC}$ for the verifier's initial commitment. This means that there is a secret information such that knowledge of this secret allows to generate a dummy commitment and to find a valid opening to any value later on. Furthermore, the dummy commitment and the fake decommitment are identically distributed to an honestly given commitment and opening to the same value. Without knowing the secret a commitment is still solidly binding. Trapdoor commitment schemes exist under standard assumptions like the intractability of

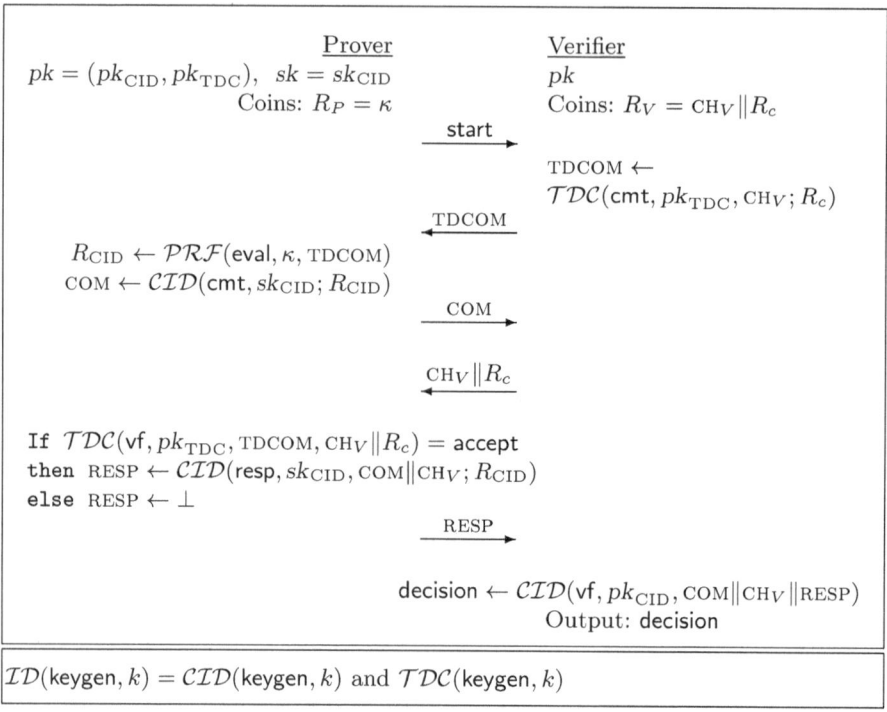

Fig. 5. Reset-secure identification protocol $\mathcal{ID}$ for the CR1 setting based on an identification scheme $\mathcal{CID}$ secure against non-resetting CR1 attacks

the discrete-log or the RSA or factoring assumption [7] and thus under the same assumptions that the aforementioned CID-identification protocols rely on.

Basically, a trapdoor commitment enables us to reduce an intrusion try of an impersonator I in the derived scheme $\mathcal{ID}$ to one for the CID-protocol. If I initiates a session with the verifier in $\mathcal{ID}$ then we can first commit to a dummy value $0^{vcl(k)}$ without having to communicate with the verifier in $\mathcal{CID}$. When I then takes the next step by sending COM, we forward this commitment to our verifier in $\mathcal{CID}$ and learn the verifier's challenge. Knowing the secret key sk_{TDC} for the trapdoor scheme we can then find a valid opening for our dummy commitment with respect to the challenge. Finally, we forward I's response in our attack.

The scheme is displayed in Figure 5. See [3] for definitions and notions. The discussion above indicates that any adversary I for $\mathcal{ID}$ does not have much more power than a non-resetting impersonator attacking $\mathcal{CID}$ and security of $\mathcal{ID}$ follows from the security of $\mathcal{CID}$.

Theorem 3. [Concrete security of the identification based scheme in the CR1 setting] *Let $\mathcal{CID}$ be an CID-identification protocol and let $vcl(\cdot)$ be a polynomially-bounded function. Also, let $\mathcal{PRF}$ be a pseudorandom function family and denote by $\mathcal{TDC}$ a trapdoor commitment scheme. Let $\mathcal{ID}$ be the associated* *identification scheme as per Figure 5. If I is an adversary of time-complexity $t(\cdot)$ and query-complexity $q(\cdot)$ attacking $\mathcal{ID}$ in the* CR1 *setting then there exists an adversary I_{CID} attacking $\mathcal{CID}$ in a non-resetting* CR1 *attack such that*

$$\mathbf{Adv}^{\mathrm{id\text{-}cr1}}_{\mathcal{ID},I}(k) \leq q(k) \cdot \mathbf{Adv}^{\mathcal{PRF}}_{(t,q)}(k) + \mathbf{Adv}^{\mathcal{TDC}}_{t}(k) + \mathbf{Adv}^{\mathrm{id\text{-}nr\text{-}cr1}}_{\mathcal{CID},I_{\mathrm{CID}}}(k) . \qquad (3)$$

Furthermore I_{CID} has time-complexity $t(k)$ and runs at most $q(k)$ sessions with the prover before trying to intrude.

As usual we have:

Corollary 4. [Polynomial-security of the identification based scheme in the CR1 setting] *Let $\mathcal{PRF}$ be a polynomially-secure pseudorandom function family and let $\mathcal{TDC}$ be a polynomially-secure trapdoor commitment scheme, set $vcl(k) = k$, and let $\mathcal{ID}$ be the associated identification scheme as per Figure 5. If $\mathcal{CID}$ is a polynomially-secure* CID-*identification protocol in the non-resetting* CR1 *model then $\mathcal{ID}$ is polynomially-secure in the* CR1 *setting.* ∎

Note that the public key in our CR1-secure identification scheme consists of two independent parts, pk_{CID} and pk_{TDC}. For concrete schemes the key generation may be combined and simplified. For instance, for Okamoto-Schnorr the public key of the identification protocol describes a group of prime order q, two generators g_1, g_2 of that group and the public key $X = g_1^{x_1} g_2^{x_2}$ for secret $x_1, x_2 \in \mathsf{Z}_q$. The prover sends $\mathrm{COM} = g_1^{r_1} g_2^{r_2}$ and replies to the challenge CH_V by transmitting $y_i = r_i + \mathrm{CH}_V x_i \bmod q$ for $i = 1, 2$. In this case, the public key for the trapdoor commitment scheme could be given by $g_1, g_3 = g_1^z$ for random trapdoor $z \in \mathsf{Z}_q$, and the commitment function maps a value c and randomness R_c to $g_1^c g_3^{R_c}$.

3.4 A Zero-Knowledge Based Protocol

As we discussed in the Introduction the idea of [11] of proving identity by employing a zero knowledge proof of knowledge has been the accepted paradigm for identification protocols in the smartcard setting. Unfortunately, as we indicated, in the resettable setting this paradigm cannot work.

RESETTABLE ZERO KNOWLEDGE BASED IDENTITY. We thus instead propose the following paradigm. Let L be a hard NP language for which there is no known efficient procedures for membership testing but for which there exists a randomized generating algorithm G which outputs pairs (x, w), where $x \in L$ and w is an NP-witness that $x \in L$. (The distribution according to which (x, w) is generated should be one for which it is hard to tell whether $x \in L$ or not). Each user Alice will run G to get a pair (x, w) and will then publish x as its public key. To prove her identity Alice will run a resettable zero-knowledge proof that $x \in L$.

PROTOCOL. To implement the above idea we need resettable zero-knowledge proofs for L. For this we turn to the work of [8]. In [8] two resettable zero-knowledge proofs for any NP language are proposed: one which takes a non-constant number of rounds and works against a computationally unbounded prover, and one which only takes a constant number of rounds and works against computationally bounded provers (i.e argument) and requires the verifiers to have published public-keys which the prover can access. We propose to utilize the latter, for efficiency sake. Thus, to implement the paradigm, we require both prover and verifier to have public-keys accessible by each other. Whereas the prover's public key is x whose membership in L it will prove to the verifier, the verifier's public key in [8] is used for specifying a perfectly private computationally binding commitment scheme which the prover must use during the protocol. (Such commitment schemes exist based for example on the strong hardness of Discrete Log Assumption.)

SECURITY. We briefly outline how to prove that the resulting ID protocol is secure in the CR1 setting. Suppose not, and that after launching a CR1 attack, an imposter can now falsely identify himself with a non-negligible probability. Then, we will construct a polynomial time algorithm A to decide membership in L. On input x, A first launches the off-line resetting attack using x as the public key and the simulator – which exists by the zero-knowledge property – to obtain views of the protocol execution. (This requires that the simulator be black-box, but this is true in the known protocols.) If $x \in L$, this view should be identical to the view obtained during the real execution, in which case a successful attack will result, which is essentially a way for A to find a language membership proof. If x not in L, then by the soundness property of a zero-knowledge proof, no matter what the simulator outputs, it will not be possible to prove membership in L.

Acknowledgments

The second author thanks Ran Canetti for discussions about resettable security. The first author is supported in part by a 1996 Packard Foundation Fellowship in Science and Engineering.

References

1. M. BELLARE, R. CANETTI, AND H. KRAWCZYK, "A modular approach to the design and analysis of authentication and key exchange protocols," *Proceedings of the 30th Annual Symposium on the Theory of Computing*, ACM, 1998.
2. M. BELLARE, A. DESAI, D. POINTCHEVAL AND P. ROGAWAY, "Relations among notions of security for public-key encryption schemes," *Advances in Cryptology – CRYPTO '98*, Lecture Notes in Computer Science Vol. 1462, H. Krawczyk ed., Springer-Verlag, 1998.
3. M. BELLARE, M. FISCHLIN, S. GOLDWASSER AND S. MICALI, "Identification protocols secure against reset attacks," Full version of this paper, available via http://www-cse.ucsd.edu/users/mihir.
4. M. BELLARE AND O. GOLDREICH, "On defining proofs of knowledge," *Advances in Cryptology – CRYPTO '92*, Lecture Notes in Computer Science Vol. 740, E. Brickell ed., Springer-Verlag, 1992.

5. M. BELLARE, D. POINTCHEVAL AND P. ROGAWAY, "Authenticated key exchange secure against dictionary attack," *Advances in Cryptology – EUROCRYPT '00*, Lecture Notes in Computer Science Vol. 1807, B. Preneel ed., Springer-Verlag, 2000.

6. M. BELLARE AND P. ROGAWAY, "Entity authentication and key distribution", *Advances in Cryptology – CRYPTO '93*, Lecture Notes in Computer Science Vol. 773, D. Stinson ed., Springer-Verlag, 1993.

7. G. BRASSARD, D. CHAUM AND C. CRÉPEAU, "Minimum Disclosure Proofs of Knowledge," *Journal of Computer and Systems Science*, Vol. 37, No. 2, 1988, pp. 156–189.

8. R. CANETTI, S. GOLDWASSER, O. GOLDREICH AND S. MICALI, "Resettable zero-knowledge," *Proceedings of the 32nd Annual Symposium on the Theory of Computing*, ACM, 2000.

9. D. DOLEV, C. DWORK AND M. NAOR, "Non-malleable cryptography", *SIAM J. on Computing*, 2001. Preliminary version in STOC 91.

10. C. DWORK, M. NAOR AND A. SAHAI, "Concurrent zero-knowledge," *Proceedings of the 30th Annual Symposium on the Theory of Computing*, ACM, 1998.

11. U. FEIGE, A. FIAT AND A. SHAMIR, "Zero-knowledge proofs of identity," *J. of Cryptology*, Vol. 1, 1988, pp. 77-94.

12. U. FEIGE AND A. SHAMIR, "Witness indistinguishable and witness hiding protocols," *Proceedings of the 22nd Annual Symposium on the Theory of Computing*, ACM, 1990.

13. A. FIAT AND A. SHAMIR, "How to prove yourself: Practical solutions to identification and signature problems," *Advances in Cryptology – CRYPTO '86*, Lecture Notes in Computer Science Vol. 263, A. Odlyzko ed., Springer-Verlag, 1986.

14. O. GOLDREICH, S. GOLDWASSER AND S. MICALI, "How to construct random functions," *Journal of the ACM*, Vol. 33, No. 4, 1986, pp. 210–217.

15. S. GOLDWASSER, S. MICALI AND C. RACKOFF, "The knowledge complexity of interactive proof systems," *SIAM J. on Computing*, Vol. 18, No. 1, pp. 186–208, February 1989.

16. S. GOLDWASSER, S. MICALI AND R. RIVEST, "A digital signature scheme secure against adaptive chosen-message attacks," *SIAM Journal of Computing*, Vol. 17, No. 2, April 1988, pp. 281–308.

17. L.C. GUILLOU AND J.-J. QUISQUATER, "A Practical Zero-Knowledge Protocol Fitted to Security Microprocessors Minimizing Both Transmission and Memory," *Advances in Cryptology EUROCRYPT '88*, Lecture Notes in Computer Science Vol. 330, C. Gunther ed., Springer-Verlag, 1988.

18. T. OKAMOTO, "Provably Secure and Practical Identification Schemes and Corresponding Signature Schemes," *Advances in Cryptology – CRYPTO '92*, Lecture Notes in Computer Science Vol. 740, E. Brickell ed., Springer-Verlag, 1992.

19. H. ONG AND C.P. SCHNORR, "Fast Signature Generation with a Fiat-Shamir Identification Scheme" *Advances in Cryptology – EUROCRYPT '90*, Lecture Notes in Computer Science Vol. 473, I. Damgård ed., Springer-Verlag, 1990.

20. C.P. SCHNORR, "Efficient Signature Generation by Smart Cards," *J. of Cryptology*, Vol. 4, 1991, pp. 161–174.

21. C.P. SCHNORR, "Security of 2^t-Root Identification and Signatures" *Advances in Cryptology – CRYPTO '96*, Lecture Notes in Computer Science Vol. 1109, N. Koblitz ed., Springer-Verlag, 1996.

22. V. SHOUP, "On the Security of a Practical Identification Scheme," *J. of Cryptology*, Vol. 12, 1999, pp. 247–260.

Does Encryption with Redundancy Provide Authenticity?

Jee Hea An and Mihir Bellare

Dept. of Computer Science & Engineering, University of California at San Diego,
9500 Gilman Drive, La Jolla, California 92093, USA.
{jeehea,mihir}@cs.ucsd.edu.
www-cse.ucsd.edu/users/{jeehea,mihir}.

Abstract. A popular paradigm for achieving privacy plus authenticity is to append some "redundancy" to the data before encrypting. We investigate the security of this paradigm at both a general and a specific level. We consider various possible notions of privacy for the base encryption scheme, and for each such notion we provide a condition on the redundancy function that is necessary and sufficient to ensure authenticity of the encryption-with-redundancy scheme. We then consider the case where the base encryption scheme is a variant of CBC called NCBC, and find sufficient conditions on the redundancy functions for NCBC encryption-with-redundancy to provide authenticity. Our results highlight an important distinction between public redundancy functions, meaning those that the adversary can compute, and secret ones, meaning those that depend on the shared key between the legitimate parties.

1 Introduction

The idea that authenticity can be easily obtained as a consequence of the privacy conferred by encryption has long attracted designers. Encryption-with-redundancy is the most popular paradigm to this end. Say that parties sharing key K are encrypting data via some encryption function $\mathcal{E}$. (Typically this is some block cipher mode of operation.) To obtain authenticity, the sender computes some function h of the data M to get a "checksum" $\tau = h(M)$. [1] It then computes a ciphertext $C \leftarrow \mathcal{E}_K(M\|\tau)$ and sends C to the receiver. The latter decrypts to get $M\|\tau$ and then checks whether $\tau = h(M)$. If not, it rejects the ciphertext as unauthentic.

The attraction of the paradigm is clear: the added cost of providing authenticity is small, amounting to computation of the checksum function plus perhaps one or two extra block cipher invocations in order to encrypt the now longer message. (Designers attempt to use simple and fast checksum functions.) However, the paradigm has a poor security record. For example, using CBC encryption with the checksum being the XOR of the message blocks (called CBCC) was proposed by the U.S. National Bureau of Standards, and was subsequently found

---

[1] Other names for the checksum include MDC —Manipulation Detection Code— and "redundancy," whence the name of the paradigm.

B. Pfitzmann (Ed.): EUROCRYPT 2001, LNCS 2045, pp. 512–528, 2001.

to not provide authenticity, as discussed in [23,16]. If the encryption algorithm is an additive stream cipher (e.g. CTR-mode encryption) where the adversary knows the plaintext, a forgery attacks by [15,16] apply. An attack attributed to Wagner on a large class of CBC-mode encryption-with-redundancy schemes is described in [24].

1.1   General Results

The many and continuing efforts to achieve authenticity via the encryption-with-redundancy paradigm point to the existence of some intuition that leads designers to think that it should work. The intuition appears to be that the privacy conveyed by the encryption makes attacks on the integrity harder. The first goal of our work is to assess the correctness of this intuition, and the security of the paradigm, at a general level. We are not concerned so much with the security of specific schemes as with trying to understand how the authenticity of the encryption-with-redundancy scheme relates to the security properties of the underlying primitives and to what extent the paradigm can be validated at a general level.

We denote the base encryption scheme by $\mathcal{SE} = (\mathcal{K}_e, \mathcal{E}, \mathcal{D})$. (It is specified by its key-generation, encryption, and decryption algorithms.) We are general with regard to the form of the redundancy computation method, allowing it to be key-based. A choice of method is given by a *redundancy code* $\mathcal{RC} = (\mathcal{K}_r, \mathcal{H})$ where $\mathcal{K}_r$ is an algorithm responsible for generating a key K_r while $\mathcal{H}$ takes K_r and the text M to return the redundancy or checksum $\tau = \mathcal{H}_{K_r}(M)$. Associated to $\mathcal{SE}$ and $\mathcal{RC}$ is the encryption-with-redundancy scheme $\mathcal{ER}$ in which one encrypts message M via $C \leftarrow \mathcal{E}_{K_e}(M \| \mathcal{H}_{K_r}(M))$. Upon receipt of ciphertext C, the receiver applies $\mathcal{D}_K$ to get back $M \| \tau$ and accepts iff $\tau = \mathcal{H}_{K_r}(M)$. Here K_e is the (secret) encryption key for $\mathcal{SE}$.

We distinguish *public redundancy* and *secret redundancy*. In the first case, K_r is public information. ($\mathcal{H}_{K_r}(\cdot)$ might be a public hash function like SHA-1, or simply return the XOR of the message blocks.) In this case, K_r is known to the adversary, who is thus capable of computing the redundancy function. In the case of secret redundancy, K_r is part of the secret key shared between the parties. (It might for example be a key for a universal hash function [11] or a message authentication code.) In this case the key K_r is not given to the adversary.

The desired authenticity property of the encryption-with-redundancy scheme $\mathcal{ER}$ is integrity of ciphertexts [7,19,8]: it should be computationally infeasible for an adversary to produce a ciphertext that is valid but different from any created by the sender.

We allow the assumed privacy attribute of the base encryption scheme to range across the various well-established notions of privacy used in the literature: IND-CPA, NM-CPA, IND-CCA. (Indistinguishability under chosen-plaintext attack [13,4], non-malleability under chosen-plaintext attack [12], and indistinguishability under chosen-ciphertext attack, respectively. Recall that non-malle-

Type of base encryption	Condition on redundancy code	
	For public redundancy	For secret redundancy
IND-CPA	None	None
NM-CPA	None	UF-NMA
IND-CCA	None	UF-NMA

Fig. 1. For each possible privacy attribute SSS-AAA of the base encryption scheme, we indicate a condition on the redundancy code that is *necessary and sufficient* for it to be integrity-providing with respect to SSS-AAA. We distinguish the cases where the redundancy is public (anyone can compute it) and secret (depends on the shared secret key). "None" means that the corresponding class of redundancy codes is empty: *No* redundancy code is integrity-providing.

---

ability under chosen-ciphertext attack is equivalent to IND-CCA [5,18] so we don't need to consider it separately.)

We say that a redundancy code $\mathcal{RC}$ is *integrity-providing with respect to se-curity notion SSS-AAA* if for *all* base encryption schemes $\mathcal{SE}$ that are SSS-AAA secure, the encryption-with-redundancy scheme $\mathcal{ER}$ obtained from $\mathcal{SE}$ and $\mathcal{RC}$ is secure in the sense of integrity of ciphertexts. (This property of a redundancy code is attractive from the design viewpoint, since a redundancy code having this property may be used in conjunction with *any* SSS-AAA-secure base encryption scheme, and authenticity of the resulting encryption-with-redundancy scheme is guaranteed.) The question we ask is the following. Given a notion of security SSS-AAA, what security attribute of the redundancy code $\mathcal{RC}$ will ensure that $\mathcal{RC}$ is integrity-providing with respect to security notion SSS-AAA?

We find that an important distinction to be made in answering this question is whether or not the redundancy computation is secret-key based. Figure 1 summarizes the results we expand on below.

ENCRYPTION WITH PUBLIC REDUNDANCY. We show that there is *no* choice of public redundancy code $\mathcal{RC}$ which is integrity-providing with respect to no-tions of security IND-CPA, NM-CPA or IND-CCA. This is a powerful indi-cation that the intuition that privacy helps provide integrity via encryption-with-redundancy is wrong in the case where the adversary can compute the redundancy function.

This conclusion is not surprising when the base encryption scheme meets only a weak notion of privacy like IND-CPA. But one might have thought that there are redundancy codes for which a condition like NM-CPA on the base encryption scheme would suffice to prove integrity of ciphertexts for the resulting encryption-with-redundancy scheme. Not only is this false, but it stays false when the base encryption scheme has even a stronger privacy attribute like IND-CCA.

Note that the most popular methods for providing redundancy are public, typically involving computing a keyless checksum of the message, and our result applies to these.

The result is proved by giving an example of a base encryption scheme meeting the notion of privacy in question such that for any redundancy code the corresponding encryption with public redundancy scheme can be attacked. (This assumes there exists some base encryption scheme meeting the notion of privacy in question, else the issue is moot.)

ENCRYPTION WITH SECRET REDUNDANCY. As Figure 1 indicates, allowing the computation of the redundancy to depend on a secret key does not help if the base encryption scheme meets only a weak notion of privacy like IND-CPA— *no* secret redundancy code is integrity-providing with respect to IND-CPA.

However secret redundancy does help if the base encryption scheme has stronger privacy attributes. We characterize the requirement on the redundancy code in this case. We say that it is UF-NMA (UnForgeable under No-Message Attack) if it is a MAC for which forgery is infeasible for an adversary that is not allowed to see the MACs of any messages before it must output its forgery. Our result is that this condition on the redundancy code is necessary and sufficient to ensure that it is integrity-providing with respect to NM-CPA and IND-CCA.

We stress that UF-NMA is a very weak security requirement, so the implication is that allowing the redundancy computation to depend on a secret key greatly increases security as long as the base encryption scheme is strong enough. We also stress that our condition on the redundancy code is both necessary and sufficient. Still in practice, the implication is largely negative because standard modes of operation do not meet notions like NM-CPA or IND-CCA.

PERSPECTIVE. The above results do not rule out obtaining secure schemes from the encryption-with-redundancy paradigm. The results refer to the ability to prove authenticity of the encryption-with-redundancy scheme *in general*, meaning based *solely* on assumed privacy attributes of the base encryption scheme and attributes of the redundancy code.

One might consider encryption with some specific redundancy code using as base encryption scheme a block cipher based mode of operation that is only IND-CPA secure, and yet be able to prove authenticity by analyzing the encryption-with-redundancy scheme directly based on the assumption that the block cipher is a pseudorandom permutation. This would not contradict the above results. What the above results do is show that the intuition that privacy helps integrity is flawed. Encryption-with-redundancy might work, but not for that reason. If a specific scheme such as the example we just mentioned works, it is not because of the privacy provided by the encryption, but, say, because of the pseudorandomness of the block cipher. In practice this tell us that to get secure encryption-with-redundancy schemes we must look at specific constructions and analyze them directly. This is what we do next.

1.2   Encryption with NCBC

We consider a variant of (random-IV) CBC mode encryption in which the enciphering corresponding to the last message block is done under a key different from that used for the other blocks. We call this mode NCBC. Here we are able to obtain positive results for both public and secret redundancy functions.

We show that if secret redundancy is used, quite simple and efficient redundancy codes suffice for the NCBC with redundancy scheme to provide authenticity. The redundancy code should satisfy the property called *AXU (Almost Xor Universal)* in [20,25]. (Any Universal-2 function [27] has this property and there are other efficient constructs as well [14,10,2].) On the other hand we show that if the redundancy is public, then authenticity of the NCBC with redundancy scheme is guaranteed if the redundancy code is *XOR-collision-resistant*. (The latter, a cryptographic property we define, can be viewed either as a variant of the standard collision-resistance property, or as an extension of the AXU property to the case where the key underlying the function is public.) These results assume the underlying block cipher is a strong pseudorandom permutation in the sense of [22].

These results should be contrasted with what we know about encryption with redundancy using the standard CBC mode as the base encryption scheme. Wagner's attack, pointed out in [24], implies that *no* public redundancy code will, in conjunction with CBC encryption, yield an encryption-with-redundancy scheme possessing integrity of ciphertexts. In the case where the redundancy is secret, Krawczyk [21] shows that it suffices for the redundancy code to be a MAC secure against chosen-message attack, but this is a strong condition on the redundancy code compared to the AXU property that suffices for NCBC. Thus, the simple modification consisting of enciphering under a different key for the last block substantially enhances CBC with regard to its ability to provide authenticity under the encryption-with-redundancy paradigm.

1.3   Related Work

Preneel gives an overview of existing authentication methods [24] that includes much relevant background. A comprehensive treatment of authenticated encryption —the goal of joint privacy and authenticity— is provided in [7]. They relate different notions of privacy and authenticity to compare their relative strengths.

Encryption-with-redundancy is one of many approaches to the design of authenticated encryption schemes. Another general approach is "generic composition:" combine an encryption scheme with a MAC in some way. This is analyzed in [7], who consider the following generic composition methods: *Encrypt-and-mac, Mac-then-encrypt, Encrypt-then-mac*. For each of these methods they consider two notions of integrity, namely integrity of ciphertexts and a weaker notion of integrity of plaintexts, and then, assuming the base encryption scheme is IND-CPA and the MAC is secure against chosen-message attack, indicate whether or not the method has the integrity property in question. Krawczyk's recent work [21] considers the same methods from the point of view of building "secure channels" over insecure networks. The drawback of the generic composition approach compared to the encryption-with-redundancy approach is that some MACs might be less efficient than redundancy codes, and that public redundancy avoids the additional independent key that is required for MACs.

Another general paradigm is "encode then encipher" [8] —add randomness and redundancy and then encipher rather than encrypt. Encode then encipher requires a variable-input length strong pseudorandom permutation, which can be relatively expensive to construct.

Let $SNCBC[F, \mathcal{RC}]$ denote NCBC encryption with block cipher F and secret redundancy provided by an efficient AXU redundancy code $\mathcal{RC}$. We compare this to other authenticated encryption schemes such as RPC mode [19], IACBC [17], and OCB [26]. RPC is computation and space inefficient compared to all the other methods. IACBC and OCB have cost comparable to that of $SNCBC[F, \mathcal{RC}]$, but OCB is parallelizable.

Encryption-with-redundancy is one of many approaches to simultaneously achieving privacy and authenticity. Our goal was to analyze and better understand this approach. We do not suggest it is superior to other approaches.

2   Definitions

A string is a member of $\{0,1\}^*$. We denote by "$\|$" an operation that combines several strings into one in such a way that the constituent strings are uniquely recoverable from the final one. (If lengths of all strings are fixed and known, concatenation will serve the purpose.) The empty string is denoted ε.

EXTENDED ENCRYPTION SCHEMES. The usual syntax of a symmetric encryption scheme (cf. [4]) is that encryption and decryption depend on a key shared between sender and receiver but not given to the adversary. We wish to consider a setting where operations depend, in addition to the shared key, on some public information, such as a hash function. The latter may be key based. (Think of the key as having been chosen at random at design time and embedded in the hash function.) All parties including the adversary have access to this key, which we call the *common key*. We need to model it explicitly because security depends on the random choice of this key even though it is public. This requires a change in encryption scheme syntax. Accordingly we define an *extended encryption scheme* which extends the usual symmetric encryption scheme by addition of another key generation algorithm. Specifically an extended encryption scheme $\mathcal{EE} = (\mathcal{K}_c, \mathcal{K}_s, \mathcal{E}, \mathcal{D})$ consists of four algorithms as follows. The randomized *common key generation* algorithm $\mathcal{K}_c$ takes input a security parameter $k \in \mathsf{N}$ and in time poly(k) returns a key K_c; we write $K_c \xleftarrow{R} \mathcal{K}_c(k)$. The randomized *secret key generation* algorithm $\mathcal{K}_s$ also takes input $k \in \mathsf{N}$ and in time poly(k) returns a key K_s; we write $K_s \xleftarrow{R} \mathcal{K}_s(k)$. We let $K = (K_c, K_s)$. The *encryption* algorithm $\mathcal{E}$ is either randomized or stateful. It takes K and a *plaintext* M and in time poly($k, |M|$) returns a *ciphertext* $C = \mathcal{E}_K(M)$; we write $C \xleftarrow{R} \mathcal{E}_K(M)$. (If randomized, it flips coins, anew upon each invocation. If stateful, it maintains a state which it updates upon each invocation.) The deterministic and stateless *decryption* algorithm $\mathcal{D}$ takes the key K and a string C and in time poly($k, |C|$) returns either the corresponding plaintext M or the distinguished symbol $\bot$; we write $x \leftarrow \mathcal{D}_K(C)$. We require that $\mathcal{D}_K(\mathcal{E}_K(M)) = M$ for all $M \in \{0,1\}^*$.

Notice that it is not apparent from the syntax why there are two keys because they are treated identically. The difference will surface when we consider security: we will view the legitimate users as possessing K_s while both they and the adversary have K_c. (It also surfaces in something we don't consider explicitly here, which is a multi-user setting. In that case, although K_s will be generated anew for each pair of users, K_c may be the same across the whole system.)

A standard symmetric encryption scheme, namely one where there is no common key, can be recovered as the special case where the common key generation algorithm $\mathcal{K}_c$ returns the empty string. Formally, we say that $\mathcal{SE} = (\mathcal{K}, \mathcal{E}, \mathcal{D})$ is a (symmetric) encryption scheme if $\mathcal{EE} = (\mathcal{K}_c, \mathcal{K}, \mathcal{E}, \mathcal{D})$ is an extended encryption scheme where $\mathcal{K}_c$ is the algorithm which on any input returns the empty string. When the common key K_c is the empty string we may also omit it in the input given to the adversary.

NOTIONS OF SECURITY. Notions of security for symmetric encryption schemes are easily adapted to extended encryption schemes by giving the adversary the common key as input. Via the formal definitions shown below and this discussion we will summarize the definitions we need.

We let $\mathcal{EE} = (\mathcal{K}_c, \mathcal{K}_s, \mathcal{E}, \mathcal{D})$ be the extended encryption scheme whose security we are defining. The formalizations, given in Definition 1 and Definition 2, associate to each notion of security and each adversary an *experiment*, and based on that, an *advantage*. The latter is a function of the security parameter that measures the success probability of the adversary. Asymptotic notions of security result by asking this function to be negligible for adversaries of time complexity polynomial in the security parameter. Concrete security assessments can be made by associating to the scheme another advantage function that for each value of the security parameter and given resources for an adversary returns the maximum, over all adversaries limited to the given resources, of the advantage of the adversary.

Note that these definitions apply to standard symmetric encryption schemes too, since as per our conventions the latter are simply the special case of extended encryption schemes in which the common key generation algorithm returns the empty string.

PRIVACY. The basic and weakest natural notion of privacy is IND-CPA. We use one of the formalizations of [4] which adapts that of [13] to the symmetric setting. A challenge bit b is chosen, the adversary is given K_c, and can query, adaptively and as often as it likes, the left-or-right encryption oracle. The adversary wins if it can guess b. For IND-CCA the adversary gets in addition a decryption oracle but loses if it queries it on any ciphertext returned by the left-or-right encryption oracle.

Non-malleability captures, intuitively, the inability of an adversary to change a ciphertext into another one such that the underlying plaintexts are meaningfully related [12]. We do not formalize it directly as per [12,5] but rather via the equivalent indistinguishability under parallel chosen-ciphertext attack characterization of [9,18]. (This facilitates our proofs.) The adversary gets the left-or-right encryption oracle and must then decide on a vector of ciphertexts **c**. (It loses if they contain an output of the left-or-right encryption oracle.) It is given their corresponding decryptions $\mathbf{p}$ and then wins if it guesses the challenge bit.

The formal definition of privacy is below with the associated experiments.

Definition 1. [Privacy] Let $\mathcal{EE} = (\mathcal{K}_c, \mathcal{K}_s, \mathcal{E}, \mathcal{D})$ be an extended encryption scheme, $b \in \{0, 1\}$ a challenge bit and $k \in \mathbb{N}$ the security parameter. Let A be an adversary that outputs a bit d. The *left-or-right* encryption oracle $\mathcal{E}_K(\mathcal{LR}(\cdot, \cdot, b))$, given to the adversary A, takes input a pair (x_0, x_1) of equal-length messages, computes ciphertext $X \leftarrow \mathcal{E}_K(x_b)$, and returns X to the adversary. (It flips coins, or updates state for the encryption function, as necessary. If the input messages are not of equal length it returns the empty string.) Now consider the following experiments each of which returns a bit.

Experiment $\mathbf{Exp}_{\mathcal{EE},A}^{\text{ind-cpa-}b}(k)$	Experiment $\mathbf{Exp}_{\mathcal{EE},A}^{\text{ind-cca-}b}(k)$
$K_c \overset{R}{\leftarrow} \mathcal{K}_c(k) \,;\; K_s \overset{R}{\leftarrow} \mathcal{K}_s(k)$	$K_c \overset{R}{\leftarrow} \mathcal{K}_c(k) \,;\; K_s \overset{R}{\leftarrow} \mathcal{K}_s(k) \,;\; K \leftarrow (K_c, K_s)$
$K \leftarrow (K_c, K_s)$	$d \leftarrow A^{\mathcal{E}_K(\mathcal{LR}(\cdot, \cdot, b)), \mathcal{D}_K(\cdot)}(k, K_c)$
$d \leftarrow A^{\mathcal{E}_K(\mathcal{LR}(\cdot, \cdot, b))}(k, K_c)$	If $\mathcal{D}_K(\cdot)$ was never queried on an output of
return d	$\mathcal{E}_K(\mathcal{LR}(\cdot, \cdot, b))$ then return d else return 0

Experiment $\mathbf{Exp}_{\mathcal{EE},A}^{\text{nm-cpa-}b}(k)$

$K_c \overset{R}{\leftarrow} \mathcal{K}_c(k) \,;\; K_s \overset{R}{\leftarrow} \mathcal{K}_s(k) \,;\; K \leftarrow (K_c, K_s)$

$(\mathbf{c}, s) \leftarrow A_1^{\mathcal{E}_K(\mathcal{LR}(\cdot, \cdot, b))}(k, K_c) \,;\; \mathbf{p} \leftarrow (\mathcal{D}_K(c_1), \cdots, \mathcal{D}_K(c_n)) \,;\; d \leftarrow A_2(\mathbf{p}, \mathbf{c}, s)$

If $\mathbf{c}$ contains no ciphertext output by $\mathcal{E}_K(\mathcal{LR}(\cdot, \cdot, b))$ then return d else return 0

For each notion of privacy sss-aaa $\in \{\text{ind-cpa, ind-cca, nm-cpa}\}$ we associate to the adversary A a corresponding advantage defined via

$$\mathbf{Adv}_{\mathcal{EE},A}^{\text{sss-aaa}}(k) = \Pr\left[\mathbf{Exp}_{\mathcal{EE},A}^{\text{sss-aaa-}1}(k) = 1\right] - \Pr\left[\mathbf{Exp}_{\mathcal{EE},A}^{\text{sss-aaa-}0}(k) = 1\right].$$

For each security notion SSS-AAA $\in \{\text{IND-CPA, IND-CCA, NM-CPA}\}$, the scheme $\mathcal{EE}$ is said to be *SSS-AAA secure* if the corresponding advantage function, $\mathbf{Adv}_{\mathcal{EE},F}^{\text{sss-aaa}}(\cdot)$ of any adversary F whose time-complexity is polynomial in k, is negligible. $\blacksquare$

INTEGRITY. The formalization of integrity follows [7]. The adversary is allowed to mount a chosen-message attack on the scheme, modeled by giving it access to an encryption oracle. Success is measured by its ability to output a "new" ciphertext that makes the decryption algorithm output a plaintext rather than reject by outputting $\perp$. Here the "new" ciphertext means that the ciphertext was never output by the encryption oracle as a response to the adversary's queries. The formal definition of integrity is below with the associated experiment.

Definition 2. [Integrity] Let $\mathcal{EE} = (\mathcal{K}_c, \mathcal{K}_s, \mathcal{E}, \mathcal{D})$ be an extended encryption scheme, and $k \in \mathbb{N}$ the security parameter. Let B be an adversary that has access to the encryption oracle and outputs a ciphertext. Now consider the following experiment.

Experiment $\mathbf{Exp}_{\mathcal{EE},B}^{\text{int-ctxt}}(k)$

$\quad K_c \stackrel{R}{\leftarrow} \mathcal{K}_c \; ; \; K_s \stackrel{R}{\leftarrow} \mathcal{K}_s \; ; \; K \leftarrow (K_c, K_s) \; ; \; C \leftarrow B^{\mathcal{E}_K(\cdot)}(k, K_c)$

$\quad$ If $\mathcal{D}_K(C) \neq \bot$ and C was never a response of $\mathcal{E}_K(\cdot)$ then return 1 else return 0

We associate to the adversary B a corresponding advantage defined via

$$\mathbf{Adv}_{\mathcal{EE},B}^{\text{int-ctxt}}(k) \;=\; \Pr\left[\, \mathbf{Exp}_{\mathcal{EE},B}^{\text{int-ctxt}}(k) = 1 \,\right] .$$

The scheme $\mathcal{EE}$ is said to be *INT-CTXT secure* if the advantage function $\mathbf{Adv}_{\mathcal{EE},F}^{\text{int-ctxt}}(\cdot)$ of any adversary F whose time-complexity is polynomial in k, is negligible. $\quad\blacksquare$

3   The Encryption-with-Redundancy Paradigm

We describe the paradigm in a general setting, as a transform that associates to any given symmetric encryption scheme and any given "redundancy code" an extended encryption scheme. We first define the syntax for redundancy codes, then detail the constructions, separating the cases of public and secret redundancy, and conclude by observing that the transform always preserves privacy. This leaves later sections to investigate the difficult issue, namely the integrity of the extended encryption scheme with redundancy.

REDUNDANCY CODES. A *redundancy code* $\mathcal{RC} = (\mathcal{K}_r, \mathcal{H})$ consists of two algorithms $\mathcal{K}_r$ and $\mathcal{H}$. The randomized key generation algorithm $\mathcal{K}_r$ takes a security parameter k and in time $\text{poly}(k)$ returns a key K_r; we write $K_r \stackrel{R}{\leftarrow} \mathcal{K}_r(k)$. The deterministic redundancy computation algorithm $\mathcal{H}$ takes K_r and a string $M \in \{0,1\}^*$ and in time $\text{poly}(k, |M|)$ returns a string τ; we write $\tau \leftarrow \mathcal{H}_{K_r}(M)$. Usually the length of τ is $\ell(k)$ where $\ell(\cdot)$, an integer valued function that depends only on the security parameter, is called the *output length* of the redundancy code. We say that the redundancy is *public* if the key K_r is public and known to the adversary. We say the redundancy is *secret* if K_r is part of the shared secret key.

EXTENDED ENCRYPTION SCHEMES WITH REDUNDANCY. Let $\mathcal{SE} = (\mathcal{K}_e, \mathcal{E}, \mathcal{D})$ be a given (symmetric) encryption scheme, which we will call the *base* encryption scheme. Let $\mathcal{RC} = (\mathcal{K}_r, \mathcal{H})$ be a given redundancy code as above. We define an associated *extended encryption scheme with public redundancy* and an associated *extended encryption scheme with secret redundancy*.

Construction 1. The extended encryption scheme with public redundancy $\mathcal{EPR} = (\mathcal{K}_c, \mathcal{K}_s, \overline{\mathcal{E}}, \overline{\mathcal{D}})$, associated to base encryption scheme $\mathcal{SE} = (\mathcal{K}_e, \mathcal{E}, \mathcal{D})$ and redundancy code $\mathcal{RC} = (\mathcal{K}_r, \mathcal{H})$, is defined as follows:

Algorithm $\mathcal{K}_c(k)$	Algorithm $\mathcal{K}_s(k)$	Algorithm $\overline{\mathcal{E}}_{\langle K_e, K_r\rangle}(M)$	Algorithm $\overline{\mathcal{D}}_{\langle K_e, K_r\rangle}(C)$
$K_r \stackrel{R}{\leftarrow} \mathcal{K}_r(k)$	$K_e \stackrel{R}{\leftarrow} \mathcal{K}_e(k)$	$\tau \leftarrow \mathcal{H}_{K_r}(M)$	$P \leftarrow \mathcal{D}_{K_e}(C)$
return K_r	return K_e	$C \stackrel{R}{\leftarrow} \mathcal{E}_{K_e}(M\|\tau)$	Parse P as $M\|\tau$
		return C	if $\tau \neq \mathcal{H}_{K_r}(M)$
			$\quad$ then return $\bot$
			$\quad$ else return M

Note that the common-key generation algorithm returns the key for the redundancy function, which is thus available to the adversary. That is why we say the redundancy is public. ∎

Construction 2. The extended encryption scheme with secret redundancy $\mathcal{ESR} = (\mathcal{K}_c, \mathcal{K}_s, \overline{\mathcal{E}}, \overline{\mathcal{D}})$, associated to base encryption scheme $\mathcal{SE} = (\mathcal{K}_e, \mathcal{E}, \mathcal{D})$ and redundancy code $\mathcal{RC} = (\mathcal{K}_r, \mathcal{H})$, is defined as follows:

Algorithm $\mathcal{K}_c(k)$	Algorithm $\mathcal{K}_s(k)$	Algorithm $\overline{\mathcal{E}}_{\langle K_e, K_r \rangle}(M)$	Algorithm $\overline{\mathcal{D}}_{\langle K_e, K_r \rangle}(C)$
return ε	$K_e \stackrel{R}{\leftarrow} \mathcal{K}_e(k)$	$\tau \leftarrow \mathcal{H}_{K_r}(M)$	$N \leftarrow \mathcal{D}_{K_e}(C)$
	$K_r \stackrel{R}{\leftarrow} \mathcal{K}_r(k)$	$C \stackrel{R}{\leftarrow} \mathcal{E}_{K_e}(M \| \tau)$	Parse N as $M \| \tau$
	return $\langle K_e, K_r \rangle$	return C	if $\tau \neq \mathcal{H}_{K_r}(M)$
			then return $\bot$
			else return M

Note that the common key generation algorithm $\mathcal{K}_c$ returns the empty string ε. We may omit the algorithm $\mathcal{K}_c$ and write $\mathcal{ESR} = (\mathcal{K}_s, \overline{\mathcal{E}}, \overline{\mathcal{D}})$. The key for the redundancy function is part of the secret key not available to the adversary. ∎

The symbol $\bot$ is a distinct symbol that indicates that the ciphertext is not valid. When we refer to an extended encryption scheme with redundancy in general we mean either of the above, and denote it by $\mathcal{ER}$.

PRIVACY IS PRESERVED. We now present a theorem regarding the privacy of an extended encryption scheme with redundancy. It applies both to the case of public and to the case of secret redundancy. The theorem below says that the encryption scheme with redundancy inherits the privacy of the base symmetric encryption scheme regardless of the redundancy code being used. This means that privacy depends only on the underlying encryption scheme, not on the redundancy code. The proof is straightforward and can be found in the full version of this paper [1].

Theorem 1. [Privacy of an extended encryption scheme with redundancy] *Let* $\mathcal{SE} = (\mathcal{K}_e, \mathcal{E}, \mathcal{D})$ *be a symmetric encryption scheme and let* $\mathcal{RC} = (\mathcal{K}_r, \mathcal{H})$ *be a redundancy code. Let* $\mathcal{ER} = (\mathcal{K}_c, \mathcal{K}_s, \overline{\mathcal{E}}, \overline{\mathcal{D}})$ *be an associated extended encryption scheme with redundancy, either public or secret. Then if* $\mathcal{SE}$ *is IND-CPA (resp. IND-CCA, NM-CPA) secure, so is* $\mathcal{ER}$. ∎

For simplicity we have stated the theorem with reference to asymptotic notions of security but we remark that the reduction in the proof is tight, and a concrete security statement reflecting this can be derived from the proof.

4   Encryption with Public Redundancy

Here we will show that in general the encryption with public redundancy paradigm fails in a strong way, meaning there is a base encryption scheme such that for *all* choices of public redundancy code, the associated extended encryption scheme with public redundancy scheme (cf. Construction 1) fails to provide integrity. This is true regardless of the security property of the base encryption scheme (i.e. IND-CPA, NM-CCA, or IND-CCA).

The result follows the paradigm of similar negative results in [4,7]. We must make the minimal assumption that some encryption scheme $\mathcal{SE}'$ secure in the given sense exists, else the question is moot. We then modify the given encryption scheme to a new scheme $\mathcal{SE}$ so that when $\mathcal{SE}$ becomes the base encryption scheme of the extended encryption scheme with public redundancy, we can provide an attack on the integrity of the latter. The proof of the following theorem can be found in the full version of this paper [1].

Theorem 2. [Encryption with public redundancy] *Suppose there exists a symmetric encryption scheme $\mathcal{SE}'$ which is IND-CCA (resp. IND-CPA, NM-CPA) secure. Then there exists a symmetric encryption scheme $\mathcal{SE}$ which is also IND-CCA (resp. IND-CPA, NM-CPA) secure but, for any redundancy code $\mathcal{RC}$, the extended encryption scheme with public redundancy $\mathcal{EPR}$ associated to $\mathcal{SE}$ and $\mathcal{RC}$ is not INT-CTXT secure.* ∎

5   Encryption with Secret Redundancy

In this section, we examine encryption schemes with secret redundancy in general so as to whether or not they provide integrity.

The following theorem states the negative result where the base encryption scheme is IND-CPA secure. The proof can be found in the full version of this paper [1].

Theorem 3. [IND-CPA encryption with secret redundancy] *Suppose there exists a symmetric encryption scheme $\mathcal{SE}'$ which is IND-CPA secure. Then there exists a symmetric encryption scheme $\mathcal{SE}$ which is also IND-CPA secure but, for any redundancy code $\mathcal{RC}$, the extended encryption scheme with secret redundancy $\mathcal{ESR}$ associated to $\mathcal{SE}$ and $\mathcal{RC}$ is not INT-CTXT secure.* ∎

For the positive result, we define below the (necessary and sufficient) security property required of the redundancy code.

We define a notion of *unforgeability under no message attack (UF-NMA)*, which is the weakest form of security required of a MAC (message authentication code) —roughly, the adversary wins if it outputs a valid message and tag pair without seeing any legitimately produced message and tag pairs. Since a MAC and a redundancy code are syntactically identical, we adopt the weakest security notion of a MAC as the security notion of a redundancy code. The formal definition is given below. Note that, in the attack model, the key to the redundancy code is not given to the adversary, indicating that the redundancy is secret.

Definition 3. [Unforgeability under no message attack (UF-NMA)] Let $\mathcal{RC} = (\mathcal{K}_r, \mathcal{H})$ be a redundancy code. Let $k \in \mathsf{N}$. Let F be an adversary. Consider the following experiment:

Algorithm $\mathcal{K}_e(k)$	Algorithm $\mathcal{E}_{a_1 \| a_2}(X)$	Algorithm $\mathcal{D}_{a_1 \| a_2}(Y)$
$a_1 \stackrel{R}{\leftarrow} \{0,1\}^\kappa$	Parse X as $x_1 \cdots x_{n+1}$	Parse Y as $y_0 y_1 \cdots y_{n+1}$
$a_2 \stackrel{R}{\leftarrow} \{0,1\}^\kappa$	$y_0 \stackrel{R}{\leftarrow} \{0,1\}^l$	For $i = 1, \cdots, n$ do
Return $(a_1 \| a_2)$	For $i = 1, \cdots, n$ do	$x_i \leftarrow F_{a_1}^{-1}(y_i) \oplus y_{i-1}$
	$y_i \leftarrow F_{a_1}(y_{i-1} \oplus x_i)$	$x_{n+1} \leftarrow F_{a_2}^{-1}(y_{n+1}) \oplus y_n$
	$y_{n+1} \leftarrow F_{a_2}(y_n \oplus x_{n+1})$	$X \leftarrow x_1 \cdots x_{n+1}$
	Return $y_0 y_1 \cdots y_{n+1}$	Return X

Fig. 2. Nested CBC encryption scheme $NCBC[F] = (\mathcal{K}_e, \mathcal{E}, \mathcal{D})$.

Experiment $\mathbf{Exp}_{\mathcal{RC},F}^{\text{uf-nma}}(k)$

$\quad K_r \stackrel{R}{\leftarrow} \mathcal{K}_r(k) \,;\, (M, \tau) \leftarrow F(k)$

$\quad$ If $\tau = \mathcal{H}_{K_r}(M)$ then return 1 else return 0

We define the *advantage* of the adversary via,

$$\mathbf{Adv}_{\mathcal{RC},F}^{\text{uf-nma}}(k) = \Pr\left[\, \mathbf{Exp}_{\mathcal{RC},F}^{\text{uf-nma}}(k) = 1 \,\right]$$

The redundancy code $\mathcal{RC}$ is said to be *UF-NMA secure* if the function $\mathbf{Adv}_{\mathcal{RC},F}^{\text{uf-nma}}(\cdot)$ is negligible for any adversary F whose time complexity is polynomial in k. ∎

The following theorem states the positive results. The proof can be found in the full version of this paper [1].

Theorem 4. [NM-CPA or IND-CCA encryption with secret redundancy] *Let $\mathcal{SE}$ be a symmetric encryption scheme which is NM-CPA or IND-CCA secure and let $\mathcal{RC}$ be a redundancy code. Then the extended encryption scheme with secret redundancy $\mathcal{ESR}$ associated to $\mathcal{SE}$ and $\mathcal{RC}$ is INT-CTXT secure if and only if the redundancy code $\mathcal{RC}$ is UF-NMA secure.* ∎

6   Nested CBC (NCBC) with Redundancy

In this section, we will consider a "natural" variant of CBC encryption, called "Nested CBC (NCBC)", designed to eliminate length-based attacks. The detailed description of NCBC is given below.

Let $F: \{0,1\}^\kappa \times \{0,1\}^l \to \{0,1\}^l$ be a family of permutations (i.e. a block cipher). We let $F_a(\cdot) = F(a, \cdot)$ and we let F_a^{-1} denote the inverse of F_a, for any key $a \in \{0,1\}^\kappa$. Our variant of CBC encryption involves the use of two keys instead of just one. The additional key is used for the last iteration of the block cipher. We call this variant of CBC the *Nested CBC (NCBC)* and denote it by $NCBC[F] = (\mathcal{K}_e, \mathcal{E}, \mathcal{D})$. The algorithms for the NCBC encryption scheme are shown in Figure 2. We assume that the messages have length a multiple of the block length l.

Given the NCBC encryption scheme, we examine what kinds of security properties for the redundancy code will provide integrity of ciphertexts for the encryption scheme with redundancy. We examine this for both public redundancy and secret redundancy. In order to facilitate the practical security analyses, we will make concrete security assessments for the schemes examined in this section.

Since the security of the NCBC scheme is based on the security of the underlying block cipher (as well as that of the redundancy code), we first define the security property of the underlying block cipher on which our security analysis will be based.

Block ciphers are usually modeled as "pseudorandom permutations" (sometimes even as "pseudorandom functions") [4]. However, we use a stronger notion called *strong pseudorandom permutation (SPRP)* [22], where the adversary gets access to both forward and inverse permutation oracles in the attack model.

Definition 4. [Strong pseudorandom permutation (SPRP)]
Let $F: \{0,1\}^{\kappa} \times \{0,1\}^{l} \to \{0,1\}^{l}$ be a block cipher with key-length κ and block-length l. Let P^{l} be the family of all permutations on l-bits. Let $k \in \mathsf{N}$ and $b \in \{0,1\}$. Let D be an adversary that has access to oracles $g(\cdot)$ and $g^{-1}(\cdot)$. Consider the following experiment:

Experiment $\mathbf{Exp}_{F,D}^{\mathrm{sprp}\text{-}b}(k)$
   If $b = 0$ then $g \xleftarrow{R} P^{l}$ else $K \xleftarrow{R} \{0,1\}^{l}$; $g \leftarrow F_K$
   $d \leftarrow D^{g(\cdot),g^{-1}(\cdot)}(k)$; return d

We define the *advantage* and the *advantage function* of the adversary as follows. For any integers $t, q \geq 0$,

$$\mathbf{Adv}_{F,D}^{\mathrm{sprp}}(k) = \Pr\left[\, \mathbf{Exp}_{F,D}^{\mathrm{sprp}\text{-}1}(k) = 1 \,\right] - \Pr\left[\, \mathbf{Exp}_{F,D}^{\mathrm{sprp}\text{-}0}(k) = 1 \,\right]$$

$$\mathbf{Adv}_{F}^{\mathrm{sprp}}(k,t,q) = \max_{D}\left\{ \mathbf{Adv}_{F,D}^{\mathrm{sprp}}(k) \right\}$$

where the maximum is over all D with time complexity t, making at most q queries to the oracles $g(\cdot)$ and $g^{-1}(\cdot)$. The block cipher F is said to be *SPRP secure* if the function $\mathbf{Adv}_{F,D}^{\mathrm{sprp}}(k)$ is negligible for any adversary D whose time complexity is polynomial in k. ∎

The "time-complexity" refers to that of the entire experiment. Here, the choice of a random permutation g is not made all at once, but rather g is simulated in the natural way.

6.1   NCBC with Secret Redundancy

Here we examine what kind of property on the redundancy code suffices to make the NCBC with secret redundancy provide integrity. We denote by $SNCBC[F, \mathcal{RC}] = (\mathcal{K}_s, \overline{\mathcal{E}}, \overline{\mathcal{D}})$ the extended encryption scheme with secret redundancy associated to the NCBC encryption scheme $NCBC[F] = (\mathcal{K}_e, \mathcal{E}, \mathcal{D})$ and a redundancy code $\mathcal{RC} = (\mathcal{K}_r, \mathcal{H})$.

It turns out that the NCBC scheme with secret redundancy provides integrity if the underlying secret redundancy meets the notion of *almost XOR universal* (AXU) introduced in [20,25].

Definition 5. [Almost XOR Universal (AXU)] Let $\mathcal{RC} = (\mathcal{K}_r, \mathcal{H})$ be a redundancy code whose output length is $\ell(k)$, where $k \in \mathsf{N}$. We define the *advantage function* of the redundancy code $\mathcal{RC}$ as follows.

$$\mathbf{Adv}_{\mathcal{RC}}^{\mathrm{axu}}(k, \mu)$$

$$= \max_{x, x' \in \{0,1\}^*, r \in \{0,1\}^{\ell(k)}} \left\{ \Pr\left[\mathcal{H}_{K_r}(x) \oplus \mathcal{H}_{K_r}(x') = r : K_r \stackrel{R}{\leftarrow} \mathcal{K}_r(k) \right] \right\}$$

where maximum is taken over all *distinct* x, x' of length at most μ each, and all $r \in \{0,1\}^{\ell(k)}$. ∎

We now state the theorem concerning the security of NCBC scheme with secret redundancy. The proof can be found in the full version of this paper [1].

Theorem 5. [Integrity of NCBC with secret redundancy] *Let $\mathcal{RC}$ be a redundancy code whose output length is l-bits. Let $F: \{0,1\}^\kappa \times \{0,1\}^l \to \{0,1\}^l$ be a block cipher, and let $NCBC[F]$ be the NCBC encryption scheme based on F. Let $SNCBC[F, \mathcal{RC}]$ be the extended encryption scheme with secret redundancy associated to $NCBC[F]$ and $\mathcal{RC}$. Let $k \in \mathsf{N}$. Then*

$$\mathbf{Adv}_{SNCBC[F, \mathcal{RC}]}^{\mathrm{int\text{-}ctxt}}(k, t, q, \mu)$$

$$\leq \left(\frac{q(q-1)}{2} + 1 \right) \cdot \mathbf{Adv}_{SRC}^{\mathrm{axu}}(k, \mu) + \frac{1}{2^l - q} + \mathbf{Adv}_F^{\mathrm{sprp}}(k, t, q + \mu/l) \quad ∎$$

6.2   NCBC with Public Redundancy

The NCBC with *public* redundancy scheme also provides authenticity if a certain condition on the underlying redundancy code is satisfied. We denote by $PNCBC[F, \mathcal{RC}] = (\mathcal{K}_c, \mathcal{K}_s, \overline{\mathcal{E}}, \overline{\mathcal{D}})$ the extended encryption scheme with public redundancy associated to the NCBC encryption scheme $NCBC[F] = (\mathcal{K}_e, \mathcal{E}, \mathcal{D})$ and a redundancy code $\mathcal{RC} = (\mathcal{K}_r, \mathcal{H})$.

   We want to examine what kind of security property for the underlying public redundancy suffices to make the NCBC scheme with public redundancy provide integrity. It turns out that, for the redundancy code, a cryptographic property called "XOR-collision-resistance" suffices to provide integrity for the NCBC scheme with public redundancy. XOR-collision-resistance is slightly stronger than "collision-resistance". Roughly, a redundancy code $\mathcal{RC} = (\mathcal{K}_r, \mathcal{H})$ is said to be *XOR-collision-resistant (XCR)* if it is "hard" to find strings x, x' where $x \neq x'$ such that $\mathcal{H}_{K_r}(x) \oplus \mathcal{H}_{K_r}(x') = r$ for any committed value r and any given key K_r. We define XOR-collision-resistance (XCR) more formally as follows.

Definition 6. [XOR-Collision-Resistance (XCR)] Let $\mathcal{RC} = (\mathcal{K}_r, \mathcal{H})$ be a redundancy code whose output length is $\ell(k)$, where $k \in \mathsf{N}$. Let $B = (B_1, B_2)$ be an adversary. Consider the following experiment:

Experiment $\mathbf{Exp}_{\mathcal{PRC}, B}^{\mathrm{xcr}}(k)$
   $(r, s) \leftarrow B_1(k)$; $K_r \stackrel{R}{\leftarrow} \mathcal{K}_r(k)$; $(x, x') \leftarrow B_2(K_r, r, s)$
   if $\mathcal{H}_{K_r}(x) \oplus \mathcal{H}_{K_r}(x') = r$ and $x \neq x'$ then return 1 else return 0

Above, the variable s denotes the state information. We define the *advantage* and the *advantage function* of the adversary via,

$$\mathbf{Adv}^{\mathrm{xcr}}_{\mathcal{PRC},B}(k) = \Pr\left[\mathbf{Exp}^{\mathrm{xcr}}_{\mathcal{PRC},B}(k) = 1\right]$$

$$\mathbf{Adv}^{\mathrm{xcr}}_{\mathcal{PRC}}(k,t) = \max_{B}\left\{\mathbf{Adv}^{\mathrm{xcr}}_{\mathcal{PRC},B}(k)\right\}$$

where the maximum is over all B with time complexity t. The scheme $\mathcal{PRC}$ is said to be *XCR secure* if the function $\mathbf{Adv}^{\mathrm{xcr}}_{\mathcal{PRC},A}(k)$ is negligible for any adversary A whose time complexity is polynomial in k. ∎

XOR-collision-resistance (XCR) as defined above is a new notion that has not been explicitly studied in the literature. In XCR, the adversary first outputs a string r and then obtains the key to the function. The adversary's goal is to find a pair of strings x, x' (called an "XOR-collision" pair) such that the XOR of their images equals r.

Given the definitions for the security properties of the underlying primitives, we now state the theorem regarding the security of the $PNCBC$ scheme. Following that we will further discuss XCR redundancy codes. The proof can be found in the full version of this paper [1].

Theorem 6. [Integrity of NCBC with public redundancy] *Let $\mathcal{RC}$ be a redundancy code whose output length is l-bits. Let $F\colon \{0,1\}^\kappa \times \{0,1\}^l \to \{0,1\}^l$ be a block cipher, and let $NCBC[F]$ be the NCBC encryption scheme based on F. Let $PNCBC[F, \mathcal{RC}]$ be the extended encryption scheme with public redundancy associated to $NCBC[F]$ and $\mathcal{RC}$. Let $k \in \mathsf{N}$. Then*

$$\mathbf{Adv}^{\mathrm{int\text{-}ctxt}}_{PNCBC[F,\mathcal{RC}]}(k,t,q,\mu)$$

$$\leq mq \cdot \mathbf{Adv}^{\mathrm{xcr}}_{\mathcal{RC}}(k,t') + \frac{2}{2^l - m} + \frac{m^2}{2(2^l - m)} + 2 \cdot \mathbf{Adv}^{\mathrm{sprp}}_{F}(k,t,q+m)$$

where $m = \mu/l$. ∎

We now further discuss XCR redundancy codes. Note that the XCR property can be thought of as a cryptographic counterpart of the AXU property described in the previous section. The combinatorial property of AXU (for secret redundancy) is weaker, and therefore, easier to implement than the cryptographic property of XCR (for public redundancy). This tells us that by adding the power of secrecy to the redundancy code, one can achieve the same security (i.e. integrity) for the NCBC with redundancy scheme under a weaker security assumption on the underlying redundancy code.

What are candidates for XCR redundancy codes? Note that an unkeyed hash function like SHA-1 does not yield an XCR redundancy code. Indeed, an adversary can choose any distinct x, x', and let $r = \mathrm{SHA\text{-}1}(x) \oplus \mathrm{SHA\text{-}1}(x')$. It can output r in its first stage, and x, x' in its second, and win the game. An XCR redundancy code must be keyed. A keyed hash function is a good candidate. Specifically, we suggest that HMAC [3] is a candidate for a XCR redundancy code. In the full version of this paper [1] we discuss other constructions including a general way to transform any collision-resistant function into an XCR redundancy code.

Acknowledgments

We thank Hugo Krawczyk for helpful comments on a previous version of this paper. We thank Daniele Micciancio for helpful discussions. The authors are supported in part by Bellare's 1996 Packard Foundation Fellowship in Science and Engineering and NSF CAREER Award CCR-9624439. The first author was also supported in part by an NSF graduate fellowship.

References

1. J. An and M. Bellare, "Does encryption with redundancy provide authenticity?" Full version of this paper, available via http://www-cse.ucsd.edu/users/mihir.
2. M. Atici and D. Stinson, "Universal Hashing and Multiple Authentication," *Advances in Cryptology – CRYPTO '96*, Lecture Notes in Computer Science Vol. 1109, N. Koblitz ed., Springer-Verlag, 1996.
3. M. Bellare, R. Canetti and H. Krawczyk, "Keying hash functions for message authentication," *Advances in Cryptology – CRYPTO '96*, Lecture Notes in Computer Science Vol. 1109, N. Koblitz ed., Springer-Verlag, 1996.
4. M. Bellare, A. Desai, E. Jokipii and P. Rogaway, "A concrete security treatment of symmetric encryption: Analysis of the DES modes of operation," *Proc. of the 38th* IEEE FOCS, IEEE, 1997.
5. M. Bellare, A. Desai, D. Pointcheval and P. Rogaway, "Relations among notions of security for public-key encryption schemes," *Advances in Cryptology – CRYPTO '98*, Lecture Notes in Computer Science Vol. 1462, H. Krawczyk ed., Springer-Verlag, 1998.
6. M. Bellare, J. Kilian and P. Rogaway, "The Security of the Cipher Block Chaining Message Authentication Code," *Journal of Computer and System Sciences*, Vol. 61, No. 3, December 2000, pp. 362–399.
7. M. Bellare and C. Namprempre, "Authenticated Encryption: Relations among notions and analysis of the generic composition paradigm," *Advances in Cryptology – ASIACRYPT '00*, Lecture Notes in Computer Science Vol. 1976, T. Okamoto ed., Springer-Verlag, 2000.
8. M. Bellare and P. Rogaway, "Encode-then-encipher encryption: How to exploit nonces or redundancy in plaintexts for efficient cryptography," *Advances in Cryptology – ASIACRYPT '00*, Lecture Notes in Computer Science Vol. 1976, T. Okamoto ed., Springer-Verlag, 2000.
9. M. Bellare and A. Sahai, "Non-Malleable Encryption: Equivalence between Two Notions, and an Indistinguishability-Based Characterization," *Advances in Cryptology – CRYPTO '99*, Lecture Notes in Computer Science Vol. 1666, M. Wiener ed., Springer-Verlag, 1999.
10. J. Black, S. Halevi, H. Krawczyk, T. Krovetz and P. Rogaway, "UMAC: Fast and secure message authentication," *Advances in Cryptology – CRYPTO '99*, Lecture Notes in Computer Science Vol. 1666, M. Wiener ed., Springer-Verlag, 1999.

11. L. CARTER AND M. WEGMAN, "Universal Classes of Hash Functions," *Journal of Computer and System Sciences*, Vol. 18, 1979, pp. 143–154.

12. D. DOLEV, C. DWORK AND M. NAOR, "Non-malleable cryptography," *Proc. of the 23rd ACM STOC*, ACM, 1991.

13. S. GOLDWASSER AND S. MICALI, "Probabilistic encryption," *Journal of Computer and System Sciences*, Vol. 28, 1984, pp. 270–299.

14. S. HALEVI AND H. KRAWCZYK, "MMH: Software Message Authentication in the Gbit/Second Rates," *Fast Software Encryption — 4th International Workshop, FSE'97 Proceedings*, Lecture Notes in Computer Science, vol. 1267, E. Biham ed., Springer, 1997.

15. R. JUENEMAN, "A high speed manipulation detection code," *Advances in Cryptology – CRYPTO '86*, Lecture Notes in Computer Science Vol. 263, A. Odlyzko ed., Springer-Verlag, 1986.

16. R. JUENEMAN, C. MEYER AND S. MATYAS, "Message Authentication with Manipulation Detection Codes," in Proceedings of the 1983 IEEE Symposium on Security and Privacy, IEEE Computer Society Press, 1984, pp.33-54.

17. C. JUTLA, "Encryption modes with almost free message integrity," Report 2000/039, *Cryptology ePrint Archive*, http://eprint.iacr.org/, August 2000.

18. J. KATZ AND M. YUNG, "Complete characterization of security notions for probabilistic private-key encryption," *Proc. of the 32nd ACM STOC*, ACM, 2000.

19. J. KATZ AND M. YUNG, "Unforgeable Encryption and Adaptively Secure Modes of Operation," *Fast Software Encryption '00*, Lecture Notes in Computer Science, B. Schneier ed., Springer-Verlag, 2000.

20. H. KRAWCZYK, "LFSR-based Hashing and Authentication," *Advances in Cryptology – CRYPTO '94*, Lecture Notes in Computer Science Vol. 839, Y. Desmedt ed., Springer-Verlag, 1994.

21. H. KRAWCZYK, "The order of encryption and authentication for protecting communications (Or: how secure is SSL?)," Manuscript, 2001.

22. M. LUBY AND C. RACKOFF, "How to Construct Pseudorandom Permutations from Pseudorandom Functions," *SIAM Journal of Computing*, Vol. 17, No. 2, pp. 373–386, April 1988.

23. A. MENEZES, P. VAN OORSHOT AND S. VANSTONE, "Handbook of applied cryptography," CRC Press LLC, 1997.

24. B. PRENEEL, "Cryptographic Primitives for Information Authentication — State of the Art," *State of the Art in Applied Cryptography*, COSIC'97, LNCS 1528, B. Preneel and V. Rijmen eds., Springer-Verlag, pp. 49-104, 1998.

25. P. ROGAWAY, "Bucket Hashing and its Application to Fast Message Authentication," *Advances in Cryptology – CRYPTO '95*, Lecture Notes in Computer Science Vol. 963, D. Coppersmith ed., Springer-Verlag, 1995.

26. P. ROGAWAY, "OCB mode: Parallelizable authenticated encryption," Presented in *NIST's workshop on modes of operations,* October, 2000. See http://csrc.nist.gov/encryption/modes/workshop1/

27. M. WEGMAN AND L. CARTER, "New hash functions and their use in authentication and set equality," *Journal of Computer and System Sciences*, Vol. 22, 1981, pp. 265–279.

Encryption Modes
with Almost Free Message Integrity

Charanjit S. Jutla

IBM T. J. Watson Research Center,
Yorktown Heights, NY 10598-704

Abstract. We define a new mode of operation for block encryption which in addition to assuring confidentiality also assures message integrity. In contrast, previously for message integrity a separate pass was required to compute a cryptographic message authentication code (MAC). The new mode of operation, called Integrity Aware CBC (IACBC) requires a total of $m + 2$ block encryptions on a plain-text of length m blocks. The well known CBC (cipher block chaining) mode requires m block encryptions. The second pass of computing the CBC-MAC essentially requires additional m block encryptions. A new highly parallelizable mode (IAPM) is also shown to be secure for both encryption and message integrity.

1 Introduction

Symmetric key encryption is an integral part of world of communication today. It refers to the schemes and algorithms used to communicate data secretly over an insecure channel between parties sharing a secret key. It is also used in other scenarios like data storage.

There are two primary aspects of any security system: *confidentiality* and *authentication*. In its most prevalent form, confidentiality is attained by encryption of bulk digital data using *block ciphers*. The block ciphers (e.g. DES [15]), which are used to encrypt fixed length data, are used in various chaining modes to encrypt bulk data. One such mode of operation is cipher block chaining (CBC) ([1,9,14]). The security of CBC has been well studied [2].

Cipher block chaining of block ciphers is also used for authentication between parties sharing a secret key. The CBC-MAC (CBC Message Authentication Code) is an international standard [10]. The security of CBC-MAC was demonstrated in [4]. Authentication in this setting is also called *Message Integrity*.

Despite similar names, the two CBC modes, one for encryption and the other for MAC are different, as in the latter the intermediate results of the computation of the MAC are kept secret. In fact in most standards (TLS, IPsec [19,17]) and proprietary security systems, two different passes with two different keys, one each of the two modes is used to achieve both confidentiality and authentication.

Nevertheless, it is enticing to combine the two passes into one so that in a single cipher block chaining pass, both confidentiality and authentication are as-

B. Pfitzmann (Ed.): EUROCRYPT 2001, LNCS 2045, pp. 529–544, 2001.

sured. Many such attempts have been made, which essentially use a simple check-sum or manipulation detection code (MDC) in the chaining mode ([16,13,6]). Unfortunately, all such previous schemes are susceptible to attacks (see e.g. [18]).

We mention here that there are two alternative approaches to authenticated encryption. The first is to generate a MAC using universal hash functions as in UMAC ([3]). UMACs on certain architectures can be generated rather fast. However, UMAC suffers from requiring too much key material or a Pseudoran-dom number generator (PRNG) to expand the key. In another scheme, block numbers are embedded into individual blocks to thwart attacks against message integrity ([11]). However, this makes the cipher-text longer.

In this paper, we present a new variant of CBC mode, which in a single pass achieves both confidentiality and authentication. To encrypt a message of length m blocks, it requires a total of $(m + \log m)$ block encryptions. All other operations are simple operations, like exclusive-or. To contrast this with the usual CBC mode, the encryption pass requires m block encryptions, and the CBC-MAC computation requires another m block encryptions.

Our new mode of operation is also simple. A simpler (though not as efficient) version of the mode just requires a usual CBC encryption of the plain-text appended with the checksum (MDC), with a random initial vector r. As already mentioned, such a scheme is susceptible to message integrity attacks. However, if one "whitens" the complete output with a random sequence, the scheme becomes secure against message integrity attacks. Whitening just refers to xor-ing the output with a random sequence. The random sequence could be generated by running the block cipher on $r + 1$, $r + 2$, ... $r + m$ (but with a different shared key). This requires m additional cryptographic operations, and hence is no more efficient than generating a MAC.

The efficiency of the new mode comes from proving that the output whiten-ing random sequence need only be pair-wise independent. In other words, if the output whitening sequence is s_1, s_2,...s_m, then each s_i is required to be random, but only pairwise-independent of the other entries. Such a sequence is easily gen-erated by performing only $\log m$ cryptographic operations like block encryption. A simple algebraic scheme can also generate such a sequence by performing only two cryptographic operations.

In fact, an even weaker condition than pair-wise independence suffices. A sequence of uniformly distributed n-bit random numbers s_1, s_2,...s_m, is called *pair-wise differentially-uniform* if for every n-bit constant c, and every pair i, j, $i \neq j$, probability that $s_i \oplus s_j$ is c is 2^{-n}. We show that the output whitening sequence need only be pair-wise differentially-uniform. A simple algebraic scheme can generate such a sequence by performing only one cryptographic operation.

The pair-wise independent sequence generated to assure message integrity can also be used to remove chaining from the encryption mode while still as-suring confidentiality. This results in a mode of operation for authenticated en-cryption which is highly parallelizable. Once again, we show that a pair-wise differentially-uniform sequence suffices to guarantee security of both confiden-tiality and authentication in this parallelizable version.

Recently and independently, Gligor and Donescu ([7]) also described a mode of operation similar to CBC (but not the parallelizable mode) which has built-in message integrity, although with a slightly weaker security bound than our construction.

The rest of the paper is organized as follows. Section 2 describes the new mode of operation. Section 3 gives definitions of random permutations, and formalizes the notions of security, for both confidentiality and message integrity. In section 4 we prove that the new (parallelizable) scheme is secure for message integrity. In section 5 we state the secrecy theorem of the new mode of operation.

2   The New Modes of Operation

We begin by defining two properties of sequence of random numbers which are slightly weaker than the well known pair-wise independence property. The first property also appeared in [8].

2.1   Pairwise Differentially-Uniform Random Numbers

Definition 2.1 (pair-wise differentially-uniform): A sequence of uniformly distributed n-bit random numbers $s_1, s_2, ..., s_z$, is called *pair-wise differentially-uniform* if for every n-bit constant c, and every pair $i, j, i \neq j$, probability that $s_i \oplus s_j$ is c is 2^{-n}.

Definition 2.2 A sequence of random numbers $s_1, s_2, ..., s_z$ uniformly distributed in GFp, is called *pair-wise differentially-uniform in* GFp if for every constant c in GFp, and every pair $i, j, i \neq j$, probability that $(s_i - s_j) \mod p$ is c is $1/p$.

2.2   The New Modes – IACBC and IAP

Now we describe the new modes of operation for encryption, which also guarantee message integrity. We will describe the parallelizable mode in more detail, as it is for this mode that we provide detailed proofs in this paper.

The mode similar to CBC is called **IACBC** for *integrity aware cipher block chaining*. It is described in Fig 1. The parallelizable mode is called **IAPM** for *integrity aware parallelizable mode*. It is described in Fig 2. We now give more details for IAPM. After reading the details for IAPM, the definition of IACBC will be clear from Fig 1.

Let n be the block size of the underlying block cipher (or pseudo-random permutation). For now we assume that if the block cipher requires keys of length k, then this mode of operation requires two keys of length k . Let these keys be called $K1$ and $K2$. From now on, we will use f_x to denote the encryption function under key x. The same notation also holds for pseudo-random permutations.

The message to be encrypted P, is divided into blocks of length n each. Let these blocks be $P_1, P_2, ...P_{z-1}$. As in CBC, a random initial vector of length n (bits) is chosen. This random vector r is expanded into $t = O(\log z)$ new random vectors $W_1, ...W_t$ using the block cipher and key $K2$ as follows:

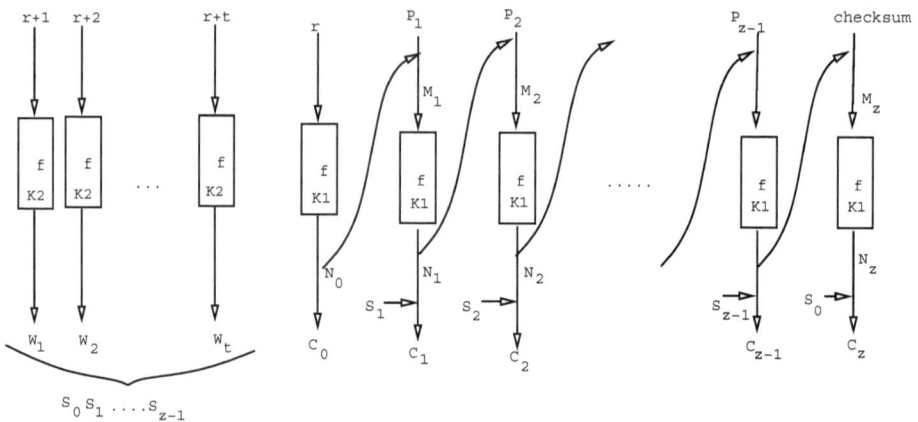

Fig. 1. Encryption with Message Integrity (IACBC)

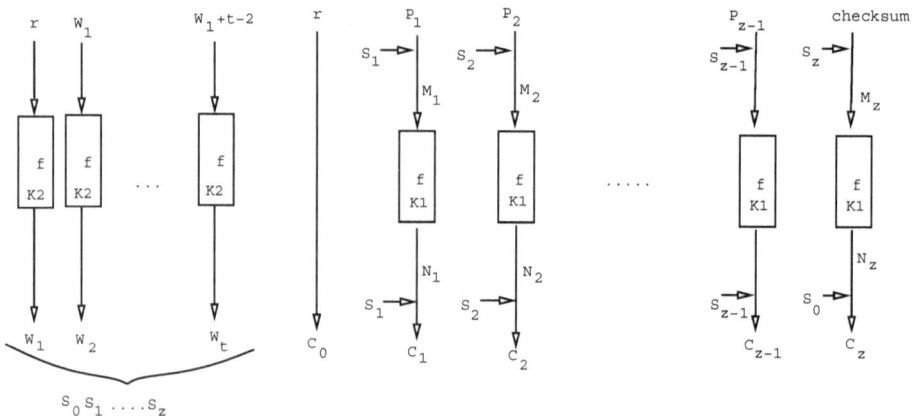

Fig. 2. Parallelizable Encryption with Message Integrity (IAPM)

$W_1 = f_{K2}(r)$
for $i = 2$ to t do
    $W_i = f_{K2}(W_1 + i - 2)$
end for

As we will show in section 4, with high probability, the t vectors are independent. The t random and independent vectors are used to prepare $z + 1$ new pair-wise differentially-uniform random vectors $S_0, S_1, ..., S_z$. There are several ways to generate such a sequence, some requiring t to be only one. Such a scheme will be described towards the end of this section. For now, consider the following method using subsets ($t = \lceil \log(z + 2) \rceil$):

    for $i = 1$ to $2^t - 1$ do
        Let $< a_1, a_2, ...a_t >$ be the binary representation of i

$$S_{i-1} = \sum_{j=1}^{t} (a_j \cdot W_j)$$
end for

The summation in the for loop above is an xor-sum.

The cipher-text message $C = < C_0, C_1, ..., C_z >$ is generated as follows (see Figure 2). The encryption pseudo-code follows:

$C_0 = r$
for $i = 1$ to $z - 1$ do
$\quad M_i = P_i \oplus S_i$
$\quad N_i = f_{K1}(M_i)$
$\quad C_i = N_i \oplus S_i$
end for
checksum $= \sum_{i=1}^{z-1} P_i$
$M_z = $ checksum $\oplus S_z$
$N_z = f_{K1}(M_z)$
$C_z = N_z \oplus S_0$

Again, the summation above is an xor-sum. Note that S_0 is used in the last step.

It is easy to see that the above scheme is invertible. The inversion process yields blocks $P_1, P_2, ..., P_z$. The decrypted plain-text is $< P_1, P_2, ..., P_{z-1} >$. Message integrity is verified by checking $P_z = P_1 \oplus P_2 \oplus ... \oplus P_{z-1}$.

The random vectors $W_1, ...W_t$ can also be generated as in Fig 1, in which case C_0 is set to $f_{K1}(r)$ (instead of r).

There are many other ways of generating the pair-wise differentially-uniform vectors $S_0, S_1, ..., S_z$ ($z < 2^n$). One could generate a sequence of pairwise differentially uniform vectors by an algebraic construction in GFp as follows: generate two random vectors W_1, and W_2, and then let $S_i = (W_1 + W_2 * i) \bmod p$, where p is a prime of appropriate size. For example, if the block cipher has block size 64 bits, p could be chosen to be $2^{64} - 257$. This leads to a fast implementation.

A sequence of $2^n - 1$ n-bit uniform random numbers, which are pair-wise differentially uniform, can also be generated by viewing the n-bit numbers as elements of $GF(2^n)$. Consider, $S_i = e(i) \cdot W$, where W is a random number in $GF(2^n)$, $e(i)$ is a one to one function from $Z_{2^n - 1}$ to non-zero elements of $GF(2^n)$, and the multiplication is in $GF(2^n)$. Then S_i is a pair-wise differentially uniform sequence of uniformly distributed random numbers. Note that this requires generation of only one W (i.e. $t = 1$).

The GFp construction with only one W, instead of two, is not pair-wise differentially uniform (as opposed to the previous construction in $GF(2^n)$). However, it is pair-wise differentially uniform in GFp (see definition 2.2). More precisely, the sequence $S_i = (W_1 * i) \bmod p$, is pair-wise differentially uniform in GFp (assuming W_1 is uniformly distributed in GFp). Such a sequence can be used securely in a slight variant of the mode described above where "whitening" now refers to addition modulo 2^n (see section 4.2).

3 Encryption Schemes: Message Security with Integrity Awareness

We give definitions of schemes which explicitly define the notion of secrecy of the input message. Of course, pseudo-random permutations can be used to build encryption schemes which guarantee such message secrecy ([2], [12]).

In addition, we also define the notion of message integrity. Moreover, we allow arbitrary length input messages (upto a certain bound).

Let Coins be the set of infinite binary strings. Let $l(n) = 2^{O(n)}$, and $w(n) = O(n)$. Let $\mathcal{N}$ be the natural numbers.

Definition A (probabilistic, symmetric, stateless) encryption scheme with message integrity consists of the following:

- **initialization:** All parties exchange information over private lines to establish a private key $x \in \{0,1\}^n$. All parties store x in their respective private memories, and $|x| = n$ is the security parameter.
- **message sending with integrity:**

$$\text{Let } E : \{0,1\}^n \times \text{Coins} \times \mathcal{N} \times \{0,1\}^{l(n)} \to \{0,1\}^{l(n)} \times \mathcal{N}$$

$$D : \{0,1\}^n \times \mathcal{N} \times \{0,1\}^{l(n)} \to \{0,1\}^{l(n)} \times \mathcal{N}$$

$$\text{MDC} : \mathcal{N} \times \{0,1\}^{l(n)} \to \{0,1\}^{w(n)}$$

be polynomial-times function ensembles. In E, the third argument is supposed to be the length of the plain-text, and E produces a pair consisting of cipher-text and its length. Similarly, in D the second argument is the length of the cipher-text. We will drop the length arguments when it is clear from context. The functions E and D have the property that for all $x \in \{0,1\}^n$, for all $P \in \{0,1\}^{l(n)}$, $c \in \text{Coins}$

$$D_x(E_x(c, P)) = P \| \text{MDC}(P)$$

We will usually drop the random argument to E as well, and just think of E as a probabilistic function ensemble. It is also conceivable that MDC may depend on Coins, cipher-text.

Definition (*Security under Find-then-Guess* [2]) Consider an adversary A that runs in two stages. During the adversary's find stage he endeavors to come up with a pair of equal length messages, P^0, P^1, whose encryptions he wants to tell apart. He also retains some state information s. In the adversary's guess stage he is given a random cipher-text y for one of the plain-texts P^0, P^1, together with s. The adversary is said to "win" if he correctly identifies the plain-text.

An Encryption Scheme is said to be (t, q, μ, ϵ)-secure in the find-then-guess sense, if for any adversary A which runs in time at most t and asks at most q queries, these totaling at most μ bits,

$$Adv_A \stackrel{\text{def}}{=} 2 \cdot Pr[(P^0, P^1, s) \leftarrow A^{E_x(\cdot)}(\text{find}); \ b \leftarrow \{0,1\}; \ y \leftarrow E_x(P^b) :$$
$$A^{E_x(\cdot)}(\text{guess}, y, s) = b] - 1 \leq \epsilon$$

The following notion of security is also called *integrity of ciphertext* ([5]).
Definition (*Message Integrity*): Consider an adversary A running in two stages.
In the first stage (*find*) A asks r queries of the oracle E_x. Let the oracle replies
be $C^1, ...C^r$. Subsequently, A produces a cipher-text C', different from each C^i,
$i \in [1..r]$. Since D has length of the cipher-text as a parameter, the breakup of
$D_x(C')$ as $P'\|P''$, where $|P''| = w(n)$, is well defined. The adversary's success
probability is given by

$$\mathrm{Succ} \overset{\mathrm{def}}{=} Pr[\mathrm{MDC}(P') = P'']$$

An encryption scheme is secure for message integrity if for any adversary A, A's
success probability is negligible.

4   Message Integrity

In this section we show that the mode of operation IAPM in Fig 2 guarantees
message integrity with high probability.

In the following theorem, we will assume that the block cipher (under a key
$K1$) is a a random permutation F. We also assume that the t W's are generated
using an independent random permutation G (for instance, using a different key
$K2$ in a block cipher).

Let the adversary's queries in the first stage be $p^1, P^2, ...P^m$. We write p^1
in lower case, as for each adversary p^1 is fixed. All random variables will be
denoted by upper case letters. Let the corresponding ciphertexts be $C^1, ..., C^m$.
We will use C to denote the sequence of ciphertext messages $C^1, ..., C^m$. For all
random variables corresponding to a block, we will use superscripts to denote
the message number, and subscripts to denote blocks in a particular message.
Thus C_j^i will be the random variable representing the jth block in ciphertext
message i. More precisely, this variable should be written $C_j^i(F, G)$, as it is a
function of the two permutations. However, we will drop the arguments when it
is clear from context.

Let the adversary's query in the second stage be cipher-text C', different
from all ciphertexts in the first stage. We will use primed variables to denote the
variables in the second stage.

We will use W to denote the set of variables $\{W_j^i : i \in [1..m], j \in [1..t]\} \cup$
$\{W_j', j \in [1..t]\}$. We will use S^i (S') to denote masks or "whitening" blocks
generated using W^i (W' resp). Any method can be used to generate S^i from
W^i, as long as S_j^i are pairwise differentially uniform. For a particular adversary,
S_j^i is a function of permutation G and the initial vector, and hence should (more
precisely) be written as $S_j^i(G, C_0^i(F, G))$ ($C_0^i(F, G)$ being the IV used to generate
W_1^i). But, we will drop the arguments as it will be clear from context. For any
constant r, we will denote by $S_j^i(r)$ the random variable $S_j^i(G, r)$.

The variables M and N are as in Fig 2. For example, $M_j^i = P_j^i \oplus S_j^i$.

We start with some informal observations to aid the reader in the eventual
formal proof. Since the new ciphertext C' is different from all old ciphertexts,

it must differ from each old ciphertext C^i in a least block number, say $d(i)$. For each C^i (except at most one C^k), the block number $d(i) = 0$, with high probability. In Lemma 3 we show that with high probability $N'_{d(k)}$ is different from all old N^i_j, and all other new N' blocks (except for a special case). Thus, $M'_{d(k)}$ is random. Then it follows (Theorem 1) that in either case the checksum is unlikely to validate.

We first prove the theorem for schemes in which the pairwise differentially uniform sequence is generated using only one W, i.e. $t = 1$. The general case is addressed in a later subsection.

Theorem 1. *Let A be an adversary attacking the message integrity of IAPM ($t = 1$) with random permutations F and G. Let A make at most m queries in the first stage, totaling at most μ blocks. Let $u = \mu + m$. Let v be the maximum number of blocks in the second stage. Then for adversary A,*

$$Succ < (2 * u^2 + m^2 + (m+1)^2 + u + v + 2 + o(1)) * 2^{-n}$$

Proof:

In the first stage the adversary makes queries with a total of at most m plaintext messages (chosen adaptively). W.l.o.g. assume that the adversary actually makes exactly m total message queries in the first stage. Let L^i be the random variable representing the length of ciphertext C^i (i.e. the checksum block has index $L^i - 1$). Similarly, L' will denote the length of C'.

We prove that either the adversary forces the following event E0, or the event E1 happens with high probability. In either case the checksum validates with low probability.

The first event E0 is called deletion attempt, as the adversary in this case just truncates an original ciphertext, but retains the last block.

Event E0 (*deletion attempt*): There is an $i \in [1..m]$, such that $2 \leq L' < L^i$, and

$$(i) \ \forall j \in [0..L' - 2] : C'_j = C^i_j$$

$$\text{and } (ii) \ C'_{L'-1} = C^i_{L^i-1}$$

Event E1 says that there is a block in the new ciphertext C', such that its N variable is different from all previous Ns (i.e. from original ciphertexts from the first stage), and also different from all other new Ns.

Event E1: there is an $x \in [1..L' - 1]$ such that

$$(i) \ \forall s \in [1..m] \forall j \in [1..L^s - 1] : N'_x \neq N^s_j$$

$$\text{and } (ii) \ \forall j \in [1..L' - 1], j \neq x : N'_x \neq N'_j$$

We next show that in both cases (i.e E0 or E1) the checksum validates with low probability.

For the case that E0 happens, we have (since $S' = S^i$ and $N'_{L'-1} = N^i_{L^i-1}$),

$$(\sum_{j=1}^{L'-1} P'_j = 0) \wedge \text{E0} \Rightarrow \sum_{j=1}^{L'-2} (P^i_j) + M^i_{L^i-1} + S^i_{L'-1} = 0$$

$$\equiv \sum_{j=1}^{L'-2} (P^i_j) + \sum_{j=1}^{L^i-2} (P^i_j) + S^i_{L^i-1} + S^i_{L'-1} = 0$$

Note that r^i can be chosen after P^i has been determined (as P^i is a deterministic function of $C^1, \ldots, C^{i-1}$), and hence the S^is are independent of P^i. Since the S^is are pairwise differentially uniform and $L' < L^i$, the above event happens with probability at most 2^{-n}.

For the case E1, by Lemma 2, the checksum validates with probability at most $1/(2^n - u - v)$

Thus the adversary's success probability is upper bounded by

$$\Pr[\neg(\text{E0} \vee \text{E1})] + \frac{1}{2^n - (u+v)} + \frac{1}{2^n}$$

which by Lemma 3 is at most

$$(u^2 + m^2 + u + v + 2) * 2^{-n} + (u^2 + (m+1)^2) * 2^{-n} + O(u+v) * 2^{-2n}$$

□

Lemma 2: $\Pr[\sum_{j=1}^{L'-1} P'_j = 0 \mid \text{E1}] \leq \frac{1}{2^n-(u+v)}$

Proof: F being a random permutation, under E1, $F^{-1}(N'_x)$ can not take values already assigned to $F^{-1}(N^s_j)$, $s \in [1..m]$, $j \in [1..L^s - 1]$. Also, $F^{-1}(N'_x)$ can be chosen after $F^{-1}(N'_j)$ have been assigned values ($j \neq x$). Thus, under the condition that event E1 has happened we have that $M'_x = F^{-1}(N'_x)$ can take any of the other values, i.e. excluding the following (at most) $(\mu + m) + L' - 2$ values, with equal probability (independently of C, C', r^i, $i \in [1..m]$, G, and hence independently of W, and independent of E1 itself):

- values already taken by $M^s_1, \ldots, M^s_{L^s-1}$, for each s, and
- the values to be taken (or already fixed) by M'_j, $j \in [1..L' - 1]$, $j \neq x$.

Now, $\sum_{j=1}^{L'-1} P'_j = 0$ iff

$$F^{-1}(N'_x) = M'_x = \sum_{j=1, j \neq x}^{L'-1} (M'_j \oplus S'_j) \oplus S'_x$$

Given any value of the RHS, since the LHS can take (at least) $2^n - (u + v - 2)$ values, the probability of LHS being equal to RHS is at most $1/(2^n - (u + v))$.
□

Lemma 3: *Let events E0,E1 be as in Theorem 1. Then,*

$$\text{Prob}[\neg(\text{E0} \vee \text{E1})] < (u^2 + m^2 + u + v) * 2^{-n} + (u^2 + (m+1)^2) * 2^{-n}$$

Proof: We first calculate the probability of event $(\text{E0} \vee \text{E1})$ happening under the assumption that F and G are random functions (instead of random permutations). Since F (and G) is invoked only u times ($(m+1)$ times resp.), a standard argument shows that the error introduced in calculating the probability of event (E0 ∨ E1) is at most $(u^2 + (m+1)^2) * 2^{-n}$.

We now consider an event, which says that all the M variables are different. The goal is to claim independence of the corresponding N variables, and hence the C variables. However, the situation is complicated by the fact that the condition that all the M_j^i variables for some i are different, may cause the variables $C_j^{i'}$, for $i' < i$, to be no more independent. However, a weaker statement can be proved by induction. To this end, consider the **event** E2(y), for $y \leq m$:

$$\forall i, i' \in [1..y], \forall j, j', j \in [1..L^i - 1], j' \in [1..L^{i'} - 1], (i,j) \neq (i',j') : (M_j^i \neq M_{j'}^{i'})$$

Event E2(m) will also be denoted by E2.

We also predicate on the event that all the initial variables C_0^i are different. Let $E3$ be the **event** that

$$\forall i, j \in [1..m], i \neq j : C_0^i \neq C_0^j$$

For $\vec{r} = r^1, ..., r^m$, all r^i different, let $E3(\vec{r})$ be the event that for all $i \in [1..m]$, $C_0^i = r^i$.

Let $l()$ be the length of the first ciphertext (determined by the adversary). We will use constant c^i to denote strings of arbitrary block length. We will use c to denote the sequence $c^1, ..., c^m$. The function $|\cdot|$ is used below to represent length of a message in blocks. Given a sequence of ciphertext messages $c^1, ..., c^i$, $i \leq m$, let $l(c^1, ..., c^i)$ be the length of the $(i+1)$th ciphertext (which is determined by the adversary, and therefore is a deterministic function of $c^1, ...c^i$). Recall that each ciphertext includes the block C_0^i, which is just r^i under $E3(\vec{r})$. Also, since C' is a deterministic function of C, given $c^1, ..., c^m$ let the ciphertext in the second stage be c' with length l'. We have

$$Pr[\neg(E0 \vee E1) \wedge E2 \mid E3(\vec{r})] = \sum_{c^1: |c^1| = l()} \cdots \sum_{c^i: |c^i| = l(c^{i-1}, ..., c^1)} \cdots$$

$$\cdots \sum_{c^m: |c^m| = l(c^{m-1}, ..., c^1)} Pr[\neg(E0 \vee E1) \wedge \bigwedge_i C^i = c^i \wedge E2 \mid E3(\vec{r})] \qquad (1)$$

In this sum, if for some i, $c_0^i \neq r^i$, then the inside expression is zero. Also, if event E0 holds for c (which determines c'), then the inside expression above for that c is zero. So, from now on, we will assume that E0 does not hold for $C = c$. Then, the inside expression above becomes:

$$Pr[\neg(E0 \vee E1) \wedge \bigwedge_i C^i = c^i \wedge E2 \mid E3(\vec{r})]$$

$$\leq min_{x \in [1..l'-1]} \left\{ \sum_{s \in [1..m], j \in [1..|c^s|-1]} Pr[(N_x' = N_j^s) \wedge \bigwedge_i C^i = c^i \wedge E2 \mid E3(\vec{r})] \right.$$

$$\left. + \sum_{j \in [1..l'-1], j \neq x} Pr[(N_x' = N_j') \wedge \bigwedge_i C^i = c^i \wedge E2 \mid E3(\vec{r})] \right\}$$

For each s, j, we have $(N'_x = N^s_j)$ iff $(S'_{x^*} \oplus S^s_{j^*}) = (C'_x \oplus C^s_j)$, where $S'_{x^*}, S^s_{j^*}$ are the masks that are used for these ciphertext blocks. That is, $j^* = j$ if $j < |c^s| - 1$ and $j^* = 0$ otherwise, and similarly $x^* = x$ if $x < l' - 1$ and $x^* = 0$ otherwise (Similarly for $j \neq x$ we have $(N'_x = N'_j)$ iff $(S'_{x^*} \oplus S'_{j^*}) = (C'_x \oplus C'_j)$).

Since each of the summands in the expression above has a conjunct $C = c$ for some constant string c (and since the forged ciphertext C' is a function of C), it follows that each of the summands in the first sum can be written as $\Pr[(S'_{x^*}(c'_0) \oplus S^s_{j^*}(c^s_0) = c'_x \oplus c^s_j) \wedge C = c \wedge E2 \mid E3(\vec{r})]$. Note that $S'_{x^*}(c'_0) \oplus S^s_{j^*}(c^s_0)$ can in some cases be identically zero. As c is some constant string, then $c'_x \oplus c^s_j$ is also constant, and recall that the variables $S(c_0)$ depend only on the choice of G. Thus, each of these summands (if $S'_{x^*}(c'_0) \oplus S^s_{j^*}(c^s_0)$ is not identically zero) can be bounded by

$$
\begin{aligned}
&\Pr[S'_{x^*}(c'_0) \oplus S^s_{j^*}(c^s_0) = c'_x \oplus c^s_j \wedge C = c \wedge E2 \mid E3(\vec{r})] \\
&= \Pr[C = c \wedge E2 \mid S'_{x^*}(c'_0) \oplus S^s_{j^*}(c^s_0) = c'_x \oplus c^s_j \wedge E3(\vec{r})] \\
&\quad * \Pr[S'_{x^*}(c'_0) \oplus S^s_{j^*}(c^s_0) = c'_x \oplus c^s_j \mid E3(\vec{r})] \\
&\leq (2^{-n})^\mu * \Pr[S'_{x^*}(c'_0) \oplus S^s_{j^*}(c^s_0) = c'_x \oplus c^s_j \mid E3(\vec{r})]
\end{aligned}
$$

where the last inequality follows by Claim 5 with $\mu = \sum_{i \in [1..m]} (l(c^{i-1}, \ldots, c^1) - 1)$. A similar inequality holds for the summands in the second sum (i.e. $N'_x = N'_j$ case). Thus, by Claim 4, the inside expression in equation (1) is at most $2^{-n\mu} * (u + v) * 2^{-n}$. Since we have $2^{n\mu}$ summands, it follows that

$$
\Pr[\neg(E0 \vee E1) \wedge E2 \mid E3(\vec{r})] \leq (u + v) * 2^{-n}
$$

Finally, we calculate $\Pr[\neg(E0 \vee E1)]$

$$
\begin{aligned}
&\Pr[\neg(E0 \vee E1)] \\
&\leq \Pr[\neg(E0 \vee E1) \wedge E2 \mid E3] + \Pr[\neg E2 \mid E3] + \Pr[\neg E3] \\
&\leq \Pr[\neg E3] + \\
&\quad \sum_{r^1, \ldots, r^m} ((\Pr[\neg(E0 \vee E1) \wedge E2 \mid E3(\vec{r})] + \Pr[\neg E2 \mid E3(\vec{r})]) * \Pr[E3(\vec{r}) \mid E3]) \\
&\leq m^2 * 2^{-n} + (u + v) * 2^{-n} + (u)^2 * 2^{-n}
\end{aligned}
$$

where the last inequality follows by Claim 6. $\square$

Claim 4: For each constant c (and its corresponding c') for which event E0 does not hold, and constant $\vec{r}$ with distinct values, there is an $x \in [1..l' - 1]$ such that

(i) $\forall s \in [1..m] \forall j \in [1..|c^s| - 1]$:
    if $S'_{x^*}(c'_0) \oplus S^s_{j^*}(c^s_0)$ is identically zero then $c'_x \oplus c^s_j \neq 0$, otherwise

$$
\Pr[S'_{x^*}(c'_0) \oplus S^s_{j^*}(c^s_0) = c'_x \oplus c^s_j \mid E3(\vec{r})] \leq 2^{-n},
$$

(ii) $\forall j \in [1..|l'-1|], j \neq x,:$

$$\Pr[S'_{x*}(c'_0) \oplus S'_{j*}(c^s_0) = c'_x \oplus c'_j \mid E3(\overrightarrow{r})] \leq 2^{-n}$$

Proof: These are the different cases (we will drop the argument from S^s and S' as it will be clear from context):

(a) (*New IV*) If for all $i \in [1..m]$, $c'_0 \neq r^i$, then we choose $x = 1$. In that case $N'_1 = N'_j$ is same as $C'_1 \oplus C'_j = S'_1 \oplus S'_{j*}$, where $j^* = j$ if $j \neq (l'-1)$, and $j^* = 0$ otherwise. Thus, for $j \in [1..l'-1], j \neq x$, since S' is pairwise differentially uniform, probability of $(S'_1 \oplus S'_{j*} = c'_1 \oplus c'_j)$ is 2^{-n} (even under $E3(\overrightarrow{r})$).

Similarly, $N'_1 = N^s_j$ is same as $C'_1 \oplus C^s_j = S'_1 \oplus S^s_{j*}$, where $j^* = j$ if $j \neq |c^s|-1$, and $j^* = 0$ otherwise. Under event $E3(\overrightarrow{r})$, and the fact that c'_0 is different from all r^i, we have that $S'_1 \oplus S^s_{j*}$ is uniformly distributed.

(b) There exists a k, $k \in [1..m]$ such that $c'_0 = r^k$. For all other $k' \in [1..m]$, $c'_0 \neq r^k$. Thus $S' = S^k$. We have several cases:

(b1) (*truncation attempt*) If c' is a truncation of c^k, then we let $x = l'-1$ which is the index of the last block of c'.

(b2) (*extension attempt*) If c' is an extension of c^k, then we let $x = |c^k|-1$ which is the index of the last block of c^k.

(b3) Otherwise, let x be the least index in which c' and c^k are different.

In all the cases (b1), (b2) and (b3), conjunct (ii) is handled as in (a).

In case (b1), $N'_x = N^s_j$ is same as $C'_{l'-1} \oplus S^s_0 = C^s_j \oplus S^s_{j*}$, where $j^* = j$ if $j \neq |c^s|-1$, and $j^* = 0$ otherwise. Now, for $s = k$, $j^* = 0$ (in which case $S^s_0 \oplus S^s_j$ is identically zero), we have $c'_x \oplus c^s_j = c'_{l'-1} \oplus c^k_{|c^k|-1}$. This quantity is not zero, since E0 (the deletion attempt) doesn't hold for c. Otherwise, $S^s_0 \oplus S^s_{j*} = S^k_0 \oplus S^s_j$ is uniformly distributed.

In case (b2), $N'_x = N^s_j$ is same as $C'_{|c^k|-1} \oplus S^k_{|c^k|-1} = C^s_j \oplus S^s_{j*}$, where $j^* = j$ if $j \neq |c^s|-1$, and $j^* = 0$ otherwise. When $s = k$, j^* is never $|c^k|-1$, and hence $S^k_{|c^k|-1} \oplus S^s_{j*}$ is uniformly distributed.

In case (b3), $N'_x = N^s_j$ is same as $C'_x \oplus S^k_{x*} = C^s_j \oplus S^s_{j*}$, where $j^* = j$ if $j \neq |c^s|-1$, and $j^* = 0$ otherwise, and $x^* = x$ if $x \neq (l'-1)$, and $x^* = 0$ otherwise. If $s = k$, and $j^* = x^*$, then either $j^* = x^* = 0$, or $j = x$. In the latter case, $c'_x \oplus c^s_j = c'_x \oplus c^k_x$, which is non-zero as x is the index in which c' and c^k differ. In the former case, $j = |c^k|-1$, and $x = (l'-1)$. In this case, $c'_x \oplus c^s_j = c'_{l'-1} \oplus c^k_{|c^k|-1}$. If this quantity is zero, then since $x (= (l'-1))$ was the least index in which c^k and c' differed, event E0 would hold for c, leading to a contradiction. In other cases, $S^k_{x*} \oplus S^s_{j*}$ is uniformly distributed.   □

Recall that $E3(\overrightarrow{r})$ is the event that all C^i_0 are distinct (and set to $\overrightarrow{r}$).

Claim 5: Let l_1 be the length of the first ciphertext. Let $y \leq m$. For any constant lengths l_i ($i \in [2..y]$) and constant strings c^i, ($i \in [1..y]$, $|c^i| = l_i$), and any function G independent of F,

$$\Pr[\bigwedge_{i \in [1..y]} C^i = c^i \wedge E2(y) \mid G \wedge E3(\overrightarrow{r})] \leq (2^{-n})^{\mu}$$

where $\mu = \Sigma_{i \in [1..y]}(l^i - 1)$.

Proof: The above probability is zero unless for all $i \in [2..y]$, $l^i = l(c^1, ..., c^{i-1})$. From now on, we will assume that the l^i are indeed such.

We do induction over y, with base case $y = 0$.
The base case is vacuously true, as $\mu = 0$ and conditional probability of TRUE is 1.

Now assume that the lemma is true for y. We prove the lemma for $y + 1$. The explanation for the inequalities is given below the sequence of inequalities.

$$\Pr[\bigwedge_{i \in [1..y+1]} C^i = c^i \wedge E2(y+1) \mid G \wedge E3(\overrightarrow{r})]$$

$$\leq \Pr[C^{y+1} = c^{y+1} \mid \bigwedge_{i \in [1..y]} C^i = c^i \wedge E2(y+1) \wedge G \wedge E3(\overrightarrow{r})]$$

$$* \Pr[\bigwedge_{i \in [1..y]} C^i = c^i \wedge E2(y+1) \mid G \wedge E3(\overrightarrow{r})]$$

$$\leq (2^{-n})^{l^{y+1}-1} * \Pr[\bigwedge_{i \in [1..y]} C^i = c^i \wedge E2(y) \mid G \wedge E3(\overrightarrow{r})]$$

$$\leq (2^{-n})^{\Sigma_{i \in [1..y]}(l^i - 1)}$$

The second inequality follows because under the condition $E2(y+1)$, all the M_j^{y+1} are different from the previous M, and hence the sequence of variables, for all $j \in [1..L^{y+1} - 1]$, $F(M_j^{y+1})$ can take all possible $(2^n)^{(L^{y+1}-1)}$ values, independently of G, and $F(M_j^{\leq y})$, and hence also all ciphertext messages till index t. Hence, the sequence $C_j^{y+1} = F(M_j^{y+1}) \oplus S_j^{y+1}$ can take all possible values. Moreover, $L^{y+1} = l(c^1, ..., c^y) = l^{y+1}$.

The last inequality follows by induction. □

Claim 6: For every fixed $\overrightarrow{r}$ with distinct values,

$$Pr[\neg E2 \mid E3(\overrightarrow{r})] < u^2 * 2^{-n}$$

Proof: Recall that Event E2 is

$$\forall i, i' \in [1..m], \forall j, j', j \in [1..L^i], j' \in [1..L^{i'}], (i,j) \neq (i',j') : (M_j^i \neq M_{j'}^{i'})$$

Under $E3(\overrightarrow{r})$, we have
(a) The set of variables $\{W_1^i\}$, $i \in [1..m]$, are uniformly random and independent variables.
(b) For each i, the variable W_1^i is independent of all ciphertext messages $C^{i'}$, $i' < i$, and hence all plaintext messages $P^{i'}$, $i' \leq i$. This follows because W_1^i can be chosen after $C^{i'}$, $i' < i$ have been chosen.

Given $E3(\overrightarrow{r})$, the probability that event E2 does not happen is at most $(\Sigma_{i \in [1..m]} L^i)^2 * 2^{-n}$, which is at most $u^2 * 2^{-n}$. This is seen as follows:

$$\Pr[M_j^i = M_{j'}^{i'}] = \Pr[P_j^i \oplus S_j^i = P_{j'}^{i'} \oplus S_{j'}^{i'}] = \Pr[S_j^i = S_{j'}^{i'} \oplus P_j^i \oplus P_{j'}^{i'}]$$

Without loss of generality, let $i \geq i'$. Then from (b) above it follows that this probability is at most 2^{-n} (if $i = i'$, then we also use the fact that the sequence S is pairwise differentially uniform). $\qquad\square$

4.1 General Case

We now prove the scheme IAPM $(t \geq 1)$ secure for message integrity. Here F and G are independent random permutations.

Theorem 4: *Let A be an adversary attacking the message integrity of IAPM $(t \geq 1)$ with random permutations F and G. Let A make at most m queries in the first stage, totaling at most μ blocks. Let $u = \mu + m$. Let v be the maximum number of blocks in the second stage. Then for adversary A,*

$$Succ < (2 * u^2 + 2tm^2 + tm + t^2(m+1)^2 + 3t(2m+1)(u+v) + 2 + o(1)) * 2^{-n}$$

Proof Sketch: We first calculate the adversary's success probability assuming that G is a random function. Then, the error introduced in the probability because of this approximation is at most $((t(m+1))^2 * 2^{-n})$.

The differences in the proof from that of Theorem 1 are (i) we can not assume a priori, that the sequence S^i is pairwise differentially uniform, (ii) E3$(\vec{r})$ as defined in Lemma 3 does not imply that S^i is independent of S^j, for $i \neq j$, (iii) in proof of Theorem 1, the case of event E0 requires S^i to be pairwise differentially uniform, and (iv) in claim 4 case (a), $S'(c_0')$ is not necessarily independent of all $S^i(r^i)$.

To this end, Event E3 is now defined to be the event that all entries in the following (multi-) set are different:

$$\{C_0^i, i \in [1..m]\} \cup \{G(C_0^i) + j - 1, i \in [1..m], j \in [1..t-1]\}$$

For $\vec{r} = r^1, ..., r^m$, all r^i different, let E3$(\vec{r})$ be the event E3 and that for all $i \in [1..m]$, $C_0^i = r^i$.

For $\vec{r} = r^1, ..., r^m$, all r^i different, $\Pr[\neg \text{ E3}(\vec{r})] \leq (2tm^2 + tm) * 2^{-n}$

Under event E3, for all $i \in [1..m]$, the sequence S^i is pairwise differentially uniform, and is independent of S^j ($j \in [1..m]$, $j \neq i$). Now (in Theorem 1) the case of event E0 is also handled under the condition E3$(\vec{r})$.

In Claim 4, case (a) (i.e. New IV) now requires showing that $S'(c_0')$ (with c_0' different from all r^i) is independent of all $S^i(r^i)$ ($i \in [1..m]$).

Consider the following events (note that $W_1^i = G(r^i)$):
Event E4: $\forall i \in [1..m], \forall j \in [1..t-1] : c_0' \neq W_1^i + j - 1$.
Event E5: $\forall i \in [1..m] : |G(c_0') - W_1^i| > t \wedge |G(c_0') - r^i| > t \wedge |G(c_0') - c_0'| > t$

Now given that, for all $k \in [1..m]$, $c_0' \neq r^k$, and under event E4, it is the case that c_0' has never been an oracle query to G, and thus $\Pr[\neg\text{E5} \mid \text{E4} \wedge \text{E3}(\vec{r})] < 2t(2m+1) * 2^{-n}$. Also, $\Pr[\neg \text{ E4} \mid \text{E3}(\vec{r})] \leq mt * 2^{-n}$.

Under events E4, E5 and E3$(\vec{r})$, and c_0' different from all r^i, $S'(c_0')$ is indeed independent of previous $S^i(r^i)$, and is also pairwise differentially uniform. $\qquad\square$

4.2   Modes Using GFp

In another variant of IACBC and IAPM, a pair-wise differentially uniform sequence in GFp is employed for "whitening" the output (and the input for parallel modes). However, now "whitening" refers to adding modulo 2^n, instead of performing an exclusive-or operation. Theorems 1 and 5 also hold for encryption schemes which employ sequences which are pair-wise differentially-uniform in GFp; the success probabilities, however are now in terms of $2/p$ instead of $1/2^n$. The condition $N_i' = N_j$ would now translate to $C_i' - S_i = C_j - S_j$, which is the same as $S_i - S_j = C_i' - C_j$ (here the subtraction is n-bit integer subtraction). It can be shown that if S_i, S_j are independent of C', C, then the probability of this event is at most $2/p$.

5   Message Secrecy

We state the theorem for security under the Find-then-Guess notion of security. The proof follows standard techniques ([2]).

Theorem 5: *Let A be an adversary attacking the encryption scheme IAPM in Figure 2 (with f being a random permutation F) in the find-then-guess sense, making at most q queries, totaling at most μ blocks. Then,*

$$Adv_A \leq (2\mu^2) \cdot \frac{1}{2^n}$$

6   Security of IACBC

Theorem 6: *Let A be an adversary attacking the message integrity of IACBC with random permutations F and G. Let A make at most m queries in the first stage, totaling at most μ blocks. Let $u = \mu + m$. Let v be the maximum number of blocks in the second stage. Then for adversary A,*

$$Succ < (2 * (u+1)^2 + 2tm^2 + t^2(m+1)^2 + 3tmu + 2(u+v+1) + 2 + o(1)) * 2^{-n}$$

Theorem 5 continues to hold for IACBC. Proofs of theorem 5, 6 and IACBC variant of theorem 5 will be given in the full version of the paper.

Acknowledgment

I am extremely grateful to Shai Halevi and Pankaj Rohatgi for help with the proof of message integrity. I would also like to thank Pankaj for helping me simplify the overall scheme, and Shai for going through the paper in excruciating detail and making numerous helpful suggestions.

I would also like to thank Don Coppersmith, Johan Hastad, Nick Howgrave-Graham, J.R. Rao, Ron Rivest, Phil Rogaway, and referees for helpful suggestions.

References

1. ANSI X3.106, "American National Standard for Information Systems - Data Encryption Algorithm - Modes of Operation", American National Standards Institute, 1983.
2. M. Bellare, A. Desai, E. Jokiph, P. Rogaway, "A Concrete Security Treatment of Symmetric Encryption: Analysis of the DES Modes of OPeration", 38th IEEE FOCS, 1997
3. J. Black, S. Halevi, H. Krawczyk, T. Krovetz and P.Rogaway, "UMAC: Fast and secure message authentication", *Advances in Cryptology-Crypto 99*, LNCS 1666, 1999
4. M. Bellare, J. Kilian, P. Rogaway, "The Security of Cipher Block Chaining", CRYPTO 94, LNCS 839, 1994
5. M. Bellare, C. Namprempre, "Authenticated Encryption: Relations among notions and analysis of the generic composition paradigm", Proc. Asiacrypt 2000, T. Okamoto ed., Springer Verlag 2000
6. V.D. Gligor, P.Donescu, "Integrity Aware PCBC Encryption Schemes", 7th Intl. Workshop on Security Protocols, Cambridge, LNCS, 1999
7. V.D. Gligor, P. Donescu, "Fast Encryption Authentication: XCBC Encryption and XECB Authentication Modes",
http://csrc.nist.gov/encryption/modes/workshop1
8. Hugo Krawczyk, "LFSR-based Hashing and Authentication", Proc. Crypto 94. LNCS 839, 1994
9. ISO 8372, " Information processing - Modes of operation for a 64-bit block cipher algorithm", International Organization for Standardization, Geneva, Switzerland, 1987
10. ISO/IEC 9797, "Data cryptographic techniques - Data integrity mechanism using a cryptographic check function employing a block cipher algorithm", 1989
11. J. Katz and M. Yung, "Unforgeable Encryption and Adaptively Secure Modes of Operation", Fast Software Encryption 2000.
12. M. Luby, "Pseudorandomness and Cryptographic Applications", Princeton Computer Science Notes, Princeton Univ. Press, 1996
13. C.H. Meyer, S. M. Matyas, "Cryptography: A New Dimension in Computer Data Security", John Wiley and Sons, New York, 1982
14. National Bureau of Standards, NBS FIPS PUB 81, "DES modes of operation", U.S. Department of Commerce, 1980.
15. National Bureau of Standards, Data Encryption Standard, U.S. Department of Commerce, FIPS 46 (1977)
16. RFC 1510,"The Kerberos network authentication service (V5)", J. Kohl and B.C. Neuman, Sept 1993
17. Security Architecture for the Internet Protocol, RFC 2401, http://www.ietf.org/rfc/rfc2401.txt
18. S.G. Stubblebine and V.D. Gligor, "On message integrity in cryptographic protocols", Proceedings of the 1992 IEEE Computer Society Symposium on Research in Security and Privacy, 1992.
19. The TLS Protocol, RFC2246, http://www.ietf.org/rfc/rfc2246.txt